FUZZY LOGIC WITH ENGINEERING APPLICATIONS

~cond Edition

FUZZY LOGIC WITH ENGINEERING APPLICATIONS

Second Edition

Timothy J. Ross
University of New Mexico, USA

John Wiley & Sons, Ltd

Other Wiley Editorial Offices

John Wiley & Sons Inc., 111 River Street, Hoboken, NJ 07030, USA

Jossey-Bass, 989 Market Street, San Francisco, CA 94103-1741, USA

Wiley-VCH Verlag GmbH, Boschstr. 12, D-69469 Weinheim, Germany

John Wiley & Sons Australia Ltd, 33 Park Road, Milton, Queensland 4064, Australia

John Wiley & Sons (Asia) Pte Ltd, 2 Clementi Loop #02-01, Jin Xing Distripark, Singapore 129809

John Wiley & Sons Canada Ltd, 22 Worcester Road, Etobicoke, Ontario, Canada M9W 1L1

Wiley also publishes its books in a variety of electronic formats. Some content that appears
in print may not be available in electronic books.

British Library Cataloguing in Publication Data

A catalogue record for this book is available from the British Library

ISBN 0-470-86074-X (Cloth)
      0-470-86075-8 (Paper)

Typeset in 10/12pt Times NewRoman by Laserwords Private Limited, Chennai, India
Printed and bound in Great Britain by TJ International, Padstow, Cornwall
This book is printed on acid-free paper responsibly manufactured from sustainable forestry
in which at least two trees are planted for each one used for paper production.

This book is dedicated to the memories of my father, Jack, and my sister, Tina – the two behavioral bookends of my life.

CONTENTS

ABOUT THE AUTHOR

Timothy J. Ross is Professor and Regents' Lecturer of Civil Engineering at the University of New Mexico. He received his PhD degree in Civil Engineering from Stanford University, his MS from Rice University, and his BS from Washington State University. Professor Ross has held previous positions as Senior Research Structural Engineer, Air Force Weapons Laboratory, from 1978 to 1986; and Vulnerability Engineer, Defense Intelligence Agency, from 1973 to 1978. Professor Ross has authored more than 120 publications and has been active in the research and teaching of fuzzy logic since 1983. He is the founding Co-Editor-in-Chief of the *International Journal of Intelligent and Fuzzy Systems* and the co-editor of *Fuzzy Logic and Control: Software and Hardware Applications*, and most recently co-editor of *Fuzzy Logic and Probability Applications: Bridging the Gap*. Professor Ross is a Fellow of the American Society of Civil Engineers. He consults for industry and such institutions as Sandia National Laboratory and the National Technological University, and is a current Faculty Affiliate with the Los Alamos National Laboratory. He was recently honored with a Senior Fulbright Fellowship for his sabbatical study at the University of Calgary, Alberta, Canada.

PREFACE TO THE
SECOND EDITION

The second edition of this text has been "on the drawing board" for quite some time. Since the first edition was published, in 1995, the technology of fuzzy set theory and its application to systems, using fuzzy logic, has moved rapidly. Developments in other theories such as possibility theory and evidence theory (both being elements of a larger collection of methods under the rubric "generalized information theories") have shed more light on the real virtues of fuzzy logic applications, and some developments in machine computation have made certain features of fuzzy logic much more useful than in the past. In fact, it would be fair to state that some developments in fuzzy systems are quite competitive with other, linear algebra-based methods in terms of computational speed and associated accuracy. To wait eight years to publish this second edition has been, perhaps, too long. On the other hand, the technology continues to move so fast that one is often caught in that uncomfortable middle-ground not wanting to miss another important development that could be included in the book. The pressures of academia and the realities of life seem to intervene at the most unexpected times, but now seems the best time for this second edition.

There are sections of the first text that have been eliminated in the second edition; I shall have more to say on this below. And there are many new sections – which are included in the second edition – to try to capture some of the newer developments; the key word here is "some" as it would be completely impossible to summarize or illustrate even a small fraction of the new developments of the last eight years. As with any book containing technical material, the first edition contained errata that have been corrected in this second edition. A new aid to students, appearing in this edition, is a section at the end of the book which contains solutions to selected end-of-chapter problems. As with the first edition, a solutions manual for all problems in the second edition can be obtained by qualified instructors by visiting http://www.wileyeurope.com/go/fuzzylogic.

One of the most important explanations I shall describe in this preface has to do with what I call the misuse of definitional terms in the past literature on uncertainty representational theories; in this edition I use these terms very cautiously. Principal among these terms is the word "coherence" and the ubiquitous use of the word "law." To begin with the latter, the axioms of a probability theory referred to as the *excluded middle* will hereinafter only be referred to as axioms – never as laws. The operations due to De Morgan also will not be referred to as a law, but as a *principle* . . . since this principle does apply to some (not all) uncertainty theories (e.g., probability and fuzzy). The *excluded middle axiom* (and its dual, the *axiom of contradiction*) are not *laws*; Newton produced *laws*, Kepler produced *laws*, Darcy, Boyle, Ohm, Kirchhoff, Bernoulli, and many others too numerous to list here all developed *laws*. *Laws* are mathematical expressions describing the immutable realizations of nature. It is perhaps a cunning, but now exposed, illusion first coined by probabilists in the last two centuries to give their established theory more legitimacy by labeling their axioms as laws. Definitions, theorems, and axioms collectively can describe a certain axiomatic foundation describing a particular kind of theory, and nothing more; in this case the *excluded middle* and other axioms (see Appendix A) can be used to describe a probability theory. Hence, if a fuzzy set theory does not happen to be *constrained* by an *excluded middle axiom*, it is not a *violation* of some immutable law of nature like Newton's laws; fuzzy set theory simply does not happen to have an axiom of the excluded middle – it does not need, nor is *constrained by*, such an axiom. In fact, as early as 1905 the famous mathematician L. E. J. Brouwer defined this excluded middle axiom as a *principle* in his writings; he showed that the *principle of the excluded middle* was inappropriate in some logics, including his own which he termed *intuitionism*. Brouwer observed that Aristotelian logic is only a part of mathematics, the special kind of mathematical thought obtained if one restricts oneself to relations of the whole and part. Brouwer had to specify in which sense the principles of logic could be considered ''laws'' because within his intuitionistic framework thought did not follow any rules, and, hence, ''law'' could no longer mean ''rule'' (see the detailed discussion on this in the summary of Chapter 5). In this regard, I shall take on the cause advocated by Brouwer almost a century ago.

In addition, the term *coherence* does not connote a *law*. It may have been a clever term used by the probabilists to describe another of their axioms (in this case a permutation of the additivity axiom) but such cleverness is now an exposed prestidigitation of the English language. Such arguments of the past like ''no uncertainty theory that is *non-coherent* can ever be considered a serious theory for describing uncertainty'' now carry literally no weight when one considers that the term *coherence* is a label and not an adjective describing the value of an axiomatic structure. I suppose that fuzzy advocates could relabel their *axiom of strong-truth functionality* to the ''law of practicability'' and then claim that any other axiomatic structure that does not use such an axiom is inadequate, to wit ''a theory that violates the practicability axiom is a violation of the law of utility,'' but we shall not resort to this hyperbole. With this edition, we will speak without the need for *linguistic slight-of-hand*. The combination of a fuzzy set theory and a probability theory is a very powerful modeling paradigm. This book is dedicated to users who are more interested in solving problems than in dealing with debates using misleading jargon.

To end my discussion on misleading definitional terms in the literature, I have made two subtle changes in the material in Chapter 15. First, following prof. Klir's lead of a couple years ago, we no longer refer to ''fuzzy measure theory'' but instead describe it now as ''monotone measure theory''. The former phrase still causes confusion when referring to fuzzy set theory; hopefully this will end that confusion. And, in Chapter 15 in describing the monotone measure, m, I have changed the phrase describing this measure from a ''basic probability assignment (bpa)'' to a ''basic evidence assignment (bea)''. Here we attempt to avoid confusion with any of the terms typically used in probability theory.

As with the first edition, this second edition is designed for the professional and academic audience interested primarily in applications of fuzzy logic in engineering and technology. Always I have found that the majority of students and practicing professionals are interested in the applications of fuzzy logic to their particular fields. Hence, the book is written for an audience primarily at the senior undergraduate and first-year graduate levels. With numerous examples throughout the text, this book is written to assist the learning process of a broad cross section of technical disciplines. The book is primarily focused on applications, but each of the book's chapters begins with the rudimentary structure of the underlying mathematics required for a fundamental understanding of the methods illustrated.

Chapter 1* introduces the basic concept of fuzziness and distinguishes fuzzy uncertainty from other forms of uncertainty. It also introduces the fundamental idea of set membership, thereby laying the foundation for all material that follows, and presents membership functions as the format used for expressing set membership. The chapter summarizes an historical review of uncertainty theories. The chapter reviews the idea of ''sets as points'' in an n-dimensional Euclidean space as a graphical analog in understanding the relationship between classical (crisp) and fuzzy sets.

Chapter 2 reviews classical set theory and develops the basic ideas of fuzzy sets. Operations, axioms, and properties of fuzzy sets are introduced by way of comparisons with the same entities for classical sets. Various normative measures to model fuzzy intersections (t-norms) and fuzzy unions (t-conorms) are summarized.

Chapter 3 develops the ideas of fuzzy relations as a means of mapping fuzziness from one universe to another. Various forms of the composition operation for relations are presented. Again, the epistemological approach in Chapter 3 uses comparisons with classical relations in developing and illustrating fuzzy relations. This chapter also illustrates methods to determine the numerical values contained within a specific class of fuzzy relations, called similarity relations.

Chapter 4 discusses the fuzzification of scalar variables and the defuzzification of membership functions. The chapter introduces the basic features of a membership function and it discusses, very briefly, the notion of interval-valued fuzzy sets. Defuzzification is necessary in dealing with the ubiquitous crisp (binary) world around us. The chapter details defuzzification of fuzzy sets and fuzzy relations into crisp sets and crisp relations, respectively, using lambda-cuts, and it describes a variety of methods to defuzzify membership functions into scalar values. Examples of all methods are given in the chapter.

Chapter 5 introduces the precepts of fuzzy logic, again through a review of the relevant features of classical, or a propositional, logic. Various logical connectives and operations are illustrated. There is a thorough discussion of the various forms of the implication operation and the composition operation provided in this chapter. Three different inference methods, popular in the literature, are illustrated. Approximate reasoning, or reasoning under imprecise (fuzzy) information, is also introduced in this chapter. Basic IF–THEN rule structures are introduced and three graphical methods for inferencing are presented.

Chapter 6 provides several classical methods of developing membership functions, including methods that make use of the technologies of neural networks, genetic algorithms, and inductive reasoning.

Chapter 7 is a new chapter which presents six new automated methods which can be used to generate rules and membership functions from observed or measured input–output

---

* Includes sections taken from Ross, T., Booker, J., and Parkinson, W., 2002, *Fuzzy Logic and Probability Applications: Bridging the Gap*, reproduced by the permission of Society for Industrial and Applied Mathematics, Philadelphia, PA.

data. The procedures are essentially computational methods of learning. Examples are provided to illustrate each method. Many of the problems at the end of the chapter will require software; this software can be downloaded from: www.wileyeurope.com/go/fuzzylogic.

Beginning the second category of chapters in the book highlighting applications, Chapter 8 continues with the rule-based format to introduce fuzzy nonlinear simulation and complex system modeling. In this context, nonlinear functions are seen as mappings of information ''patches'' from the input space to information ''patches'' of the output space, instead of the ''point-to-point'' idea taught in classical engineering courses. Fidelity of the simulation is illustrated with standard functions, but the power of the idea can be seen in systems too complex for an algorithmic description. This chapter formalizes fuzzy associative memories (FAMs) as generalized mappings.

Chapter 9 is a new chapter covering the area of rule-base reduction. Fuzzy systems are becoming popular, but they can also present computational challenges as the rule-bases, especially those derived from automated methods, can become large in an exponential sense as the number of inputs and their dimensionality grows. This chapter summarizes two relatively new reduction techniques and provides examples of each.

Chapter 10 develops fuzzy decision making by introducing some simple concepts in ordering, preference and consensus, and multiobjective decisions. It introduces the powerful concept of Bayesian decision methods by fuzzifying this classic probabilistic approach. This chapter illustrates the power of combining fuzzy set theory with probability to handle random and nonrandom uncertainty in the decision-making process.

Chapter 11 discusses a few fuzzy classification methods by contrasting them with classical methods of classification, and develops a simple metric to assess the goodness of the classification, or misclassification. This chapter also summarizes classification using equivalence relations. The algebra of fuzzy vectors is summarized here. Classification is used as a springboard to introduce fuzzy pattern recognition. A single-feature and a multiple-feature procedure are summarized. Some simple ideas in image processing and syntactic pattern recognition are also illustrated.

Chapter 12 summarizes some typical operations in fuzzy arithmetic and fuzzy numbers. The extension of fuzziness to nonfuzzy mathematical forms using Zadeh's extension principle and several approximate methods to implement this principle are illustrated.

Chapter 13 introduces the field of fuzzy control systems. A brief review of control system design and control surfaces is provided. Some example problems in control are provided. Two new sections have been added to this book: fuzzy engineering process control, and fuzzy statistical process control. Examples of these are provided in the chapter.

Chapter 14 briefly addresses some important ideas embodied in fuzzy optimization, fuzzy cognitive mapping, fuzzy system identification, and fuzzy regression.

Finally, Chapter 15 enlarges the reader's understanding of the relationship between fuzzy uncertainty and random uncertainty (and other general forms of uncertainty, for that matter) by illustrating the foundations of monotone measures. The chapter discusses monotone measures in the context of evidence theory and probability theory. Because this chapter is an expansion of ideas relating to other disciplines (Dempster–Shafer evidence theory and probability theory), it can be omitted without impact on the material preceding it.

Appendix A of the book shows the axiomatic similarity of fuzzy set theory and probability theory and Appendix B provides answers to selected problems from each chapter.

Most of the text can be covered in a one-semester course at the senior undergraduate level. In fact, most science disciplines and virtually all math and engineering disciplines contain the basic ideas of set theory, mathematics, and deductive logic, which form the only knowledge necessary for a complete understanding of the text. For an introductory class, instructors may want to exclude some or all of the material covered in the last section of Chapter 6 (neural networks, genetic algorithms, and inductive reasoning), Chapter 7 (automated methods of generation), Chapter 9 on rule-base reduction methods, and any of the final three chapters: Chapter 13 (fuzzy control), Chapter 14 (miscellaneous fuzzy applications), and Chapter 15 on alternative measures of uncertainty. I consider the applications in Chapter 8 on simulations, Chapter 10 on decision making, Chapter 11 on classification, and Chapter 12 on fuzzy arithmetic to be important in the first course on this subject. The other topics could be used either as introductory material for a graduate-level course or for additional coverage for graduate students taking the undergraduate course for graduate credit.

The book is organized a bit differently from the first edition. I have moved most of the information for rule-based deductive systems closer to the front of the book, and have moved fuzzy arithmetic toward the end of the book; the latter does not disturb the flow of the book to get quickly into fuzzy systems development. A significant amount of new material has been added in the area of automated methods of generating fuzzy systems (Chapter 7); a new section has been added on additional methods of inference in Chapter 5; and a new chapter has been added on the growing importance of rule-base reduction methods (Chapter 9). Two new sections in fuzzy control have been added in Chapter 13. I have also deleted materials that either did not prove useful in the pedagogy of fuzzy systems, or were subjects of considerable depth which are introduced in other, more focused texts. Many of the rather lengthy example problems from the first edition have been reduced for brevity. In terms of organization, the first eight chapters of the book develop the foundational material necessary to get students to a position where they can generate their own fuzzy systems. The last seven chapters use the foundation material from the first eight chapters to present specific applications.

The problems in this text are typically based on current and potential applications, case studies, and education in intelligent and fuzzy systems in engineering and related technical fields. The problems address the disciplines of computer science, electrical engineering, manufacturing engineering, industrial engineering, chemical engineering, petroleum engineering, mechanical engineering, civil engineering, environmental engineering, engineering management, and a few related fields such as mathematics, medicine, operations research, technology management, the hard and soft sciences, and some technical business issues. The references cited in the chapters are listed toward the end of each chapter. These references provide sufficient detail for those readers interested in learning more about particular applications using fuzzy sets or fuzzy logic. The large number of problems provided in the text at the end of each chapter allows instructors a sizable problem base to afford instruction using this text on a multisemester or multiyear basis, without having to assign the same problems term after term.

I was most fortunate this past year to have co-edited a text with Drs. Jane Booker and Jerry Parkinson, entitled *Fuzzy Logic and Probability Applications: Bridging the Gap*, published by the Society for Industrial and Applied Mathematics (SIAM), in which many of my current thoughts on the matter of the differences between fuzzy logic and probability theory were noted; some of this appears in Chapters 1 and 15 of this edition. Moreover, I am also grateful to Prof. Kevin Passino whose text, *Fuzzy Control*, published by Prentice Hall, illustrated some very recent developments in the automated generation of membership functions and rules in fuzzy systems. The algorithms discussed in his book, while being developed by others earlier, are collected in one chapter in his book; some of these are illustrated here in Chapter 7, on automated methods. The added value to Dr. Passino's material and methods is that I have expanded their explanation and have added some simple numerical examples of these methods to aid first-time students in this field.

Again I wish to give credit either to some of the individuals who have shaped my thinking about this subject since the first edition of 1995, or to others who by their simple association with me have caused me to be more circumspect about the use of the material contained in the book. In addition to the previously mentioned colleagues Jane Booker and Jerry Parkinson, who both overwhelm me with their knowledge and enthusiasm, my other colleagues at Los Alamos National Laboratory have shaped or altered my thinking critically and positively: Scott Doebling, Ed Rodriquez, and John Birely for their steadfast support over the years to investigate alternative uncertainty paradigms, Jason Pepin for his useful statistical work in mechanics, Cliff Joslyn for his attention to detail in the axiomatic structure of random sets, Brian Reardon for his critical questions of relevance, François Hemez and Mark Anderson for their expertise in applying uncertainty theory to validation methods, Kari Sentz for her humor and her perspective in linguistic uncertainty, Ron Smith and Karen Hench for their collaborations in process control, and Steve Eisenhawer and Terry Bott for their early and continuing applications of fuzzy logic in risk assessment.

Some of the newer sections of the second edition were first taught to a group of faculty and students at the University of Calgary, Alberta, during my most recent sabbatical leave. My host, Prof. Gopal Achari, was instrumental in giving me this exposure and outreach to these individuals and I shall remain indebted to him. Among this group, faculty members Drs. Brent Young, William Svrcek, and Tom Brown, and students Jeff Macisaac, Rachel Mintz, and Rodolfo Tellez, all showed leadership and critical inquiry in adopting many fuzzy skills into their own research programs. Discussions with Prof. Mihaela Ulieru, already a fuzzy advocate, and her students proved useful. Finally, paper collaborations with Ms. Sumita Fons, Messrs. Glen Hay and James Vanderlee all gave me a feeling of accomplishment on my "mission to Canada."

Collaborations, discussions, or readings from Drs. Lotfi Zadeh, George Klir, and Vladik Kreinovich over the past few years have enriched my understanding in this field immeasurably. In particular, Dr. Klir's book of 1995 (*Fuzzy Sets and Fuzzy Logic*) and his writings in various journals collectively have helped me deepen my understanding of some of the nuances in the mathematics of fuzzy logic; his book is referenced in many places in this second edition. I wish to thank some of my recent graduate students who have undertaken projects, MS theses, or PhD dissertations related to this field and whose hard work for me and alongside me has given me a sense of pride in their own remarkable tenacity and productive efforts: Drs. Sunil Donald and Jonathan Lucero and Mr. Greg Chavez, and Mss. Terese Gabocy Anderson and Rhonda Young. There have been numerous students over the past eight years who have contributed many example problems for updating the text; unfortunately too numerous to mention in this brief preface. I want to thank them all again for their contributions.

Four individuals need specific mention because they have contributed some sections to this text. I would like to thank specifically Dr. Jerry Parkinson for his contributions to Chapter 13 in the areas of chemical process control and fuzzy statistical process control, Dr. Jonathan Lucero for his contributions in developing the material in Chapter 9 for rule-reduction methods (which form the core of his PhD dissertation), Greg Chavez for his text preparation of many of the new, contributed problems in this text and of the material in Chapter 7, and Dr. Sunil Donald for one new section in Chapter 15 on empirical methods to generate possibility distributions.

I am most grateful for financial support over the past three years while I have generated most of the background material in my own research for some of the newer material in the book. I would like to thank the Los Alamos National Laboratory, Engineering and Science Applications Division, the University of New Mexico, and the US–Canadian Fulbright Foundation for their generous support during this period of time.

With so many texts covering specific niches of fuzzy logic it is not possible to summarize all these important facets of fuzzy set theory and fuzzy logic in a single textbook. The hundreds of edited works and tens of thousands of archival papers show clearly that this is a rapidly growing technology, where new discoveries are being published every month. It remains my fervent hope that this introductory textbook will assist students and practising professionals to learn, to apply, and to be comfortable with fuzzy set theory and fuzzy logic. I welcome comments from all readers to improve this textbook as a useful guide for the community of engineers and technologists who will become knowledgeable about the potential of fuzzy system tools for their use in solving the problems that challenge us each day.

Timothy J. Ross
Santa Fe, New Mexico

CHAPTER
1

INTRODUCTION

It is the mark of an instructed mind to rest satisfied with that degree of precision which the nature of the subject admits, and not to seek exactness where only an approximation of the truth is possible.

Aristotle, 384–322 BC
Ancient Greek philosopher

Precision is not truth.

Henri E. B. Matisse, 1869–1954
Impressionist painter

All traditional logic habitually assumes that precise symbols are being employed. It is therefore not applicable to this terrestrial life but only to an imagined celestial existence.

Bertrand Russell, 1923
British philosopher and Nobel Laureate

We must exploit our tolerance for imprecision.

Lotfi Zadeh
Professor, Systems Engineering, UC Berkeley, 1973

The quotes above, all of them legendary, have a common thread. That thread represents the relationship between precision and uncertainty. The more uncertainty in a problem, the less precise we can be in our understanding of that problem. It is ironic that the oldest quote, above, is due to the philosopher who is credited with the establishment of Western logic – a binary logic that only admits the opposites of true and false, a logic which does not admit degrees of truth in between these two extremes. In other words, Aristotelian logic does not admit imprecision in truth. However, Aristotle's quote is so appropriate today; it is a quote that admits uncertainty. It is an admonishment that we should heed; we should balance the precision we seek with the uncertainty that exists. Most engineering texts do not address the uncertainty in the information, models, and solutions that are conveyed

Fuzzy Logic with Engineering Applications, Second Edition   T. J. Ross
© 2004 John Wiley & Sons, Ltd   ISBNs: 0-470-86074-X (HB); 0-470-86075-8 (PB)

within the problems addressed therein. This text is dedicated to the characterization and quantification of uncertainty within engineering problems such that an appropriate level of precision can be expressed. When we ask ourselves why we should engage in this pursuit, one reason should be obvious: achieving high levels of precision costs significantly in time or money or both. Are we solving problems that require precision? The more complex a system is, the more imprecise or inexact is the information that we have to characterize that system. It seems, then, that precision and information and complexity are inextricably related in the problems we pose for eventual solution. However, for most of the problems that we face, the quote above due to Professor Zadeh suggests that we can do a better job in accepting some level of imprecision.

It seems intuitive that we should balance the degree of precision in a problem with the associated uncertainty in that problem. Hence, this book recognizes that uncertainty of various forms permeates all scientific endeavors and it exists as an integral feature of all abstractions, models, and solutions. It is the intent of this book to introduce methods to handle one of these forms of uncertainty in our technical problems, the form we have come to call fuzziness.

THE CASE FOR IMPRECISION

Our understanding of most physical processes is based largely on imprecise human reasoning. This imprecision (when compared to the precise quantities required by computers) is nonetheless a form of information that can be quite useful to humans. The ability to embed such reasoning in hitherto intractable and complex problems is the criterion by which the efficacy of fuzzy logic is judged. Undoubtedly this ability cannot solve problems that require precision – problems such as shooting precision laser beams over tens of kilometers in space; milling machine components to accuracies of parts per billion; or focusing a microscopic electron beam on a specimen the size of a nanometer. The impact of fuzzy logic in these areas might be years away, if ever. But not many human problems require such precision – problems such as parking a car, backing up a trailer, navigating a car among others on a freeway, washing clothes, controlling traffic at intersections, judging beauty contestants, and a preliminary understanding of a complex system.

Requiring precision in engineering models and products translates to requiring high cost and long lead times in production and development. For other than simple systems, expense is proportional to precision: more precision entails higher cost. When considering the use of fuzzy logic for a given problem, an engineer or scientist should ponder the need for *exploiting the tolerance for imprecision*. Not only does high precision dictate high costs but also it entails low tractability in a problem. Articles in the popular media illustrate the need to exploit imprecision. Take the "traveling salesrep" problem, for example. In this classic optimization problem a sales representative wants to minimize total distance traveled by considering various itineraries and schedules between a series of cities on a particular trip. For a small number of cities, the problem is a trivial exercise in enumerating all the possibilities and choosing the shortest route. As the number of cities continues to grow, the problem quickly approaches a combinatorial explosion impossible to solve through an exhaustive search, even with a computer. For example, for 100 cities there are $100 \times 99 \times 98 \times 97 \times \cdots \times 2 \times 1$, or about 10^{200}, possible routes to consider! No computers exist today that can solve this problem through a brute-force enumeration of all the possible routes. There are real, practical problems analogous to the traveling salesrep problem. For example, such problems arise in the fabrication of circuit boards, where precise lasers drill hundreds of thousands of holes in the board. Deciding in which order to drill the holes (where the board moves under a stationary laser) so as to minimize drilling time is a traveling salesrep problem [Kolata, 1991].

Thus, algorithms have been developed to solve the traveling salesrep problem in an optimal sense; that is, the exact answer is not guaranteed but an optimum answer is achievable – the optimality is measured as a percent accuracy, with 0% representing the exact answer and accuracies larger than zero representing answers of lesser accuracy. Suppose we consider a signal routing problem analogous to the traveling salesrep problem where we want to find the optimum path (i.e., minimum travel time) between 100,000 nodes in a network to an accuracy within 1% of the exact solution; this requires significant CPU time on a supercomputer. If we take the same problem and increase the precision requirement a modest amount to an accuracy of 0.75%, the computing time approaches a few months! Now suppose we can live with an accuracy of 3.5% (quite a bit more accurate than most problems we deal with), and we want to consider an order-of-magnitude more nodes in the network, say 1,000,000; the computing time for this problem is on the order of several minutes [Kolata, 1991]. This remarkable reduction in cost (translating time to dollars) is due solely to the acceptance of a lesser degree of precision in the optimum solution. Can humans live with a little less precision? The answer to this question depends on the situation, but for the vast majority of problems we deal with every day the answer is a resounding yes.

AN HISTORICAL PERSPECTIVE

From an historical point of view the issue of uncertainty has not always been embraced within the scientific community [Klir and Yuan, 1995]. In the traditional view of science, uncertainty represents an undesirable state, a state that must be avoided at all costs. This was the state of science until the late nineteenth century when physicists realized that Newtonian mechanics did not address problems at the molecular level. Newer methods, associated with statistical mechanics, were developed which recognized that statistical averages could replace the specific manifestations of microscopic entities. These statistical quantities, which summarized the activity of large numbers of microscopic entities, could then be connected in a model with appropriate macroscopic variables [Klir and Yuan, 1995]. Now, the role of Newtonian mechanics and its underlying calculus which considered no uncertainty was replaced with statistical mechanics which could be described by a probability theory – a theory which could capture a form of uncertainty, the type generally referred to as random uncertainty. After the development of statistical mechanics there has been a gradual trend in science during the past century to consider the influence of uncertainty on problems, and to do so in an attempt to make our models more robust, in the sense that we achieve credible solutions and at the same time quantify the amount of uncertainty.

Of course, the leading theory in quantifying uncertainty in scientific models from the late nineteenth century until the late twentieth century had been probability theory. However, the gradual evolution of the expression of uncertainty using probability theory was challenged, first in 1937 by Max Black, with his studies in vagueness, then with the introduction of fuzzy sets by Lotfi Zadeh in 1965. Zadeh's work [1965] had a profound influence on the thinking about uncertainty because it challenged not only probability theory as the sole representation for uncertainty, but the very foundations upon which probability theory was based: classical binary (two-valued) logic [Klir and Yuan, 1995].

Probability theory dominated the mathematics of uncertainty for over five centuries. Probability concepts date back to the 1500s, to the time of Cardano when gamblers recognized the rules of probability in games of chance. The concepts were still very much in the limelight in 1685, when the Bishop of Wells wrote a paper that discussed a problem in determining the truth of statements made by two witnesses who were both known to be unreliable to the extent that they only tell the truth with probabilities p_1 and p_2, respectively. The Bishop's answer to this was based on his assumption that the two witnesses were independent sources of information [Lindley, 1987].

Probability theory was initially developed in the eighteenth century in such landmark treatises as Jacob Bernoulli's *Ars Conjectandi* (1713) and Abraham DeMoiver's *Doctrine of Chances* (1718, 2nd edition 1738). Later in that century a small number of articles appeared in the periodical literature that would have a profound effect on the field. Most notable of these were Thomas Bayes's "An essay towards solving a problem in the doctrine of chances" (1763) and Pierre Simon Laplace's formulation of the axioms relating to games of chance, "Memoire sur la probabilite des causes par les evenemens" (1774). Laplace, only 25 years old at the time he began his work in 1772, wrote the first substantial article in mathematical statistics prior to the nineteenth century. Despite the fact that Laplace, at the same time, was heavily engaged in mathematical astronomy, his memoir was an explosion of ideas that provided the roots for modern decision theory, Bayesian inference with nuisance parameters (historians claim that Laplace did not know of Bayes's earlier work), and the asymptotic approximations of posterior distributions [Stigler, 1986].

By the time of Newton, physicists and mathematicians were formulating different theories of probability. The most popular ones remaining today are the relative frequency theory and the subjectivist or personalistic theory. The later development was initiated by Thomas Bayes (1763), who articulated his very powerful theorem for the assessment of subjective probabilities. The theorem specified that a human's degree of belief could be subjected to an objective, coherent, and measurable mathematical framework within subjective probability theory. In the early days of the twentieth century Rescher developed a formal framework for a conditional probability theory.

The twentieth century saw the first developments of alternatives to probability theory and to classical Aristotelian logic as paradigms to address more kinds of uncertainty than just the random kind. Jan Lukasiewicz developed a multivalued, discrete logic (*circa* 1930). In the 1960's Arthur Dempster developed a theory of evidence which, for the first time, included an assessment of ignorance, or the absence of information. In 1965 Lotfi Zadeh introduced his seminal idea in a continuous-valued logic that he called fuzzy set theory. In the 1970s Glenn Shafer extended Dempster's work to produce a complete theory of evidence dealing with information from more than one source, and Lotfi Zadeh illustrated a possibility theory resulting from special cases of fuzzy sets. Later in the 1980s other investigators showed a strong relationship between evidence theory, probability theory, and possibility theory with the use of what was called fuzzy measures [Klir and Wierman, 1996], and what is now being termed monotone measures.

Uncertainty can be thought of in an epistemological sense as being the inverse of information. Information about a particular engineering or scientific problem may be incomplete, imprecise, fragmentary, unreliable, vague, contradictory, or deficient in some other way [Klir and Yuan, 1995]. When we acquire more and more information about a problem, we become less and less uncertain about its formulation and solution. Problems that are characterized by very little information are said to be ill-posed, complex, or not sufficiently known. These problems are imbued with a high degree of uncertainty. Uncertainty can be manifested in many forms: it can be fuzzy (not sharp, unclear, imprecise, approximate), it can be vague (not specific, amorphous), it can be ambiguous (too many choices, contradictory), it can be of the form of ignorance (dissonant, not knowing something), or it can be a form due to natural variability (conflicting, random, chaotic, unpredictable). Many other linguistic labels have been applied to these various forms, but for now these shall suffice. Zadeh [2002] posed some simple examples of these forms in terms of a person's statements about when they shall return to a current place in time. The statement "I shall return soon" is vague, whereas the statement "I shall return in a few minutes" is fuzzy; the former is not known to be associated with any unit of time (seconds, hours, days), and the latter is associated with an uncertainty that is at least known to be on the order of minutes. The phrase, "I shall return within 2 minutes of 6pm" involves an uncertainty which has a quantifiable imprecision; probability theory could address this form.

Vagueness can be used to describe certain kinds of uncertainty associated with linguistic information or intuitive information. Examples of vague information are that the data quality is "good," or that the transparency of an optical element is "acceptable." Moreover, in terms of semantics, even the terms vague and fuzzy cannot be generally considered synonyms, as explained by Zadeh [1995]: "usually a vague proposition is fuzzy, but the converse is not generally true."

Discussions about vagueness started with a famous work by the philosopher Max Black. Black [1937] defined a vague proposition as a proposition where the possible states (of the proposition) are not clearly defined with regard to inclusion. For example, consider the proposition that a person is young. Since the term "young" has different interpretations to different individuals, we cannot decisively determine the age(s) at which an individual is young versus the age(s) at which an individual is not considered to be young. Thus, the proposition is vaguely defined. Classical (binary) logic does not hold under these circumstances, therefore we must establish a different method of interpretation.

Max Black, in writing his 1937 essay "Vagueness: An exercise in logical analysis" first cites remarks made by the ancient philosopher Plato about uncertainty in geometry, then embellishes on the writings of Bertrand Russell (1923) who emphasized that "all traditional logic habitually assumes that precise symbols are being employed." With these great thoughts as a prelude to his own arguments, he proceeded to produce his own, now-famous quote:

> It is a paradox, whose importance familiarity fails to diminish, that the most highly developed and useful scientific theories are ostensibly expressed in terms of objects never encountered in experience. The line traced by a draftsman, no matter how accurate, is seen beneath the microscope as a kind of corrugated trench, far removed from the ideal line of pure geometry. And the "point-planet" of astronomy, the "perfect gas" of thermodynamics, or the "pure-species" of genetics are equally remote from exact realization. Indeed the unintelligibility at the atomic or subatomic level of the notion of a rigidly demarcated boundary shows that such objects not merely are not but could not be encountered. While the mathematician constructs a theory in terms of "perfect" objects, the experimental scientist observes objects of which

the properties demanded by theory are and can, in the very nature of measurement, be only approximately true.

More recently, in support of Black's work, Quine [1981] states:

Diminish a table, conceptually, molecule by molecule: when is a table not a table? No stipulations will avail us here, however arbitrary. If the term 'table' is to be reconciled with bivalence, we must posit an exact demarcation, exact to the last molecule, even though we cannot specify it. We must hold that there are physical objects, coincident except for one molecule, such that one is a table and the other is not.

Bruno de Finetti [1974], publishing in his landmark book *Theory of Probability*, gets his readers' attention quickly by proclaiming, "Probability does not exist; it is a subjective description of a person's uncertainty. We should be normative about uncertainty and not descriptive." He further emphasizes that the frequentist view of probability (objectivist view) "requires individual trials to be equally probable and stochastically independent." In discussing the difference between possibility and probability he states, "The logic of certainty furnishes us with the range of possibility (and the possible has no gradations); probability is an additional notion that one applies within the range of possibility, thus giving rise to graduations ('more or less' probable) that are meaningless in the logic of uncertainty." In his book, de Finetti gives us warnings: "The calculus of probability can say absolutely nothing about reality," and in referring to the dangers implicit in attempts to confuse certainty with high probability, he states

We have to stress this point because these attempts assume many forms and are always dangerous. In one sentence: to make a mistake of this kind leaves one inevitably faced with all sorts of fallacious arguments and contradictions whenever an attempt is made to state, on the basis of probabilistic considerations, that something must occur, or that its occurrence confirms or disproves some probabilistic assumptions.

In a discussion about the use of such vague terms as "very probable" or "practically certain," or "almost impossible," de Finetti states:

The field of probability and statistics is then transformed into a Tower of Babel, in which only the most naive amateur claims to understand what he says and hears, and this because, in a language devoid of convention, the fundamental distinctions between what is certain and what is not, and between what is impossible and what is not, are abolished. Certainty and impossibility then become confused with high or low degrees of a subjective probability, which is itself denied precisely by this falsification of the language. On the contrary, the preservation of a clear, terse distinction between certainty and uncertainty, impossibility and possibility, is the unique and essential precondition for making meaningful statements (which could be either right or wrong), whereas the alternative transforms every sentence into a nonsense.

THE UTILITY OF FUZZY SYSTEMS

Several sources have shown and proven that fuzzy systems are universal approximators [Kosko, 1994; Ying et al., 1999]. These proofs stem from the isomorphism between two algebras: an abstract algebra (one dealing with groups, fields, and rings) and a linear algebra (one dealing with vector spaces, state vectors, and transition matrices) and the structure of a fuzzy system, which is comprised of an implication between actions and conclusions (antecedents and consequents). The reason for this isomorphism is that both entities (algebra and fuzzy systems) involve a mapping between elements of two or more domains. Just as an algebraic function maps an input variable to an output variable, a fuzzy system maps an input group to an output group; in the latter these groups can be linguistic propositions or other forms of fuzzy information. The foundation on which fuzzy systems theory rests is a fundamental theorem from real analysis in algebra known as the Stone–Weierstrass theorem, first developed in the late nineteenth century by Weierstrass [1885], then simplified by Stone [1937].

In the coming years it will be the consequence of this isomorphism that will make fuzzy systems more and more popular as solution schemes, and it will make fuzzy systems theory a routine offering in the classroom as opposed to its previous status as a "new, but curious technology." Fuzzy systems, or whatever label scientists eventually come to call it in the future, will be a standard course in any science or engineering curriculum. It contains all of what algebra has to offer, plus more, because it can handle all kinds of information not just numerical quantities. More on this similarity between abstract or linear algebras and fuzzy systems is discussed in Chapter 9 on rule-reduction methods.

While fuzzy systems are shown to be universal approximators to algebraic functions, it is not this attribute that actually makes them valuable to us in understanding new or evolving problems. Rather, the primary benefit of fuzzy systems theory is to approximate system behavior where analytic functions or numerical relations do not exist. Hence, fuzzy systems have high potential to understand the very systems that are devoid of analytic formulations: complex systems. Complex systems can be new systems that have not been tested, they can be systems involved with the human condition such as biological or medical systems, or they can be social, economic, or political systems, where the vast arrays of inputs and outputs could not all possibly be captured analytically or controlled in any conventional sense. Moreover, the relationship between the causes and effects of these systems is generally not understood, but often can be observed.

Alternatively, fuzzy systems theory can have utility in assessing some of our more conventional, less complex systems. For example, for some problems exact solutions are not always necessary. An approximate, but fast, solution can be useful in making preliminary design decisions, or as an initial estimate in a more accurate numerical technique to save computational costs, or in the myriad of situations where the inputs to a problem are vague, ambiguous, or not known at all. For example, suppose we need a controller to bring an aircraft out of a vertical dive. Conventional controllers cannot handle this scenario as they are restricted to linear ranges of variables; a dive situation is highly nonlinear. In this case, we could use a fuzzy controller, which is adept at handling nonlinear situations albeit in an imprecise fashion, to bring the plane out of the dive into a more linear range, then hand off the control of the aircraft to a conventional, linear, highly accurate controller. Examples of other situations where exact solutions are not warranted abound in our daily lives. For example, in the following quote from a popular science fiction movie,

C-3PO: Sir, the possibility of successfully navigating an asteroid field is approximately 3,720 to 1!

Han Solo: Never tell me the odds!

Characters in the movie *Star Wars: The Empire Strikes Back* (Episode V), 1980 we have an illustration of where the input information (the odds of navigating through an asteroid field) is useless, so how does one make a decision in the presence of this information?

Hence, fuzzy systems are very useful in two general contexts: (1) in situations involving highly complex systems whose behaviors are not well understood, and (2) in situations where an approximate, but fast, solution is warranted.

As pointed out by Ben-Haim [2001], there is a distinction between models of systems and models of uncertainty. A fuzzy system can be thought of as an aggregation of both because it attempts to understand a system for which no model exists, and it does so with information that can be uncertain in a sense of being vague, or fuzzy, or imprecise, or altogether lacking. Systems whose behaviors are both understood and controllable are of the kind which exhibit a certain robustness to spurious changes. In this sense, robust systems are ones whose output (such as a decision system) does not change significantly under the influence of changes in the inputs, because the system has been designed to operate within some window of uncertain conditions. It is maintained that fuzzy systems too are robust. They are robust because the uncertainties contained in both the inputs and outputs of the system are used in formulating the system structure itself, unlike conventional systems analysis which first poses a model, based on a collective set of assumptions needed to formulate a mathematical form, then uncertainties in each of the parameters of that mathematical abstraction are considered.

The positing of a mathematical form for our system can be our first mistake, and any subsequent uncertainty analysis of this mathematical abstraction could be misleading. We call this the Optimist's dilemma: find out how a chicken clucks, by first "assuming a spherical chicken." Once the sphericity of the chicken has been assumed, there are all kinds of elegant solutions that can be found; we can predict any number of sophisticated clucking sounds with our model. Unfortunately when we monitor a real chicken it does not cluck the way we predict. The point being made here is that there are few physical and no mathematical abstractions that can be made to solve some of our complex problems, so we need new tools to deal with complexity; fuzzy systems and their associated developments can be one of these newer tools.

LIMITATIONS OF FUZZY SYSTEMS

However, this is not to suggest that we can now stop looking for additional tools. Realistically, even fuzzy systems, as they are posed now, can be described as shallow models in the sense that they are primarily used in deductive reasoning. This is the kind of reasoning where we infer the specific from the general. For example, in the game of tic-tac-toe there are only a few moves for the entire game; we can deduce our next move from the previous move, and our knowledge of the game. It is this kind of reasoning that we also called shallow reasoning, since our knowledge, as expressed linguistically, is of a shallow and meager kind. In contrast to this is the kind of reasoning that is inductive, where we infer the general from the particular; this method of inference is called deep, because our knowledge is of a deep and substantial kind – a game of chess would be closer to an inductive kind of model.

We should understand the distinction between using mathematical models to account for observed data, and using mathematical models to describe the underlying process by which the observed data are generated or produced by nature [Arciszewski et al., 2003]. Models of systems where the behavior can be observed, and whose predictions can only account for these observed data, are said to be shallow, as they do not account for the underlying realities. Deep models, those of the inductive kind, are alleged to capture the physical process by which nature has produced the results we have observed. In his *Republic* (360 BC), Plato suggests the idea that things that are perceived are only imperfect copies of the true reality that can only be comprehended by pure thought. Plato was fond of mathematics, and he saw in its very precise structure of logic idealized abstraction and separation from the material world. He thought of these things being so important, that above the doorway to his Academy was placed the inscription "Let no one ignorant of mathematics enter here." In Plato's doctrine of forms, he argued that the phenomenal world was a mere shadowy image of the eternal, immutable real world, and that matter was docile and disorderly, governed by a Mind that was the source of coherence, harmony, and orderliness. He argued that if man was occupied with the things of the senses, then he could never gain true knowledge. In his work the *Phaedo* he declares that as mere mortals we cannot expect to attain absolute truth about the universe, but instead must be content with developing a descriptive picture – a model [Barrow, 2000].

Centuries later, Galileo was advised by his inquisitors that he must not say that his mathematical models were describing the realities of nature, but rather that they simply were adequate models of the observations he made with his telescope [Drake, 1957]; hence, that they were solely deductive. In this regard, models that only attempt to replicate some phenomenological behavior are considered shallow models, or models of the deductive kind, and they lack the knowledge needed for true understanding of a physical process. The system that emerges under inductive reasoning will have connections with both evolution and complexity. How do humans reason in situations that are complicated or ill-defined? Modern psychology tells us that as humans we are only moderately good at deductive logic, and we make only moderate use of it. But we are superb at seeing or recognizing or matching patterns – behaviors that confer obvious evolutionary benefits. In problems of complication then, we look for patterns; and we simplify the problem by using these to construct temporary internal models or hypotheses or schemata to work with [Bower and Hilgard, 1981]. We carry out localized deductions based on our current hypotheses and we act on these deductions. Then, as feedback from the environment comes in, we may strengthen or weaken our beliefs in our current hypotheses, discarding some when they cease to perform, and replacing them as needed with new ones. In other words, where we cannot fully reason or lack full definition of the problem, we use simple models to fill the gaps in our understanding; such behavior is inductive.

Some sophisticated models may, in fact, be a complex weave of deductive and inductive steps. But, even our so-called "deep models" may not be deep enough. An illustration of this comes from a recent popular decision problem, articulated as the El Farol problem by W. Brian Arthur [1994]. This problem involves a decision-making scenario in which inductive reasoning is assumed and modeled, and its implications are examined. El Farol is a bar in Santa Fe, New Mexico, where on one night of the week in particular there is popular Irish music offered. Suppose N bar patrons decide independently each week whether to go to El Farol on this certain night. For simplicity, we set $N = 100$. Space in the bar is limited, and the evening is enjoyable if things are not too crowded – specifically, if fewer than 60% of the possible 100 are present. There is no way to tell the number coming for sure in advance, therefore a bar patron goes – deems it worth going – if he expects fewer than 60 to show up, or stays home if he expects more than 60 to go; there is no need that utilities differ much above and below 60. Choices are unaffected by previous visits; there is no collusion or prior communication among the bar patrons; and the only information available is the numbers who came in past weeks. Of interest is the dynamics of the number of bar patrons attending from week to week.

There are two interesting features of this problem. First, if there were an obvious model that all bar patrons could use to forecast attendance and on which to base their decisions, then a deductive solution would be possible. But no such model exists in this case. Given the numbers attending in the recent past, a large number of expectational models might be reasonable and defensible. Thus, not knowing which model other patrons might choose, a reference patron cannot choose his in a well-defined way. There is no deductively rational solution – no ''correct'' expectational model. From the patrons' viewpoint, the problem is ill-defined and they are propelled into a realm of induction. Second, any commonality of expectations gets disintegrated: if everyone believes few will go, then all will go. But this would invalidate that belief. Similarly, if all believe most will go, nobody will go, invalidating that belief. Expectations will be forced to differ, but not in a methodical, predictive way.

Scientists have long been uneasy with the assumption of perfect, deductive rationality in decision contexts that are complicated and potentially ill-defined. The level at which humans can apply perfect rationality is surprisingly modest. Yet it has not been clear how to deal with imperfect or bounded rationality. From the inductive example given above (El Farol problem), it would be easy to suggest that as humans in these contexts we use inductive reasoning: we induce a variety of working hypotheses, act upon the most credible, and replace hypotheses with new ones if they cease to work. Such reasoning can be modeled in a variety of ways. Usually this leads to a rich psychological world in which peoples' ideas or mental models compete for survival against other peoples' ideas or mental models – a world that is both evolutionary and complex. And, while this seems the best course of action for modeling complex questions and problems, this text stops short of that longer term goal with only a presentation of simple deductive models, of the rule-based kind, that are introduced and illustrated in Chapters 5–8.

THE ALLUSION: STATISTICS AND RANDOM PROCESSES

The uninitiated often claim that fuzzy set theory is just another form of probability theory in disguise. This statement, of course, is simply not true (Appendix A formally rejects this claim with an axiomatic discussion of both probability theory and fuzzy logic). Basic statistical analysis is founded on probability theory or stationary random processes, whereas most experimental results contain both random (typically noise) and nonrandom processes. One class of random processes, stationary random processes, exhibits the following three characteristics: (1) The sample space on which the processes are defined cannot change from one experiment to another; that is, the outcome space cannot change. (2) The frequency of occurrence, or probability, of an event within that sample space is constant and cannot change from trial to trial or experiment to experiment. (3) The outcomes must be repeatable from experiment to experiment. The outcome of one trial does not influence the outcome of a previous or future trial. There are more general classes of random processes than the class mentioned here. However, fuzzy sets are not governed by these characteristics.

Stationary random processes are those that arise out of chance, where the chances represent frequencies of occurrence that can be measured. Problems like picking colored balls out of an urn, coin and dice tossing, and many card games are good examples of stationary random processes. How many of the decisions that humans must make every day could be categorized as random? How about the uncertainty in the weather – is this random? How about your uncertainty in choosing clothes for the next day, or which car to buy, or your preference in colors – are these random uncertainties? How about your ability to park a car; is this a random process? How about the risk in whether a substance consumed by an individual now will cause cancer in that individual 15 years from now; is this a form of random uncertainty? Although it is possible to model all of these forms of uncertainty with various classes of random processes, the solutions may not be reliable. Treatment of these forms of uncertainty using fuzzy set theory should also be done with caution. One needs to study the character of the uncertainty, then choose an appropriate approach to develop a model of the process. Features of a problem that vary in time and space should be considered. For example, when the weather report suggests that there is a 60% chance of rain tomorrow, does this mean that there has been rain on tomorrow's date for 60 of the last 100 years? Does it mean that somewhere in your community 60% of the land area will receive rain? Does it mean that 60% of the time it will be raining and 40% of the time it will not be raining? Humans often deal with these forms of uncertainty linguistically, such as, "It will likely rain tomorrow." And with this crude assessment of the possibility of rain, humans can still make appropriately accurate decisions about the weather.

Random errors will generally average out over time, or space. Nonrandom errors, such as some unknown form of bias (often called a systematic error) in an experiment, will not generally average out and will likely grow larger with time. The systematic errors generally arise from causes about which we are ignorant, for which we lack information, or that we cannot control. Distinguishing between random and nonrandom errors is a difficult problem in many situations, and to quantify this distinction often results in the illusion that the analyst knows the extent and character of each type of error. In all likelihood nonrandom errors can increase without bounds. Moreover, variability of the random kind cannot be reduced with additional information, although it can be quantified. By contrast, nonrandom uncertainty, which too can be quantified with various theories, can be reduced with the acquisition of additional information.

It is historically interesting that the word *statistics* is derived from the now obsolete term *statist*, which means *an expert in statesmanship*. Statistics were the numerical facts that statists used to describe the operations of states. To many people, statistics, and other recent methods to represent uncertainty like evidence theory and fuzzy set theory, are still the facts by which politicians, newspapers, insurance sellers, and other broker occupations approach us as potential customers for their services or products! The air of sophistication that these methods provide to an issue should not be the basis for making a decision; it should be made only after a good balance has been achieved between the information content in a problem and the proper representation tool to assess it.

Popular lore suggests that the various uncertainty theories allow engineers to fool themselves in a highly sophisticated way when looking at relatively incoherent heaps of data (computational or experimental), as if this form of deception is any more palatable than just plain ignorance. All too often, scientists and engineers are led to use these theories as a crutch to explain vagaries in their models or in their data. For example, in probability applications the assumption of independent random variables is often assumed to provide a simpler method to prescribe joint probability distribution functions. An analogous assumption, called noninteractive sets, is used in fuzzy applications to develop joint membership functions from individual membership functions for sets from different universes of discourse. Should one ignore apparently aberrant information, or consider all information in the model whether or not it conforms to the engineers' preconceptions? Additional experiments to increase understanding cost money, and yet, they might increase the uncertainty by revealing conflicting information. It could best be said that statistics alone, or fuzzy sets alone, or evidence theory alone, are individually insufficient to explain many of the imponderables that people face every day. Collectively they could be very powerful. A poem by J. V. Cunningham [1971] titled ''Meditation on Statistical Method'' provides a good lesson in caution for any technologist pondering the thought that ignoring uncertainty (again, using statistics because of the era of the poem) in a problem will somehow make its solution seem more certain.

Plato despair!
We prove by norms
How numbers bear
Empiric forms,

How random wrongs
Will average right
If time be long
And error slight;

But in our hearts
Hyperbole
Curves and departs
To infinity.

Error is boundless.
Nor hope nor doubt,
Though both be groundless,
Will average out.

UNCERTAINTY AND INFORMATION

Only a small portion of the knowledge (information) for a typical problem might be regarded as certain, or deterministic. Unfortunately, the vast majority of the material taught in engineering classes is based on the presumption that the knowledge involved is deterministic. Most processes are neatly and surreptitiously reduced to closed-form algorithms – equations and formulas. When students graduate, it seems that their biggest fear upon entering the real world is ''forgetting the correct formula.'' These formulas typically describe a deterministic process, one where there is no uncertainty in the physics of the process (i.e., the right formula) and there is no uncertainty in the parameters of the process (i.e., the coefficients are known with impunity). It is only after we leave the university, it seems, that we realize we were duped in academe, and that the information we have for a particular problem virtually always contains uncertainty. For how many of our problems can we say that the information content is known absolutely, i.e., with no ignorance, no vagueness, no imprecision, no element of chance? Uncertain information can take on many different forms. There is uncertainty that arises because of complexity; for example, the complexity in the reliability network of a nuclear reactor. There is uncertainty that arises from ignorance, from various classes of randomness, from the inability to perform adequate measurements, from lack of knowledge, or from vagueness, like the fuzziness inherent in our natural language.

The nature of uncertainty in a problem is a very important point that engineers should ponder prior to their selection of an appropriate method to express the uncertainty. Fuzzy sets provide a mathematical way to represent vagueness and fuzziness in humanistic systems. For example, suppose you are teaching your child to bake cookies and you want to give instructions about when to take the cookies out of the oven. You could say to take them out when the temperature inside the cookie dough reaches 375°F, or you could advise your child to take them out when the tops of the cookies turn *light brown*. Which instruction would you give? Most likely, you would use the second of the two instructions. The first instruction is too precise to implement practically; in this case precision is not useful. The vague term *light brown* is useful in this context and can be acted upon even by a child. We all use vague terms, imprecise information, and other fuzzy data just as easily as we deal with situations governed by chance, where probability techniques are warranted and very useful. Hence, our sophisticated computational methods should be able to represent and manipulate a variety of uncertainties. Other representations of uncertainties due to ambiguity, nonspecificity, beliefs, and ignorance are introduced in Chapter 15.

FUZZY SETS AND MEMBERSHIP

The foregoing sections discuss the various elements of uncertainty. Making decisions about processes that contain nonrandom uncertainty, such as the uncertainty in natural language, has been shown to be less than perfect. The idea proposed by Lotfi Zadeh suggested that *set membership* is the key to decision making when faced with uncertainty. In fact, Zadeh made the following statement in his seminal paper of 1965:

> The notion of a fuzzy set provides a convenient point of departure for the construction of a conceptual framework which parallels in many respects the framework used in the case of ordinary sets, but is more general than the latter and, potentially, may prove to have a much wider scope of applicability, particularly in the fields of pattern classification and information processing. Essentially, such a framework provides a natural way of dealing with problems in which the source of imprecision is the absence of sharply defined criteria of class membership rather than the presence of random variables.

As an example, we can easily assess whether someone is over 6 feet tall. In a binary sense, the person either is or is not, based on the accuracy, or imprecision, of our measuring device. For example, if "tall" is a set defined as heights equal to or greater than 6 feet, a computer would not recognize an individual of height 5′11.999″ as being a member of the set "tall." But how do we assess the uncertainty in the following question: Is the person *nearly* 6 feet tall? The uncertainty in this case is due to the vagueness or ambiguity of the adjective *nearly*. A 5′11″ person could clearly be a member of the set of "nearly 6 feet tall" people. In the first situation, the uncertainty of whether a person, whose height is unknown, is 6 feet or not is binary; the person either is or is not, and we can produce a probability assessment of that prospect based on height data from many people. But the uncertainty of whether a person is nearly 6 feet is nonrandom. The degree to which the person approaches a height of 6 feet is fuzzy. In reality, "tallness" is a matter of degree and is relative. Among peoples of the Tutsi tribe in Rwanda and Burundi a height for a male of 6 feet is considered short. So, 6 feet can be tall in one context and short in another. In the real (fuzzy) world, the set of tall people can overlap with the set of not-tall people, an impossibility when one follows the precepts of classical binary logic (this is discussed in Chapter 5).

This notion of set membership, then, is central to the representation of objects within a universe by sets defined on the universe. Classical sets contain objects that satisfy precise properties of membership; fuzzy sets contain objects that satisfy imprecise properties of membership, i.e., membership of an object in a fuzzy set can be approximate. For example, the set of heights *from 5 to 7 feet* is precise (crisp); the set of heights in the region *around 6 feet* is imprecise, or fuzzy. To elaborate, suppose we have an exhaustive collection of individual elements (singletons) x, which make up a universe of information (discourse), X. Further, various combinations of these individual elements make up sets, say A, on the universe. For crisp sets an element x in the universe X is either a member of some crisp set A or not. This binary issue of membership can be represented mathematically with the indicator function,

$$\chi_A(x) = \begin{cases} 1, & x \in A \\ 0, & x \notin A \end{cases} \tag{1.1}$$

where the symbol $\chi_A(x)$ gives the indication of an unambiguous membership of element x in set A, and the symbols $\in$ and $\notin$ denote contained in and not contained in, respectively. For our example of the universe of heights of people, suppose set A is the crisp set of all people with $5.0 \leq x \leq 7.0$ feet, shown in Fig. 1.1a. A particular individual, x_1, has a height of 6.0 feet. The membership of this individual in crisp set A is equal to 1, or full membership, given symbolically as $\chi_A(x_1) = 1$. Another individual, say, x_2, has a height of 4.99 feet. The membership of this individual in set A is equal to 0, or no membership, hence $\chi_A(x_2) = 0$, also seen in Fig. 1.1a. In these cases the membership in a set is binary, either an element is a member of a set or it is not.

Zadeh extended the notion of binary membership to accommodate various "degrees of membership" on the real continuous interval [0, 1], where the endpoints of 0 and 1

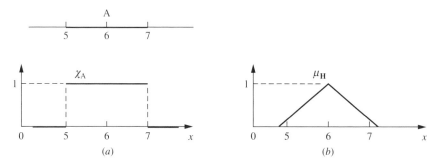

FIGURE 1.1
Height membership functions for (*a*) a crisp set A and (*b*) a fuzzy set H.

conform to no membership and full membership, respectively, just as the indicator function does for crisp sets, but where the infinite number of values in between the endpoints can represent various degrees of membership for an element x in some set on the universe. The sets on the universe X that can accommodate ''degrees of membership'' were termed by Zadeh as ''fuzzy sets.'' Continuing further on the example on heights, consider a set H consisting of heights *near 6 feet*. Since the property *near 6 feet* is fuzzy, there is not a unique membership function for H. Rather, the analyst must decide what the membership function, denoted μ_H, should look like. Plausible properties of this function might be (1) normality ($\mu_H(6) = 1$), (2) monotonicity (the closer H is to 6, the closer μ_H is to 1), and (3) symmetry (numbers equidistant from 6 should have the same value of μ_H) [Bezdek, 1993]. Such a membership function is illustrated in Fig. 1.1*b*. A key difference between crisp and fuzzy sets is their membership function; a crisp set has a unique membership function, whereas a fuzzy set can have an infinite number of membership functions to represent it. For fuzzy sets, the uniqueness is sacrificed, but flexibility is gained because the membership function can be adjusted to maximize the utility for a particular application.

James Bezdek provided one of the most lucid comparisons between crisp and fuzzy sets [Bezdek, 1993]. It bears repeating here. Crisp sets of real objects are equivalent to, and isomorphically described by, a unique membership function, such as χ_A in Fig. 1.1*a*. But there is no set-theoretic equivalent of ''real objects'' corresponding to χ_A. Fuzzy sets are always *functions*, which map a universe of objects, say X, onto the unit interval [0, 1]; that is, the fuzzy set H is the *function* μ_H that carries X into [0, 1]. Hence, *every* function that maps X onto [0, 1] is a fuzzy set. Although this statement is true in a formal mathematical sense, many functions that qualify on the basis of this definition cannot be suitable fuzzy sets. But they *become* fuzzy sets when, and only when, they match some intuitively plausible semantic description of imprecise properties of the objects in X.

The membership function embodies the mathematical representation of membership in a set, and the notation used throughout this text for a fuzzy set is a set symbol with a tilde underscore, say A̰, where the functional mapping is given by

$$\mu_{\underline{A}}(x) \in [0, 1] \tag{1.2}$$

and the symbol $\mu_{\underline{A}}(x)$ is the degree of membership of element x in fuzzy set A̰. Therefore, $\mu_{\underline{A}}(x)$ is a value on the unit interval that measures the degree to which element x belongs to fuzzy set A̰; equivalently, $\mu_{\underline{A}}(x) =$ degree to which $x \in$ A̰.

CHANCE VERSUS FUZZINESS

Suppose you are a basketball recruiter and are looking for a ''very tall'' player for the center position on a men's team. One of your information sources tells you that a hot prospect in Oregon has a 95% chance of being over 7 feet tall. Another of your sources tells you that a good player in Louisiana has a high membership in the set of ''very tall'' people. The problem with the information from the first source is that it is a probabilistic quantity. There is a 5% chance that the Oregon player is not over 7 feet tall and could, conceivably, be someone of extremely short stature. The second source of information would, in this case, contain a different kind of uncertainty for the recruiter; it is a fuzziness due to the linguistic qualifier ''very tall'' because if the player turned out to be less than 7 feet tall there is still a high likelihood that he would be quite tall.

Another example involves a personal choice. Suppose you are seated at a table on which rest two glasses of liquid. The liquid in the first glass is described to you as having a 95% chance of being healthful and good. The liquid in the second glass is described as having a 0.95 membership in the class of ''healthful and good'' liquids. Which glass would you select, keeping in mind that the first glass has a 5% chance of being filled with nonhealthful liquids, including poisons [Bezdek, 1993]?

What philosophical distinction can be made regarding these two forms of information? Suppose we are allowed to measure the basketball players' heights and test the liquids in the glasses. The prior probability of 0.95 in each case becomes a posterior probability of 1.0 or 0; that is, either the player is or is not over 7 feet tall and the liquid is either benign or not. However, the membership value of 0.95, which measures the extent to which the player's height is over 7 feet, or the drinkability of the liquid is ''healthful and good,'' remains 0.95 after measuring or testing. These two examples illustrate very clearly the difference in the information content between chance and fuzziness.

This brings us to the clearest distinction between fuzziness and chance. *Fuzziness describes the lack of distinction of an event, whereas chance describes the uncertainty in the occurrence of the event.* The event will occur or not occur; but is the description of the event clear enough to measure its occurrence or nonoccurrence? Consider the following geometric questions, which serve to illustrate our ability to address fuzziness (lack of distinctiveness) with certain mathematical relations. The geometric shape in Fig. 1.2a can resemble a disk, a cylinder, or a rod, depending on the aspect ratio of d/h. For $d/h \ll 1$ the shape of the object approaches a long rod; in fact, as $d/h \to 0$ the shape approaches a line. For $d/h \gg 1$ the object approaches the shape of a flat disk; as $d/h \to \infty$ the object approaches a circular area. For other values of this aspect ratio, e.g., for $d/h \approx 1$, the shape is typical of what we would call a ''right circular cylinder.'' See Fig. 1.2b.

The geometric shape in Fig. 1.3a is an ellipse, with parameters a and b. Under what conditions of these two parameters will a general elliptic shape become a circle? Mathematically, we know that a circle results when $a/b = 1$, and hence this is a specific, crisp geometric shape. We know that when $a/b \ll 1$ or $a/b \gg 1$ we clearly have an elliptic shape; and as $a/b \to \infty$, a line segment results. Using this knowledge, we can

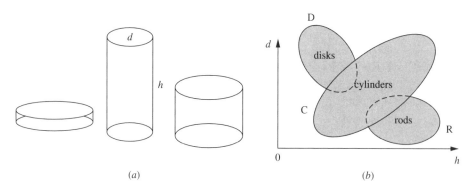

(a)   (b)

FIGURE 1.2
Relationship between (a) mathematical terms and (b) fuzzy linguistic terms.

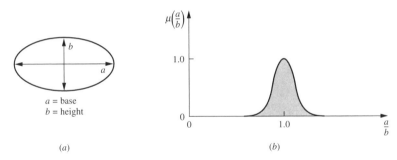

FIGURE 1.3

Figure 1.3 The (*a*) geometric shape and (*b*) membership function for an approximate circle.

develop a description of the membership function to describe the geometric set we call an "approximate circle." Without a theoretical development, the following expression describing a Gaussian curve (for this membership function all points on the real line have nonzero membership; this can be an advantage or disadvantage depending on the nature of the problem) offers a good approximation for the membership function of the fuzzy set "approximate circle," denoted $\underset{\sim}{C}$:

$$\mu_{\underset{\sim}{C}}\left(\frac{a}{b}\right) = \exp\left[-3\left(\frac{a}{b} - 1\right)^2\right] \tag{1.3}$$

Figure 1.3*b* is a plot of the membership function given in Eq. (1.3). As the elliptic ratio a/b approaches a value of unity, the membership value approaches unity; for $a/b = 1$ we have an unambiguous circle. As $a/b \to \infty$ or $a/b \to 0$, we get a line segment; hence, the membership of the shape in the fuzzy set $\underset{\sim}{C}$ approaches zero, because a line segment is not very similar in shape to a circle. In Fig. 1.3*b* we see that as we get farther from $a/b = 1$ our membership in the set " approximate circle" gets smaller and smaller. All values of a/b that have a membership value of unity are called the prototypes; in this case $a/b = 1$ is the only prototype for the set "approximate circle," because at this value it is exactly a circle.

Suppose we were to place in a bag a large number of generally elliptical two-dimensional shapes and ask the question: What is the probability of randomly selecting an "approximate circle" from the bag? We could not answer this question without first assessing the two different kinds of uncertainty. First, we would have to address the issue of fuzziness in the meaning of the term "approximate circle" by selecting a value of membership, above which we would be willing to call the shape an approximate circle; for example, any shape with a membership value above 0.9 in the fuzzy set "approximate circle" would be considered a circle. Second, we would have to know the proportion of the shapes in the bag that have membership values above 0.9. The first issue is one of assessing fuzziness and the second relates to the frequencies required to address questions of chance.

SETS AS POINTS IN HYPERCUBES

There is an interesting geometric analog for illustrating the idea of set membership [Kosko, 1992]. Heretofore we have described a fuzzy set $\underset{\sim}{A}$ defined on a universe X. For a universe with only one element, the membership function is defined on the unit interval [0,1]; for a two-element universe, the membership function is defined on the unit square; and for a three-element universe, the membership function is defined on the unit cube. All of these situations are shown in Fig. 1.4. For a universe of n elements we define the membership on the unit hypercube, $I^n = [0, 1]^n$.

The endpoints on the unit interval in Fig. 1.4a, and the vertices of the unit square and the unit cube in Figs. 1.4b and 1.4c, respectively, represent the possible crisp subsets, or collections, of the elements of the universe in each figure. This collection of possible crisp (nonfuzzy) subsets of elements in a universe constitutes the power set of the universe. For example, in Fig. 1.4c the universe comprises three elements, $X = \{x_1, x_2, x_3\}$. The point $(0, 0, 1)$ represents the crisp subset in 3-space, where x_1 and x_2 have no membership and

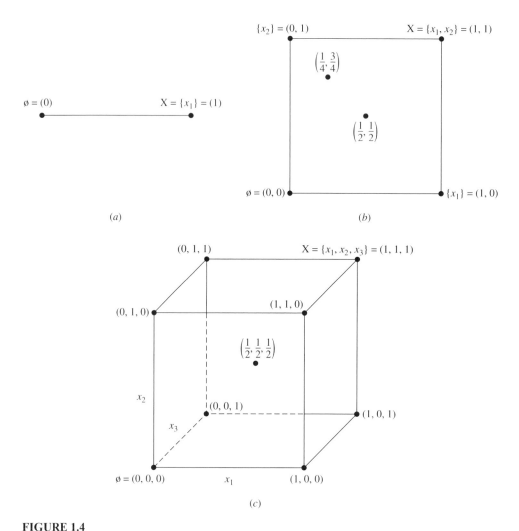

FIGURE 1.4
''Sets as points'' [Kosko, 1992]: (a) one-element universe, (b) two-element universe, (c) three-element universe.

element x_3 has full membership, i.e., the subset $\{x_3\}$; the point $(1, 1, 0)$ is the crisp subset where x_1 and x_2 have full membership and element x_3 has no membership, i.e., the subset $\{x_1, x_2\}$; and so on for the other six vertices in Fig. 1.4c. In general, there are 2^n subsets in the power set of a universe with n elements; geometrically, this universe is represented by a hypercube in n-space, where the 2^n vertices represent the collection of sets constituting the power set. Two points in the diagrams bear special note, as illustrated in Fig. 1.4c. In this figure the point $(1, 1, 1)$, where all elements in the universe have full membership, is called the whole set, X, and the point $(0, 0, 0)$, where all elements in the universe have no membership, is called the null set, $\emptyset$.

The centroids of each of the diagrams in Fig. 1.4 represent single points where the membership value for each element in the universe equals $\frac{1}{2}$. For example, the point $(\frac{1}{2}, \frac{1}{2})$ in Fig. 1.4b is in the midpoint of the square. This midpoint in each of the three figures is a special point – it is the set of maximum "fuzziness." A membership value of $\frac{1}{2}$ indicates that the element belongs to the fuzzy set as much as it does not – that is, it holds equal membership in both the fuzzy set and its complement. In a geometric sense, this point is the location in the space that is farthest from any of the vertices and yet equidistant from all of them. In fact, all points interior to the vertices of the spaces represented in Fig. 1.4 represent fuzzy sets, where the membership value of each variable is a number between 0 and 1. For example, in Fig. 1.4b, the point $(\frac{1}{4}, \frac{3}{4})$ represents a fuzzy set where variable x_1 has a 0.25 degree of membership in the set and variable x_2 has a 0.75 degree of membership in the set. It is obvious by inspection of the diagrams in Fig. 1.4 that, although the number of subsets in the power set is enumerated by the 2^n vertices, the number of fuzzy sets on the universe is infinite, as represented by the infinite number of points on the interior of each space.

Finally, the vertices of the cube in Fig. 1.4c are the identical coordinates found in the value set, V{P(X)}, developed in Example 2.4 of the next chapter.

SUMMARY

This chapter has discussed models with essentially two different kinds of information: fuzzy membership functions, which represent similarities of objects to nondistinct properties, and probabilities, which provide knowledge about relative frequencies. The value of either of these kinds of information in making decisions is a matter of preference; popular, but controversial, contrary views have been offered [Ross et al., 2002]. Fuzzy models are *not* replacements for probability models. As seen in Fig. 1.1, every crisp set is fuzzy, but the converse does not hold. The idea that crisp sets are special forms of fuzzy sets was illustrated graphically in the section on sets as points, where crisp sets are represented by the vertices of a unit hypercube. All other points within the unit hypercube, or along its edges, are graphically analogous to a fuzzy set. Fuzzy models are not that different from more familiar models. Sometimes they work better, and sometimes they do not. After all, the efficacy of a model in solving a problem should be the only criterion used to judge that model. Lately, a growing body of evidence suggests that fuzzy approaches to real problems are an effective alternative to previous, traditional methods.

REFERENCES

Arciszewski, T, Sauer, T., and Schum, D. (2003). "Conceptual designing: Chaos-based approach," *J. Intell. Fuzzy Syst.*, vol. 13, pp. 45–60.

Arthur, W. B. (1994). ''Inductive reasoning and bounded rationality,'' *Am. Econ. Rev.*, vol. 84, pp. 406–411.

Barrow, J. (2000). *The Universe That Discovered Itself*, Oxford University Press, Oxford, UK.

Ben-Haim, Y. (2001). *Information Gap Decision Theory: Decisions Under Severe Uncertainty*, Series on Decision and Risk, Academic Press, London.

Bezdek, J. (1993). ''Editorial: Fuzzy models – What are they, and why?'' *IEEE Trans. Fuzzy Syst.*, vol. 1, pp. 1–5.

Black, M. (1937). ''Vagueness: An exercise in logical analysis,'' *Int. J. Gen. Syst.*, vol. 17, pp. 107–128.

Bower, G. and Hilgard, E. (1981). *Theories of Learning*, Prentice Hall, Englewood Cliffs, NJ.

Cunningham, J. (1971). *The collected poems and epigrams of J. V. Cunningham*, Swallow Press, Chicago.

de Finetti, B. (1974). *Theory of Probability*, John Wiley & Sons, New York.

Drake, S. (1957). *Discoveries and Opinions of Galileo*, Anchor Books, New York, pp. 162–169.

Klir, G. and Wierman, M. (1996). *Uncertainty-Based Information*, Physica-Verlag, Heidelberg.

Klir, G. and Yuan, B. (1995). *Fuzzy Sets and Fuzzy Logic: Theory and Applications*, Prentice Hall, Upper Saddle River, NJ.

Kolata, G. (1991). ''Math problem, long baffling, slowly yields,'' *New York Times*, March 12, p. C1.

Kosko, B. (1992). *Neural Networks and Fuzzy Systems*, Prentice Hall, Englewood Cliffs, NJ.

Kosko, B. (1994). ''Fuzzy systems as universal approximators,'' *IEEE Trans. Comput.*, vol. 43, no. 11.

Lindley, D. (1987). ''Comment: A tale of two wells,'' *Stat. Sci.*, vol. 2, pp. 38–40.

Quine, W. (1981). *Theories and Things*, Harvard University Press, Cambridge, MA.

Ross, T., Booker, J., and Parkinson, W. J. (2002). *Fuzzy Logic and Probability Applications: Bridging the Gap*, Society for Industrial and Applied Mathematics, Philadelphia, PA.

Stigler, S. (1986). ''Laplace's 1774 memoir on inverse probability,'' *Stat. Sci.*, vol. 1, pp. 359–378.

Stone, M. H. (1937). ''Applications of the theory of Boolean rings to general topology,'' *Trans. Am. Math. Soc.*, vol. 41, pp. 375–481, esp. pp. 453–481.

Weierstrass, K. (1885). ''Mathematische Werke, Band 3, Abhandlungen III,'' pp. 1–37, esp. p. 5, *Sitzungsber. koniglichen preuss. Akad. Wiss.*, July 9 and July 30.

Ying, H., Ding, Y., Li, S., and Shao, S. (1999). ''Fuzzy systems as universal approximators,'' *IEEE Trans. Syst., Man, Cybern – Part A: Syst. Hum.*, vol. 29, no. 5.

Zadeh, L. (1965). ''Fuzzy sets,'' *Inf. Control*, vol. 8, pp. 338–353.

Zadeh, L. (1973). ''Outline of a new approach to the analysis of complex systems and decision processes,'' *IEEE Trans. Syst., Man, Cybern.*, vol. SMC-3, pp. 28–44.

Zadeh, L. (1995). ''Discussion: Probability theory and fuzzy logic are complementary rather than competitive,'' *Technometrics*, vol. 37, pp. 271–276.

Zadeh, L. (2002). *Forward to Fuzzy Logic and Probability Applications: Bridging the Gap*, Society for Industrial and Applied Mathematics, Philadelphia, PA.

PROBLEMS

1.1. Develop a reasonable membership function for the following fuzzy sets based on height measured in centimeters:

    (*a*) ''Tall''

    (*b*) ''Short''

    (*c*) ''Not short''

1.2. Develop a membership function for laminar and turbulent flow for a typical flat plate with a sharp leading edge in a typical air stream. Transition usually takes place between Reynolds numbers (Re) of 2×10^5 and 3×10^6. An Re of 5×10^5 is usually considered the point of turbulent flow for this situation.

1.3. Develop a reasonable membership function for a square, based on the geometric properties of a rectangle. For this problem use L as the length of the longer side and l as the length of the smaller side.

1.4. For the cylindrical shapes shown in Fig. 1.2, develop a membership function for each of the following shapes using the ratio d/h, and discuss the reason for any overlapping among the three membership functions:

(a) Rod

(b) Cylinder

(c) Disk

1.5. The question of whether a glass of water is half-full or half-empty is an age-old philosophical issue. Such descriptions of the volume of liquid in a glass depend on the state of mind of the person asked the question. Develop membership functions for the fuzzy sets ''half-full,'' ''full,'' ''empty,'' and ''half-empty'' using percent volume as the element of information. Assume the maximum volume of water in the glass is V_0. Discuss whether the terms ''half-full'' and ''half-empty'' should have identical membership functions. Does your answer solve this ageless riddle?

1.6. Landfills are a primary source of methane, a greenhouse gas. Landfill caps, called biocaps, are designed to minimize methane emission by maximizing methane oxidation; these caps are classified as ''best'' if they are capable of oxidizing 80% of the methane that originates in the landfill's interior. Complete oxidation was found to be difficult to establish. Develop a reasonable membership function of the percent methane oxidation to show the performance of the biocap and emissions of methane.

1.7. Industry A discharges wastewater into a nearby river. Wastewater contains high biological oxygen demand (BOD) and other inorganic contaminants. The discharge rate of rivers and wastewater is constant through the year. From research, it has been found that BOD values not exceeding 250 mg/L do not cause any harmful effect to aquatic ecosystems. However, BOD values higher than 250 mg/L have significant impact. Draw both a crisp and fuzzy membership function to show the effects of the BOD value on aquatic ecosystems.

1.8. A fuzzy set for a major storm event in Calgary, Alberta, could be described as a rainstorm in a subdivision that raised the level of the storm-water pond to within 70% of its design capacity. The membership function for a major storm set could be described as having full membership when 70% of the pond volume has been reached but varying from zero membership to full membership at 40% capacity and 70% capacity, respectively. Draw a typical membership function as it is described.

1.9. In Alberta a waste is orally toxic if it has an oral toxicity (LD_{50}) of less than 5000 mg/kg. Develop and draw a crisp and a fuzzy membership function for the oral toxicity.

1.10. Using the ratios of internal angles or sides of a hexagon, draw the membership diagrams for ''regular'' and ''irregular'' hexagons.

1.11. Develop algorithms for the following membership function shapes:

(a) Triangular

(b) Gamma function

(c) Quadratic S-function

(d) Trapezoid

(e) Gaussian

(f) Exponential-wire function

1.12. In soil mechanics soils are classified based on the size of their particles as clay, silt, or sand (clays having the smallest particles and sands having the largest particles). Though silts have larger particles than clays, it is often difficult to distinguish between these two soil types; silts and sands present the same problem. Develop membership functions for these three soil types, in terms of their grain size, S.

1.13. A sour natural gas stream is contacted with a lean amine solution in an absorber; this allows the amine to remove the sour component in the natural gas producing a rich amine solution and a "sales gas" which is the natural gas with a much lower sour gas concentration than the feed gas as shown in Fig. P1.13. Concentrations above C_2, which is the pipeline specification for sour gas concentration, are considered to have full membership in the set of "high concentrations." A concentration below C_1, which is the lower limit of sour gas concentration that can be detected by analysis instrumentation, is considered to have full membership in the set of "low concentrations." Sketch a membership function for the absorber "sales gas" sour gas concentration as a function of concentration, C; show the points C_1 and C_2.

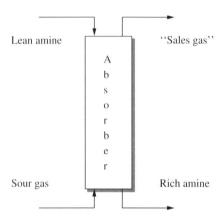

FIGURE P1.13

1.14. A circular column loaded axially is assumed to be eccentric when the load is acting at 5% of the axis, depending on the diameter of the column, d as shown in Fig. P1.14. We have the following conditions: $e/d = 0.05$ eccentric; $e/d < 0.05$ not-very-eccentric; $e/d > 0.05$, very eccentric. Develop a membership function for "eccentricity" on the scale of e/d ratios.

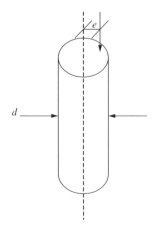

FIGURE P1.14

1.15. A rectangular sheet of perimeter $2L + 2h$ is to be rolled into a cylinder with height, h. Classify the cylinder as a function of the rectangular sheet as being a rod, cylinder, or disk by developing membership functions for these shapes.

1.16. Enumerate the nonfuzzy subsets of the power set for a universe with $n = 4$ elements, i.e., $X = \{x_1, x_2, x_3, x_4\}$, and indicate their coordinates as the vertices on a 4-cube.

1.17. Probability distributions can be shown to exist on certain planes that intersect the regions shown in Fig. 1.4. Draw the points, lines, and planes on which probability distributions exist for the one-, two-, and three-parameter cases shown in Fig. 1.4.

CHAPTER
2

CLASSICAL SETS AND FUZZY SETS

Philosophical objections may be raised by the logical implications of building a mathematical structure on the premise of fuzziness, since it seems (at least superficially) necessary to require that an object be or not be an element of a given set. From an aesthetic viewpoint, this may be the most satisfactory state of affairs, but to the extent that mathematical structures are used to model physical actualities, it is often an unrealistic requirement.... Fuzzy sets have an intuitively plausible philosophical basis. Once this is accepted, analytical and practical considerations concerning fuzzy sets are in most respects quite orthodox.

James Bezdek
Professor, Computer Science, 1981

As alluded to in Chapter 1, the *universe of discourse* is the universe of all available information on a given problem. Once this universe is defined we are able to define certain events on this information space. We will describe sets as mathematical abstractions of these events and of the universe itself. Figure 2.1*a* shows an abstraction of a universe of discourse, say X, and a crisp (classical) set A somewhere in this universe. A classical set is defined by *crisp* boundaries, i.e., there is no uncertainty in the prescription or location of the boundaries of the set, as shown in Fig. 2.1*a* where the boundary of crisp set A is an unambiguous line. A fuzzy set, on the other hand, is prescribed by vague or ambiguous properties; hence its boundaries are ambiguously specified, as shown by the fuzzy boundary for set A̰ in Fig. 2.1*b*.

In Chapter 1 we introduced the notion of set membership, from a one-dimensional viewpoint. Figure 2.1 again helps to explain this idea, but from a two-dimensional perspective. Point *a* in Fig. 2.1*a* is clearly a member of crisp set A; point *b* is unambiguously *not* a member of set A. Figure 2.1*b* shows the vague, ambiguous boundary of a fuzzy set A̰ on the same universe X: the shaded boundary represents the boundary region of A̰. In the central (unshaded) region of the fuzzy set, point *a* is clearly a full member of the set.

Fuzzy Logic with Engineering Applications, Second Edition   T. J. Ross
© 2004 John Wiley & Sons, Ltd   ISBNs: 0-470-86074-X (HB); 0-470-86075-8 (PB)

X (Universe of discourse)   X (Universe of discourse)

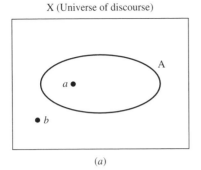
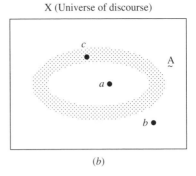

(a)   (b)

FIGURE 2.1
Diagrams for (a) crisp set boundary and (b) fuzzy set boundary.

Outside the boundary region of the fuzzy set, point b is clearly not a member of the fuzzy set. However, the membership of point c, which is on the boundary region, is ambiguous. If complete membership in a set (such as point a in Fig. 2.1b) is represented by the number 1, and no-membership in a set (such as point b in Fig. 2.1b) is represented by 0, then point c in Fig. 2.1b must have some intermediate value of membership (partial membership in fuzzy set A) on the interval [0,1]. Presumably the membership of point c in A approaches a value of 1 as it moves closer to the central (unshaded) region in Fig. 2.1b of A, and the membership of point c in A approaches a value of 0 as it moves closer to leaving the boundary region of A.

In this chapter, the precepts and operations of fuzzy sets are compared with those of classical sets. Several good books are available for reviewing this basic material [see for example, Dubois and Prade, 1980; Klir and Folger, 1988; Zimmermann, 1991; Klir and Yuan, 1995]. Fuzzy sets embrace virtually all (with one exception, as will be seen) of the definitions, precepts, and axioms that define classical sets. As indicated in Chapter 1, crisp sets are a special form of fuzzy sets; they are sets without ambiguity in their membership (i.e., they are sets with unambiguous boundaries). It will be shown that fuzzy set theory is a mathematically rigorous and comprehensive set theory useful in characterizing concepts (sets) with natural ambiguity. It is instructive to introduce fuzzy sets by first reviewing the elements of classical (crisp) set theory.

CLASSICAL SETS

Define a universe of discourse, X, as a collection of objects all having the same characteristics. The individual elements in the universe X will be denoted as x. The features of the elements in X can be discrete, countable integers or continuous valued quantities on the real line. Examples of elements of various universes might be as follows:

The clock speeds of computer CPUs
The operating currents of an electronic motor
The operating temperature of a heat pump (in degrees Celsius)
The Richter magnitudes of an earthquake
The integers 1 to 10

Most real-world engineering processes contain elements that are real and non-negative. The first four items just named are examples of such elements. However, for purposes of modeling, most engineering problems are simplified to consider only integer values of the elements in a universe of discourse. So, for example, computer clock speeds might be measured in integer values of megahertz and heat pump temperatures might be measured in integer values of degrees Celsius. Further, most engineering processes are simplified to consider only finite-sized universes. Although Richter magnitudes may not have a theoretical limit, we have not historically measured earthquake magnitudes much above 9; this value might be the upper bound in a structural engineering design problem. As another example, suppose you are interested in the stress under one leg of the chair in which you are sitting. You might argue that it is possible to get an infinite stress on one leg of the chair by sitting in the chair in such a manner that only one leg is supporting you and by letting the area of the tip of that leg approach zero. Although this is theoretically possible, in reality the chair leg will either buckle elastically as the tip area becomes very small or yield plastically and fail because materials that have infinite strength have not yet been developed. Hence, choosing a universe that is discrete and finite or one that is continuous and infinite is a modeling choice; the choice does not alter the characterization of sets defined on the universe. If elements of a universe are continuous, then sets defined on the universe will be composed of continuous elements. For example, if the universe of discourse is defined as all Richter magnitudes up to a value of 9, then we can define a set of "destructive magnitudes," which might be composed (1) of all magnitudes greater than or equal to a value of 6 in the crisp case or (2) of all magnitudes "approximately 6 and higher" in the fuzzy case.

A useful attribute of sets and the universes on which they are defined is a metric known as the cardinality, or the cardinal number. The total number of elements in a universe X is called its cardinal number, denoted n_x, where x again is a label for individual elements in the universe. Discrete universes that are composed of a countably finite collection of elements will have a finite cardinal number; continuous universes comprised of an infinite collection of elements will have an infinite cardinality. Collections of elements within a universe are called sets, and collections of elements within sets are called subsets. Sets and subsets are terms that are often used synonymously, since any set is also a subset of the universal set X. The collection of all possible sets in the universe is called the *whole set*.

For crisp sets A and B consisting of collections of some elements in X, the following notation is defined:

$x \in X$	$\Rightarrow$	x belongs to X
$x \in A$	$\Rightarrow$	x belongs to A
$x \notin A$	$\Rightarrow$	x does not belong to A

For sets A and B on X, we also have

$A \subset B$	$\Rightarrow$	A is fully contained in B (if $x \in A$, then $x \in B$)
$A \subseteq B$	$\Rightarrow$	A is contained in or is equivalent to B
$(A \leftrightarrow B)$	$\Rightarrow$	$A \subseteq B$ and $B \subseteq A$ (A is equivalent to B)

We define the null set, $\emptyset$, as the set containing no elements, and the whole set, X, as the set of all elements in the universe. The null set is analogous to an impossible event, and the whole set is analogous to a certain event. All possible sets of X constitute a special set called the power set, P(X). For a specific universe X, the power set P(X) is enumerated in the following example.

Example 2.1. We have a universe comprised of three elements, $X = \{a, b, c\}$, so the cardinal number is $n_x = 3$. The power set is

$$P(X) = \{\emptyset, \{a\}, \{b\}, \{c\}, \{a, b\}, \{a, c\}, \{b, c\}, \{a, b, c\}\}$$

The cardinality of the power set, denoted $n_{P(X)}$, is found as

$$n_{P(X)} = 2^{n_X} = 2^3 = 8$$

Note that if the cardinality of the universe is infinite, then the cardinality of the power set is also infinity, i.e., $n_X = \infty \Rightarrow n_{P(X)} = \infty$.

Operations on Classical Sets

Let A and B be two sets on the universe X. The union between the two sets, denoted $A \cup B$, represents all those elements in the universe that reside in (or belong to) the set A, the set B, or both sets A and B. (This operation is also called the *logical or*; another form of the union is the *exclusive or* operation. The *exclusive or* will be described in Chapter 5.) The intersection of the two sets, denoted $A \cap B$, represents all those elements in the universe X that simultaneously reside in (or belong to) both sets A and B. The complement of a set A, denoted $\overline{A}$, is defined as the collection of all elements in the universe that do not reside in the set A. The difference of a set A with respect to B, denoted $A \mid B$, is defined as the collection of all elements in the universe that reside in A and that do not reside in B simultaneously. These operations are shown below in set-theoretic terms.

Union	$A \cup B = \{x \mid x \in A \text{ or } x \in B\}$	(2.1)
Intersection	$A \cap B = \{x \mid x \in A \text{ and } x \in B\}$	(2.2)
Complement	$\overline{A} = \{x \mid x \notin A, x \in X\}$	(2.3)
Difference	$A \mid B = \{x \mid x \in A \text{ and } x \notin B\}$	(2.4)

These four operations are shown in terms of Venn diagrams in Figs. 2.2–2.5.

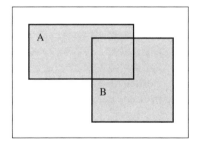

FIGURE 2.2
Union of sets A and B (logical or).

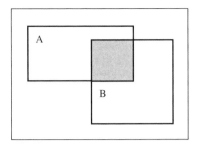

FIGURE 2.3
Intersection of sets A and B.

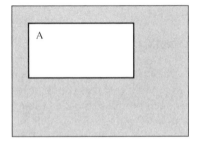

FIGURE 2.4
Complement of set A.

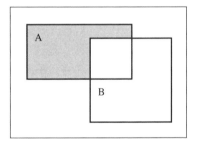

FIGURE 2.5
Difference operation A | B.

Properties of Classical (Crisp) Sets

Certain properties of sets are important because of their influence on the mathematical manipulation of sets. The most appropriate properties for defining classical sets and showing their similarity to fuzzy sets are as follows:

Commutativity
$$A \cup B = B \cup A$$
$$A \cap B = B \cap A \tag{2.5}$$

Associativity	$A \cup (B \cup C) = (A \cup B) \cup C$	
	$A \cap (B \cap C) = (A \cap B) \cap C$	(2.6)
Distributivity	$A \cup (B \cap C) = (A \cup B) \cap (A \cup C)$	
	$A \cap (B \cup C) = (A \cap B) \cup (A \cap C)$	(2.7)
Idempotency	$A \cup A = A$	
	$A \cap A = A$	(2.8)
Identity	$A \cup \emptyset = A$	
	$A \cap X = A$	
	$A \cap \emptyset = \emptyset$	(2.9)
	$A \cup X = X$	
Transitivity	If $A \subseteq B$ and $B \subseteq C$, then $A \subseteq C$	(2.10)
Involution	$\overline{\overline{A}} = A$	(2.11)

The double-cross-hatched area in Fig. 2.6 is a Venn diagram example of the associativity property for intersection, and the double-cross-hatched areas in Figs. 2.7 and 2.8

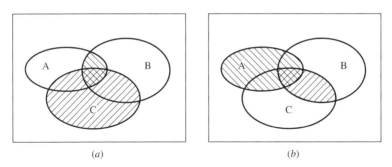

(a)   (b)

FIGURE 2.6
Venn diagrams for (a) $(A \cap B) \cap C$ and (b) $A \cap (B \cap C)$.

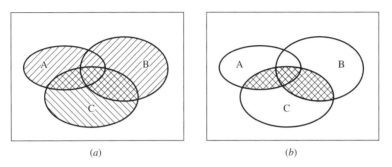

(a)   (b)

FIGURE 2.7
Venn diagrams for (a) $(A \cup B) \cap C$ and (b) $(A \cap C) \cup (B \cap C)$.

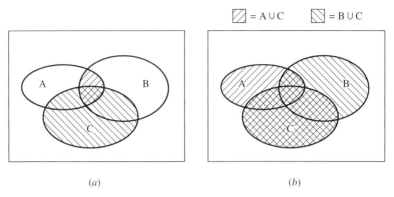

(a)   (b)

FIGURE 2.8
Venn diagrams for (a) $(A \cap B) \cup C$ and (b) $(A \cup C) \cap (B \cup C)$.

are Venn diagram examples of the distributivity property for various combinations of the intersection and union properties.

Two special properties of set operations are known as the *excluded middle axioms* and *De Morgan's principles*. These properties are enumerated here for two sets A and B. The *excluded middle axioms* are very important because these are the only set operations described here that are *not* valid for both classical sets and fuzzy sets. There are two excluded middle axioms (given in Eqs. (2.12)). The first, called the *axiom of the excluded middle,* deals with the union of a set A and its complement; the second, called the *axiom of contradiction*, represents the intersection of a set A and its complement.

Axiom of the excluded middle $\qquad\qquad A \cup \overline{A} = X$ $\qquad\qquad\qquad\qquad$ (2.12a)

Axiom of the contradiction $\qquad\qquad A \cap \overline{A} = \emptyset$ $\qquad\qquad\qquad\qquad$ (2.12b)

De Morgan's principles are important because of their usefulness in proving tautologies and contradictions in logic, as well as in a host of other set operations and proofs. De Morgan's principles are displayed in the shaded areas of the Venn diagrams in Figs. 2.9 and 2.10 and described mathematically in Eq. (2.13).

$$\overline{A \cap B} = \overline{A} \cup \overline{B} \qquad\qquad\qquad\qquad (2.13a)$$

$$\overline{A \cup B} = \overline{A} \cap \overline{B} \qquad\qquad\qquad\qquad (2.13b)$$

In general, De Morgan's principles can be stated for n sets, as provided here for events, E_i:

$$\overline{E_1 \cup E_2 \cup \cdots \cup E_n} = \overline{E_1} \cap \overline{E_2} \cap \cdots \cap \overline{E_n} \qquad\qquad (2.14a)$$

$$\overline{E_1 \cap E_2 \cap \cdots \cap E_n} = \overline{E_1} \cup \overline{E_2} \cup \cdots \cup \overline{E_n} \qquad\qquad (2.14b)$$

From the general equations, Eqs. (2.14), for De Morgan's principles we get a duality relation: the complement of a union or an intersection is equal to the intersection or union, respectively, of the respective complements. This result is very powerful in dealing with

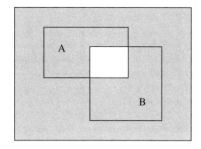

FIGURE 2.9
De Morgan's principle $(\overline{A \cap B})$.

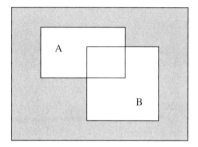

FIGURE 2.10
De Morgan's principle $(\overline{A \cup B})$.

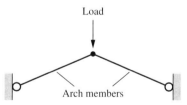

FIGURE 2.11
A two-member arch.

set structures since we often have information about the complement of a set (or event), or the complement of combinations of sets (or events), rather than information about the sets themselves.

Example 2.2. A shallow arch consists of two slender members as shown in Fig. 2.11. If either member fails, then the arch will collapse. If E_1 = survival of member 1 and E_2 = survival of member 2, then survival of the arch = $E_1 \cap E_2$, and, conversely, collapse of the arch = $\overline{E_1 \cap E_2}$. Logically, collapse of the arch will occur if either of the members fails, i.e., when $\overline{E_1} \cup \overline{E_2}$. Therefore,

$$\overline{E_1 \cap E_2} = \overline{E_1} \cup \overline{E_2}$$

which is an illustration of De Morgan's principle.

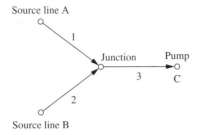

FIGURE 2.12
Hydraulic hose system.

As Eq. (2.14) suggests, De Morgan's principles are very useful for compound events, as illustrated in the following example.

Example 2.3. For purposes of safety, the fluid supply for a hydraulic pump C in an airplane comes from two redundant source lines, A and B. The fluid is transported by high-pressure hoses consisting of branches 1, 2, and 3, as shown in Fig. 2.12. Operating specifications for the pump indicate that either source line alone is capable of supplying the necessary fluid pressure to the pump. Denote E_1 = failure of branch 1, E_2 = failure of branch 2, and E_3 = failure of branch 3. Then insufficient pressure to operate the pump would be caused by $(E_1 \cap E_2) \cup E_3$, and sufficient pressure would be the complement of this event. Using De Morgan's principles, we can calculate the condition of sufficient pressure to be

$$\overline{(E_1 \cap E_2) \cup E_3} = (\overline{E_1} \cup \overline{E_2}) \cap \overline{E_3}$$

in which $(\overline{E_1} \cup \overline{E_2})$ means the availability of pressure at the junction, and $\overline{E_3}$ means the absence of failure in branch 3.

Mapping of Classical Sets to Functions

Mapping is an important concept in relating set-theoretic forms to function-theoretic representations of information. In its most general form it can be used to map elements or subsets on one universe of discourse to elements or sets in another universe. Suppose X and Y are two different universes of discourse (information). If an element x is contained in X and corresponds to an element y contained in Y, it is generally termed a mapping from X to Y, or $f : X \rightarrow Y$. As a mapping, the characteristic (indicator) function χ_A is defined by

$$\chi_A(x) = \begin{cases} 1, & x \in A \\ 0, & x \notin A \end{cases} \qquad (2.15)$$

where χ_A expresses "membership" in set A for the element x in the universe. This membership idea is a mapping from an element x in universe X to one of the two elements in universe Y, i.e., to the elements 0 or 1, as shown in Fig. 2.13.

For any set A defined on the universe X, there exists a function-theoretic set, called a value set, denoted V(A), under the mapping of the characteristic function, χ. By convention, the null set $\emptyset$ is assigned the membership value 0 and the whole set X is assigned the membership value 1.

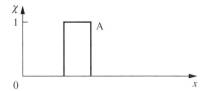

FIGURE 2.13
Membership function is a mapping for crisp set A.

Example 2.4. Continuing with the example (Example 2.1) of a universe with three elements, $X = \{a, b, c\}$, we desire to map the elements of the power set of X, i.e., P(X), to a universe, Y, consisting of only two elements (the characteristic function),

$$Y = \{0, 1\}$$

As before, the elements of the power set are enumerated.

$$P(X) = \{\emptyset, \{a\}, \{b\}, \{c\}, \{a, b\}, \{b, c\}, \{a, c\}, \{a, b, c\}\}$$

Thus, the elements in the value set V(A) as determined from the mapping are

$$V\{P(X)\} = \{\{0, 0, 0\}, \{1, 0, 0\}, \{0, 1, 0\}, \{0, 0, 1\}, \{1, 1, 0\}, \{0, 1, 1\}, \{1, 0, 1\}, \{1, 1, 1\}\}$$

For example, the third subset in the power set P(X) is the element b. For this subset there is no a, so a value of 0 goes in the first position of the data triplet; there is a b, so a value of 1 goes in the second position of the data triplet; and there is no c, so a value of 0 goes in the third position of the data triplet. Hence, the third subset of the value set is the data triplet, $\{0, 1, 0\}$, as already seen. The value set has a graphical analog that is described in Chapter 1 in the section "Sets as Points in Hypercubes."

Now, define two sets, A and B, on the universe X. The union of these two sets in terms of function-theoretic terms is given as follows (the symbol $\vee$ is the maximum operator and $\wedge$ is the minimum operator):

Union $\qquad A \cup B \longrightarrow \chi_{A \cup B}(x) = \chi_A(x) \vee \chi_B(x) = \max(\chi_A(x), \chi_B(x))$ $\qquad$ (2.16)

The intersection of these two sets in function-theoretic terms is given by

Intersection $\qquad A \cap B \longrightarrow \chi_{A \cap B}(x) = \chi_A(x) \wedge \chi_B(x) = \min(\chi_A(x), \chi_B(x))$ $\qquad$ (2.17)

The complement of a single set on universe X, say A, is given by

Complement $\qquad\qquad\qquad \overline{A} \longrightarrow \chi_{\overline{A}}(x) = 1 - \chi_A(x)$ $\qquad\qquad\qquad$ (2.18)

For two sets on the same universe, say A and B, if one set (A) is contained in another set (B), then

Containment $\qquad\qquad\qquad A \subseteq B \longrightarrow \chi_A(x) \leq \chi_B(x)$ $\qquad\qquad\qquad$ (2.19)

Function-theoretic operators for union and intersection (other than maximum and minimum, respectively) are discussed in the literature [Gupta and Qi, 1991].

FUZZY SETS

In classical, or crisp, sets the transition for an element in the universe between membership and nonmembership in a given set is abrupt and well-defined (said to be "crisp"). For an element in a universe that contains fuzzy sets, this transition can be gradual. This transition among various degrees of membership can be thought of as conforming to the fact that the boundaries of the fuzzy sets are vague and ambiguous. Hence, membership of an element from the universe in this set is measured by a function that attempts to describe vagueness and ambiguity.

A fuzzy set, then, is a set containing elements that have varying degrees of membership in the set. This idea is in contrast with classical, or crisp, sets because members of a crisp set would not be members unless their membership was full, or complete, in that set (i.e., their membership is assigned a value of 1). Elements in a fuzzy set, because their membership need not be complete, can also be members of other fuzzy sets on the same universe.

Elements of a fuzzy set are mapped to a universe of *membership values* using a function-theoretic form. As mentioned in Chapter 1 (Eq. (1.2)), fuzzy sets are denoted in this text by a set symbol with a tilde understrike; so, for example, $\underset{\sim}{A}$ would be the *fuzzy set A*. This function maps elements of a fuzzy set $\underset{\sim}{A}$ to a real numbered value on the interval 0 to 1. If an element in the universe, say x, is a member of fuzzy set $\underset{\sim}{A}$, then this mapping is given by Eq. (1.2), or $\mu_{\underset{\sim}{A}}(x) \in [0,1]$. This mapping is shown in Fig. 2.14 for a typical fuzzy set.

A notation convention for fuzzy sets when the universe of discourse, X, is discrete and finite, is as follows for a fuzzy set $\underset{\sim}{A}$:

$$\underset{\sim}{A} = \left\{ \frac{\mu_{\underset{\sim}{A}}(x_1)}{x_1} + \frac{\mu_{\underset{\sim}{A}}(x_2)}{x_2} + \cdots \right\} = \left\{ \sum_i \frac{\mu_{\underset{\sim}{A}}(x_i)}{x_i} \right\} \tag{2.20}$$

When the universe, X, is continuous and infinite, the fuzzy set $\underset{\sim}{A}$ is denoted by

$$\underset{\sim}{A} = \left\{ \int \frac{\mu_{\underset{\sim}{A}}(x)}{x} \right\} \tag{2.21}$$

In both notations, the horizontal bar is not a quotient but rather a delimiter. The numerator in each term is the membership value in set $\underset{\sim}{A}$ associated with the element of the universe

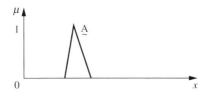

FIGURE 2.14
Membership function for fuzzy set $\underset{\sim}{A}$.

indicated in the denominator. In the first notation, the summation symbol is not for algebraic summation but rather denotes the collection or aggregation of each element; hence the "+" signs in the first notation are not the algebraic "add" but are an aggregation or collection operator. In the second notation the integral sign is not an algebraic integral but a continuous function-theoretic aggregation operator for continuous variables. Both notations are due to Zadeh [1965].

Fuzzy Set Operations

Define three fuzzy sets A, B, and C on the universe X. For a given element x of the universe, the following function-theoretic operations for the set-theoretic operations of union, intersection, and complement are defined for A, B, and C on X:

Union $$\mu_{A \cup B}(x) = \mu_A(x) \vee \mu_B(x) \tag{2.22}$$

Intersection $$\mu_{A \cap B}(x) = \mu_A(x) \wedge \mu_B(x) \tag{2.23}$$

Complement $$\mu_{\bar{A}}(x) = 1 - \mu_A(x) \tag{2.24}$$

Venn diagrams for these operations, extended to consider fuzzy sets, are shown in Figs. 2.15–2.17. The operations given in Eqs. (2.22)–(2.24) are known as the *standard fuzzy operations*. There are many other fuzzy operations, and a discussion of these is given later in this chapter.

Any fuzzy set A defined on a universe X is a subset of that universe. Also by definition, just as with classical sets, the membership value of any element x in the null set $\emptyset$ is 0,

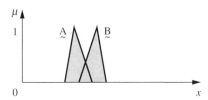

FIGURE 2.15
Union of fuzzy sets A and B.

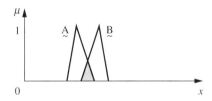

FIGURE 2.16
Intersection of fuzzy sets A and B.

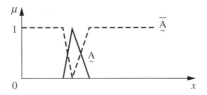

FIGURE 2.17
Complement of fuzzy set $\underset{\sim}{A}$.

and the membership value of any element x in the whole set X is 1. Note that the null set and the whole set are not fuzzy sets in this context (no tilde understrike). The appropriate notation for these ideas is as follows:

$$\underset{\sim}{A} \subseteq X \Rightarrow \mu_{\underset{\sim}{A}}(x) \leq \mu_X(x) \qquad (2.25a)$$

$$\text{For all } x \in X, \ \mu_\emptyset(x) = 0 \qquad (2.25b)$$

$$\text{For all } x \in X, \ \mu_X(x) = 1 \qquad (2.25c)$$

The collection of all fuzzy sets and fuzzy subsets on X is denoted as the fuzzy power set P(X). It should be obvious, based on the fact that all fuzzy sets can overlap, that the cardinality, $n_{P(X)}$, of the fuzzy power set is infinite; that is, $n_{P(X)} = \infty$.

De Morgan's principles for classical sets also hold for fuzzy sets, as denoted by these expressions:

$$\overline{\underset{\sim}{A} \cap \underset{\sim}{B}} = \overline{\underset{\sim}{A}} \cup \overline{\underset{\sim}{B}} \qquad (2.26a)$$

$$\overline{\underset{\sim}{A} \cup \underset{\sim}{B}} = \overline{\underset{\sim}{A}} \cap \overline{\underset{\sim}{B}} \qquad (2.26b)$$

As enumerated before, all other operations on classical sets also hold for fuzzy sets, except for the excluded middle axioms. These two axioms do not hold for fuzzy sets since they do not form part of the basic axiomatic structure of fuzzy sets (see Appendix A); since fuzzy sets can overlap, a set and its complement can also overlap. The *excluded middle axioms,* extended for fuzzy sets, are expressed by

$$\underset{\sim}{A} \cup \overline{\underset{\sim}{A}} \neq X \qquad (2.27a)$$

$$\underset{\sim}{A} \cap \overline{\underset{\sim}{A}} \neq \emptyset \qquad (2.27b)$$

Extended Venn diagrams comparing the *excluded middle axioms* for classical (crisp) sets and fuzzy sets are shown in Figs. 2.18 and 2.19, respectively.

Properties of Fuzzy Sets

Fuzzy sets follow the same properties as crisp sets. Because of this fact and because the membership values of a crisp set are a subset of the interval [0,1], classical sets can be

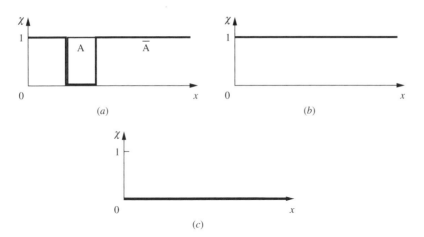

FIGURE 2.18
Excluded middle axioms for crisp sets. (*a*) Crisp set A and its complement; (*b*) crisp $A \cup \overline{A} = X$ (axiom of excluded middle); (*c*) crisp $A \cap \overline{A} = \emptyset$ (axiom of contradiction).

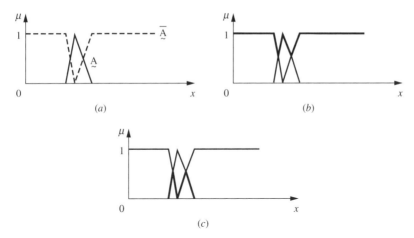

FIGURE 2.19
Excluded middle axioms for fuzzy sets. (*a*) Fuzzy set $\underset{\sim}{A}$ and its complement; (*b*) fuzzy $\underset{\sim}{A} \cup \overline{\underset{\sim}{A}} \neq X$ (axiom of excluded middle); (*c*) fuzzy $A \cap \overline{A} \neq \emptyset$ (axiom of contradiction).

thought of as a special case of fuzzy sets. Frequently used properties of fuzzy sets are listed below.

Commutativity
$$\underset{\sim}{A} \cup \underset{\sim}{B} = \underset{\sim}{B} \cup \underset{\sim}{A}$$
$$\underset{\sim}{A} \cap \underset{\sim}{B} = \underset{\sim}{B} \cap \underset{\sim}{A} \tag{2.28}$$

Associativity
$$\underset{\sim}{A} \cup \left(\underset{\sim}{B} \cup \underset{\sim}{C} \right) = \left(\underset{\sim}{A} \cup \underset{\sim}{B} \right) \cup \underset{\sim}{C}$$
$$\underset{\sim}{A} \cap \left(\underset{\sim}{B} \cap \underset{\sim}{C} \right) = \left(\underset{\sim}{A} \cap \underset{\sim}{B} \right) \cap \underset{\sim}{C} \tag{2.29}$$

Distributivity
$$A \cup \left(B \cap C \right) = \left(A \cup B \right) \cap \left(A \cup C \right)$$

$$A \cap \left(B \cup C \right) = \left(A \cap B \right) \cup \left(A \cap C \right) \tag{2.30}$$

Idempotency
$$A \cup A = A \quad \text{and} \quad A \cap A = A \tag{2.31}$$

Identity
$$A \cup \emptyset = A \quad \text{and} \quad A \cap X = A$$

$$A \cap \emptyset = \emptyset \quad \text{and} \quad A \cup X = X \tag{2.32}$$

Transitivity
$$\text{If } A \subseteq B \text{ and } B \subseteq C, \text{ then } A \subseteq C \tag{2.33}$$

Involution
$$\overline{\overline{A}} = A \tag{2.34}$$

Example 2.5. Consider a simple hollow shaft of approximately 1 m radius and wall thickness $1/(2\pi)$ m. The shaft is built by stacking a ductile section, D, of the appropriate cross section over a brittle section, B, as shown in Fig. 2.20. A downward force P and a torque T are simultaneously applied to the shaft. Because of the dimensions chosen, the nominal shear stress on any element in the shaft is T (pascals) and the nominal vertical component of stress in the shaft is P (pascals). We also assume that the failure properties of both B and D are not known with any certainty.

   We define the fuzzy set A to be the region in (P,T) space for which material D is "safe" using as a metric the failure function $\mu_A = f([P^2 + 4T^2]^{1/2})$. Similarly, we define the set B to be the region in (P,T) space for which material B is "safe," using as a metric the failure function $\mu_B = g(P - \beta|T|)$, where β is an assumed material parameter. The functions f and g will, of course, be membership functions on the interval $[0, 1]$. Their exact specification is not important at this point. What is useful, however, prior to specifying f and g, is to discuss the basic set operations in the context of this problem. This discussion is summarized below:

1. $A \cup B$ is the set of loadings for which one expects that either material B or material D will be "safe."
2. $A \cap B$ is the set of loadings for which one expects that both material B and material D are "safe."
3. $\overline{A}$ and $\overline{B}$ are the sets of loadings for which material D and material B are unsafe, respectively.

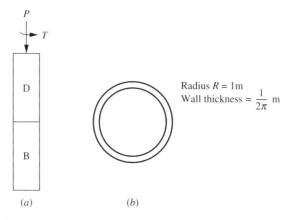

(a)   (b)

FIGURE 2.20
(*a*) Axial view and (*b*) cross-sectional view of example hollow shaft.

4. $A \mid B$ is the set of loadings for which the ductile material is safe but the brittle material is in jeopardy.

5. $B \mid A$ is the set of loadings for which the brittle material is safe but the ductile material is in jeopardy.

6. De Morgan's principle $\overline{A \cap B} = \overline{A} \cup \overline{B}$ asserts that the loadings that are not safe with respect to both materials are the union of those that are unsafe with respect to the brittle material with those that are unsafe with respect to the ductile material.

7. De Morgan's principle $\overline{A \cup B} = \overline{A} \cap \overline{B}$ asserts that the loads that are safe for neither material D nor material B are the intersection of those that are unsafe for material D with those that are unsafe for material B.

To illustrate these ideas numerically, let's say we have two discrete fuzzy sets, namely,

$$A = \left\{ \frac{1}{2} + \frac{0.5}{3} + \frac{0.3}{4} + \frac{0.2}{5} \right\} \quad \text{and} \quad B = \left\{ \frac{0.5}{2} + \frac{0.7}{3} + \frac{0.2}{4} + \frac{0.4}{5} \right\}$$

We can now calculate several of the operations just discussed (membership for element 1 in both A and B is implicitly 0):

Complement $$\overline{A} = \left\{ \frac{1}{1} + \frac{0}{2} + \frac{0.5}{3} + \frac{0.7}{4} + \frac{0.8}{5} \right\}$$

$$\overline{B} = \left\{ \frac{1}{1} + \frac{0.5}{2} + \frac{0.3}{3} + \frac{0.8}{4} + \frac{0.6}{5} \right\}$$

Union $$A \cup B = \left\{ \frac{1}{2} + \frac{0.7}{3} + \frac{0.3}{4} + \frac{0.4}{5} \right\}$$

Intersection $$A \cap B = \left\{ \frac{0.5}{2} + \frac{0.5}{3} + \frac{0.2}{4} + \frac{0.2}{5} \right\}$$

Difference $$A \mid B = A \cap \overline{B} = \left\{ \frac{0.5}{2} + \frac{0.3}{3} + \frac{0.3}{4} + \frac{0.2}{5} \right\}$$

$$B \mid A = B \cap \overline{A} = \left\{ \frac{0}{2} + \frac{0.5}{3} + \frac{0.2}{4} + \frac{0.4}{5} \right\}$$

De Morgan's principles $$\overline{A \cup B} = \overline{A} \cap \overline{B} = \left\{ \frac{1}{1} + \frac{0}{2} + \frac{0.3}{3} + \frac{0.7}{4} + \frac{0.6}{5} \right\}$$

$$\overline{A \cap B} = \overline{A} \cup \overline{B} = \left\{ \frac{1}{1} + \frac{0.5}{2} + \frac{0.5}{3} + \frac{0.8}{4} + \frac{0.8}{5} \right\}$$

Example 2.6. Continuing from the chemical engineering case described in Problem 1.13 of Chapter 1, suppose the selection of an appropriate analyzer to monitor the "sales gas" sour gas concentration is important. This selection process can be complicated by the fact that one type of analyzer, say A, does not provide an average suitable pressure range but it does give a borderline value of instrument dead time; in contrast another analyzer, say B, may give a good value of process dead time but a poor pressure range. Suppose for this problem we consider three analyzers: A, B and C.
Let

$$P = \left\{ \frac{0.7}{A} + \frac{0.3}{B} + \frac{0.9}{C} \right\}$$

represent the fuzzy set showing the pressure range suitability of analyzers A, B, and C (a membership of 0 is not suitable, a value of 1 is excellent).

Also let

$$\underset{\sim}{OT} = \left\{ \frac{0.5}{A} + \frac{0.9}{B} + \frac{0.4}{C} \right\}$$

represent the fuzzy set showing the instrument dead time suitability of analyzers A, B, and C (again, 0 is not suitable and 1 is excellent).

$\overline{\underset{\sim}{P}}$ and $\overline{\underset{\sim}{OT}}$ will show the analyzers that are not suitable for pressure range and instrument dead time, respectively:

$$\overline{\underset{\sim}{P}} = \left\{ \frac{0.3}{A} + \frac{0.7}{B} + \frac{0.1}{C} \right\} \text{ and } \overline{\underset{\sim}{OT}} = \left\{ \frac{0.5}{A} + \frac{0.1}{B} + \frac{0.6}{C} \right\},$$

$$\text{therefore } \overline{\underset{\sim}{P}} \cap \overline{\underset{\sim}{OT}} = \left\{ \frac{0.3}{A} + \frac{0.1}{B} + \frac{0.1}{C} \right\}$$

$\underset{\sim}{P} \cup \underset{\sim}{OT}$ will show which analyzer is most suitable in either category:

$$\underset{\sim}{P} \cup \underset{\sim}{OT} = \left\{ \frac{0.7}{A} + \frac{0.9}{B} + \frac{0.9}{C} \right\}$$

$\underset{\sim}{P} \cap \underset{\sim}{OT}$ will show which analyzer is suitable in both categories:

$$\underset{\sim}{P} \cap \underset{\sim}{OT} = \left\{ \frac{0.5}{A} + \frac{0.3}{B} + \frac{0.4}{C} \right\}$$

Example 2.7. One of the crucial manufacturing operations associated with building the external fuel tank for the Space Shuttle involves the spray-on foam insulation (SOFI) process, which combines two critical component chemicals in a spray gun under high pressure and a precise temperature and flow rate. Control of these parameters to *near* setpoint values is crucial for satisfying a number of important specification requirements. Specification requirements consist of aerodynamic, mechanical, chemical, and thermodynamic properties.

Fuzzy characterization techniques could be employed to enhance initial screening experiments; for example, to determine the critical values of both flow and temperature. The true levels can only be approximated in the real world. If we target a low flow rate for 48 lb/min, it may be 38 to 58 lb/min. Also, if we target a high temperature for 135°F, it may be 133 to 137°F.

How the imprecision of the experimental setup influences the variabilities of key process end results could be modeled using fuzzy set methods, e.g., high flow with high temperature, low flow with low temperature, etc. Examples are shown in Fig. 2.21, for low flow rate and high temperature.

Suppose we have a fuzzy set for flow, normalized on a universe of integers [1, 2, 3, 4, 5] and a fuzzy set for temperature, normalized on a universe of integers [1, 2, 3, 4], as follows:

$$\underset{\sim}{F} = \left\{ \frac{0}{1} + \frac{0.5}{2} + \frac{1}{3} + \frac{0.5}{4} + \frac{0}{5} \right\} \text{ and } \underset{\sim}{D} = \left\{ \frac{0}{2} + \frac{1}{3} + \frac{0}{4} \right\}$$

Further suppose that we are interested in how flow and temperature are related in a pairwise sense; we could take the intersection of these two sets. A three-dimensional image should be constructed when we take the union or intersection of sets from two different universes. For example, the intersection of $\underset{\sim}{F}$ and $\underset{\sim}{D}$ is given in Fig. 2.22. The idea of combining membership functions from two different universes in an orthogonal form, as indicated in Fig. 2.22, is associated with what is termed *noninteractive fuzzy sets*, and this will be described below.

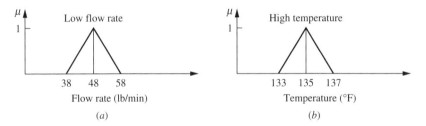

FIGURE 2.21
Foam insulation membership function for (*a*) low flow rate and (*b*) high temperature.

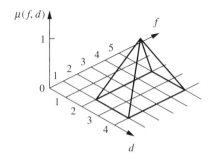

FIGURE 2.22
Three-dimensional image of the intersection of two fuzzy sets, i.e., $\underset{\sim}{F} \cap \underset{\sim}{D}$.

Noninteractive Fuzzy Sets

Later in the text, in Chapter 8 on simulation, we will make reference to noninteractive fuzzy sets. Noninteractive sets in fuzzy set theory can be thought of as being analogous to independent events in probability theory. They always arise in the context of relations or in *n*-dimensional mappings [Zadeh, 1975; Bandemer and Näther, 1992]. A noninteractive fuzzy set can be defined as follows. Suppose we define a fuzzy set $\underset{\sim}{A}$ on the Cartesian space $X = X_1 \times X_2$. The set $\underset{\sim}{A}$ is separable into two *noninteractive* fuzzy sets, called its orthogonal projections, if and only if

$$\underset{\sim}{A} = Pr_{X_1}(\underset{\sim}{A}) \times Pr_{X_2}(\underset{\sim}{A}) \tag{2.35a}$$

where

$$\mu_{Pr_{X_1}(\underset{\sim}{A})}(x_1) = \max_{x_2 \in X_2} \mu_{\underset{\sim}{A}}(x_1, x_2), \quad \forall x_1 \in X_1 \tag{2.35b}$$

$$\mu_{Pr_{X_2}(\underset{\sim}{A})}(x_2) = \max_{x_1 \in X_1} \mu_{\underset{\sim}{A}}(x_1, x_2), \quad \forall x_2 \in X_2 \tag{2.35c}$$

are the membership functions for the projections of $\underset{\sim}{A}$ on universes X_1 and X_2, respectively. Hence, if Eq. (2.35a) holds for a fuzzy set, the membership functions $\mu_{Pr_{X_1}(\underset{\sim}{A})}(x_1)$

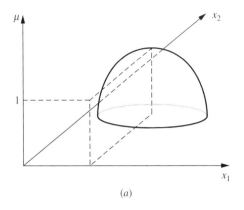
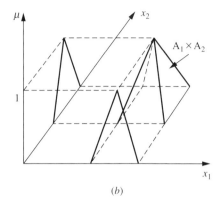

(a) (b)

FIGURE 2.23
Fuzzy sets: (a) interactive and (b) noninteractive.

and $\mu_{\mathrm{Pr}_{x_2}(A)}(x_2)$ describe noninteractive fuzzy sets, i.e., the projections are noninteractive fuzzy sets.

Separability or noninteractivity of fuzzy set $\underset{\sim}{A}$ describes a kind of independence of the components (x_1 and x_2): $\underset{\sim}{A}$ can be uniquely reconstructed by its projections; the components of the fuzzy set $\underset{\sim}{A}$ can vary without consideration of the other components. As an example, the two-dimensional planar fuzzy set shown in Fig. 2.22 comprises noninteractive fuzzy sets ($\underset{\sim}{F}$ and $\underset{\sim}{D}$), because it was constructed by the Cartesian product (intersection in this case) of the two fuzzy sets $\underset{\sim}{F}$ and $\underset{\sim}{D}$, whereas a two-dimensional fuzzy set comprising curved surfaces will be nonseparable, i.e., its components will be *interactive*. Interactive components are characterized by the fact that variation of one component depends on the values of the other components. See Fig. 2.23.

Alternative Fuzzy Set Operations

The operations on fuzzy sets listed as Eqs. (2.22–2.24) are called the *standard fuzzy operations*. These operations are the same as those for classical sets, when the range of membership values is restricted to the unit interval. However, these standard fuzzy operations are not the only operations that can be applied to fuzzy sets. For each of the three standard operations, there exists a broad class of functions whose members can be considered fuzzy generalizations of the standard operations. Functions that qualify as fuzzy intersections and fuzzy unions are usually referred to in the literature as *t-norms* and *t-conorms* (or *s-norms*), respectively [e.g., Klir and Yuan, 1995; Klement et al., 2000]. These t-norms and t-conorms are so named because they were originally introduced as *triangular norms* and *triangular conorms*, respectively, by Menger [1942] in his study of statistical metric spaces.

The standard fuzzy operations have special significance when compared to all of the other *t-norms* and *t-conorms*. The standard fuzzy intersection, min operator, when applied to a fuzzy set produces the largest membership value of all the t-norms, and the standard fuzzy union, max operator, when applied to a fuzzy set produces the smallest membership value of all the t-conorms. These features of the standard fuzzy intersection and union are significant because they both prevent the compounding of errors in the operands [Klir and Yuan, 1995]. Most of the alternative norms lack this significance.

Aggregation operations on fuzzy sets are operations by which several fuzzy sets are combined in a desirable way to produce a single fuzzy set. For example, suppose a computer's performance in three test trials is described as excellent, very good, and nominal, and each of these linguistic labels is represented by a fuzzy set on the universe [0, 100]. Then, a useful aggregation operation would produce a meaningful expression, in terms of a single fuzzy set, of the overall performance of the computer. The standard fuzzy intersections and unions qualify as aggregation operations on fuzzy sets and, although they are defined for only two arguments, the fact that they have a property of associativity provides a mechanism for extending their definitions to three or more arguments. Other common aggregation operations, such as *averaging operations* and *ordered weighted averaging operations*, can be found in the literature [see Klir and Yuan, 1995]. The averaging operations have their own range that happens to fill the gap between the largest intersection (the min operator) and the smallest union (the max operator). These averaging operations on fuzzy sets have no counterparts in classical set theory and, because of this, extensions of fuzzy sets into fuzzy logic allows for the latter to be much more expressive in natural categories revealed by empirical data or required by intuition [Belohlavek et al., 2002].

SUMMARY

In this chapter we have developed the basic definitions for, properties of, and operations on crisp sets and fuzzy sets. It has been shown that the only basic axioms not common to both crisp and fuzzy sets are the two excluded middle axioms; however, these axioms are not part of the axiomatic structure of fuzzy set theory (see Appendix A). All other operations detailed here are common to both crisp and fuzzy sets; however, other operations such as aggregation and averaging operators that are allowed in fuzzy sets have no counterparts in classical set theory. For many situations in reasoning, the excluded middle axioms present constraints on reasoning (see Chapters 5 and 15). Aside from the difference of set membership being an infinite-valued idea as opposed to a binary-valued quantity, fuzzy sets are handled and treated in the same mathematical form as are crisp sets. The principle of *noninteractivity* between sets was introduced and is analogous to the assumption of independence in probability modeling. Noninteractive fuzzy sets will become a necessary idea in fuzzy systems simulation when inputs from a variety of universes are aggregated in a collective sense to propagate an output; Chapters 5 and 8 will discuss this propagation process in more detail. Finally, it was pointed out that there are many other operations, called *norms*, that can be used to extend fuzzy intersections, unions, and complements, but such extensions are beyond the scope of this text.

REFERENCES

Bandemer, H. and Näther, W. (1992). *Fuzzy data analysis*, Kluwer Academic, Dordrecht.
Belohlavek, R., Klir, G., Lewis, H., and Way, E. (2002). ''On the capability of fuzzy set theory to represent concepts'', *Int. J. Gen. Syst.*, vol. 31, pp. 569–585.

Bezdek, J. (1981). *Pattern recognition with fuzzy objective function algorithms*, Plenum Press, New York.

Dubois, D. and Prade, H. (1980). *Fuzzy sets and systems, theory and applications*, Academic Press, New York.

Gupta, M. and Qi, J. (1991). "Theory of T-norms and fuzzy inference methods," *Fuzzy Sets Syst.*, vol. 40, pp. 431–450.

Klement, E., Mesiar, R., and Pap, E. (2000). *Triangular Norms*, Kluwer Academic, Boston.

Klir, G. and Folger, T. (1988). *Fuzzy sets, uncertainty, and information*, Prentice Hall, Englewood Cliffs, NJ.

Klir, G. and Yuan, B. (1995). *Fuzzy sets and fuzzy logic: theory and applications*, Prentice Hall, Upper Saddle River, NJ.

Kosko, B. (1992). *Neural networks and fuzzy systems*, Prentice Hall, Englewood Cliffs, NJ.

Menger, K. (1942). "Statistical metrics," *Proc. Natl. Acad. Sci.*, vol. 28, pp. 535–537.

Zadeh, L. (1965). "Fuzzy sets", *Inf. Control*, vol. 8, pp. 338–353.

Zadeh, L. (1975). "The concept of a linguistic variable and its application to approximate reasoning, Parts 1, 2, and 3," *Inf. Sci.*, vol. 8, pp. 199–249, 301–357; vol. 9, pp. 43–80.

Zimmermann, H. (1991). *Fuzzy set theory and its applications*, 2nd ed., Kluwer Academic, Dordrecht, Germany.

PROBLEMS

2.1. Typical membership functions for laminar and turbulent flow for a flat plate with a sharp leading edge in a typical air stream are shown in Fig. P2.1. Transition between laminar and turbulent flow usually takes place between Reynolds numbers of 2×10^5 and 3×10^6. An $Re = 5 \times 10^5$ is usually considered the point of turbulent flow for this situation. Find the intersection and union for the two flows.

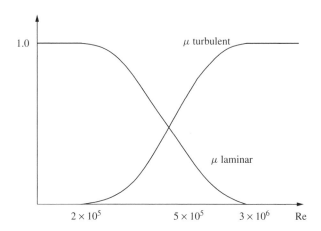

FIGURE P2.1

2.2. In neighborhoods there may be several storm-water ponds draining to a single downstream trunk sewer. In this neighborhood the city monitors all ponds for height of water caused by storm events. For two storms (labeled A and B) identified as being significant based on rainfall data collected at the airport, determine the corresponding performance of the neighborhood storm-water ponds.

Suppose the neighborhood has five ponds, i.e., $X = [1, 2, 3, 4, 5]$, and suppose that significant pond storage membership is 1.0 for any pond that is 70% or more to full depth. For storm $\underset{\sim}{A}$ the pond performance set is

$$\underset{\sim}{A} = \left\{ \frac{0.6}{1} + \frac{0.3}{2} + \frac{0.9}{3} + \frac{1}{4} + \frac{1}{5} \right\}$$

For storm $\underset{\sim}{B}$ the pond performance set is

$$\underset{\sim}{B} = \left\{ \frac{0.8}{1} + \frac{0.4}{2} + \frac{0.9}{3} + \frac{0.7}{4} + \frac{1}{5} \right\}$$

(a) To assess the impacts on pond performance suppose only two ponds can be monitored due to budget constraints. Moreover, data from the storms indicate that there may be a difference in thunderburst locations around this neighborhood. Which two of the five ponds should be monitored?

(b) Determine the most conservative estimate of pond performance (i.e., find $\underset{\sim}{A} \cup \underset{\sim}{B}$).

2.3. Methane biofilters can be used to oxidize methane using biological activities. It has become necessary to compare performance of two test columns, A and B. The methane outflow level at the surface, in nondimensional units of $X = \{50, 100, 150, 200\}$, was detected and is tabulated below against the respective methane inflow into each test column. The following fuzzy sets represent the test columns:

$$\underset{\sim}{A} = \left\{ \frac{0.15}{50} + \frac{0.25}{100} + \frac{0.5}{150} + \frac{0.7}{200} \right\} \quad \underset{\sim}{B} = \left\{ \frac{0.2}{50} + \frac{0.3}{100} + \frac{0.6}{150} + \frac{0.65}{200} \right\}$$

Calculate the union, intersection, and the difference for the test columns.

2.4. Given a set of measurements of the magnetic field near the surface of a person's head, we want to locate the electrical activity in the person's brain that would give rise to the measured magnetic field. This is called the inverse problem, and it has no unique solution. One approach is to model the electrical activity as dipoles and attempt to find one to four dipoles that would produce a magnetic field closely resembling the measured field. For this problem we will model the procedure a neuroscientist would use in attempting to fit a measured magnetic field using either one or two dipoles. The scientist uses a reduced chi-square statistic to determine how good the fit is. If $R = 1.0$, the fit is exact. If $R \geq 3$, the fit is bad. Also a two-dipole model must have a lower R than a one-dipole model to give the same amount of confidence in the model. The range of R will be taken as $R = \{1.0, 1.5, 2.0, 2.5, 3.0\}$ and we define the following fuzzy sets for D_1 = the one-dipole model and D_2 = the two-dipole model:

$$\underset{\sim}{D}_1 = \left\{ \frac{1}{1.0} + \frac{0.75}{1.5} + \frac{0.3}{2.0} + \frac{0.15}{2.5} + \frac{0}{3.0} \right\}$$

$$\underset{\sim}{D}_2 = \left\{ \frac{1}{1.0} + \frac{0.6}{1.5} + \frac{0.2}{2.0} + \frac{0.1}{2.5} + \frac{0}{3.0} \right\}$$

For these two fuzzy sets, find the following:

(a) $\underset{\sim}{D}_1 \cup \underset{\sim}{D}_2$

(b) $\underset{\sim}{D}_1 \cap \underset{\sim}{D}_2$

(c) $\overline{\underset{\sim}{D}_1}$

(d) $\overline{\underset{\sim}{D}_2}$

(e) $\underset{\sim}{D_1} \mid \underset{\sim}{D_2}$

(f) $\overline{\underset{\sim}{D_1} \cup \underset{\sim}{D_2}}$

2.5. In determining corporate profitability, many construction companies must make decisions based upon the particular client's spending habits, such as the amount the client spends and their capacity for spending. Many of these attributes are fuzzy. A client which spends a "large amount" is considered to be "profitable" to the construction company. A "large" amount of spending is a fuzzy variable, as is a "profitable" return. These two fuzzy sets should have some overlap, but they should not be defined on an identical range.

$$\underset{\sim}{A} = \{\text{"large" spenders}\}$$

$$\underset{\sim}{B} = \{\text{"profitable" clients}\}$$

For the two fuzzy sets shown in Fig. P2.5, find the following properties graphically:

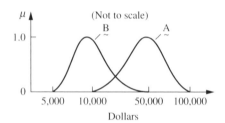

FIGURE P2.5

(a) $\underset{\sim}{A} \cup \underset{\sim}{B}$: all clients deemed profitable or who are large spenders.

(b) $\underset{\sim}{A} \cap \underset{\sim}{B}$: all clients deemed profitable and large spenders.

(c) $\overline{\underset{\sim}{A}}$ and $\overline{\underset{\sim}{B}}$: those clients (i) deemed not profitable, and (ii) deemed not large spenders (separately).

(d) $\overline{\underset{\sim}{A}} \mid \overline{\underset{\sim}{B}}$: entities deemed profitable clients, but not large spenders.

(e) $\overline{\underset{\sim}{A} \cup \underset{\sim}{B}} = \overline{\underset{\sim}{A}} \cap \overline{\underset{\sim}{B}}$ (De Morgan's principle).

2.6. Suppose you are a soils engineer. You wish to track the movement of soil particles under strain in an experimental apparatus that allows viewing of the soil motion. You are building pattern recognition software to allow a computer to monitor and detect the motions. However, there are two difficulties in "teaching" your software to view the motion: (1) the tracked particle can be occluded by another particle; (2) your segmentation algorithm can be inadequate. One way to handle the occlusion is to assume that the area of the occluded particle is smaller than the area of the unoccluded particle. Therefore, when the area is changing you know that the particle is occluded. However, the segmentation algorithm also makes the area of the particle shrink if the edge detection scheme in the algorithm cannot do a good job because of poor illumination in the experimental apparatus. In other words, the area of the particle becomes small as a result of either occlusion or bad segmentation. You define two fuzzy sets on a universe of nondimensional particle areas, $X = [0, 1, 2, 3, 4]$: $\underset{\sim}{A}$ is a fuzzy set whose elements belong to the occlusion, and $\underset{\sim}{B}$ is a fuzzy set whose elements belong to inadequate segmentation. Let

$$\underset{\sim}{A} = \left\{ \frac{0.1}{0} + \frac{0.4}{1} + \frac{1}{2} + \frac{0.3}{3} + \frac{0.2}{4} \right\}$$

$$\underset{\sim}{B} = \left\{ \frac{0.2}{0} + \frac{0.5}{1} + \frac{1}{2} + \frac{0.4}{3} + \frac{0.1}{4} \right\}$$

Find the following:

(a) $\underset{\sim}{A} \cup \underset{\sim}{B}$

(b) $\underset{\sim}{A} \cap \underset{\sim}{B}$

(c) $\overline{\underset{\sim}{A}}$

(d) $\overline{\underset{\sim}{B}}$

(e) $\overline{\underset{\sim}{A} \cap \underset{\sim}{B}}$

(f) $\overline{\underset{\sim}{A} \cup \underset{\sim}{B}}$

2.7. You are asked to select an implementation technology for a numerical processor. Computation throughput is directly related to clock speed. Assume that all implementations will be in the same family (e.g., CMOS). You are considering whether the design should be implemented using medium-scale integration (MSI) with discrete parts, field-programmable array parts (FPGA), or multichip modules (MCM). Define the universe of potential clock frequencies as $X = \{1, 10, 20, 40, 80, 100\}$ MHz; and define MSI, FPGA, and MCM as fuzzy sets of clock frequencies that should be implemented in each of these technologies, where the following table defines their membership values:

Clock frequency, MHz	MSI	FPGA	MCM
1	1	0.3	0
10	0.7	1	0
20	0.4	1	0.5
40	0	0.5	0.7
80	0	0.2	1
100	0	0	1

Representing the three sets as MSI $= \underset{\sim}{M}$, FPGA $= \underset{\sim}{F}$, and MCM $= \underset{\sim}{C}$, find the following:

(a) $\underset{\sim}{M} \cup \underset{\sim}{F}$

(b) $\underset{\sim}{M} \cap \underset{\sim}{F}$

(c) $\overline{\underset{\sim}{M}}$

(d) $\overline{\underset{\sim}{F}}$

(e) $\underset{\sim}{C} \cap \overline{\underset{\sim}{F}}$

(f) $\overline{\underset{\sim}{M} \cap \underset{\sim}{C}}$

2.8. We want to compare two sensors based upon their detection levels and gain settings. For a universe of discourse of gain settings, $X = \{0, 20, 40, 60, 80, 100\}$, the sensor detection levels for the monitoring of a standard item provides typical membership functions to represent the detection levels for each of the sensors; these are given below in standard discrete form:

$$\underset{\sim}{S_1} = \left\{ \frac{0}{0} + \frac{0.5}{20} + \frac{0.65}{40} + \frac{0.85}{60} + \frac{1.0}{80} + \frac{1.0}{100} \right\}$$

$$\underset{\sim}{S_2} = \left\{ \frac{0}{0} + \frac{0.45}{20} + \frac{0.6}{40} + \frac{0.8}{60} + \frac{0.95}{80} + \frac{1.0}{100} \right\}$$

Find the following membership functions using standard fuzzy operations:

(a) $\mu_{\underset{\sim}{S_1} \cup \underset{\sim}{S_2}}(x)$

(b) $\mu_{\underset{\sim}{S_1} \cap \underset{\sim}{S_2}}(x)$

(c) $\mu_{\overline{\underset{\sim}{S_1}}}(x)$

(d) $\mu_{\overline{\underset{\sim}{S_2}}}(x)$

(e) $\mu_{\overline{\underset{\sim}{S_1}} \cup \underset{\sim}{S_1}}(x)$

(f) $\mu_{\overline{\underset{\sim}{S_1}} \cap \underset{\sim}{S_1}}(x)$

2.9. For flight simulator data the determination of certain changes in operating conditions of the aircraft is made on the basis of hard breakpoints in the Mach region. Let us define a fuzzy set to represent the condition of "near" a Mach number of 0.74. Further, define a second fuzzy set to represent the condition of "in the region of" a Mach number of 0.74. In typical simulation data a Mach number of 0.74 is a hard breakpoint.

$$\underset{\sim}{A} = \text{near Mach } 0.74 = \left\{ \frac{0}{0.730} + \frac{0.8}{0.735} + \frac{1}{0.740} + \frac{0.6}{0.745} + \frac{0}{0.750} \right\}$$

$$\underset{\sim}{B} = \text{in the region of Mach } 0.74 = \left\{ \frac{0}{0.730} + \frac{0.4}{0.735} + \frac{0.8}{0.740} + \frac{1}{0.745} + \frac{0.6}{0.750} \right\}$$

For these two fuzzy sets find the following:

(a) $\underset{\sim}{A} \cup \underset{\sim}{B}$

(b) $\underset{\sim}{A} \cap \underset{\sim}{B}$

(c) $\overline{\underset{\sim}{A}}$

(d) $\underset{\sim}{A} \mid \underset{\sim}{B}$

(e) $\overline{\underset{\sim}{A} \cup \underset{\sim}{B}}$

(f) $\overline{\underset{\sim}{A} \cap \underset{\sim}{B}}$

2.10. A system component is tested on a drop table in the time domain, t, to shock loads of haversine pulses of various acceleration amplitudes, $\ddot{x}$, as shown in Fig. P2.10a. After the test the component is evaluated for damage. Define two fuzzy sets, "Passed" $= \underset{\sim}{P}$ and "Failed" $= \underset{\sim}{F}$. Of course, failed and passed are fuzzy notions, since failure to the component might be some partial level between the extremes of pass and fail. These sets are defined on a linear scale of accelerations, $|\ddot{x}|$, which is the magnitude of the input pulse (see Fig. P2.10b). We define the following set operations:

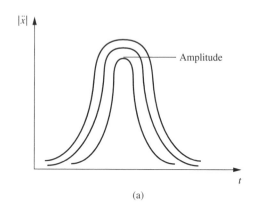

(a)

FIGURE P2.10a

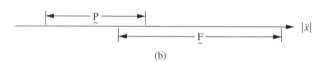

(b)

FIGURE P2.10b

(a) $\underset{\sim}{F} \cup \underset{\sim}{P} = \{|\ddot{x}|\}$: the universe of input shock level results.

(b) $\underset{\sim}{F} \cap \underset{\sim}{P}$: the portion of the universe where the component could both fail and pass.

(c) $\overline{\underset{\sim}{F}}$: portion of universe that definitely passed.

(d) $\overline{\underset{\sim}{P}}$: portion of universe that definitely failed.

(e) $\underset{\sim}{F} \mid \underset{\sim}{P}$: the portion of the failed set that definitely failed.

Define suitable membership functions for the two fuzzy sets $\underset{\sim}{F}$ and $\underset{\sim}{P}$ and determine the operations just described.

2.11. Suppose an engineer is addressing a problem in the power control of a mobile cellular telephone transmitting to its base station. Let MP be the medium-power fuzzy set, and HP be the high-power set. Let the universe of discourse be comprised of discrete units of dB · m, i.e., $X = \{0, 1, 2, \ldots, 10\}$. The membership functions for these two fuzzy sets are shown in Fig. P2.11. For these two fuzzy sets, demonstrate union, intersection, complement, and the difference.

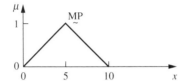
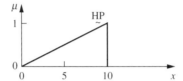

FIGURE P2.11

2.12. Consider a local area network (LAN) of interconnected workstations that communicate using Ethernet protocols at a maximum rate of 10 Mbit/s. Traffic rates on the network can be expressed as the peak value of the total bandwidth (BW) used; and the two fuzzy variables, "Quiet" and "Congested," can be used to describe the perceived loading of the LAN. If the discrete universal set $X = \{0, 1, 2, 5, 7, 9, 10\}$ represents bandwidth usage, in Mbit/s, then the membership functions of the fuzzy sets Quiet $\underset{\sim}{Q}$ and Congested $\underset{\sim}{C}$ are as shown in Fig. P2.12.

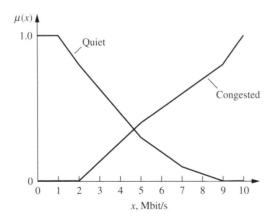

FIGURE P2.12

For these two fuzzy sets, graphically determine the union, intersection, complement of each, difference $\underset{\sim}{Q} \mid \underset{\sim}{C}$, and both De Morgan's principles.

2.13. An engineer is asked to develop a glass break detector/discriminator for use with residential alarm systems. The detector should be able to distinguish between the breaking of a pane of a glass (a window) and a drinking glass. From analysis it has been determined that the sound of a shattering window pane contains most of its energy at frequencies centered about 4 kHz whereas the sound of a shattering drinking glass contains most of its energy at frequencies centered about 8 kHz. The spectra of the two shattering sounds overlap. The membership functions for the window pane and the glass are given as $\mu_{\underset{\sim}{A}}(x)$ and $\mu_{\underset{\sim}{B}}(x)$, respectively. Illustrate the basic operations of union, intersection, complement, and difference for the following membership functions:

$$x = 0, 1, \ldots, 10 \quad \sigma = 2 \quad \mu_{\underset{\sim}{A}} = 4 \quad \mu_{\underset{\sim}{B}} = 8 \quad \mu_{\underset{\sim}{A}}(x) = \exp\left[\frac{-(x - \mu_A)^2}{2\sigma^2}\right]$$

$$\mu_{\underset{\sim}{B}}(x) = \exp\left[\frac{-(x - \mu_B)^2}{2\sigma^2}\right]$$

2.14. Samples of a new microprocessor IC chip are to be sent to several customers for beta testing. The chips are sorted to meet certain maximum electrical characteristics, say frequency and temperature rating, so that the "best" chips are distributed to preferred customer 1. Suppose that each sample chip is screened and all chips are found to have a maximum operating frequency in the range 7–15 MHz at 20°C. Also, the maximum operating temperature range (20°C $\pm \Delta T$) at 8 MHz is determined. Suppose there are eight sample chips with the following electrical characteristics:

				Chip number				
	1	2	3	4	5	6	7	8
f_{max}, MHz	6	7	8	9	10	11	12	13
ΔT_{max}, °C	0	0	20	40	30	50	40	60

The following fuzzy sets are defined:

$\underset{\sim}{A}$ = set of "fast" chips = chips with $f_{max} \geq 12$ MHz

$$= \left\{\frac{0}{1} + \frac{0}{2} + \frac{0}{3} + \frac{0}{4} + \frac{0.2}{5} + \frac{0.6}{6} + \frac{1}{7} + \frac{1}{8}\right\}$$

$\underset{\sim}{B}$ = set of "slow" chips = chips with $f_{max} \geq 8$ MHz

$$= \left\{\frac{0.1}{1} + \frac{0.5}{2} + \frac{1}{3} + \frac{1}{4} + \frac{1}{5} + \frac{1}{6} + \frac{1}{7} + \frac{1}{8}\right\}$$

$\underset{\sim}{C}$ = set of "cold" chips = chips with $\Delta T_{max} \geq 10$°C

$$= \left\{\frac{0}{1} + \frac{0}{2} + \frac{1}{3} + \frac{1}{4} + \frac{1}{5} + \frac{1}{6} + \frac{1}{7} + \frac{1}{8}\right\}$$

$\underset{\sim}{D}$ = set of "hot" chips = chips with $\Delta T_{max} \geq 50$°C

$$= \left\{\frac{0}{1} + \frac{0}{2} + \frac{0}{3} + \frac{0.5}{4} + \frac{0.1}{5} + \frac{1}{6} + \frac{0.5}{7} + \frac{1}{8}\right\}$$

It is seen that the units for operating frequencies and temperatures are different; hence, the associated fuzzy sets could be considered from different universes and operations on combinations of them would involve the Cartesian product. However, both sets of universes have been transformed to a different universe, simply the universe of countable integers from 1 to 8. Based on a single universe, use these four fuzzy sets to illustrate various set operations. For example, the following operations relate the sets of ''fast'' and ''hot'' chips:

(a) $\underset{\sim}{A} \cup \underset{\sim}{D}$

(b) $\underset{\sim}{A} \cap \underset{\sim}{D}$

(c) $\overline{\underset{\sim}{A}}$

(d) $\underset{\sim}{A} \mid \underset{\sim}{D}$

(e) $\overline{\underset{\sim}{A} \cup \underset{\sim}{D}}$

(f) $\overline{\underset{\sim}{A} \cap \underset{\sim}{D}}$

CHAPTER
3

CLASSICAL RELATIONS AND FUZZY RELATIONS

. . . assonance means getting the rhyme wrong.

Michael Caine as Professor Bryant in the movie
Educating Rita, 1983

This chapter introduces the notion of a relation as the basic idea behind numerous operations on sets such as Cartesian products, composition of relations, and equivalence properties. Like a set, a relation is of fundamental importance in all engineering, science, and mathematically based fields. It is also associated with graph theory, a subject of wide impact in design and data manipulation. Relations can be also be used to represent *similarity*, a notion that is important to many different technologies and, as expressed in the humorous metaphorical quote above, a concept that is a key ingredient in our natural language and its many uses, e.g., its use in poems. The *American Heritage Dictionary* defines assonance as "approximate agreement or partial similarity"; assonance is an example of a prototypical fuzzy concept.

Understanding relations is central to the understanding of a great many areas addressed in this textbook. Relations are intimately involved in logic, approximate reasoning, rule-based systems, nonlinear simulation, synthetic evaluation, classification, pattern recognition, and control. Relations will be referred to repeatedly in this text in many different applications areas. Relations represent mappings for sets just as mathematical functions do; relations are also very useful in representing connectives in logic (see Chapter 5).

This chapter begins by describing Cartesian products as a means of producing ordered relationships among sets. Following this is an introduction to classical (crisp) relations – structures that represent the presence or absence of correlation, interaction, or

Fuzzy Logic with Engineering Applications, Second Edition T. J. Ross
© 2004 John Wiley & Sons, Ltd ISBNs: 0-470-86074-X (HB); 0-470-86075-8 (PB)

propinquity between the elements of two or more crisp sets; in this case, a set could also be the universe. There are only two degrees of relationship between elements of the sets in a crisp relation: the relationships ''completely related'' and ''not related,'' in a binary sense. Basic operations, properties, and the cardinality of relations are explained and illustrated. Two composition methods to relate elements of three or more universes are illustrated.

Fuzzy relations are then developed by allowing the relationship between elements of two or more sets to take on an infinite number of degrees of relationship between the extremes of ''completely related'' and ''not related.'' In this sense, fuzzy relations are to crisp relations as fuzzy sets are to crisp sets; crisp sets and relations are constrained realizations of fuzzy sets and relations. Operations, properties, and cardinality of fuzzy relations are introduced and illustrated, as are Cartesian products and compositions of fuzzy relations. Some engineering examples are given to illustrate various issues associated with relations. The reader can consult the literature for more details on relations [e.g., Gill, 1976; Dubois and Prade, 1980; Kandel, 1985; Klir and Folger, 1988; Zadeh, 1971].

This chapter contains a section on tolerance and equivalence relations – both classical and fuzzy – which is introduced for use in later chapters of the book. Both tolerance and equivalence relations are illustrated with some examples. Finally, the chapter concludes with a section on value assignments, which discusses various methods to develop the elements of relations, and a list of additional composition operators. These assignment methods are discussed, and a few examples are given in the area of similarity methods.

CARTESIAN PRODUCT

An ordered sequence of r elements, written in the form $(a_1, a_2, a_3, \ldots, a_r)$, is called an ordered r-tuple; an unordered r-tuple is simply a collection of r elements without restrictions on order. In a ubiquitous special case where $r = 2$, the r-tuple is referred to as an ordered *pair*. For crisp sets $A_1, A_2, \ldots, A_r$, the set of all r-tuples $(a_1, a_2, a_3, \ldots, a_r)$, where $a_1 \in A_1$, $a_2 \in A_2$, and $a_r \in A_r$, is called the *Cartesian product* of $A_1, A_2, \ldots, A_r$, and is denoted by $A_1 \times A_2 \times \cdots \times A_r$. The Cartesian product of two or more sets is *not* the same thing as the arithmetic product of two or more sets. The latter will be dealt with in Chapter 12, when the extension principle is introduced.

When all the A_r are identical and equal to A, the Cartesian product $A_1 \times A_2 \times \cdots \times A_r$ can be denoted as A^r.

Example 3.1. The elements in two sets A and B are given as $A = \{0, 1\}$ and $B = \{a, b, c\}$. Various Cartesian products of these two sets can be written as shown:

$$A \times B = \{(0, a), (0, b), (0, c), (1, a), (1, b), (1, c)\}$$

$$B \times A = \{(a, 0), (a, 1), (b, 0), (b, 1), (c, 0), (c, 1)\}$$

$$A \times A = A^2 = \{(0, 0), (0, 1), (1, 0), (1, 1)\}$$

$$B \times B = B^2 = \{(a, a), (a, b), (a, c), (b, a), (b, b), (b, c), (c, a), (c, b), (c, c)\}$$

CRISP RELATIONS

A subset of the Cartesian product $A_1 \times A_2 \times \cdots \times A_r$ is called an *r-ary relation* over $A_1, A_2, \ldots, A_r$. Again, the most common case is for $r = 2$; in this situation the relation is a subset of the Cartesian product $A_1 \times A_2$ (i.e., a set of pairs, the first coordinate of which is from A_1 and the second from A_2). This subset of the full Cartesian product is called a *binary relation from* A_1 *into* A_2. If three, four, or five sets are involved in a subset of the full Cartesian product, the relations are called ternary, quaternary, and quinary, respectively. In this text, whenever the term *relation* is used without qualification, it is taken to mean a *binary relation*.

The Cartesian product of two universes X and Y is determined as

$$X \times Y = \{(x, y) \mid x \in X, y \in Y\} \tag{3.1}$$

which forms an ordered pair of every $x \in X$ with every $y \in Y$, forming *unconstrained* matches between X and Y. That is, every element in universe X is related completely to every element in universe Y. The *strength* of this relationship between ordered pairs of elements in each universe is measured by the characteristic function, denoted χ, where a value of unity is associated with *complete relationship* and a value of zero is associated with *no relationship*, i.e.,

$$\chi_{X \times Y}(x, y) = \begin{cases} 1, & (x, y) \in X \times Y \\ 0, & (x, y) \notin X \times Y \end{cases} \tag{3.2}$$

One can think of this strength of relation as a mapping from ordered pairs of the universe, or ordered pairs of sets defined on the universes, to the characteristic function. When the universes, or sets, are finite the relation can be conveniently represented by a matrix, called a *relation matrix*. An r-ary relation can be represented by an r-dimensional relation matrix. Hence, binary relations can be represented by two-dimensional matrices (used throughout this text).

An example of the strength of relation for the unconstrained case is given in the Sagittal diagram shown in Fig. 3.1 (a Sagittal diagram is simply a schematic depicting points as elements of universes, and lines as relationships between points, or it can be a pictorial of the elements as nodes which are connected by directional lines, as seen in Fig. 3.8). Lines in the Sagittal diagram and values of unity in the *relation matrix*

$$R = \begin{matrix} & \begin{matrix} a & b & c \end{matrix} \\ \begin{matrix} 1 \\ 2 \\ 3 \end{matrix} & \begin{bmatrix} 1 & 1 & 1 \\ 1 & 1 & 1 \\ 1 & 1 & 1 \end{bmatrix} \end{matrix}$$

correspond to the ordered pairs of mappings in the relation. Here, the elements in the two universes are defined as $X = \{1, 2, 3\}$ and $Y = \{a, b, c\}$.

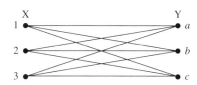

FIGURE 3.1
Sagittal diagram of an unconstrained relation.

A more general crisp relation, R, exists when matches between elements in two universes are *constrained*. Again, the characteristic function is used to assign values of relationship in the mapping of the Cartesian space X × Y to the binary values of (0, 1):

$$\chi_R(x, y) = \begin{cases} 1, & (x, y) \in R \\ 0, & (x, y) \notin R \end{cases} \tag{3.3}$$

Example 3.2. In many biological models, members of certain species can reproduce only with certain members of another species. Hence, only some elements in two or more universes have a relationship (nonzero) in the Cartesian product. An example is shown in Fig. 3.2 for two two-member species, i.e., for X = {1, 2} and for Y = {a, b}. In this case the locations of zeros in the relation matrix

$$R = \{(1, a), (2, b)\} \quad R \subset X \times Y$$

and the absence of lines in the Sagittal diagram correspond to pairs of elements between the two universes where there is "no relation"; that is, the strength of the relationship is zero.

Special cases of the constrained and the unconstrained Cartesian product for sets where $r = 2$ (i.e., for A^2) are called the *identity relation* and the *universal relation,* respectively. For example, for A = {0, 1, 2} the universal relation, denoted U_A, and the identity relation, denoted I_A, are found to be

$$U_A = \{(0, 0), (0, 1), (0, 2), (1, 0), (1, 1), (1, 2), (2, 0), (2, 1), (2, 2)\}$$

$$I_A = \{(0, 0), (1, 1), (2, 2)\}$$

Example 3.3. Relations can also be defined for continuous universes. Consider, for example, the continuous relation defined by the following expression:

$$R = \{(x, y) \mid y \geq 2x, x \in X, y \in Y\}$$

which is also given in function-theoretic form using the characteristic function as

$$\chi_R(x, y) = \begin{cases} 1, & y \geq 2x \\ 0, & y < 2x \end{cases}$$

Graphically, this relation is equivalent to the shaded region shown in Fig. 3.3.

Cardinality of Crisp Relations

Suppose n elements of the universe X are related (paired) to m elements of the universe Y. If the cardinality of X is n_X and the cardinality of Y is n_Y, then the cardinality of the

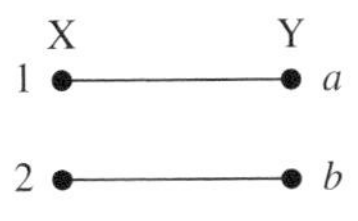

FIGURE 3.2
Relation matrix and Sagittal diagram for a constrained relation.

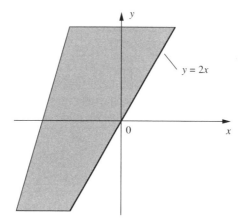

FIGURE 3.3
Relation corresponding to the expression $y \geq 2x$.

relation, R, between these two universes is $n_{X \times Y} = n_X * n_Y$. The cardinality of the power set describing this relation, $P(X \times Y)$, is then $n_{P(X \times Y)} = 2^{(n_X n_Y)}$.

Operations on Crisp Relations

Define R and S as two separate relations on the Cartesian universe $X \times Y$, and define the null relation and the complete relation as the relation matrices $\underset{\sim}{\mathbf{O}}$ and $\underset{\sim}{\mathbf{E}}$, respectively. An example of a 4×4 form of the $\underset{\sim}{\mathbf{O}}$ and $\underset{\sim}{\mathbf{E}}$ matrices is given here:

$$\mathbf{O} = \begin{bmatrix} 0 & 0 & 0 & 0 \\ 0 & 0 & 0 & 0 \\ 0 & 0 & 0 & 0 \\ 0 & 0 & 0 & 0 \end{bmatrix} \qquad \mathbf{E} = \begin{bmatrix} 1 & 1 & 1 & 1 \\ 1 & 1 & 1 & 1 \\ 1 & 1 & 1 & 1 \\ 1 & 1 & 1 & 1 \end{bmatrix}$$

The following function-theoretic operations for the two crisp relations (R, S) can now be defined.

Union $\qquad R \cup S \longrightarrow \chi_{R \cup S}(x, y) : \chi_{R \cup S}(x, y) = \max[\chi_R(x, y), \chi_S(x, y)]$ $\qquad$ (3.4)

Intersection $\quad R \cap S \longrightarrow \chi_{R \cap S}(x, y) : \chi_{R \cap S}(x, y) = \min[\chi_R(x, y), \chi_S(x, y)]$ $\qquad$ (3.5)

Complement $\qquad \overline{R} \longrightarrow \chi_{\overline{R}}(x, y) : \chi_{\overline{R}}(x, y) = 1 - \chi_R(x, y)$ $\qquad$ (3.6)

Containment $\qquad R \subset S \longrightarrow \chi_R(x, y) : \chi_R(x, y) \leq \chi_S(x, y)$ $\qquad$ (3.7)

Identity $\qquad\qquad\qquad \emptyset \longrightarrow \mathbf{O} \text{ and } X \longrightarrow \mathbf{E}$ $\qquad$ (3.8)

Properties of Crisp Relations

The properties of commutativity, associativity, distributivity, involution, and idempotency all hold for crisp relations just as they do for classical set operations. Moreover, *De Morgan's principles* and the *excluded middle axioms* also hold for crisp (classical) relations just as they do for crisp (classical) sets. The null relation, **O**, and the complete relation, **E**, are analogous to the null set, Ø, and the whole set, X, respectively, in the set-theoretic case (see Chapter 2).

Composition

Let R be a relation that relates, or maps, elements from universe X to universe Y, and let S be a relation that relates, or maps, elements from universe Y to universe Z.

A useful question we seek to answer is whether we can find a relation, T, that relates the same elements in universe X that R contains to the same elements in universe Z that S contains. It turns out we can find such a relation using an operation known as *composition*. For the Sagittal diagram in Fig. 3.4, we see that the only "path" between relation R and relation S is the two routes that start at x_1 and end at z_2 (i.e., $x_1 - y_1 - z_2$ and $x_1 - y_3 - z_2$). Hence, we wish to find a relation T that relates the ordered pair (x_1, z_2), i.e., $(x_1, z_2) \in T$. In this example,

$$R = \{(x_1, y_1), (x_1, y_3), (x_2, y_4)\}$$

$$S = \{(y_1, z_2), (y_3, z_2)\}$$

There are two common forms of the composition operation; one is called the max−min composition and the other the max−product composition. (Five other forms of the composition operator are available for certain logic issues; these are described at the end of this chapter.) The max−min composition is defined by the set-theoretic and membership function-theoretic expressions

$$T = R \circ S$$

$$\chi_T(x, z) = \bigvee_{y \in Y} (\chi_R(x, y) \wedge \chi_S(y, z)) \tag{3.9}$$

and the max−product (sometimes called max−dot) composition is defined by the set-theoretic and membership function-theoretic expressions

$$T = R \circ S$$

$$\chi_T(x, z) = \bigvee_{y \in Y} (\chi_R(x, y) \bullet \chi_S(y, z)) \tag{3.10}$$

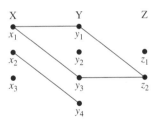

FIGURE 3.4
Sagittal diagram relating elements of three universes.

Tensile force ← ... → Tensile force

FIGURE 3.5
Chain strength analogy for max–min composition.

There is a very interesting physical analogy for the max–min composition operator. Figure 3.5 illustrates a system comprising several chains placed together in a parallel fashion. In the system, each chain comprises a number of chain links. If we were to take one of the chains out of the system, place it in a tensile test machine, and exert a large tensile force on the chain, we would find that the chain would break at its weakest link. Hence, the strength of one chain is equal to the strength of its weakest link; in other words, the *minimum* ($\wedge$) strength of all the links in the chain governs the strength of the overall chain. Now, if we were to place the entire chain system in a tensile device and exert a tensile force on the chain system, we would find that the chain system would continue to carry increasing loads until the last chain in the system broke. That is, weaker chains would break with an increasing load until the strongest chain was left alone, and eventually it would break; in other words, the *maximum* ($\vee$) strength of all the chains in the chain system would govern the overall strength of the chain system. Each chain in the system is analogous to the min operation in the max–min composition, and the overall chain system strength is analogous to the max operation in the max–min composition.

> **Example 3.4.** The matrix expression for the crisp relations shown in Fig. 3.4 can be found using the max–min composition operation. Relation matrices for R and S would be expressed as

$$R = \begin{array}{c} \\ x_1 \\ x_2 \\ x_3 \end{array} \begin{array}{cccc} y_1 & y_2 & y_3 & y_4 \\ \left[\begin{array}{cccc} 1 & 0 & 1 & 0 \\ 0 & 0 & 0 & 1 \\ 0 & 0 & 0 & 0 \end{array}\right] \end{array} \quad \text{and} \quad S = \begin{array}{c} \\ y_1 \\ y_2 \\ y_3 \\ y_4 \end{array} \begin{array}{cc} z_1 & z_2 \\ \left[\begin{array}{cc} 0 & 1 \\ 0 & 0 \\ 0 & 1 \\ 0 & 0 \end{array}\right] \end{array}$$

> The resulting relation T would then be determined by max–min composition, Eq. (3.9), or max–product composition, Eq. (3.10). (In the crisp case these forms of the composition operators produce identical results; other forms of this operator, such as those listed at the end of this chapter, will not produce identical results.) For example,

$$\mu_T(x_1, z_1) = \max[\min(1, 0), \min(0, 0), \min(1, 0), \min(0, 0)] = 0$$

$$\mu_T(x_1, z_2) = \max[\min(1, 1), \min(0, 0), \min(1, 1), \min(0, 0)] = 1$$

and for the rest,

$$T = \begin{array}{c} \\ x_1 \\ x_2 \\ x_3 \end{array} \begin{array}{cc} z_1 & z_2 \\ \left[\begin{array}{cc} 0 & 1 \\ 0 & 0 \\ 0 & 0 \end{array}\right] \end{array}$$

FUZZY RELATIONS

Fuzzy relations also map elements of one universe, say X, to those of another universe, say Y, through the Cartesian product of the two universes. However, the ''strength'' of the relation between ordered pairs of the two universes is not measured with the characteristic function, but rather with a membership function expressing various "degrees" of strength of the relation on the unit interval [0,1]. Hence, a fuzzy relation R is a mapping from the Cartesian space $X \times Y$ to the interval [0,1], where the strength of the mapping is expressed by the membership function of the relation for ordered pairs from the two universes, or $\mu_R(x, y)$.

Cardinality of Fuzzy Relations

Since the cardinality of fuzzy sets on any universe is infinity, the cardinality of a fuzzy relation between two or more universes is also infinity.

Operations on Fuzzy Relations

Let R and S be fuzzy relations on the Cartesian space $X \times Y$. Then the following operations apply for the membership values for various set operations:

Union	$\mu_{R \cup S}(x, y) = \max(\mu_R(x, y), \mu_S(x, y))$	(3.11)
Intersection	$\mu_{R \cap S}(x, y) = \min(\mu_R(x, y), \mu_S(x, y))$	(3.12)
Complement	$\mu_{\overline{R}}(x, y) = 1 - \mu_R(x, y)$	(3.13)
Containment	$R \subset S \Rightarrow \mu_R(x, y) \leq \mu_S(x, y)$	(3.14)

Properties of Fuzzy Relations

Just as for crisp relations, the properties of commutativity, associativity, distributivity, involution, and idempotency all hold for fuzzy relations. Moreover, De Morgan's principles hold for fuzzy relations just as they do for crisp (classical) relations, and the null relation, **O**, and the complete relation, **E**, are analogous to the null set and the whole set in set-theoretic form, respectively. Fuzzy relations are not constrained, as is the case for fuzzy sets in general, by the excluded middle axioms. Since a fuzzy relation R is also a fuzzy set, there is overlap between a relation and its complement; hence,

$$R \cup \overline{R} \neq E$$

$$R \cap \overline{R} \neq O$$

As seen in the foregoing expressions, the *excluded middle axioms* for relations do not result, in general, in the null relation, **O**, or the complete relation, **E**.

Fuzzy Cartesian Product and Composition

Because fuzzy relations in general are fuzzy sets, we can define the Cartesian product to be a relation between two or more fuzzy sets. Let A be a fuzzy set on universe X and B be a fuzzy set on universe Y; then the Cartesian product between fuzzy sets A and B will result in a fuzzy relation R, which is contained within the full Cartesian product space, or

$$A \times B = R \subset X \times Y \tag{3.15}$$

where the fuzzy relation $\underset{\sim}{R}$ has membership function

$$\mu_{\underset{\sim}{R}}(x, y) = \mu_{\underset{\sim}{A} \times \underset{\sim}{B}}(x, y) = \min\left(\mu_{\underset{\sim}{A}}(x), \mu_{\underset{\sim}{B}}(y)\right) \tag{3.16}$$

The Cartesian product defined by $\underset{\sim}{A} \times \underset{\sim}{B} = \underset{\sim}{R}$, Eq. (3.15), is implemented in the same fashion as is the cross product of two vectors. Again, the Cartesian product is *not* the same operation as the arithmetic product. In the case of two-dimensional relations ($r = 2$), the former employs the idea of pairing of elements among sets, whereas the latter uses actual arithmetic products between elements of sets. Each of the fuzzy sets could be thought of as a vector of membership values; each value is associated with a particular element in each set. For example, for a fuzzy set (vector) $\underset{\sim}{A}$ that has four elements, hence column vector of size 4×1, and for a fuzzy set (vector) $\underset{\sim}{B}$ that has five elements, hence a row vector size of 1×5, the resulting fuzzy relation, $\underset{\sim}{R}$, will be represented by a matrix of size 4×5, i.e., $\underset{\sim}{R}$ will have four rows and five columns. This result is illustrated in the following example.

> **Example 3.5.** Suppose we have two fuzzy sets, $\underset{\sim}{A}$ defined on a universe of three discrete temperatures, $X = \{x_1, x_2, x_3\}$, and $\underset{\sim}{B}$ defined on a universe of two discrete pressures, $Y = \{y_1, y_2\}$, and we want to find the fuzzy Cartesian product between them. Fuzzy set $\underset{\sim}{A}$ could represent the "ambient" temperature and fuzzy set $\underset{\sim}{B}$ the "near optimum" pressure for a certain heat exchanger, and the Cartesian product might represent the conditions (temperature–pressure pairs) of the exchanger that are associated with "efficient" operations. For example, let
>
> $$\underset{\sim}{A} = \frac{0.2}{x_1} + \frac{0.5}{x_2} + \frac{1}{x_3} \quad \text{and} \quad \underset{\sim}{B} = \frac{0.3}{y_1} + \frac{0.9}{y_2}$$
>
> Note that $\underset{\sim}{A}$ can be represented as a column vector of size 3×1 and $\underset{\sim}{B}$ can be represented by a row vector of 1×2. Then the fuzzy Cartesian product, using Eq. (3.16), results in a fuzzy relation $\underset{\sim}{R}$ (of size 3×2) representing "efficient" conditions, or
>
> $$\underset{\sim}{A} \times \underset{\sim}{B} = \underset{\sim}{R} = \begin{array}{c} x_1 \\ x_2 \\ x_3 \end{array} \begin{array}{cc} y_1 & y_2 \\ \left[\begin{array}{cc} 0.2 & 0.2 \\ 0.3 & 0.5 \\ 0.3 & 0.9 \end{array} \right] \end{array}$$

Fuzzy composition can be defined just as it is for crisp (binary) relations. Suppose $\underset{\sim}{R}$ is a fuzzy relation on the Cartesian space $X \times Y$, $\underset{\sim}{S}$ is a fuzzy relation on $Y \times Z$, and $\underset{\sim}{T}$ is a fuzzy relation on $X \times Z$; then fuzzy max–min composition is defined in terms of the set-theoretic notation and membership function-theoretic notation in the following manner:

$$\underset{\sim}{T} = \underset{\sim}{R} \circ \underset{\sim}{S}$$

$$\mu_{\underset{\sim}{T}}(x, z) = \bigvee_{y \in Y}(\mu_{\underset{\sim}{R}}(x, y) \wedge \mu_{\underset{\sim}{S}}(y, z)) \tag{3.17a}$$

and fuzzy max–product composition is defined in terms of the membership function-theoretic notation as

$$\mu_{\underset{\sim}{T}}(x, z) = \bigvee_{y \in Y}(\mu_{\underset{\sim}{R}}(x, y) \bullet \mu_{\underset{\sim}{S}}(y, z)) \tag{3.17b}$$

It should be pointed out that neither crisp nor fuzzy compositions are commutative in general; that is,

$$R \circ S \neq S \circ R \tag{3.18}$$

Equation (3.18) is general for any matrix operation, fuzzy or otherwise, that must satisfy consistency between the cardinal counts of elements in respective universes. Even for the case of square matrices, the composition converse, represented by Eq. (3.18), is not guaranteed.

Example 3.6. Let us extend the information contained in the Sagittal diagram shown in Fig. 3.4 to include fuzzy relationships for $X \times Y$ (denoted by the fuzzy relation R) and $Y \times Z$ (denoted by the fuzzy relation S). In this case we change the elements of the universes to,

$$X = \{x_1, x_2\}, \quad Y = \{y_1, y_2\}, \quad \text{and} \quad Z = \{z_1, z_2, z_3\}$$

Consider the following fuzzy relations:

$$R = \begin{array}{c} x_1 \\ x_2 \end{array} \begin{bmatrix} \overset{y_1}{0.7} & \overset{y_2}{0.5} \\ 0.8 & 0.4 \end{bmatrix} \quad \text{and} \quad S = \begin{array}{c} y_1 \\ y_2 \end{array} \begin{bmatrix} \overset{z_1}{0.9} & \overset{z_2}{0.6} & \overset{z_3}{0.2} \\ 0.1 & 0.7 & 0.5 \end{bmatrix}$$

Then the resulting relation, T, which relates elements of universe X to elements of universe Z, i.e., defined on Cartesian space $X \times Z$, can be found by max–min composition, Eq. (3.17a), to be, for example,

$$\mu_T(x_1, z_1) = \max[\min(0.7, 0.9), \min(0.5, 0.1)] = 0.7$$

and the rest,

$$T = \begin{array}{c} x_1 \\ x_2 \end{array} \begin{bmatrix} \overset{z_1}{0.7} & \overset{z_2}{0.6} & \overset{z_3}{0.5} \\ 0.8 & 0.6 & 0.4 \end{bmatrix}$$

and by max–product composition, Eq. (3.17b), to be, for example,

$$\mu_T(x_2, z_2) = \max[(0.8 \cdot 0.6), (0.4 \cdot 0.7)] = 0.48$$

and the rest,

$$T = \begin{array}{c} x_1 \\ x_2 \end{array} \begin{bmatrix} \overset{z_1}{0.63} & \overset{z_2}{0.42} & \overset{z_3}{0.25} \\ 0.72 & 0.48 & 0.20 \end{bmatrix}$$

We now illustrate the use of relations with fuzzy sets for three examples from the fields of medicine, electrical, and civil engineering.

Example 3.7. A certain type of virus attacks cells of the human body. The infected cells can be visualized using a special microscope. The microscope generates digital images that medical doctors can analyze and identify the infected cells. The virus causes the infected cells to have a black spot, within a darker grey region (Fig. 3.6).

A digital image process can be applied to the image. This processing generates two variables: the first variable, P, is related to black spot quantity (black pixels), and the second variable, S, is related to the shape of the black spot, i.e., if they are circular or elliptic. In these images it is often difficult to actually count the number of black pixels, or to identify a perfect circular cluster of pixels; hence, both these variables must be estimated in a linguistic way.

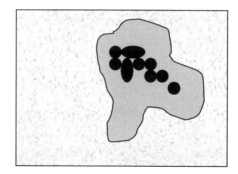

FIGURE 3.6
An infected cell shows black spots with different shapes in a micrograph.

Suppose that we have two fuzzy sets, P which represents the number of black pixels (e.g., none with black pixels, C_1, a few with black pixels, C_2, and a lot of black pixels, C_3), and S which represents the shape of the black pixel clusters, e.g., S_1 is an ellipse and S_2 is a circle. So we have

$$P = \left\{ \frac{0.1}{C_1} + \frac{0.5}{C_2} + \frac{1.0}{C_3} \right\} \quad \text{and} \quad S = \left\{ \frac{0.3}{S_1} + \frac{0.8}{S_2} \right\}$$

and we want to find the relationship between quantity of black pixels in the virus and the shape of the black pixel clusters. Using a Cartesian product between P and S gives

$$R = P \times S = \begin{array}{c} \\ C_1 \\ C_2 \\ C_3 \end{array} \begin{array}{cc} S_1 & S_2 \\ \left[\begin{array}{cc} 0.1 & 0.1 \\ 0.3 & 0.5 \\ 0.3 & 0.8 \end{array} \right] \end{array}$$

Now, suppose another microscope image is taken and the number of black pixels is slightly different; let the new black pixel quantity be represented by a fuzzy set, P':

$$P' = \left\{ \frac{0.4}{C_1} + \frac{0.7}{C_2} + \frac{1.0}{C_3} \right\}$$

Using max–min composition with the relation R will yield a new value for the fuzzy set of pixel cluster shapes that are associated with the new black pixel quantity:

$$S' = P' \circ R = \begin{bmatrix} 0.4 & 0.7 & 1.0 \end{bmatrix} \circ \begin{bmatrix} 0.1 & 0.1 \\ 0.3 & 0.5 \\ 0.3 & 0.8 \end{bmatrix} = \begin{bmatrix} 0.3 & 0.8 \end{bmatrix}$$

Example 3.8. Suppose we are interested in understanding the speed control of the DC (direct current) shunt motor under no-load condition, as shown diagrammatically in Fig. 3.7. Initially, the series resistance R_{se} in Fig. 3.7 should be kept in the cut-in position for the following reasons:

1. The back electromagnetic force, given by $E_b = kN\phi$, where k is a constant of proportionality, N is the motor speed, and ϕ is the flux (which is proportional to input voltage, V), is equal to zero because the motor speed is equal to zero initially.
2. We have $V = E_b + I_a(R_a + R_{se})$, therefore $I_a = (V - E_b)/(R_a + R_{se})$, where I_a is the armature current and R_a is the armature resistance. Since E_b is equal to zero initially, the armature current will be $I_a = V/(R_a + R_{se})$, which is going to be quite large initially and may destroy the armature.

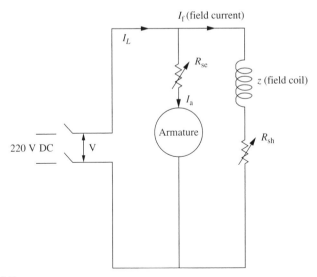

FIGURE 3.7
A DC shunt motor system.

On the basis of both cases 1 and 2, keeping the series resistance R_{se} in the cut-in position will restrict the speed to a very low value. Hence, if the rated no-load speed of the motor is 1500 rpm, then the resistance in series with the armature, or the shunt resistance R_{sh}, has to be varied.

Two methods provide this type of control: armature control and field control. For example, in armature control, suppose that ϕ (flux) is maintained at some constant value; then motor speed N is proportional to E_b.

If R_{se} is decreased step by step from its high value, I_a (armature current) increases. Hence, this method increases I_a. On the other hand, as I_a is increased the motor speed N increases. These two possible approaches to control could have been done manually or automatically. Either way, however, results in at least two problems, presuming we do not want to change the design of the armature:

What should be the minimum and maximum level of R_{se}?

What should be the minimum and maximum value of I_a?

Now let us suppose that load on the motor is taken into consideration. Then the problem of control becomes two-fold. First, owing to fluctuations in the load, the armature current may change, resulting in change in the motor speed. Second, as a result of changes in speed, the armature resistance control must be accomplished in order to maintain the motor's rated speed. Such control issues become very important in applications involving electric trains and a large number of consumer appliances making use of small batteries to run their motors.

We wish to use concepts of fuzzy sets to address this problem. Let $\underset{\sim}{R}_{se}$ be a fuzzy set representing a number of possible values for series resistance, say s_n values, given by

$$\underset{\sim}{R}_{se} = \{R_{s1}, R_{s2}, R_{s3}, \ldots, R_{s_n}\}$$

and let $\underset{\sim}{I}_a$ be a fuzzy set having a number of possible values of the armature current, say m values, given by

$$\underset{\sim}{I}_a = \{I_1, I_2, I_3, \ldots, I_m\}$$

The fuzzy sets $\underset{\sim}{R}_{se}$ and $\underset{\sim}{I}_a$ can be related through a fuzzy relation, say $\underset{\sim}{R}$, which would allow for the establishment of various degrees of relationship between pairs of resistance and current. In this way, the resistance–current pairings could conform to the modeler's intuition about the trade-offs involved in control of the armature.

Let N be another fuzzy set having numerous values for the motor speed, say v values, given by

$$N = \{N_1, N_2, N_3, \ldots, N_v\}$$

Now, we can determine another fuzzy relation, say S, to relate current to motor speed, i.e., I_a to N.

Using the operation of composition, we could then compute a relation, say T, to be used to relate series resistance to motor speed, i.e., R_{se} to N. The operations needed to develop these relations are as follows – two fuzzy cartesian products and one composition:

$$R = R_{se} \times I_a$$

$$S = I_a \times N$$

$$T = R \circ S$$

Suppose the membership functions for both series resistance R_{se} and armature current I_a are given in terms of percentages *of their respective rated values*, i.e.,

$$\mu_{R_{se}}(\%se) = \frac{0.3}{30} + \frac{0.7}{60} + \frac{1.0}{100} + \frac{0.2}{120}$$

and

$$\mu_{I_a}(\%a) = \frac{0.2}{20} + \frac{0.4}{40} + \frac{0.6}{60} + \frac{0.8}{80} + \frac{1.0}{100} + \frac{0.1}{120}$$

and the membership value for N is given in units of motor speed in rpm,

$$\mu_N(\text{rpm}) = \frac{0.33}{500} + \frac{0.67}{1000} + \frac{1.0}{1500} + \frac{0.15}{1800}$$

The following relations then result from use of the Cartesian product to determine R and S:

$$R = \begin{array}{c} \\ 30 \\ 60 \\ 100 \\ 120 \end{array} \begin{array}{cccccc} 20 & 40 & 60 & 80 & 100 & 120 \\ \left[\begin{array}{cccccc} 0.2 & 0.3 & 0.3 & 0.3 & 0.3 & 0.1 \\ 0.2 & 0.4 & 0.6 & 0.7 & 0.7 & 0.1 \\ 0.2 & 0.4 & 0.6 & 0.8 & 1 & 0.1 \\ 0.2 & 0.2 & 0.2 & 0.2 & 0.2 & 0.1 \end{array} \right] \end{array}$$

and

$$S = \begin{array}{c} \\ 20 \\ 40 \\ 60 \\ 80 \\ 100 \\ 120 \end{array} \begin{array}{cccc} 500 & 1000 & 1500 & 1800 \\ \left[\begin{array}{cccc} 0.2 & 0.2 & 0.2 & 0.15 \\ 0.33 & 0.4 & 0.4 & 0.15 \\ 0.33 & 0.6 & 0.6 & 0.15 \\ 0.33 & 0.67 & 0.8 & 0.15 \\ 0.33 & 0.67 & 1 & 0.15 \\ 0.1 & 0.1 & 0.1 & 0.1 \end{array} \right] \end{array}$$

For example, $\mu_R(60, 40) = \min(0.7, 0.4) = 0.4$, $\mu_R(100, 80) = \min(1.0, 0.8) = 0.8$, and $\mu_S(80, 1000) = \min(0.8, 0.67) = 0.67$.

The following relation results from a max–min composition for T:

$$T = R \circ S = \begin{array}{c} \\ 30 \\ 60 \\ 100 \\ 120 \end{array} \begin{array}{cccc} 500 & 1000 & 1500 & 1800 \\ \left[\begin{array}{cccc} 0.3 & 0.3 & 0.3 & 0.15 \\ 0.33 & 0.67 & 0.7 & 0.15 \\ 0.33 & 0.67 & 1 & 0.15 \\ 0.2 & 0.2 & 0.2 & 0.15 \end{array} \right] \end{array}$$

For instance,

$$\mu_{\underset{\sim}{T}}(60, 1500) = \max[\min(0.2, 0.2), \min(0.4, 0.4), \min(0.6, 0.6),$$

$$\min(0.7, 0.8), \min(0.7, 1.0), \min(0.1, 0.1)]$$

$$= \max[0.2, 0.4, 0.6, 0.7, 0.7, 0.1] = 0.7$$

Example 3.9. In the city of Calgary, Alberta, there are a significant number of neighborhood ponds that store overland flow from rainstorms and release the water downstream at a controlled rate to reduce or eliminate flooding in downstream areas. To illustrate a relation using the Cartesian product let us compare the level in the neighborhood pond system based on a 1-in-100 year storm volume capacity with the closest three rain gauge stations that measure total rainfall.

Let $\underset{\sim}{A}$ = Pond system relative depths based on 1-in-100 year capacity (assume the capacities of four ponds are p_1, p_2, p_3, and p_4, and all combine to form one outfall to the trunk sewer). Let $\underset{\sim}{B}$ = Total rainfall for event based on 1-in-100 year values from three different rain gage stations, g_1, g_2, and g_3. Suppose we have the following specific fuzzy sets:

$$\underset{\sim}{A} = \frac{0.2}{p_1} + \frac{0.6}{p_2} + \frac{0.5}{p_3} + \frac{0.9}{p_4}$$

$$\underset{\sim}{B} = \frac{0.4}{g_1} + \frac{0.7}{g_2} + \frac{0.8}{g_3}$$

The Cartesian product of these two fuzzy sets could then be formed:

$$\underset{\sim}{A} \times \underset{\sim}{B} = \underset{\sim}{C} = \begin{array}{c} \\ p_1 \\ p_2 \\ p_3 \\ p_4 \end{array} \begin{array}{ccc} g_1 & g_2 & g_3 \\ \left[\begin{array}{ccc} 0.2 & 0.2 & 0.2 \\ 0.4 & 0.6 & 0.6 \\ 0.4 & 0.5 & 0.5 \\ 0.4 & 0.7 & 0.8 \end{array}\right] \end{array}$$

The meaning of this Cartesian product would be to relate the rain gauge's prediction of large storms to the actual pond performance during rain events. Higher values indicate designs and station information that could model and control flooding in a reasonable way. Lower relative values may indicate a design problem or a nonrepresentative gauge location.

To illustrate composition for the same situation let us try to determine if the rainstorms are widespread or localized. Let us compare the results from a pond system well removed from the previous location during the same storm.

Suppose we have a relationship between the capacity of five more ponds within a new pond system ($p_5, \ldots, p_9$) and the rainfall data from the original rainfall gauges (g_1, g_2, and g_3). This relation is given by

$$\underset{\sim}{D} = \begin{array}{c} \\ g_1 \\ g_2 \\ g_3 \end{array} \begin{array}{ccccc} p_5 & p_6 & p_7 & p_8 & p_9 \\ \left[\begin{array}{ccccc} 0.3 & 0.6 & 0.5 & 0.2 & 0.1 \\ 0.4 & 0.7 & 0.5 & 0.3 & 0.3 \\ 0.2 & 0.6 & 0.8 & 0.9 & 0.8 \end{array}\right] \end{array}$$

Let $\underset{\sim}{E}$ be a fuzzy max$-$min composition for the two ponding systems:

$$\underset{\sim}{E} = \underset{\sim}{C} \circ \underset{\sim}{D} = \begin{array}{c} \\ p_1 \\ p_2 \\ p_3 \\ p_4 \end{array} \begin{array}{ccccc} p_5 & p_6 & p_7 & p_8 & p_9 \\ \left[\begin{array}{ccccc} 0.2 & 0.2 & 0.2 & 0.2 & 0.1 \\ 0.4 & 0.6 & 0.6 & 0.6 & 0.6 \\ 0.4 & 0.5 & 0.5 & 0.5 & 0.5 \\ 0.4 & 0.7 & 0.8 & 0.8 & 0.8 \end{array}\right] \end{array}$$

For example,

$$\mu_{\underset{\sim}{E}}(p_2, p_7) = \max[\min(0.4, 0.5), \min(0.6, 0.5), \min(0.6, 0.8)] = 0.6$$

This new relation, $\underset{\sim}{E}$, actually represents the character of the rainstorm for the two geographically separated pond systems: the first system from the four ponds, $p_1, \ldots, p_4$ and the second system from the ponds $p_5, \ldots, p_9$. If the numbers in this relation are large, it means that the rainstorm was widespread, whereas if the numbers are closer to zero, then the rainstorm is more localized and the original rain gauges are not a good predictor for both systems.

TOLERANCE AND EQUIVALENCE RELATIONS

Relations can exhibit various useful properties, a few of which will be discussed here. As mentioned in the introduction of this chapter, relations can be used in graph theory [Gill, 1976; Zadeh, 1971]. Consider the simple graphs in Fig. 3.8. This figure describes a universe of three elements, which are labeled as the vertices of this graph, 1, 2, and 3, or in set notation, X = {1, 2, 3}. The useful properties we wish to discuss are reflexivity, symmetry, and transitivity (there are other properties of relations that are the antonyms of these three, i.e., irreflexivity, asymmetry, and nontransitivity; these, and an additional property of antisymmetry, will not be discussed in this text). When a relation is reflexive every vertex in the graph originates a single loop, as shown in Fig. 3.8a. If a relation is symmetric, then in the graph for every edge pointing (the arrows on the edge lines in Fig. 3.8b) from vertex i to vertex j ($i, j = 1, 2, 3$), there is an edge pointing in the opposite direction, i.e., from vertex j to vertex i. When a relation is transitive, then for every pair of edges in the graph, one pointing from vertex i to vertex j and the other from vertex j to vertex k ($i, j, k = 1, 2, 3$), there is an edge pointing from vertex i directly to vertex k, as seen in Fig. 3.8c (e.g., an arrow from vertex 1 to vertex 2, an arrow from vertex 2 to vertex 3, and an arrow from vertex 1 to vertex 3).

Crisp Equivalence Relation

A relation R on a universe X can also be thought of as a relation from X to X. The relation R is an equivalence relation if it has the following three properties: (1) reflexivity, (2) symmetry, and (3) transitivity. For example, for a matrix relation the following properties will hold:

Reflexivity $(x_i, x_i) \in R$ or $\chi_R(x_i, x_i) = 1$ (3.19a)

Symmetry $(x_i, x_j) \in R \longrightarrow (x_j, x_i) \in R$ (3.19b)

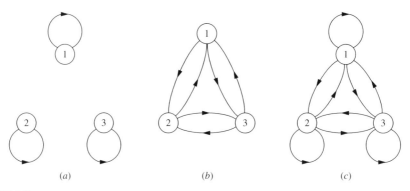

(a)   (b)   (c)

FIGURE 3.8
Three-vertex graphs for properties of (a) reflexivity, (b) symmetry, (c) transitivity [Gill, 1976].

or $\chi_R(x_i, x_j) = \chi_R(x_j, x_i)$

Transitivity $(x_i, x_j) \in R$ and $(x_j, x_k) \in R \longrightarrow (x_i, x_k) \in R$ (3.19c)

or $\chi_R(x_i, x_j)$ and $\chi_R(x_j, x_k) = 1 \longrightarrow \chi_R(x_i, x_k) = 1$

The most familiar equivalence relation is that of equality among elements of a set. Other examples of equivalence relations include the relation of parallelism among lines in plane geometry, the relation of similarity among triangles, the relation "works in the same building as" among workers of a given city, and others.

Crisp Tolerance Relation

A tolerance relation R (also called a *proximity* relation) on a universe X is a relation that exhibits only the properties of reflexivity and symmetry. A tolerance relation, R, can be reformed into an equivalence relation by at most $(n - 1)$ compositions with itself, where n is the cardinal number of the set defining R, in this case X, i.e.,

$$R_1^{n-1} = R_1 \circ R_1 \circ \cdots \circ R_1 = R$$ (3.20)

Example 3.10. Suppose in an airline transportation system we have a universe composed of five elements: the cities Omaha, Chicago, Rome, London, and Detroit. The airline is studying locations of potential hubs in various countries and must consider air mileage between cities and takeoff and landing policies in the various countries. These cities can be enumerated as the elements of a set, i.e.,

$$X = \{x_1, x_2, x_3, x_4, x_5\} = \{\text{Omaha, Chicago, Rome, London, Detroit}\}$$

Further, suppose we have a tolerance relation, R_1, that expresses relationships among these cities:

$$R_1 = \begin{bmatrix} 1 & 1 & 0 & 0 & 0 \\ 1 & 1 & 0 & 0 & 1 \\ 0 & 0 & 1 & 0 & 0 \\ 0 & 0 & 0 & 1 & 0 \\ 0 & 1 & 0 & 0 & 1 \end{bmatrix}$$

This relation is reflexive and symmetric. The graph for this tolerance relation would involve five vertices (five elements in the relation), as shown in Fig. 3.9. The property of reflexivity

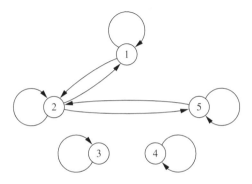

FIGURE 3.9
Five-vertex graph of tolerance relation (reflexive and symmetric) in Example 3.10.

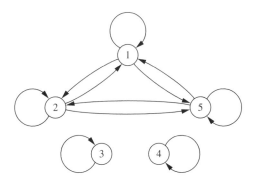

FIGURE 3.10
Five-vertex graph of equivalence relation (reflexive, symmetric, transitive) in Example 3.10.

(diagonal elements equal unity) simply indicates that a city is totally related to itself. The property of symmetry might represent proximity: Omaha and Chicago (x_1 and x_2) are close (in a binary sense) geographically, and Chicago and Detroit (x_2 and x_5) are close geographically. This relation, R_1, does not have properties of transitivity, e.g.,

$$(x_1, x_2) \in R_1 \quad (x_2, x_5) \in R_1 \quad \text{but} \quad (x_1, x_5) \notin R_1$$

R_1 can become an equivalence relation through one ($1 \le n$, where $n = 5$) composition. Using Eq. (3.20), we get

$$R_1 \circ R_1 = \begin{bmatrix} 1 & 1 & 0 & 0 & 1 \\ 1 & 1 & 0 & 0 & 1 \\ 0 & 0 & 1 & 0 & 0 \\ 0 & 0 & 0 & 1 & 0 \\ 1 & 1 & 0 & 0 & 1 \end{bmatrix} = R$$

Now, we see in this matrix that transitivity holds, i.e., $(x_1, x_5) \in R_1$, and R is an equivalence relation. Although the point is not important here, we will see in Chapter 11 that equivalence relations also have certain special properties useful in classification. For instance, in this example the equivalence relation expressed in the foregoing R matrix could represent cities in separate countries. Inspection of the matrix shows that the first, second, and fifth columns are identical, i.e., Omaha, Chicago, and Detroit are in the same class; and columns the third and fourth are unique, indicating that Rome and London are cities each in their own class; these three different classes could represent distinct countries. The graph for this equivalence relation would involve five vertices (five elements in the relation), as shown in Fig. 3.10.

FUZZY TOLERANCE AND EQUIVALENCE RELATIONS

A fuzzy relation, $\underset{\sim}{R}$, on a single universe X is also a relation from X to X. It is a fuzzy equivalence relation if all three of the following properties for matrix relations define it:

Reflexivity $\mu_R(x_i, x_i) = 1$ (3.21a)

Symmetry $\mu_R(x_i, x_j) = \mu_R(x_j, x_i)$ (3.21b)

Transitivity $\mu_R(x_i, x_j) = \lambda_1$ and $\mu_R(x_j, x_k) = \lambda_2 \longrightarrow \mu_R(x_i, x_k) = \lambda$ (3.21c)

where $\lambda \ge \min[\lambda_1, \lambda_2]$.

Looking at the physical analog (see Fig. 3.5) of a composition operation, we see it comprises a parallel system of chains, where each chain represents a particular path through the chain system. The physical analogy behind transitivity is that the shorter the chain, the stronger the relation (the stronger is the chain system). In particular, the strength of the link between two elements must be greater than or equal to the strength of any indirect chain involving other elements, i.e., Eq. (3.21c) [Dubois and Prade, 1980].

It can be shown that any fuzzy tolerance relation, R_1, that has properties of reflexivity and symmetry can be reformed into a fuzzy equivalence relation by at most $(n-1)$ compositions, just as a crisp tolerance relation can be reformed into a crisp equivalence relation. That is,

$$R_1^{n-1} = R_1 \circ R_1 \circ \cdots \circ R_1 = R \tag{3.22}$$

Example 3.11. Suppose, in a biotechnology experiment, five potentially new strains of bacteria have been detected in the area around an anaerobic corrosion pit on a new aluminum–lithium alloy used in the fuel tanks of a new experimental aircraft. In order to propose methods to eliminate the biocorrosion caused by these bacteria, the five strains must first be categorized. One way to categorize them is to compare them to one another. In a pairwise comparison, the following "similarity" relation, R_1, is developed. For example, the first strain (column 1) has a strength of similarity to the second strain of 0.8, to the third strain a strength of 0 (i.e., no relation), to the fourth strain a strength of 0.1, and so on. Because the relation is for pairwise similarity it will be reflexive and symmetric. Hence,

$$R_1 = \begin{bmatrix} 1 & 0.8 & 0 & 0.1 & 0.2 \\ 0.8 & 1 & 0.4 & 0 & 0.9 \\ 0 & 0.4 & 1 & 0 & 0 \\ 0.1 & 0 & 0 & 1 & 0.5 \\ 0.2 & 0.9 & 0 & 0.5 & 1 \end{bmatrix}$$

is reflexive and symmetric. However, it is not transitive, e.g.,

$$\mu_R(x_1, x_2) = 0.8, \quad \mu_R(x_2, x_5) = 0.9 \geq 0.8$$

but

$$\mu_R(x_1, x_5) = 0.2 \leq \min(0.8, 0.9)$$

One composition results in the following relation:

$$R_1^2 = R_1 \circ R_1 = \begin{bmatrix} 1 & 0.8 & 0.4 & 0.2 & 0.8 \\ 0.8 & 1 & 0.4 & 0.5 & 0.9 \\ 0.4 & 0.4 & 1 & 0 & 0.4 \\ 0.2 & 0.5 & 0 & 1 & 0.5 \\ 0.8 & 0.9 & 0.4 & 0.5 & 1 \end{bmatrix}$$

where transitivity still does not result; for example,

$$\mu_{R^2}(x_1, x_2) = 0.8 \geq 0.5 \quad \text{and} \quad \mu_{R^2}(x_2, x_4) = 0.5$$

but

$$\mu_{R^2}(x_1, x_4) = 0.2 \leq \min(0.8, 0.5)$$

Finally, after one or two more compositions, transitivity results:

$$\underset{\sim}{R}_1^3 = \underset{\sim}{R}_1^4 = \underset{\sim}{R} = \begin{bmatrix} 1 & 0.8 & 0.4 & 0.5 & 0.8 \\ 0.8 & 1 & 0.4 & 0.5 & 0.9 \\ 0.4 & 0.4 & 1 & 0.4 & 0.4 \\ 0.5 & 0.5 & 0.4 & 1 & 0.5 \\ 0.8 & 0.9 & 0.4 & 0.5 & 1 \end{bmatrix}$$

$\underset{\sim}{R}_1^3(x_1, x_2) = 0.8 \geq 0.5$

$\underset{\sim}{R}_1^3(x_2, x_4) = 0.5 \geq 0.5$

$\underset{\sim}{R}_1^3(x_1, x_4) = 0.5 \geq 0.5$

Graphs can be drawn for fuzzy equivalence relations, but the arrows in the graphs between vertices will have various ''strengths,'' i.e., values on the interval [0, 1]. Once the fuzzy relation $\underset{\sim}{R}$ in Example 3.11 is an equivalence relation, it can be used in categorizing the various bacteria according to preestablished levels of confidence. These levels of confidence will be illustrated with a method called ''alpha cuts'' in Chapter 4, and the categorization idea will be illustrated using classification in Chapter 11.

There is an interesting graphical analog for fuzzy equivalence relations. An inspection of a three-dimensional plot of the preceding equivalence relation, $\underset{\sim}{R}_1^3$, is shown in Fig. 3.11. In this graph, which is a plot of the membership values of the equivalence relation, we can see that, if it were a watershed, there would be no location where water would *pool*, or be trapped. In fact, every equivalence relation will produce a surface on which water cannot be trapped; the converse is not true in general, however. That is, there can be relations that are not equivalence relations but whose three-dimensional surface representations will not trap water. An example of the latter is given in the original tolerance relation, $\underset{\sim}{R}_1$, of Example 3.11.

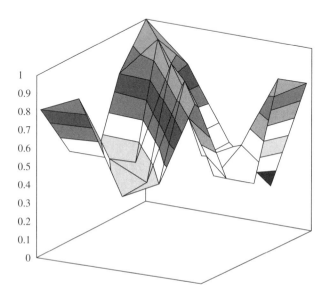

FIGURE 3.11
Three-dimensional representation of an equivalence relation.

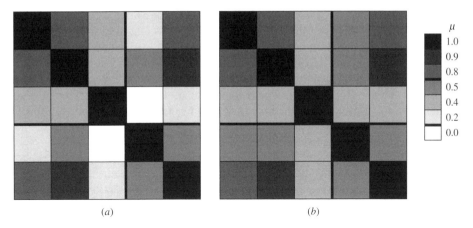

$$\mu$$
1.0
0.9
0.8
0.5
0.4
0.2
0.0

(a)   (b)

FIGURE 3.12
Two-dimensional contours of (a) tolerance relation $\underset{\sim}{R}_1^2$ and (b) equivalence relation $\underset{\sim}{R}_1^3$.

Another way to show this same information is to construct a two-dimensional contour of the relation using various contour levels; these correspond to different degrees of membership. The graphic in Fig. 3.12 shows contour plots for the tolerance relation, $\underset{\sim}{R}_1^2$, and the equivalence level, $\underset{\sim}{R}_1^3$. The contours in Fig. 3.12a would trap water (the lightest areas are inside other darker areas), whereas water would not be trapped in the contour diagram of Fig. 3.12b.

VALUE ASSIGNMENTS

An appropriate question regarding relations is: Where do the membership values that are contained in a relation come from? The answer to this question is that there are at least seven different ways to develop the numerical values that characterize a relation:

1. Cartesian product
2. Closed-form expression
3. Lookup table
4. Linguistic rules of knowledge
5. Classification
6. Automated methods from input/output data
7. Similarity methods in data manipulation

The first way is the one that has been illustrated so far in this chapter – to calculate relations from the Cartesian product of two or more fuzzy sets. A second way is through simple observation of a physical process. For a given set of inputs we observe a process yielding a set of outputs. If there is no variation between specific input – output pairs we may be led to model the process with a crisp relation. Moreover, if no variability exists, one might be able to express the relation as a closed-form algorithm of the form $\mathbf{Y} = f(\mathbf{X})$, where $\mathbf{X}$ is a vector of inputs and $\mathbf{Y}$ is a vector of outputs. If some variability exists, membership values on the interval [0, 1] may lead us to develop a fuzzy relation from a third approach – the use of a lookup table. Fuzzy relations can also be assembled from linguistic knowledge, expressed as if – then rules. Such knowledge may come from experts, from polls, or from consensus building. This fourth method is illustrated in more detail in Chapters 5 and 8. Relations also arise from notions of classification where issues associated with similarity are central to determining relationships among patterns or clusters of data. The ability to develop relations in classification, the fifth method, is developed in more detail in Chapter 11. The sixth method involves the development of membership functions from procedures used on input and output data which could be observed and measured from some complex process; this method is the subject of Chapter 7.

One of the most prevalent forms of determining the values in relations, and which is simpler than the sixth method, is through manipulations of data, the seventh method mentioned. The more robust a data set, the more accurate the relational entities are in establishing relationships among elements of two or more data sets. This seventh way for determining value assignments for relations is actually a family of procedures termed similarity methods [see Zadeh, 1971; or Dubois and Prade, 1980]. All of these methods attempt to determine some sort of similar pattern or structure in data through various metrics. There are many of these methods available, but the two most prevalent will be discussed here.

Cosine Amplitude

A useful method is the cosine amplitude method. As with all the following methods, this similarity metric makes use of a collection of data samples, n data samples in particular. If these data samples are collected they form a data array, X,

$$X = \{x_1, x_2, \ldots, x_n\}$$

Each of the elements, x_i, in the data array X is itself a vector of length m, i.e.,

$$x_i = \{x_{i_1}, x_{i_2}, \ldots, x_{i_m}\}$$

Hence, each of the data samples can be thought of as a point in m-dimensional space, where each point needs m coordinates for a complete description. Each element of a relation, r_{ij}, results from a pairwise comparison of two data samples, say x_i and x_j, where the strength of the relationship between data sample x_i and data sample x_j is given by the membership value expressing that strength, i.e., $r_{ij} = \mu_R(x_i, y_j)$. The relation matrix will be of size $n \times n$ and, as will be the case for all similarity relations, the matrix will be reflexive and symmetric – hence a tolerance relation. The cosine amplitude method calculates r_{ij} in the following manner, and guarantees, as do all the similarity methods, that $0 \le r_{ij} \le 1$:

$$r_{ij} = \frac{\left| \sum_{k=1}^{m} x_{ik} x_{jk} \right|}{\sqrt{\left(\sum_{k=1}^{m} x_{ik}^2 \right) \left(\sum_{k=1}^{m} x_{jk}^2 \right)}}, \quad \text{where } i, j = 1, 2, \ldots, n \qquad (3.23)$$

Close inspection of Eq. (3.23) reveals that this method is related to the dot product for the cosine function. When two vectors are colinear (most similar), their dot product is unity;

when the two vectors are at right angles to one another (most dissimilar), their dot product is zero.

Example 3.12 [Ross, 1995]. Five separate regions along the San Andreas fault in California have suffered damage from a recent earthquake. For purposes of assessing payouts from insurance companies to building owners, the five regions must be classified as to their damage levels. Expression of the damage in terms of relations will prove helpful.

Surveys are conducted of the buildings in each region. All of the buildings in each region are described as being in one of three damage states: no damage, medium damage, and serious damage. Each region has each of these three damage states expressed as a percentage (ratio) of the total number of buildings. Hence, for this problem $n = 5$ and $m = 3$. The following table summarizes the findings of the survey team:

Regions	x_1	x_2	x_3	x_4	x_5
x_{i1}–Ratio with no damage	0.3	0.2	0.1	0.7	0.4
x_{i2}–Ratio with medium damage	0.6	0.4	0.6	0.2	0.6
x_{i3}–Ratio with serious damage	0.1	0.4	0.3	0.1	0.0

We wish to use the cosine amplitude method to express these data as a fuzzy relation. Equation (3.23) for an element in the fuzzy relation, r_{ij}, thus takes on the specific form

$$r_{ij} = \frac{\left| \sum_{k=1}^{3} x_{ik} x_{jk} \right|}{\sqrt{\left(\sum_{k=1}^{3} x_{ik}^2 \right) \left(\sum_{k=1}^{3} x_{jk}^2 \right)}}$$

For example, for $i = 1$ and $j = 2$ we get

$$r_{12} = \frac{0.3 \times 0.2 + 0.6 \times 0.4 + 0.1 \times 0.4}{\left[(0.3^2 + 0.6^2 + 0.1^2)(0.2^2 + 0.4^2 + 0.4^2) \right]^{1/2}} = \frac{0.34}{[0.46 \times 0.36]^{1/2}} = 0.836$$

Computing the other elements of the relation results in the following tolerance relation:

$$R_1 = \begin{bmatrix} 1 & & & & \\ 0.836 & 1 & & \text{sym} & \\ 0.914 & 0.934 & 1 & & \\ 0.682 & 0.6 & 0.441 & 1 & \\ 0.982 & 0.74 & 0.818 & 0.774 & 1 \end{bmatrix}$$

and two compositions of R_1 produce the equivalence relation, R:

$$R = R_1^3 = \begin{bmatrix} 1 & & & & \\ 0.914 & 1 & & \text{sym} & \\ 0.914 & 0.934 & 1 & & \\ 0.774 & 0.774 & 0.774 & 1 & \\ 0.982 & 0.914 & 0.914 & 0.774 & 1 \end{bmatrix}$$

The tolerance relation, R_1, expressed the pairwise similarity of damage for each of the regions; the equivalence relation, R, also expresses this same information but additionally can be used to classify the regions into categories with *like properties* (see Chapter 11).

Max–Min Method

Another popular method, which is computationally simpler than the cosine amplitude method, is known as the max–min method. Although the name sounds similar to the max–min composition method, this similarity method is different from composition. It is found through simple min and max operations on pairs of the data points, x_{ij}, and is given by

$$r_{ij} = \frac{\sum_{k=1}^{m} \min(x_{ik}, x_{jk})}{\sum_{k=1}^{m} \max(x_{ik}, x_{jk})}, \quad \text{where } i, j = 1, 2, \ldots, n \tag{3.24}$$

Example 3.13. If we reconsider Example 3.12, the min–max method will produce the following result for $i = 1$, $j = 2$:

$$r_{12} = \frac{\sum_{k=1}^{3} [\min(0.3, 0.2), \min(0.6, 0.4), \min(0.1, 0.4)]}{\sum_{k=1}^{3} [\max(0.3, 0.2), \max(0.6, 0.4), \max(0.1, 0.4)]} = \frac{0.2 + 0.4 + 0.1}{0.3 + 0.6 + 0.4} = 0.538$$

Computing the other elements of the relation results in the following tolerance relation:

$$R_1 = \begin{bmatrix} 1 & & & & \\ 0.538 & 1 & & \text{sym} & \\ 0.667 & 0.667 & 1 & & \\ 0.429 & 0.333 & 0.250 & 1 & \\ 0.818 & 0.429 & 0.538 & 0.429 & 1 \end{bmatrix}$$

Other Similarity Methods

The list of other similarity methods is quite lengthy. Ross [1995] presents nine additional similarity methods, and others can be found in the literature.

OTHER FORMS OF THE COMPOSITION OPERATION

Max–min and max–product (also referred to as max–dot) methods of composition of fuzzy relations are the two most commonly used techniques. Many other techniques are mentioned in the literature. Each method of composition of fuzzy relations reflects a special inference machine and has its own significance and applications. The max–min method is the one used by Zadeh in his original paper on approximate reasoning using natural language if–then rules. Many have claimed, since Zadeh's introduction, that this method of composition effectively expresses the approximate and interpolative reasoning used by humans when they employ linguistic propositions for deductive reasoning [Ross, 1995].

The following additional methods are among those proposed in the literature for the composition operation $B = A \circ R$, where A is the input, or antecedent defined on the universe X, B is the output, or consequent defined on universe Y, and R is a fuzzy relation characterizing the relationship between specific inputs (x) and specific outputs (y):

$$\text{min--max} \qquad \mu_B(y) = \min_{x \in X}\{\max[\mu_A(x), \mu_R(x, y)]\} \qquad (3.25)$$

$$\text{max--max} \qquad \mu_B(y) = \max_{x \in X}\{\max[\mu_A(x), \mu_R(x, y)]\} \qquad (3.26)$$

$$\text{min--min} \qquad \mu_B(y) = \min_{x \in X}\{\min[\mu_A(x), \mu_R(x, y)]\} \qquad (3.27)$$

$$\text{max--average} \qquad \mu_B(y) = \tfrac{1}{2}\max_{x \in X}[\mu_A(x) + \mu_R(x, y)] \qquad (3.28)$$

$$\text{sum--product} \qquad \mu_B(y) = f\left\{\sum_{x \in X}[\mu_A(x) \cdot \mu_R(x, y)]\right\} \qquad (3.29)$$

where $f(\cdot)$ is a logistic function (like a sigmoid or a step function) that limits the value of the function within the interval $[0, 1]$. This composition method is commonly used in applications of artificial neural networks for mapping between parallel layers in a multilayer network.

It is left as an exercise for the reader (see Problems 3.26 and 3.27) to determine the relationship among these additional forms of the composition operator for various combinations of the membership values for $\mu_A(x)$ and $\mu_R(x, y)$.

SUMMARY

This chapter has shown some of the properties and operations of crisp and fuzzy relations. There are many more, but these will provide a sufficient foundation for the rest of the material in the text. The idea of a relation is most powerful; this modeling power will be shown in subsequent chapters dealing with such issues as logic, nonlinear simulation, classification, and control. The idea of composition was introduced, and it will be seen in Chapter 12 that the composition of a relation is similar to a method used to extend fuzziness into functions, called the *extension principle*. Tolerant and equivalent relations hold some special properties, as will be illustrated in Chapter 11, when they are used in similarity applications and classification applications, respectively. There are some very interesting graphical analogies for relations as seen in some of the example problems (also see Problem 3.9 at the end of the chapter). Finally, several similarity metrics were shown to be useful in developing the relational *strengths,* or distances, within fuzzy relations from data sets.

REFERENCES

Dubois, D. and Prade, H. (1980). *Fuzzy sets and systems: Theory and applications*, Academic Press, New York.

Gill, A. (1976). *Applied algebra for the computer sciences*, Prentice Hall, Englewood Cliffs, NJ.

Kandel, A. (1985). *Fuzzy mathematical techniques with applications*, Addison-Wesley, Menlo Park, CA.

Klir, G. and Folger, T. (1988). *Fuzzy sets, uncertainty, and information*, Prentice Hall, Englewood Cliffs, NJ.

Ross, T. (1995). *Fuzzy Logic and Engineering Applications*, 1st ed., McGraw-Hill, New York.
Zadeh, L. (1971). "Similarity relations and fuzzy orderings," *Inf. Sci.*, vol. 3, pp. 177–200.

PROBLEMS

General Relations

3.1. The provision of high-quality drinking water remains one of the greatest environmental challenges for public health officials and water utilities worldwide. In order to ensure maximum water quality in distribution networks, proper attention must be given to the pressure head at nodal points (measured by pressure probes) and the demand consumption pattern (measured by telemetry meters) along the whole length of the distribution network. Suppose we have two fuzzy sets, P defined on a universe of three discrete pressures $\{x_1, x_2, x_3\}$, and D defined on a universe of two discrete demand consumptions $\{y_1, y_2\}$, where fuzzy set P represents the near optimum pressure for high-quality drinking water and P represents the instantaneous demand (water demand) obtained from a time-series demand forecasting. Thus, P and D represent the variable inputs and water quality represents the output.

The Cartesian product represents the conditions (pressure–demand consumption) of the distribution system that are associated with near maximum water quality.

Let

$$P = \left\{ \frac{0.1}{x_1} + \frac{0.4}{x_2} + \frac{1}{x_3} \right\} \quad \text{and} \quad D = \left\{ \frac{0.5}{y_1} + \frac{0.8}{y_2} \right\}$$

Calculate the Cartesian product $T = P \times D$.

3.2. In a water treatment process, we use a biological process to remove biodegradable organic matter. The organic matter is measured as the biological oxygen demand (BOD), where the optimal BOD of effluent should be less them 20 mg/L. Let B represent a fuzzy set "good effluent" on the universe of optical BOD values (20, 40, 60) as defined by the membership function

$$\mu_B = \frac{0.5}{60} + \frac{0.7}{40} + \frac{1.0}{20}$$

The retention time is critical to a bioreactor; we try to find the retention time, measured in days. Let T represent a fuzzy set called "optimal retention time" on the universe of days (6, 8, 10) as given by the membership function

$$\mu_T = \frac{0.9}{10} + \frac{0.7}{8} + \frac{0.5}{6}$$

The utilization rate of organic food indicates the level of the treatment in the biological process, and this rate is measured on a universe of fractions from 0 to 1, where 1 is optimal. Fuzzy set U will represent "high utilization rates," as defined by the membership function

$$\mu_U = \frac{1}{0.9} + \frac{0.8}{0.8} + \frac{0.6}{0.7} + \frac{0.4}{0.6}$$

We can define the following relations:
$R = B \times T$, which reflects how retention time affects BOD removal;
$S = T \times U$, which relates how retention time affects organic food consumption; and
$W = R \circ S$, which represents the BOD removal and food utilization.

(a) Find R and S using Cartesian products.

(b) Find W using max–min composition.

(c) Find W using max–product composition.

3.3. Assume storm magnitudes are recorded on a rain gauge station within a 24 h period. We will represent our assessment of the size of a storm on the universe of rainfall depths, h_i, $i = 1, 2, 3$, where $h_3 > h_2 > h_1$. The data on depths are based on statistical estimates acquired from numerous rainfall records. The membership function representing the confidence in the rainfall depth of a particular "moderate storm" F is given by

$$F = \left\{ \frac{0.4}{h_1} + \frac{0.9}{h_2} + \frac{0.6}{h_3} \right\}$$

Suppose D is a fuzzy set which represents the rainfall duration, t_i ($t_i < 24$ h), where $t_2 > t_1$ and the duration can again be derived from statistics. The membership function of a "long duration storm" might be

$$D = \left\{ \frac{0.1}{t_1} + \frac{1.0}{t_2} \right\}$$

(a) Find the Cartesian product $F \times D = G$, which provides a relation between rainfall depth and duration.

(b) Then assume you have a fuzzy set of confidence in the measurement of the rainfall depth due to factors such as wind, human error, instrument type, etc. Such a fuzzy set on the universe of depths, say "high confidence in depth h_2", could be

$$E = \left\{ \frac{0.2}{h_1} + \frac{1.0}{h_2} + \frac{0.3}{h_3} \right\}$$

Using a max–min composition find $C = E \circ G$, which represents the best strength of the estimate with respect to the storm duration.

3.4. A company sells a product called a video multiplexer, which multiplexes the video from 16 video cameras into a single video cassette recorder (VCR). The product has a motion detection feature that can increase the frequency with which a given camera's video is recorded to tape depending on the amount of motion that is present. It does this by recording more information from that camera at the expense of the amount of video that is recorded from the other 15 cameras. Define a universe X to be the speed of the objects that are present in the video of camera 1 (there are 16 cameras). For example, let X = {Low Speed, Medium Speed, High Speed} = {LS, MS, HS}. Now, define a universe Y to represent the frequency with which the video from camera 1 is recorded to a VCR tape, i.e., the record rate of camera 1. Suppose Y = {Slow Record Rate, Medium Record Rate, Fast Record Rate} = {SRR, MRR, FRR}. Let us now define a fuzzy set A on X and a fuzzy set B on Y, where A represents a fuzzy slow-moving object present in video camera 1, and B represents a fuzzy slow record rate, biased to the slow side. For example,

$$A = \left\{ \frac{1}{LS} + \frac{0.4}{MS} + \frac{0.2}{HS} \right\}$$

$$B = \left\{ \frac{1}{SRR} + \frac{0.5}{MRR} + \frac{0.25}{FRR} \right\}$$

(a) Find the fuzzy relation for the Cartesian product of A and B, i.e., find $R = A \times B$.

(b) Suppose we introduce another fuzzy set, $\underset{\sim}{C}$, which represents a fuzzy fast-moving object present in video camera 1, say, for example, the following:

$$\underset{\sim}{C} = \left\{ \frac{0.1}{\text{LS}} + \frac{0.3}{\text{MS}} + \frac{1}{\text{HS}} \right\}$$

Find the relation between $\underset{\sim}{C}$ and $\underset{\sim}{B}$ using a Cartesian product, i.e., find $\underset{\sim}{S} = \underset{\sim}{C} \times \underset{\sim}{B}$.

(c) Find $\underset{\sim}{C} \circ \underset{\sim}{R}$ using max–min composition.

(d) Find $\underset{\sim}{C} \circ \underset{\sim}{R}$ using max–product composition.

(e) Comment on the differences between the results of parts (c) and (d).

3.5. Three variables of interest in power transistors are the amount of current that can be switched, the voltage that can be switched, and the cost. The following membership functions for power transistors were developed from a hypothetical components catalog:

$$\text{Average current (in amps)} = \underset{\sim}{I} = \left\{ \frac{0.4}{0.8} + \frac{0.7}{0.9} + \frac{1}{1} + \frac{0.8}{1.1} + \frac{0.6}{1.2} \right\}$$

$$\text{Average voltage (in volts)} = \underset{\sim}{V} = \left\{ \frac{0.2}{30} + \frac{0.8}{45} + \frac{1}{60} + \frac{0.9}{75} + \frac{0.7}{90} \right\}$$

Note how the membership values in each set taper off faster toward the lower voltage and currents. These two fuzzy sets are related to the "power" of the transistor. Power in electronics is defined by an algebraic operation, $P = VI$, but let us deal with a general Cartesian relationship between voltage and current, i.e., simply with $\underset{\sim}{P} = \underset{\sim}{V} \times \underset{\sim}{I}$. Keep in mind that the Cartesian product is different from the arithmetic product. The Cartesian product expresses the relationship between V_i and I_j, where V_i and I_j are individual elements in the fuzzy sets $\underset{\sim}{V}$ and $\underset{\sim}{I}$.

(a) Find the fuzzy Cartesian product $\underset{\sim}{P} = \underset{\sim}{V} \times \underset{\sim}{I}$.

Now let us define a fuzzy set for the cost $\underset{\sim}{C}$, in dollars, of a transistor, e.g.,

$$\underset{\sim}{C} = \left\{ \frac{0.4}{0.5} + \frac{1}{0.6} + \frac{0.5}{0.7} \right\}$$

(b) Using a fuzzy Cartesian product, find $\underset{\sim}{T} = \underset{\sim}{I} \times \underset{\sim}{C}$. What would this relation, $\underset{\sim}{T}$, represent physically?

(c) Using max–min composition, find $\underset{\sim}{E} = \underset{\sim}{P} \circ \underset{\sim}{T}$. What would this relation, $\underset{\sim}{E}$, represent physically?

(d) Using max–product composition, find $\underset{\sim}{E} = \underset{\sim}{P} \circ \underset{\sim}{T}$.

3.6. The relationship between temperature and maximum operating frequency R depends on various factors for a given electronic circuit. Let $\underset{\sim}{T}$ be a temperature fuzzy set (in degrees Fahrenheit) and $\underset{\sim}{F}$ represent a frequency fuzzy set (in MHz) on the following universes of discourse:

$$\underset{\sim}{T} = \{-100, -50, 0, 50, 100\} \quad \text{and} \quad \underset{\sim}{F} = \{8, 16, 25, 33\}$$

Suppose a Cartesian product between $\underset{\sim}{T}$ and $\underset{\sim}{F}$ is formed that results in the following relation:

$$\underset{\sim}{R} = \begin{array}{c} \\ 8 \\ 16 \\ 25 \\ 33 \end{array} \begin{array}{ccccc} -100 & -50 & 0 & 50 & 100 \\ \left[\begin{array}{ccccc} 0.2 & 0.5 & 0.7 & 1 & 0.9 \\ 0.3 & 0.5 & 0.7 & 1 & 0.8 \\ 0.4 & 0.6 & 0.8 & 0.9 & 0.4 \\ 0.9 & 1 & 0.8 & 0.6 & 0.4 \end{array} \right] \end{array}$$

The reliability of the electronic circuit is related to the maximum operating temperature. Such a relation S can be expressed as a Cartesian product between the reliability index, $M = \{1, 2, 4, 8, 16\}$ (in dimensionless units), and the temperature, as in the following example:

$$S = \begin{array}{c@{\,}c} & \begin{array}{ccccc} 1 & 2 & 4 & 8 & 16 \end{array} \\ \begin{array}{r} -100 \\ -50 \\ 0 \\ 50 \\ 100 \end{array} & \left[\begin{array}{ccccc} 1 & 0.8 & 0.6 & 0.3 & 0.1 \\ 0.7 & 1 & 0.7 & 0.5 & 0.4 \\ 0.5 & 0.6 & 1 & 0.8 & 0.8 \\ 0.3 & 0.4 & 0.6 & 1 & 0.9 \\ 0.9 & 0.3 & 0.5 & 0.7 & 1 \end{array}\right] \end{array}$$

Composition can be performed on any two or more relations with compatible row–column consistency. To find a relationship between frequency and the reliability index, use

(a) max–min composition

(b) max–product composition

3.7. The formation of algal solutions and other biological colonies in surface waters is strongly dependent on such factors as the pH of the water, the temperature, and oxygen content. Relationships among these various factors enable environmental engineers to study issues involving bioremediation using the algae. Suppose we define a set T of water temperatures from a lake on the following discrete universe of temperatures in degrees Fahrenheit:

$$T = \{50, 55, 60\}$$

And suppose we define a universe O of oxygen content values in the water, as percent by volume:

$$O = \{1, 2, 6\}$$

Suppose a Cartesian product is performed between specific fuzzy sets T and O defined on T and Q to produce the following relation:

$$R = T \times O = \begin{array}{c@{\,}c} & \begin{array}{ccc} 1 & 2 & 6 \end{array} \\ \begin{array}{r} 50 \\ 55 \\ 60 \end{array} & \left[\begin{array}{ccc} 0.1 & 0.2 & 0.9 \\ 0.1 & 1 & 0.7 \\ 0.8 & 0.7 & 0.1 \end{array}\right] \end{array}$$

Now suppose we define another fuzzy set of temperatures, "about 55°F," with the following membership values:

$$I_T = \left\{ \frac{0.5}{50} + \frac{1}{55} + \frac{0.7}{60} \right\}$$

(a) Using max–min composition, find $S = I_T \circ (T \times O)$.

(b) Using max–product composition, find $S = I_T \circ R$.

3.8. Relating earthquake intensity to ground acceleration is an imprecise science. Suppose we have a universe of earthquake intensities (on the Mercalli scale), $I = \{5, 6, 7, 8, 9\}$, and a universe of accelerations, $A = \{0.2, 0.4, 0.6, 0.8, 1.0, 1.2\}$, in g. The following fuzzy relation, R, exists on the Cartesian space $I \times A$:

$$R = \begin{array}{c@{\,}c} & \begin{array}{cccccc} 0.2 & 0.4 & 0.6 & 0.8 & 1.0 & 1.2 \end{array} \\ \begin{array}{r} 5 \\ 6 \\ 7 \\ 8 \\ 9 \end{array} & \left[\begin{array}{cccccc} 0.75 & 1 & 0.85 & 0.5 & 0.2 & 0 \\ 0.5 & 0.8 & 1 & 0.7 & 0.3 & 0 \\ 0.1 & 0.5 & 0.8 & 1 & 0.7 & 0.1 \\ 0 & 0.2 & 0.5 & 0.85 & 1 & 0.6 \\ 0 & 0 & 0.2 & 0.5 & 0.9 & 1 \end{array}\right] \end{array}$$

If the fuzzy set "intensity about 7" is defined as

$$I_7 = \left\{ \frac{0.1}{5} + \frac{0.6}{6} + \frac{1}{7} + \frac{0.8}{8} + \frac{0.2}{9} \right\}$$

determine the fuzzy membership of I_7 on the universe of accelerations, A.

3.9. Given the continuous, noninteractive fuzzy sets A and B on universes X and Y, using Zadeh's notation for continuous fuzzy variables,

$$A = \left\{ \int \frac{1 - 0.1|x|}{x} \right\} \quad \text{for } x \in [0, +10]$$

$$B = \left\{ \int \frac{0.2|y|}{y} \right\} \quad \text{for } y \in [0, +5]$$

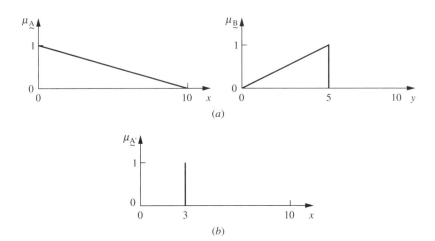

(a)

(b)

FIGURE P3.9

as seen in Fig. P3.9a:

(a) Construct a fuzzy relation R for the Cartesian product of A and B.

(b) Use max–min composition to find B', given the fuzzy singleton $A' = \frac{1}{3}$ (see Fig. P3.9b). *Hint:* You can solve this problem graphically by segregating the Cartesian space into various regions according to the min and max operations, or you can approximate the continuous fuzzy variables as discrete variables and use matrix operations. In any case, sketch the solution.

3.10. Risk assessment of hazardous waste situations requires the assimilation of a great deal of linguistic information. In this context, consider risk as being defined as "consequence of a hazard" multiplied by "possibility of the hazard" (instead of the conventional "probability of hazard"). Consequence is the result of an unintended action on humans, equipment, or facilities, or the environment. Possibility is the estimate of the likelihood that the unintended action will occur. Consequence and possibility are dependent on several factors and therefore cannot be determined with precision. We will use composition, then, to define risk; hence risk = consequence ∘ possibility, or

$$R = C \circ P$$

We will consider that the consequence is the logical intersection of the hazard mitigation and the hazard source term (the source term defines the type of initiation, such as a smokestack emitting a toxic gas, or a truck spill of a toxic organic); hence we define consequence = mitigation ∩ source term, or

$$C = M \cap ST$$

Since humans and their systems are ultimately responsible for preventing or causing non-natural hazards, we define the possibility of a hazard as the logical intersection of human errors and system vulnerabilities; hence possibility = human factors ∩ system reliabilities,

$$P = H \cap S$$

From these postulates, show that the membership form of the risk, R, is given by the expression

$$\mu_R(x, y) = \max\{\min(\min[\mu_M(x, y), \mu_{ST}(x, y)], \min[\mu_H(x, y), \mu_S(x, y)])\}$$

3.11. A new optical microscope camera uses a lookup table to relate voltage readings (which are related to illuminance) to exposure time. To aid in the creation of this lookup table, we need to determine how much time the camera should expose the pictures at a certain light level. Define a fuzzy set "around 3 volts" on a universe of voltage readings in volts

$$V_{1 \times 5} = \left\{ \frac{0.1}{2.98} + \frac{0.3}{2.99} + \frac{0.7}{3} + \frac{0.4}{3.01} + \frac{0.2}{3.02} \right\} \quad \text{(volts)}$$

and a fuzzy set "around 1/10 second" on a universe of exposure time in seconds

$$T_{1 \times 6} = \left\{ \frac{0.1}{0.05} + \frac{0.3}{0.06} + \frac{0.3}{0.07} + \frac{0.4}{0.08} + \frac{0.5}{0.09} + \frac{0.2}{0.1} \right\} \quad \text{(seconds)}$$

(a) Find $R = V \times T$.

Now define a third universe of "stops." In photography, stops are related to making the picture some degree lighter or darker than the "average" exposed picture. Therefore, let Universe of Stops = $\{-2, -1.5, -1, 0, .5, 1, 1.5, 2\}$ (stops). We will define a fuzzy set on this universe as

$$Z = \text{a little bit lighter} = \left\{ \frac{0.1}{0} + \frac{0.7}{0.5} + \frac{0.3}{1} \right\}$$

(b) Find $S = T \times Z$.

(c) Find $M = R \circ S$ by max–min composition.

(d) Find $M = R \circ S$ by max–product composition.

3.12. Music is not a precise science. Tactile movements by musicians on various instruments come from years of practice, and such movements are very subjective and imprecise. When a guitar player changes from an A chord to a C chord (major), his or her fingers have to move some distance, which can be measured in terms of frets (e.g., 1 fret = 0.1). This change in finger movement is described by the relation given in the following table. The table is for a six-string guitar: x_i is the string number, for $i = 1, 2, \ldots, 6$. For example, -0.2 is two frets down and 0.3 is three frets up, where 0 is located at the top of the guitar fingerboard.

A chord	C chord					
	x_6	x_5	x_4	x_3	x_2	x_1

The finger positions on the guitar strings for the two chords can be given in terms of the following membership functions:

$$\underset{\sim}{C} \text{ chord} = \left\{ \frac{0}{x_6} + \frac{0.3}{x_5} + \frac{0.2}{x_4} + \frac{0}{x_3} + \frac{0.1}{x_2} + \frac{0}{x_1} \right\}$$

$$\underset{\sim}{A} \text{ chord} = \left\{ \frac{0}{x_6} + \frac{0}{x_5} + \frac{0.2}{x_4} + \frac{0.2}{x_3} + \frac{0.2}{x_2} + \frac{0}{x_1} \right\}$$

Suppose the placement of fingers on the six strings for a G chord is given as

$$\underset{\sim}{G} \text{ chord} = \left\{ \frac{0.3}{x_6} + \frac{0.2}{x_5} + \frac{0}{x_4} + \frac{0}{x_3} + \frac{0}{xx_2} + \frac{0.3}{x_1} \right\}$$

(a) Find the relation that expresses moving from an A chord to a G chord; call this $\underset{\sim}{R}$.
(b) Use max–product composition to determine $\underset{\sim}{C} \circ \underset{\sim}{R}$.

3.13. In neuroscience research, it is often necessary to relate functional information to anatomical information. One source of anatomical information for a subject is a series of magnetic resonance imaging (MRI) pictures, composed of gray-level pixels, of the subject's head. For some applications, it is useful to segment (label) the brain into MRI slices (images along different virtual planes through the brain). This procedure can be difficult to do using gray-level values alone. A standard – or model – brain, combined with a distance criterion, can be used in conjunction with the gray-level information to improve the segmentation process. Define the following elements for the problem:

1. Normalized distance from the model (D)

$$\underset{\sim}{D} = \left\{ \frac{1}{0} + \frac{0.7}{1} + \frac{0.3}{2} \right\}$$

2. Intensity range for the cerebral cortex $(\underset{\sim}{I_C})$

$$\underset{\sim}{I_C} = \left\{ \frac{0.5}{20} + \frac{1}{30} + \frac{0.6}{40} \right\}$$

3. Intensity range for the medulla $(\underset{\sim}{I_M})$

$$\underset{\sim}{I_M} = \left\{ \frac{0.7}{20} + \frac{0.9}{30} + \frac{0.4}{40} \right\}$$

Based on these membership functions, find the following:
(a) $\underset{\sim}{R} = \underset{\sim}{I_C} \times \underset{\sim}{D}$
(b) Max–min composition of $\underset{\sim}{I_M} \circ \underset{\sim}{R}$
(c) Max–product composition of $\underset{\sim}{I_M} \circ \underset{\sim}{R}$

3.14. In the field of computer networking there is an imprecise relationship between the level of use of a network communication bandwidth and the latency experienced in peer-to-peer communications. Let $\underset{\sim}{X}$ be a fuzzy set of use levels (in terms of the percentage of full bandwidth used) and $\underset{\sim}{Y}$ be a fuzzy set of latencies (in milliseconds) with the following membership functions:

$$\underset{\sim}{X} = \left\{ \frac{0.2}{10} + \frac{0.5}{20} + \frac{0.8}{40} + \frac{1.0}{60} + \frac{0.6}{80} + \frac{0.1}{100} \right\}$$

$$\underset{\sim}{Y} = \left\{ \frac{0.3}{0.5} + \frac{0.6}{1} + \frac{0.9}{1.5} + \frac{1.0}{4} + \frac{0.6}{8} + \frac{0.3}{20} \right\}$$

(a) Find the Cartesian product represented by the relation $R = X \times Y$.
Now, suppose we have a second fuzzy set of bandwidth usage given by

$$Z = \left\{ \frac{0.3}{10} + \frac{0.6}{20} + \frac{0.7}{40} + \frac{0.9}{60} + \frac{1}{80} + \frac{0.5}{100} \right\}$$

Find $S = Z_{1\times6} \circ R_{6\times6}$
(b) using max–min composition;
(c) using max–product composition.

3.15. High-speed rail monitoring devices sometimes make use of sensitive sensors to measure the deflection of the earth when a rail car passes. These deflections are measured with respect to some distance from the rail car and, hence, are actually very small angles measured in microradians. Let a universe of deflections be $A = \{1, 2, 3, 4\}$ where A is the angle in microradians, and let a universe of distances be $D = \{1, 2, 5, 7\}$ where D is distance in feet. Suppose a relation between these two parameters has been determined as follows:

$$R = \begin{array}{c} \\ A_1 \\ A_2 \\ A_3 \\ A_4 \end{array} \begin{array}{cccc} D_1 & D_2 & D_3 & D_4 \\ \left[\begin{array}{cccc} 1 & 0.3 & 0.1 & 0 \\ 0.2 & 1 & 0.3 & 0.1 \\ 0 & 0.7 & 1 & 0.2 \\ 0 & 0.1 & 0.4 & 1 \end{array}\right] \end{array}$$

Now let a universe of rail car weights be $W = \{1, 2\}$, where W is the weight in units of 100,000 pounds. Suppose the fuzzy relation of W to A is given by

$$S = \begin{array}{c} \\ A_1 \\ A_2 \\ A_3 \\ A_4 \end{array} \begin{array}{cc} W_1 & W_2 \\ \left[\begin{array}{cc} 1 & 0.4 \\ 0.5 & 1 \\ 0.3 & 0.1 \\ 0 & 0 \end{array}\right] \end{array}$$

Using these two relations, find the relation, $R^T \circ S = T$ (note the matrix transposition here)
(a) using max–min composition;
(b) using max–product composition.

3.16. In the field of soil mechanics new research methods involving vision recognition systems are being applied to soil masses to watch individual grains of soil as they translate and rotate under confining pressures. In tracking the motion of the soil particles some problems arise with the vision recognition software. One problem is called ''occlusion,'' whereby a soil particle that is being tracked becomes partially or completely occluded from view by passing behind other soil particles. Occlusion can also occur when a tracked particle is behind a mark on the camera's lens, or the tracked particle is partly out of sight of the camera. In this problem we will consider only occlusions for the first two problems mentioned. Let us define a universe of parameters for particle occlusion, say

$$X = \{x_1, x_2, x_3\}$$

and a universe of parameters for lens mark occlusion, say

$$Y = \{y_1, y_2, y_3\}$$

Then, define

$$A = \left\{ \frac{0.1}{x_1} + \frac{0.9}{x_2} + \frac{0.0}{x_3} \right\}$$

as a specific fuzzy set for a tracked particle behind another particle, and let

$$\underset{\sim}{B} = \left\{ \frac{0}{y_1} + \frac{1}{y_2} + \frac{0}{y_3} \right\}$$

be a particular fuzzy set for an occlusion behind a lens mark.

(a) Find the relation, $\underset{\sim}{R} = \underset{\sim}{A} \times \underset{\sim}{B}$, using a Cartesian product.

Let $\underset{\sim}{C}$ be another fuzzy set in which a tracked particle is behind a particle, e.g.,

$$\underset{\sim}{C} = \left\{ \frac{0.3}{x_1} + \frac{1.0}{x_2} + \frac{0.0}{x_3} \right\}$$

(b) Using max–min composition, find $\underset{\sim}{S} = \underset{\sim}{C} \circ \underset{\sim}{R}$.

3.17. A common way to control the outlet composition of a distillation column is by controlling the temperature of one of the trays. In order for the system to be controlled properly, the set point of the chosen tray must guarantee that the outlet composition of the distillation column (could be the distillate, the bottoms, or a side stream) is under specifications for the expected range of disturbances. Tray temperature is measured instead of outlet composition because it is hard to measure online the fraction of a compound without dead time (e.g., common gas chromatographs take minutes to make a reading), whereas temperature is easy, fast, and inexpensive to measure. Steady state simulation runs are made for the expected disturbances in order to pick a tray from the distillation column and a temperature value for the control scheme. The problem with this methodology is that there is never a "best" tray temperature for all the disturbances.

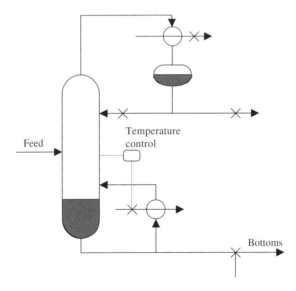

FIGURE P3.17
Distillation column.

For this problem, we define a distillation column (see Fig. P3.17) that separates light hydrocarbons, propane, butane, and iso-butane. We also state that the composition of propane in the bottoms must be ≤ 0.01 mole fraction, and that the distillation column is made of 20 trays counted from top to bottom.

We define the set $\underset{\sim}{F}$ to represent three types of feed flows = {High flow, Medium flow, Low flow}. Besides the bulk amount of the feed, another important factor is the comparison in the feed of the key component. We will use propane (C_3) as the key component. Define the fuzzy set $\underset{\sim}{P}$ = {High C_3, Low C_3}. Finally, let us say that from previous raw studies, the prospective trays to be chosen for control are reduced to the set T = {Tray$_{14}$, Tray$_{15}$, Tray$_{16}$, Tray$_{17}$}, where fuzzy set $\underset{\sim}{T}$ refers to how well it keeps the column under specifications.

If we have the fuzzy sets

$$\underset{\sim}{F} = \begin{array}{c} \text{High flow} \\ \text{Med flow} \\ \text{Low flow} \end{array} \begin{bmatrix} 0.1 \\ 0.7 \\ 0.4 \end{bmatrix} \qquad \underset{\sim}{P} = \begin{array}{cc} \text{High } C_3 & \text{Low } C_3 \\ [\quad 0.3 & 0.8 \quad] \end{array}$$

(a) Find the Cartesian product $\underset{\sim}{R} = \underset{\sim}{F} \times \underset{\sim}{P}$.

(b) $\underset{\sim}{P}$ and $\underset{\sim}{T}$ are highly related. If we have

$$\underset{\sim}{T} = \begin{array}{cccc} \text{Tray}_{14} & \text{Tray}_{15} & \text{Tray}_{16} & \text{Tray}_{17} \\ [\, 0.1 & 0.3 & 0.8 & 0.7 \,] \end{array}$$

then find the Cartesian product $\underset{\sim}{S} = \underset{\sim}{P}^T \times \underset{\sim}{T}$.

(c) Since outlet composition is not easily measured, it is still important to know how $\underset{\sim}{F}$ and $\underset{\sim}{T}$ are related. This knowledge can be acquired with a max–min composition; find $\underset{\sim}{C} = \underset{\sim}{R} \circ \underset{\sim}{S}$.

Value Assignments and Similarity

3.18. Pyrolysis is a widely used high-temperature reaction to produce ethylene and propylene. When this product leaves the furnace reactor it is necessary to reduce their temperature as quickly as possible in order to stop the reaction and avoid producing undesired products. This "quenching" of the products is made in equipment that works and looks just like any heat exchanger, with the difference that a reaction is actually happening in the tubes. Very good heat transfer is obtained when the tubes are clean, but coke deposition on the inner surface of the tubes reduces the heat transfer coefficient by a considerable amount and it also increases the pressure drop. Therefore, coke deposition in the pyrolysis of light hydrocarbons becomes an important factor in the design of the equipment (usually, the equipment needs cleaning every four months). An experiment was set in order to determine the coke deposition in the exchangers built with 10 tubes for different conditions. The different conditions were:

$$X_1 = \text{low tube diameter}$$

$$X_2 = \text{big tube diameter}$$

$$X_3 = \text{tubes made of material 1}$$

$$X_4 = \text{tubes made of material 2}$$

$$X_5 = \text{high pressure}$$

$$X_6 = \text{very high pressure}$$

For every run, all of the 10 tubes were examined and distributed in four categories by percentage. Not all the tubes were expected to get exactly the same amount of coke deposition because there is never perfect distribution of the feed into the 10 tubes.

The examination categories are: High deposition, Med–High deposition, Med deposition, and Moderate deposition. The results are as follows:

	X_1	X_2	X_3	X_4	X_5	X_6
High deposition	0.05	0.01	0.7	0.03	0	0.6
Med–High deposition	0.8	0.5	0.2	0.9	0	0.3
Med deposition	0.1	0.4	0.1	0.04	0.2	0.1
Moderate deposition	0.05	0.09	0	0.03	0.8	0

(a) It is desired to find the similarity among the six pyrolysis conditions; use the max–min method to find the similarity.

(b) Use the cosine amplitude method to find the similarity.

3.19. A structural designer is considering four different kinds of structural beams ($S_1, \ldots, S_4$) for a new building. Laboratory experiments on the deflection resistance for these four different kinds of beams have been performed, and the engineer wants to determine their suitability in the new structure. The following data have been observed, based on the overall deflection capacity of each bean type:

		S_1	S_2	S_3	S_4
No deflection	x_1	0.4	0.6	0.5	0.9
Some deflection	x_2	0.5	0.3	0.5	0.1
Excessive deflection	x_3	0.1	0.1	0	0

Using the cosine amplitude method determine the similarity of the four beam types.

3.20. Given a particular stream gauge record from three gauging stations, g_1-g_3, during major rainstorms, we can count the years that have had storms with equal magnitude over different periods to get a quantity used in the design of storm sewers: the number of years with similar storms over a specified time period.

	Gauge		
	g_1	g_2	g_3
$T_1 = 10$ years	0.00	0.10	0.10
$T_2 = 25$ years	0.04	0.04	0.08
$T_3 = 50$ years	0.02	0.04	0.06

Find the similarity relation, $\underset{\sim}{R}$, among the three gauges using the max–min method.

3.21. A certain type of virus attacks cells of human body, as explained in Example 3.7. The infected cells can be visualized using a special microscope; the virus causes the infected cells to have a black spot in the image as depicted in Fig. 3.6. The microscope generates digital images that medical doctors can analyze to identify the infected cells. A physician analyzes different samples of an infected organ (S) and classifies it as Not infected, Moderately infected, and Seriously infected. Five samples were analyzed as given in the table below:

Samples	S_1	S_2	S_3	S_4	S_5
Not infected	0.6	0.3	0.1	0.9	0.8
Moderately infected	0.4	0.5	0.3	0.1	0.1
Seriously infected	0.0	0.2	0.6	0.0	0.1

Use the min–max method to find the similarity relation.

3.22. In the statistical characterization of fractured reservoirs, the goal is to classify the geology according to different kinds of fractures, which are mainly tectonic and regional fractures. The purpose of this classification is to do critical simulation based on well data, seismic data, and fracture pattern. After pattern recognition (using Cauchy–Euler detection algorithms or other methods) and classification of the fracture images derived from the outcrops of fractured reservoirs, a geological engineer can get different patterns corresponding to different fracture morphologies. Suppose the engineer obtains five images ($I_1 \ldots I_5$) from five different outcrops of fractured reservoirs, and their percentage values corresponding to three kinds of fractures (Tectonic fracture, Regional fracture, and Other fracture), as given below:

	I_1	I_2	I_3	I_4	I_5
Tectonic fracture	0.6	0.6	0.3	0.5	0.2
Regional fracture	0.3	0.1	0.2	0.2	0.6
Other fracture	0.1	0.3	0.5	0.3	0.2

Develop a similarity relation
(*a*) using the cosine amplitude method and
(*b*) using the max–min method.
(*c*) Since the similarity relation will be a tolerance relation, fund the associated equivalence relation.

3.23. The latitude of the receiver can affect the positional accuracies of geo-positioning-system (GPS) results because the GPS was designed primarily for operation close to the equator. The accuracy of a GPS degrades as the distance from the equator increases. We define the following fuzzy sets:
discrete set for position accuracies ($\underset{\sim}{P}$)
discrete set for latitude ($\underset{\sim}{L}$)
(*a*) Discuss the physical meaning of the relation $R_1 = \underset{\sim}{P} \times \underset{\sim}{L}$.
  Position accuracies of GPS are also affected by ionospheric effects. Ionospheric effects are strongest at the magnetic equator and they weaken as the GPS gets closer to the poles. We define a fuzzy discrete set for ionospheric effect ($\underset{\sim}{I}$).
(*b*) Discuss the physical meaning of the relation $R_2 = \underset{\sim}{L} \times \underset{\sim}{I}$.
(*c*) Through composition we can relate the ionospheric effect with the position accuracies, i.e., by calculating $\underset{\sim}{IP} = R_1 \circ R_2$. Discuss the physical meaning of this composition.
(*d*) Five sets of GPS measurements were made and the quality of each range to a single satellite was graded based on its elevation angle, in degrees. The quality ranged from poor to acceptable and to high quality. The following table shows the distribution of GPS ranges for each of the five sets of measurements. Using the max–min similarity method find the fuzzy similarity relation for the five sets.

| | GPS measurement | | | | |
	1	2	3	4	5
Poor quality ($0°–15°$)	0.2	0.3	0.4	0.5	0.4
Acceptable quality ($10°–60°$)	0.6	0.6	0.5	0.5	0.6
High quality ($50°–90°$)	0.2	0.1	0.1	0.0	0.0

(*e*) Find the equivalence relation of the matrix developed in part (*d*).

3.24. Over the last decade Calgary, Alberta, has made use of a large number of storm-water ponds to mitigate flooding during extreme rainfall events by reducing the peak flows in trunk sewers or receiving waterways. The ponds have control structures that have data recorders monitoring water levels during the storm events. More ponds are built as urban development occurs. To determine the similarity of the pond performances within the city quadrants would be useful in determining if citywide drainage policies work effectively. Let the regions of the city be represented by the four quadrants (NW, NE, SW, SE). The ponds in the area can have performance based on design capacities broken down into three categories of stormwater filling levels: low, moderate, and high values based on expectations and percentage of ponds in the category. The following table represents a recent storm event summary:

Location	NW	NE	SW	SE
x_{i1}–Ratio with low filling	0.3	0.2	0.1	0.7
x_{i2}–Ratio with moderate filling	0.6	0.4	0.6	0.2
x_{i3}–Ratio with high filling	0.1	0.4	0.3	0.1

(a) Use the cosine amplitude method to express these data as a fuzzy relation.
(b) Comment on the similarity of the ponds in the four quadrants of the city.
(c) Find the equivalence relation associated with the matrix calculated in part (b). How does the equivalence relation differ, in terms of the physical significance, from the original tolerance relation?

Equivalence Relations

3.25. The accompanying Sagittal diagrams (Fig. P3.25) show two relations on the universe, $X = \{1, 2, 3\}$. Are these relations equivalence relations [Gill, 1976]?

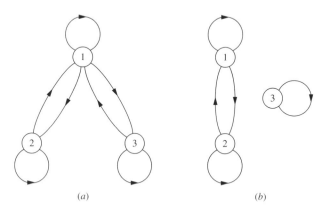

(a)   (b)

FIGURE P3.25
From [Gill, 1976].

Other Composition Operations

3.26. For Example 3.6 in this chapter, recalculate the fuzzy relation $\underset{\sim}{T}$ using
(i) Equation (3.25)
(ii) Equation (3.26)

(iii) Equation (3.27)
(iv) Equation (3.28)
(v) Equation (3.29), where

$$f(\bullet) = \begin{cases} 1 - e^{-x}, & \text{for } x \geq 0 \\ e^x - 1, & \text{for } x \leq 0 \end{cases}$$

3.27. Fill in the following table using Eqs. (3.25)–(3.29) to determine values of the composition $\underset{\sim}{B} = \underset{\sim}{A} \circ \underset{\sim}{R}$ for the fuzzy relation

$$\underset{\sim}{R} = \begin{array}{c} \\ x_1 \\ x_2 \\ x_3 \end{array} \begin{array}{ccc} y_1 & y_2 & y_3 \\ \left[\begin{array}{ccc} 0.1 & 0.2 & 0.3 \\ 0.4 & 0.5 & 0.6 \\ 0.7 & 0.8 & 0.9 \end{array}\right] \end{array}$$

Comment on the similarities and dissimilarities of the various composition methods with respect to the various antecedents, $\underset{\sim}{A}$.

$\underset{\sim}{A}$	$\underset{\sim}{B}$
[0.1 0.5 1.0]	
[1.0 0.6 0.1]	
[0.2 0.6 0.4]	
[0.7 0.9 0.8]	

CHAPTER
4

PROPERTIES OF MEMBERSHIP FUNCTIONS, FUZZIFICATION, AND DEFUZZIFICATION

*"Let's consider your age, to begin with – how old are you?" "I'm seven and a half, exactly."
"You needn't say 'exactually,'" the Queen remarked; "I can believe it without that. Now I'll
give you something to believe. I'm just one hundred and one, five months, and a day." "I can't
believe that!" said Alice.*

*"Can't you?" the Queen said in a pitying tone. "Try again; draw a long breath, and shut
your eyes." Alice laughed. "There's no use trying," she said; "one can't believe impossible
things."*

Lewis Carroll
Through the Looking Glass, 1871

It is one thing to compute, to reason, and to model with fuzzy information; it is another
to apply the fuzzy results to the world around us. Despite the fact that the bulk of the
information we assimilate every day is fuzzy, like the age of people in the Lewis Carroll
example above, most of the actions or decisions implemented by humans or machines are
crisp or binary. The decisions we make that require an action are binary, the hardware we
use is binary, and certainly the computers we use are based on binary digital instructions.
For example, in making a decision about developing a new engineering product the eventual
decision is to go forward with development or not; the fuzzy choice to "maybe go forward"
might be acceptable in planning stages, but eventually funds are released for development
or they are not. In giving instructions to an aircraft autopilot, it is not possible to turn the
plane "slightly to the west"; an autopilot device does not understand the natural language

Fuzzy Logic with Engineering Applications, Second Edition   T. J. Ross
© 2004 John Wiley & Sons, Ltd   ISBNs: 0-470-86074-X (HB); 0-470-86075-8 (PB)

of a human. We have to turn the plane by 15°, for example, a crisp number. An electrical circuit typically is either on or off, not partially on.

The bulk of this textbook illustrates procedures to "fuzzify" the mathematical and engineering principles we have so long considered to be deterministic. But in various applications and engineering scenarios there will be a need to "defuzzify" the fuzzy results we generate through a fuzzy systems analysis. In other words, we may eventually find a need to convert the fuzzy results to crisp results. For example, in classification and pattern recognition (see Chapter 11) we may want to transform a fuzzy partition or pattern into a crisp partition or pattern; in control (see Chapter 13) we may want to give a single-valued input to a semiconductor device instead of a fuzzy input command. This "defuzzification" has the result of reducing a fuzzy set to a crisp single-valued quantity, or to a crisp set; of converting a fuzzy matrix to a crisp matrix; or of making a fuzzy number a crisp number.

Mathematically, the defuzzification of a fuzzy set is the process of "rounding it off" from its location in the unit hypercube to the nearest (in a geometric sense) vertex (see Chapter 1). If one thinks of a fuzzy set as a collection of membership values, or a vector of values on the unit interval, defuzzification reduces this vector to a single scalar quantity – presumably to the most typical (prototype) or representative value. Various popular forms of converting fuzzy sets to crisp sets or to single scalar values are introduced later in this chapter.

FEATURES OF THE MEMBERSHIP FUNCTION

Since all information contained in a fuzzy set is described by its membership function, it is useful to develop a lexicon of terms to describe various special features of this function. For purposes of simplicity, the functions shown in the following figures will all be continuous, but the terms apply equally for both discrete and continuous fuzzy sets. Figure 4.1 assists in this description.

The *core* of a membership function for some fuzzy set A is defined as that region of the universe that is characterized by complete and full membership in the set A. That is, the core comprises those elements x of the universe such that $\mu_A(x) = 1$.

The *support* of a membership function for some fuzzy set A is defined as that region of the universe that is characterized by nonzero membership in the set A. That is, the support comprises those elements x of the universe such that $\mu_A(x) > 0$.

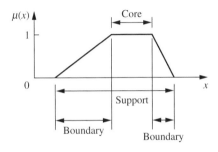

FIGURE 4.1
Core, support, and boundaries of a fuzzy set.

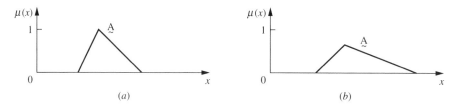

FIGURE 4.2
Fuzzy sets that are normal (a) and subnormal (b).

The *boundaries* of a membership function for some fuzzy set $\underset{\sim}{A}$ are defined as that region of the universe containing elements that have a nonzero membership but not complete membership. That is, the boundaries comprise those elements x of the universe such that $0 < \mu_{\underset{\sim}{A}}(x) < 1$. These elements of the universe are those with some *degree* of fuzziness, or only partial membership in the fuzzy set $\underset{\sim}{A}$. Figure 4.1 illustrates the regions in the universe comprising the core, support, and boundaries of a typical fuzzy set.

A *normal* fuzzy set is one whose membership function has at least one element x in the universe whose membership value is unity. For fuzzy sets where one and only one element has a membership equal to one, this element is typically referred to as the *prototype* of the set, or the prototypical element. Figure 4.2 illustrates typical normal and subnormal fuzzy sets.

A *convex* fuzzy set is described by a membership function whose membership values are strictly monotonically increasing, or whose membership values are strictly monotonically decreasing, or whose membership values are strictly monotonically increasing then strictly monotonically decreasing with increasing values for elements in the universe. Said another way, if, for any elements x, y, and z in a fuzzy set $\underset{\sim}{A}$, the relation $x < y < z$ implies that

$$\mu_{\underset{\sim}{A}}(y) \geq \min[\mu_{\underset{\sim}{A}}(x), \mu_{\underset{\sim}{A}}(z)] \tag{4.1}$$

then $\underset{\sim}{A}$ is said to be a convex fuzzy set [Ross, 1995]. Figure 4.3 shows a typical convex fuzzy set and a typical nonconvex fuzzy set. It is important to remark here that this definition of convexity is *different* from some definitions of the same term in mathematics. In some areas of mathematics, convexity of shape has to do with whether a straight line through any part of the shape goes outside the boundaries of that shape. This definition of convexity is *not* used here; Eq. (4.1) succinctly summarizes our definition of convexity.

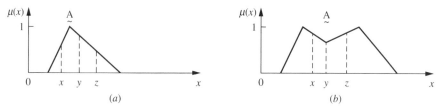

FIGURE 4.3
Convex, normal fuzzy set (a) and nonconvex, normal fuzzy set (b).

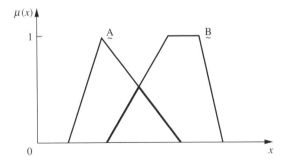

FIGURE 4.4
The intersection of two convex fuzzy sets produces a convex fuzzy set.

A special property of two convex fuzzy sets, say $\underset{\sim}{A}$ and $\underset{\sim}{B}$, is that the intersection of these two convex fuzzy sets is also a convex fuzzy set, as shown in Fig. 4.4. That is, for $\underset{\sim}{A}$ and $\underset{\sim}{B}$, which are both convex, $\underset{\sim}{A} \cap \underset{\sim}{B}$ is also convex.

The *crossover points* of a membership function are defined as the elements in the universe for which a particular fuzzy set $\underset{\sim}{A}$ has values equal to 0.5, i.e., for which $\mu_{\underset{\sim}{A}}(x) = 0.5$.

The *height* of a fuzzy set $\underset{\sim}{A}$ is the maximum value of the membership function, i.e., $\mathrm{hgt}(\underset{\sim}{A}) = \max\{\mu_{\underset{\sim}{A}}(x)\}$. If the $\mathrm{hgt}(\underset{\sim}{A}) < 1$, the fuzzy set is said to be subnormal. The $\mathrm{hgt}(\underset{\sim}{A})$ may be viewed as the degree of validity or credibility of information expressed by $\underset{\sim}{A}$ [Klir and Yuan, 1995].

If $\underset{\sim}{A}$ is a convex single-point normal fuzzy set defined on the real line, then $\underset{\sim}{A}$ is often termed a *fuzzy number*.

VARIOUS FORMS

The most common forms of membership functions are those that are normal and convex. However, many operations on fuzzy sets, hence operations on membership functions, result in fuzzy sets that are subnormal and nonconvex. For example, the extension principle to be discussed in Chapter 12 and the union operator both can produce subnormal or nonconvex fuzzy sets.

Membership functions can be symmetrical or asymmetrical. They are typically defined on one-dimensional universes, but they certainly can be described on multidimensional (or n-dimensional) universes. For example, the membership functions shown in this chapter are one-dimensional curves. In two dimensions these curves become surfaces and for three or more dimensions these surfaces become hypersurfaces. These hypersurfaces, or curves, are simple mappings from combinations of the parameters in the n-dimensional space to a membership value on the interval [0, 1]. Again, this membership value expresses the degree of membership that the specific combination of parameters in the n-dimensional space has in a particular fuzzy set defined on the n-dimensional universe of discourse. The hypersurfaces for an n-dimensional universe are analogous to joint probability density functions; but, of course, the mapping for the membership function is to membership in a particular set and not to relative frequencies, as it is for probability density functions.

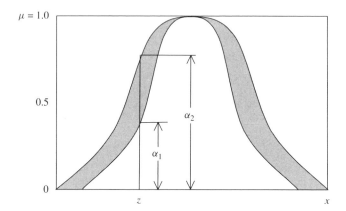

FIGURE 4.5
An interval-valued membership function.

Fuzzy sets of the types depicted in Fig. 4.2 are by far the most common ones encountered in practice; they are described by *ordinary membership functions*. However, several other types of fuzzy membership functions have been proposed [Klir and Yuan, 1995] as *generalized membership functions*. The primary reason for considering other types of membership functions is that the values used in developing ordinary membership functions are often overly precise. They require that each element of the universe x on which the fuzzy set $\underset{\sim}{A}$ is defined be assigned a specific membership value, $\mu_A(x)$. Suppose the level of information is not adequate to specify membership functions with this precision. For example, we may only know the upper and lower bounds of membership grades for each element of the universe for a fuzzy set. Such a fuzzy set would be described by an *interval-valued membership function*, such as the one shown in Fig. 4.5. In this figure, for a particular element, $x = z$, the membership in a fuzzy set $\underset{\sim}{A}$, i.e., $\mu_{\underset{\sim}{A}}(z)$, would be expressed by the membership interval $[\alpha_1, \alpha_2]$. Interval-valued fuzzy sets can be generalized further by allowing their intervals to become fuzzy. Each membership interval then becomes an ordinary fuzzy set. This type of membership function is referred to in the literature as a *type-2 fuzzy set*. Other generalizations of the fuzzy membership functions are available as well [see Klir and Yuan, 1995].

FUZZIFICATION

Fuzzification is the process of making a crisp quantity fuzzy. We do this by simply recognizing that many of the quantities that we consider to be crisp and deterministic are actually not deterministic at all: They carry considerable uncertainty. If the form of uncertainty happens to arise because of imprecision, ambiguity, or vagueness, then the variable is probably fuzzy and can be represented by a membership function.

In the real world, hardware such as a digital voltmeter generates crisp data, but these data are subject to experimental error. The information shown in Fig. 4.6 shows one possible range of errors for a typical voltage reading and the associated membership function that might represent such imprecision.

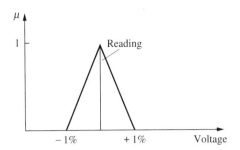

FIGURE 4.6
Membership function representing imprecision in "crisp voltage reading."

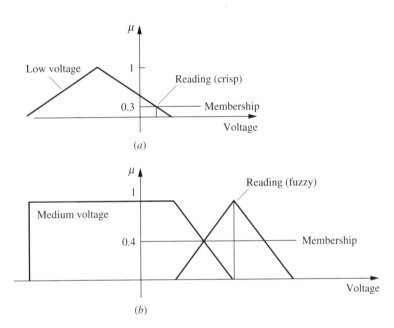

FIGURE 4.7
Comparisons of fuzzy sets and crisp or fuzzy readings: (a) fuzzy set and crisp reading; (b) fuzzy set and fuzzy reading.

The representation of imprecise data as fuzzy sets is a useful but not mandatory step when those data are used in fuzzy systems. This idea is shown in Fig. 4.7, where we consider the data as a crisp reading, Fig. 4.7a, or as a fuzzy reading, as shown in Fig. 4.7b. In Fig. 4.7a we might want to compare a crisp voltage reading to a fuzzy set, say "low voltage." In the figure we see that the crisp reading intersects the fuzzy set "low voltage" at a membership of 0.3, i.e., the fuzzy set and the reading can be said to agree at a membership value of 0.3. In Fig. 4.7b the intersection of the fuzzy set "medium voltage" and a fuzzified voltage reading occurs at a membership of 0.4. We can see in Fig. 4.7b that the set intersection of the two fuzzy sets is a small triangle, whose largest membership occurs at the membership value of 0.4.

We will say more about the importance of fuzzification of crisp variables in Chapters 8 and 13 of this text. In Chapter 8 the topic is simulation, and the inputs for any nonlinear or complex simulation will be expressed as fuzzy sets. If the process is inherently quantitative or the inputs derive from sensor measurements, then these crisp numerical inputs could be fuzzified in order for them to be used in a fuzzy inference system (to be discussed in Chapter 5). In Chapter 13 the topic is fuzzy control, and, again, this is a discipline where the inputs generally originate from a piece of hardware, or a sensor and the measured input could be fuzzified for utility in the rule-based system which describes the fuzzy controller. If the system to be controlled is not hardware based, e.g., the control of an economic system or the control of an ecosystem subjected to a toxic chemical, then the inputs could be scalar quantities arising from statistical sampling, or other derived numerical quantities. Again, for utility in fuzzy systems, these scalar quantities could first be fuzzified, i.e., translated into a membership function, and then used to form the input structure necessary for a fuzzy system.

DEFUZZIFICATION TO CRISP SETS

We begin by considering a fuzzy set $\underset{\sim}{A}$, then define a lambda-cut set, A_λ, where $0 \leq \lambda \leq 1$. The set A_λ is a crisp set called the lambda (λ)-cut (or alpha-cut) set of the fuzzy set $\underset{\sim}{A}$, where $A_\lambda = \{x | \mu_A(x) \geq \lambda\}$. Note that the λ-cut set A_λ does not have a tilde underscore; it is a crisp set derived from its parent fuzzy set, $\underset{\sim}{A}$. Any particular fuzzy set $\underset{\sim}{A}$ can be transformed into an infinite number of λ-cut sets, because there are an infinite number of values λ on the interval $[0, 1]$.

Any element $x \in A_\lambda$ belongs to $\underset{\sim}{A}$ with a grade of membership that is greater than or equal to the value λ. The following example illustrates this idea.

Example 4.1. Let us consider the discrete fuzzy set, using Zadeh's notation, defined on universe $X = \{a, b, c, d, e, f\}$,

$$\underset{\sim}{A} = \left\{ \frac{1}{a} + \frac{0.9}{b} + \frac{0.6}{c} + \frac{0.3}{d} + \frac{0.01}{e} + \frac{0}{f} \right\}$$

This fuzzy set is shown schematically in Fig. 4.8. We can reduce this fuzzy set into several λ-cut sets, all of which are crisp. For example, we can define λ-cut sets for the values of $\lambda = 1$, 0.9, 0.6, 0.3, 0^+, and 0.

$$A_1 = \{a\}, \quad A_{0.9} = \{a, b\}$$

$$A_{0.6} = \{a, b, c\}, \quad A_{0.3} = \{a, b, c, d\}$$

$$A_{0^+} = \{a, b, c, d, e\}, \quad A_0 = X$$

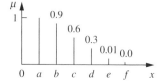

FIGURE 4.8
A discrete fuzzy set $\underset{\sim}{A}$.

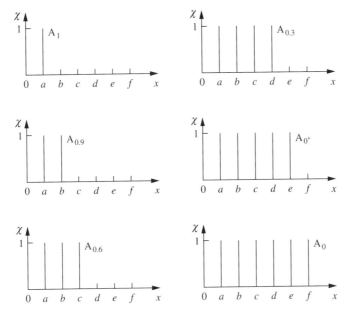

FIGURE 4.9
Lambda-cut sets for $\lambda = 1, 0.9, 0.6, 0.3, 0^+, 0$.

The quantity $\lambda = 0^+$ is defined as a small "ε" value >0, i.e., a value just greater than zero. By definition, $\lambda = 0$ produces the universe X, since all elements in the universe have at least a 0 membership value in any set on the universe. Since all A_λ are crisp sets, all the elements just shown in the example λ-cut sets have unit membership in the particular λ-cut set. For example, for $\lambda = 0.3$, the elements a, b, c, and d of the universe have a membership of 1 in the λ-cut set, $A_{0.3}$, and the elements e and f of the universe have a membership of 0 in the λ-cut set, $A_{0.3}$. Figure 4.9 shows schematically the crisp λ-cut sets for the values $\lambda = 1$, 0.9, 0.6, 0.3, 0^+, and 0. Notice in these plots of membership value versus the universe X that the effect of a λ-cut is to rescale the membership values: to one for all elements of the fuzzy set $\underset{\sim}{A}$ having membership values greater than or equal to λ, and to zero for all elements of the fuzzy set $\underset{\sim}{A}$ having membership values less than λ.

We can express λ-cut sets using Zadeh's notation. For the example, λ-cut sets for the values $\lambda = 0.9$ and 0.25 are given here:

$$A_{0.9} = \left\{ \frac{1}{a} + \frac{1}{b} + \frac{0}{c} + \frac{0}{d} + \frac{0}{e} + \frac{0}{f} \right\} \qquad A_{0.25} = \left\{ \frac{1}{a} + \frac{1}{b} + \frac{1}{c} + \frac{1}{d} + \frac{0}{e} + \frac{0}{f} \right\}$$

λ-cut sets obey the following four very special properties:

1. $(\underset{\sim}{A} \cup \underset{\sim}{B})_\lambda = A_\lambda \cup B_\lambda$ (4.1a)

2. $(\underset{\sim}{A} \cap \underset{\sim}{B})_\lambda = A_\lambda \cap B_\lambda$ (4.1b)

3. $\overline{(\underset{\sim}{A})}_\lambda \neq \overline{A}_\lambda$ except for a value of $\lambda = 0.5$ (4.1c)

4. For any $\lambda \leq \alpha$, where $0 \leq \alpha \leq 1$, it is true that $A_\alpha \subseteq A_\lambda$, where $A_0 = X$ (4.1d)

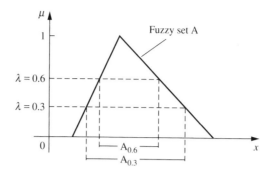

FIGURE 4.10
Two different λ-cut sets for a continuous-valued fuzzy set.

These properties show that λ-cuts on standard operations on fuzzy sets are equivalent with standard set operations on λ-cut sets. The last operation, Eq. (4.1d), can be shown more conveniently using graphics. Figure 4.10 shows a continuous-valued fuzzy set with two λ-cut values. Notice in the graphic that for $\lambda = 0.3$ and $\alpha = 0.6$, $A_{0.3}$ has a greater domain than $A_{0.6}$, i.e., for $\lambda \leq \alpha$ $(0.3 \leq 0.6)$, $A_{0.6} \subseteq A_{0.3}$.

In this chapter, various definitions of a membership function are discussed and illustrated. Many of these same definitions arise through the use of λ-cut sets. As seen in Fig. 4.1, we can provide the following definitions for a convex fuzzy set $\underset{\sim}{A}$. The core of $\underset{\sim}{A}$ is the $\lambda = 1$ cut set, A_1. The support of $\underset{\sim}{A}$ is the λ-cut set A_{0^+}, where $\lambda = 0^+$, or symbolically, $A_{0^+} = \{x \mid \mu_{\underset{\sim}{A}(x)} > 0\}$. The intervals $[A_{0^+}, A_1]$ form the boundaries of the fuzzy set $\underset{\sim}{A}$, i.e., those regions that have membership values between 0 and 1 (exclusive of 0 and 1): that is, for $0 < \lambda < 1$.

λ-CUTS FOR FUZZY RELATIONS

In Chapter 3, a biotechnology example, Example 3.11, was developed using a fuzzy relation that was reflexive and symmetric. Recall this matrix,

$$\underset{\sim}{R} = \begin{bmatrix} 1 & 0.8 & 0 & 0.1 & 0.2 \\ 0.8 & 1 & 0.4 & 0 & 0.9 \\ 0 & 0.4 & 1 & 0 & 0 \\ 0.1 & 0 & 0 & 1 & 0.5 \\ 0.2 & 0.9 & 0 & 0.5 & 1 \end{bmatrix}$$

We can define a λ-cut procedure for relations similar to the one developed for sets. Consider a fuzzy relation $\underset{\sim}{R}$, where each row of the relational matrix is considered a fuzzy set, i.e., the jth row in $\underset{\sim}{R}$ represents a discrete membership function for a fuzzy set, R_j. Hence, a fuzzy relation can be converted to a crisp relation in the following manner. Let us define $R_\lambda = \{(x, y) \mid \mu_R(x, y) \geq \lambda\}$ as a λ-cut relation of the fuzzy relation, $\underset{\sim}{R}$. Since in this case $\underset{\sim}{R}$ is a two-dimensional array defined on the universes X and Y, then any pair $(x, y) \in R_\lambda$ belongs to $\underset{\sim}{R}$ with a "strength" of relation greater than or equal to λ. These ideas for relations can be illustrated with an example.

Example 4.2. Suppose we take the fuzzy relation from the biotechnology example in Chapter 3 (Example 3.11), and perform λ-cut operations for the values of $\lambda = 1, 0.9, 0$. These crisp relations are given below:

$$\lambda = 1, \; R_1 = \begin{bmatrix} 1 & 0 & 0 & 0 & 0 \\ 0 & 1 & 0 & 0 & 0 \\ 0 & 0 & 1 & 0 & 0 \\ 0 & 0 & 0 & 1 & 0 \\ 0 & 0 & 0 & 0 & 1 \end{bmatrix}$$

$$\lambda = 0.9, \; R_{0.9} = \begin{bmatrix} 1 & 0 & 0 & 0 & 0 \\ 0 & 1 & 0 & 0 & 1 \\ 0 & 0 & 1 & 0 & 0 \\ 0 & 0 & 0 & 1 & 0 \\ 0 & 1 & 0 & 0 & 1 \end{bmatrix}$$

$$\lambda = 0, \; R_0 = \underset{\sim}{E} \text{ (whole relation; see Chapter 3)}$$

λ-cuts on fuzzy relations obey certain properties, just as λ-cuts on fuzzy sets do (see Eqs. (4.1)), as given in Eqs. (4.2):

1. $(\underset{\sim}{R} \cup \underset{\sim}{S})_\lambda = R_\lambda \cup S_\lambda$ (4.2a)

2. $(\underset{\sim}{R} \cap \underset{\sim}{S})_\lambda = R_\lambda \cap S_\lambda$ (4.2b)

3. $(\overline{\underset{\sim}{R}})_\lambda \neq \overline{R_\lambda}$ (4.2c)

4. For any $\lambda \le \alpha, 0 \le \alpha \le 1$, then $R_\alpha \subseteq R_\lambda$ (4.2d)

DEFUZZIFICATION TO SCALARS

As mentioned in the introduction, there may be situations where the output of a fuzzy process needs to be a single scalar quantity as opposed to a fuzzy set. Defuzzification is the conversion of a fuzzy quantity to a precise quantity, just as fuzzification is the conversion of a precise quantity to a fuzzy quantity. The output of a fuzzy process can be the logical union of two or more fuzzy membership functions defined on the universe of discourse of the output variable. For example, suppose a fuzzy output is comprised of two parts: the first part, $\underset{\sim}{C_1}$, a trapezoidal shape, shown in Fig. 4.11a, and the second part, $\underset{\sim}{C_2}$, a triangular membership shape, shown in Fig. 4.11b. The union of these two membership functions, i.e., $\underset{\sim}{C} = \underset{\sim}{C_1} \cup \underset{\sim}{C_2}$, involves the max operator, which graphically is the outer envelope of the two shapes shown in Figs. 4.11a and b; the resulting shape is shown in Fig. 4.11c. Of course, a general fuzzy output process can involve many output parts (more than two), and the membership function representing each part of the output can have shapes other than triangles and trapezoids. Further, as Fig. 4.11a shows, the membership functions may not always be normal. In general, we can have

$$\underset{\sim}{C_k} = \bigcup_{i=1}^{k} \underset{\sim}{C_i} = \underset{\sim}{C} \qquad (4.3)$$

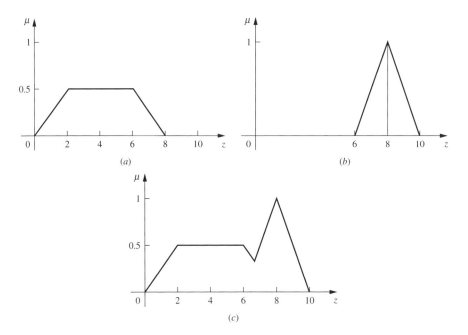

FIGURE 4.11
Typical fuzzy process output: (*a*) first part of fuzzy output; (*b*) second part of fuzzy output; (*c*) union of both parts.

Among the many methods that have been proposed in the literature in recent years, seven are described here for defuzzifying fuzzy output functions (membership functions) [Hellendoorn and Thomas, 1993]. Four of these methods are first summarized and then illustrated in two examples; then the additional three methods are described, then illustrated in two other examples.

1. *Max membership principle*: Also known as the *height* method, this scheme is limited to peaked output functions. This method is given by the algebraic expression

$$\mu_{\underset{\sim}{C}}(z^*) \geq \mu_{\underset{\sim}{C}}(z) \qquad \text{for all } z \in Z \tag{4.4}$$

where z^* is the defuzzified value, and is shown graphically in Fig. 4.12.

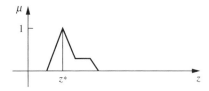

FIGURE 4.12
Max membership defuzzification method.

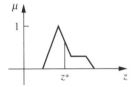

FIGURE 4.13
Centroid defuzzification method.

2. *Centroid method*: This procedure (also called center of area, center of gravity) is the most prevalent and physically appealing of all the defuzzification methods [Sugeno, 1985; Lee, 1990]; it is given by the algebraic expression

$$z^* = \frac{\displaystyle\int \mu_{\underset{\sim}{C}}(z) \cdot z \, dz}{\displaystyle\int \mu_{\underset{\sim}{C}}(z) \, dz} \qquad (4.5)$$

where $\int$ denotes an algebraic integration. This method is shown in Fig. 4.13.
3. *Weighted average method*: The weighted average method is the most frequently used in fuzzy applications since it is one of the more computationally efficient methods. Unfortunately it is *usually* restricted to symmetrical output membership functions. It is given by the algebraic expression

$$z^* = \frac{\displaystyle\sum \mu_{\underset{\sim}{C}}(\bar{z}) \cdot \bar{z}}{\displaystyle\sum \mu_{\underset{\sim}{C}}(\bar{z})} \qquad (4.6)$$

where $\sum$ denotes the algebraic sum and where $\bar{z}$ is the centroid of each symmetric membership function. This method is shown in Fig. 4.14. The weighted average method is formed by weighting each membership function in the output by its respective maximum membership value. As an example, the two functions shown in Fig. 4.14 would result in the following general form for the defuzzified value:

$$z^* = \frac{a(0.5) + b(0.9)}{0.5 + 0.9}$$

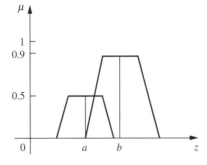

FIGURE 4.14
Weighted average method of defuzzification.

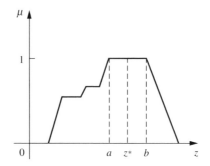

FIGURE 4.15
Mean max membership defuzzification method.

Since the method is limited to symmetrical membership functions, the values a and b are the means (centroids) of their respective shapes.

4. *Mean max membership*: This method (also called middle-of-maxima) is closely related to the first method, except that the locations of the maximum membership can be nonunique (i.e., the maximum membership can be a plateau rather than a single point). This method is given by the expression [Sugeno, 1985; Lee, 1990]

$$z^* = \frac{a + b}{2} \tag{4.7}$$

where a and b are as defined in Fig. 4.15.

> **Example 4.3.** A railroad company intends to lay a new rail line in a particular part of a county. The whole area through which the new line is passing must be purchased for right-of-way considerations. It is surveyed in three stretches, and the data are collected for analysis. The surveyed data for the road are given by the sets, B_1, B_2, and B_3, where the sets are defined on the universe of right-of-way widths, in meters. For the railroad to purchase the land, it must have an assessment of the amount of land to be bought. The three surveys on right-of-way width are ambiguous, however, because some of the land along the proposed railway route is already public domain and will not need to be purchased. Additionally, the original surveys are so old (*circa* 1860) that some ambiguity exists on boundaries and public right-of-way for old utility lines and old roads. The three fuzzy sets, B_1, B_2, and B_3, shown in Figs. 4.16, 4.17, and 4.18, respectively, represent the uncertainty in each survey as to the membership of right-of-way width, in meters, in privately owned land.
>
> We now want to aggregate these three survey results to find the single most nearly representative right-of-way width (z) to allow the railroad to make its initial estimate of the

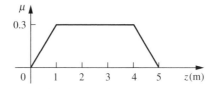

FIGURE 4.16
Fuzzy set B_1: public right-of-way width (z) for survey 1.

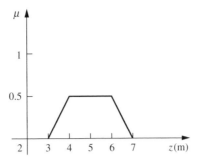

FIGURE 4.17
Fuzzy set B_2: public right-of-way width (z) for survey 2.

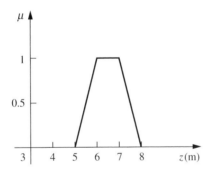

FIGURE 4.18
Fuzzy set B_3: public right-of-way width (z) for survey 3.

right-of-way purchasing cost. Using Eqs. (4.5)–(4.7) and the preceding three fuzzy sets, we want to find z^*.

According to the centroid method, Eq. (4.5), z^* can be found using

$$z^* = \frac{\int \mu_B(z) \cdot z \, dz}{\int \mu_B(z) \, dz}$$

$$= \left[\int_0^1 (0.3z)z \, dz + \int_1^{3.6} (0.3z) \, dz + \int_{3.6}^4 \left(\frac{z - 3.6}{2} \right) z \, dz + \int_4^{5.5} (0.5)z \, dz \right.$$

$$\left. + \int_{5.5}^6 (z - 5.5)z \, dz + \int_6^7 z \, dz + \int_7^8 \left(\frac{7 - z}{2} \right) z \, dz \right]$$

$$\div \left[\int_0^1 (0.3z) \, dz + \int_1^{3.6} (0.3) \, dz + \int_{3.6}^4 \left(\frac{z - 3.6}{2} \right) dz + \int_4^{5.5} (0.5) \, dz \right.$$

$$\left. + \int_{5.5}^6 \left(\frac{z - 5.5}{2} \right) dz + \int_6^7 dz + \int_7^8 \left(\frac{7 - z}{2} \right) dz \right]$$

$$= 4.9 \text{ meters}$$

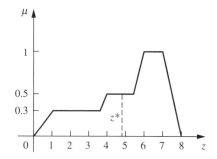

FIGURE 4.19
The centroid method for finding z^*.

where z^* is shown in Fig. 4.19. According to the weighted average method, Eq. (4.6),

$$z^* = \frac{(0.3 \times 2.5) + (0.5 \times 5) + (1 \times 6.5)}{0.3 + 0.5 + 1} = 5.41 \text{ meters}$$

and is shown in Fig. 4.20. According to the mean max membership method, Eq. (4.7), z^* is given by $(6 + 7)/2 = 6.5$ meters, and is shown in Fig. 4.21.

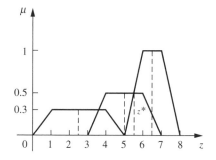

FIGURE 4.20
The weighted average method for finding z^*.

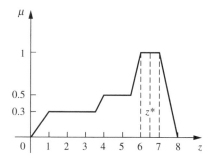

FIGURE 4.21
The mean max membership method for finding z^*.

Example 4.4. Many products, such as tar, petroleum jelly, and petroleum, are extracted from crude oil. In a newly drilled oil well, three sets of oil samples are taken and tested for their viscosity. The results are given in the form of the three fuzzy sets B_1, B_2, and B_3, all defined on a universe of normalized viscosity, as shown in Figs. 4.22–4.24. Using Eqs. (4.4)–(4.6), we want to find the most nearly representative viscosity value for all three oil samples, and hence find z^* for the three fuzzy viscosity sets.

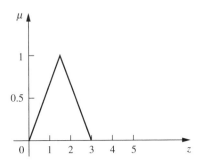

FIGURE 4.22
Membership in viscosity of oil sample 1, B_1.

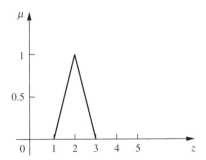

FIGURE 4.23
Membership in viscosity of oil sample 2, B_2.

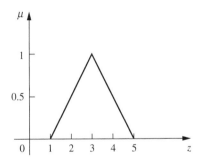

FIGURE 4.24
Membership in viscosity of oil sample 3, B_3.

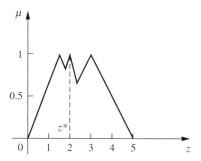

FIGURE 4.25
Logical union of three fuzzy sets B_1, B_2, and B_3.

To find z^* using the centroid method, we first need to find the logical union of the three fuzzy sets. This is shown in Fig. 4.25. Also shown in Fig. 4.25 is the result of the max membership method, Eq. (4.4). For this method, we see that $\mu_B(z^*)$ has three locations where the membership equals unity. This result is ambiguous and, in this case, the selection of the intermediate point is arbitrary, but it is closer to the centroid of the area shown in Fig. 4.25. There could be other compelling reasons to select another value in this case; perhaps max membership is not a good metric for this problem.

According to the centroid method, Eq. (4.5),

$$z^* = \frac{\displaystyle\int \mu_B(z)z\,dz}{\displaystyle\int \mu_B(z)\,dz}$$

$$= \left[\int_0^{1.5} (0.67z)z\,dz + \int_{1.5}^{1.8} (2 - 0.67z)z\,dz + \int_{1.8}^2 (z - 1)z\,dz + \int_2^{2.33} (3 - z)z\,dz \right.$$

$$\left. + \int_{2.33}^3 (0.5z - 0.5)z\,dz + \int_3^5 (2.5 - 0.5z)z\,dz \right]$$

$$\div \left[\int_0^{1.5} (0.67z)\,dz + \int_{1.5}^{1.8} (2 - 0.67z)\,dz + \int_{1.8}^2 (z - 1)\,dz + \int_2^{2.33} (3 - z)\,dz \right.$$

$$\left. + \int_{2.33}^3 (0.5z - 0.5)\,dz + \int_3^5 (2.5 - 0.5z)\,dz \right]$$

$$= 2.5 \text{ meters}$$

The centroid value obtained, z^*, is shown in Fig. 4.26.

According to the weighted average method, Eq. (4.6),

$$z^* = \frac{(1 \times 1.5) + (1 \times 2) + (1 \times 3)}{1 + 1 + 1} = 2.25 \text{ meters}$$

and is shown in Fig. 4.27.

Three other popular methods, which are worthy of discussion because of their appearance in some applications, are the center of sums, center of largest area, and first of maxima methods [Hellendoorn and Thomas, 1993]. These methods are now developed.

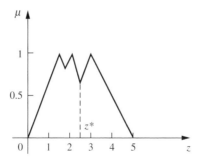

FIGURE 4.26
Centroid value z^* for three fuzzy oil samples.

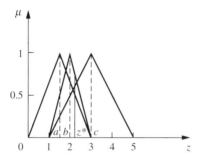

FIGURE 4.27
Weighted average method for z^*.

5. *Center of sums*: This is faster than many defuzzification methods that are presently in use, and the method is not restricted to symmetric membership functions. This process involves the algebraic sum of individual output fuzzy sets, say $\underset{\sim}{C}_1$ and $\underset{\sim}{C}_2$, instead of their union. Two drawbacks to this method are that the intersecting areas are added twice, and the method also involves finding the centroids of the individual membership functions. The defuzzified value z^* is given by the following equation:

$$z^* = \frac{\int_z \bar{z} \sum_{k=1}^{n} \mu_{\underset{\sim}{C}_k}(z)\, dz}{\int_z \sum_{k=1}^{n} \mu_{\underset{\sim}{C}_k}(z)\, dz} \tag{4.8}$$

where the symbol $\bar{z}$ is the distance to the centroid of each of the respective membership functions.

This method is similar to the weighted average method, Eq. (4.6), except in the center of sums method the weights are the areas of the respective membership functions whereas in the weighted average method the weights are individual membership values. Figure 4.28 is an illustration of the center of sums method.

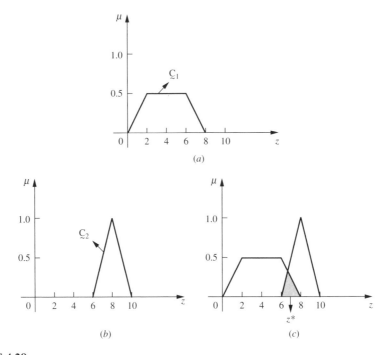

FIGURE 4.28
Center of sums method: (a) first membership function; (b) second membership function; and (c) defuzzification step.

6. *Center of largest area*: If the output fuzzy set has at least two convex subregions, then the center of gravity (i.e., z^* is calculated using the centroid method, Eq. (4.5)) of the convex fuzzy subregion with the largest area is used to obtain the defuzzified value z^* of the output. This is shown graphically in Fig. 4.29, and given algebraically here:

$$z^* = \frac{\displaystyle\int \mu_{\underset{\sim}{C}_m}(z)z\,dz}{\displaystyle\int \mu_{\underset{\sim}{C}_m}(z)\,dz} \tag{4.9}$$

where $\underset{\sim}{C}_m$ is the convex subregion that has the largest area making up $\underset{\sim}{C}_k$. This condition applies in the case when the overall output $\underset{\sim}{C}_k$ is nonconvex; and in the case when $\underset{\sim}{C}_k$ is convex, z^* is the same quantity as determined by the centroid method or the center of largest area method (because then there is only one convex region).

7. *First (or last) of maxima*: This method uses the overall output or union of all individual output fuzzy sets $\underset{\sim}{C}_k$ to determine the smallest value of the domain with maximized membership degree in $\underset{\sim}{C}_k$. The equations for z^* are as follows.
   First, the largest height in the union (denoted hgt($\underset{\sim}{C}_k$)) is determined,

$$\text{hgt}(\underset{\sim}{C}_k) = \sup_{z \in Z} \mu_{\underset{\sim}{C}_k}(z) \tag{4.10}$$

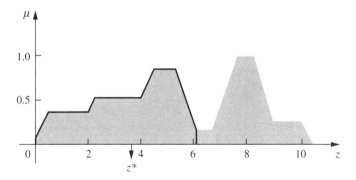

FIGURE 4.29
Center of largest area method (outlined with bold lines), shown for a nonconvex $\underset{\sim}{C_k}$.

Then the first of the maxima is found,

$$z* = \inf_{z \in Z} \left\{ z \in Z \mid \mu_{C_k}(z) = \text{hgt}(\underset{\sim}{C_k}) \right\} \tag{4.11}$$

An alternative to this method is called the last of maxima, and it is given by

$$z^* = \sup_{z \in Z} \left\{ z \in Z \mid \mu_{C_k}(z) = \text{hgt}(\underset{\sim}{C_k}) \right\} \tag{4.12}$$

In Eqs. (4.10)–(4.12) the supremum (sup) is the least upper bound and the infimum (inf) is the greatest lower bound. Graphically, this method is shown in Fig. 4.30, where, in the case illustrated in the figure, the first max is also the last max and, because it is a distinct max, is also the mean max. Hence, the methods presented in Eqs. (4.4) (max or height), (4.7) (mean max), (4.11) (first max), and (4.12) (last max) all provide the same defuzzified value, z^*, for the particular situation illustrated in Fig. 4.30.

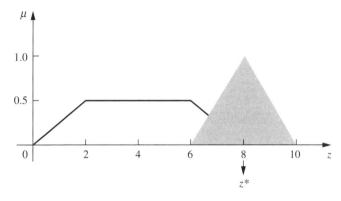

FIGURE 4.30
First of max (and last of max) method.

The problems illustrated in Examples 4.3 and 4.4 are now continued, to illustrate the last three methods presented.

Example 4.5. Continuing with Example 4.3 on the railroad company planning to lay a new rail line, we will calculate the defuzzified values using the (1) center of sums method, (2) center of largest area, and (3) first maxima and last maxima.

According to the center of sums method, Eq. (4.8), z^* will be as follows:

$$z^* = \frac{[2.5 \times 0.5 \times 0.3(3+5) + 5 \times 0.5 \times 0.5(2+4) + 6.5 \times 0.5 \times 1(3+1)]}{[0.5 \times 0.3(3+5) + 0.5 \times 0.5(2+4) + 0.5 \times 1(3+1)]}$$

$$= 5.0 \text{ m}$$

with the result shown in Fig. 4.31. The center of largest area method, Eq. (4.9), provides the same result (i.e., $z^* = 4.9$) as the centroid method, Eq. (4.5), because the complete output fuzzy set is convex, as seen in Fig. 4.32. According to the first of maxima and last of maxima methods, Eqs. (4.11) and (4.12), z^* is shown as z_1^* and z_2^*, respectively, in Fig. 4.33.

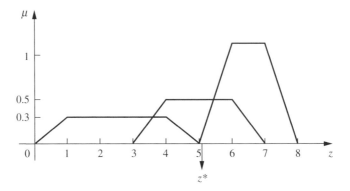

FIGURE 4.31
Center of sums result for Example 4.5.

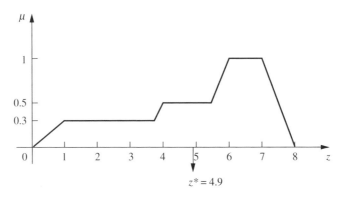

FIGURE 4.32
Output fuzzy set for Example 4.5 is convex.

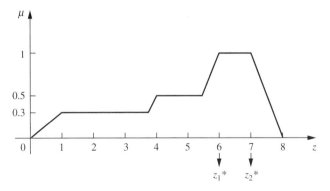

FIGURE 4.33
First of maxima solution ($z_1^* = 6$) and last of maxima solution ($z_2^* = 7$).

Example 4.6. Continuing with Example 4.4 on the crude oil problem, the center of sums method, Eq. (4.8), produces a defuzzified value for z^* of

$$z^* = \frac{(0.5 \times 3 \times 1 \times 1.5 + 0.5 \times 2 \times 1 \times 2 + 0.5 \times 4 \times 1 \times 3)}{(0.5 \times 3 \times 1 + 0.5 \times 2 \times 1 + 0.5 \times 4 \times 1)} = 2.3 \text{ m}$$

which is shown in Fig. 4.34. In the center of largest area method we first determine the areas of the three individual convex fuzzy output sets, as seen in Fig. 4.35. These areas are 1.02, 0.46, and 1.56 square units, respectively. Among them, the third area is largest, so the centroid of that area will be the center of the largest area. The defuzzified value is calculated to be $z^* = 3.3$:

$$z^* = \frac{\left[\left(\dfrac{0.67}{2} + 2.33\right)[0.5 \times 0.67(1 + 0.67)]\right] + 3.66(0.5 \times 2 \times 1)}{[0.5 \times 0.67(1 + 0.67)] + (0.5 \times 2 \times 1)} = 3.3 \text{ m}$$

Finally, one can see graphically in Fig. 4.36 that the first of maxima and last of maxima, Eqs. (4.11)–(4.12), give different values for z^*, namely, $z^* = 1.5$ and 3.0, respectively.

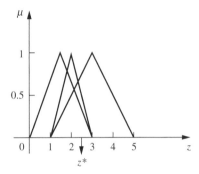

FIGURE 4.34
Center of sums solution for Example 4.6.

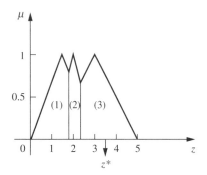

FIGURE 4.35
Center of largest area method for Example 4.6.

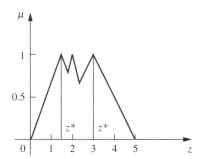

FIGURE 4.36
First of maxima gives $z^* = 1.5$ and last of maxima gives $z^* = 3$.

SUMMARY

This chapter has introduced the various features and forms of a membership function and the idea of fuzzyifying scalar quantities to make them fuzzy sets. The primary focus of the chapter, however, has been to explain the process of converting from fuzzy membership functions to crisp formats – a process called defuzzification. Defuzzification is necessary because, for example, we cannot instruct the voltage going into a machine to increase "slightly," even if this instruction comes from a fuzzy controller – we must alter its voltage by a specific amount. Defuzzification is a natural and necessary process. In fact, there is an analogous form of defuzzification in mathematics where we solve a complicated problem in the complex plane, find the real and imaginary parts of the solution, then *decomplexify* the imaginary solution back to the real numbers space [Bezdek, 1993]. There are numerous other methods for defuzzification that have not been presented here. A review of the literature will provide the details on some of these [see, for example, Filev and Yager, 1991; Yager and Filev, 1993].

A natural question to ask is: Of the seven defuzzification methods presented, which is the best? One obvious answer to the question is that it is context- or problem-dependent.

To answer this question in more depth, Hellendoorn and Thomas [1993] have specified five criteria against which to measure the methods. These criteria will be repeated here for the benefit of the reader who also ponders the question just given in terms of the advantages and disadvantages of the various methods. The first criterion is *continuity*. A small change in the input of a fuzzy process should not produce a large change in the output. Second, a criterion known as *disambiguity* simply points out that a defuzzification method should always result in a unique value for z^*, i.e., no ambiguity in the defuzzified value. This criterion is not satisfied by the center of largest area method, Eq. (4.9), because, as seen in Fig. 4.31, when the largest membership functions have equal area, there is ambiguity in selecting a z^*. The third criterion is called *plausibility*. To be plausible, z^* should lie approximately in the middle of the support region of C_k and have a high degree of membership in C_k. The centroid method, Eq. (4.5), does not exhibit plausibility in the situation illustrated in Fig. 4.31 because, although z^* lies in the middle of the support of C_k, it does not have a high degree of membership (also seen in the darkened area of Fig. 4.28c). The fourth criterion is that of *computational simplicity*, which suggests that the more time consuming a method is, the less value it should have in a computation system. The height method, Eq. (4.4), the mean max method, Eq. (4.7), and the first of maxima method are faster than the centroid, Eq. (4.5), or center of sum, Eq. (4.8), methods, for example. The fifth criterion is called the *weighting method*, which weights the output fuzzy sets. This criterion constitutes the difference between the centroid method, Eq. (4.5), the weighted average method, Eq. (4.6), and center of sum methods, Eq. (4.8). The problem with the fifth criterion is that it is problem-dependent, as there is little by which to judge the best weighting method; the weighted average method involves less computation than the center of sums, but that attribute falls under the fourth criterion, computational simplicity.

As with many issues in fuzzy logic, the method of defuzzification should be assessed in terms of the goodness of the answer in the context of the data available. Other methods are available that purport to be superior to the simple methods presented here [Hellendoorn and Thomas, 1993].

REFERENCES

Bezdek, J. (1993). ''Editorial: Fuzzy models – what are they, and why?'' *IEEE Trans. Fuzzy Syst.*, vol. 1, pp. 1–5.

Filev, D. and Yager, R. (1991). ''A generalized defuzzification method under BAD distributions,'' *Int. J. Intell. Syst.*, vol. 6, pp. 689–697.

Hellendoorn, H. and Thomas, C. (1993). ''Defuzzification in fuzzy controllers,'' *Intell. Fuzzy Syst.*, vol. 1, pp. 109–123.

Klir, G. and Folger, T. (1988). *Fuzzy sets, uncertainty, and information*, Prentice Hall, Englewood Cliffs, NJ.

Klir, G. and Yuan, B. (1995). *Fuzzy sets and fuzzy logic: theory and applications*, Prentice Hall, Upper Saddle River, NJ.

Lee, C. (1990). ''Fuzzy logic in control systems: fuzzy logic controller, Parts I and II,'' *IEEE Trans. Syst., Man, Cybern.*, vol. 20, pp. 404–435.

Sugeno, M. (1985). ''An introductory survey of fuzzy control,'' *Inf. Sci.*, vol. 36, pp. 59–83.

Yager, R. and Filev, D. (1993). ''SLIDE: A simple adaptive defuzzification method,'' *IEEE Trans. Fuzzy Syst.*, vol. 1, pp. 69–78.

PROBLEMS

4.1. Two fuzzy sets A and B, both defined on X, are as follows:

$\mu(x_i)$	x_1	x_2	x_3	x_4	x_5	x_6
A	0.1	0.6	0.8	0.9	0.7	0.1
B	0.9	0.7	0.5	0.2	0.1	0

Express the following λ-cut sets using Zadeh's notation:

(a) $(\overline{A})_{0.7}$ (e) $(A \cup \overline{A})_{0.7}$
(b) $(B)_{0.4}$ (f) $(B \cap \overline{B})_{0.5}$
(c) $(A \cup B)_{0.7}$ (g) $(\overline{A \cap B})_{0.7}$
(d) $(A \cap B)_{0.6}$ (h) $(\overline{A \cup B})_{0.7}$

4.2. [Klir and Folger, 1988] Show that all λ-cuts of any fuzzy set A defined in R^n space ($n \geq 1$) are convex if and only if

$$\mu_A[\lambda r + (1 - \lambda)s] \geq \min[\mu_A(r), \mu_A(s)]$$

for all $r, s \in R^n$, and all $\lambda \in [0, 1]$.

4.3. The fuzzy sets A, B, and C are all defined on the universe $X = [0, 5]$ with the following membership functions:

$$\mu_A(x) = \frac{1}{1 + 5(x - 5)^2} \qquad \mu_B(x) = 2^{-x} \qquad \mu_C(x) = \frac{2x}{x + 5}$$

(a) Sketch the membership functions.
(b) Define the intervals along the x axis corresponding to the λ-cut sets for each of the fuzzy sets A, B, and C for the following values of λ:
   (i) $\lambda = 0.2$
   (ii) $\lambda = 0.4$
   (iii) $\lambda = 0.7$
   (iv) $\lambda = 0.9$
   (v) $\lambda = 1.0$

4.4. Determine the crisp λ-cut relations for $\lambda = 0.1j$, for $j = 0,1,\ldots,10$, for the following fuzzy relation matrix R:

$$R = \begin{bmatrix} 0.2 & 0.7 & 0.8 & 1 \\ 1 & 0.9 & 0.5 & 0.1 \\ 0 & 0.8 & 1 & 0.6 \\ 0.2 & 0.4 & 1 & 0.3 \end{bmatrix}$$

4.5. For the fuzzy relation R_4 in Example 3.11 find the λ-cut relations for the following values of λ:
(a) $\lambda = 0^+$
(b) $\lambda = 0.1$
(c) $\lambda = 0.4$
(d) $\lambda = 0.7$

4.6. For the fuzzy relation R in Problem 3.9(a) sketch (in 3D) the λ-cut relations for the following values of λ:

(a) $\lambda = 0^+$
(b) $\lambda = 0.3$
(c) $\lambda = 0.5$
(d) $\lambda = 0.9$
(e) $\lambda = 1$

4.7. Show that any λ-cut relation (for $\lambda > 0$) of a fuzzy tolerance relation results in a crisp tolerance relation.

4.8. Show that any λ-cut relation (for $\lambda > 0$) of a fuzzy equivalence relation results in a crisp equivalence relation.

4.9. In metallurgy materials are made with mixtures of various metals and other elements to achieve certain desirable properties. In a particular preparation of steel, three elements, namely iron, manganese, and carbon, are mixed in two different proportions. The samples obtained from these two different proportions are placed on a normalized scale, as shown in Fig. P4.9 and are represented as fuzzy sets A_1 and A_2. You are interested in finding some sort of "average" steel proportion. For the logical union of the membership functions shown we want to find the defuzzified quantity. For each of the seven methods presented in this chapter assess (a) whether each is applicable and, if so, (b) calculate the defuzzified value, z^*.

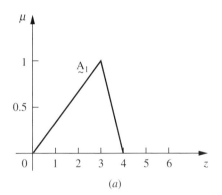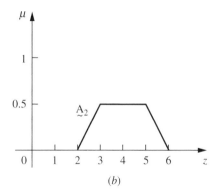

FIGURE P4.9

4.10. Two companies bid for a contract. A committee has to review the estimates of those companies and give reports to its chairperson. The reviewed reports are evaluated on a nondimensional scale and assigned a weighted score that is represented by a fuzzy membership function, as illustrated by the two fuzzy sets, B_1 and B_2, in Fig. P4.10. The chairperson is interested in the lowest bid, as well as a metric to measure the combined "best" score. For the logical union of the membership functions shown we want to find the defuzzified quantity. For each of the seven methods presented in this chapter assess (a) whether each is applicable and, if so, (b) calculate the defuzzified value, z^*.

4.11. A landfill is the cheapest method of solid waste treatment and disposal. Once disposed into a landfill, solid waste can be a major source of energy due to its potential to produce methane. However, all the solid waste disposed cannot generate methane at the same rate and in the same quantities. Based on its biodegradability, solid waste is classified into three distinct groups, namely: rapidly biodegradable, moderately biodegradable, and slowly biodegradable. Design of a landfill gas extraction system is based on gas production through the first two groups; both have different gas production patterns. The data collected from experiments and experiences are presented by the sets A_1 and A_2 as shown in Fig. P4.11, where A_1 and A_2 are defined as the fuzzy

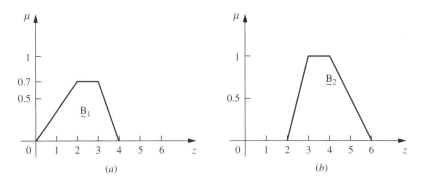

FIGURE P4.10

sets rapidly biodegradable and slowly biodegradable, respectively, in units of years. In order to properly design the gas extraction system we need a single representative gas production value. For the logical union of the membership functions shown we want to find the defuzzified quantity. For each of the seven methods presented in this chapter assess (*a*) whether each is applicable and, if so, (*b*) calculate the defuzzified value, z^*.

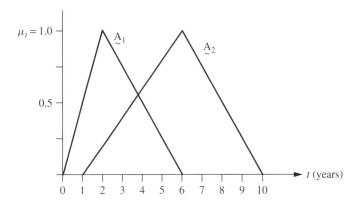

FIGURE P4.11

4.12. Uniaxial compressive strength is easily performed on cylindrical or prismatic ice samples and can vary with strain rate, temperature, porosity, grain orientation, and grain size ratio. While strain rate and temperature can be controlled easily, the other variables cannot. This lack of control yields an uncertainty in the uniaxial test results.

A test was conducted on each type of sample at a constant strain rate of 10^{-4} s^{-1}, and a temperature of $-5°$C. Upon inspection of the results the exact yield point could not be determined; however, there was enough information to form fuzzy sets for the failure of the cylindrical and prismatic samples $\underset{\sim}{A}$ and $\underset{\sim}{B}$, respectively, as shown in Fig. P4.12. Once the union of $\underset{\sim}{A}$ and $\underset{\sim}{B}$ has been identified (the universe of compressive strengths, megapascals N/m^2 × 10^6) we can obtain a defuzzified value for the yield strength of this ice under a compressive axial load. For each of the seven methods presented in this chapter assess (*a*) whether each is applicable and, if so, (*b*) calculate the defuzzified value, z^*.

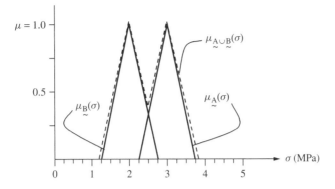

$\mu = 1.0$

$\mu_{A \cup B}(\sigma)$

0.5

$\mu_B(\sigma)$

$\mu_A(\sigma)$

σ (MPa)

0   1   2   3   4   5

FIGURE P4.12

4.13. In the field of heat exchanger network (HEN) synthesis, a chemical plant can have two different sets of independent HENs. The term "optimum cost" is considered fuzzy, because for the design and construction of the HENs we have to consider other parameters in addition to the capital cost. The membership function of fuzzy sets HEN1 and HEN2 is shown in Figs.P4.13(a) and P4.13(b), respectively.

    We wish to determine the optimum capital cost of a project to optimize the plant using both independent networks (HEN1 and HEN2); hence, the logical union of their membership functions, as shown in Fig. P4.13(c). For each of the seven methods presented in this chapter assess (*a*) whether each is applicable and, if so, (*b*) calculate the defuzzified value, z^*.

4.14. In reactor design, it is often best to simplify a reactor by assuming ideal conditions. For a continuous stirred tank reactor (CSTR), the concentration inside the reactor is the same as the concentration of the effluent stream. In a plug flow reactor (PFR), the concentrations of the inlet and outlet streams are different as the concentrations of the reactants change along the length of the tube. For a fluidized bed in which catalyst is removed from the top of the reactor, there exists both characteristics of a CSTR and PFR. The difference between inside reactor concentration (C_i) and effluent concentration (C_e) gives the membership of either CSTR or PFR, as seen in Fig. P4.14.

    Find the difference in concentration that represents the optimum design, i.e., find the most representative value for the union of PFR and CSTR. For each of the seven methods presented in this chapter assess (*a*) whether each is applicable and, if so, (*b*) calculate the defuzzified value, z^*.

4.15. Often in chemical processing plants there will be more than one type of instrumentation measuring the same variable at the same instance during the process. Due to the nature of measurements they are almost never exact, and hence can be represented as a fuzzy set. Due to the differences in instrumentation the measurements will usually not be the same. Take for example two types of temperature sensors, namely a thermocouple (TC) and an RTD (Resistance Temperature Detector) measuring the same stream temperature. The membership function of the two types of temperature sensors may look as in Fig. P4.15.

    When an operator who prefers one measuring device ends his or her shift, and then is replaced by another operator with a different preference in measuring device, there may be a problem in determining the actual value of a variable. To avoid this problem it was decided to plot the membership functions of the two types of sensors, take their union, and employ defuzzification to select one temperature for the operator to use. To find this temperature, for each of the seven methods presented in this chapter, assess (*a*) whether each is applicable and, if so, (*b*) calculate the defuzzified value, z^*.

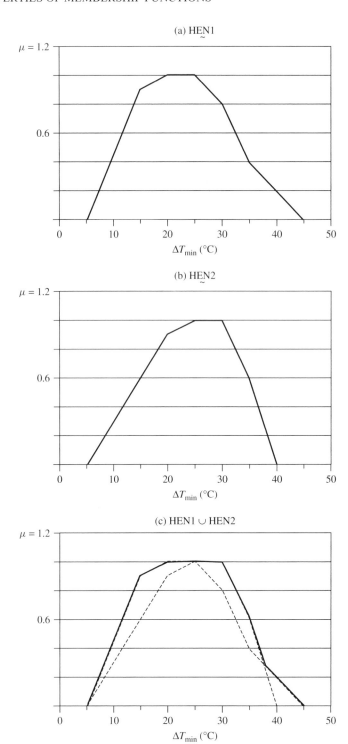

FIGURE P4.13

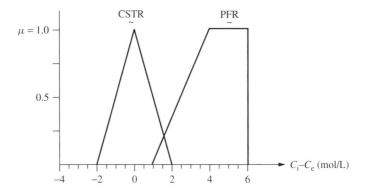

FIGURE P4.14

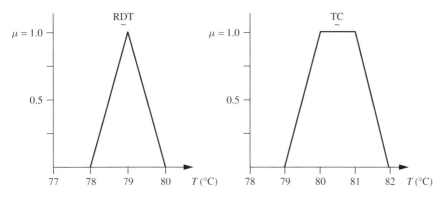

FIGURE P4.15

CHAPTER
5

LOGIC AND FUZZY SYSTEMS

PART I LOGIC

"I know what you're thinking about," said Tweedledum; *"but it isn't so, nohow."* *"Contrariwise,"* continued Tweedledee, *"if it was so, it might be; and if it were so, it would be; but as it isn't, it ain't. That's logic."*

Lewis Carroll
Through the Looking Glass, 1871

Logic is but a small part of the human capacity to reason. Logic can be a means to compel us to infer correct answers, but it cannot by itself be responsible for our creativity or for our ability to remember. In other words, logic can assist us in organizing words to make clear sentences, but it cannot help us determine what sentences to use in various contexts. Consider the passage above from the nineteenth-century mathematician Lewis Carroll in his classic *Through the Looking Glass*. How many of us can see the logical context in the discourse of these fictional characters? Logic for humans is a way quantitatively to develop a reasoning process that can be replicated and manipulated with mathematical precepts. The interest in logic is the study of truth in logical propositions; in classical logic this truth is binary – a proposition is either true or false.

From this perspective, fuzzy logic is a method to formalize the human capacity of imprecise reasoning, or – later in this chapter – approximate reasoning. Such reasoning represents the human ability to reason approximately and judge under uncertainty. In fuzzy logic all truths are partial or approximate. In this sense this reasoning has also been termed interpolative reasoning, where the process of interpolating between the binary extremes of true and false is represented by the ability of fuzzy logic to encapsulate partial truths.

Fuzzy Logic with Engineering Applications, Second Edition   T. J. Ross
© 2004 John Wiley & Sons, Ltd   ISBNs: 0-470-86074-X (HB); 0-470-86075-8 (PB)

Part-I of this chapter introduces the reader to fuzzy logic with a review of classical logic and its operations, logical implications, and certain classical inference mechanisms such as tautologies. The concept of a proposition is introduced as are associated concepts of truth sets, tautologies, and contradictions. The operations of disjunction, conjunction, and negation are introduced as well as classical implication and equivalence; all of these are useful tools to construct compound propositions from single propositions. Operations on propositions are shown to be isomorphic with operations on sets; hence an algebra of propositions is developed by using the algebra of sets discussed in Chapter 2. Fuzzy logic is then shown to be an extension of classical logic when partial truths are included to extend bivalued logic (true or false) to a multivalued logic (degrees of truth between true and not true).

In Part-II of this chapter we introduce the use of fuzzy sets as a calculus for the interpretation of natural language. Natural language, despite its vagueness and ambiguity, is the vehicle for human communication, and it seems appropriate that a mathematical theory that deals with fuzziness and ambiguity is also the same tool used to express and interpret the linguistic character of our language. The chapter continues with the use of natural language in the expression of a knowledge form known as rule-based systems, which shall be referred to generally as *fuzzy systems*. The chapter concludes with a simple graphical interpretation of inference, which is illustrated with some examples.

CLASSICAL LOGIC

In classical logic, a simple proposition P is a linguistic, or declarative, statement contained within a universe of elements, say X, that can be identified as being a collection of elements in X that are strictly true or strictly false. Hence, a proposition P is a collection of elements, i.e., a set, where the truth values for all elements in the set are either all true or all false. The veracity (truth) of an element in the proposition P can be assigned a binary truth value, called $T(P)$, just as an element in a universe is assigned a binary quantity to measure its membership in a particular set. For binary (Boolean) classical logic, $T(P)$ is assigned a value of 1 (truth) or 0 (false). If U is the universe of all propositions, then T is a mapping of the elements, u, in these propositions (sets) to the binary quantities (0, 1), or

$$T : u \in U \longrightarrow (0, 1)$$

All elements u in the universe U that are true for proposition P are called the truth set of P, denoted $T(P)$. Those elements u in the universe U that are false for proposition P are called the falsity set of P.

In logic we need to postulate the boundary conditions of truth values just as we do for sets; that is, in function-theoretic terms we need to define the truth value of a universe of discourse. For a universe Y and the null set Ø, we define the following truth values:

$$T(Y) = 1 \quad \text{and} \quad T(\emptyset) = 0$$

Now let P and Q be two simple propositions on the same universe of discourse that can be combined using the following five logical connectives

Disjunction	$(\vee)$
Conjunction	$(\wedge)$
Negation	$(-)$
Implication	$(\rightarrow)$
Equivalence	$(\leftrightarrow)$

to form logical expressions involving the two simple propositions. These connectives can be used to form new propositions from simple propositions.

The disjunction connective, the logical *or*, is the term used to represent what is commonly referred to as the *inclusive or*. The natural language term *or* and the logical *or* differ in that the former implies exclusion (denoted in the literature as the *exclusive or*; further details are given in this chapter). For example, ''soup or salad'' on a restaurant menu implies the choice of one or the other option, but not both. The *inclusive or* is the one most often employed in logic; the inclusive or (*logical or* as used here) implies that a compound proposition is true if either of the simple propositions is true or both are true.

The equivalence connective arises from dual implication; that is, for some propositions P and Q, if P $\rightarrow$ Q and Q $\rightarrow$ P, then P $\leftrightarrow$ Q.

Now define sets A and B from universe X (universe X is isomorphic with universe U), where these sets might represent linguistic ideas or thoughts. A *propositional calculus* (sometimes called the *algebra of propositions*) will exist for the case where proposition P measures the truth of the statement that an element, x, from the universe X is contained in set A and the truth of the statement Q that this element, x, is contained in set B, or more conventionally,

$$P : \text{truth that } x \in A$$

$$Q : \text{truth that } x \in B$$

where truth is measured in terms of the truth value, i.e.,

$$\text{if } x \in A, \ T(P) = 1; \ \text{otherwise}, \ T(P) = 0$$

$$\text{if } x \in B, \ T(Q) = 1; \ \text{otherwise}, \ T(Q) = 0$$

or, using the characteristic function to represent truth (1) and falsity (0), the following notation results:

$$\chi_A(x) = \begin{cases} 1, & x \in A \\ 0, & x \notin A \end{cases}$$

A notion of *mutual exclusivity* arises in this calculus. For the situation involving two propositions P and Q, where $T(P) \cap T(Q) = \emptyset$, we have that the truth of P always implies the falsity of Q and vice versa; hence, P and Q are mutually exclusive propositions.

Example 5.1. Let P be the proposition ''The structural beam is an 18WF45'' and let Q be the proposition ''The structural beam is made of steel.'' Let X be the universe of structural members comprised of girders, beams, and columns; x is an element (beam), A is the set of all wide-flange (WF) beams, and B is the set of all steel beams. Hence,

$$P : x \text{ is in A}$$

$$Q : x \text{ is in B}$$

The five logical connectives already defined can be used to create compound propositions, where a compound proposition is defined as a logical proposition formed by logically connecting two or more simple propositions. Just as we are interested in the truth of a simple proposition, classical logic also involves the assessment of the truth of compound propositions. For the case of two simple propositions, the resulting compound propositions are defined next in terms of their binary truth values.

Given a proposition $P : x \in A$, $\overline{P} : x \notin A$, we have the following for the logical connectives:

Disjunction

$$P \vee Q : x \in A \text{ or } x \in B$$

$$\text{Hence, } T(P \vee Q) = \max(T(P), T(Q)) \tag{5.1a}$$

Conjunction

$$P \wedge Q : x \in A \text{ and } x \in B$$

$$\text{Hence, } T(P \wedge Q) = \min(T(P), T(Q)) \tag{5.1b}$$

Negation

$$\text{If } T(P) = 1, \text{ then } T(\overline{P}) = 0; \text{ if } T(P) = 0, \text{ then } T(\overline{P}) = 1. \tag{5.1c}$$

Implication

$$(P \longrightarrow Q) : x \notin A \text{ or } x \in B$$

$$\text{Hence, } T(P \longrightarrow Q) = T(\overline{P} \cup Q) \tag{5.1d}$$

Equivalence

$$(P \longleftrightarrow Q) : T(P \longleftrightarrow Q) = \begin{cases} 1, & \text{for } T(P) = T(Q) \\ 0, & \text{for } T(P) \neq T(Q) \end{cases} \tag{5.1e}$$

The logical connective *implication*, i.e., $P \to Q$ (P implies Q), presented here is also known as the classical implication, to distinguish it from an alternative form devised in the 1930s by Lukasiewicz, a Polish mathematician, who was first credited with exploring logics other than Aristotelian (classical or binary logic) [Rescher, 1969], and from several other forms (see end of this chapter). In this implication the proposition P is also referred to as the *hypothesis* or the *antecedent*, and the proposition Q is also referred to as the *conclusion* or the *consequent*. The compound proposition $P \to Q$ is true in all cases except where a true antecedent P appears with a false consequent, Q, i.e., a true hypothesis cannot imply a false conclusion.

Example 5.2 **[Similar to Gill, 1976].** Consider the following four propositions:

1. If $1 + 1 = 2$, then $4 > 0$.
2. If $1 + 1 = 3$, then $4 > 0$.
3. If $1 + 1 = 3$, then $4 < 0$.
4. If $1 + 1 = 2$, then $4 < 0$.

The first three propositions are all true; the fourth is false. In the first two, the conclusion $4 > 0$ is true regardless of the truth of the hypothesis; in the third case both propositions are false, but this does not disprove the implication; finally, in the fourth case, a true hypothesis cannot produce a false conclusion.

Hence, the classical form of the implication is true for all propositions of P and Q except for those propositions that are in both the truth set of P and the false set of Q, i.e.,

$$T(P \longrightarrow Q) = \overline{T(P) \cap T(\overline{Q})} \qquad (5.2)$$

This classical form of the implication operation requires some explanation. For a proposition P defined on set A and a proposition Q defined on set B, the implication "P implies Q" is equivalent to taking the union of elements in the complement of set A with the elements in the set B (this result can also be derived by using De Morgan's principles on Eq. (5.2)). That is, the logical implication is analogous to the set-theoretic form

$$(P \longrightarrow Q) \equiv (\overline{A} \cup B \text{ is true}) \equiv (\text{either "not in A" or "in B"})$$

so that

$$T(P \longrightarrow Q) = T(\overline{P} \vee Q) = \max(T(\overline{P}), T(Q)) \qquad (5.3)$$

This expression is linguistically equivalent to the statement, "P → Q is true" when either "not A" or "B" is true (logical or). Graphically, this implication and the analogous set operation are represented by the Venn diagram in Fig. 5.1. As noted in the diagram, the region represented by the difference A | B is the set region where the implication P → Q is false (the implication "fails"). The shaded region in Fig. 5.1 represents the collection of elements in the universe where the implication is true; that is, the set

$$\overline{A \mid B} = \overline{\overline{A} \cup \overline{B}} = \overline{A \cap \overline{B}}$$

If x is in A *and* x is not in B, then

$$A \longrightarrow B \text{ fails} \equiv A \mid B \text{ (difference)}$$

Now, with two propositions (P and Q) each being able to take on one of two truth values (true or false, 1 or 0), there will be a total of $2^2 = 4$ propositional situations. These situations are illustrated, along with the appropriate truth values, for the propositions P and

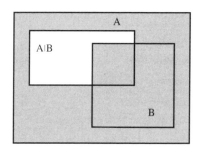

FIGURE 5.1
Graphical analog of the classical implication operation; gray area is where implication holds.

TABLE 5.1
Truth table for various compound propositions

P	Q	$\overline{P}$	P ∨ Q	P ∧ Q	P → Q	P ↔ Q
T (1)	T (1)	F (0)	T (1)	T (1)	T (1)	T (1)
T (1)	F (0)	F (0)	T (1)	F (0)	F (0)	F (0)
F (0)	T (1)	T (1)	T (1)	F (0)	T (1)	F (0)
F (0)	F (0)	T (1)	F (0)	F (0)	T (1)	T (1)

Q and the various logical connectives between them in Table 5.1. The values in the last five columns of the table are calculated using the expressions in Eqs. (5.1) and (5.3). In Table 5.1 T (or 1) denotes true and F (or 0) denotes false.

Suppose the implication operation involves two different universes of discourse; P is a proposition described by set A, which is defined on universe X, and Q is a proposition described by set B, which is defined on universe Y. Then the implication P → Q can be represented in set-theoretic terms by the relation R, where R is defined by

$$R = (A \times B) \cup (\overline{A} \times Y) \equiv \text{IF A, THEN B}$$

$$\text{IF } x \in A \text{ where } x \in X \text{ and } A \subset X \tag{5.4}$$

$$\text{THEN } y \in B \text{ where } y \in Y \text{ and } B \subset Y$$

This implication, Eq. (5.4), is also equivalent to the linguistic rule form, IF A, THEN B. The graphic shown in Fig. 5.2 represents the space of the Cartesian product X × Y, showing typical sets A and B; and superposed on this space is the set-theoretic equivalent of the implication. That is,

$$P \longrightarrow Q : \text{IF } x \in A, \text{ THEN } y \in B, \quad \text{or} \quad P \longrightarrow Q \equiv \overline{A} \cup B$$

The shaded regions of the compound Venn diagram in Fig. 5.2 represent the truth domain of the implication, IF A, THEN B (P → Q).

Another compound proposition in linguistic rule form is the expression

$$\text{IF A, THEN B, ELSE C}$$

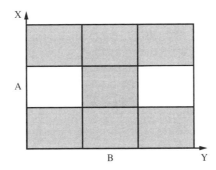

FIGURE 5.2
The Cartesian space showing the implication IF A, THEN B.

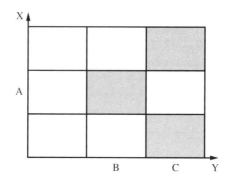

FIGURE 5.3
Truth domain for IF A, THEN B, ELSE C.

Linguistically, this compound proposition could be expressed as

$$\text{IF A, THEN B,} \quad \text{and} \quad \text{IF } \overline{\text{A}}, \text{ THEN C}$$

In classical logic this rule has the form

$$(P \longrightarrow Q) \wedge (\overline{P} \longrightarrow S) \tag{5.5}$$

$$P: \quad x \in A, A \subset X$$

$$Q: \quad y \in B, B \subset Y$$

$$S: \quad y \in C, C \subset Y$$

The set-theoretic equivalent of this compound proposition is given by

$$\text{IF A, THEN B, ELSE C} \equiv (A \times B) \cup (\overline{A} \times C) = R = \text{ relation on } X \times Y \tag{5.6}$$

The graphic in Fig. 5.3 illustrates the shaded region representing the truth domain for this compound proposition for the particular case where $B \cap C = \emptyset$.

Tautologies

In classical logic it is useful to consider compound propositions that are always true, irrespective of the truth values of the individual simple propositions. Classical logical compound propositions with this property are called *tautologies*. Tautologies are useful for deductive reasoning, for proving theorems, and for making deductive inferences. So, if a compound proposition can be expressed in the form of a tautology, the truth value of that compound proposition is known to be true. Inference schemes in expert systems often employ tautologies because tautologies are formulas that are true on logical grounds alone. For example, if A is the set of all prime numbers ($A_1 = 1, A_2 = 2, A_3 = 3, A_4 = 5,\ldots$) on the real line universe, X, then the proposition "A_i is not divisible by 6" is a tautology.

One tautology, known as *modus ponens* deduction, is a very common inference scheme used in forward-chaining rule-based expert systems. It is an operation whose task is to find the truth value of a consequent in a production rule, given the truth value of the antecedent in the rule. *Modus ponens* deduction concludes that, given two propositions, P and P → Q, both of which are true, then the truth of the simple proposition Q is automatically inferred. Another useful tautology is the *modus tollens* inference, which is used in backward-chaining expert systems. In *modus tollens* an implication between two propositions is combined with a second proposition and both are used to imply a third proposition. Some common tautologies follow:

$$\overline{B} \cup B \longleftrightarrow X$$

$$A \cup X; \quad \overline{A} \cup X \longleftrightarrow X$$

$$(A \wedge (A \longrightarrow B)) \longrightarrow B \quad (\textit{modus ponens}) \tag{5.7}$$

$$(\overline{B} \wedge (A \longrightarrow B)) \longrightarrow \overline{A} \quad (\textit{modus tollens}) \tag{5.8}$$

A simple proof of the truth value of the *modus ponens* deduction is provided here, along with the various properties for each step of the proof, for purposes of illustrating the utility of a tautology in classical reasoning.

Proof

$$(A \wedge (A \longrightarrow B)) \longrightarrow B$$

$$(A \wedge (\overline{A} \cup B)) \longrightarrow B \qquad \textit{Implication}$$

$$((A \wedge \overline{A}) \cup (A \wedge B)) \longrightarrow B \qquad \textit{Distributivity}$$

$$(\emptyset \cup (A \wedge B)) \longrightarrow B \qquad \textit{Excluded middle axioms}$$

$$(A \wedge B) \longrightarrow B \qquad \textit{Identity}$$

$$\overline{(A \wedge B)} \cup B \qquad \textit{Implication}$$

$$(\overline{A} \vee \overline{B}) \cup B \qquad \textit{De Morgan's principles}$$

$$\overline{A} \vee (\overline{B} \cup B) \qquad \textit{Associativity}$$

$$\overline{A} \cup X \qquad \textit{Excluded middle axioms}$$

$$X \Longrightarrow T(X) = 1 \qquad \textit{Identity; QED}$$

A simpler manifestation of the truth value of this tautology is shown in Table 5.2 in truth table form, where a column of all ones for the result shows a tautology.

TABLE 5.2
Truth table (*modus ponens*)

A	B	A → B	(A ∧ (A → B))	(A ∧ (A → B)) → B
0	0	1	0	1
0	1	1	0	1 Tautology
1	0	0	0	1
1	1	1	1	1

TABLE 5.3
Truth table (*modus tollens*)

A	B	$\overline{A}$	$\overline{B}$	A → B	$(\overline{B} \wedge (A \to B))$	$(\overline{B} \wedge (A \to B)) \to \overline{A}$	
0	0	1	1	1	1	1	
0	1	1	0	1	0	1	Tautology
1	0	0	1	0	0	1	
1	1	0	0	1	0	1	

Similarly, a simple proof of the truth value of the *modus tollens* inference is listed here.

Proof

$$(\overline{B} \wedge (A \longrightarrow B)) \longrightarrow \overline{A}$$

$$(\overline{B} \wedge (\overline{A} \cup B)) \longrightarrow \overline{A}$$

$$((\overline{B} \wedge \overline{A}) \cup (\overline{B} \wedge B)) \longrightarrow \overline{A}$$

$$((\overline{B} \wedge \overline{A}) \cup \emptyset) \longrightarrow \overline{A}$$

$$(\overline{B} \wedge \overline{A}) \longrightarrow \overline{A}$$

$$\overline{(\overline{B} \wedge \overline{A})} \cup \overline{A}$$

$$(\overline{\overline{B}} \vee \overline{\overline{A}}) \cup \overline{A}$$

$$B \cup (A \cup \overline{A})$$

$$B \cup X = X \Longrightarrow T(X) = 1 \qquad \text{QED}$$

The truth table form of this result is shown in Table 5.3.

Contradictions

Compound propositions that are always false, regardless of the truth value of the individual simple propositions constituting the compound proposition, are called contradictions. For example, if A is the set of all prime numbers ($A_1 = 1$, $A_2 = 2$, $A_3 = 3$, $A_4 = 5$, ...) on the real line universe, X, then the proposition "A_i is a multiple of 4" is a contradiction. Some simple contradictions are listed here:

$$\overline{B} \cap B$$

$$A \cap \emptyset; \quad \overline{A} \cap \emptyset$$

Equivalence

As mentioned, propositions P and Q are equivalent, i.e., P ↔ Q, is true only when both P and Q are true or when both P and Q are false. For example, the propositions P: "triangle

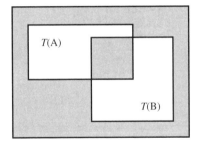

FIGURE 5.4
Venn diagram for equivalence (darkened areas), i.e., for $T(A \leftrightarrow B)$.

is equilateral'' and Q: "triangle is equiangular" are equivalent because they are either both true or both false for some triangle. This condition of equivalence is shown in Fig. 5.4, where the shaded region is the region of equivalence.

It can be easily proved that the statement P $\leftrightarrow$ Q is a tautology if P is identical to Q, i.e., if and only if $T(P) = T(Q)$.

Example 5.3. Suppose we consider the universe of positive integers, $X = \{1 \leq n \leq 8\}$. Let P = "n is an even number" and let Q = "$(3 \leq n \leq 7) \wedge (n \neq 6)$." Then $T(P) = \{2, 4, 6, 8\}$ and $T(Q) = \{3, 4, 5, 7\}$. The equivalence P $\leftrightarrow$ Q has the truth set

$$T(P \longleftrightarrow Q) = (T(P) \cap T(Q)) \cup (\overline{T(P)} \cap \overline{T(Q)}) = \{4\} \cup \{1\} = \{1, 4\}$$

One can see that "1 is an even number" and "$(3 \leq 1 \leq 7) \wedge (1 \neq 6)$" are both false, and "4 is an even number" and "$(3 \leq 4 \leq 7) \wedge (4 \neq 6)$" are both true.

Example 5.4. Prove that P $\leftrightarrow$ Q if P = "n is an integer power of 2 less than 7 and greater than zero" and Q = "$n^2 - 6n + 8 = 0$." Since $T(P) = \{2, 4\}$ and $T(Q) = \{2, 4\}$, it follows that P $\leftrightarrow$ Q is an equivalence.

Suppose a proposition R has the form P $\rightarrow$ Q. Then the proposition $\overline{Q} \rightarrow \overline{P}$ is called the *contrapositive* of R; the proposition Q $\rightarrow$ P is called the *converse* of R; and the proposition $\overline{P} \rightarrow \overline{Q}$ is called the *inverse* of R. Interesting properties of these propositions can be shown (see Problem 5.3 at the end of this chapter).

The *dual* of a compound proposition *that does not involve implication* is the same proposition with false (0) replacing true (1) (i.e., a set being replaced by its complement), true replacing false, conjunction ($\wedge$) replacing disjunction ($\vee$), and disjunction replacing conjunction. If a proposition is true, then its *dual* is also true (see Problems 5.4 and 5.5).

Exclusive Or and Exclusive Nor

Two more interesting compound propositions are worthy of discussion. These are the *exclusive or* and the *exclusive nor*. The exclusive or is of interest because it arises in many situations involving natural language and human reasoning. For example, when you are going to travel by plane or boat to some destination, the implication is that you can travel by air or sea, but not both, i.e., one or the other. This situation involves the exclusive or; it

TABLE 5.4
Truth table for exclusive
or, *XOR*

P	Q	P *XOR* Q
1	1	0
1	0	1
0	1	1
0	0	0

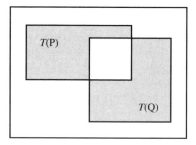

FIGURE 5.5
Exclusive or shown in gray areas.

TABLE 5.5
Truth table for exclusive nor

P	Q	P *XOR* Q
1	1	1
1	0	0
0	1	0
0	0	1

does not involve the intersection, as does the logical or (union in Eq. (2.1) and Fig. 2.2 and disjunction in Eq. (5.1*a*)). For two propositions, P and Q, the exclusive or, denoted here as *XOR*, is given in Table 5.4 and Fig. 5.5.

The *exclusive nor* is the complement of the *exclusive or* [Mano, 1988]. A look at its truth table, Table 5.5, shows that it is an equivalence operation, i.e.,

$$\overline{P\,XOR\,Q} \longleftrightarrow (P \longleftrightarrow Q)$$

and, hence, it is graphically equivalent to the Venn diagram in Fig. 5.4.

Logical Proofs

Logic involves the use of inference in everyday life, as well as in mathematics. In the latter, we often want to prove theorems to form foundations for solution procedures. In natural language, if we are given some hypotheses it is often useful to make certain conclusions from them – the so-called process of inference (inferring new facts from established facts). In the terminology we have been using, we want to know if the proposition $(P_1 \wedge P_2 \wedge \cdots \wedge P_n) \rightarrow Q$ is true. That is, is the statement a tautology?

The process works as follows. First, the linguistic statement (compound proposition) is made. Second, the statement is decomposed into its respective single propositions. Third, the statement is expressed algebraically with all pertinent logical connectives in place. Fourth, a truth table is used to establish the veracity of the statement.

Example 5.5.

Hypotheses: Engineers are mathematicians. Logical thinkers do not believe in magic. Mathematicians are logical thinkers.

Conclusion: Engineers do not believe in magic.

Let us decompose this information into individual propositions.

$$P : \text{a person is an engineer}$$

$$Q : \text{a person is a mathematician}$$

$$R : \text{a person is a logical thinker}$$

$$S : \text{a person believes in magic}$$

The statements can now be expressed as algebraic propositions as

$$((P \longrightarrow Q) \wedge (R \longrightarrow \overline{S}) \wedge (Q \longrightarrow R)) \longrightarrow (P \longrightarrow \overline{S})$$

It can be shown that this compound proposition is a tautology (see Problem 5.6).

Sometimes it might be difficult to prove a proposition by a direct proof (i.e., verify that it is true), so an alternative is to use an indirect proof. For example, the popular *proof by contradiction* (*reductio ad absurdum*) exploits the fact that $P \rightarrow Q$ is true if and only if $P \wedge \overline{Q}$ is false. Hence, if we want to prove that the compound statement $(P_1 \wedge P_2 \wedge \cdots \wedge P_n) \rightarrow Q$ is a tautology, we can alternatively show that the alternative statement $P_1 \wedge P_2 \wedge \cdots \wedge P_n \wedge \overline{Q}$ is a contradiction.

Example 5.6.

Hypotheses: If an arch-dam fails, the failure is due to a poor subgrade. An arch-dam fails.

Conclusion: The arch-dam failed because of a poor subgrade.

This information can be shown to be algebraically equivalent to the expression

$$((P \longrightarrow Q) \wedge P) \longrightarrow Q$$

To prove this by contradiction, we need to show that the algebraic expression

$$((P \longrightarrow Q) \wedge P \wedge \overline{Q})$$

is a contradiction. We can do this by constructing the truth table in Table 5.6. Recall that a contradiction is indicated when the last column of a truth table is filled with zeros.

TABLE 5.6
Truth table for dam failure problem

P	Q	$\bar{P}$	$\bar{Q}$	$\bar{P} \vee Q$	$(\bar{P} \vee Q) \wedge P \wedge \bar{Q}$
0	0	1	1	1	0
0	1	1	0	1	0
1	0	0	1	0	0
1	1	0	0	1	0

Deductive Inferences

The *modus ponens* deduction is used as a tool for making inferences in rule-based systems. A typical if–then rule is used to determine whether an antecedent (cause or action) infers a consequent (effect or reaction). Suppose we have a rule of the form IF A, THEN B, where A is a set defined on universe X and B is a set defined on universe Y. As discussed before, this rule can be translated into a relation between sets A and B; that is, recalling Eq. (5.4), $R = (A \times B) \cup (\bar{A} \times Y)$. Now suppose a new antecedent, say A′, is known. Can we use *modus ponens* deduction, Eq. (5.7), to infer a new consequent, say B′, resulting from the new antecedent? That is, can we deduce, in rule form, IF A′, THEN B′? The answer, of course, is yes, through the use of the composition operation (defined initially in Chapter 3). Since "A implies B" is defined on the Cartesian space $X \times Y$, B′ can be found through the following set-theoretic formulation, again from Eq. (5.4):

$$B' = A' \circ R = A' \circ ((A \times B) \cup (\bar{A} \times Y))$$

where the symbol ∘ denotes the composition operation. *Modus ponens* deduction can also be used for the compound rule IF A, THEN B, ELSE C, where this compound rule is equivalent to the relation defined in Eq. (5.6) as $R = (A \times B) \cup (\bar{A} \times C)$. For this compound rule, if we define another antecedent A′, the following possibilities exist, depending on (1) whether A′ is fully contained in the original antecedent A, (2) whether A′ is contained only in the complement of A, or (3) whether A′ and A overlap to some extent as described next:

$$\text{IF } A' \subset A, \text{ THEN } y = B$$

$$\text{IF } A' \subset \bar{A}, \text{ THEN } y = C$$

$$\text{IF } A' \cap A \neq \emptyset, \ A' \cap \bar{A} \neq \emptyset, \text{ THEN } y = B \cup C$$

The rule IF A, THEN B (proposition P is defined on set A in universe X, and proposition Q is defined on set B in universe Y), i.e., $(P \rightarrow Q) = R = (A \times B) \cup (\bar{A} \times Y)$, is then defined in function-theoretic terms as

$$\chi_R(x, y) = \max[(\chi_A(x) \wedge \chi_B(y)), ((1 - \chi_A(x)) \wedge 1)] \tag{5.9}$$

where $\chi()$ is the characteristic function as defined before.

Example 5.7. Suppose we have two universes of discourse for a heat exchanger problem described by the following collection of elements, $X = \{1, 2, 3, 4\}$ and $Y = \{1, 2, 3, 4, 5, 6\}$.

Suppose X is a universe of normalized temperatures and Y is a universe of normalized pressures. Define crisp set A on universe X and crisp set B on universe Y as follows: A = {2, 3} and B = {3, 4}. The deductive inference IF A, THEN B (i.e., IF temperature is A, THEN pressure is B) will yield a matrix describing the membership values of the relation R, i.e., $\chi_R(x, y)$ through the use of Eq. (5.9). That is, the matrix R represents the rule IF A, THEN B as a matrix of characteristic (crisp membership) values.

Crisp sets A and B can be written using Zadeh's notation,

$$A = \left\{ \tfrac{0}{1} + \tfrac{1}{2} + \tfrac{1}{3} + \tfrac{0}{4} \right\}$$

$$B = \left\{ \tfrac{0}{1} + \tfrac{0}{2} + \tfrac{1}{3} + \tfrac{1}{4} + \tfrac{0}{5} + \tfrac{0}{6} \right\}$$

If we treat set A as a column vector and set B as a row vector, the following matrix results from the Cartesian product of A × B, using Eq. (3.16):

$$A \times B = \begin{bmatrix} 0 & 0 & 0 & 0 & 0 & 0 \\ 0 & 0 & 1 & 1 & 0 & 0 \\ 0 & 0 & 1 & 1 & 0 & 0 \\ 0 & 0 & 0 & 0 & 0 & 0 \end{bmatrix}$$

The Cartesian product $\overline{A} \times Y$ can be determined using Eq. (3.16) by arranging $\overline{A}$ as a column vector and the universe Y as a row vector (sets $\overline{A}$ and Y can be written using Zadeh's notation),

$$\overline{A} = \left\{ \tfrac{1}{1} + \tfrac{0}{2} + \tfrac{0}{3} + \tfrac{1}{4} \right\}$$

$$Y = \left\{ \tfrac{1}{1} + \tfrac{1}{2} + \tfrac{1}{3} + \tfrac{1}{4} + \tfrac{1}{5} + \tfrac{1}{6} \right\}$$

$$\overline{A} \times Y = \begin{bmatrix} 1 & 1 & 1 & 1 & 1 & 1 \\ 0 & 0 & 0 & 0 & 0 & 0 \\ 0 & 0 & 0 & 0 & 0 & 0 \\ 1 & 1 & 1 & 1 & 1 & 1 \end{bmatrix}$$

Then the full relation R describing the implication IF A, THEN B is the maximum of the two matrices A × B and $\overline{A} \times Y$, or, using Eq. (5.9),

$$R = \begin{array}{c} \\ 1 \\ 2 \\ 3 \\ 4 \end{array} \begin{array}{cccccc} 1 & 2 & 3 & 4 & 5 & 6 \\ \begin{bmatrix} 1 & 1 & 1 & 1 & 1 & 1 \\ 0 & 0 & 1 & 1 & 0 & 0 \\ 0 & 0 & 1 & 1 & 0 & 0 \\ 1 & 1 & 1 & 1 & 1 & 1 \end{bmatrix} \end{array}$$

The compound rule IF A, THEN B, ELSE C can also be defined in terms of a matrix relation as $R = (A \times B) \cup (\overline{A} \times C) \Rightarrow (P \rightarrow Q) \wedge (\overline{P} \rightarrow S)$, as given by Eqs. (5.5) and (5.6), where the membership function is determined as

$$\chi_R(x, y) = \max[(\chi_A(x) \wedge \chi_B(y)), ((1 - \chi_A(x)) \wedge \chi_C(y))] \tag{5.10}$$

Example 5.8. Continuing with the previous heat exchanger example, suppose we define a crisp set C on the universe of normalized temperatures Y as C = {5, 6}, or, using Zadeh's notation,

$$C = \left\{ \tfrac{0}{1} + \tfrac{0}{2} + \tfrac{0}{3} + \tfrac{0}{4} + \tfrac{1}{5} + \tfrac{1}{6} \right\}$$

The deductive inference IF A, THEN B, ELSE C (i.e., IF pressure is A, THEN temperature is B, ELSE temperature is C) will yield a relational matrix R, with characteristic values $\chi_R(x, y)$

obtained using Eq. (5.10). The first half of the expression in Eq. (5.10) (i.e., A × B) has already been determined in the previous example. The Cartesian product $\overline{A}$ × C can be determined using Eq. (3.16) by arranging the set $\overline{A}$ as a column vector and the set C as a row vector (see set $\overline{A}$ in Example 5.7), or

$$
\overline{A} \times C = \begin{bmatrix} 0 & 0 & 0 & 0 & 1 & 1 \\ 0 & 0 & 0 & 0 & 0 & 0 \\ 0 & 0 & 0 & 0 & 0 & 0 \\ 0 & 0 & 0 & 0 & 1 & 1 \end{bmatrix}
$$

Then the full relation R describing the implication IF A, THEN B, ELSE C is the maximum of the two matrices A × B and $\overline{A}$ × C (see Eq. (5.10)),

$$
R = \begin{matrix} & 1 & 2 & 3 & 4 & 5 & 6 \\ 1 & 0 & 0 & 0 & 0 & 1 & 1 \\ 2 & 0 & 0 & 1 & 1 & 0 & 0 \\ 3 & 0 & 0 & 1 & 1 & 0 & 0 \\ 4 & 0 & 0 & 0 & 0 & 1 & 1 \end{matrix}
$$

FUZZY LOGIC

The restriction of classical propositional calculus to a two-valued logic has created many interesting paradoxes over the ages. For example, the Barber of Seville is a classic paradox (also termed Russell's barber). In the small Spanish town of Seville, there is a rule that all and only those men who do not shave themselves are shaved by the barber. Who shaves the barber? Another example comes from ancient Greece. Does the liar from Crete lie when he claims, ''All Cretians are liars?'' If he is telling the truth, his statement is false. But if his statement is false, he is not telling the truth. A simpler form of this paradox is the two-word proposition, ''I lie.'' The statement can not be both true and false.

Returning to the Barber of Seville, we conclude that the only way for this paradox (or any classic paradox for that matter) to work is if the statement is both true and false simultaneously. This can be shown, using set notation [Kosko, 1992]. Let S be the proposition that the barber shaves himself and $\overline{S}$ (not S) that he does not. Then since $S \rightarrow \overline{S}$ (S implies not S), and $\overline{S} \rightarrow S$, the two propositions are logically equivalent: $S \leftrightarrow \overline{S}$. Equivalent propositions have the same truth value; hence,

$$
T(S) = T(\overline{S}) = 1 - T(S)
$$

which yields the expression

$$
T(S) = \tfrac{1}{2}
$$

As seen, paradoxes reduce to half-truths (or half-falsities) mathematically. In classical binary (bivalued) logic, however, such conditions are not allowed, i.e., only $T(S) = 1$ or 0 is valid; this is a manifestation of the constraints placed on classical logic by the excluded middle axioms.

A more subtle form of paradox can also be addressed by a multivalued logic. Consider the paradoxes represented by the classical *sorites* (literally, a heap of syllogisms); for example, the case of a liter-full glass of water. Often this example is called the Optimist's conclusion (is the glass half-full or half-empty when the volume is at 500 milliliters?). Is the liter-full glass still full if we remove 1 milliliter of water? Is the glass still full if we remove

2 milliliters of water, 3, 4, or 100 milliliters? If we continue to answer yes, then eventually we will have removed all the water, and an empty glass will still be characterized as full! At what point did the liter-full glass of water become empty? Perhaps at 500 milliliters full? Unfortunately no single milliliter of liquid provides for a transition between full and empty. This transition is gradual, so that as each milliliter of water is removed, the truth value of the glass being full gradually diminishes from a value of 1 at 1000 milliliters to 0 at 0 milliliters. Hence, for many problems we have need for a multivalued logic other than the classic binary logic that is so prevalent today.

A relatively recent debate involving similar ideas to those in paradoxes stems from a paper by psychologists Osherson and Smith [1981], in which they claim (incorrectly) that fuzzy set theory is not expressive enough to represent strong intuitionistic concepts. This idea can be described as the logically empty and logically universal concepts. The authors argued that the concept *apple that is not an apple is logically empty*, and that the concept *fruit that either is or is not an apple is logically universal*. These concepts are correct for classical logic; the logically empty idea and the logically universal idea are the axiom of contradiction and the axiom of the excluded middle, respectively. The authors argued that fuzzy logic also should adhere to these axioms to correctly represent concepts in natural language but, of course, there is a compelling reason why they should not. Several authorities have disputed this argument (see Belohlavek et al., 2002). While the *standard fuzzy operations* do not obey the excluded middle axioms, there are other fuzzy operations for intersection, union, and complement that do conform to these axioms if such confirmation is required by empirical evidence. More to the point, however, is that the concepts of *apple* and *fruit* are fuzzy and, as fruit geneticists will point out, there are some fruits that can appear to be an apple that genetically are not an apple.

A fuzzy logic proposition, $\underset{\sim}{P}$, is a statement involving some concept without clearly defined boundaries. Linguistic statements that tend to express subjective ideas and that can be interpreted slightly differently by various individuals typically involve fuzzy propositions. Most natural language is fuzzy, in that it involves vague and imprecise terms. Statements describing a person's height or weight or assessments of people's preferences about colors or menus can be used as examples of fuzzy propositions. The truth value assigned to $\underset{\sim}{P}$ can be any value on the interval [0, 1]. The assignment of the truth value to a proposition is actually a mapping from the interval [0, 1] to the universe U of truth values, T, as indicated in Eq. (5.11),

$$T : u \in U \longrightarrow (0, 1) \tag{5.11}$$

As in classical binary logic, we assign a logical proposition to a set in the universe of discourse. Fuzzy propositions are assigned to fuzzy sets. Suppose proposition $\underset{\sim}{P}$ is assigned to fuzzy set $\underset{\sim}{A}$; then the truth value of a proposition, denoted $T(\underset{\sim}{P})$, is given by

$$T(\underset{\sim}{P}) = \mu_{\underset{\sim}{A}}(x) \quad \text{where } 0 \le \mu_{\underset{\sim}{A}} \le 1 \tag{5.12}$$

Equation (5.12) indicates that the degree of truth for the proposition $\underset{\sim}{P} : x \in \underset{\sim}{A}$ is equal to the membership grade of x in the fuzzy set $\underset{\sim}{A}$.

The logical connectives of negation, disjunction, conjunction, and implication are also defined for a fuzzy logic. These connectives are given in Eqs. (5.13)–(5.16) for two simple propositions: proposition $\underset{\sim}{P}$ defined on fuzzy set $\underset{\sim}{A}$ and proposition $\underset{\sim}{Q}$ defined on fuzzy set $\underset{\sim}{B}$.

Negation

$$T(\overline{\underset{\sim}{P}}) = 1 - T(\underset{\sim}{P}) \tag{5.13}$$

Disjunction

$$\underset{\sim}{P} \vee \underset{\sim}{Q} : x \text{ is } \underset{\sim}{A} \text{ or } \underset{\sim}{B} \quad T(\underset{\sim}{P} \vee \underset{\sim}{Q}) = \max(T(\underset{\sim}{P}), T(\underset{\sim}{Q})) \tag{5.14}$$

Conjunction

$$\underset{\sim}{P} \wedge \underset{\sim}{Q} : x \text{ is } \underset{\sim}{A} \text{ and } \underset{\sim}{B} \quad T(\underset{\sim}{P} \wedge \underset{\sim}{Q}) = \min(T(\underset{\sim}{P}), T(\underset{\sim}{Q})) \tag{5.15}$$

Implication [Zadeh, 1973]

$$\underset{\sim}{P} \longrightarrow \underset{\sim}{Q} : x \text{ is } \underset{\sim}{A}, \text{ then x is } \underset{\sim}{B}$$

$$T(\underset{\sim}{P} \longrightarrow \underset{\sim}{Q}) = T(\overline{\underset{\sim}{P}} \vee \underset{\sim}{Q}) = \max(T(\overline{\underset{\sim}{P}}), T(\underset{\sim}{Q})) \tag{5.16}$$

As before in binary logic, the implication connective can be modeled in rule-based form; $\underset{\sim}{P} \to \underset{\sim}{Q}$ is, IF x is $\underset{\sim}{A}$, THEN y is $\underset{\sim}{B}$ and it is equivalent to the following fuzzy relation, $\underset{\sim}{R} = (\underset{\sim}{A} \times \underset{\sim}{B}) \cup (\overline{\underset{\sim}{A}} \times Y)$ (recall Eq. (5.4)), just as it is in classical logic. The membership function of $\underset{\sim}{R}$ is expressed by the following formula:

$$\mu_{\underset{\sim}{R}}(x, y) = \max[(\mu_{\underset{\sim}{A}}(x) \wedge \mu_{\underset{\sim}{B}}(y)), (1 - \mu_{\underset{\sim}{A}}(x))] \tag{5.17}$$

Example 5.9. Suppose we are evaluating a new invention to determine its commercial potential. We will use two metrics to make our decisions regarding the innovation of the idea. Our metrics are the "uniqueness" of the invention, denoted by a universe of novelty scales, $X = \{1, 2, 3, 4\}$, and the "market size" of the invention's commercial market, denoted on a universe of scaled market sizes, $Y = \{1, 2, 3, 4, 5, 6\}$. In both universes the lowest numbers are the "highest uniqueness" and the "largest market," respectively. A new invention in your group, say a compressible liquid of very useful temperature and viscosity conditions, has just received scores of "medium uniqueness," denoted by fuzzy set $\underset{\sim}{A}$, and "medium market size," denoted fuzzy set $\underset{\sim}{B}$. We wish to determine the implication of such a result, i.e., IF $\underset{\sim}{A}$, THEN $\underset{\sim}{B}$. We assign the invention the following fuzzy sets to represent its ratings:

$$\underset{\sim}{A} = \text{medium uniqueness} = \left\{ \frac{0.6}{2} + \frac{1}{3} + \frac{0.2}{4} \right\}$$

$$\underset{\sim}{B} = \text{medium market size} = \left\{ \frac{0.4}{2} + \frac{1}{3} + \frac{0.8}{4} + \frac{0.3}{5} \right\}$$

$$\underset{\sim}{C} = \text{diffuse market size} = \left\{ \frac{0.3}{1} + \frac{0.5}{2} + \frac{0.6}{3} + \frac{0.6}{4} + \frac{0.5}{5} + \frac{0.3}{6} \right\}$$

The following matrices are then determined in developing the membership function of the implication, $\mu_{\underset{\sim}{R}}(x, y)$, illustrated in Eq. (5.17),

$$\underset{\sim}{A} \times \underset{\sim}{B} = \begin{array}{c} 1 \\ 2 \\ 3 \\ 4 \end{array} \begin{array}{cccccc} 1 & 2 & 3 & 4 & 5 & 6 \\ \left[\begin{array}{cccccc} 0 & 0 & 0 & 0 & 0 & 0 \\ 0 & 0.4 & 0.6 & 0.6 & 0.3 & 0 \\ 0 & 0.4 & 1 & 0.8 & 0.3 & 0 \\ 0 & 0.2 & 0.2 & 0.2 & 0.2 & 0 \end{array}\right] \end{array}$$

$$\overline{A} \times Y = \begin{array}{c} \\ 1 \\ 2 \\ 3 \\ 4 \end{array} \begin{array}{cccccc} 1 & 2 & 3 & 4 & 5 & 6 \\ \left[\begin{array}{cccccc} 1 & 1 & 1 & 1 & 1 & 1 \\ 0.4 & 0.4 & 0.4 & 0.4 & 0.4 & 0.4 \\ 0 & 0 & 0 & 0 & 0 & 0 \\ 0.8 & 0.8 & 0.8 & 0.8 & 0.8 & 0.8 \end{array}\right] \end{array}$$

and finally, $R = \max(A \times B, \overline{A} \times Y)$

$$R = \begin{array}{c} \\ 1 \\ 2 \\ 3 \\ 4 \end{array} \begin{array}{cccccc} 1 & 2 & 3 & 4 & 5 & 6 \\ \left[\begin{array}{cccccc} 1 & 1 & 1 & 1 & 1 & 1 \\ 0.4 & 0.4 & 0.6 & 0.6 & 0.4 & 0.4 \\ 0 & 0.4 & 1 & 0.8 & 0.3 & 0 \\ 0.8 & 0.8 & 0.8 & 0.8 & 0.8 & 0.8 \end{array}\right] \end{array}$$

When the logical conditional implication is of the compound form

IF x is A, THEN y is B, ELSE y is C

then the equivalent fuzzy relation, R, is expressed as $R = (A \times B) \cup (\overline{A} \times C)$, in a form just as Eq. (5.6), whose membership function is expressed by the following formula:

$$\mu_R(x, y) = \max\left[(\mu_A(x) \wedge \mu_B(y)), ((1 - \mu_A(x)) \wedge \mu_C(y))\right] \tag{5.18}$$

Hence, using the result of Eq. 5.18, the new relation is

$$R = (A \times B) \cup (\overline{A} \times C)(\overline{A} \times C): \ \overline{A} \times C = \begin{array}{c} \\ 1 \\ 2 \\ 3 \\ 4 \end{array} \begin{array}{cccccc} 1 & 2 & 3 & 4 & 5 & 6 \\ \left[\begin{array}{cccccc} 0.3 & 0.5 & 0.6 & 0.6 & 0.5 & 0.3 \\ 0.3 & 0.4 & 0.4 & 0.4 & 0.4 & 0.3 \\ 0 & 0 & 0 & 0 & 0 & 0 \\ 0.3 & 0.5 & 0.6 & 0.6 & 0.5 & 0.3 \end{array}\right] \end{array}$$

and finally,

$$R = \begin{array}{c} \\ 1 \\ 2 \\ 3 \\ 4 \end{array} \begin{array}{cccccc} 1 & 2 & 3 & 4 & 5 & 6 \\ \left[\begin{array}{cccccc} 0.3 & 0.5 & 0.6 & 0.6 & 0.5 & 0.3 \\ 0.3 & 0.4 & 0.6 & 0.6 & 0.4 & 0.3 \\ 0 & 0.4 & 1 & 0.8 & 0.3 & 0 \\ 0.3 & 0.5 & 0.6 & 0.6 & 0.5 & 0.3 \end{array}\right] \end{array}$$

APPROXIMATE REASONING

The ultimate goal of fuzzy logic is to form the theoretical foundation for reasoning about imprecise propositions; such reasoning has been referred to as approximate reasoning [Zadeh, 1976, 1979]. Approximate reasoning is analogous to classical logic for reasoning with precise propositions, and hence is an extension of classical propositional calculus that deals with partial truths.

Suppose we have a rule-based format to represent fuzzy information. These rules are expressed in conventional antecedent-consequent form, such as

Rule 1: IF x is $\underset{\sim}{A}$, THEN y is $\underset{\sim}{B}$, where $\underset{\sim}{A}$ and $\underset{\sim}{B}$ represent fuzzy propositions (sets).

Now suppose we introduce a new antecedent, say $\underset{\sim}{A}'$, and we consider the following rule:

Rule 2: IF x is $\underset{\sim}{A}'$, THEN y is $\underset{\sim}{B}'$.

From information derived from Rule 1, is it possible to derive the consequent in Rule 2, $\underset{\sim}{B}'$? The answer is yes, and the procedure is fuzzy composition. The consequent $\underset{\sim}{B}'$ can be found from the composition operation, $\underset{\sim}{B}' = \underset{\sim}{A}' \circ \underset{\sim}{R}$.

The two most common forms of the composition operator are the max–min and the max–product compositions, as initially defined in Chapter 3.

Example 5.10. Continuing with the invention example, Example 5.9, suppose that the fuzzy relation just developed, i.e., $\underset{\sim}{R}$, describes the invention's commercial potential. We wish to know what market size would be associated with a uniqueness score of "almost high uniqueness." That is, with a new antecedent, $\underset{\sim}{A}'$, the following consequent, $\underset{\sim}{B}'$, can be determined using composition. Let

$$\underset{\sim}{A}' = \text{almost high uniqueness} = \left\{ \frac{0.5}{1} + \frac{1}{2} + \frac{0.3}{3} + \frac{0}{4} \right\}$$

Then, using the following max–min composition,

$$\underset{\sim}{B}' = \underset{\sim}{A}' \circ \underset{\sim}{R} = \left\{ \frac{0.5}{1} + \frac{0.5}{2} + \frac{0.6}{3} + \frac{0.6}{4} + \frac{0.5}{5} + \frac{0.5}{6} \right\}$$

we get the fuzzy set describing the associated market size. In other words, the consequent is fairly diffuse, where there is no strong (or weak) membership value for any of the market size scores (i.e., no membership values near 0 or 1).

This power of fuzzy logic and approximate reasoning to assess qualitative knowledge can be illustrated in more familiar terms to engineers in the context of the following example in the field of biophysics.

Example 5.11. For research on the human visual system, it is sometimes necessary to characterize the strength of response to a visual stimulus based on a magnetic field measurement or on an electrical potential measurement. When using magnetic field measurements, a typical experiment will require nearly 100 off/on presentations of the stimulus at one location to obtain useful data. If the researcher is attempting to map the visual cortex of the brain, several stimulus locations must be used in the experiments. When working with a new subject, a researcher will make preliminary measurements to determine if the type of stimulus being used evokes a good response in the subject. The magnetic measurements are in units of femtotesla (10^{-15} tesla). Therefore, the inputs and outputs are both measured in terms of magnetic units.

We will define inputs on the universe $X = [0, 50, 100, 150, 200]$ femtotesla, and outputs on the universe $Y = [0, 50, 100, 150, 200]$ femtotesla. We will define two fuzzy sets, two different stimuli, on universe X:

$$\underset{\sim}{W} = \text{"weak stimulus"} = \left\{ \frac{1}{0} + \frac{0.9}{50} + \frac{0.3}{100} + \frac{0}{150} + \frac{0}{200} \right\} \subset X$$

$$\underset{\sim}{M} = \text{"medium stimulus"} = \left\{ \frac{0}{0} + \frac{0.4}{50} + \frac{1}{100} + \frac{0.4}{150} + \frac{0}{200} \right\} \subset X$$

and one fuzzy set on the output universe Y,

$$\underset{\sim}{S} = \text{``severe response''} = \left\{ \frac{0}{0} + \frac{0}{50} + \frac{0.5}{100} + \frac{0.9}{150} + \frac{1}{200} \right\} \subset Y$$

The complement of $\underset{\sim}{S}$ will then be

$$\overline{\underset{\sim}{S}} = \left\{ \frac{1}{0} + \frac{1}{50} + \frac{0.5}{100} + \frac{0.1}{150} + \frac{0}{200} \right\}$$

We will construct the proposition: IF ``weak stimulus'' THEN not ``severe response,'' using classical implication.

IF $\underset{\sim}{W}$ THEN $\overline{\underset{\sim}{S}} = \underset{\sim}{W} \longrightarrow \overline{\underset{\sim}{S}} = (\underset{\sim}{W} \times \overline{\underset{\sim}{S}}) \cup (\overline{\underset{\sim}{W}} \times Y)$

$$
\underset{\sim}{W} \times \overline{\underset{\sim}{S}} =
\begin{bmatrix} 1 \\ 0.9 \\ 0.3 \\ 0 \\ 0 \end{bmatrix}
[1\ 1\ 0.5\ 0.1\ 0] =
\begin{matrix}
 & & 0 & 50 & 100 & 150 & 200 \\
0 & \\
50 & \\
100 & \\
150 & \\
200 &
\end{matrix}
\begin{bmatrix}
1 & 1 & 0.5 & 0.1 & 0 \\
0.9 & 0.9 & 0.5 & 0.1 & 0 \\
0.3 & 0.3 & 0.3 & 0.1 & 0 \\
0 & 0 & 0 & 0 & 0 \\
0 & 0 & 0 & 0 & 0
\end{bmatrix}
$$

$$
\overline{\underset{\sim}{W}} \times Y =
\begin{bmatrix} 0 \\ .1 \\ .7 \\ 1 \\ 1 \end{bmatrix}
[1\ 1\ 1\ 1\ 1] =
\begin{matrix}
 & & 0 & 50 & 100 & 150 & 200 \\
0 & \\
50 & \\
100 & \\
150 & \\
200 &
\end{matrix}
\begin{bmatrix}
0 & 0 & 0 & 0 & 0 \\
0.1 & 0.1 & 0.1 & 0.1 & 0.1 \\
0.7 & 0.7 & 0.7 & 0.7 & 0.7 \\
1 & 1 & 1 & 1 & 1 \\
1 & 1 & 1 & 1 & 1
\end{bmatrix}
$$

$$
\underset{\sim}{R} = (\underset{\sim}{W} \times \overline{\underset{\sim}{S}}) \cup (\overline{\underset{\sim}{W}} \times Y) =
\begin{matrix}
 & 0 & 50 & 100 & 150 & 200 \\
0 & \\
50 & \\
100 & \\
150 & \\
200 &
\end{matrix}
\begin{bmatrix}
1 & 1 & 0.5 & 0.1 & 0 \\
0.9 & 0.9 & 0.5 & 0.1 & 0.1 \\
0.7 & 0.7 & 0.7 & 0.7 & 0.7 \\
1 & 1 & 1 & 1 & 1 \\
1 & 1 & 1 & 1 & 1
\end{bmatrix}
$$

This relation $\underset{\sim}{R}$, then, expresses the knowledge embedded in the rule: IF ``weak stimuli'' THEN not ``severe response.'' Now, using a new antecedent (IF part) for the input, $\underset{\sim}{M} = $ ``medium stimuli,'' and a max–min composition we can find another response on the Y universe to relate approximately to the new stimulus $\underset{\sim}{M}$, i.e., to find $\underset{\sim}{M} \circ \underset{\sim}{R}$:

$$
\underset{\sim}{M} \circ \underset{\sim}{R} = [0\ 0.4\ 1\ 0.4\ 0]
\begin{matrix}
0 & 50 & 100 & 150 & 200 \\
\end{matrix}
\begin{bmatrix}
1 & 1 & 0.5 & 0.1 & 0 \\
0.9 & 0.9 & 0.5 & 0.1 & 0.1 \\
0.7 & 0.7 & 0.7 & 0.7 & 0.7 \\
1 & 1 & 1 & 1 & 1 \\
1 & 1 & 1 & 1 & 1
\end{bmatrix}
= [0.7\ 0.7\ 0.7\ 0.7\ 0.7]
$$

This result might be labeled linguistically as ``no measurable response.''

An interesting issue in approximate reasoning is the idea of an inverse relationship between fuzzy antecedents and fuzzy consequences arising from the composition operation.

Consider the following problem. Suppose we use the original antecedent, $\underset{\sim}{A}$, in the fuzzy composition. Do we get the original fuzzy consequent, $\underset{\sim}{B}$, as a result of the operation? That is, does the composition operation have a unique inverse, i.e., $\underset{\sim}{B} = \underset{\sim}{A} \circ \underset{\sim}{R}$? The answer is an unqualified no, and one should not expect an inverse to exist for fuzzy composition.

Example 5.12. Again, continuing with the invention example, Examples 5.9 and 5.10, suppose that $\underset{\sim}{A}' = \underset{\sim}{A} =$ ''medium uniqueness.'' Then

$$\underset{\sim}{B}' = \underset{\sim}{A}' \circ \underset{\sim}{R} = \underset{\sim}{A} \circ \underset{\sim}{R} = \left\{ \frac{0.4}{1} + \frac{0.4}{2} + \frac{1}{3} + \frac{0.8}{4} + \frac{0.4}{5} + \frac{0.4}{6} \right\} \neq \underset{\sim}{B}$$

That is, the new consequent does not yield the original consequent ($\underset{\sim}{B} =$ medium market size) because the inverse is not guaranteed with fuzzy composition.

In classical binary logic this inverse does exist; that is, crisp *modus ponens* would give

$$B' = A' \circ R = A \circ R = B$$

where the sets A and B are crisp, and the relation R is also crisp. In the case of approximate reasoning, the fuzzy inference is not precise but rather is approximate. However, the inference does represent an approximate linguistic characteristic of the relation between two universes of discourse, X and Y.

Example 5.13. Suppose you are a soils engineer and you wish to track the movement of soil particles under applied loading in an experimental apparatus that allows viewing of the soil motion. You are building pattern recognition software to enable a computer to monitor and detect the motions. However, there are some difficulties in ''teaching'' your software to view the motion. The tracked particle can be occluded by another particle. The occlusion can occur when a tracked particle is behind another particle, behind a mark on the camera's lens, or partially out of sight of the camera. We want to establish a relationship between particle occlusion, which is a poorly known phenomenon, and lens occlusion, which is quite well-known in photography. Let these membership functions,

$$\underset{\sim}{A} = \left\{ \frac{0.1}{x_1} + \frac{0.9}{x_2} + \frac{0.0}{x_3} \right\} \quad \text{and} \quad \underset{\sim}{B} = \left\{ \frac{0}{y_1} + \frac{1}{y_2} + \frac{0}{y_3} \right\}$$

describe fuzzy sets for a *tracked particle moderately occluded* behind another particle and a *lens mark associated with moderate image quality*, respectively. Fuzzy set $\underset{\sim}{A}$ is defined on a universe $X = \{x_1, x_2, x_3\}$ of tracked particle indicators, and fuzzy set $\underset{\sim}{B}$ (note in this case that B is a crisp singleton) is defined on a universe $Y = \{y_1, y_2, y_3\}$ of lens obstruction indices. A typical rule might be: IF occlusion due to particle occlusion is moderate, THEN image quality will be similar to a moderate lens obstruction, or symbolically,

$$\text{IF } x \text{ is } \underset{\sim}{A}, \text{ THEN } y \text{ is } \underset{\sim}{B} \text{ or } (\underset{\sim}{A} \times \underset{\sim}{B}) \cup (\overline{\underset{\sim}{A}} \times Y) = \underset{\sim}{R}$$

We can find the relation, $\underset{\sim}{R}$, as follows:

$$\underset{\sim}{A} \times \underset{\sim}{B} = \begin{array}{c} x_1 \\ x_2 \\ x_3 \end{array} \begin{bmatrix} \overset{y_1}{0} & \overset{y_2}{0.1} & \overset{y_3}{0} \\ 0 & 0.9 & 0 \\ 0 & 0 & 0 \end{bmatrix} \qquad \overline{\underset{\sim}{A}} \times Y = \begin{array}{c} x_1 \\ x_2 \\ x_3 \end{array} \begin{bmatrix} \overset{y_1}{0.9} & \overset{y_2}{0.9} & \overset{y_3}{0.9} \\ 0.1 & 0.1 & 0.1 \\ 1 & 1 & 1 \end{bmatrix}$$

$$R = (\underset{\sim}{A} \times \underset{\sim}{B}) \cup (\overline{\underset{\sim}{A}} \times \underset{\sim}{Y}) = \begin{bmatrix} 0.9 & 0.9 & 0.9 \\ 0.1 & 0.9 & 0.1 \\ 1 & 1 & 1 \end{bmatrix}$$

This relation expresses in matrix form all the knowledge embedded in the implication. Let $\underset{\sim}{A}'$ be a fuzzy set, in which a tracked particle is behind a particle with *slightly more occlusion* than the particle expressed in the original antecedent $\underset{\sim}{A}$, given by

$$\underset{\sim}{A}' = \left\{ \frac{0.3}{x_1} + \frac{1.0}{x_2} + \frac{0.0}{x_3} \right\}$$

We can find the associated membership of the image quality using max–min composition. For example, approximate reasoning will provide

$$\text{IF } x \text{ is } \underset{\sim}{A}', \text{ THEN } \underset{\sim}{B}' = \underset{\sim}{A}' \circ \underset{\sim}{R}$$

and we get

$$\underset{\sim}{B}' = [0.3 \ 1 \ 0] \circ \begin{bmatrix} 0.9 & 0.9 & 0.9 \\ 0.1 & 0.9 & 0.1 \\ 1 & 1 & 1 \end{bmatrix} = \left\{ \frac{0.3}{y_1} + \frac{0.9}{y_2} + \frac{0.3}{y_3} \right\}$$

This image quality, $\underset{\sim}{B}'$, is more fuzzy than $\underset{\sim}{B}$, as indicated by the former's membership function.

OTHER FORMS OF THE IMPLICATION OPERATION

There are other techniques for obtaining the fuzzy relation $\underset{\sim}{R}$ based on the IF $\underset{\sim}{A}$, THEN $\underset{\sim}{B}$, or $\underset{\sim}{R} = \underset{\sim}{A} \rightarrow \underset{\sim}{B}$. These are known as fuzzy implication operations, and they are valid for all values of $x \in X$ and $y \in Y$. The following forms of the implication operator show different techniques for obtaining the membership function values of fuzzy relation $\underset{\sim}{R}$ defined on the Cartesian product space $X \times Y$:

$$\mu_{\underset{\sim}{R}}(x, y) = \max[\mu_{\underset{\sim}{B}}(y), 1 - \mu_{\underset{\sim}{A}}(x)] \qquad (5.19)$$

$$\mu_{\underset{\sim}{R}}(x, y) = \min[\mu_{\underset{\sim}{A}}(x), \mu_{\underset{\sim}{B}}(y)] \qquad (5.20)$$

$$\mu_{\underset{\sim}{R}}(x, y) = \min\{1, [1 - \mu_{\underset{\sim}{A}}(x) + \mu_{\underset{\sim}{B}}(y)]\} \qquad (5.21)$$

$$\mu_{\underset{\sim}{R}}(x, y) = \mu_{\underset{\sim}{A}}(x) \cdot \mu_{\underset{\sim}{B}}(y) \qquad (5.22)$$

$$\mu_{\underset{\sim}{R}}(x, y) = \begin{cases} 1, & \text{for } \mu_{\underset{\sim}{A}}(x) \leq \mu_{\underset{\sim}{B}}(y) \\ \mu_{\underset{\sim}{B}}(y), & \text{otherwise} \end{cases} \qquad (5.23)$$

In situations where the universes are represented by discrete elements the fuzzy relation $\underset{\sim}{R}$ is a matrix.

Equation (5.19) is equivalent to classical implication (Eq. 5.16) for $\mu_{\underset{\sim}{B}}(y) \leq \mu_{\underset{\sim}{A}}(x)$. Equation (5.20) has been given various terms in the literature; it has been referred to as *correlation-minimum* and as *Mamdani's implication*, after British Prof. Mamdani's work in the area of system control [Mamdani, 1976]. This formulation for the implication is also equivalent to the fuzzy cross product of fuzzy sets $\underset{\sim}{A}$ and $\underset{\sim}{B}$, i.e., $\underset{\sim}{R} = \underset{\sim}{A} \times \underset{\sim}{B}$. For

$\mu_{\underset{\sim}{A}}(x) \geq 0.5$ and $\mu_{\underset{\sim}{B}}(y) \geq 0.5$ classical implication reduces to Mamdani's implication. The implication defined by Eq. (5.21) is known as *Lukasiewicz's implication*, after the Polish logician Jan Lukasiewicz [Rescher, 1969]. Equation (5.22) describes a form of *correlation-product implication* and is based on the notions of conditioning and reinforcement. This product form tends to dilute the influence of joint membership values that are small and, as such, are related to Hebbian-type learning algorithms in neuropsychology when used in artificial neural network computations. Equation (5.23) is sometimes called *Brouwerian* implication and is discussed in Sanchez [1976]. Although the classical implication continues to be the most popular and is valid for fuzzy and crisp applications, these other methods have been introduced as computationally effective under certain conditions of the membership values, $\mu_{\underset{\sim}{A}}(x)$ and $\mu_{\underset{\sim}{B}}(y)$. The appropriate choice of an implication operator is a matter left to the analyst, since it is typically context-dependent (see Problems 5.20 and 5.21 for comparisons). Ross [1995] gives a few other fuzzy implications.

PART II   FUZZY SYSTEMS

> *It was the best of times, it was the worst of times, it was the age of wisdom, it was the age of foolishness, it was the epoch of belief, it was the epoch of incredulity, it was the season of Light, it was the season of Darkness, it was the spring of hope, it was the winter of despair, we had everything before us . . . .*
>
> **Charles Dickens**
> *A Tale of Two Cities,* Chapter 1, 1859

Natural language is perhaps the most powerful form of conveying information that humans possess for any given problem or situation that requires solving or reasoning. This power has largely remained untapped in today's mathematical paradigms; not so anymore with the utility of fuzzy logic. Consider the information contained in the passage above from Charles Dickens' *A Tale of Two Cities*. Imagine reducing this passage to a more precise form such that it could be assimilated by a binary computer. First, we will have to remove the fuzziness inherent in the passage, limiting the statements to precise, either–or, Aristotelian logic. Consider the following crisp version of the first few words of the Dickens passage:

> The time interval x was the period exhibiting a 100 percent maximum of possible values as measured along some arbitrary social scale, [and] the interval x was also the period of time exhibiting a 100 percent minimum of these values as measured along the same scale. [Clark, 1992]

The crisp version of this passage has established an untenable paradox, identical to that posed by the excluded middle axioms in probability theory. Another example is available from the same classic, the last sentence in Dickens' *A Tale of Two Cities*: ''It is a far, far better thing that I do, than I have ever done; it is a far, far better rest that I go to, than I have ever known.'' It would also be difficult to address this original fuzzy phrase by an intelligent machine using binary logic. Both of these examples demonstrate the power of communication inherent in natural language, and they demonstrate how far we are from enabling intelligent machines to reason the way humans do – a long way!

NATURAL LANGUAGE

Cognitive scientists tell us that humans base their thinking primarily on conceptual patterns and mental images rather than on any numerical quantities. In fact the expert system paradigm known as "frames" is based on the notion of a cognitive picture in one's mind. Furthermore, humans communicate with their own natural language by referring to previous mental images with rather vague but simple terms. Despite the vagueness and ambiguity in natural language, humans communicating in a common language have very little trouble in basic understanding. Our language has been termed the *shell of our thoughts* [Zadeh, 1975a]. Hence, any attempts to model the human thought process as expressed in our communications with one another must be preceded by models that attempt to emulate our natural language.

Our natural language consists of fundamental terms characterized as atoms in the literature. A collection of these atoms will form the molecules, or phrases, of our natural language. The fundamental terms can be called *atomic* terms. Examples of some atomic terms are *slow, medium, young, beautiful*, etc. A collection of atomic terms is called a composite, or simply a set of terms. Examples of composite terms are *very slow horse, medium-weight female, young tree, fairly beautiful painting*, etc. Suppose we define the atomic terms and sets of atomic terms to exist as elements and sets on a universe of natural language terms, say universe X. Furthermore, let us define another universe, called Y, as a universe of cognitive interpretations, or meanings. Although it may seem straightforward to envision a universe of terms, it may be difficult to ponder a universe of *interpretations*. Consider this universe, however, to be a collection of individual elements and sets that represent the cognitive patterns and mental images referred to earlier in this chapter. Clearly, then, these interpretations would be rather vague, and they might best be represented as fuzzy sets. Hence, an atomic term, or as Zadeh [1975a] defines it, a linguistic variable, can be interpreted using fuzzy sets.

The need for expressing linguistic variables using the precepts of mathematics is quite well established. Leibniz, who was an early developer of calculus, once claimed, "If we could find characters or signs appropriate for expressing all our thoughts as definitely and as exactly as arithmetic expresses numbers or geometric analysis expresses lines, we could in all subjects, in so far as they are amenable to reasoning, accomplish what is done in arithmetic and geometry." Fuzzy sets are a relatively new quantitative method to accomplish just what Leibniz had suggested.

With these definitions and foundations, we are now in a position to establish a formal model of linguistics using fuzzy sets. Suppose we define a specific atomic term in the universe of natural language, X, as element α, and we define a fuzzy set $\underset{\sim}{A}$ in the universe of interpretations, or meanings, Y, as a specific meaning for the term α. Then natural language can be expressed as a mapping $\underset{\sim}{M}$ from a set of atomic terms in X to a corresponding set of interpretations defined on universe Y. Each atomic term α in X corresponds to a fuzzy set $\underset{\sim}{A}$ in Y, which is the "interpretation" of α. This mapping, which can be denoted $\underset{\sim}{M}(\alpha, \underset{\sim}{A})$, is shown schematically in Fig. 5.6.

The fuzzy set $\underset{\sim}{A}$ represents the fuzziness in the mapping between an atomic term and its interpretation, and can be denoted by the membership function $\mu_{\underset{\sim}{M}}(\alpha, y)$, or more simply by

$$\mu_{\underset{\sim}{M}}(\alpha, y) = \mu_{\underset{\sim}{A}}(y) \tag{5.24}$$

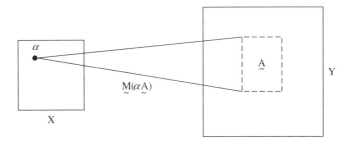

FIGURE 5.6
Mapping of a linguistic atom, α, to a cognitive interpretation, $\underset{\sim}{A}$.

As an example, suppose we have the atomic term "young" (α) and we want to interpret this linguistic atom in terms of age, y, by a membership function that expresses the term "young." The membership function given here in the notation of Zadeh [1975b], and labeled $\underset{\sim}{A}$, might be one interpretation of the term young expressed as a function of age,

$$\underset{\sim}{A} = \text{"young"} = \int_0^{25} \frac{1}{y} + \int_{25}^{100} \frac{1}{y}\left[1 + \left(\frac{y-25}{5}\right)^2\right]^{-1}$$

or alternatively,

$$\mu_{\underset{\sim}{M}}(\text{young}, y) = \begin{cases} \left[1 + \left(\dfrac{y-25}{5}\right)^2\right]^{-1} & y > 25 \text{ years} \\ 1 & y \leq 25 \text{ years} \end{cases}$$

Similarly, the atomic term "old" might be expressed as another fuzzy set, $\underset{\sim}{O}$, on the universe of interpretation, Y, as

$$\mu_{\underset{\sim}{M}}(\text{old}, y) = 1 - \left[1 + \left(\frac{y-50}{5}\right)^2\right]^{-1} \quad \text{for } 50 \leq y \leq 100$$

On the basis of the foregoing, we can call α a natural language variable whose "value" is defined by the fuzzy set $\mu_\alpha(y)$. Hereinafter, the "value" of a linguistic variable will be synonymous with its *interpretation*.

As suggested before, a composite is a collection, or set, of atomic terms combined by various linguistic connectives such as *and*, *or*, and *not*. Define two atomic terms, α and β, on the universe X. The *interpretation* of the composite, defined on universe Y, can be defined by the following set-theoretic operations [Zadeh, 1975b],

$$\alpha \text{ or } \beta : \mu_{\alpha \text{ or } \beta}(y) = \max(\mu_\alpha(y), \mu_\beta(y))$$

$$\alpha \text{ and } \beta : \mu_{\alpha \text{ and } \beta}(y) = \min(\mu_\alpha(y), \mu_\beta(y)) \tag{5.25}$$

$$\text{Not } \alpha = \bar{\alpha} : \mu_{\bar{\alpha}}(y) = 1 - \mu_\alpha(y)$$

These operations are analogous to those proposed earlier in this chapter (*standard fuzzy operations*), where the natural language connectives *and*, *or*, and *not* were logical connectives.

LINGUISTIC HEDGES

In linguistics, fundamental atomic terms are often modified with adjectives (nouns) or adverbs (verbs) like *very, low, slight, more or less, fairly, slightly, almost, barely, mostly, roughly, approximately*, and so many more that it would be difficult to list them all. We will call these modifiers "linguistic hedges": that is, the singular meaning of an atomic term is modified, or hedged, from its original interpretation. Using fuzzy sets as the calculus of interpretation, these linguistic hedges have the effect of modifying the membership function for a basic atomic term [Zadeh, 1972]. As an example, let us look at the basic linguistic atom, α, and subject it to some hedges. Define $\alpha = \int_Y \mu_\alpha(y)/y$; then

$$\text{"Very"} \; \alpha = \alpha^2 = \int_Y \frac{[\mu_\alpha(y)]^2}{y} \tag{5.26}$$

$$\text{"Very, very"} \; \alpha = \alpha^4 \tag{5.27}$$

$$\text{"Plus"} \; \alpha = \alpha^{1.25} \tag{5.28}$$

$$\text{"Slightly"} \; \alpha = \sqrt{\alpha} = \int_Y \frac{[\mu_\alpha(y)]^{0.5}}{y} \tag{5.29}$$

$$\text{"Minus"} \; \alpha = \alpha^{0.75} \tag{5.30}$$

The expressions shown in Eqs. (5.26)–(5.28) are linguistic hedges known as *concentrations* [Zadeh, 1972]. Concentrations tend to concentrate the elements of a fuzzy set by reducing the degree of membership of all elements that are only "partly" in the set. The less an element is in a set (i.e., the lower its original membership value), the more it is reduced in membership through concentration. For example, by using Eq. (5.26) for the hedge *very*, a membership value of 0.9 is reduced by 10% to a value of 0.81, but a membership value of 0.1 is reduced by an order of magnitude to 0.01. This decrease is simply a manifestation of the properties of the membership value itself; for $0 \le \mu \le 1$, then $\mu \ge \mu^2$. Alternatively, the expressions given in Eqs. (5.29) and (5.30) are linguistic hedges known as *dilations* (or dilutions in some publications). Dilations stretch or dilate a fuzzy set by increasing the membership of elements that are "partly" in the set [Zadeh, 1972]. For example, using Eq. (5.29) for the hedge *slightly*, a membership value of 0.81 is increased by 11% to a value of 0.9, whereas a membership value of 0.01 is increased by an order of magnitude to 0.1.

Another operation on linguistic fuzzy sets is known as *intensification*. This operation acts in a combination of concentration and dilation. It increases the degree of membership of those elements in the set with original membership values greater than 0.5, and it decreases the degree of membership of those elements in the set with original membership values less than 0.5. This also has the effect of making the boundaries of the membership function (see Fig. 4.1) steeper. *Intensification* can be expressed by numerous algorithms, one of which, proposed by Zadeh [1972], is

$$\text{"intensify"} \; \alpha = \begin{cases} 2\mu_\alpha^2(y) & \text{for } 0 \le \mu_\alpha(y) \le 0.5 \\ 1 - 2[1 - \mu_\alpha(y)]^2 & \text{for } 0.5 \le \mu_\alpha(y) \le 1 \end{cases} \tag{5.31}$$

Intensification increases the contrast between the elements of the set that have more than half-membership and those that have less than half-membership. Figures 5.7, 5.8,

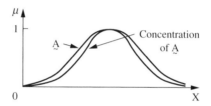

FIGURE 5.7
Fuzzy concentration.

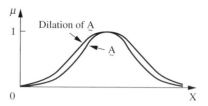

FIGURE 5.8
Fuzzy dilation.

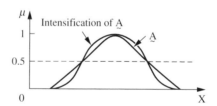

FIGURE 5.9
Fuzzy intensification.

and 5.9 illustrate the operations of concentration, dilation, and intensification, respectively, for fuzzy linguistic hedges on a typical fuzzy set A.

Composite terms can be formed from one or more combinations of atomic terms, logical connectives, and linguistic hedges. Since an atomic term is essentially a fuzzy mapping from the universe of terms to a universe of fuzzy sets represented by membership functions, the implementation of linguistic hedges and logical connectives is manifested as function-theoretic operations on the values of the membership functions. In order to conduct the function-theoretic operations, a precedence order must be established. For example, suppose we have two atomic terms "small" and "red," and their associated membership functions, and we pose the following linguistic expression: a "not small" *and* "very red" fruit. Which of the operations, i.e., not, and, very, would we perform first, which would we perform second, and so on? In the literature, the following preference table (Table 5.7) has been suggested for standard Boolean operations.

Parentheses may be used to change the precedence order and ambiguities may be resolved by the use of association-to-the-right. For example, "plus very minus very small"

TABLE 5.7
Precedence for linguistic hedges
and logical operations

Precedence	Operation
First	Hedge. not
Second	And
Third	Or

Source: Zadeh [1973]

should be interpreted as

$$\text{plus (very (minus (very (small))))}$$

Every atomic term and every composite term has a syntax represented by its linguistic label and a semantics, or meaning (interpretation), which is given by a membership function. The use of a membership function gives the flexibility of an elastic meaning to a linguistic term. On the basis of this elasticity and flexibility, it is possible to incorporate subjectivity and bias into the meaning of a linguistic term. These are some of the most important benefits of using fuzzy mathematics in the modeling of linguistic variables. This capability allows us to encode and automate human knowledge, which is often expressed in natural language propositions.

In our example, a "not small" *and* "very red" fruit, we would perform the hedges "not small" and "very red" first, then we would perform the logical operation *and* on the two phrases as suggested in Table 5.7. To further illustrate Table 5.7 consider the following numerical example.

Example 5.14. Suppose we have a universe of integers, $Y = \{1, 2, 3, 4, 5\}$. We define the following linguistic terms as a mapping onto Y:

$$\text{``Small''} = \left\{ \frac{1}{1} + \frac{0.8}{2} + \frac{0.6}{3} + \frac{0.4}{4} + \frac{0.2}{5} \right\}$$

$$\text{``Large''} = \left\{ \frac{0.2}{1} + \frac{0.4}{2} + \frac{0.6}{3} + \frac{0.8}{4} + \frac{1}{5} \right\}$$

Now we modify these two linguistic terms with hedges,

$$\text{``Very small''} = \text{``small''}^2 \text{ (Eq. (5.26))} = \left\{ \frac{1}{1} + \frac{0.64}{2} + \frac{0.36}{3} + \frac{0.16}{4} + \frac{0.04}{5} \right\}$$

$$\text{``Not very small''} = 1 - \text{``very small''} = \left\{ \frac{0}{1} + \frac{0.36}{2} + \frac{0.64}{3} + \frac{0.84}{4} + \frac{0.96}{5} \right\}$$

Then we construct a phrase, or a composite term:

$$\alpha = \text{``not very small and not very, very large''}$$

which involves the following set-theoretic operations:

$$\alpha = \left(\frac{0.36}{2} + \frac{0.64}{3} + \frac{0.84}{4} + \frac{0.96}{5} \right) \cap \left(\frac{1}{1} + \frac{1}{2} + \frac{0.9}{3} + \frac{0.6}{4} \right) = \left(\frac{0.36}{2} + \frac{0.64}{3} + \frac{0.6}{4} \right)$$

Suppose we want to construct a linguistic variable "intensely small" (extremely small); we will make use of Eq. (5.31) to modify "small" as follows:

$$\text{"Intensely small"} = \left\{ \frac{1 - 2[1 - 1]^2}{1} + \frac{1 - 2[1 - 0.8]^2}{2} \right.$$

$$+ \left. \frac{1 - 2[1 - 0.6]^2}{3} + \frac{2[0.4]^2}{4} + \frac{2[0.2]^2}{5} \right\}$$

$$= \left\{ \frac{1}{1} + \frac{0.92}{2} + \frac{0.68}{3} + \frac{0.32}{4} + \frac{0.08}{5} \right\}$$

In summary, the foregoing material introduces the idea of a *linguistic variable* (atomic term), which is a variable whose values (interpretation) are natural language expressions referring to the contextual semantics of the variable. Zadeh [1975b] described this notion quite well:

A linguistic variable differs from a numerical variable in that its values are not numbers but words or sentences in a natural or artificial language. Since words, in general, are less precise than numbers, the concept of a linguistic variable serves the purpose of providing a means of approximate characterization of phenomena which are too complex or too ill-defined to be amenable to description in conventional quantitative terms. More specifically, the fuzzy sets which represent the restrictions associated with the values of a linguistic variable may be viewed as summaries of various subclasses of elements in a universe of discourse. This, of course, is analogous to the role played by words and sentences in a natural language. For example, the adjective handsome is a summary of a complex of characteristics of the appearance of an individual. It may also be viewed as a label for a fuzzy set which represents a restriction imposed by a fuzzy variable named handsome. From this point of view, then, the terms *very handsome, not handsome, extremely handsome, quite handsome*, etc., are names of fuzzy sets which result from operating on the fuzzy set handsome with the modifiers named *very, not, extremely, quite*, etc. In effect, these fuzzy sets, together with the fuzzy set labeled *handsome*, play the role of values of the linguistic variable *Appearance*.

FUZZY (RULE-BASED) SYSTEMS

In the field of artificial intelligence (machine intelligence) there are various ways to represent knowledge. Perhaps the most common way to represent human knowledge is to form it into natural language expressions of the type

$$\text{IF premise (antecedent), THEN conclusion (consequent)} \qquad (5.32)$$

The form in Expression (5.32) is commonly referred to as the IF–THEN *rule-based* form; this form generally is referred to as the *deductive* form. It typically expresses an inference such that if we know a fact (premise, hypothesis, antecedent), then we can infer, or derive, another fact called a conclusion (consequent). This form of knowledge

TABLE 5.8
The canonical form for a fuzzy rule-based system

Rule 1:	IF condition C^1, THEN restriction R^1
Rule 2:	IF condition C^2, THEN restriction R^2
$\vdots$	
Rule r:	IF condition C^r, THEN restriction R^r

representation, characterized as *shallow knowledge*, is quite appropriate in the context of linguistics because it expresses human empirical and heuristic knowledge in our own language of communication. It does not, however, capture the *deeper* forms of knowledge usually associated with intuition, structure, function, and behavior of the objects around us simply because these latter forms of knowledge are not readily reduced to linguistic phrases or representations; this deeper form, as described in Chapter 1, is referred to as *inductive*. The fuzzy rule-based system is most useful in modeling some complex systems that can be observed by humans because it makes use of linguistic variables as its antecedents and consequents; as described here these linguistic variables can be naturally represented by fuzzy sets and logical connectives of these sets.

By using the basic properties and operations defined for fuzzy sets (see Chapter 2), any compound rule structure may be decomposed and reduced to a number of simple canonical rules as given in Table 5.8. These rules are based on natural language representations and models, which are themselves based on fuzzy sets and fuzzy logic. The fuzzy level of understanding and describing a complex system is expressed in the form of a set of restrictions on the output based on certain conditions of the input (see Table 5.8). Restrictions are generally modeled by fuzzy sets and relations. These restriction statements are usually connected by linguistic connectives such as "and," "or," or "else." The restrictions $R^1, R^2, \ldots, R^r$ apply to the output actions, or consequents of the rules. The following illustrates a couple of the most common techniques [Ross, 1995] for decomposition of linguistic rules involving multiple antecedents into the simple canonical form illustrated in Table 5.8.

Multiple conjunctive antecedents

$$\text{IF } x \text{ is } \underset{\sim}{A}^1 \text{ and } \underset{\sim}{A}^2 \ldots \text{ and } \underset{\sim}{A}^L \text{ THEN } y \text{ is } \underset{\sim}{B}^s \tag{5.33}$$

Assuming a new fuzzy subset A^s as

$$\underset{\sim}{A}^s = \underset{\sim}{A}^1 \cap \underset{\sim}{A}^2 \cap \cdots \cap \underset{\sim}{A}^L$$

expressed by means of membership function

$$\mu_{\underset{\sim}{A}^s}(x) = \min[\mu_{\underset{\sim}{A}^1}(x), \mu_{\underset{\sim}{A}^2}(x), \ldots, \mu_{\underset{\sim}{A}^L}(x)]$$

based on the definition of the standard fuzzy intersection operation, the compound rule may be rewritten as

$$\text{IF } \underset{\sim}{A}^s \text{ THEN } \underset{\sim}{B}^s$$

Multiple disjunctive antecedents

$$\text{IF } x \text{ is } \underset{\sim}{A}^1 \text{ OR } x \text{ is } \underset{\sim}{A}^2 \dots \text{OR } x \text{ is } \underset{\sim}{A}^L \text{ THEN } y \text{ is } \underset{\sim}{B}^s \tag{5.34}$$

could be rewritten as

$$\text{IF } x \text{ is } \underset{\sim}{A}^s \text{ THEN } y \text{ is } \underset{\sim}{B}^s$$

where the fuzzy set $\underset{\sim}{A}^s$ is defined as

$$\underset{\sim}{A}^s = \underset{\sim}{A}^1 \cup \underset{\sim}{A}^2 \cup \dots \cup \underset{\sim}{A}^L$$

$$\mu_{\underset{\sim}{A}^s}(x) = \max\left[\mu_{\underset{\sim}{A}^1}(x), \mu_{\underset{\sim}{A}^2}(x), \dots, \mu_{\underset{\sim}{A}^L}(x)\right]$$

which is based on the definition of the standard fuzzy union operation.

Aggregation of fuzzy rules

Most rule-based systems involve more than one rule. The process of obtaining the overall consequent (conclusion) from the individual consequents contributed by each rule in the rule-base is known as aggregation of rules. In determining an aggregation strategy, two simple extreme cases exist [Ross, 1995]:

1. *Conjunctive system of rules.* In the case of a system of rules that must be jointly satisfied, the rules are connected by ''and'' connectives. In this case the aggregated output (consequent), y, is found by the fuzzy intersection of all individual rule consequents, y^i, where $i = 1, 2, \dots r$ (see Table 5.8), as

$$y = y^1 \text{ and } y^2 \text{ and } \dots \text{ and } y^r$$

   or $\qquad\qquad\qquad\qquad\qquad\qquad\qquad\qquad\qquad\qquad\qquad\qquad (5.35)$

$$y = y^1 \cap y^2 \cap \dots \cap y^r$$

   which is defined by the membership function

$$\mu_y(y) = \min(\mu_{y^1}(y), \mu_{y^2}(y), \dots, \mu_{y^r}(y)) \text{ for } y \in Y \tag{5.36}$$

2. *Disjunctive system of rules.* For the case of a disjunctive system of rules where the satisfaction of at least one rule is required, the rules are connected by the ''or'' connectives. In this case the aggregated output is found by the fuzzy union of all individual rule contributions, as

$$y = y^1 \text{ or } y^2 \text{ or } \dots \text{ or } y^r$$

   or $\qquad\qquad\qquad\qquad\qquad\qquad\qquad\qquad\qquad\qquad\qquad\qquad (5.37)$

$$y = y^1 \cup y^2 \cup \dots \cup y^r$$

   which is defined by the membership function

$$\mu_y(y) = \max(\mu_{y^1}(y), \mu_{y^2}(y), \dots, \mu_{y^r}(y)) \text{ for } y \in Y \tag{5.38}$$

GRAPHICAL TECHNIQUES OF INFERENCE

Part I of this chapter illustrates mathematical procedures to conduct deductive inferencing of IF–THEN rules. These procedures can be implemented on a computer for processing speed. Sometimes, however, it is useful to be able to conduct the inference computation manually with a few rules to check computer programs or to verify the inference operations. Conducting the matrix operations illustrated in this chapter, Part I, for a few rule sets can quickly become quite onerous. Graphical methods that emulate the inference process and that make manual computations involving a few simple rules straightforward have been proposed [see Jang et al., 1997]. This section will describe three common methods of deductive inference for fuzzy systems based on linguistic rules: (1) Mamdani systems, (2) Sugeno models, and (3) Tsukamoto models.

The first inference method, due to Mamdani and Assilian [1975], is the most common in practice and in the literature. To begin the general illustration of this idea, we consider a simple two-rule system where each rule comprises two antecedents and one consequent. This is analogous to a dual-input and single-output fuzzy system. The graphical procedures illustrated here can be easily extended and will hold for fuzzy rule-bases (or fuzzy systems) with any number of antecedents (inputs) and consequents (outputs). A fuzzy system with two noninteractive inputs x_1 and x_2 (antecedents) and a single output y (consequent) is described by a collection of r linguistic IF–THEN propositions in the Mamdani form:

$$\text{IF } x_1 \text{ is } \underset{\sim}{A_1^k} \text{ and } x_2 \text{ is } \underset{\sim}{A_2^k} \text{ THEN } y^k \text{ is } \underset{\sim}{B^k} \qquad \text{for } k = 1, 2, \ldots, r \qquad (5.39)$$

where $\underset{\sim}{A_1^k}$ and $\underset{\sim}{A_2^k}$ are the fuzzy sets representing the kth antecedent pairs, and $\underset{\sim}{B^k}$ is the fuzzy set representing the kth consequent.

In the following presentation, we consider two different cases of two-input Mamdani systems: (1) the inputs to the system are scalar values, and we use a max–min inference method, and (2) the inputs to the system are scalar values, and we use a max–product inference method. Of course the inputs to any fuzzy system can also be a membership function, such as a gauge reading that has been fuzzified, but we shall lose no generality in describing the method by employing fuzzy singletons (scalar values) as the input.

Case 1

Inputs x_1 and x_2 are crisp values, i.e., delta functions. The rule-based system is described by Eq. (5.39), so membership for the inputs x_1 and x_2 will be described by

$$\mu(x_1) = \delta(x_1 - \text{input }(i)) = \begin{cases} 1, & x_1 = \text{input}(i) \\ 0, & \text{otherwise} \end{cases} \qquad (5.40)$$

$$\mu(x_2) = \delta(x_2 - \text{input }(j)) = \begin{cases} 1, & x_2 = \text{input}(j) \\ 0, & \text{otherwise} \end{cases} \qquad (5.41)$$

Based on the Mamdani implication method of inference given in this chapter, Eq. (5.20), and for a set of disjunctive rules, the aggregated output for the r rules will be given by

$$\mu_{\underset{\sim}{B^k}}(y) = \max_k [\min[\mu_{\underset{\sim}{A_1^k}}(\text{input}(i)), \mu_{\underset{\sim}{A_2^k}}(\text{input}(j))]] \quad k = 1, 2, \ldots, r \qquad (5.42)$$

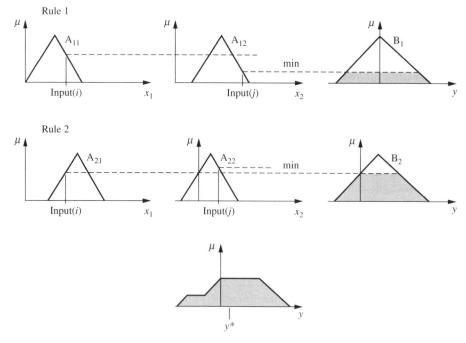

FIGURE 5.10
Graphical Mamdani (max–min) inference method with crisp inputs.

Equation (5.42) has a very simple graphical interpretation, as seen in Fig. 5.10. Figure 5.10 illustrates the graphical analysis of two rules, where the symbols A_{11} and A_{12} refer to the first and second fuzzy antecedents of the first rule, respectively, and the symbol B_1 refers to the fuzzy consequent of the first rule; the symbols A_{21} and A_{22} refer to the first and second fuzzy antecedents, respectively, of the second rule, and the symbol B_2 refers to the fuzzy consequent of the second rule. The minimum function in Eq. (5.42) is illustrated in Fig. 5.10 and arises because the antecedent pairs given in the general rule structure for this system are connected by a logical "and" connective, as seen in Eq. (5.39). The minimum membership value for the antecedents propagates through to the consequent and truncates the membership function for the consequent of each rule. This graphical inference is done for each rule. Then the truncated membership functions for each rule are aggregated, using the graphical equivalent of either Eq. (5.36), for conjunction rules, or Eq. (5.38), for disjunctive rules; in Fig. 5.10 the rules are disjunctive, so the aggregation operation *max* results in an aggregated membership function comprised of the outer envelope of the individual truncated membership forms from each rule. If one wishes to find a crisp value for the aggregated output, some appropriate defuzzification technique (see Chapter 4) could be employed to the aggregated membership function, and a value such as y^* shown in Fig. 5.10 would result.

Case 2

In the preceding example, if we were to use a max–product (or correlation-product) implication technique (see Eq. (5.22)) for a set of disjunctive rules, the aggregated output for the r rules would be given by

$$\mu_{\underset{\sim}{B}^k}(y) = \max_k[\mu_{\underset{\sim}{A}_1^k}(\text{input}(i)) \cdot \mu_{\underset{\sim}{A}_2^k}(\text{input}(j))] \quad k = 1, 2, \ldots, r \tag{5.43}$$

and the resulting graphical equivalent of Eq. (5.43) would be as shown in Fig. 5.11. In Fig. 5.11 the effect of the max–product implication is shown by the consequent membership functions remaining as scaled triangles (instead of truncated triangles as in case 1). Again, Fig. 5.11 shows the aggregated consequent resulting from a disjunctive set of rules (the outer envelope of the individual scaled consequents) and a defuzzified value, y^*, resulting from some defuzzification method (see Chapter 4).

Example 5.15. In mechanics, the energy of a moving body is called kinetic energy. If an object of mass m (kilograms) is moving with a velocity v (meters per second), then the kinetic energy k (in joules) is given by the equation $k = \frac{1}{2}mv^2$. Suppose we model the mass and velocity as inputs to a system (moving body) and the energy as output, then observe the system for a while and deduce the following two disjunctive rules of inference based on our observations:

Rule 1 : IF x_1 is $\underset{\sim}{A}_1^1$ (small mass) *and* x_2 is $\underset{\sim}{A}_2^1$ (high velocity),

THEN y is $\underset{\sim}{B}^1$ (medium energy).

Rule 2 : IF x_1 is $\underset{\sim}{A}_1^2$ (large mass) *or* x_2 is $\underset{\sim}{A}_2^2$ (medium velocity),

THEN y is $\underset{\sim}{B}^2$ (high energy).

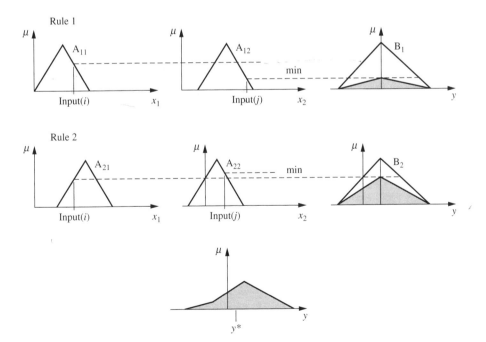

FIGURE 5.11
Graphical Mamdani (max–product) implication method with crisp inputs.

We now proceed to describe these two rules in a graphical form and illustrate the two cases of graphical inference presented earlier in this section.

Suppose we have made some observations of the system (moving body) and we estimate the values of the two inputs, mass and velocity, as crisp values. For example, let input(i) = 0.35 kg (mass) and input(j) = 55 m/s (velocity). Case 1 models the inputs as delta functions, Eqs. (5.40) – (5.41), and uses a Mamdani implication, Eq. (5.42). Graphically, this is illustrated in Fig. 5.12, where the output fuzzy membership function is defuzzified using a centroid method.

In Figs. 5.12 and 5.13, the two rules governing the behavior of the moving body system are illustrated graphically. The antecedents, mass (kg) and velocity (m/s), for each rule are shown as fuzzy membership functions corresponding to the linguistic values for each antecedent. Moreover, the consequent, energy (joules), for each rule is also shown as a fuzzy membership function corresponding to the linguistic label for that consequent. The inputs for mass and velocity intersect the antecedent membership functions at some membership level. The minimum or maximum of the two membership values is propagated to the consequent depending on whether the ''and'' or ''or'' connective, respectively, is used between the two antecedents in the rule. The propagated membership value from operations on the antecedents then truncates (for Mamdani implication) or scales (for max–product implication) the membership function for the consequent for that rule. This truncation or scaling is conducted for each rule, and then the truncated or scaled membership functions from each rule are aggregated according to Eq. (5.36) (conjunctive) or (5.38) (disjunctive). In this example we are using two disjunctive rules.

In case 2 we only change the method of implication from the first case. Now using a max–product implication method, Eq. (5.43), and a centroidal defuzzification method, the graphical result is shown in Figure 5.13.

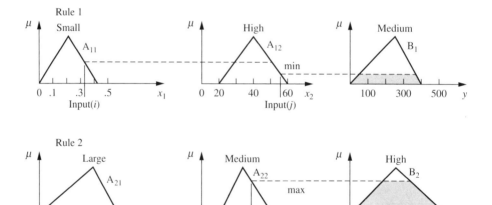

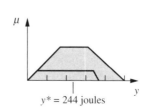

$y^* = 244$ joules

FIGURE 5.12
Fuzzy inference method using the case 1 graphical approach.

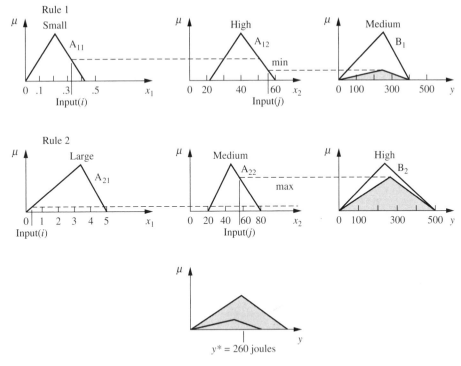

FIGURE 5.13
Fuzzy inference method using the case 2 graphical approach.

The Mamdani method has several variations. There are different t-norms to use for the connectives of the antecedents, different aggregation operators for the rules, and numerous defuzzification methods that could be used. As the foregoing example illustrates, the two Mamdani methods yield different shapes for the aggregated fuzzy consequents for the two rules used. However, the defuzzified values for the output energy are both fairly consistent: 244 joules and 260 joules. The power of fuzzy rule-based systems is their ability to yield "good" results with reasonably simple mathematical operations.

The second inference method, generally referred to as the Sugeno method, or the TSK method (Takagi, Sugeno, and Kang) [Takagi and Sugeno, 1985; Sugeno and Kang, 1988], was proposed in an effort to develop a systematic approach to generating fuzzy rules from a given input–output data set. A typical rule in a Sugeno model, which has two-inputs x and y, and output z, has the form

$$\text{IF } x \text{ is } \underset{\sim}{A} \textit{ and } y \text{ is } \underset{\sim}{B}, \text{ THEN } z \text{ is } z = f(x, y)$$

where $z = f(x, y)$ is a crisp function in the consequent. Usually $f(x, y)$ is a polynomial function in the inputs x and y, but it can be any general function as long as it describes the output of the system within the fuzzy region specified in the antecedent of the rule to which it is applied. When $f(x, y)$ is a constant the inference system is called a *zero-order Sugeno model*, which is a special case of the Mamdani system in which each rule's consequent is specified as a fuzzy singleton. When $f(x, y)$ is a linear function of x and y, the inference

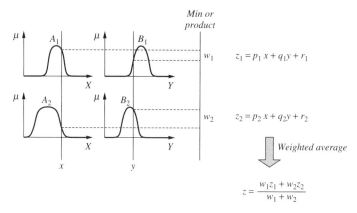

FIGURE 5.14
The Sugeno fuzzy model (Jang, Jyh-Shing Roger; Sun, Chuen-Tsai; Mizutani, Eiji, *Neuro-Fuzzy and Soft Computing: A Computational Approach to Learning and Machine Intelligence, 1st Edition*, © 1997. Reprinted by permission of Pearson Education Inc., Upper Saddle River, NJ).

system is called a *first-order Sugeno model.* Jang et al. [1997] point out that the output of a zero-order Sugeno model is a smooth function of its input variables as long as the neighboring membership functions in the antecedent have enough overlap. By contrast, the overlap of the membership functions in the consequent of a Mamdani model does not have a decisive effect on the smoothness; it is the overlap of the antecedent membership functions that determines the smoothness of the resulting system behavior.

In a Sugeno model each rule has a crisp output, given by a function; because of this the overall output is obtained via a weighted average defuzzification (Eq. (4.6)), as shown in Fig. 5.14. This process avoids the time-consuming methods of defuzzification necessary in the Mamdani model.

> **Example 5.16.** An example of a two-input single-output Sugeno model with four rules is repeated from Jang et al. [1997]:
>
> $$\text{IF X is small and Y is small, THEN } z = -x + y + 1$$
> $$\text{IF X is small and Y is large, THEN } z = -y + 3$$
> $$\text{IF X is large and Y is small, THEN } z = -x + 3$$
> $$\text{IF X is large and Y is large, THEN } z = x + y + 2$$
>
> Figure 5.15*a* plots the membership function of inputs X and Y, and Fig. 5.15*b* is the resulting input – output surface of the system. The surface is complex, but it is still obvious that the surface is comprised of four planes, each of which is specified by the output function of each of the four rules. Figure 5.15*b* shows that there is a smooth transition between the four output planes. Without the mathematically difficult process of a defuzzification operation, the Sugeno model is a very popular method for sample-based fuzzy systems modeling.

The third inference method is due to Tsukamoto [1979]. In this method the consequent of each fuzzy rule is represented by a fuzzy set with a monotonic membership function, as shown in Fig. 5.16. In a monotonic membership function, sometimes called a *shoulder function*, the inferred output of each rule is defined as a crisp value induced by the membership value coming from the antecedent clause of the rule. The overall output is

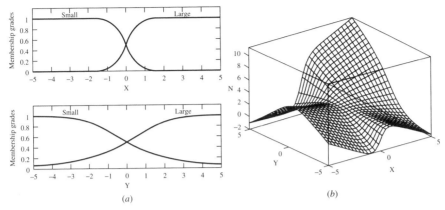

FIGURE 5.15
Sugeno Model for Example 5.16 (Jang, Jyh-Shing Roger; Sun, Chuen-Tsai; Mizutani, Eiji, *Neuro-Fuzzy and Soft Computing: A Computational Approach to Learning and Machine Intelligence, 1st Edition*, © 1997. Reprinted by permission of Pearson Education Inc., Upper Saddle River, NJ): (*a*) antecedent and consequent membership functions; (*b*) overall system response surface.

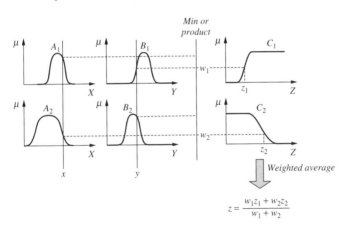

FIGURE 5.16
The Tsukamoto fuzzy model (Jang, Jyh-Shing Roger; Sun, Chuen-Tsai; Mizutani, Eiji, *Neuro-Fuzzy and Soft Computing: A Computational Approach to Learning and Machine Intelligence, 1st Edition*, © 1997. Reprinted by permission of Pearson Education Inc., Upper Saddle River, NJ).

calculated by the weighted average of each rule's output, as seen in Fig. 5.16. Since each rule infers a crisp output, the Tsukamoto model's aggregation of the overall output also avoids the time-consuming process of defuzzification. Because of the special nature of the output membership functions required by the method, it is not as useful as a general approach, and must be employed in specific situations.

Example 5.17. An example of a single input, single-output Tsukamoto fuzzy model is given by the following rules:

$$IF\ X\ is\ small,\ THEN\ Y\ is\ C_1$$
$$IF\ X\ is\ medium,\ THEN\ Y\ is\ C_2$$
$$IF\ X\ is\ large,\ THEN\ Y\ is\ C_3$$

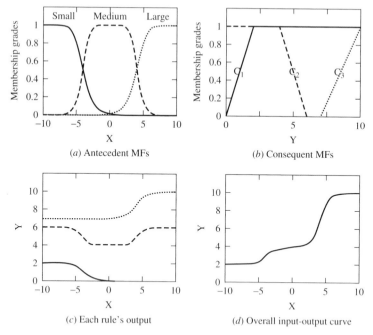

FIGURE 5.17
Tsukamoto model for Example 5.17 (Jang, Jyh-Shing Roger; Sun, Chuen-Tsai; Mizutani, Eiji, *Neuro-Fuzzy and Soft Computing: A Computational Approach to Learning and Machine Intelligence, 1st Edition*, © 1997. Reprinted by permission of Pearson Education Inc., Upper Saddle River, NJ): (*a*) antecedent membership functions; (*b*) consequent membership functions; (*c*) each rule's output curve; (*d*) overall system response curve.

where the antecedent and consequent fuzzy sets are as shown in Fig. 5.17*a* and Fig. 5.17*b*, respectively. If we plot the output of each of the three rules as a function of the input, X, we get the three curves shown in Fig. 5.17*c*. The overall output of the three-rule system is shown in Fig. 5.17*d*. Since the reasoning mechanism of the Tsukamoto fuzzy model does not strictly follow a composition operation in its inference it always generates a crisp output even when the input and output membership functions are fuzzy membership functions.

Example 5.18. In heat exchanger design, a flexibility analysis requires the designer to determine if the size of the heat exchanger is either small or large. In order to quantify this linguistic vagueness of size, we form the general design equation for a heat exchanger, $Q = AU \Delta T_{\log \text{ mean}}$, where the heat transfer coefficient U and area A need to be determined. Figure 5.18 show a schematic of this exchanger.

We want to determine the sizes for a heat exchanger in which a stream of benzene is heated using saturated steam at pressure 68.95 kPa and temperature 362.7 K. The initial temperature of the benzene steam is 17°C, and the model used to determine the size of the heat exchanger is the following:

$$AU = wC_p \ln \left(\frac{T_s - T_1}{\Delta T_{\text{app}}} \right)$$

where C_p is the heat capacity of the benzene (1.7543 kJ/K kg) and $T_s - T_1 = 72.55$ K.

We will model the benzene flow rate, w, in kg/s, and temperature approach $\left(\Delta T_{\text{app}} \right)$ in kelvin, as the inputs, and we will model the size of the heat exchanger as output. We will

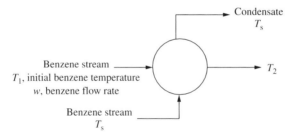

FIGURE 5.18
Heat exchanger design.

deduce the following disjunctive rules of inference based on observations of the model:

Rule 1 :   IF w is $\underset{\sim}{A}_1^1$ (large flow rate) and ΔT_{app} is $\underset{\sim}{A}_2^1$ (small approach),

THEN AU is $\underset{\sim}{B}^1$ (large heat exchanger).

Rule 2 :   IF w is $\underset{\sim}{A}_1^2$ (small flow rate) or ΔT_{app} is $\underset{\sim}{A}_2^2$ (large approach),

THEN AU is $\underset{\sim}{B}^1$ (small heat exchanger).

Rule 3 :   IF w is $\underset{\sim}{A}_1^2$ (small flow rate) and ΔT_{app} is $\underset{\sim}{A}_2^1$ (small approach),

THEN AU is $\underset{\sim}{B}^1$ (large heat exchanger).

The graphical equivalent of these rules is shown in Fig. 5.19. A weighted average defuzzification method will be employed to compare the results from one input pair for each of the three following inference methods: Mamdani, Sugeno and Tsukamoto.
    We will input two crisp values of benzene flow rate and temperature approach:

$$w = 1300 \text{ kg/s} \quad \text{and} \quad \Delta T_{app} = 6.5 \text{ K}$$

1. Using the max–min Mamdani implication method of inference, we know that

$$\mu_{B^k}(AU) = \max_k \left\{ \min \left[\mu_{A_1^k}(w), \mu_{A_2^k}(\Delta T_{app}) \right] \right\}$$

Using the graphical approach we get the rules shown in Fig. 5.19.
And using a weighted average defuzzification, we get

$$AU^* = \frac{(4500 \ \text{m}^2 \ \text{kW/m}^2 \ \text{K})(0.5) + (10,000 \ \text{m}^2 \ \text{kW/m}^2 \ \text{K})(0.25)}{0.5 + 0.25} = 6333.3 \ \text{m}^2 \ \text{kW/m}^2 \ \text{K}$$

which is also shown in Fig. 5.20.
2. for the Sugeno fuzzy method of inference, we have experience in heat exchanger design that gives the following expressions in a polynomial form for our two consequents (small and large heat exchangers):

$$AU_{small} = 3.4765w - 210.5\Delta T_{app} + 2103$$

$$AU_{large} = 4.6925w - 52.62\Delta T_{app} + 2631$$

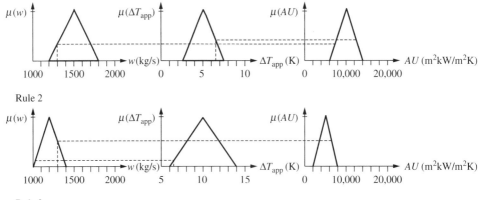

Rule 1

Rule 2

Rule 3

FIGURE 5.19
Graphical inference using the Mamdani method for three rules.

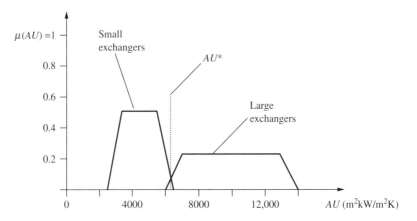

FIGURE 5.20
Result of the defuzzification step in the Mamdani method.

Taking the minimum value for the conjunction and the maximum value for the disjunction of the membership values of the inputs, for each rule we will have:

$$\text{Rule 1}: \quad w_1 = 0.25$$

$$\text{Rule 2}: \quad w_2 = 0.5$$

$$\text{Rule 3}: \quad w_1 = 0.25$$

Then

$$AU_{small} = 5256 \text{ m}^2 \text{ kW/m}^2 \text{ K} \quad \text{and} \quad AU_{large} = 5311 \text{ m}^2 \text{ kW/m}^2 \text{ K}$$

Finally, the defuzzified value of the heat exchange size is

$$AU^* = \frac{(5311 \text{ m}^2 \text{ kW/m}^2 \text{ K})(0.25) + (5256 \text{ m}^2 \text{ kW/m}^2 \text{ K})(0.5) + (5311 \text{ m}^2 \text{ kW/m}^2 \text{ K}(0.25)}{0.25 + 0.5 + 0.25}$$

$$= 5283.5 \text{ m}^2 \text{ kW/m}^2 \text{ K}$$

3. For the Tsukamoto fuzzy method of inference, we modify the output membership functions from the Mamdani case (see Fig. 5.19), but we added shoulders to them for Tsukamoto. Using a graphical approach, we get the rules shown in Fig. 5.21.
   Using the minimum value of the membership values, the defuzzified value of the heat exchanger size is

$$AU^* = \frac{(7000 \text{ m}^2 \text{ kW/m}^2 \text{ K})(0.25) + (5500 \text{ m}^2 \text{ kW/m}^2 \text{ K})(0.5) + (7000 \text{ m}^2 \text{ kW/m}^2 \text{ K}(0.25)}{0.25 + 0.5 + 0.25}$$

$$= 6250 \text{ m}^2 \text{ kW/m}^2 \text{ K}$$

The Mamdani and Tsukamoto methods yield similar values of AU, since they are based on similar membership functions for the output. The difference with the Sugeno method is a function of the accuracy of the polynomials that model the output.

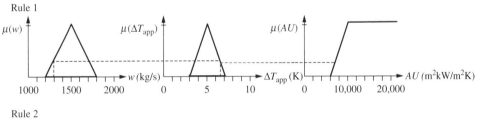

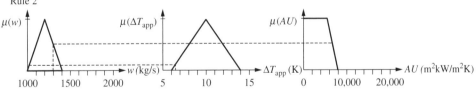

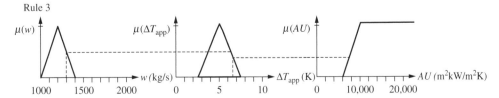

FIGURE 5.21
Tsukamoto method of inference for the three rules.

SUMMARY

This chapter has presented the basic axioms, operations, and properties of binary logic and fuzzy logic. Just as in Chapter 2, we find that the only significant difference between a binary logic and a fuzzy logic stems from the logical equivalent of the excluded middle axioms. Examples are provided that illustrate the various operations of a fuzzy logic. An approximate reasoning, proposed by Zadeh [1976, 1979], is presented to illustrate the power of using fuzzy sets in the reasoning process. Other works in the area of fuzzy reasoning and approximate reasoning have been helpful in explaining the theory; for example, a useful comparison study [Mizumoto and Zimmerman, 1982] and a work defining the mathematical foundations [Yager, 1985]. From a general point of view, other multivalued logics have been developed [Dubois and Prade, 1980; Klir and Folger, 1988], and these other logics may be viewed as *fuzzy logics* in the sense that they represent more than just the crisp truth values of 0 and 1. In fact, Gaines [1976] has shown that some forms of multivalued logics result from fuzzifying, in the sense of the extension principle, the standard propositional calculus. The illustration of approximate reasoning given here is conducted using fuzzy relations to represent the rules of inference. The chapter concludes by pointing out the rich variety in reasoning possible with fuzzy logic when one considers the vast array of implication and composition operations; an example of this can be found in Yager [1983]. The implications can be interpreted as specific chains of reasoning. Giles [1976] gives a very nice interpretation of these chains of reasoning in terms of risk: every chain of reasoning is analogous to a dialogue between speakers whose assertions entail a commitment about their truth.

The subjectivity that exists in fuzzy modeling is a blessing rather than a curse. The vagueness present in the definition of terms is consistent with the information contained in the conditional rules developed by the engineer when observing some complex process. Even though the set of linguistic variables and their meanings is compatible and consistent with the set of conditional rules used, the overall outcome of the qualitative process is translated into objective and quantifiable results. Fuzzy mathematical tools and the calculus of fuzzy IF–THEN rules provide a most useful paradigm for the automation and implementation of an extensive body of human knowledge heretofore not embodied in the quantitative modeling process; we call this paradigm *fuzzy systems*. These mathematical tools provide a means of sharing, communicating, and transferring this human subjective knowledge of systems and processes.

This chapter has also summarized the seminal works of Zadeh [1972, 1973, 1975a,b] in the area of linguistic modeling. Modeling in the area of linguistics has reached far beyond the boundaries of engineering. For example, Kickert [1979] used fuzzy linguistic modeling to adapt a factual prototype of Mulder's power theory to a numerical simulation. This is a marvelous illustration of the power of fuzzy sets in a situation where ideals of the *rational man* run contrary to the satisfaction gained simply through the exercise of power. The material developed in this chapter provides a good foundation for discussions in Chapter 8 on nonlinear simulation and in Chapter 13 on fuzzy control.

This chapter has focused on two popular forms of logic – classical and fuzzy. There are other forms, of course. For example, almost a century ago (1905), L. E. J. Brouwer posed a form of logic known as intuitionism. This logic has been subject to debate for all of this time [Franchella, 1995]. Intuitionism is a branch of logic which stresses that mathematics has priority over logic, the objects of mathematics are constructed and operated upon in the mind by the mathematician, and it is impossible to define the properties of mathematical objects simply by establishing a number of axioms. In particular, intuitionists reject, as they called it then, the *principle of the excluded middle* (we refer to this as the excluded middle axiom in this text) which allows proof by contradiction.

Brouwer rejected in mathematical proofs the *principle of the excluded middle*, which states that any mathematical statement is either true or false. In 1918 he published a set theory, in 1919 a measure theory, and in 1923 a theory of functions all developed without using this principle.

In 1928 Brouwer's paper, "Reflections on Formalism," identifies and discusses four key differences between formalism and intuitionism, all having to do either with the role of the *principle of the excluded middle* or with the relation between mathematics and language. Brouwer emphasizes, as he had done in his dissertation in 1907, that formalism presupposes contextual mathematics at the metalevel. In this paper Brouwer presents his first strong counterexample of the *principle of the excluded middle*, by showing that it is false that every real number is either rational or irrational. An illustration of this is the following: A is a statement: "π has infinitely many 7s in its expansion" and $\overline{A}$ is a statement: "π has only finitely many 7s in its expansion." We do not know whether A is true or false, so we cannot claim that A or $\overline{A}$ is true, because that would imply that we either know A or we know $\overline{A}$ [Kreinovich, 2003].

Brouwer also considered weak counterexamples to the *principle of the excluded middle*. A still open problem in mathematics, known as Goldbach's conjecture (the conjecture that every even number equal to or greater than 4 is the sum of two prime numbers), is one such counterexample. The conjecture Brouwer states: "we have at present experienced neither its truth nor its falsity, so intuitionistically speaking, it is at present neither true nor false, and hence we cannot assert 'Goldbach's conjecture is true, or it is false'" [Franchella, 1995].

Another form of logic is termed *linear logic*, where we have two different versions of conjunction. For example, in the phrase *I can buy a snack and a drink* we can mean that we can only buy one, not both, or that we can buy both. Both forms of the conjunction are allowed [Kreinovich, 2003].

In Brouwer's intuitionism there is a single description of the connective, "or." In intuitionism (also termed constructive logic) the meaning of "or" is as follows: the statement "A or B" means that either we know A or we know B. In a nonconstructive logic the statement "A or B" means that we know that one or the other (A or B) is true, but we do not know which one is true. In classical logic we have both types of "or." What Brouwer pointed out is that if we interpret the "or" as a *constructive or*, then the excluded middle axiom is not valid.

The significance of other forms of logic is that we oftentimes intertwine our human intuition with formal logic structures that are likely *layered* in our minds, just like the laws of nature are *layered* in reality. For example, sometimes we use Newton's laws to describe behavior in mechanics that we can see visually, yet the phenomena might better be described by quantum mechanics laws at a scale that is not known to us through simple observation.

REFERENCES

Belohlavek, R., Klir, G., Lewis, H., and Way, E. (2002). "On the capability of fuzzy set theory to represent concepts," *Int. J. Gen. Syst.*, vol. 31, pp. 569–585.

Clark, D. W. (1992). "Computer illogic...," *Mirage Mag.*, University of New Mexico Alumni Association, Fall, pp. 12–13.

Dubois, D. and Prade, H. (1980). *Fuzzy sets and systems: Theory and applications*, Academic Press, New York.

Franchella, M. (1995). "L. E. J. Brouwer: Toward intuitionistic logic," *Hist. Math.*, vol. 22, pp. 304–322.

Gaines, B. (1976). "Foundations of fuzzy reasoning," *Int. J. Man Mach. Stud.*, vol. 8, pp. 623–688.

Giles, R. (1976). "Lukasiewicz logic and fuzzy theory," *Int. J. Man Mach. Stud.*, vol. 8, pp. 313–327.

Gill, A. (1976). *Applied algebra for the computer sciences*, Prentice Hall, Englewood Cliffs, NJ.

Jang, J., Sun, C. and Mizutani, E. (1997). *Neuro-Fuzzy and Soft Computing: A Computational Approach to Learning and Machine Intelligence, 1st Edition*, Pearson Education, Inc., Upper Saddle River, NJ.

Kickert, W. (1979). "An example of linguistic modelling: The case of Mulder's theory of power," in M. Gupta, R. Ragade, and R. Yager (eds.), *Advances in fuzzy set theory and applications*, Elsevier, Amsterdam, pp. 519–540.

Klir, G. and Folger, T. (1988). *Fuzzy sets, uncertainty, and information*, Prentice Hall, Englewood Cliffs, NJ.

Kosko, B. (1992). *Neural networks and fuzzy systems*, Prentice Hall, Englewood Cliffs, NJ.

Kreinovich, V. (2003). Personal discussion.

Mamdani, E. and Assilian, S. (1975). "An experiment in linguistic synthesis with a fuzzy logic controller," *Int. J. Man Mach. Syst.*, vol. 7, pp. 1–13.

Mamdani, E. H. (1976). "Advances in linguistic synthesis of fuzzy controllers," *Int. J. Man Mach. Stud.*, vol. 8, pp. 669–678.

Mano, M. (1988). *Computer engineering: Hardware design*, Prentice Hall, Englewood Cliffs, NJ, p. 59.

Mizumoto, M. and Zimmerman, H.-J. (1982). "Comparison of fuzzy reasoning methods," *Fuzzy Sets Syst.*, vol. 8, pp. 253–283.

Osherson, D. and Smith, E. (1981). "On the adequacy of prototype theory as a theory of concepts," *Cognition*, vol. 9, pp. 35–58.

Rescher, N. (1969). *Many-valued logic*, McGraw-Hill, New York.

Ross, T. (1995). *Fuzzy logic with engineering applications*, McGraw-Hill, New York.

Sanchez, E. (1976). "Resolution of composite fuzzy relation equations," *Inf. Control*, vol. 30, pp. 38–48.

Sugeno, M. and Kang, G. (1988). "Structure identification of fuzzy model," *Fuzzy Sets Syst.*, vol. 28, pp. 15–33.

Takagi, T. and Sugeno, M. (1985). "Fuzzy identification of systems and its applications to modeling and control," *IEEE Trans. Syst., Man, Cybern.*, vol. 15, pp. 116–132.

Tsukamoto, Y. (1979). "An approach to fuzzy reasoning method," in M. Gupta, R. Ragade, and R. Yager, (eds.), *Advances in fuzzy set theory and applications*, Elsevier, Amsterdam, pp. 137–149.

Yager, R. R. (1983). "On the implication operator in fuzzy logic," *Inf. Sci.*, vol. 31, pp. 141–164.

Yager, R. R. (1985). "Strong truth and rules of inference in fuzzy logic and approximate reasoning," *Cybern. Syst.*, vol. 16, pp. 23–63.

Zadeh, L. (1972) "A fuzzy-set-theoretic interpretation of linguistic hedges," *J. Cybern.*, vol. 2, pp. 4–34.

Zadeh, L. (1973). "Outline of a new approach to the analysis of complex systems and decision processes," *IEEE Trans. Syst. Man, Cybern.*, vol. 3, pp. 28–44.

Zadeh, L. (1975a). "The concept of a linguistic variable and its application to approximate reasoning – I," *Inf. Sci.*, vol. 8, pp. 199–249.

Zadeh, L. (1975b). "The concept of a linguistic variable and its application to approximate reasoning – II," *Inf. Sci.*, vol. 8, pp. 301–357.

Zadeh, L. (1976). "The concept of a linguistic variable and its application to approximate reasoning – Part 3," *Inf. Sci.*, vol. 9, pp. 43–80.

Zadeh, L. (1979). ''A theory of approximate reasoning,'' in J. Hayes, D. Michie, and L. Mikulich (eds.), *Machine Intelligence*, Halstead Press, New York, pp. 149–194.

PROBLEMS

5.1. Under what conditions of P and Q is the implication $P \rightarrow Q$ a tautology?

5.2. The exclusive-or is given by the expression $P \, XOR \, Q = (\bar{P} \wedge Q) \vee (P \wedge \bar{Q})$. Show that the logical-or, given by $P \vee Q$, gives a different result from the exclusive-or and comment on this difference using an example in your own field.

5.3. For a proposition R of the form $P \rightarrow Q$, show the following:

(a) R and its contrapositive are equivalent, i.e., prove that $(P \rightarrow Q) \leftrightarrow (\bar{Q} \rightarrow \bar{P})$.

(b) The converse of R and the inverse of R are equivalent, i.e., prove that $(Q \rightarrow P) \leftrightarrow (\bar{P} \rightarrow \bar{Q})$.

5.4. Show that the dual of the equivalence $((P \vee Q) \vee ((\bar{P}) \wedge (\bar{Q}))) \leftrightarrow X$ is also true.

5.5. Show that De Morgan's principles are duals.

5.6. Show that the compound proposition $((P \rightarrow Q) \wedge (R \rightarrow \bar{S}) \wedge (Q \rightarrow R)) \rightarrow (P \rightarrow \bar{S})$ is a tautology.

5.7. Show that the following propositions from Lewis Carroll are tautologies [Gill, 1976]:

(a) No ducks waltz; no officers ever decline to waltz; all my poultry are ducks. Therefore, none of my poultry are officers.

(b) Babies are illogical; despised persons cannot manage crocodiles; illogical persons are despised; therefore, babies cannot manage crocodiles.

(c) Promise-breakers are untrustworthy; wine-drinkers are very communicative; a man who keeps his promise is honest; all pawnbrokers are wine-drinkers; we can always trust a very communicative person; therefore, all pawnbrokers are honest. (This problem requires $2^6 = 64$ lines of a truth table; perhaps it should be tackled with a computer.)

5.8. Prove the following statements by contradiction.

(a) $((P \rightarrow Q) \wedge P) \rightarrow Q$

(b) $((P \rightarrow \bar{Q}) \wedge (Q \vee \bar{R}) \wedge (R \wedge \bar{S})) \rightarrow \bar{P}$

5.9. Prove that $((P \rightarrow \bar{Q}) \wedge (R \rightarrow \bar{Q}) \wedge (P \vee R)) \rightarrow R$ is not a tautology (i.e., a fallacy) by developing a counterexample.

5.10. Prove that the following statements are tautologies.

(a) $((P \rightarrow Q) \wedge P) \rightarrow Q$

(b) $P \rightarrow (P \vee Q)$

(c) $(P \wedge Q) \rightarrow P$

(d) $((P \rightarrow Q) \wedge (Q \rightarrow R)) \rightarrow (P \rightarrow R)$

(e) $((P \vee Q) \wedge \bar{P}) \rightarrow Q$

5.11. For this inference rule,

$$[(A \rightarrow B) \wedge (B \rightarrow C)] \rightarrow (A \rightarrow C)$$

Prove that the rule is a tautology.

5.12. Consider the following two discrete fuzzy sets, which are defined on universe $X = \{-5, 5\}$:

$$\underset{\sim}{A} = \text{``zero''} = \left\{ \frac{0}{-2} + \frac{0.5}{-1} + \frac{1.0}{0} + \frac{0.5}{1} + \frac{0}{2} \right\}$$

$$\underset{\sim}{B} = \text{``positive medium''} = \left\{ \frac{0}{0} + \frac{0.5}{1} + \frac{1.0}{2} + \frac{0.5}{3} + \frac{0}{4} \right\}$$

(a) Construct the relation for the rule IF $\underset{\sim}{A}$, THEN $\underset{\sim}{B}$ (i.e., IF x is "zero" THEN y is "positive medium") using the Mamdani implication, Eq. (5.20), and the product implication, Eq. (5.22), or

$$\mu_R(x, y) = \min[\mu_A(x), \mu_B(y)]$$

and

$$\mu_R(x, y) = \mu_A(x) \cdot \mu_B(y)$$

(b) If we introduce a new antecedent,

$$\underset{\sim}{A}' = \text{"positive small"} = \left\{ \frac{0}{-1} + \frac{0.5}{0} + \frac{1.0}{1} + \frac{0.5}{2} + \frac{0}{3} \right\}$$

find the new consequent $\underset{\sim}{B}'$, using max–min composition, i.e., $\underset{\sim}{B}' = \underset{\sim}{A}' \circ \underset{\sim}{R}$, for both relations from part (a).

5.13. Given the fuzzy sets $\underset{\sim}{A}$ and $\underset{\sim}{B}$ on X and Y, respectively,

$$\underset{\sim}{A} = \int \left\{ \frac{1 - 0.1x}{x} \right\}, \qquad \text{for } x \in [0, +10]$$

$$\underset{\sim}{B} = \int \left\{ \frac{0.2y}{y} \right\}, \qquad \text{for } y \in [0, +5]$$

$$\mu_A(x) = 0 \quad \text{outside the } [0, 10] \text{ interval}$$

$$\mu_B(y) = 0 \quad \text{outside the } [0, 5] \text{ interval}$$

(a) Construct a fuzzy relation $\underset{\sim}{R}$ for the implication $\underset{\sim}{A} \to \underset{\sim}{B}$ using the classical implication operation, i.e., construct $\underset{\sim}{R} = (\underset{\sim}{A} \times \underset{\sim}{B}) \cup (\overline{\underset{\sim}{A}} \times Y)$.

(b) Use max–min composition to find $\underset{\sim}{B}'$, given

$$\underset{\sim}{A}' = \left\{ \frac{1}{3} \right\}$$

Note: $\underset{\sim}{A}'$ is a crisp singleton, i.e., the number 3 has a membership of 1, and all other numbers in the universe X have a membership of 0.

Hint: You can solve this problem graphically by segregating the Cartesian space into various regions according to the min and max operations, or you can approximate the continuous fuzzy variables as discrete variables and use matrix operations. In either case, "sketch" the solutions for part (a) in 3D space (x, y, μ) and (b) in 2D space (y, μ).

5.14. Suppose we have a distillation process where the objective is to separate components of a mixture in the input stream. The process is pictured in Fig. P5.14. The relationship between the input variable, temperature, and the output variable, distillate fractions, is not precise but the human operator of this process has developed an intuitive understanding of this relationship. The universe for each of these variables is

$$X = \text{universe of temperatures } (^\circ F) = \{160, 165, 170, 175, 180, 185, 190, 195\}$$

$$Y = \text{universe of distillate fractions (percentages)} = \{77, 80, 83, 86, 89, 92, 95, 98\}$$

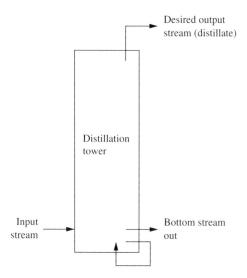

FIGURE P5.14

Now we define fuzzy sets $\underset{\sim}{A}$ and $\underset{\sim}{B}$ on X and Y, respectively:

$$\underset{\sim}{A} = \text{temperature of input steam is hot} = \left\{ \frac{0}{175} + \frac{0.7}{180} + \frac{1}{185} + \frac{0.4}{190} \right\}$$

$$\underset{\sim}{B} = \text{separation of mixture is good} = \left\{ \frac{0}{89} + \frac{0.5}{92} + \frac{0.8}{95} + \frac{1}{98} \right\}$$

We wish to determine the proposition, IF "temperature is hot" THEN "separation of mixture is good," or symbolically, $\underset{\sim}{A} \to \underset{\sim}{B}$. From this,

(a) Find $\underset{\sim}{R} = (\underset{\sim}{A} \times \underset{\sim}{B}) \cup (\overline{\underset{\sim}{A}} \times Y)$.

(b) Now define another fuzzy linguistic variable as

$$\underset{\sim}{A}' = \left\{ \frac{1}{170} + \frac{0.8}{175} + \frac{0.5}{180} + \frac{0.2}{185} \right\}$$

and for the "new" rule IF $\underset{\sim}{A}'$ THEN $\underset{\sim}{B}'$, find $\underset{\sim}{B}'$ using max–min composition, i.e., find $\underset{\sim}{B}' = \underset{\sim}{A}' \circ \underset{\sim}{R}$.

5.15. The calculation of the vibration of an elastic structure depends on knowing the material properties of the structure as well as its support conditions. Suppose we have an elastic structure, such as a bar of known material, with properties like wave speed (C), modulus of elasticity (E), and cross-sectional area (A). However, the support stiffness is not well-known; hence the fundamental natural frequency of the system is not precise either. A relationship does exist between them, though, as illustrated in Fig. P5.15.

FIGURE P5.15

Define two fuzzy sets,

$$\underset{\sim}{K} = \text{``support stiffness,''} \text{ in pounds per square inch}$$

$$\underset{\sim}{f_1} = \text{``first natural frequency of the system,''} \text{ in hertz}$$

with membership functions

$$\underset{\sim}{K} = \left\{ \frac{0}{1e+3} + \frac{0.2}{1e+4} + \frac{0.5}{1e+5} + \frac{0.8}{5e+5} + \frac{1}{1e+6} + \frac{0.8}{5e+6} + \frac{0.2}{1e+7} \right\}$$

$$\underset{\sim}{f_1} = \left\{ \frac{0}{100} + \frac{0}{200} + \frac{0.2}{500} + \frac{0.5}{800} + \frac{1}{1000} + \frac{0.8}{2000} + \frac{0.2}{5000} \right\}$$

(a) Using the proposition, IF x is $\underset{\sim}{K}$, THEN y is $\underset{\sim}{f_1}$, find this relation using the following forms of the implication $\underset{\sim}{K} \rightarrow \underset{\sim}{f_1}$:
  (i) Classical $\mu_R = \max[\min(\mu_K, \mu_{f_1}), (1 - \mu_K)]$
  (ii) Mamdani $\mu_R = \min(\mu_K, \mu_{f_1})$
  (iii) Product $\mu_R = \mu_K \cdot \mu_{f_1}$
(b) Now define another antecedent, say $\underset{\sim}{K'} = \text{``damaged support,''}$

$$\underset{\sim}{K'} = \left\{ \frac{0}{1e+3} + \frac{0.8}{1e+4} + \frac{0.1}{1e+5} \right\}$$

Find the system's fundamental (first) natural frequency due to the change in the support conditions, i.e., find $\underset{\sim}{f_1} = \text{``first natural frequency due to damaged support''}$ using classical implication from part (a), sub-part (i) preceding, and
  (i) max–min composition
  (ii) max–product composition

5.16. When gyros are calibrated for axis bias, they are matched with a temperature. Thus, we can have a relation of gyro bias (GB) vs. temperature ($\underset{\sim}{T}$). Suppose we have fuzzy sets for a given gyro bias and a given Fahrenheit temperature, as follows:

$$\mu_{GB}(x) = \left\{ \frac{0.2}{3} + \frac{0.4}{4} + \frac{1}{5} + \frac{0.4}{6} + \frac{0.2}{7} \right\} \quad \text{bias in degrees Fahrenheit per hour}$$

$$\mu_T(y) = \left\{ \frac{0.4}{66} + \frac{0.6}{68} + \frac{1}{70} + \frac{0.6}{72} + \frac{0.4}{74} \right\} \quad \text{temperature in degrees Fahrenheit}$$

(a) Use a Mamdani implication to find the relation IF gyro bias, THEN temperature.
(b) Suppose we are given a new gyro bias (GB') as follows:

$$\mu_{GB'}(x) = \left\{ \frac{0.6}{3} + \frac{1}{4} + \frac{0.6}{5} \right\}$$

Using max–min composition, find the temperature associated with this new bias.

5.17. You are asked to develop a controller to regulate the temperature of a room. Knowledge of the system allows you to construct a simple rule of thumb: when the temperature is HOT then cool room down by turning the fan at the fast speed, or, expressed in rule form, IF temperature is HOT, THEN fan should turn FAST. Fuzzy sets for hot temperature and fast fan speed can be developed: for example,

$$\underset{\sim}{H} = \text{``hot''} = \left\{ \frac{0}{60} + \frac{0.1}{70} + \frac{0.7}{80} + \frac{0.9}{90} + \frac{1}{100} \right\}$$

represents universe X in °F, and

$$\underset{\sim}{F} = \text{``fast''} = \left\{ \frac{0}{0} + \frac{0.2}{1} + \frac{0.5}{2} + \frac{0.9}{3} + \frac{1}{4} \right\}$$

represents universe Y in 1000 rpm.

(a) From these two fuzzy sets construct a relation for the rule using classical implication.

(b) Suppose a new rule uses a slightly different temperature, say "moderately hot," and is expressed by the fuzzy membership function for "moderately hot," or

$$\underset{\sim}{H'} = \left\{ \frac{0}{60} + \frac{0.2}{70} + \frac{1}{80} + \frac{1}{90} + \frac{1}{100} \right\}$$

Using max–product composition, find the resulting fuzzy fan speed.

5.18. In public transportation systems there often is a significant need for speed control. For subway systems, for example, the train speed cannot go too far beyond a certain target speed or the trains will have trouble stopping at a desired location in the station. Set up a fuzzy set

$$\underset{\sim}{A} = \text{``speed way over target''} = \left\{ \frac{0}{T_0} + \frac{0.8}{T_0 + 5} + \frac{1}{T_0 + 10} + \frac{0.8}{T_0 + 15} \right\}$$

on a universe of target speeds, say $T = [T_0, T_0 + 15]$, where T_0 is a lower bound on speed. Define another fuzzy set,

$$\underset{\sim}{B} = \text{``apply brakes with high force''} = \left\{ \frac{0.3}{10} + \frac{0.8}{20} + \frac{0.9}{30} + \frac{1}{40} \right\}$$

on a universe of braking pressures, say $S = [10, 40]$.

(a) For the compound proposition, IF speed is "way over target," THEN "apply brakes with high force," find a fuzzy relation using classical implication.

(b) For a new antecedent,

$$\underset{\sim}{A'} = \text{``speed moderately over target''} = \left\{ \frac{0.2}{T_0} + \frac{0.6}{T_0 + 5} + \frac{0.8}{T_0 + 10} + \frac{0.3}{T_0 + 15} \right\}$$

find the fuzzy brake pressure using max–min composition.

5.19. We want to consider the engineering of amplifiers. Here, the amplifier is a simple voltage-measuring input and current output, as shown in Fig. P5.19. We define two fuzzy linguistic variables for a fuzzy relation: $\underset{\sim}{V}_{in}$, the input voltage, and $\underset{\sim}{I}_{out}$, the output current:

$$\underset{\sim}{V}_{in} = \text{``small''} = \left\{ \frac{0.5}{0.10} + \frac{1}{0.20} + \frac{0.8}{0.30} + \frac{0.2}{0.40} \right\} \quad \text{volts}$$

$$\underset{\sim}{I}_{out} = \text{``big''} = \left\{ \frac{0.3}{0.6} + \frac{1}{1} + \frac{0.5}{1.4} \right\} \quad \text{amps}$$

where $\underset{\sim}{V}_{in}$ is defined on a universe of voltages, and $\underset{\sim}{I}_{out}$ is defined on a universe of currents.

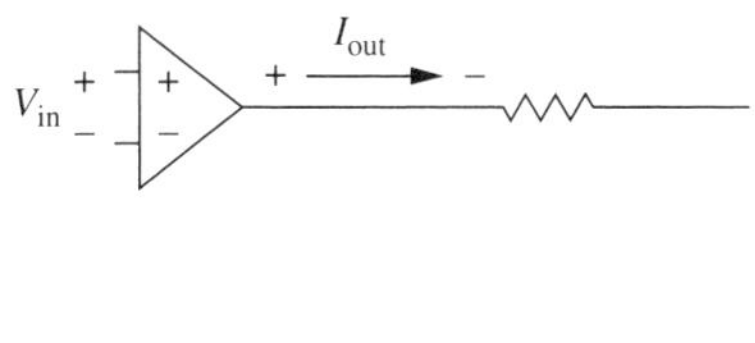

FIGURE P5.19

(a) Find the relation, IF $\underset{\sim}{V}_{in}$, THEN $\underset{\sim}{I}_{out}$, using classical implication.

(b) Another fuzzy linguistic variable in this problem is input impedance, $\underset{\sim}{Z}$. The higher the impedance, generally the better the amplifier. For the following impedance defined on a universe of resistances,

$$\underset{\sim}{Z} = \text{``high impedance''} = \left\{ \frac{0}{10^4} + \frac{0.3}{10^5} + \frac{1}{10^6} + \frac{0.6}{10^7} \right\} \quad \text{ohms}$$

find the relation, IF $\underset{\sim}{Z}$, THEN $\underset{\sim}{I}_{out}$, using Mamdani implication.

5.20 For Example 5.9 in this chapter, recalculate the fuzzy relation $\underset{\sim}{R}$ using

(a) Equation (5.19)

(b) Equation (5.20)

(c) Equation (5.21)

(d) Equation (5.22)

(e) Equation (5.23)

5.21. Fill in the following table using Eqs. (5.19) – (5.23) to determine the values of the implication $\underset{\sim}{A} \rightarrow \underset{\sim}{B}$. Comment on the similarities and dissimilarities of the various implication methods with respect to the various values for $\underset{\sim}{A}$ and $\underset{\sim}{B}$.

$\underset{\sim}{A}$	$\underset{\sim}{B}$	$\underset{\sim}{A} \rightarrow \underset{\sim}{B}$
0	0	
0	1	
1	0	
1	1	
0.2	0.3	
0.2	0.7	
0.8	0.3	
0.8	0.7	

5.22. A factory process control operation involves two linguistic (atomic) parameters consisting of pressure and temperature in a fluid delivery system. Nominal pressure limits range from 400 psi minimum to 1000 psi maximum. Nominal temperature limits are 130 to 140°F. We characterize each parameter in fuzzy linguistic terms as follows:

$$\text{``Low temperature''} = \left\{ \frac{1}{131} + \frac{0.8}{132} + \frac{0.6}{133} + \frac{0.4}{134} + \frac{0.2}{135} + \frac{0}{136} \right\}$$

$$\text{``High temperature''} = \left\{ \frac{0}{134} + \frac{0.2}{135} + \frac{0.4}{136} + \frac{0.6}{137} + \frac{0.8}{138} + \frac{1}{139} \right\}$$

$$\text{``High pressure''} = \left\{ \frac{0}{400} + \frac{0.2}{600} + \frac{0.4}{700} + \frac{0.6}{800} + \frac{0.8}{900} + \frac{1}{1000} \right\}$$

$$\text{``Low pressure''} = \left\{ \frac{1}{400} + \frac{0.8}{600} + \frac{0.6}{700} + \frac{0.4}{800} + \frac{0.2}{900} + \frac{0}{1000} \right\}$$

(a) Find the following membership functions:

(i) Temperature not very low

(ii) Temperature not very high

(iii) Temperature not very low and not very high

(b) Find the following membership functions:
   (i) Pressure slightly high
   (ii) Pressure fairly high ($[\text{high}]^{2/3}$)
   (iii) Pressure not very low or fairly low

5.23. In information retrieval, having fuzzy information about the size of a document helps when trying to scale the word frequencies of the document (i.e., how often a word occurs in a document is important when determining relevance). So on the universe of document sizes, we define two fuzzy sets:

$$\text{``Small'' document''} = \begin{cases} 1 - e^{-k(a-x)} & \text{for } x \leq a \\ 0 & \text{for } x > a \end{cases}$$

$$\text{``Large document''} = \begin{cases} 1 - e^{-k(x-b)} & \text{for } x \geq b \\ 0 & \text{for } x < b \end{cases}$$

where the parameters k, a, and b change from database to database. Graphically the parameters a and b look as shown in Fig. P5.23. Develop a graphical solution to the following linguistic phrases, for the specific values of $a = 2$, $b = 4$, and $k = 0.5$:

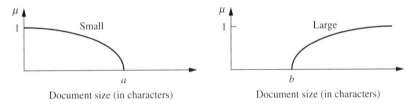

FIGURE P5.23

(a) "Not very large" document
(b) "Large and small" documents
(c) "Not very large or small" documents

5.24. In a problem related to the computer tracking of soil particles as they move under stress, the program displays desired particles on the screen. Particles can be small and large. Because of segmentation problems in computer imaging, the particles can become too large and obscure particles of interest or become too small and be obscured. To solve this problem linguistically, suppose we define the following atomic terms on a scale of sizes [0, 50] in units of mm^2:

$$\text{``Large''} = \left\{ \frac{0}{0} + \frac{0.1}{10} + \frac{0.3}{20} + \frac{0.5}{30} + \frac{0.6}{40} + \frac{0.7}{50} \right\}$$

$$\text{``Small''} = \left\{ \frac{1}{0} + \frac{0.8}{10} + \frac{0.5}{20} + \frac{0.3}{30} + \frac{0.1}{40} + \frac{0}{50} \right\}$$

For these atomic terms find membership functions for the following phrases:
(a) Very small or very large
(b) Not small and not large
(c) Large or not small

5.25. In vehicle navigation the mapping source of information uses *shape points* to define the curvature of a turning maneuver. A segment is a length of road between two points. If the segment is linear, it has no or very few shape points. If the road is winding or circular, the segment can have many shape points. Figure P5.25 shows the relationship of curvature and shape points. Assume that up to nine shape points can define any curvature in a typical road segment. The universe of discourse of shape points then varies from 0 (linear road) to 9 (extremely curved). Define the following membership functions:

$$\text{``Somewhat straight''} = \left\{ \frac{1}{0} + \frac{0.9}{1} + \frac{0.8}{2} + \frac{0.7}{3} + \frac{0.6}{4} + \frac{0.5}{5} + \frac{0.4}{6} + \frac{0.3}{7} + \frac{0.2}{8} + \frac{0.1}{9} \right\}$$

$$\text{``Curved''} = \left\{ \frac{0}{0} + \frac{0.1}{1} + \frac{0.2}{2} + \frac{0.3}{3} + \frac{0.4}{4} + \frac{0.5}{5} + \frac{0.6}{6} + \frac{0.7}{7} + \frac{0.8}{8} - \frac{0.9}{9} \right\}$$

(x_0, y_0) (x_i, y_i)

Linear road segment

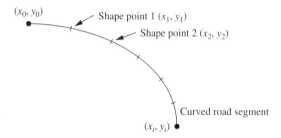
(x_0, y_0)

Shape point 1 (x_1, y_1)

Shape point 2 (x_2, y_2)

Curved road segment

(x_i, y_i)

FIGURE P5.25

Calculate the membership functions for the following phrases:
(a) Very curved
(b) Fairly curved ($= [\text{curved}]^{2/3}$)
(c) Very, very somewhat straight
(d) Not fairly curved and very, very somewhat straight

5.26. This problems deals with the voltages generated internally in switching power supplies. Embedded systems are often supplied 120 V AC for power. A "power supply" is required to convert this to a useful voltage (quite often +5 V DC). Some power supply designs employ a technique called "switching." This technique generates the appropriate voltages by storing and releasing the energy between inductors and capacitors. This problem characterizes two linguistic variables, high and low voltage, on the voltage range of 0 to 200 V AC:

$$\text{``High''} = \left\{ \frac{0}{0} + \frac{0}{25} + \frac{0}{50} + \frac{0.1}{75} + \frac{0.2}{100} + \frac{0.4}{125} + \frac{0.6}{150} + \frac{0.8}{175} + \frac{1}{200} \right\}$$

$$\text{``Medium''} = \left\{ \frac{0.2}{0} + \frac{0.4}{25} + \frac{0.6}{50} + \frac{0.8}{75} + \frac{1}{100} + \frac{0.8}{125} + \frac{0.6}{150} + \frac{0.4}{175} + \frac{0.2}{200} \right\}$$

Find the membership functions for the following phrases:
(a) Not very high
(b) Slightly medium and very high
(c) Very, very high or very, very medium

5.27. In risk assessment we deal with characterizing uncertainty in assessing the hazard to human health posed by various toxic chemicals. Because the pharmacokinetics of the human body are very difficult to explain for long-term chemical hazards, such as chronic exposure to lead or to cigarette smoke, hazards can sometimes be uncertain because of scarce data or uncertainty in the exposure patterns. Let us characterize hazard linguistically with two terms: "low" hazard and "high" hazard:

$$\text{"Low" hazard} = \left\{ \frac{0}{1} + \frac{0.3}{2} + \frac{0.8}{3} + \frac{0.1}{4} + \frac{0}{5} \right\}$$

$$\text{"High" hazard} = \left\{ \frac{0}{1} + \frac{0.1}{2} + \frac{0.2}{3} + \frac{0.8}{4} + \frac{0}{5} \right\}$$

Find the membership functions for the following linguistic expressions:

(a) Low hazard and not high hazard

(b) Very high hazard and not low hazard

(c) Low hazard or high hazard

5.28. In reference to car speeds we have the linguistic variables "fast" and "slow" for speed:

$$\text{"Fast"} = \left\{ \frac{0}{0} + \frac{0.1}{10} + \frac{0.2}{20} + \frac{0.3}{30} + \frac{0.4}{40} + \frac{0.5}{50} + \frac{0.6}{60} + \frac{0.7}{70} + \frac{0.8}{80} + \frac{0.9}{90} + \frac{1}{100} \right\}$$

$$\text{"Slow"} = \left\{ \frac{1}{0} + \frac{0.9}{10} + \frac{0.8}{20} + \frac{0.7}{30} + \frac{0.6}{40} + \frac{0.5}{50} + \frac{0.4}{60} + \frac{0.3}{70} + \frac{0.2}{80} + \frac{0.1}{90} + \frac{0}{100} \right\}$$

Using these variables, compute the membership function for the following linguistic terms:

(a) Very fast

(b) Very, very fast

(c) Highly fast (= minus very, very fast)

(d) Plus very fast

(e) Fairly fast (= $[\text{fast}]^{2/3}$)

(f) Not very slow and not very fast

(g) Slow or not very slow

5.29. For finding the volume of a cylinder, we need two parameters, namely, radius and height of the cylinder. When the radius is 7 centimeters and height is 12 centimeters, then the volume equals 1847.26 cubic centimeters (using volume $= \pi r^2 h$). Reduce the following rule to canonical form: IF x_1 is radius AND x_2 is height, THEN y is volume.

5.30. According to Boyle's law, for an ideal gas at constant temperature t, pressure is inversely proportional to volume, or volume is inversely proportional to pressure. When we consider different sets of pressures and volumes under the same temperature, we can apply the following rule: IF x_1 is $p_1 v_1$ AND x_2 is $p_2 v_2$, THEN t is a constant. Here p is pressure and v is volume of the gas considered. Reduce this rule to canonical form.

5.31. In Example 5.16 recalculate the response function shown in Fig. 5.15b using the following membership function shapes:

(a) two triangles for the input and two triangles for the output;

(b) two trapezoids for the input and two trapezoids for the output

5.32. In Example 5.17 recalculate the response function shown in Fig. 5.17d using the following membership function shapes for the inputs: (a) triangles for small, medium, large; (b) trapezoids for small, medium, large.

5.33. Repeat Example 5.16 using a *weighted sum defuzzifier* instead of the weighted average defuzzification, i.e., use $z = w_1 z_1 + w_2 z_2$ in Fig. 5.14. Do you get the same response surface as in Example 5.16? Why?

5.34. From thermodynamics it is known that for an ideal gas in an adiabatic reversible process

$$\frac{T_2}{T_1} = \left(\frac{P_2}{P_1}\right)^{\frac{\gamma - 1}{\gamma}}$$

where T_1 and T_2 are temperatures in kelvin (k) and P_1 and P_2 are pressures in bars and, for an ideal gas, the constant γ is

$$\gamma = 1 + \frac{R}{Cv} = 1.4$$

For this problem, T_1 will be fixed at 300 K and the fuzzy model will predict P_2 for the given input variables P_1 and T_2. In other words, we are interested in finding the final pressure, P_2, of the system if the temperature of the system is changed to T_2 from an original pressure equal to P_1. A real application could use a similar model built from experimental data to do a prediction on nonideal gases.

The rules used are

Rule 1 :   IF $P_1 = $ atmP AND $T_2 = $ lowT THEN $P_2 = $ lowP

Rule 2 :   IF $P_1 = $ atmP AND $T_2 = $ midT THEN $P_2 = $ lowP

Rule 1 :   IF $P_1 = $ lowP AND $T_2 = $ lowT THEN $P_2 = $ very highP

The rules and membership functions are based on values obtained from the known formula where

IF $P_1 = 1$ bar AND $T_2 = 410$ K THEN $P_2 = 3$ bar

IF $P_1 = 1$ bar AND $T_2 = 430$ K THEN $P_2 = 3.5$ bar

IF $P_1 = 2$ bar AND $T_2 = 420$ K THEN $P_2 = 6$ bar

Given the rule-base, the membership functions shown in Fig. P5.34, and the following pair of input values, $P_1 = 1.6$ bar and $T_2 = 415$ K, conduct a simulation to determine P_2 for the three inference methods of Mamdani, Sugeno, and Tsukamoto. For the Sugeno consequents use the ideal gas formula, given above.

For Mamdani and Sugeno, use the input membership functions in Fig. P5.34a.

For the Tsukamoto method, use the output membership functions shown in Fig. P5.34b with the same inputs as used in Fig. P5.34a.

5.35. In finding the Nusselt number (a dimensionless number for determining heat transfer) for an hexagonal cylinder in cross flow, there are two correlations (which are to be used as the consequent terms in a Sugeno inference method):

$$Nu_1 = 0.16Re^{0.638}Pr^{1/3} \quad 5000 < Re < 19,650$$

$$Nu_2 = 0.0385Re^{0.728}Pr^{1/3} \ Re > 19,650$$

Re is the Reynolds number and Pr is the Prandtl number.

The Nusselt number is a function of convective heat transfer (h), diameter of the hexagonal cylinder (D) over which cooling fluid travels, and the conductivity of the material (K):

$$Nu = \frac{hD}{K}$$

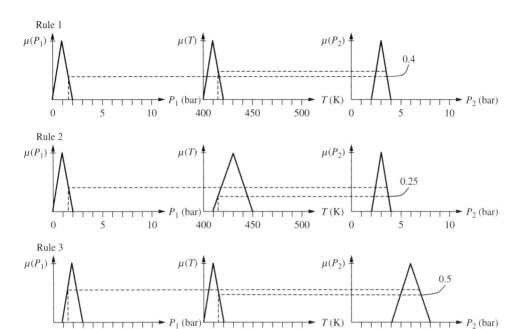

FIGURE P5.34*a*

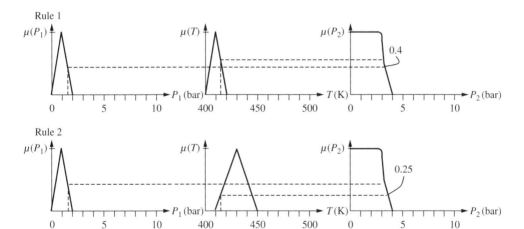

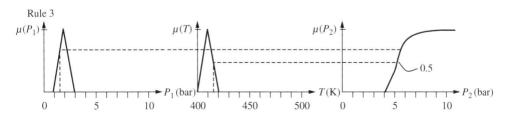

FIGURE P5.34*b*

Both Re and Pr can be fuzzy due to uncertainty in the variables in velocity. It would be convenient to find Nu (output) based on Re and Pr (inputs) without having to do all the calculations. More specifically, there is uncertainty in calculating the Reynolds number because velocity is not known exactly:

$$Re = \frac{\rho V D}{\mu}$$

where ρ is the density, V is the velocity, D is the characteristic length (or pipe diameter), and μ is the dynamic viscosity. And there is also uncertainty in the value for the Prandtl number due to its constituents

$$Pr = \frac{v}{\alpha}$$

where v is the kinematic viscosity and α is the specific gravity.

Calculation of Nu is very involved and the incorporation of a rule-base can be used to bypass these calculations; we have the following rules to govern this process:

If Re is high and Pr is low $\longrightarrow$ Then Nu is low

If Re is low and Pr is low $\longrightarrow$ Then Nu is low

If Re is high and Pr is high $\longrightarrow$ Then Nu is medium

If Re is low and Pr is high $\longrightarrow$ Then Nu is medium

For this problem, conduct a Mamdani and a Sugeno inference, based on the membership functions given in Figs P5.35a, b, and c, and use the following inputs:

$$Re = 19.65 \times 10^3$$

$$Pr = 275$$

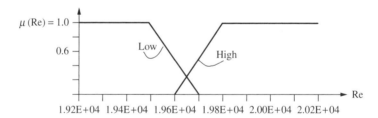

FIGURE P5.35a
Input for Reynolds number.

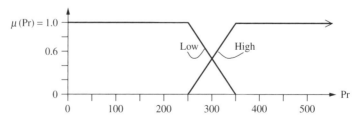

FIGURE P5.35b
Input for Prandtl number.

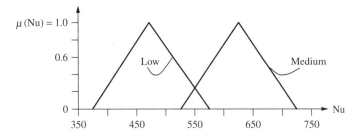

FIGURE P5.35c
Output for Mamdani inference.

Comment on the differences in the results.

CHAPTER
6

DEVELOPMENT OF MEMBERSHIP FUNCTIONS

So far as the laws of mathematics refer to reality, they are not certain. And so far as they are certain, they do not refer to reality.

Albert Einstein Theoretical physicist and Nobel Laureate
"Geometrie und Erfahrung," Lecture to Prussian Academy, 1921

The statement above, from Albert Einstein, attests to the fact that few things in real life are certain or can be conveniently reduced to the axioms of mathematical theories and models. A metaphorical expression that represents this idea is known as the "Law of Probable Dispersal," to wit, "Whatever it is that hits the fan will not be evenly distributed." As this enlightened law implies, most things in nature cannot be characterized with simple or convenient shapes or distributions. Membership functions characterize the fuzziness in a fuzzy set – whether the elements in the set are discrete or continuous – in a graphical form for eventual use in the mathematical formalisms of fuzzy set theory. But the shapes used to describe the fuzziness have very few restrictions indeed; some of these have been described in Chapters 1 and 4. Just as there are an infinite number of ways to characterize fuzziness, there are an infinite number of ways to graphically depict the membership functions that describe this fuzziness. This chapter describes a few procedures to develop these membership functions based on deductive intuition or numerical data; Chapter 7 develops this idea further with an explanation of additional procedures which build membership functions and deductive rules from measured observations of systems.

Since the membership function essentially embodies all fuzziness for a particular fuzzy set, its description is the essence of a fuzzy property or operation. Because of the importance of the "shape" of the membership function, a great deal of attention has been focused on development of these functions. This chapter describes, then illustrates, six

Fuzzy Logic with Engineering Applications, Second Edition   T. J. Ross
© 2004 John Wiley & Sons, Ltd   ISBNs: 0-470-86074-X (HB); 0-470-86075-8 (PB)

procedures that have been used to build membership functions. There are many more; references at the end of this chapter can be consulted on this topic.

MEMBERSHIP VALUE ASSIGNMENTS

There are possibly more ways to assign membership values or functions to fuzzy variables than there are to assign probability density functions to random variables [see Dubois and Prade, 1980]. This assignment process can be intuitive or it can be based on some algorithmic or logical operations. The following is a list of six straightforward methods described in the literature to assign membership values or functions to fuzzy variables. Each of these methods will be illustrated in simple examples in this chapter. The literature on this topic is rich with references, and a short list of those consulted is provided in the summary of this chapter.

1. Intuition
2. Inference
3. Rank ordering
4. Neural networks
5. Genetic algorithms
6. Inductive reasoning

Intuition

This method needs little or no introduction. It is simply derived from the capacity of humans to develop membership functions through their own innate intelligence and understanding. Intuition involves contextual and semantic knowledge about an issue; it can also involve linguistic truth values about this knowledge [see Zadeh, 1972]. As an example, consider the membership functions for the fuzzy variable temperature. Figure 6.1 shows various shapes on the universe of temperature as measured in units of degrees Celsius. Each curve is a membership function corresponding to various fuzzy variables, such as very cold, cold, normal, hot, and very hot. Of course, these curves are a function of context and the analyst developing them. For example, if the temperatures are referred to the range of human comfort we get one set of curves, and if they are referred to the range of safe operating temperatures for a steam turbine we get another set. However, the important character of these curves for purposes of use in fuzzy operations is the fact that they overlap. In numerous examples throughout the rest of this text we shall see that the precise shapes of

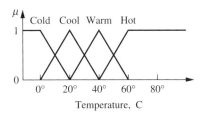

FIGURE 6.1
Membership functions for the fuzzy variable "temperature."

these curves are not so important in their utility. Rather, it is the approximate placement of the curves on the universe of discourse, the number of curves (partitions) used, and the overlapping character that are the most important ideas.

Inference

In the inference method we use knowledge to perform deductive reasoning. That is, we wish to deduce or infer a conclusion, given a body of facts and knowledge. There are many forms of this method documented in the literature, but the one we will illustrate here relates to our formal knowledge of geometry and geometric shapes, similar to ideas posed in Chapter 1.

In the identification of a triangle, let A, B, and C be the inner angles of a triangle, in the order $A \geq B \geq C \geq 0$, and let U be the universe of triangles, i.e.,

$$U = \{(A, B, C) \mid A \geq B \geq C \geq 0; \; A + B + C = 180°\} \tag{6.1}$$

We define a number of geometric shapes that we wish to be able to identify for any collection of angles fulfilling the constraints given in Eq. (6.1). For this purpose we will define the following five types of triangles:

$\underset{\sim}{I}$	Approximate isosceles triangle
$\underset{\sim}{R}$	Approximate right triangle
$\underset{\sim}{IR}$	Approximate isosceles *and* right triangle
$\underset{\sim}{E}$	Approximate equilateral triangle
$\underset{\sim}{T}$	Other triangles

We can infer membership values for all of these triangle types through the method of inference, because we possess knowledge about geometry that helps us to make the membership assignments. So we shall list this knowledge here to develop an algorithm to assist us in making these membership assignments for any collection of angles meeting the constraints of Eq. (6.1).

For the approximate isosceles triangle we have the following algorithm for the membership, again for the situation of $A \geq B \geq C \geq 0$ and $A + B + C = 180°$:

$$\mu_{\underset{\sim}{I}}(A, B, C) = 1 - \frac{1}{60°} \min(A - B, B - C) \tag{6.2}$$

So, for example, if $A = B$ or $B = C$, the membership value in the approximate isosceles triangle is $\mu_I = 1$; if $A = 120°$, $B = 60°$, and $C = 0°$, then $\mu_I = 0$. For a fuzzy right triangle, we have

$$\mu_{\underset{\sim}{R}}(A, B, C) = 1 - \frac{1}{90°} |A - 90°| \tag{6.3}$$

For instance, when $A = 90°$, the membership value in the fuzzy right triangle, $\mu_R = 1$, or when $A = 180°$, this membership vanishes, i.e., $\mu_R = 0$. For the case of an approximate isosceles *and* right triangle (there is only one of these in the crisp domain), we can find this membership function by taking the logical intersection (*and* operator) of the isosceles and right triangle membership functions, or

$$\underset{\sim}{IR} = \underset{\sim}{I} \cap \underset{\sim}{R}$$

which results in

$$\mu_{\underset{\sim}{IR}}(A, B, C) = \min[\mu_{\underset{\sim}{I}}(A, B, C), \mu_{\underset{\sim}{R}}(A, B, C)]$$

$$= 1 - \max\left[\frac{1}{60°}\min(A - B, B - C), \frac{1}{90°}|A - 90°|\right] \qquad (6.4)$$

For the case of a fuzzy equilateral triangle, the membership function is given by

$$\mu_{\underset{\sim}{E}}(A, B, C) = 1 - \frac{1}{180°}(A - C) \qquad (6.5)$$

For example, when $A = B = C$, the membership value is $\mu_{\underset{\sim}{E}}(A, B, C) = 1$; when $A = 180°$, the membership value vanishes, or $\mu_{\underset{\sim}{E}} = 0$. Finally, for the set of "all other triangles" (all triangular shapes other than $\underset{\sim}{I}$, $\underset{\sim}{R}$, and $\underset{\sim}{E}$) we simply invoke the complement of the logical union of the three previous cases (or, from De Morgan's principles (Eq. (2.13)), the intersection of the complements of the triangular shapes),

$$\underset{\sim}{T} = (\overline{\underset{\sim}{I} \cup \underset{\sim}{R} \cup \underset{\sim}{E}}) = \overline{\underset{\sim}{I}} \cap \overline{\underset{\sim}{R}} \cap \overline{\underset{\sim}{E}}$$

which results in

$$\mu_{\underset{\sim}{T}}(A, B, C) = \min\{1 - \mu_{\underset{\sim}{I}}(A, B, C), 1 - \mu_{\underset{\sim}{E}}(A, B, C), 1 - \mu_{\underset{\sim}{R}}(A, B, C)\}$$

$$= \frac{1}{180°}\min\{3(A - B), 3(B - C), 2|A - 90°|, A - C\} \qquad (6.6)$$

Example 6.1   **[Ross, 1995].** Define a specific triangle, as shown in Fig. 6.2, with these three ordered angles:

$$\{X : A = 85° \geq B = 50° \geq C = 45°, \text{where} A + B + C = 180°\}$$

The membership values for the fuzzy triangle shown in Fig. 6.2 for each of the fuzzy triangles types are determined from Eqs. (6.2)–(6.6), as listed here:

$$\mu_{\underset{\sim}{R}}(x) = 0.94$$

$$\mu_{\underset{\sim}{I}}(x) = 0.916$$

$$\mu_{\underset{\sim}{IR}}(x) = 0.916$$

$$\mu_{\underset{\sim}{E}}(x) = 0.7$$

$$\mu_{\underset{\sim}{T}}(x) = 0.05$$

Hence, it appears that the triangle given in Fig. 6.2 has the highest membership in the set of fuzzy right triangles, i.e., in $\underset{\sim}{R}$. Notice, however, that the triangle in Fig. 6.2 also has high membership in the isosceles triangle fuzzy set, and reasonably high membership in the equilateral fuzzy triangle set.

Rank Ordering

Assessing preferences by a single individual, a committee, a poll, and other opinion methods can be used to assign membership values to a fuzzy variable. Preference is determined

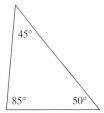

FIGURE 6.2
A specific triangle.

by pairwise comparisons, and these determine the ordering of the membership. More is provided on the area of rank ordering in Chapter 10, "Fuzzy Decision Making." This method is very similar to a relative preferences method developed by Saaty [1974].

> **Example 6.2 [Ross, 1995].** Suppose 1000 people respond to a questionnaire about their pairwise preferences among five colors, X = {red, orange, yellow, green, blue}. Define a fuzzy set as $\underset{\sim}{A}$ on the universe of colors "best color." Table 6.1 is a summary of the opinion survey. In this table, for example, out of 1000 people 517 preferred the color red to the color orange, 841 preferred the color orange to the color yellow, etc. Note that the color columns in the table represent an "antisymmetric" matrix. Such a matrix will be seen to relate to a *reciprocal relation,* which is introduced in Chapter 10. The total number of responses is 10,000 (10 comparisons). If the sum of the preferences of each color (row sum) is normalized to the total number of responses, a rank ordering can be determined as shown in the last two columns of the table.
>
> If the percentage preference (the percentage column of Table 6.1) of these colors is plotted to a normalized scale on the universe of colors in an ascending order on the color universe, the membership function for "best color" shown in Fig. 6.3 would result. Alternatively, the membership function could be formed based on the rank order developed (last column of Table 6.1).

Neural Networks

In this section we explain how a neural network can be used to determine membership functions. We first present a brief introduction to neural networks and then show how they can be used to determine membership functions.

A neural network is a technique that seeks to build an intelligent program (to implement intelligence) using models that simulate the working network of the neurons in

TABLE 6.1
Example in Rank Ordering

| | Number who preferred | | | | | | | |
	Red	Orange	Yellow	Green	Blue	Total	Percentage	Rank order
Red	–	517	525	545	661	2248	22.5	2
Orange	483	–	841	477	576	2377	23.8	1
Yellow	475	159	–	534	614	1782	17.8	4
Green	455	523	466	–	643	2087	20.9	3
Blue	339	424	386	357	–	1506	15	5
Total						10,000		

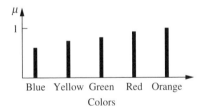

FIGURE 6.3
Membership function for best color.

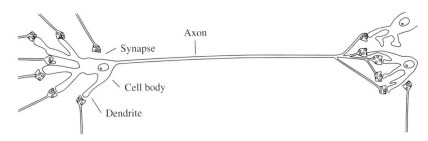

FIGURE 6.4
A simple schematic of a human neuron.

the human brain [Yamakawa, 1992; Hopfield, 1982; Hopfield and Tank, 1986]. A neuron, Fig. 6.4, is made up of several protrusions called dendrites and a long branch called the axon. A neuron is joined to other neurons through the dendrites. The dendrites of different neurons meet to form synapses, the areas where messages pass. The neurons receive the impulses via the synapses. If the total of the impulses received exceeds a certain threshold value, then the neuron sends an impulse down the axon where the axon is connected to other neurons through more synapses. The synapses may be excitatory or inhibitory in nature. An excitatory synapse adds to the total of the impulses reaching the neuron, whereas an inhibitory neuron reduces the total of the impulses reaching the neuron. In a global sense, a neuron receives a set of input pulses and sends out another pulse that is a function of the input pulses.

This concept of how neurons work in the human brain is utilized in performing computations on computers. Researchers have long felt that the neurons are responsible for the human capacity to learn, and it is in this sense that the physical structure is being emulated by a neural network to accomplish machine learning. Each computational unit computes some function of its inputs and passes the result to connected units in the network. The knowledge of the system comes out of the entire network of the neurons.

Figure 6.5 shows the analog of a neuron as a threshold element. The variables x_1, $x_2, \ldots, x_i, \ldots, x_n$ are the n inputs to the threshold element. These are analogous to impulses arriving from several different neurons to one neuron. The variables $w_1, w_2, \ldots, w_i, \ldots, w_n$ are the weights associated with the impulses/inputs, signifying the relative importance that is associated with the path from which the input is coming. When w_i is positive, input x_i acts as an excitatory signal for the element. When w_i is negative, input x_i acts as an inhibitory signal for the element. The threshold element sums the product of these inputs

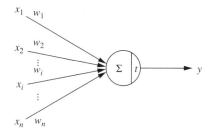

FIGURE 6.5
A threshold element as an analog to a neuron.

and their associated weights ($\sum w_i x_i$), compares it to a prescribed threshold value, and, if the summation is greater than the threshold value, computes an output using a nonlinear function (F). The signal output y (Fig. 6.5) is a nonlinear function (F) of the difference between the preceding computed summation and the threshold value and is expressed as

$$y = F\left(\sum w_i x_i - t\right) \tag{6.7}$$

where x_i signal input ($i = 1, 2, \ldots, n$)
         w_i weight associated with the signal input x_i
          t threshold level prescribed by user

         $F(s)$ is a nonlinear function, e.g., a sigmoid function $F(s) = \dfrac{1}{1 + e^{-s}}$

The nonlinear function, F, is a modeling choice and is a function of the type of output signal desired in the neural network model. Popular choices for this function are a sigmoid function, a step function, and a ramp function on the unit interval.

Figure 6.6 shows a simple neural network for a system with single-input signal x and a corresponding single-output signal $f(x)$. The first layer has only one element that has a single input, but the element sends its output to four other elements in the second layer. Elements shown in the second layer are all single-input, single-output elements. The third layer has only one element that has four inputs, and it computes the output for the system. This neural network is termed a ($1 \times 4 \times 1$) neural network. The numbers represent the number of elements in each layer of the network. The layers other than the first (input layer) and the last (output layer) layers constitute the set of hidden layers. (Systems can have more than three layers, in which case we would have more than one hidden layer.)

Neural systems solve problems by adapting to the nature of the data (signals) they receive. One of the ways to accomplish this is to use a training data set and a checking data set of input and output data/signals (x, y) (for a multiple-input, multiple-output system using a neural network, we may use input–output sets comprised of vectors ($\mathbf{x}_1, \mathbf{x}_2, \ldots, \mathbf{x}_n, \mathbf{y}_1, \mathbf{y}_2, \ldots, \mathbf{y}_n$)). We start with a random assignment of weights w^i_{jk} to the paths joining the elements in the different layers (Fig. 6.6). Then an input x from the training data set is passed through the neural network. The neural network computes a value ($f(x)_{\text{output}}$), which is compared with the actual value ($f(x)_{\text{actual}} = y$). The error measure E is computed from these two output values as

$$E = f(x)_{\text{actual}} - f(x)_{\text{output}} \tag{6.8}$$

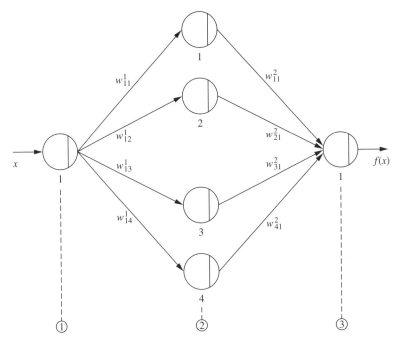

FIGURE 6.6
A simple $1 \times 4 \times 1$ neural network, where w^i_{jk} represents the weight associated with the path connecting the jth element of the ith layer to the kth element of the $(i + 1)$th layer.

This is the error measure associated with the last layer of the neural network (for Fig. 6.6); in this case the error measure E would be associated with the third layer in the neural network. Next we try to distribute this error to the elements in the hidden layers using a technique called back-propagation.

The error measure associated with the different elements in the hidden layers is computed as follows. Let E_j be the error associated with the jth element (Fig. 6.7). Let w_{nj} be the weight associated with the line from element n to element j and let I be the input to unit n. The error for element n is computed as

$$E_n = F'(I)w_{nj}E_j \tag{6.9}$$

where, for $F(I) = 1/(1 + e^{-I})$, the sigmoid function, we have

$$F'(I) = F(I)(1 - F(I)) \tag{6.10}$$

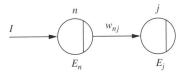

FIGURE 6.7
Distribution of error to different elements.

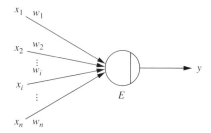

FIGURE 6.8
A threshold element with an error E associated with it.

Next the different weights w_{jk}^i connecting different elements in the network are corrected so that they can approximate the final output more closely. For updating the weights, the error measure on the elements is used to update the weights on the lines joining the elements.

For an element with an error E associated with it, as shown in Fig. 6.8, the associated weights may be updated as

$$w_i \text{ (new)} = w_i \text{ (old)} + \alpha E x_i \tag{6.11}$$

where $\quad \alpha =$ learning constant
$\quad E =$ associated error measure
$\quad x_i =$ input to the element

The input value x_i is passed through the neural network (now having the updated weights) again, and the errors, if any, are computed again. This technique is iterated until the error value of the final output is within some user-prescribed limits.

The neural network then uses the next set of input–output data. This method is continued for all data in the training data set. This technique makes the neural network simulate the nonlinear relation between the input–output data sets. Finally a checking data set is used to verify how well the neural network can simulate the nonlinear relationship.

For systems where we may have data sets of inputs and corresponding outputs, and where the relationship between the input and output may be highly nonlinear or not known at all, we may want to use fuzzy logic to classify the input and the output data sets broadly into different fuzzy classes. Furthermore, for systems that are dynamic in nature (the system parameters may change in a nondeterministic fashion) the fuzzy membership functions would have to be repeatedly updated. For these types of systems it is advantageous to use a neural network since the network can modify itself (by changing the weight assignments in the neural network) to accommodate the changes. Unlike symbolic learning algorithms, e.g., conventional expert systems [Luger and Stubblefield, 1989], neural networks do not learn by adding new rules to their knowledge base; they learn by modifying their overall structure. The lack of intuitive knowledge in the learning process is one of the major drawbacks of neural networks for use in cognitive learning.

Generation of membership functions using a neural network

We consider here a method by which fuzzy membership functions may be created for fuzzy classes of an input data set [Takagi and Hayashi, 1991]. We select a number of input data

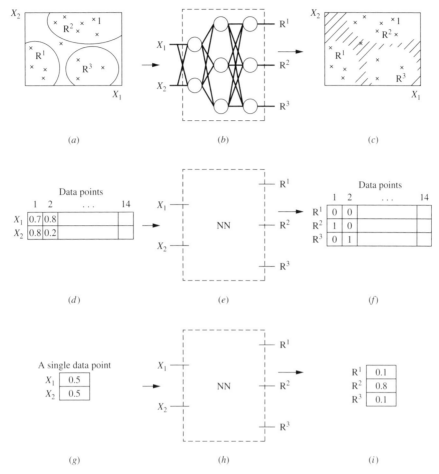

FIGURE 6.9
Using a neural network to determine membership functions [Takagi and Hayashi, 1991].

TABLE 6.2
Variables describing the data points to be used as a training data set

Data point	1	2	3	4	5	6	7	8	9	10
x_1	0.05	0.09	0.12	0.15	0.20	0.75	0.80	0.82	0.90	0.95
x_2	0.02	0.11	0.20	0.22	0.25	0.75	0.83	0.80	0.89	0.89

values and divide them into a training data set and a checking data set. The training data set is used to train the neural network. Let us consider an input training data set as shown in Fig. 6.9a. Table 6.2 shows the coordinate values of the different data points considered (e.g., crosses in Fig. 6.9a). The data points are expressed with two coordinates each, since the data shown in Fig. 6.9a represent a two-dimensional problem. The data points are first divided into different classes (Fig. 6.9a) by conventional clustering techniques (these are explained in Chapter 11).

As shown in Fig. 6.9*a* the data points have been divided into three regions, or classes, R^1, R^2, and R^3. Let us consider data point 1, which has input coordinate values of $x_1 = 0.7$ and $x_2 = 0.8$ (Fig. 6.9*d*). As this is in region R_2, we assign to it a complete membership of one in class R_2 and zero membership in classes R_1 and R_3 (Fig. 6.9*f*). Similarly, the other data points are assigned membership values of unity for the classes they belong to initially. A neural network is created (Figs. 6.9*b, e, h*) that uses the data point marked 1 and the corresponding membership values in different classes for training itself to simulate the relationship between coordinate locations and the membership values. Figure 6.9*c* represents the output of the neural network, which classifies data points into one of the three regions. The neural network then uses the next set of data values (e.g., point 2) and membership values to train itself further as seen in Fig. 6.9*d*. This repetitive process is continued until the neural network can simulate the entire set of input–output (coordinate location–membership value) values. The performance of the neural network is then checked using the checking data set. Once the neural network is ready, its final version (Fig. 6.9*h*) can be used to determine the membership values (function) of any input data (Fig. 6.9*g*) in the different regions (Fig. 6.9*i*).

Notice that the points shown in the table in Fig. 6.9*i* are actually the membership values in each region for the data point shown in Fig. 6.9*g*. These could be plotted as a membership function, as shown in Fig. 6.10. A complete mapping of the membership of different data points in the different fuzzy classes can be derived to determine the overlap of the different classes (the hatched portion in Fig. 6.9*c* shows the overlap of the three fuzzy classes). These steps will become clearer as we go through the computations in the following example.

Example 6.3. Let us consider a system that has 20 data points described in two-dimensional format (two variables) as shown in Tables 6.2 and 6.3. We have placed these data points in two fuzzy classes, R_1 and R_2, using a clustering technique (see Chapter 11). We would like to form a neural network that can determine the membership values of any data point in the two classes. We would use the data points in Table 6.2 to train the neural network and the

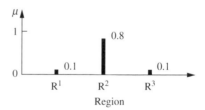

FIGURE 6.10
Membership function for data point $(X_1, X_2 = (0.5, 0.5))$.

TABLE 6.3
Variables describing the data points to be used as a checking data set

Data point	11	12	13	14	15	16	17	18	19	20
x_1	0.09	0.10	0.14	0.18	0.22	0.77	0.79	0.84	0.94	0.98
x_2	0.04	0.10	0.21	0.24	0.28	0.78	0.81	0.82	0.93	0.99

TABLE 6.4
Membership values of the data points in the training and checking data sets to be used for training
and checking the performance of the neural network

Data points	1 & 11	2 & 12	3 & 13	4 & 14	5 & 15	6 & 16
R_1	1.0	1.0	1.0	1.0	1.0	0.0
R_2	0.0	0.0	0.0	0.0	0.0	1.0
	7 & 17	8 & 18	9 & 19	10 & 20		
	0.0	0.0	0.0	0.0		
	1.0	1.0	1.0	1.0		

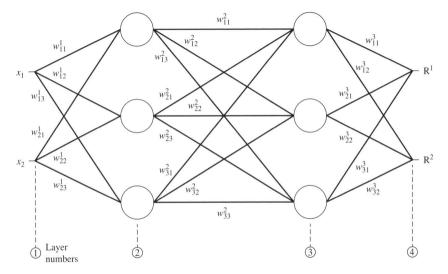

FIGURE 6.11
The $[2 \times 3 \times 3 \times 2]$ neural network to be trained for the data set of Example 6.3.

data points in Table 6.3 to check its performance. The membership values in Table 6.4 are to
be used to train and check the performance of the neural network. The data points that are to
be used for training and checking the performance of the neural network have been assigned
membership values of unity for the classes into which they have been originally assigned, as
seen in Table 6.4.

   We select a $2 \times 3 \times 3 \times 2$ neural network to simulate the relationship between the data
points and their membership in the two fuzzy sets, R_1 and R_2 (Fig. 6.11). The coordinates x_1
and x_2 for each data point are used as the input values, and the corresponding membership
values in the two fuzzy classes for each data point are the output values for the neural
network.

   Table 6.5 shows the initial quasi-random values that have been assigned to the different
weights connecting the paths between the elements in the layers in the network shown in
Fig. 6.11. We take the first data point ($x_1 = 0.05$, $x_2 = 0.02$) as the input to the neural network.
We will use Eq. (6.7) in the form

$$O = \frac{1}{1 + \exp[-\left(\sum x_i w_i - t\right)]} \qquad (6.12)$$

TABLE 6.5
The initial quasi-random values that have been
assigned to the different weights connecting
the paths between the elements in the layers in
the network of Fig. 6.11

$w_{11}^1 = 0.5$	$w_{11}^2 = 0.10$	$w_{11}^3 = 0.30$
$w_{12}^1 = 0.4$	$w_{12}^2 = 0.55$	$w_{12}^3 = 0.35$
$w_{13}^1 = 0.1$	$w_{13}^2 = 0.35$	$w_{21}^3 = 0.35$
$w_{21}^1 = 0.2$	$w_{21}^2 = 0.20$	$w_{22}^3 = 0.25$
$w_{22}^1 = 0.6$	$w_{22}^2 = 0.45$	$w_{31}^3 = 0.45$
$w_{23}^1 = 0.2$	$w_{23}^2 = 0.35$	$w_{32}^3 = 0.30$
	$w_{31}^2 = 0.25$	
	$w_{32}^2 = 0.15$	
	$w_{33}^2 = 0.60$	

where O = output of the threshold element computed using the sigmoidal function
$\quad\quad\quad x_i$ = inputs to the threshold element ($i = 1, 2, \ldots, n$)
$\quad\quad\quad w_i$ = weights attached to the inputs
$\quad\quad\quad t$ = threshold for the element

First iteration: We start off with the first iteration in training the neural network using
Eq. (6.12) to determine the outputs of the different elements by calculating the outputs for each
of the neural network layers. We select a threshold value of $t = 0$.
Outputs for the second layer:

$$O_1^2 = \frac{1}{1 + \exp\{-[(0.05 \times 0.50) + (0.02 \times 0.20) - 0.0]\}} = 0.507249$$

$$O_2^2 = \frac{1}{1 + \exp\{-[(0.05 \times 0.40) + (0.02 \times 0.60) - 0.0]\}} = 0.507999$$

$$O_3^2 = \frac{1}{1 + \exp\{-[(0.05 \times 0.10) + (0.02 \times 0.20) - 0.0]\}} = 0.502250$$

Outputs for the third layer:

$$O_1^3 = \frac{1}{1 + \exp\{-[(0.507249 \times 0.10) + (0.507999 \times 0.20) + (0.502250 \times 0.25) - 0.0]\}}$$
$$= 0.569028$$

$$O_2^3 = \frac{1}{1 + \exp\{-[(0.507249 \times 0.55) + (0.507999 \times 0.45) + (0.502250 \times 0.15) - 0.0]\}}$$
$$= 0.641740$$

$$O_3^3 = \frac{1}{1 + \exp\{-[(0.507249 \times 0.35) + (0.507999 \times 0.35) + (0.502250 \times 0.60) - 0.0]\}}$$
$$= 0.658516$$

Outputs for the fourth layer:

$$O_1^4 = \frac{1}{1 + \exp\{-[(0.569028 \times 0.30) + (0.641740 \times 0.35) + (0.658516 \times 0.45) - 0.0]\}}$$
$$= 0.666334$$

$$O_2^4 = \frac{1}{1 + \exp\{-[(0.569028 \times 0.35) + (0.641740 \times 0.25) + (0.658516 \times 0.30) - 0.0]\}}$$
$$= 0.635793$$

Determining errors:

$$R_1 : E_1^4 = O_1^4{}_{\text{actual}} - O_1^4 = 1.0 - 0.666334 = 0.333666$$
$$R_2 : E_2^4 = O_2^4{}_{\text{actual}} - O_2^4 = 0.0 - 0.635793 = -0.635793$$

Now that we know the final errors for the neural network for the first iteration, we distribute this error to the other nodes (elements) in the network using Eqs. (6.9)–(6.10) in the form

$$E_n = O_n(1 - O_n) \sum_j w_{nj} E_j \tag{6.13}$$

Assigning errors: First, we assign errors to the elements in the third layer,

$$E_1^3 = 0.569028(1.0 - 0.569028)[(0.30 \times 0.333666) + (0.35 \times (-0.635793))] = -0.030024$$
$$E_2^3 = 0.641740(1.0 - 0.641740)[(0.35 \times 0.333666) + (0.25 \times (-0.635793))] = -0.009694$$
$$E_3^3 = 0.658516(1.0 - 0.658516)[(0.45 \times 0.333666) + (0.30 \times (-0.635793))] = -0.009127$$

and then assign errors to the elements in the second layer,

$$E_1^2 = 0.507249(1.0 - 0.507249)[(0.10 \times (-0.030024)) + (0.55 \times (-0.009694))$$
$$+ (0.35 \times (-0.009127))] = -0.002882$$
$$E_2^2 = 0.507999(1.0 - 0.507999)[(0.20 \times (-0.030024)) + (0.45 \times (-0.009694))$$
$$+ (0.35 \times (-0.009127))] = -0.003390$$
$$E_3^2 = 0.502250(1.0 - 0.502250)[(0.25 \times (-0.030024)) + (0.15 \times (-0.009694))$$
$$+ (0.60 \times (-0.009127))] = -0.003609$$

Now that we know the errors associated with each element in the network we can update the weights associated with these elements so that the network approximates the output more closely. To update the weights we use Eq. (6.11) in the form

$$w_{jk}^i(\text{new}) = w_{jk}^i(\text{old}) + \alpha E_k^{i+1} x_{jk} \tag{6.14}$$

where w_{jk}^i = represents the weight associated with the path connecting the jth element of the ith layer to the kth element of the $(i + 1)$th layer

α = learning constant, which we will take as 0.3 for this example

E_k^{i+1} = error associated with the kth element of the $(i + 1)$th layer

x_{jk} = input from the jth element in the ith layer to the kth element in the $(i + 1)$th layer (O_j^i)

Updating weights: We will update the weights connecting elements in the third and the fourth layers,

$$w_{11}^3 = 0.30 + 0.3 \times 0.333666 \times 0.569028 = 0.356960$$

$$w_{21}^3 = 0.35 + 0.3 \times 0.333666 \times 0.641740 = 0.414238$$

$$w_{31}^3 = 0.45 + 0.3 \times 0.333666 \times 0.658516 = 0.515917$$

$$w_{12}^3 = 0.35 + 0.3 \times (-0.635793) \times 0.569028 = 0.241465$$

$$w_{22}^3 = 0.25 + 0.3 \times (-0.635793) \times 0.641740 = 0.127596$$

$$w_{32}^3 = 0.30 + 0.3 \times (-0.635793) \times 0.658516 = 0.174396$$

then update weights connecting elements in the second and the third layers,

$$w_{11}^2 = 0.10 + 0.3 \times (-0.030024) \times 0.507249 = 0.095431$$

$$w_{21}^2 = 0.20 + 0.3 \times (-0.030024) \times 0.507999 = 0.195424$$

$$w_{31}^2 = 0.25 + 0.3 \times (-0.030024) \times 0.502250 = 0.245476$$

$$w_{12}^2 = 0.55 + 0.3 \times (-0.009694) \times 0.507249 = 0.548525$$

$$w_{22}^2 = 0.45 + 0.3 \times (-0.009694) \times 0.507999 = 0.448523$$

$$w_{32}^2 = 0.15 + 0.3 \times (-0.009694) \times 0.502250 = 0.148540$$

$$w_{13}^2 = 0.35 + 0.3 \times (-0.009127) \times 0.507249 = 0.348611$$

$$w_{23}^2 = 0.35 + 0.3 \times (-0.009127) \times 0.507999 = 0.348609$$

$$w_{33}^2 = 0.60 + 0.3 \times (-0.009127) \times 0.502250 = 0.598625$$

and then, finally, update weights connecting elements in the first and the second layers,

$$w_{11}^1 = 0.50 + 0.3 \times (-0.002882) \times 0.05 = 0.499957$$

$$w_{12}^1 = 0.40 + 0.3 \times (-0.003390) \times 0.05 = 0.399949$$

$$w_{13}^1 = 0.10 + 0.3 \times (-0.003609) \times 0.05 = 0.099946$$

$$w_{21}^1 = 0.20 + 0.3 \times (-0.002882) \times 0.02 = 0.199983$$

$$w_{22}^1 = 0.60 + 0.3 \times (-0.003390) \times 0.02 = 0.599980$$

$$w_{23}^1 = 0.20 + 0.3 \times (-0.003609) \times 0.02 = 0.199978$$

Now that all the weights in the neural network have been updated, the input data point $(x_1 = 0.05, x_2 = 0.02)$ is again passed through the neural network. The errors in approximating the output are computed again and redistributed as before. This process is continued until the errors are within acceptable limits. Next, the second data point $(x_1 = 0.09, x_2 = 0.11,$ Table 6.2) and the corresponding membership values $(R^1 = 1, R^2 = 0,$ Table 6.4) are used to train the network. This process is continued until all the data points in the *training* data set (Table 6.2) are used. The performance of the neural network (how closely it can predict the value of the membership of the data point) is then checked using the data points in the *checking* data set (Table 6.3).

Once the neural network is trained and verified to be performing satisfactorily, it can be used to find the membership of any other data points in the two fuzzy classes. A complete mapping of the membership of different data points in the different fuzzy classes can be derived to determine the overlap of the different classes (R_1 and R_2).

Genetic Algorithms

As in the previous section we will first provide a brief introduction to genetic algorithms and then show how these can be used to determine membership functions. In the previous section we introduced the concept of a neural network. In implementing a neural network algorithm, we try to recreate the working of neurons in the human brain. In this section we introduce another class of algorithms, which use the concept of Darwin's theory of evolution. Darwin's theory basically stressed the fact that the existence of all living things is based on the rule of "survival of the fittest." Darwin also postulated that new breeds or classes of living things come into existence through the processes of reproduction, crossover, and mutation among existing organisms [Forrest, 1993].

These concepts in the theory of evolution have been translated into algorithms to search for solutions to problems in a more "natural" way. First, different possible solutions to a problem are created. These solutions are then tested for their performance (i.e., how good a solution they provide). Among all possible solutions, a fraction of the good solutions is selected, and the others are eliminated (survival of the fittest). The selected solutions undergo the processes of reproduction, crossover, and mutation to create a new generation of possible solutions (which are expected to perform better than the previous generation). This process of production of a new generation and its evaluation is repeated until there is convergence within a generation. The benefit of this technique is that it searches for a solution from a broad spectrum of possible solutions, rather than restrict the search to a narrow domain where the results would be normally expected. Genetic algorithms try to perform an intelligent search for a solution from a nearly infinite number of possible solutions.

In the following material we show how the concepts of genetics are translated into a search algorithm [Goldberg, 1989]. In a genetic algorithm, the parameter set of the problem is coded as a finite string of bits. For example, given a set of two-dimensional data $((x, y)$ data points), we want to fit a linear curve (straight line) through the data. To get a linear fit, we encode the parameter set for a line ($y = C_1 x + C_2$) by creating independent bit strings for the two unknown constants C_1 and C_2 (parameter set describing the line) and then join them (concatenate the strings). The bit strings are combinations of zeros and ones, which represent the value of a number in binary form. An n-bit string can accommodate all integers up to the value $2^n - 1$. For example, the number 7 requires a 3-bit string, i.e., $2^3 - 1 = 7$, and the bit string would look like "111," where the first unit digit is in the 2^2 place $(= 4)$, the second unit digit is in the 2^1 place $(= 2)$, and the last unit digit is in the 2^0 place $(= 1)$; hence, $4 + 2 + 1 = 7$. The number 10 would look like "1010," i.e., $2^3 + 2^1 = 10$, from a 4-bit string. This bit string may be mapped to the value of a parameter, say $C_i, i = 1, 2$, by the mapping

$$C_i = C_{\min} + \frac{b}{2^L - 1} \left(C_{\max_i} - C_{\min_i} \right) \tag{6.15}$$

where "b" is the number in decimal form that is being represented in binary form (e.g., 152 may be represented in binary form as 10011000), L is the length of the bit string (i.e., the number of bits in each string), and $C_{\max}$ and $C_{\min}$ are user-defined constants between which C_1 and C_2 vary linearly. The parameters C_1 and C_2 depend on the problem. The length of the bit strings is based on the handling capacity of the computer being used, i.e., on how long a string (strings of each parameter are concatenated to make one long string representing the whole parameter set) the computer can manipulate at an optimum speed.

All genetic algorithms contain three basic operators: reproduction, crossover, and mutation, where all three are analogous to their namesakes in genetics. Let us consider the overall process of a genetic algorithm before trying to understand the basic processes.

First, an initial population of n strings (for n parameters) of length L is created. The strings are created in a random fashion, i.e., the values of the parameters that are coded in the strings are random values (created by randomly placing the zeros and ones in the strings). Each of the strings is decoded into a set of parameters that it represents. This set of parameters is passed through a numerical model of the problem space. The numerical model gives out a solution based on the input set of parameters. On the basis of the quality of this solution, the string is assigned a fitness value. The fitness values are determined for each string in the entire population of strings. With these fitness values, the three genetic operators are used to create a new generation of strings, which is expected to perform better than the previous generations (better fitness values). The new set of strings is again decoded and evaluated, and a new generation is created using the three basic genetic operators. This process is continued until convergence is achieved within a population.

Among the three genetic operators, reproduction is the process by which strings with better fitness values receive correspondingly *better copies* in the new generation, i.e., we try to ensure that better solutions persist and contribute to better offspring (new strings) during successive generations. This is a way of ensuring the ''survival of the fittest'' strings. Because the total number of strings in each generation is kept a constant (for computational economy and efficiency), strings with lower fitness values are eliminated.

The second operator, crossover, is the process in which the strings are able to mix and match their desirable qualities in a random fashion. After reproduction, crossover proceeds in three simple steps. First, two new strings are selected at random (Fig. 6.12a). Second, a random location in both strings is selected (Fig. 6.12b). Third, the portions of the strings to the right of the randomly selected location in the two strings are exchanged (Fig. 6.13c). In this way information is exchanged between strings, and portions of high-quality solutions are exchanged and combined.

Reproduction and crossover together give genetic algorithms most of their searching power. The third genetic operator, mutation, helps to increase the searching power. In order to understand the need for mutation, let us consider the case where reproduction or crossover may not be able to find an optimum solution to a problem. During the creation of a generation it is possible that the entire population of strings is missing a vital bit of information (e.g., none of the strings has a one at the fourth location) that is important for determining the correct or the most nearly optimum solution. Future generations that would be created using reproduction and crossover would not be able to alleviate this problem. Here mutation becomes important. Occasionally, the value at a certain string location is changed, i.e., if there is a one originally at a location in the bit string, it is changed to a zero, or vice versa. Mutation thus ensures that the vital bit of information is introduced into the generation. Mutation, as it does in nature, takes place very rarely, on the order of once in a thousand bit string locations (a suggested mutation rate is 0.005/bit/generation [Forrest, 1993]).

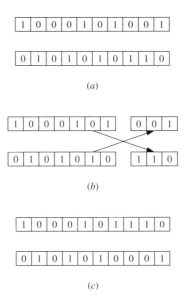

(a)

(b)

(c)

FIGURE 6.12
Crossover in strings. (a) Two strings are selected at random to be mated; (b) a random location in the strings is located (here the location is before the last three bit locations); and (c) the string portions following the selected location are exchanged.

Let us now consider an example that shows how a line may be fit through a given data set using a genetic algorithm.

Example 6.4. Let us consider the data set in Table 6.6. For performing a line ($y = C_1 x + C_2$) fit, as mentioned earlier, we first encode the parameter set (C_1, C_2) in the form of bit strings. Bit strings are created with random assignment of ones and zeros at different bit locations. We start with an initial population of four strings (Table 6.7a, column 2). The strings are 12 bits in length. The first 6 bits encode the parameter C_1, and the next 6 bits encode the parameter C_2. Table 6.7a, columns 3 and 5, shows the decimal equivalent of their binary coding. These binary values for C_1 and C_2 are then mapped into values relevant to the problem using Eq. (6.15). We assume that the minimum value to which we would expect C_1 or C_2 to go would be -2 and the maximum would be 5 (these are arbitrary values – any other values could just as easily have been chosen). Therefore, for Eq. (6.15), $C_{\min i} = -2$ and $C_{\max i} = 5$. Using these values, we compute C_1 and C_2 (Table 6.7a, columns 4 and 6). The values shown in Table 6.7a, columns 7, 8, 9, and 10, are the values computed using the equation $y = C_1 x + C_2$, using the

TABLE 6.6
Data set through which a line fit is required

Data number	x	y'
1	1.0	1.0
2	2.0	2.0
3	4.0	4.0
4	6.0	6.0

TABLE 6.7a
First iteration using a genetic algorithm, Example 6.4

(1) String number	(2) String	(3) C_1 (bin.)	(4) C_1	(5) C_2 (bin.)	(6) C_2	(7) y_1	(8) y_2	(9) y_3	(10) y_4	(11) $f(x) = 400 - \sum(y_i - y_r)^2$	(12) Expected count = f/f_{av}	(13) Actual count
1	000111 010100	7	−1.22	20	0.22	−1.00	−2.22	−3.44	−7.11	147.49	0.48	0
2	010010 001100	18	0.00	12	−0.67	−0.67	−0.67	−0.67	−0.67	332.22	1.08	1
3	010101 101010	21	0.33	42	2.67	3.00	3.33	5.00	4.67	391.44	1.27	2
4	100100 001001	36	2.00	9	−1.00	1.00	3.00	3.67	11.00	358.00	1.17	1
—	—	—	—	—	—	—	—	—	—	Sum 1229.15	—	—
—	—	—	—	—	—	—	—	—	—	Average 307.29	—	—
—	—	—	—	—	—	—	—	—	—	Maximum 391.44	—	—

TABLE 6.7b
Second iteration using a genetic algorithm, Example 6.4

(1) Selected strings	(2) New strings	(3) C_1 (bin.)	(4) C_1	(5) C_2 (bin.)	(6) C_2	(7) y_1	(8) y_2	(9) y_3	(10) y_4	(11) $f(x) = 400 - \sum(y_i - y_r)^2$	(12) Expected count = f/f_{av}	(13) Actual count
0101|01 101010	010110 001100	22	0.44	12	−0.67	−0.22	0.22	1.11	2.00	375.78	1.15	1
0100|10 001100	010001 101010	17	−0.11	42	2.67	2.56	2.44	2.22	2.00	380.78	1.17	2
010101 101|010	010101 101001	21	0.33	41	2.56	2.89	3.22	3.89	4.56	292.06	0.90	1
100100 001|001	100100 001010	36	2.0	10	−0.89	1.11	3.11	7.11	11.11	255.73	0.78	0
—	—	—	—	—	—	—	—	—	—	Sum 1304.35	—	—
—	—	—	—	—	—	—	—	—	—	Average 326.09	—	—
—	—	—	—	—	—	—	—	—	—	Maximum 380.78	—	—

values of C_1 and C_2 from columns 4 and 6, respectively, for different values of x as given in Table 6.6. These computed values for the y are compared with the correct values (Table 6.6), and the square of the errors in estimating the y is estimated for each string. This summation is subtracted from a large number (400 in this problem) (Table 6.7a, column 11) to convert the problem into a maximization problem. The values in Table 6.7a, column 11, are the fitness values for the four strings. These fitness values are added. Their average is also computed. The fitness value of each string is divided by the average fitness value of the whole population of strings to give an estimate of the relative fitness of each string (Table 6.7a, column 12). This measure also acts as a guide as to which strings are eliminated from consideration for the next generation and which string "gets reproduced" in the next generation. In this problem a cutoff value of 0.80 (relative fitness) has been used for the acceptability of a string succeeding into the next generation. Table 6.7a, column 13, shows the number of copies of each of the four strings that would be used to create the next generation of strings.

   Table 6.7b is a continuation of Table 6.7a. The first column in Table 6.7b shows the four strings selected from the previous generation aligned for crossover at the locations shown in the strings in the column. After crossover, the new strings generated are shown in Table 6.7b, column 2. These strings undergo the same process of decoding and evaluation as the previous generation. This process is shown in Table 6.7b, columns 3–13. We notice that the average fitness of the second generation is greater than that of the first generation of strings.

   The process of generation of strings and their evaluation is continued until we get a convergence to the solution within a generation.

Computing membership functions using genetic algorithms

Genetic algorithms as just described can be used to compute membership functions [Karr and Gentry, 1993]. Given some functional mapping for a system, some membership functions and their shapes are assumed for the various fuzzy variables defined for a problem. These membership functions are then coded as bit strings that are then concatenated. An evaluation (fitness) function is used to evaluate the fitness of each set of membership functions (parameters that define the functional mapping). This procedure is illustrated for a simple problem in the next example.

Example 6.5. Let us consider that we have a single-input (x), single-output (y) system with input–output values as shown in Table 6.8. Table 6.9 shows a functional mapping for this system between the input (x) and the output (y).

   In Table 6.9 we see that each of the variables x and y makes use of two fuzzy classes (x uses S (small) and L (large); y uses L (large) and VL (very large)). The functional mapping tells us that a *small x* maps to a *small y*, and a *large x* maps to a *very large y*. We assume that the range of the variable x is [0, 5] and that that of y is [0, 25]. We assume that each membership function has the shape of a right triangle, as shown in Fig. 6.13.

   The membership function on the right side of Fig. 6.13 is constrained to have the right-angle wedge at the upper limit of the range of the fuzzy variable. The membership function on the left side is constrained to have the right-angle wedge on the lower limit of the range of the fuzzy variable. It is intuitively obvious that under the foregoing constraints the only thing needed to describe the shape and position of the membership function fully is the length of the base of the right-triangle membership functions. We use this fact in encoding the membership functions as bit strings.

   The unknown variables in this problem are the lengths of the bases of the four membership functions (x(S, L) and y(S, VL)). We use 6-bit binary strings to define the base of each of the membership functions. (The binary values are later mapped to decimal values using Eq. (6.15).) These strings are then concatenated to give us a 24-bit (6×4) string. As shown in Table 6.10a, column 1, we start with an initial population of four strings. These are decoded to the binary values of the variables as shown in Table 6.10a, columns 2, 3, 4, and 5. The binary values are mapped to decimal values for the fuzzy variables using Eq. (6.15)(Table 6.10a,

TABLE 6.8
Data for a single-input, single-output system

x	1	2	3	4	5
y	1	4	9	16	25

TABLE 6.9
Functional mapping for the system

x	S	L
y	S	VL

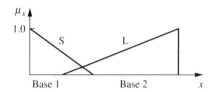

FIGURE 6.13
Membership functions for the input variables are assumed to be right triangles.

columns 6, 7, 8, and 9). For the fuzzy variable x (range $x = 0, 5$) we use $C_{min} = 0$ and $C_{max} = 5$ for both the membership functions S (Small) and L (Large). For the fuzzy variable y (range $y = 0, 25$) we use $C_{min} = 0$ and $C_{max} = 25$.

The physical representation of the first string is shown in Fig. 6.14. In this figure the base values are obtained from Table 6.10a, columns 6, 7, 8, and 9. So, for example, the base values for the x variable for string number 1 are 0.56 and $5 - 1.59 = 3.41$, and the base values for the y variable are 8.73 and $25 - 20.24 = 4.76$. To determine the fitness of the combination of membership functions in each of the strings, we want a measure of the square of the errors that are produced in estimating the value of the outputs y, given the inputs x from Table 6.8. Figure 6.14 shows how the value of the output y can be computed graphically from the membership functions for string number 1 in Table 6.10a. For example, for $x = 4$ we see that the membership of x in the fuzzy class Large is 0.37. Referring to the rules in Table 6.9, we see that if x is Large then y is Very Large. Therefore, we look for the value in the fuzzy class Very Large (VL) of fuzzy variable y that has a membership of 0.37. We determine this to be equal to 12.25. The corresponding actual value for y is 16 (Table 6.8). Therefore, the squared error is $(16 - 12.25)^2 = 14.06$. Columns 10, 11, 12, 13, and 14 of Table 6.10a show the values computed for y using the respective membership functions. Table 6.10a, column 15, shows the sum of the squared errors subtracted from 1000 (this is done to convert the fitness function from a minimization problem to a maximization problem). Table 6.10a, column 15, thus shows the fitness values for the four strings. We find the sum of all the fitness values in the generation and the average fitness of the generation. The average fitness of the generation is used to determine the relative fitness of the strings in the generation, as seen in Table 6.10a, column 16. These relative fitness values are used to determine which strings are to be eliminated and which string gets how many copies to make the next generation of strings. In this problem a cutoff value of 0.75 (relative fitness) has been used for the acceptability of a string propagating into the next generation. Table 6.10a, column 17, shows the number of copies of each of the four strings that would be used to create the next generation of strings.

Table 6.10b is a continuation of Table 6.10a. The first column in Table 6.10b shows the four strings selected from the previous generation aligned for crossover at the locations

TABLE 6.10a

First iteration using a genetic algorithm for determining optimal membership functions

String number	(1) String	(2) base 1 (bin)	(3) base 2 (bin)	(4) base 3 (bin)	(5) base 4 (bin)	(6) base 1	(7) base 2	(8) base 3	(9) base 4	(10) y' ($x=1$)	(11) y' ($x=2$)	(12) y' ($x=3$)	(13) y' ($x=4$)	(14) y' ($x=5$)	(15) $1000 - \sum(y_i - y_r)^2$	(16) Expected count = f/f_{av}	(17) Actual count
1	000111 010100 010110 110011	7	20	22	51	0.56	1.59	8.73	20.24	0	0	0	12.25	25	887.94	1.24	1
2	010010 001100 101100 100110	18	12	44	38	1.43	0.95	17.46	15.08	12.22	0	0	0	25	521.11	0.73	0
3	010101 101010 001101 101000	21	42	13	40	1.67	3.33	5.16	15.87	3.1	10.72	15.48	20.24	25	890.46	1.25	2
4	100100 001001 101100 100011	36	9	44	35	2.86	0.71	17.46	13.89	6.98	12.22	0	0	25	559.67	0.78	1
														Sum	2859.18		
														Average	714.80		
														Maximum	890.46		

TABLE 6.10b

Second iteration using a genetic algorithm for determining optimal membership functions

(1) Selected strings	(2) New Strings	(3) base 1 (bin)	(4) base 2 (bin)	(5) base 3 (bin)	(6) base 4 (bin)	(7) base 1	(8) base 2	(9) base 3	(10) base 4	(11) y' ($x=1$)	(12) y' ($x=2$)	(13) y' ($x=3$)	(14) y' ($x=4$)	(15) y' ($x=5$)	(16) $1000 - \sum(y_i - y_r)^2$	(17) Expected count = f/f_{av}	(18) Actual count
000111 0101|00 010110 110011	000111 010110 001101 101000	7	22	13	40	0.56	1.75	5.16	15.87	0	0	0	15.93	25	902.00	1.10	1
010101 1010|10 001101 101000	010101 101000 010110 110011	21	40	22	51	1.67	3.17	8.73	20.24	5.24	5.85	12.23	18.62	25	961.30	1.18	2
010101 101010 0011|01 1|01000	010101 101010 001101 100011	21	42	13	35	1.67	3.33	5.16	13.89	3.1	12.51	16.68	20.84	25	840.78	1.03	1
100100 001001 101100 1|00011	100100 001001 101100 101000	36	9	44	40	2.86	0.71	17.46	15.87	6.11	12.22	0	0	25	569.32	0.70	0
													Sum		3273.40		
													Average		818.35		
													Maximum		961.30		

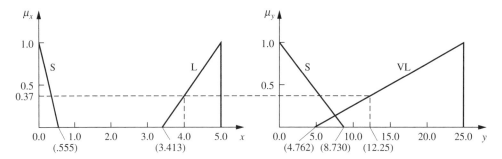

FIGURE 6.14
Physical representation of the first string in Table 4.12*a* and the graphical determination of *y* for a given *x*.

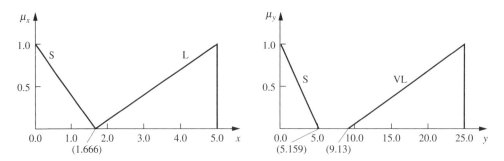

FIGURE 6.15
Physical mapping of the best string in the first generation of strings in the genetic algorithm.

shown in the strings in the column. After crossover, the new strings generated are shown in Table 6.10*b*, column 2. These strings undergo the same process of decoding and evaluation as the previous generation. This process is shown in Table 6.10*b*, columns 3–18. We notice that the average fitness of the second generation is greater than that of the first generation of strings. Also, the fitness of the best string in the second generation is greater than the fitness of the best string in the first generation. Figure 6.15 shows the physical mapping of the best string in the first generation. Figure 6.16 shows the physical mapping of the best string in the second generation; notice that the membership values for the *y* variable in Fig. 6.16 show overlap, which is a very desirable property of membership functions.

The process of generating and evaluating strings is continued until we get a convergence to the solution within a generation, i.e., we get the membership functions with the best fitness value.

Inductive Reasoning

An automatic generation of membership functions can also be accommodated by using the essential characteristic of *inductive reasoning,* which derives a general consensus from the particular (derives the generic from the specific). The induction is performed by the entropy minimization principle, which clusters most optimally the parameters corresponding to the output classes [De Luca and Termini, 1972].

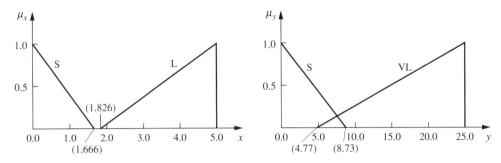

FIGURE 6.16
Physical mapping of the best string in the second generation of strings in the genetic algorithm.

This method is based on an ideal scheme that describes the input and output relationships for a well-established database, i.e., the method generates membership functions based solely on the data provided. The method can be quite useful for complex systems where the data are abundant and static. In situations where the data are dynamic, the method may not be useful, since the membership functions will continually change with time (see the chapter summary for a discussion on the merits of this method).

The intent of induction is to discover a law having objective validity and universal application. Beginning with the particular, induction concludes with the general. The essential principles of induction have been known for centuries. Three laws of induction are summarized here [Christensen, 1980]:

1. Given a set of irreducible outcomes of an experiment, the induced probabilities are those probabilities consistent with all available information that maximize the entropy of the set.
2. The induced probability of a set of independent observations is proportional to the probability density of the induced probability of a single observation.
3. The induced rule is that rule consistent with all available information of which the entropy is minimum.

Among the three laws above, the third one is appropriate for classification (or, for our purposes, membership function development) and the second one for calculating the mean probability of each step of separation (or partitioning). In classification, the probability aspects of the problem are completely disregarded since the issue is simply a binary one: a data point is either in a class or not.

A key goal of entropy minimization analysis is to determine the quantity of information in a given data set. The entropy of a probability distribution is a measure of the uncertainty of the distribution [Yager and Filev, 1994]. This information measure compares the contents of data to a prior probability for the same data. The higher the prior estimate of the probability for an outcome to occur, the lower will be the information gained by observing it to occur. The entropy on a set of possible outcomes of a trial where one and only one outcome is true is defined by the summation of probability and the logarithm of the probability for all outcomes. In other words, the entropy is the expected value of information.

For a simple one-dimensional (one uncertain variable) case, let us assume that the probability of the ith sample w_i to be true is $\{p(w_i)\}$. If we actually observe the sample w_i

in the future and discover that it is true, then we gain the following information, $I(w_i)$:

$$I(w_i) = -k \ln p(w_i) \tag{6.16}$$

where k is a normalizing parameter. If we discover that it is false, we still gain this information:

$$I(\overline{w}_i) = -k \ln[1 - p(w_i)] \tag{6.17}$$

Then the entropy of the inner product of all the samples (N) is

$$S = -k \sum_{i=1}^{N} [p_i \ln p_i + (1 - p_i) \ln(1 - p_i)] \tag{6.18}$$

where $p_i = p(w_i)$. The minus sign before parameter k in Eq. (6.18) ensures that $S \geq 0$, because $\ln x \leq 0$ for $0 \leq x \leq 1$.

The third law of induction, which is typical in pattern classification, says that the entropy of a rule should be minimized. Minimum entropy (S) is associated with all the p_i being as close to ones or zeros as possible, which in turn implies that they have a very high probability of either happening or not happening, respectively. Note in Eq. (6.18) that if $p_i = 1$ then $S = 0$. This result makes sense since p_i is the probability measure of whether a value belongs to a partition or not.

Membership function generation

To subdivide our data set into membership functions we need some procedure to establish fuzzy thresholds between classes of data. We can determine a threshold line with an entropy minimization screening method, then start the segmentation process, first into two classes. By partitioning the first two classes one more time, we can have three different classes. Therefore, a repeated partitioning with threshold value calculations will allow us to partition the data set into a number of classes, or fuzzy sets, depending on the shape used to describe membership in each set.

Membership function generation is based on a partitioning or analog screening concept, which draws a threshold line between two classes of sample data. The main idea behind drawing the threshold line is to classify the samples while minimizing the entropy for an optimum partitioning. The following is a brief review of the threshold value calculation using the induction principle for a two-class problem. First, we assume that we are seeking a threshold value for a sample in the range between x_1 and x_2. Considering this sample alone, we write an entropy equation for the regions $[x_1, x]$ and $[x, x_2]$. We denote the first region p and the second region q, as is shown in Fig. 6.17. By moving an imaginary threshold value x between x_1 and x_2, we calculate entropy for each value of x.

An entropy with each value of x in the region x_1 and x_2 is expressed by Christensen [1980] as

$$S(x) = p(x)S_p(x) + q(x)S_q(x) \tag{6.19}$$

where

$$S_p(x) = -[p_1(x) \ln p_1(x) + p_2(x) \ln p_2(x)] \tag{6.20}$$

$$S_q(x) = -[q_1(x) \ln q_1(x) + q_2(x) \ln q_2(x)] \tag{6.21}$$

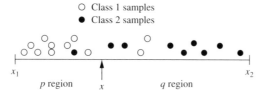

FIGURE 6.17
Illustration of threshold value idea.

where $p_k(x)$ and $q_k(x)$ = conditional probabilities that the class k sample is in the region $[x_1, x_1 + x]$ and $[x_1 + x, x_2]$, respectively
$p(x)$ and $q(x)$ = probabilities that all samples are in the region $[x_1, x_1 + x]$ and $[x_1 + x, x_2]$, respectively
$p(x) + q(x) = 1$

A value of x that gives the minimum entropy is the optimum threshold value. We calculate entropy estimates of $p_k(x)$, $q_k(x)$, $p(x)$, and $q(x)$, as follows [Christensen, 1980]:

$$p_k(x) = \frac{n_k(x) + 1}{n(x) + 1} \tag{6.22}$$

$$q_k(x) = \frac{N_k(x) + 1}{N(x) + 1} \tag{6.23}$$

$$p(x) = \frac{n(x)}{n} \tag{6.24}$$

$$q(x) = 1 - p(x) \tag{6.25}$$

where $n_k(x)$ = number of class k samples located in $[x_l, x_l + x]$
$n(x)$ = the total number of samples located in $[x_l, x_l + x]$
$N_k(x)$ = number of class k samples located in $[x_l + x, x_2]$
$N(x)$ = the total number of samples located in $[x_l + x, x_2]$
n = total number of samples in $[x_1, x_2]$
l = a general length along the interval $[x_1, x_2]$

While moving x in the region $[x_1, x_2]$ we calculate the values of entropy for each position of x. The value of x that holds the minimum entropy we will call the primary threshold (PRI) value. With this PRI value, we divide the region $[x_1, x_2]$ in two. We may say that the left side of the primary threshold is the *negative* side and the right the *positive* side; these labels are purely arbitrary but should hold some contextual meaning for the particular problem. With this first PRI value we can choose a shape for the two membership functions; one such shape uses two trapezoids, as seen in Fig. 6.18a. But the particular choice of shape is arbitrary; we could just as well have chosen to make the threshold crisp and use two rectangles as membership functions. However, we do want to employ some amount of overlap since this develops the power of a membership function. As we get more and more subdivisions of the region $[x_1, x_2]$, the choice of shape for the membership function becomes less and less important as long as there is overlap between sets. Therefore, selection of simple shapes like triangles and trapezoids, which exhibit some degree of overlap, is judicious. In the next sequence we conduct the segmentation again, on each of the regions

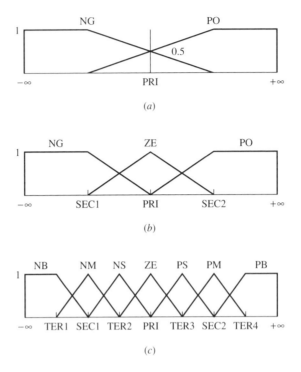

FIGURE 6.18
Repeated partitions and corresponding fuzzy set labels: (*a*) the first partition, (*b*) the second partition, and (*c*) the third partition.

shown in Fig. 6.18*a;* this process will determine *secondary* threshold values. The same procedure is applied to calculate these secondary threshold values. If we denote a secondary threshold in the negative area as SEC1 and the other secondary threshold in the positive area SEC2, we now have three threshold lines in the sample space. The thresholds SEC1 and SEC2 are the minimum entropy points that divide the respective areas into two classes. Then we can use three labels of PO (positive), ZE (zero), and NG (negative) for each of the classes, and the three threshold values (PRI, SEC1, SEC2) are used as the *toes* of the three separate membership shapes shown in Fig. 6.18*b*. In fuzzy logic applications we often use an odd number of membership functions to partition a region, say five labels or seven. To develop seven partitions we would need *tertiary* threshold values in each of the three classes of Fig. 6.18*b*. Each threshold level, in turn, gradually separates the region into more and more classes. We have four tertiary threshold values: TER1, TER2, TER3, and TER4. Two of the tertiary thresholds lie between primary and secondary thresholds, and the other two lie between secondary thresholds and the ends of the sample space; this arrangement is shown in Fig. 6.18*c*. In this figure we use labels such as NB, NM, NS, ZE, PS, PM, and PB.

Example 6.6. The shape of an ellipse may be characterized by the ratio of the length of two chords *a* and *b,* as shown in Fig. 6.19 (a similar problem was originally posed in Chapter 1; see Fig. 1.3).
    Let $x = a/b$; then as the ratio $a/b \to \infty$, the shape of the ellipse tends to a horizontal line, whereas as $a/b \to 0$, the shape tends to a vertical line. For $a/b = 1$ the shape is a

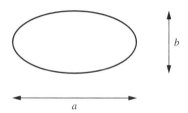

FIGURE 6.19
Geometry of an ellipse.

TABLE 6.11
Segmentation of x into two arbitrary classes (from raw data)

$x = a/b$	0	0.1	0.15	0.2	0.2	0.5	0.9	1.1	1.9	5	50	100
Class	1	1	1	1	1	2	1	1	2	2	2	2

TABLE 6.12
Calculations for selection of partition point PRI

x	0.7	1.0	1.5	3.45
p_1	$\dfrac{5+1}{6+1} = \dfrac{6}{7}$	$\dfrac{6+1}{7+1} = \dfrac{7}{8}$	$\dfrac{7+1}{8+1} = \dfrac{8}{9}$	$\dfrac{7+1}{9+1} = \dfrac{8}{10}$
p_2	$\dfrac{1+1}{6+1} = \dfrac{2}{7}$	$\dfrac{1+1}{7+1} = \dfrac{2}{8}$	$\dfrac{1+1}{8+1} = \dfrac{2}{9}$	$\dfrac{2+1}{9+1} = \dfrac{3}{10}$
q_1	$\dfrac{2+1}{6+1} = \dfrac{3}{7}$	$\dfrac{1+1}{5+1} = \dfrac{2}{6}$	$\dfrac{0+1}{4+1} = \dfrac{1}{5}$	$\dfrac{0+1}{3+1} = \dfrac{1}{4}$
q_2	$\dfrac{4+1}{6+1} = \dfrac{5}{7}$	$\dfrac{4+1}{5+1} = \dfrac{5}{6}$	$\dfrac{4+1}{4+1} = 1.0$	$\dfrac{3+1}{3+1} = 1.0$
$p(x)$	$\dfrac{6}{12}$	$\dfrac{7}{12}$	$\dfrac{8}{12}$	$\dfrac{9}{12}$
$q(x)$	$\dfrac{6}{12}$	$\dfrac{5}{12}$	$\dfrac{4}{12}$	$\dfrac{3}{12}$
$S_p(x)$	0.49	0.463	0.439	0.54
$S_q(x)$	0.603	0.518	0.32	0.347
S	0.547	0.486	0.4✓	0.49

circle. Given a set of a/b values that have been classified into two classes (class division is not necessarily based on the value of x alone; other properties like line thickness, shading of the ellipse, etc., may also be criteria), divide the variable $x = a/b$ into fuzzy partitions, as illustrated in Table 6.11.

First we determine the entropy for different values of x. The value of x is selected as approximately the midvalue between any two adjacent values. Equations (6.19)–(6.25) are then used to compute p_1, p_2, q_1, q_2, $p(x)$, $q(x)$, $S_p(x)$, $S_q(x)$, and S; and the results are displayed in Table 6.12. The value of x that gives the minimum value of the entropy (S) is selected as the first threshold partition point, PRI. From Table 6.12 (see checkmark at $S = 0.4$) we see that the first partition point is selected at $x = 1.5$, and its location for determining membership function selection is shown in Fig. 6.20.

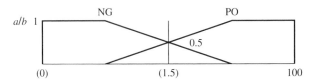

FIGURE 6.20
Partitioning of the variable $x = a/b$ into positive (PO) and negative (NG) partitions.

TABLE 6.13
Calculations to determine secondary threshold value: NG side

x	0.175	0.35	0.7
p_1	$\dfrac{3+1}{3+1} = 1.0$	$\dfrac{5+1}{5+1} = 1.0$	$\dfrac{5+1}{6+1} = \dfrac{6}{7}$
p_2	$\dfrac{0+1}{3+1} = \dfrac{1}{4}$	$\dfrac{0+1}{5+1} = \dfrac{1}{6}$	$\dfrac{1+1}{6+1} = \dfrac{2}{7}$
q_1	$\dfrac{4+1}{5+1} = \dfrac{5}{6}$	$\dfrac{2+1}{3+1} = \dfrac{3}{4}$	$\dfrac{2+1}{2+1} = 1.0$
q_2	$\dfrac{1+1}{5+1} = \dfrac{2}{6}$	$\dfrac{1+1}{3+1} = \dfrac{2}{4}$	$\dfrac{0+1}{2+1} = \dfrac{1}{3}$
$p(x)$	$\dfrac{3}{8}$	$\dfrac{5}{8}$	$\dfrac{6}{8}$
$q(x)$	$\dfrac{5}{8}$	$\dfrac{3}{8}$	$\dfrac{2}{8}$
$S_p(x)$	0.347	0.299	0.49
$S_q(x)$	0.518	0.562	0.366
S	0.454	0.398$\checkmark$	0.459

The same process as displayed in Table 6.12 is repeated for the negative and positive partitions for different values of x. For example, in determining the threshold value to partition the negative (NG) side of Fig. 6.20, Table 6.13 displays the appropriate calculations.

Table 6.14 illustrates the calculations to determine the threshold value to partition the positive side of Fig. 6.20.

The partitions are selected based on the minimum entropy principle; the S values with a checkmark in Tables 6.13 and 6.14 are those selected. The resulting fuzzy partitions are as shown in Fig. 6.21. If required, these partitions can be further subdivided into more fuzzy subpartitions of the variable x.

SUMMARY

This chapter attempts to summarize several methods – classical and modern – that have been and are being used to develop membership functions. This field is rapidly developing, and this chapter is simply an introduction. Many methods for developing membership functions have not been discussed in this chapter. Ideas like deformable prototypes [Bremermann, 1976], implicit analytical definition [Kochen and Badre, 1976], relative preferences [Saaty, 1974], and various uses of statistics [Dubois and Prade, 1980] are just a few of the many omitted here for brevity.

TABLE 6.14
Calculations to determine secondary threshold value: PO side

x	27.5
p_1	$\dfrac{0+1}{2+1}=\dfrac{1}{3}$
p_2	$\dfrac{2+1}{2+1}=1.0$
q_1	$\dfrac{0+1}{2+1}=\dfrac{1}{3}$
q_2	$\dfrac{2+1}{2+1}=1.0$
$p(x)$	$\dfrac{2}{4}$
$q(x)$	$\dfrac{2}{4}$
$S_p(x)$	0.366
$S_q(x)$	0.366
S	0.366 $\checkmark$

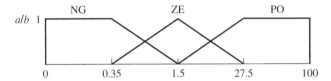

FIGURE 6.21
Secondary partitioning for Example 6.6.

This chapter has dealt at length with only six of the methods currently used in developing membership functions. There are a growing number of papers in the area of cognitive systems, where learning methods like neural networks and reasoning systems like fuzzy systems are being combined to form powerful problem solvers. In these cases, the membership functions are generally tuned in a cyclic fashion and are inextricably tied to their associated rule structure [e.g., see Hayashi et al., 1992].

In the case of genetic algorithms a number of works have appeared [see Karr and Gentry, 1993; Lee and Takagi, 1993]. Vast improvements have been made in the ability of genetic algorithms to find optimum solutions: for example, the *best* shape for a membership function. One of these improvements makes use of gray codes in solving a traditional binary coding problem, where sometimes all the bits used to map a decimal number had to be changed to increase that number by 1 [Forrest, 1993]. This problem had made it difficult for some algorithms to find an optimum solution from a point in the solution space that was already close to the optimum. Both neural network and genetic algorithm approaches to determining membership functions generally make use of associated rules in the knowledge base.

In inductive reasoning, as long as the database is not dynamic the method will produce good results; when the database changes, the partitioning must be reaccomplished. Compared to neural networks and genetic algorithms, inductive reasoning has an advantage in the fact that the method may not require a convergence analysis, which in the case of genetic algorithms and neural networks is computationally very expensive. On the other hand, the inductive reasoning method uses the entire database to formulate rules and membership functions and, if the database is large, this method can also be computationally expensive. The choice of which of the three methods to use depends entirely on the problem size and problem type.

REFERENCES

Bremermann, H. (1976). "Pattern recognition," in H. Bossel,S. Klaszko, and N. Müller (eds.), *Systems theory in the social sciences*, Birkhaeuser, Basel, pp. 116–159.

Christensen, R. (1980). *Entropy minimax sourcebook*, vols. 1–4, and Fundamentals of inductive reasoning, Entropy Ltd., Lincoln, MA.

De Luca, A. and Termini, S. (1972). "A definition of a non-probabilistic entropy in the setting of fuzzy sets theory," *Inf. Control*, vol. 20, pp. 301–312.

Dubois, D., and Prade, H. (1980). *Fuzzy sets and systems: Theory and applications*, Academic Press, New York.

Einstein, A. (1922). "Geometry and experience," in *Sidelights of relativity*, Methuen, London (English translation of 1921 speech to Prussian Academy).

Forrest, S. (1993). "Genetic algorithms: Principles of natural selection applied to computation," *Science*, vol. 261, pp. 872–878.

Goldberg, D. (1989). *Genetic algorithms*, Addison-Wesley, New York.

Hayashi, I., Nomura, H., Yamasaki, H. and Wakami, N. (1992). "Construction of fuzzy inference rules by NDF and NDFL," *Int. J. Approximate Reasoning*, vol. 6, pp. 241–266.

Hopfield, J. (1982). "Neural networks and physical systems with emergent collective computational abilities," *Proc. Natl. Acad. Sci. USA*, vol. 79, pp. 2554–2558.

Hopfield, J. and Tank, D. (1986). "Computing with neural circuits: A model," *Science*, vol. 233, pp. 625–633.

Karr, C. L., and Gentry, E. J. (1993). "Fuzzy control of pH using genetic algorithms," *IEEE Trans. Fuzzy Syst.*, vol. 1, no. 1, pp. 46–53.

Kim, C. J. and Russel, B. D. (1993). "Automatic generation of membership function and fuzzy rule using inductive reasoning," *IEEE Trans.*, Paper 0-7803-1485-9/93.

Kochen, M. and Badre, A. (1976). "On the precision of adjectives which denote fuzzy sets," *J. Cybern.*, vol. 4, no. 1, pp. 49–59.

Lee, M., and Takagi, H. (1993). "Integrating design stages of fuzzy systems using genetic algorithms," *IEEE Trans.*, Paper 0-7803-0614-7/93.

Luger, G., and Stubblefield, W. (1989). *Artificial intelligence and the design of expert systems*, Benjamin-Cummings, Redwood City, CA.

Ross, T. (1995). *Fuzzy logic with engineering applications*, McGraw-Hill, New York.

Saaty, T. (1974). "Measuring the fuzziness of sets," *J. Cybern.*, vol. 4, no. 4, pp. 53–61.

Takagi, H. and Hayashi, I. (1991). "NN-driven fuzzy reasoning," *Int. J. Approximate Reasoning*, vol. 5, pp. 191–212.

Yager, R. and Filev, D. (1994). "Template-based fuzzy systems modeling," *Intell. Fuzzy Syst.*, vol. 2, no. 1, pp. 39–54.

Yamakawa, T. (1992). "A fuzzy logic controller," *J. Biotechnol.*, vol. 24, pp. 1–32.

Zadeh, L. (1972). "A rationale for fuzzy control," *J. Dyn. Syst. Meas. Control Trans. ASME*, vol. 94, pp. 3–4.

PROBLEMS

6.1. Using your own intuition, develop fuzzy membership functions on the real line for the fuzzy number 3, using the following function shapes:

(*a*) Symmetric triangle

(*b*) Trapezoid

(*c*) Gaussian function

6.2. Using your own intuition, develop fuzzy membership functions on the real line for the fuzzy number "approximately 2 *or* approximately 8" using the following function shapes:

(*a*) Symmetric triangles

(*b*) Trapezoids

(*c*) Gaussian functions

6.3. Using your own intuition, develop fuzzy membership functions on the real line for the fuzzy number "approximately 6 *to* approximately 8" using the following function shapes:

(*a*) Symmetric triangles

(*b*) Trapezoids

(*c*) Gaussian functions

6.4. Using your own intuition and your own definitions of the universe of discourse, plot fuzzy membership functions for the following variables:

(*a*) Age of people

  (*i*) Very young

  (*ii*) Young

  (*iii*) Middle-aged

  (*iv*) Old

  (*v*) Very old

(*b*) Education of people

  (*i*) Fairly educated

  (*ii*) Educated

  (*iii*) Highly educated

  (*iv*) Not highly educated

  (*v*) More or less educated

6.5. Using the inference approach outlined in this chapter, find the membership values for each of the triangular shapes (I, R, IR, E, T) for each of the following triangles:

(*a*) $80°, 75°, 25°$

(*b*) $55°, 65°, 60°$

(*c*) $45°, 75°, 60°$

6.6. Develop a membership function for rectangles that is similar to the algorithm on triangles in this chapter. This function should have two independent variables; hence, it can be plotted.

6.7. The following raw data were determined in a pairwise comparison of new premium car preferences in a poll of 100 people. When it was compared with a Porsche (P), 79 of those polled preferred a BMW (B), 85 preferred a Mercedes (M), 59 preferred a Lexus (L), and 67 preferred an Infinity (I). When a BMW was compared, the preferences were 21–P, 23–M, 37–L, and 45–I. When a Mercedes was compared, the preferences were 15–P, 77–B, 35–L, and 48–I. When a Lexus was compared, the preferences were 41–P, 63–B, 65–M, and 51–I. Finally, when an Infinity was compared, the preferences were 33–P, 55–B, 52–M, and 49–L. Using rank ordering, plot the membership function for "most preferred car."

6.8. For the data shown in the accompanying table, show the first iteration in trying to compute the membership values for the input variables x_1, x_2, and x_3 in the output regions R^1 and R^2. Assume a random set of weights for your neural network.

x_1	x_2	x_3	R^1	R^2
1.0	0.5	2.3	1.0	0.0

(a) Use a $3 \times 3 \times 1$ neural network.

(b) Use a $3 \times 3 \times 2$ neural network.

(c) Explain the difference in results when using (a) and (b).

6.9. For the data shown in the following table, show the first iteration in trying to compute the membership values for the input variables x_1, x_2, x_3, and x_4 in the regions R^1, R^2, and R^3.

x_1	x_2	x_3	x_4	R^1	R^2	R^3
10	0	−4	2	0	1	0

Use a $4 \times 3 \times 3$ neural network with a random set of weights.

6.10. For the data shown in the accompanying Table A, show the first two iterations using a genetic algorithm in trying to find the optimum membership functions (use right-triangle functions) for the input variable x and output variable y in the rule table, Table B.

Table A Data

x	0	45	90
y	0	0.71	1

Table B Rules

x	SM	MD
y	SM	LG

For the rule table, the symbols SM, MD, and LG mean small, medium, and large, respectively.

6.11. For the data shown in the following Table A, show the first two iterations using a genetic algorithm in trying to find the optimum membership functions (use right-triangle functions) for the input variable x and output variable y in the rule table, Table B. For the rule table, Table B, the symbols ZE, S, and LG mean zero, small, and large, respectively.

Table A Data

x	0	0.3	0.6	1.0	100
y	1	0.74	0.55	0.37	0

Table B Rules

x	LG	S
y	ZE	S

6.12. The results of a price survey for 30 automobiles is presented here:

Class	Automobile prices (in units of $1,000)
Economy	5.5, 5.8, 7.5, 7.9, 8.2, 8.5, 9.2, 10.4, 11.2, 13.5
Midsize	11.9, 12.5, 13.2, 14.9, 15.6, 17.8, 18.2, 19.5, 20.5, 24.0
Luxury	22.0, 23.5, 25.0, 26.0, 27.5, 29.0, 32.0, 37.0, 43.0, 47.5

Consider the automobile prices as a variable and the classes as economy, midsize, and luxury automobiles. Develop three membership function envelopes for car prices using the method of inductive reasoning.

CHAPTER
7

AUTOMATED METHODS FOR FUZZY SYSTEMS

Measure what is measurable, and make measurable what is not so.

Galileo Galilei,
circa 1630

It is often difficult or impossible to accurately model complicated natural processes or engineered systems using a conventional nonlinear mathematical approach with limited prior knowledge. Ideally, the analyst uses the information and knowledge gained from prior experiments or trials with the system to develop a model and predict the outcome, but for new systems where little is known or where experimental analyses are too costly to perform, prior knowledge and information is often unavailable. This lack of data on, or extensive knowledge of, the system makes developing a model using conventional means extremely difficult and often impossible. Furthermore, forming a linguistic rule-base of the system may be impractical without conducting additional observations. Fortunately, for situations such as these, fuzzy modeling is very practical and can be used to develop a model for the system using the "limited" available information. *Batch least squares*, *recursive least squares*, *gradient method*, *learning from example* (LFE), *modified learning from example* (MLFE), and *clustering method* are some of the algorithms available for developing a fuzzy model [Passino and Yurkovich, 1998]. The choice of which method to implement depends on such factors as the amount of prior knowledge of the system to be modeled. These methods, which are referred to as automated methods, are provided as additional procedures to develop membership functions, like those of Chapter 6, but also to provide rules as well.

Fuzzy Logic with Engineering Applications, Second Edition T. J. Ross
© 2004 John Wiley & Sons, Ltd ISBNs: 0-470-86074-X (HB); 0-470-86075-8 (PB)

DEFINITIONS

The description of these methods provided by Passino and Yurkovich [1998] is expanded in this chapter with a detailed example for developing a fuzzy model using each of the algorithms mentioned above. The explanation is given in such a manner that allows the reader to easily prepare a MATLAB code for other applications (see the preface for instructions on accessing this software). Only two-input, single-output systems are illustrated here but the algorithms can be extended to multiple-input, single-output systems and even multiple-input, multiple-output systems. An example of a two-input, single-output system is illustrated in Fig. 7.1, where the information is provided by three points and where the inputs are x_1 and x_2 and the output is y. Most of the algorithms used in the examples of this chapter incorporate Gaussian membership functions for the inputs $\mu(x)$,

$$\mu(x) = \exp\left[-\frac{1}{2}\left(\frac{x_i - c_i}{\sigma_i}\right)^2\right] \tag{7.1}$$

where x_i is the ith input variable, c_i is the ith center of the membership function (i.e., where the membership function achieves a maximum value), and σ_i is a constant related to the spread of the ith membership function). Figure 7.2 illustrates a typical Gaussian membership function and these parameters.

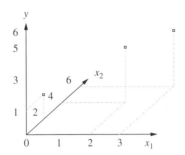

FIGURE 7.1
Example of two-input, single-output system for three data points (Reproduced by permission of Kevin M. Passino and Stephen Yurkovich, from: Kevin M. Passino and Stephen Yurkovich, *Fuzzy Control*, Addison Wesley Longman, Menlo Park, CA, 1998).

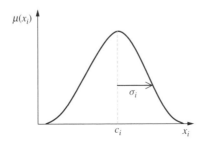

FIGURE 7.2
Typical Gaussian membership function.

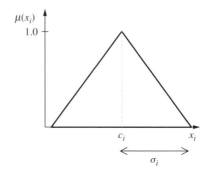

FIGURE 7.3
Typical triangular membership function.

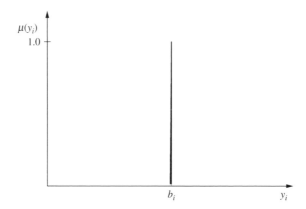

FIGURE 7.4
Delta membership function.

For demonstrative purposes triangular membership functions are used in the example given for the *learning from example* algorithm; Eq. (7.2) shows the formula used for the triangular membership function while Fig. 7.3 illustrates the membership function. In fact any type of membership function may be used for the input and output functions but only the Gaussian and triangular membership functions are illustrated here. In most of the examples provided in this chapter, the output membership function is a delta function, which is an impulse function of zero width with only one value with full membership located at b_i and all other values have zero (see Fig. 7.4). However, the algorithms provided here accommodate any type of output membership function.

$$\mu(x) = \begin{cases} \max\left\{0, 1 + \dfrac{x_i - c_i}{1.0\sigma_i}\right\} & \text{if } x_i \leq c_i \\[3mm] \max\left\{0, 1 + \dfrac{c_i - x_i}{1.0\sigma_i}\right\} & \text{otherwise} \end{cases} \tag{7.2}$$

The six automated methods presented here either develop a rule-base or use a predetermined rule-base (such as the LFE method, batch, and recursive least squares algorithms) to model the system and predict outputs given the inputs; in any case the rules are comprised of a premise clause and a consequence. A typical example of a rule for a multiple-input, single-output system is as follows:

IF *premise*$_1$ and *premise*$_2$ THEN *consequence*

These rules are developed by the algorithms to predict and/or govern an output for the system with given inputs. Most importantly, the algorithms incorporate the use of fuzzy values rather than fuzzy linguistic terms in these rules (hence the membership functions). In other words, the premise and consequence are fuzzy values. In the *batch least squares* (BLS), *recursive least squares* (RLS), and *gradient method* (GM) algorithms this rule-base must be specified by the user of the algorithm from other automated procedures (e.g., MLFE); however, the gradient method has the capability to update the parameters of the rule-base (i.e., the parameters of the membership functions). The *clustering method* (CM) and *modified learning from example* (MLFE) form a rule-base from the input–output which is then used to model the system. The LFE algorithm relies on complete specification of the membership functions and only constructs the rules of the rule-base. Because of this, some algorithms can be used together to develop a refined model for the system. For instance, the MLFE can be used in conjunction with the RLS to develop a more effective model. Once the parameters of the membership functions of the rule-base have been specified they are used by the algorithms to predict an output given the inputs. A detailed description of this process is provided later in the chapter.

The examples to follow all employ a center-average defuzzification, and a product t-norm for the premise, and a product implication, as given by Eq. (7.3). A Takagi–Sugeno or other inference mechanism may be used instead but their application and respective discussion are not included in this chapter [see Passino and Yurkovich, 1998]. As mentioned, most of our examples use Gaussian membership functions for the premise and delta functions for the output, resulting in the following equation to predict the output given an input data-tuple x_j:

$$f(x|\theta) = \frac{\sum\limits_{i=1}^{R} b_i \prod\limits_{j=1}^{n} \exp\left[-\frac{1}{2}\left(\frac{x_j - c_j^i}{\sigma_j^i}\right)^2\right]}{\sum\limits_{i=1}^{R} \prod\limits_{j=1}^{n} \exp\left[-\frac{1}{2}\left(\frac{x_j - c_j^i}{\sigma_j^i}\right)^2\right]} \tag{7.3}$$

where R is the number of rules in the rule-base and n is the number of inputs per data data-tuple. For instance, the system of Fig. 7.1 has two inputs (x_1 and x_2); thus $n = 2$ and if there were two rules in the rule-base, $R = 2$. The parameter R is not known a priori for some methods, but is determined by the algorithms. The symbol θ is a vector that includes the membership function parameters for the rule-base, c_i, σ_i, and b_i.

The data-tuples we shall use for our examples are the same as those used in Passino and Yurkovich [1998]. Table 7.1 and Fig. 7.1 contain these data, which are presumably a representative portion of a larger nonlinear data set, $Z = \{([x_1, x_2], y)\}$. The data of

TABLE 7.1
Training data set,
$Z = \{([x_1, x_2], y)\}$

x_1	x_2	y
0	2	1
2	4	5
3	6	6

Table 7.1 are used to train the fuzzy system to model the output y given the two inputs x_1 and x_2. The BLS and RLS methods are presented first followed by the GM, CM, and LFE the and finally MLFE methods. In consideration for space, only the training of the fuzzy set is demonstrated in each example. However, at the end of this chapter the result of a more thorough application of fuzzy modeling is presented for a system described by numerous data-tuples.

To illustrate the various algorithms, we will use input–output data from an experiment on a new, crushable foam material called *syntactic foam*; the strength of the foam will be verified by triaxial compression tests. Unfortunately, due to the costs involved in preparing the foam only a few cylindrical specimens (2.800 inches in length and 1.400 inches in diameter) are tested under various triaxial compressive loads to determine the longitudinal deformations associated with these loads (see Fig. 7.11). The collected input–output data consist of the longitudinal stress (major principal stress) and the lateral stress (minor principal stress), and their respective longitudinal deformation is the output. These input–output data are then used by various fuzzy algorithms to develop a model to predict the longitudinal deformation given the lateral and longitudinal stress.

BATCH LEAST SQUARES ALGORITHM

The following example demonstrates the development of a nonlinear fuzzy model for the data in Table 7.1 using the BLS algorithm. The algorithm constructs a fuzzy model from numerical data which can then be used to predict outputs given any input. Thus, the data set Z can be thought of as a training set used to model the system. When using the BLS algorithm to develop a fuzzy model it is helpful to have knowledge about the behavior of the data set in order to form a rule-base. In the cases where this knowledge is not available another algorithm with rule-forming capabilities (such as MLFE or CM) may be used to develop a rule-base.

To begin with, let us denote the number of input data-tuples, $m = 3$, where there are two inputs for each data-tuple, $n = 2$ (i.e., x_1, x_2). As required by the algorithm we must designate the number of rules (two rules, $R = 2$) and the rule parameters. The consequence in each rule is denoted by the output membership function centers b_1 and b_2. Recall that there are no other parameters needed for the consequence. The two premises of each rule are defined by the input membership function centers (c_i) and their respective spread (σ_i):

IF *premise*$_1$ and *premise*$_2$ THEN *consequence*

Say we designate values for the premise and consequence of the rule-base that are close to the first two data-tuples of Z (presume that Z is a good representation of the data contained in a larger data set). This way the premise and consequence capture as much of the data set as possible thus improving the predictability of the model. We have the following for the input membership functions centers c_j^i, where i is the rule number and j denotes input number:

$$c_1^1 = 1.5 \qquad c_1^2 = 3$$
$$c_2^1 = 3 \qquad c_2^2 = 5$$

This places the peaks of the membership functions between the input portions of the training data pairs. We could make a conjecture as to the whereabouts of the respective output centers for these input centers as well, but for demonstrative purposes we use the output from the first two data sets for now. Thus, we have the following two rules in our rule-base:

Rule 1:   If x_1 is "about 1.5" and x_2 is "about 3" then b_1 is 1.

Rule 2:   If x_1 is "about 3" and x_2 is "about 5" then b_2 is 5.

Next we pick the spreads, σ_j^i, for the input membership functions we selected. As a good start we select $\sigma_j^i = 2$, for $i = 1$, 2 and $j = 1$, 2, to provide reasonable overlap between membership functions. We may have to increase or decrease the overlap among the input membership functions in the rule-base to improve the output of the fuzzy model. The input membership functions for Rules 1 and 2 are Gaussian membership functions and are displayed in Figs. 7.5 and 7.6. The output membership functions for the rules are delta functions that are displayed in Fig. 7.7.

The objective is to determine the predicted output using Eq. (7.3) when given an input data-tuple. Up to now we have only defined the rule-base but have not developed an output mapping function; we do this next using the training data set.

We calculate the membership value that each input data-tuple has in the specified rules of the rule-base and multiply these two values by one another, resulting in the membership value that the input data-tuple has in a particular rule. This is accomplished by

$$\mu_i(x) = \prod_{j=1}^n \exp\left(-\frac{1}{2}\left(\frac{x - c_j^i}{\sigma_j^i}\right)^2\right) \tag{7.4}$$

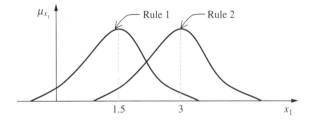

FIGURE 7.5
Input membership functions for x_1.

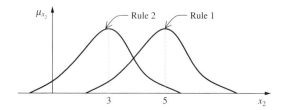

FIGURE 7.6
Input membership functions for x_2.

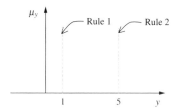

FIGURE 7.7
Output membership functions for y.

where $n = 2$, x are the input data-tuples, and c_j^i and μ_j^i are the rule-base parameters. This reduces Eq. (7.3) to the following form:

$$f(x|\theta) = \frac{\sum_{i=1}^{R} b_i \mu_i(x)}{\sum_{i=1}^{R} \mu_i(x)} \tag{7.5}$$

Passino and Yurkovich [1998] point out that this is also equal to

$$f(x|\theta) = \frac{\sum_{i=1}^{R} b_i \mu_i(x)}{\sum_{i=1}^{R} \mu_i(x)} = \frac{b_1 \mu_1(x)}{\sum_{i=1}^{R} \mu_i(x)} + \frac{b_2 \mu_2(x)}{\sum_{i=1}^{R} \mu_i(x)} + \cdots + \frac{b_R \mu_R(x)}{\sum_{i=1}^{R} \mu_i(x)}$$

and if we define the regression vector ξ as

$$\xi_i(x) = \frac{\mu_i(x)}{\sum_{i=1}^{R} \mu_i(x)} = \frac{\prod_{j=1}^{n} \exp\left[-\frac{1}{2}\left(\frac{x_j - c_j^i}{\sigma_j^i}\right)^2\right]}{\sum_{i=1}^{R} \prod_{j=1}^{n} \exp\left[-\frac{1}{2}\left(\frac{x_j - c_j^i}{\sigma_j^i}\right)^2\right]} \qquad \text{for } i = 1, 2 \tag{7.6}$$

we can use Eq. (7.7) to calculate the output. However, since we are using the BLS approach the output is calculated a little differently [for theory see Passino and Yurkovich; 1998]. The resulting mapping, for the BLS approach, is

$$f(x|\hat{\theta}) = \hat{\theta}^{\mathrm{T}}\xi(x) \tag{7.7}$$

where $\hat{\theta}$ is the least squares estimate vector from the training set, and $\hat{\theta}^{\mathrm{T}}$ is the transpose. The calculation of the least squares estimate and ξ_i are explained below.

For each input data-tuple we have two ($i = 2$) values for ξ, one for Rule 1 and another for Rule 2, resulting in a total of six values:

$$\xi_1(x^1) \quad \xi_2(x^1)$$
$$\xi_1(x^2) \quad \xi_2(x^2)$$
$$\xi_1(x^3) \quad \xi_2(x^3)$$

Using Eq. (7.6), we get

$$\xi_1(x^1) = \frac{\exp\left[-\frac{1}{2}\left(\frac{x_1 - c_1^1}{\sigma_1^1}\right)^2\right] * \exp\left[-\frac{1}{2}\left(\frac{x_2 - c_2^1}{\sigma_2^1}\right)^2\right]}{\exp\left[-\frac{1}{2}\left(\frac{x_1 - c_1^1}{\sigma_1^1}\right)^2\right] * \exp\left[-\frac{1}{2}\left(\frac{x_2 - c_2^1}{\sigma_2^1}\right)^2\right] + \exp\left[-\frac{1}{2}\left(\frac{x_1 - c_1^2}{\sigma_1^2}\right)^2\right] * \exp\left[-\frac{1}{2}\left(\frac{x_2 - c_2^2}{\sigma_2^2}\right)^2\right]}$$

$$\xi_1(x^1) = \frac{\exp\left[-\frac{1}{2}\left(\frac{0 - 1.5}{2}\right)^2\right] * \exp\left[-\frac{1}{2}\left(\frac{2 - 3}{2}\right)^2\right]}{\exp\left[-\frac{1}{2}\left(\frac{0 - 1.5}{2}\right)^2\right] * \exp\left[-\frac{1}{2}\left(\frac{2 - 3}{2}\right)^2\right] + \exp\left[-\frac{1}{2}\left(\frac{0 - 3}{2}\right)^2\right] * \exp\left[-\frac{1}{2}\left(\frac{2 - 5}{2}\right)^2\right]}$$

$$\xi_1(x^1) = \frac{0.66614}{0.77154} = 0.8634$$

and

$$\xi_2(x^1) = \frac{\exp\left[-\frac{1}{2}\left(\frac{x_1 - c_1^2}{\sigma_1^2}\right)^2\right] * \exp\left[-\frac{1}{2}\left(\frac{x_2 - c_2^2}{\sigma_2^2}\right)^2\right]}{\exp\left[-\frac{1}{2}\left(\frac{x_1 - c_1^1}{\sigma_1^1}\right)^2\right] * \exp\left[-\frac{1}{2}\left(\frac{x_2 - c_2^1}{\sigma_2^1}\right)^2\right] + \exp\left[-\frac{1}{2}\left(\frac{x_1 - c_1^2}{\sigma_1^2}\right)^2\right] * \exp\left[-\frac{1}{2}\left(\frac{x_2 - c_2^2}{\sigma_2^2}\right)^2\right]}$$

$$\xi_2(x^1) = \frac{\exp\left[-\frac{1}{2}\left(\frac{0 - 3}{2}\right)^2\right] * \exp\left[-\frac{1}{2}\left(\frac{2 - 5}{2}\right)^2\right]}{\exp\left[-\frac{1}{2}\left(\frac{0 - 1.5}{2}\right)^2\right] * \exp\left[-\frac{1}{2}\left(\frac{2 - 3}{2}\right)^2\right] + \exp\left[-\frac{1}{2}\left(\frac{0 - 3}{2}\right)^2\right] * \exp\left[-\frac{1}{2}\left(\frac{2 - 5}{2}\right)^2\right]}$$

$$\xi_2(x^1) = 0.13661$$

For x^2 and x^3 of data set Z we obtain the following values of $\xi_i(x)$:

$$\xi_1(x^2) = 0.5234 \quad \xi_2(x^2) = 0.4766$$
$$\xi_1(x^3) = 0.2173 \quad \xi_2(x^3) = 0.7827$$

With $\xi_i(x)$ completely specified the transpose of $\xi_i(x)$ is determined and placed into a matrix, Φ:

$$\Phi = \begin{bmatrix} \xi^{\mathrm{T}}(x^1) \\ \xi^{\mathrm{T}}(x^2) \\ \xi^{\mathrm{T}}(x^3) \end{bmatrix} = \begin{bmatrix} 0.8634 & 0.1366 \\ 0.5234 & 0.4766 \\ 0.2173 & 0.7827 \end{bmatrix}$$

And from Z we have the following outputs placed in vector $\mathbf{Y}$:

$$\mathbf{Y} = [y^1 \ y^2 \ y^3]^{\mathrm{T}} = [1 \ 5 \ 6]^{\mathrm{T}}$$

Using $\mathbf{Y}$ and Φ we determine $\hat{\theta}$,

$$\hat{\theta} = (\Phi^{\mathrm{T}}\Phi)^{-1}\Phi^{\mathrm{T}}\mathbf{Y} \tag{7.8}$$

thus producing

$$\hat{\theta} = \left\{ \begin{bmatrix} 0.8634 & 0.5234 & 0.2173 \\ 0.1366 & 0.4766 & 0.7827 \end{bmatrix} * \begin{bmatrix} 0.8634 & 0.1366 \\ 0.5234 & 0.4766 \\ 0.2173 & 0.7827 \end{bmatrix} \right\}^{-1} * \begin{bmatrix} 0.8634 & 0.5234 & 0.2173 \\ 0.1366 & 0.4766 & 0.7827 \end{bmatrix} * \begin{bmatrix} 1 \\ 5 \\ 6 \end{bmatrix}$$

$$\hat{\theta} = \begin{bmatrix} 0.3647 \\ 8.1775 \end{bmatrix}$$

Using Eq. (7.7) we calculate the output for the training data set:

$$f(x|\hat{\theta}) = \hat{\theta}^{\mathrm{T}}\xi(x)$$

$$f(x^1|\hat{\theta}) = [0.3647 \quad 8.1775] * \begin{bmatrix} 0.8634 \\ 0.1366 \end{bmatrix}$$

$$f(x^1|\hat{\theta}) = 1.4319$$

$$f(x^2|\hat{\theta}) = 4.0883$$

$$f(x^3|\hat{\theta}) = 6.4798$$

As seen, the fuzzy system maps the training data set reasonably accurately and if we use additional points not in the training set, as a test set, to see how the system interpolates, we find, for example,

$$f([1, \ 2]^{\mathrm{T}}|\hat{\theta}) = 1.8267; \ f([2.5, \ 5]^{\mathrm{T}}|\hat{\theta}) = 5.3981; \ f([4, \ 7]^{\mathrm{T}}|\hat{\theta}) = 7.3673$$

The accuracy of the fuzzy model developed using BLS primarily depends on the rules specified in the rule-base and the data set used to train the fuzzy model.

RECURSIVE LEAST SQUARES ALGORITHM

The RLS algorithm is very similar to the BLS algorithm; however, the RLS algorithm makes updating $\hat{\theta}$ much easier. The algorithm is a recursive version of the BLS method (the theory behind this algorithm is available [see Passino and Yurkovich, 1998]). It operates without using all the training data and most importantly without having to compute the inverse of $\Phi^T\Phi$ each time the $\hat{\theta}$ in updated. RLS calculates $\hat{\theta}(k)$ at each time step k from the past estimate $\hat{\theta}(k-1)$ and the latest data pair that is received, x^k, y^k. The following example demonstrates the training of a fuzzy model using the RLS algorithm given data set Z (see Table 7.1).

As before, we use Gaussian membership functions for the input and a delta function for the output in the rule-base. Recall that b_i is the point in the output space at which the output membership function for the ith rule is a delta function, and c_j^i is the point in the jth input universe of discourse where the membership function for the ith rule achieves a maximum. The relative width, σ_j^i, of the jth input membership function for the ith rule is always greater than zero.

The RLS algorithm requires that the rule-base be specified, (i.e., number of rules, input membership function centers, input membership function relative widths, and the output centers). The training data set should include a good representative subset of the data set. If the analyst does not have enough knowledge of the system to specify the parameters needed to define the rule-base he or she can do so by using another algorithm first, such as the MLFE. In this example we are able to specify these parameters for the rule-base. We decide on using two rules to model the system and make an educated guess as to where to set the input membership function centers based on some type of regular spacing so as to lie in the middle of the training data, just as we did in the BLS example.

Like the BLS we can vary the spread for each premise of the rules and thus achieve greater or lesser overlap among the input membership functions $\mu_{x_j^i}$. This is very useful when dealing with inputs of different ranges where we would like the spreads of the inputs to reflect this variability. Again, we select $\sigma_j^i = 2$, for $i = 1, 2$ and $j = 1, 2$, which should provide sufficient overlap between membership functions for the data in Table 7.1. We tune $f(x|\theta)$ to interpolate the data set Z by selecting two rules ($R = 2$). If we choose the same values for c_j^i that we used in the BLS example we have

$$c_1^1 = 1.5 \quad c_1^2 = 3$$
$$c_2^1 = 3 \quad c_2^2 = 5$$

We have two inputs for each data-tuple, $n = 2$, and three input data-tuples in our training set, $m = 3$. We assume that the training set may be increased by one each time step, so we let the time index $k = m$. In the RLS we can cycle through the training data a number of times to develop the fuzzy model but in this example we elect to cycle through the data only once for demonstrative purpose.

Now we calculate the regression vector based on the training data set and obtain the same regression vector ξ, using Eq. (7.6) as we did for the BLS example. Recall that in the least squares algorithm the training data x^i are mapped into $\xi(x^i)$ which is then used to develop an output $f(x^i)$ for the model. We get the identical results for $\xi(x^i)$ as for the BLS approach, i.e.,

$$\xi_1(x^1) = 0.8634 \quad \xi_2(x^1) = 0.13661$$
$$\xi_1(x^2) = 0.5234 \quad \xi_2(x^2) = 0.4766$$
$$\xi_1(x^3) = 0.2173 \quad \xi_2(x^3) = 0.7827$$

If we decide to use a *weighted recursive least squares* (WRLS) algorithm because the parameters of the physical system θ vary slowly we employ a "forgetting factor," λ, which gives the more recent data more weight. The forgetting factor varies from 0 to 1 where $\lambda = 1$ results in a standard RLS. For our example we choose to use $\lambda = 1$ in order to weight all the training data equally. Before proceeding to find an estimate of the output, we need to decide in what order to have the RLS process the data pairs (x^i, y^i). There are many possibilities, but in this example we choose to cycle through the data just once beginning with the first input pair and ending with the last (note that all the data-tuples are weighted equally). As mentioned above, we could repeat this a number of times which may improve our results; however, for illustrative purposes we decide to cycle through just once.

For the RLS algorithm we use a covariance matrix to determine $\hat{\theta}$, which is calculated using the regression vector and a previous covariant (see Eq. (7.11)). To do this, we must first calculate an initial covariance matrix P_0 using a parameter α and the identity matrix, I (see Eq. (7.9)). P_0 is the covariance matrix at time step 0 ($k = 0$) and is used to update the covariance matrix, P, in the next time step. A recursive relation is established to calculate values of the P matrix for each time step (see Eq. (7.10)). The value of the parameter α should be greater than 0. Here we use a value of $\alpha = 2000$; I is an $R \times R$ matrix. Next we set our initial conditions for $\hat{\theta}$, at time step 0 ($k = 0$); a good starting point for this would be to use the results from our BLS example, thus

$$\hat{\theta}(0) = \begin{bmatrix} 0.3647 \\ 8.1775 \end{bmatrix}$$

If these values are not readily available, another set of values may be used but more cycles may be needed to arrive at good values. As mentioned previously, this example only demonstrates the training of the fuzzy model using one cycle.

$$P_0 = \alpha I \tag{7.9}$$

$$P_0 = P(0) = 2000 * \begin{bmatrix} 1 & 0 \\ 0 & 1 \end{bmatrix} = \begin{bmatrix} 2000 & 0 \\ 0 & 2000 \end{bmatrix}$$

Once P_0 is determined we use it along with $\xi_i(x^{k=1})$ to calculate the next P and $\hat{\theta}$ for the next step, $P(k = 1)$ and $\hat{\theta}(k = 1)$. This is accomplished using Eqs. (7.10) and (7.11):

$$P(k) = \frac{1}{\lambda}\{I - P(k-1)\xi(x^k)[\lambda I + (\xi(x^k))^{\mathrm{T}}P(k-1)\xi(x^k)]^{-1}(\xi(x^k))^{\mathrm{T}}\}P(k-1) \tag{7.10}$$

$$\hat{\theta}(k) = \hat{\theta}(k-1) + P(k)\xi(x^k)[y^k - (\xi(x^k))^{\mathrm{T}}\hat{\theta}(k-1)] \tag{7.11}$$

For $k = 1$ and $\xi_1(x^1) = 0.8634$, $\quad \xi_2(x^1) = 0.1366$,

$$P(1) = \frac{1}{1}\left\{ \begin{bmatrix} 1 & 0 \\ 0 & 1 \end{bmatrix} - \left(\begin{bmatrix} 2000 & 0 \\ 0 & 2000 \end{bmatrix} * \begin{bmatrix} 0.8634 \\ 0.1366 \end{bmatrix} \right) * \left(1 + [0.8634 \quad 0.1366] * \begin{bmatrix} 2000 & 0 \\ 0 & 2000 \end{bmatrix} \right.\right.$$

$$\left.\left. * \begin{bmatrix} 0.8634 \\ 0.1366 \end{bmatrix} \right)^{-1} * [0.8634 \quad 0.1366] \right\} * \begin{bmatrix} 2000 & 0 \\ 0 & 2000 \end{bmatrix}$$

$$P(1) = \left(\begin{bmatrix} 1 & 0 \\ 0 & 1 \end{bmatrix} - \begin{bmatrix} 0.9749 & 0.1543 \\ 0.1543 & 0.0244 \end{bmatrix} \right) * \begin{bmatrix} 2000 & 0 \\ 0 & 2000 \end{bmatrix}$$

$$P(1) = \begin{bmatrix} 50.12 & -308.5 \\ -308.5 & 1951 \end{bmatrix}$$

$$\hat{\theta}(1) = \begin{bmatrix} 0.3467 \\ 8.1775 \end{bmatrix} + \begin{bmatrix} 50.12 & -308.5 \\ -308.5 & 1951 \end{bmatrix} * \begin{bmatrix} 0.8634 \\ 0.1366 \end{bmatrix} * \left(1 - \begin{bmatrix} 0.8634 & 0.1366 \end{bmatrix} * \begin{bmatrix} 0.3467 \\ 8.1775 \end{bmatrix} \right)$$

$$\hat{\theta}(1) = \begin{bmatrix} -0.1232 \\ 8.1007 \end{bmatrix}$$

The next time step, with $k = 2$ and $\xi_1(x^2) = 0.5234$, $\xi_2(x^2) = 0.4766$, results in the following:

$$P(2) = \begin{bmatrix} 2.1193 & -3.1614 \\ -3.1614 & 8.7762 \end{bmatrix}$$

$$\hat{\theta}(2) = \begin{bmatrix} -0.6016 \\ 11.1438 \end{bmatrix}$$

Finally for the third time step of the cycle, $k = 3$ and $\xi_1(x^3) = 0.2173$, $\xi_2(x^3) = 0.7827$, results are

$$P(3) = \begin{bmatrix} 1.3684 & -0.8564 \\ -0.8564 & 1.7003 \end{bmatrix}$$

$$\hat{\theta}(3) = \begin{bmatrix} 0.3646 \\ 8.1779 \end{bmatrix}$$

We have now calculated the vector parameters $\hat{\theta}$, based on the three inputs needed to model the system. For this example, performing another cycle with the training data set changes $\hat{\theta}$ very little; this is left as an exercise at the end of the chapter. Now that $\hat{\theta}$ has been determined it is used in conjunction with ξ in Eq. (7.7) to calculate the resulting output values for the training data-tuples:

$$f(x^1 | \hat{\theta}) = \begin{bmatrix} 0.3646 & 8.1779 \end{bmatrix} * \begin{bmatrix} 0.8634 \\ 0.1366 \end{bmatrix} = 1.432$$

$$f(x^2 | \hat{\theta}) = \begin{bmatrix} 0.3646 & 8.1779 \end{bmatrix} * \begin{bmatrix} 0.5234 \\ 0.4766 \end{bmatrix} = 4.088$$

$$f(x^3 | \hat{\theta}) = \begin{bmatrix} 0.3646 & 8.1779 \end{bmatrix} * \begin{bmatrix} 0.2173 \\ 0.7827 \end{bmatrix} = 6.480$$

This compares well with the original output (Table 7.1). Modifying the input membership function parameters may improve the predicted output; this is left as an exercise at the end of this chapter.

GRADIENT METHOD

In the RLS method we noticed that the predicted output for the training data set could have been improved and recommended modifying the input parameters. The gradient method (GM) does just that and provides a means for tuning the parameters of the fuzzy model, i.e., the parameters of the rule-base. Recall that for the input membership function we have the membership function centers and the spread of the membership functions. In addition to the input parameters the GM provides a method to tune the output membership function.

Using the training data set Z of Table 7.1 we illustrate the development of a fuzzy model using the GM. Like the least squares algorithms we must specify the rules; however, the GM has the ability to tune the parameters associated with the rules based on the training set. Thus the data used in the training set are of utmost importance in achieving a good approximation. We shall illustrate the method with two rules ($R = 2$). The GM's goal is to minimize the error between the predicted output value, $f(x^m|\theta)$, and the actual output value y^m through the quadratic function e_m, which we call the "error surface." The equation for this error surface is

$$e_m = \tfrac{1}{2}[f(x^m|\theta) - y^m]^2 \tag{7.12}$$

Here m denotes the input–output data-tuple from the training data set. We want to find the minimum value on this error surface which may be used to determine when the model has achieved desired predictability. In this example we demonstrate how cycling through the training data updates the rule-base parameters thus reducing the difference between the predicted output and the actual output as provided here,

$$\varepsilon_m = f(x^m|\theta) - y^m \tag{7.13}$$

We can keep cycling through the training data set each time step (k) modifying the rule parameters, thus decreasing ε_m and obtaining an improved fuzzy system. In this example we update the rule-base using the first data-tuple of Table 7.1 which is then used in the first time step. The second and third time steps, for the remaining data-tuples in Table 7.1, are reserved for an exercise at the end of the chapter.

The GM requires that a step size λ be specified for each of the three parameters being tuned (b_i, c^i_j, and σ^i_j) which are used by the algorithm to determine the updated rule parameters and decrease the error value. Selecting a large step size will converge faster but may risk overstepping the minimum value of e_m, and selecting a small step size means the parameter converges very slowly [Passino and Yurkovich, 1998]. Here we designate the step size for the error surface of the output membership centers, input membership centers, and input membership spreads equal to 1, λ_1, λ_2, and $\lambda_3 = 1$, respectively. In this example the step size values were selected primarily to simplify the calculations.

The algorithm requires that initial values for the rules be designated, but these rules are updated through the iterations with each time step (i.e., the next data-tuple). Thus, to initiate the algorithm for the first rule we choose x^1, y^1 for the input and output membership function centers and select the input spreads to be equal to 1. For the second rule we choose x^2, y^2 as the input and output membership function centers and select the input spreads to be equal to 1. It is important to note that these values initiate the algorithm and are updated by the process to obtain a better model to predict the output in the first time step, i.e., $k = 1$. These initiating values correspond to the zero time step ($k = 0$):

$$\begin{bmatrix} c^1_1(0) \\ c^1_2(0) \end{bmatrix} = \begin{bmatrix} 0 \\ 2 \end{bmatrix} \qquad \begin{bmatrix} \sigma^1_1(0) \\ \sigma^1_2(0) \end{bmatrix} = \begin{bmatrix} 1 \\ 1 \end{bmatrix} \qquad b_1(0) = 1$$

$$\begin{bmatrix} c^2_1(0) \\ c^2_2(0) \end{bmatrix} = \begin{bmatrix} 2 \\ 4 \end{bmatrix} \qquad \begin{bmatrix} \sigma^2_1(0) \\ \sigma^2_2(0) \end{bmatrix} = \begin{bmatrix} 1 \\ 1 \end{bmatrix} \qquad b_2(0) = 5$$

Let us calculate the predicted outputs for the current fuzzy model. First we need to calculate the membership values for data-tuples of Table 7.1, using

$$\mu_i(x^m, k = 0) = \prod_{j=1}^{n} \exp\left[-\frac{1}{2}\left(\frac{x_j^m - c_j^i(k=0)}{\sigma_j^i(k=0)}\right)^2\right] \tag{7.14}$$

$$\mu_1(x^1,\ 0) = \exp\left[-\frac{1}{2}\left(\frac{0-0}{1}\right)^2\right] * \exp\left[-\frac{1}{2}\left(\frac{2-2}{1}\right)^2\right] = 1$$

$$\mu_1(x^2,\ 0) = \exp\left[-\frac{1}{2}\left(\frac{2-0}{1}\right)^2\right] * \exp\left[-\frac{1}{2}\left(\frac{4-2}{1}\right)^2\right] = 0.0183156$$

$$\mu_1(x^3,\ 0) = \exp\left[-\frac{1}{2}\left(\frac{3-0}{1}\right)^2\right] * \exp\left[-\frac{1}{2}\left(\frac{6-2}{1}\right)^2\right] = 3.72665 \times 10^{-6}$$

$$\mu_2(x^1,\ 0) = \exp\left[-\frac{1}{2}\left(\frac{0-2}{1}\right)^2\right] * \exp\left[-\frac{1}{2}\left(\frac{2-4}{1}\right)^2\right] = 0.0183156$$

$$\mu_2(x^2,\ 0) = \exp\left[-\frac{1}{2}\left(\frac{2-2}{1}\right)^2\right] * \exp\left[-\frac{1}{2}\left(\frac{4-4}{1}\right)^2\right] = 1$$

$$\mu_2(x^3,\ 0) = \exp\left[-\frac{1}{2}\left(\frac{3-2}{1}\right)^2\right] * \exp\left[-\frac{1}{2}\left(\frac{6-4}{1}\right)^2\right] = 0.082085$$

From the membership values the training data set has in the current rule-base we obtain the fuzzy output from Eq. (7.3) as follows:

$$f(x^m|\theta(k=0)) = \frac{\sum_{i=1}^{R} b_i(0) \prod_{j=1}^{n} \exp\left[-\frac{1}{2}\left(\frac{x_j^m - c_j^i(k=0)}{\sigma_j^i(k=0)}\right)^2\right]}{\sum_{i=1}^{R}\prod_{j=1}^{R}\exp\left[-\frac{1}{2}\left(\frac{x_j^m - c_j^i(k=0)}{\sigma_j^i(k=0)}\right)^2\right]}$$

$$f(x^2|\theta(0)) = \frac{1 * \exp\left[-\frac{1}{2}\left(\frac{2-0}{1}\right)^2\right] * \exp\left[-\frac{1}{2}\left(\frac{4-2}{1}\right)^2\right] + 5 * \exp\left[-\frac{1}{2}\left(\frac{2-2}{1}\right)^2\right] * \exp\left[-\frac{1}{2}\left(\frac{4-4}{1}\right)^2\right]}{\exp\left[-\frac{1}{2}\left(\frac{2-0}{1}\right)^2\right] * \exp\left[-\frac{1}{2}\left(\frac{4-2}{1}\right)^2\right] + \exp\left[-\frac{1}{2}\left(\frac{2-2}{1}\right)^2\right] * \exp\left[-\frac{1}{2}\left(\frac{4-4}{1}\right)^2\right]}$$

$$f(x^2|\theta(0)) = \frac{1 * \mu_1(x^2, 0) + 5 * \mu_2(x^2, 0)}{\mu_1(x^2, 0) + \mu_2(x^2, 0)}$$

$$f(x^2|\theta(0)) = \frac{1 * 0.0183156 + 5 * 1}{1.083156} = 4.92805$$

$$f(x^3|\theta(0)) = \frac{1 * \mu_1(x^3, 0) + 5 * \mu_2(x^3, 0)}{\mu_1(x^3, 0) + \mu_2(x^3, 0)}$$

$$f(x^3|\theta(0)) = \frac{1 * 0.00000372665 + 5 * 0.082085}{0.00000372665 + 0.082085} = 4.999818$$

$$f(x^1|\theta(0)) = \frac{1 * \mu_1(x^1, 0) + 5 * \mu_2(x^1, 0)}{\mu_1(x^1, 0) + \mu_2(x^1, 0)}$$

$$f(x^1|\theta(0)) = \frac{1 * 1 + 5 * 0.0183156}{1 + 0.0183156} = 1.0719447$$

To compute the approximate error between the predicted output values and the actual output values, we use Eq. (7.12):

$$e_m = \tfrac{1}{2}[f(x^m|\theta(k = 0)) - y^m]^2$$

$$e_1 = \tfrac{1}{2}[1.0719447 - 1]^2 = 2.58802 \times 10^{-3}$$

$$e_2 = \tfrac{1}{2}[4.928055 - 5]^2 = 2.58802 \times 10^{-3}$$

$$e_3 = \tfrac{1}{2}[4.999818 - 6]^2 = 0.500182$$

From the above results it can be seen that the algorithm maps the first two data points much better than the third. The predicted output is improved by cycling through the model with the training data set. The rule-base parameters are modified and improved after each time step; through this process the algorithm will learn to map the third data pair but does not forget how to map the first two data pairs.

Now we demonstrate how the GM updates the rule-base parameters b_i, c_j^i, and σ_j^i using the first time step, $k = 1$. Note that the first time step uses the first data-tuple of the training set, the second time step uses the second data-tuple, and the third time step uses the third data-tuple. We could cycle through the training data set repeatedly to improve the algorithm's predictability, which is what the GM does. We start by calculating the difference between the predicted fuzzy output and the actual output for the first data-tuple of the training data set, using Eq. (7.15). We then use this value to update the parameters of our rule-base using Eqs. (7.16), (7.17), and (7.18).

$$\varepsilon_m(k = 0) = f(x^m|\theta(k = 0)) - y^m \tag{7.15}$$

$$\varepsilon_1(0) = 1.0719447 - 1 = 0.0719447$$

We begin with the output membership function centers using

$$b_i(k) = b_i(k - 1) - \lambda_1(\varepsilon_k(k - 1))\frac{\mu_i(x^k, k - 1)}{\sum_{i=1}^{R} \mu_i(x^k, k - 1)} \tag{7.16}$$

$$b_1(1) = b_1(0) - \lambda_1 * (\varepsilon_1(0))\frac{\mu_1(x^1, 0)}{\mu_1(x^1, 0) + \mu_2(x^1, 0)}$$

$$b_1(1) = 1 - 1 * (0.0719447)\left(\frac{1}{1 + 0.0183156}\right) = 0.964354$$

$$b_2(1) = b_2(0) - \lambda_1 * (\varepsilon_1(0))\frac{\mu_2(x^1, 0)}{\mu_1(x^1, 0) + \mu_2(x^1, 0)}$$

$$b_2(1) = 5 - 1 * (0.0719447) \left(\frac{0.0183156}{1 + .0183156} \right) = 4.998706$$

Then the input membership function centers for the rule-base are updated based on the first time step $k = 1$:

$$c_j^i(k) = c_j^i(k-1) - \lambda_2 \varepsilon_k(k-1) \left(\frac{b_i(k-1) - f(x^k|\theta(k-1))}{\displaystyle\sum_{i=1}^{R} \mu_i(x^k, k-1)} \right)$$

$$* \mu_i(x^k, k-1) \left(\frac{x_j^k - c_j^i(k-1)}{(\sigma_j^i(k-1))^2} \right) \tag{7.17}$$

$$c_1^1(1) = c_1^1(0) - 1 * \varepsilon_1(0) \left(\frac{b_1(0) - f(x^1|\theta(0))}{\mu_1(x^1, 0) + \mu_2(x^1, 0)} \right) * \mu_1(x^1, 0) \left(\frac{x_1^1 - c_1^1(0)}{(\sigma_1^1(0))^2} \right)$$

$$c_1^1(1) = 0 - 1 * (0.0719447) * \left(\frac{1 - 1.0719447}{1 + 0.0183156} \right) * 1 * \left(\frac{0 - 0}{(1)^2} \right) = 0$$

$$c_2^1(1) = c_2^1(0) - 1 * \varepsilon_1(0) \left(\frac{b_1(0) - f(x^1|\theta(0))}{\mu_1(x^1, 0) + \mu_2(x^1, 0)} \right) * \mu_2(x^1, 0) \left(\frac{x_2^1 - c_2^1(0)}{(\sigma_2^1(0))^2} \right)$$

$$c_2^1(1) = 2 - 1 * (0.0719447) * \left(\frac{1 - 1.0719447}{1 + 0.0183156} \right) * 0.0183156 * \left(\frac{2 - 2}{(1)^2} \right) = 2$$

Since the input membership functions for the first rule are the first data-tuples, the updated centers do not change. The time step will affect the second rule because the rule's parameters are based on the second data-tuple of the training data set:

$$c_1^2(1) = c_1^2(0) - 1 * \varepsilon_1(0) \left(\frac{b_2(0) - f(x^1|\theta(0))}{\mu_1(x^1, 0) + \mu_2(x^1, 0)} \right) * \mu_2(x^1, 0) \left(\frac{x_1^1 - c_1^2(0)}{(\sigma_1^2(0))^2} \right)$$

$$c_2^1(1) = 2 - 1 * (0.0719447) * \left(\frac{5 - 1.0719447}{1 + 0.0183156} \right) * 0.0183156 * \left(\frac{0 - 2}{(1)^2} \right) = 2.010166$$

$$c_2^2(1) = c_2^2(0) - 1 * \varepsilon_1(0) \left(\frac{b_2(0) - f(x^1|\theta(0))}{\mu_1(x^1, 0) + \mu_2(x^1, 0)} \right) * \mu_2(x^1, 0) \left(\frac{x_2^1 - c_2^2(0)}{(\sigma_2^2(0))^2} \right)$$

$$c_2^2(1) = 4 - 1 * (0.0719447) * \left(\frac{5 - 1.0719447}{1 + 0.0183156} \right) * 0.0183156 * \left(\frac{2 - 4}{(1)^2} \right) = 4.010166$$

As expected, the first time step would have an effect on the input membership functions for the second rule. This is an iterative process and may take several iterations (time steps) to obtain a desired fuzzy system model.

Finally we update the input membership function spreads, using the following equation:

$$\sigma_j^i(k) = \sigma_j^i(k-1) - \lambda_3 * \varepsilon_k(k-1) * \left(\frac{b_i(k-1) - f(x^k|\theta(k-1))}{\displaystyle\sum_{i=1}^{R} \mu_i(x^k, k-1)} \right.$$

$$\left. * \mu_i(x^k, k-1) * \left(\frac{(x_j^k - c_j^i(k-1))^2}{(\sigma_j^i(k-1))^3} \right) \right) \tag{7.18}$$

$$\sigma_1^1(1) = \sigma_1^1(0) - 1 * \varepsilon_1(0) \left(\frac{b_1(0) - f(x^1|\theta(0))}{\mu_1(x^1,0) + \mu_2(x^1,0)} \right) * \mu_1(x^1,0) * \left(\frac{(x_1^1 - c_1^1(0))^2}{(\sigma_1^1(0))^3} \right)$$

$$\sigma_1^1(1) = 1 - 1 * (0.0719447) \left(\frac{1 - 1.0719447}{1.0183156} \right) * 1 * \left(\frac{(0-0)^2}{(1)^3} \right) = 1$$

$$\sigma_2^1(1) = \sigma_2^1(0) - 1 * (0.0719447) \left(\frac{b_1(0) - f(x^1|\theta(0))}{\mu_1(x^1,0) + \mu_2(x^1,0)} \right) * \mu_1(x^1,0) * \left(\frac{(x_2^1 - c_2^1(0))^2}{(\sigma_2^1(0))^3} \right)$$

$$\sigma_2^1(1) = 1 - 1 * (0.0719447) \left(\frac{1 - 1.0719447}{1.0183156} \right) * 1 * \left(\frac{(2-2)^2}{(1)^3} \right) = 1$$

$$\sigma_1^2(1) = \sigma_1^2(0) - 1 * (0.0719447) \left(\frac{b_2(0) - f(x^3|\theta(0))}{\mu_1(x^1,0) + \mu_2(x^1,0)} \right) * \mu_2(x^1,0) * \left(\frac{(x_1^1 - c_1^2(0))^2}{(\sigma_1^2(0))^3} \right)$$

$$\sigma_1^2(1) = 1 - 1 * (0.0719447) \left(\frac{5 - 1.0719447}{1.0183156} \right) * 0.0183156 * \left(\frac{(0-2)^2}{(1)^3} \right) = 0.979668$$

$$\sigma_2^2(1) = \sigma_2^2(0) - 1 * (0.0719447) \left(\frac{b_2(0) - f(x^1|\theta(0))}{\mu_1(x^1,0) + \mu_2(x^1,0)} \right) * \mu_2(x^1,0) * \left(\frac{(x_2^1 - c_2^2(0))^2}{(\sigma_2^2(0))^3} \right)$$

$$\sigma_2^2(1) = 1 - 1 * (0.0719447) \left(\frac{5 - 1.0719447}{1.0183156} \right) * 0.0183156 * \left(\frac{(2-4)^2}{(1)^3} \right) = 0.979668$$

We now have completed one iteration (using one time step), which updated the parameters of the rule-base. Further iterations with the training set will improve the predictive power of the rule-base.

CLUSTERING METHOD

Fuzzy clustering is the partitioning of a collection of data into fuzzy subsets or clusters based on similarities between the data [Passino and Yurkovich, 1998]. The *clustering method* (CM), like the other methods described previously, develops a fuzzy estimation model, to predict the output given the input. The algorithm forms rules (or clusters) with training data using a nearest neighbor approach for the fuzzy system. This is demonstrated in the following example where the same training data set used in the previous examples is again used here (see Table 7.1).

Recall that these data consist of two inputs ($n = 2$) and one output for each data-tuple. Again we employ Gaussian membership functions for the input fuzzy sets, and delta functions for the output functions. In addition, we make use of center-average defuzzification and product premise for developing our fuzzy model which is given by $f(x|\theta)$ in Eq. (7.3); however, for the clustering method we employ slightly different variables as shown below:

$$
f(x|\theta) = \frac{\sum\limits_{i=1}^{R} A_i \prod\limits_{j=1}^{n} \exp\left[-\left(\frac{x_j - v_j^i}{2\sigma}\right)^2\right]}{\sum\limits_{i=1}^{R} B_i \prod\limits_{j=1}^{n} \exp\left[-\left(\frac{x_j - v_j^i}{2\sigma}\right)^2\right]}
\tag{7.19}
$$

In the above equation, R is the total number of rules, v_j^i are the input membership function centers, x_j is the input, and σ is spread for the input membership functions.

In this example we initiate the parameters A_i and B_i which are then updated or optimized by the algorithm during training of the fuzzy model to predict the output. This is clarified later on in this section. Passino and Yurkovich [1998] make the following recommendations on σ:

- A small σ provides narrow membership functions that may yield a less smooth fuzzy system, which may cause the fuzzy system mapping not to generalize well for the data points in the training set.
- Increasing the parameter σ will result in a smoother fuzzy system mapping.

The use of only one value for the spread may pose a problem when developing a model for the inputs of different ranges. For instance, suppose it is desired to use five input membership function centers for each input variable. Presume that the first input has a range of 0 to 10, that the second input has a range of 20 to 200, and that the resulting width of the membership functions for the first input would have to be much smaller than that of the second. The use of only one spread to develop a model for systems such as this would not work very well. This could be remedied by increasing the number of input membership function centers for the second input.

The above parameters make up the vector θ, shown below, which is developed during the training of the fuzzy model $f(x|\theta)$. The dimensions of θ are determined by the number of inputs n and the number of rules, R, in the rule-base.

$$
\theta = [A_1, \ldots, A_R, B_1, \ldots, B_R, v_1^1, \ldots, v_n^1, \ldots, v_1^R, \ldots, v_n^R, \sigma]^T
$$

The CM develops its rules by first forming cluster centers $v^j = [v_1^j, v_2^j, \ldots, v_n^j]^T$ for the input data. These cluster centers are formed by measuring the distance between the existing R cluster centers and the input training data; if the distance is greater than a specified maximum ε_f we add another cluster center (i.e., rule), otherwise the available cluster center will suffice and we update the parameters A and B for that rule.

Let us begin the training of the fuzzy model by specifying the above parameters using the first data-tuple in Table 7.1 to initiate the process. For this we use $A_1 = y^1 = 1$, $B_1 = 1$, $v_1^1 = x_1^1 = 0$, $v_2^1 = x_2^1 = 2$, and $\sigma = 0.3$. We also specify the maximum distance between our cluster centers and the input data as $\varepsilon_f = 3.0$. Our fuzzy model $f(x|\theta)$ now has one rule ($R = 1$) and the cluster centers for the input of this rule are

$$\underline{v}^1 = \begin{bmatrix} 0 \\ 2 \end{bmatrix}$$

Now using the second data-tuple

$$(x^2, y^2) = \left(\begin{bmatrix} 2 \\ 4 \end{bmatrix}, 5 \right)$$

we check if the clusters are applicable for this data-tuple or if another rule is needed. We accomplish this by measuring the distance ε for each data-tuple and compare this value to the specified ε_f:

$$\varepsilon_{ij} = \left| x^i - v^l \right| \tag{7.20}$$

$$\varepsilon_{21} = \left| x_1^2 - v_1^1 \right|$$

$$\varepsilon_{21} = |2 - 0| = 2 < \varepsilon_f = 3$$

$$\varepsilon_{22} = \left| x_2^2 - v_2^1 \right|$$

$$\varepsilon_{22} = |4 - 2| = 2 < \varepsilon_f = 3$$

Thus there is no need to incorporate an additional cluster, as both ε_{11} and ε_{12} are less than ε_f. However, we must update the existing parameters A_1 and B_1 to account for the output of the current input data-tuple. Updating A_1 modifies the numerator to better predict the output value for the current input data-tuple while the updated B_1 normalizes this predicted output value. This is accomplished using the following equations:

$$A_l = A_l^{\text{old}} + y^i \tag{7.21}$$

$$B_l = B_l^{\text{old}} + 1 \tag{7.22}$$

$$A_1 = A_1^{\text{old}} + y^1 = 1 + 5 = 6$$

$$B_1 = B_1^{\text{old}} + 1 = 1 + 1 = 2$$

This results in the following parameters for our fuzzy model:

$$\theta = [A_1 = 6, B_1 = 2, v_1^1 = 0, v_2^1 = 2, \sigma = 0.3]^{\text{T}}$$

Let us continue training our fuzzy model using the third data-tuple,

$$(x^3, y^3) = \left(\begin{bmatrix} 3 \\ 6 \end{bmatrix}, 6 \right)$$

$$\varepsilon_{31} = \left| x_1^3 - v_1^1 \right| = |3 - 0| = 3 = \varepsilon_f$$

$$\varepsilon_{32} = \left| x_2^3 - v_2^1 \right| = |6 - 0| = 6 > \varepsilon_f = 3$$

$$\varepsilon_{32} = 6 > \varepsilon_f = 3$$

Since both the calculated distances are not less than the specified maximum distance ($\varepsilon_f = 3$) we include another two additional clusters for the next rule. We now have two rules, the second of which is based on the third data-tuple. The cluster centers for the second rule are assigned the values equivalent to the input of the third data-tuple and parameter A_2 is assigned a value equivalent to the output of the third data-tuple while B_2 is assigned a value of 1. The parameters needed for the fuzzy model are displayed below in the updated parameter vector θ:

$$\theta = [A_1 = 6, A_2 = 6, B_1 = 2, B_2 = 1, v_1^1 = 0, v_2^1 = 2, v_1^2 = 3, v_2^2 = 6, \sigma = 0.3]^T$$

LEARNING FROM EXAMPLE

The LFE training procedure relies entirely on a complete specification of the membership functions by the analyst and it only constructs the rules. Again we use the data set Z illustrated in Table 7.1 as the training data set for the fuzzy model. Like the other examples we initiate the algorithm by designating two rules for this data set, $R = 2$. For this method we use triangular membership functions rather than Gaussian for both the input and output. This is done to demonstrate the use of other types of membership functions.

We define the expected range of variation in the input and output variables:

$$X_i = [x_i^-, \quad x_i^+], \ i = 1, 2 \quad Y = [y^-, \quad y^+]$$

We designate $x_1^- = 0, x_1^+ = 4, x_2^- = 0, x_2^+ = 8, y^- = -1$ and $y^+ = 9$ as a choice for known regions within which all data points lie. Recall that a triangular membership function was defined mathematically in Eq. (7.2), which is rewritten here as

$$\mu^c(u) = \begin{cases} \max\left\{0, 1 + \dfrac{u-c}{0.5w}\right\} & \text{if } u \leq c \\[2mm] \max\left\{0, 1 + \dfrac{c-u}{0.5w}\right\} & \text{if } u > c \end{cases}$$

In the above equation, u is the point of interest, c is the membership function center, and w is the base width of the membership function (see Fig. 7.3).

Next the membership functions are defined for each input and output universe of discourse. Our system has two inputs and one output, thus we have X_1^j, X_2^k, and Y^l. It is important to recognize that the number of membership functions on each universe of discourse affects the accuracy of the function approximations and in this example X_1 has fewer membership functions than does X_2. This may occur if the first input is limited to a few values and the second input has more variability. Here X_1^j, X_2^k, and Y^l denote the fuzzy sets with associated membership functions $\mu_{X_i^j}(x_i)$ and $\mu_{Y^j}(y)$ respectively. These fuzzy sets and their membership functions are illustrated in Figs. 7.8, 7.9, and 7.10.

In both Figs. 7.8 and 7.9 the membership functions are saturated (membership value for the fuzzy set is equal to 1) at the far left and right extremes. In this case we use Eqs. (7.23) and (7.24) to determine the membership values for the leftmost and rightmost

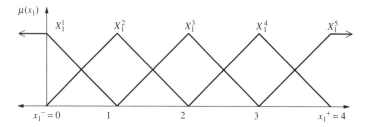

FIGURE 7.8
Specified triangular input membership functions for x_1.

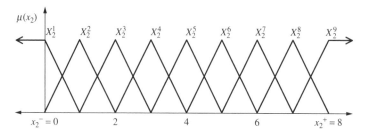

FIGURE 7.9
Specified triangular input membership functions for x_2.

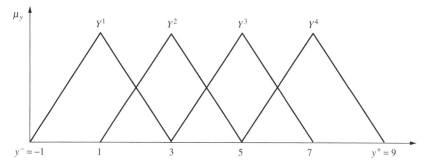

FIGURE 7.10
Output membership functions for y.

membership functions, respectively:

$$
\mu^{L}(u) = \begin{cases} 1, & \text{if } u \leq c^{L} \\ \max\left\{0, 1 + \dfrac{c^{L} - u}{0.5w^{L}}\right\}, & \text{otherwise} \end{cases} \tag{7.23}
$$

$$
\mu^{R}(u) = \begin{cases} \max\left\{0, 1 + \dfrac{u - c^{R}}{0.5w^{R}}\right\}, & \text{if } u \leq c^{R} \\ 1, & \text{otherwise} \end{cases} \tag{7.24}
$$

In the above equations c^L and c^R specify the point on the horizontal axis when full membership begins for the leftmost and rightmost membership functions (μ^L and μ^R), respectively, while w^L and w^R specify two times the width of the nonunity and nonzero part of μ^L and μ^R, respectively.

Now for the first training data (d) point for the fuzzy system we have

$$d^{(1)} = (x^1, y^1) = \left(\begin{bmatrix} 0 \\ 2 \end{bmatrix}, 1 \right)$$

which we add to the rule-base since there are currently no other rules in the set:

$$R_1 := \text{ if } x_1 \text{ is } X_1^1 \text{ and } x_2 \text{ is } X_2^3 \text{ then } y \text{ is } Y^1.$$

Using Eqs. (7.2), (7.23), or (7.24) we can determine the membership values that the first training data point has in the two existing rules. These values are then used to calculate the "degree of attainment" using Eq. (7.25). The data-tuple in $d^{(1)}$ has full membership in the leftmost membership of Fig. 7.8 (fuzzy value 0 or X_1^1), full membership in the third membership function from the left in Fig. 7.9 (fuzzy value 2 or X_2^3) and membership equal to 1 in the output membership function for the determined rule-base:

$$\mu_{X_1^1}(x_1 = 0) = 1$$
$$\mu_{X_2^3}(x_2 = 2) = 1$$
$$\mu_{Y^1}(y = 1) = 1$$
$$\text{degree}(R_i) = \mu_{X_1^j}(x_1)^* \mu_{X_2^k}(x_2)^* \mu_{Y^l}(y) \qquad (7.25)$$

Then,

$$\text{degree}(R_1) = \mu_{X_1^1}(x_1)^* \mu_{X_2^3}(x_2)^* \mu_{Y^1}(y) = 1^*1^*1 = 1$$

We now move on to the next data-tuple, $d^{(2)}$:

$$d^{(2)} = \left(\begin{bmatrix} 2 \\ 4 \end{bmatrix}, 5 \right)$$

Since the existing rule does not model $d^{(2)}$ we add another rule to the rule-base, which is as follows:

$$R_2 := \text{ if } x_1 \text{ is } X_1^3 \text{ and } x_2 \text{ is } X_2^5 \text{ then } y \text{ is } Y^3.$$

The data tuple in $d^{(2)}$ has full membership in the center membership function of Fig. 7.8 (fuzzy value 2 or X_1^3), full membership in the fifth membership function from the left in Fig. 7.9 (fuzzy value 4 or X_2^5), and a membership value of 1 in the fuzzy set of 4:

$$\mu_{X_1^3}(x_1 = 2) = 1$$
$$\mu_{X_2^5}(x_2 = 4) = 1$$
$$\mu_{Y^5}(y = 5) = 1$$

where we get

$$\text{degree}(R_1) = \mu_{X_1^1}(x_1)^* \mu_{X_2^3}(x_2)^* \mu_{Y^1}(y) = 0^*0^*0 = 0$$

$$\text{degree}(R_2) = \mu_{X_1^3}(x_1)^* \mu_{X_2^5}(x_2)^* \mu_{Y^3}(y) = 1^*1^*1 = 1$$

Of the rules specified, the second training data-tuple has zero membership in Rule 1, R_1, and full membership in Rule 2, R_2. Thus, it was beneficial to add Rule 2 to the rule-base, since it represents $d^{(2)}$ better than any other existing rule.

Now we move on to the third data-tuple in the training data set:

$$d^{(3)} = \left(\begin{bmatrix} 3 \\ 6 \end{bmatrix}, 6 \right)$$

Again the existing rules in the rule-base do not model $d^{(3)}$, thus another rule should be added. Since the data set is so small we find that we are including an additional rule for each data-tuple in the data set. Ideally we do not prefer that a rule be specified for every data-tuple in the training set; although, earlier, it was mentioned that Z is a representative portion of a larger data set, so in this example there would be more data points in the entire set. If we were to train a fuzzy system with a much larger data set Z, we would find that there will not be a rule for each of the m data-tuples in Z. Some rules will adequately represent more than one data pair. Nevertheless, the system can be improved by designating the fuzzy sets X_1^j, X_2^k, and Y^l differently and perhaps avoiding the addition of unnecessary rules to the rule-base. We do not attempt this correction here but continue with the addition of another rule to the rule-base:

$$R_3 = \text{ if } x_1 \text{ is } X_1^4 \text{ and } x_2 \text{ is } X_2^7 \text{ then } y \text{ is } Y^3.$$

$$\text{degree}(R_1) = \mu_{X_1^1}(x_1)^* \mu_{X_2^3}(x_2)^* \mu_{Y^1}(y) = 0^*0^*0 = 0$$

$$\text{degree}(R_2) = \mu_{X_1^3}(x_1)^* \mu_{X_2^5}(x_2)^* \mu_{Y^3}(y) = 0^*0^*0.5 = 0$$

$$\text{degree}(R_3) = \mu_{X_1^4}(x_1)^* \mu_{X_2^7}(x_2)^* \mu_{Y^3}(y) = 1^*1^*0.5 = 0.5$$

The third data-tuple has 0.5 membership in Rule 3 but has zero membership in the other rules. We now have a complete rule-base and, thus, a fuzzy model for the system. LFE is a very useful algorithm for developing a fuzzy model because it can be used as a basis in other algorithms for developing a stronger model. For instance, it can be used to form a rule-base which can then be improved in the RLS algorithm to refine the fuzzy model.

MODIFIED LEARNING FROM EXAMPLE

Unlike the LFE algorithm which relies entirely on user-specified membership functions, the MLFE calculates both membership functions and rules.

In this example we again employ delta functions for the outputs and Gaussian membership functions for the inputs. Again, b_i is the output value for the ith rule, c_j^i is the point in the jth input universe of discourse where the membership function for the ith rule achieves a maximum, and σ_j^i ($\sigma_j^i > 0$) is the spread of the membership function for the jth input and the ith rule. Recall that θ is a vector composed of the above parameters, where the dimensions of θ are determined by the number of inputs n and the number of rules R in the rule-base.

Again we use the data in Table 7.1 to develop a fuzzy model. Let us begin by specifying an "initial fuzzy system" that the MLFE procedure will use to initialize the parameters in θ. We initiate the process by setting the number of rules equal to 1, $R = 1$, and for b_1, c_1^1, and c_2^1 we use the first training data-tuple in Z and the spreads will be assumed to be equal to 0.5. It is important to note that the spreads cannot be set at zero, to avoid a division by zero error in the algorithm.

$$b_1 = y^1 = 1 \quad c_1^1 = 0 \quad c_2^1 = 2 \quad \sigma_1^1 = \sigma_2^1 = 0.5$$

For this example we would like the fuzzy system to approximate the output to within a tolerance of 0.25, thus we set $\varepsilon_f = 0.25$. We also introduce a weighting factor W which is used to calculate the spreads for the membership functions, as given later in Eq. (7.28). The weighting factor W is used to determine the amount of overlap between the membership function of the new rule and that of its nearest neighbor. This is demonstrated later in this section. For this example we will set the value of W equal to 2.

Following the initial procedure for the first data-tuple, we use the second data-tuple,

$$(x^2, y^2) = \left(\begin{bmatrix} 2 \\ 4 \end{bmatrix}, 5 \right)$$

and compare the data-tuple output portion y^2 with the existing fuzzy system output value, $f(x^2|\theta)$. The existing fuzzy system contains only the one rule which was previously added to the rule-base:

$$f(x^2|\theta) = \frac{1 * \exp\left[-\frac{1}{2} \left(\frac{2-0}{0.5} \right)^2 \right] * \exp\left[-\frac{1}{2} \left(\frac{4-2}{0.5} \right)^2 \right]}{\exp\left[-\frac{1}{2} \left(\frac{2-0}{0.5} \right)^2 \right] * \exp\left[-\frac{1}{2} \left(\frac{4-2}{0.5} \right)^2 \right]} = 1$$

Next we determine how accurate or adequate our fuzzy system is at mapping the information. To do this we take the absolute value of Eq. (7.15), where m denotes the data-tuple $m = 2$:

$$|f(x^2|\theta) - y^2| = |1 - 5| = 4$$

This value is much greater than the tolerance specified, $\varepsilon_f = 0.25$. Thus we add a rule to the rule-base to represent (x^2, y^2) by modifying the current parameters θ by letting $R = 2, b_2 = y^2$, and $c_j^2 = x_j^2$ for $j = 1, 2$:

$$b_2 = 5 \quad c_1^2 = 2 \quad c_2^2 = 4$$

If the calculated ε_f had been less than the specified tolerance there would have been no need to add another rule.

Next we develop the spreads σ_j^i to adjust the spacing between membership functions for the new rule. MLFE does this in a manner that does not distort what has already been learned, so there is a smooth interpolation between training points. The development of σ_j^i

for $i = R$ is accomplished through a series of steps, the first step being the calculation of the distances between the new membership function center and the existing membership function centers for each input,

$$|c_j^{i'} - c_j^i| \tag{7.26}$$

These values are then placed in the vector $\overline{h}_j$ below:

$$\overline{h}_j^{i'} = \{|c_j^{i'} - c_j^i| : i' = 1, 2, \ldots, R, \quad i' \neq i\}$$

For each input j the smallest nonzero element of this vector of distances $\overline{h}_j^{i'}$ is the nearest neighbor to the newly developed input membership function center for any added rule. This nearest neighbor is then used to create the relative width of the membership function for the new rule. If all the distances are zero in this vector, which means all the membership functions defined before are at the same center value, do not modify the relative width of the new rule but keep the same width defined as for the other input membership functions. The determination of this minimum is accomplished using

$$k_j = \min(\overline{h}_j) \tag{7.27}$$

where $j = 1, 2, \ldots, n$ and c_j^i are fixed.

For instance, in our example we have the following:

$$i' = 1 \text{ and } j = 1 \quad |c_j^{i'} - c_j^i| : i = 2$$
$$|c_1^1 - c_1^2| = |2 - 0| = 2$$
$$i' = 1 \text{ and } j = 2 \quad |c_j^{i'} - c_j^i| : i = 2$$
$$|c_2^1 - c_2^2| = |4 - 2| = 2$$

and our vector of distances for each input consists of only one nonzero value:

$$\overline{h}_1^2 = \{2\} \quad \overline{h}_2^2 = \{2\}$$

Therefore, c_1^1 is closest to c_1^2 and is used to develop the relative width of the membership function for the first input of the new rule and c_2^1 is the nearest neighbor to c_2^2. The weighting factor W is used to determine the amount of overlap between the membership function of the new rule with that of its nearest neighbor. This is accomplished using

$$\sigma_j^{i'} = \frac{1}{W}|c_j^{i'} - c_j^{\min}| \tag{7.28}$$

where $c_j^{\min}$ is the nearest membership function center to the new membership function center $c_j^{i'}$.

As mentioned above, we select a value of $W = 2$ which results in the following relative widths for the membership functions of the new rule:

$$\sigma_1^2 = \frac{1}{W}|c_1^2 - c_1^1| = \frac{1}{2}|2 - 0| = 1$$

$$\sigma_2^2 = \frac{1}{W}|c_2^2 - c_2^1| = \frac{1}{2}|4 - 2| = 1$$

The spread (relative width of the membership function) could be altered by increasing or decreasing W, which results in more or less, respectively, overlapping of the input membership functions.

We now have a rule-base consisting of two rules and θ consists of

$$\theta = [b_1, b_2, c_1^1, c_2^1, c_1^2, c_2^2, \sigma_1^1, \sigma_2^1, \sigma_1^2, \sigma_2^2]^T$$

where

$$b_1 = 1 \qquad b_2 = 5$$
$$c_1^1 = 0 \qquad c_1^2 = 2$$
$$c_2^1 = 2 \qquad c_2^2 = 4$$
$$\sigma_1^1 = 0.5 \qquad \sigma_1^2 = 1$$
$$\sigma_2^1 = 0.5 \qquad \sigma_2^2 = 1$$

We now move on to the third data-tuple in the training set to determine if our rule-base is adequate and if we have to include any additional rules:

$$f(x^3|\theta) = \frac{1 * \exp\left[-\frac{1}{2}\left(\frac{3-0}{1}\right)^2\right] * \exp\left[-\frac{1}{2}\left(\frac{6-2}{1}\right)^2\right] + 5 * \exp\left[-\frac{1}{2}\left(\frac{3-2}{1}\right)^2\right] * \exp\left[-\frac{1}{2}\left(\frac{6-4}{1}\right)^2\right]}{\exp\left[-\frac{1}{2}\left(\frac{3-0}{1}\right)^2\right] * \exp\left[-\frac{1}{2}\left(\frac{6-2}{1}\right)^2\right] + \exp\left[-\frac{1}{2}\left(\frac{3-2}{1}\right)^2\right] * \exp\left[-\frac{1}{2}\left(\frac{6-4}{1}\right)^2\right]}$$

$$f(x^3|\theta) = \frac{0.410428}{0.08208872} = 5$$

Again we determine how accurate or adequate our fuzzy system is at mapping the information:

$$|f(x^3|\theta) - y^3| = |5 - 6| = 1 > \varepsilon_f = 0.25$$

According to our specified tolerance, another rule is necessary to adequately model our training data:

$$b_3 = 6 \quad c_1^3 = 2 \quad c_2^3 = 4$$

We now can determine the spreads for the third rule. We begin with Eq. (7.26) to form the vector of distances:

$$i' = 3 \text{ and } j = 1 \quad \{|c_j^{i'} - c_j^i| : i = 1, 2\}$$
$$\{|c_1^3 - c_1^1|, |c_1^3 - c_1^2|\}$$
$$\{3, 1\}$$
$$i' = 3 \text{ and } j = 2 \quad \{|c_j^{i'} - c_j^i| : i = 1, 2\}$$
$$\{|c_2^3 - c_2^1|, |c_2^3 - c_2^1|\}$$
$$\{4, 2\}$$
$$\overline{h}_1^3 = \left\{\begin{matrix} 3 \\ 1 \end{matrix}\right\} \quad \overline{h}_2^3 = \left\{\begin{matrix} 4 \\ 2 \end{matrix}\right\}$$

In the distance vector $\overline{h}_1^3$ the minimum nonzero value is 1 and in the distance vector $\overline{h}_2^3$ the minimum nonzero value is 2. Therefore, the nearest neighbor to the first input membership function center is c_1^2 and the nearest neighbor to the second input membership function center is c_2^2, which are both used with the weighting factor and Eq. (7.28) to calculate the spreads for each membership function:

$$\sigma_1^3 = \frac{1}{W}|c_1^3 - c_1^2| = \frac{1}{2}|3 - 2| = \frac{1}{2}$$

$$\sigma_2^3 = \frac{1}{W}|c_2^3 - c_2^2| = \frac{1}{2}|6 - 4| = 1$$

The resulting rule-base is the following:

$$\theta = [b_1, b_2, b_3, c_1^1, c_2^1, c_1^2, c_2^2, c_1^3, c_2^3, \sigma_1^1, \sigma_2^1, \sigma_1^2, \sigma_2^2, \sigma_1^3, \sigma_2^3]^{\mathrm{T}}$$

where $b_1 = 1$ $b_2 = 5$ $b_3 = 6$

$c_1^1 = 0$ $c_1^2 = 2$ $c_1^3 = 2$

$c_2^1 = 2$ $c_2^2 = 4$ $c_2^3 = 4$

$\sigma_1^1 = 0.5$ $\sigma_1^2 = 1$ $\sigma_1^3 = 0.5$

$\sigma_2^1 = 0.5$ $\sigma_2^2 = 1$ $\sigma_2^3 = 1$

Example 7.1. Fuzzy modeling has been employed to generate a rule-base to model a "syntactic foam" from only limited input–output data obtained through various compressive tests. Due to their light weight and high compressive strength, foams have been incorporated into the design of many engineering systems. This is especially true in the aircraft industry where these characteristics are extremely important. For instance, aluminum syntactic foam is theorized to have a shear strength that is three times greater than existing aluminum [Erikson, 1999]. Syntactic foams are composite materials formed by mechanically combining manufactured material bubbles or microspheres with resin. They are referred to as syntactic because the microspheres are arranged together, unlike blown foams which are created by injecting gas into a liquid slurry causing the bubbles to solidify and producing foam [Erikson, 1999]. As is often the case with newly developed materials, the cost of preparing the material is high; thus, only limited information is available on the material.

A newly developed syntactic foam has been selected as an encasing material due to its light weight, high compressive strength, isotropic behavior, low porosity, and because the material is noncombustible and can be machined to a specific shape and tolerance. Due to the high costs involved in preparing the syntactic foam, only a limited amount has been made. Fortunately, in addition to the two pieces of syntactic foam specimens prepared for use as the encasing material, four specimens were prepared for conducting triaxial compression tests. The four specimens each had a height of 2.800 ± 0.001 inches and a diameter of 1.400 ± 0.001 inches, as shown in Fig. 7.11. The triaxial compression test is capable of applying two different compressive stresses to the sample, a radial (minor principal) and a longitudinal (major principal) stress. In each test performed the compressive pressure (stress) was gradually increased causing the sample to deform. The first test applied a continuous equivalent major and minor principal compressive stress to the specimen (hydrostatic compression) and the yield stress was observed at about 10,000 psi. In the second test the major and minor principal stresses were gradually increased to a value of 3750 psi; the minor principal stress was then held constant at this value while the major principal stress was continuously increased. In order to probe a good portion of stress space the same procedure was followed for the third and fourth tests; however, the minor principal stress was held constant at values of 6500 and 9000 psi, respectively, for these.

In each of the previous tests, the experimentalist found that maintaining the minor principal stress constant was difficult and the minor principal stress was noted to fluctuate by as much as ± 200 psi. The experimentalist also noted that at about the yielding stress the syntactic

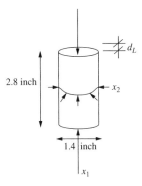

FIGURE 7.11
Syntactic foam cylinders.

foam began to harden and exhibit nonlinear behavior. A portion of the nonlinear data collected from the triaxial compression tests is provided in Table 7.2. Using the information provided by the experimentalist and the portion of the input–output data we develop a fuzzy model using the modified learning from example (MLFE) algorithm to obtain a general rule-base for governing the system, followed by the recursive least squares (RLS) algorithm to fine-tune the rule-base and the model parameters.

We begin by specifying a rule and its parameters that is used by the MLFE algorithm to develop the remainder of the rules in the rule-base. To do this we use one of the data-tuples from

TABLE 7.2
Major and minor principal stress and resulting longitudinal deformation

Training set			Testing set		
Major (x_1 psi)	Minor (x_2 psi)	δ_L (inch)	Major (x_1 psi)	Minor (x_2 psi)	δ_L (inch)
12,250	3750	3.92176E-2	12,911.1	12,927	2.0273E-2
11,500	6500	2.90297E-2	11,092.4	10,966.4	1.59737E-2
11,250	9000	2.51901E-2	14,545.8	14,487.1	2.40827E-2
11,000	11,000	1.63300E-2	13,012.1	12,963.8	2.02157E-2
11,960	9000	2.50463E-2	12,150	3750	3.8150E-2
12,140	6510	3.22360E-2	12,904	3744.8	4.28953E-2
13,000	3750	4.37127E-2	14,000	3770.4	5.05537E-2
13,800	3750	4.91016E-2	11,406	6520.3	2.83967E-2
12,950	6500	3.60650E-2	12,100	6535.5	3.2120E-2
12,600	9000	2.80437E-2	13,109	6525.3	3.69773E-2
11,170	11,110	1.65160E-2	11,017.6	9000.7	2.11967E-2
12,930	13,000	2.02390E-2	12,105.9	8975.1	2.57443E-2
13,000	12,500	2.36600E-2	12,700	900.0	2.8504E-2
11,130	9000	2.19333E-2			
11,250	6500	2.77190E-2			
12,000	3750	3.88243E-2			
12,900	3750	4.28953E-2			
11,990	6500	3.15040E-2			
12,010	9000	2.58113E-2			
12,000	12,000	1.95610E-2			
14,490	14,450	2.41490E-2			
12,900	9000	2.97153E-2			
13,220	6500	3.81633E-2			
14,000	3750	5.05537E-2			

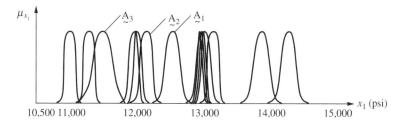

FIGURE 7.12
Membership functions for major principal stress.

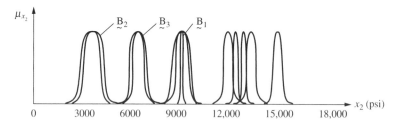

FIGURE 7.13
Membership functions for minor principal stress.

the training set: say we use $x_1 = 12,600$ psi and $x_2 = 9000$ psi as initial input membership function centers and $\delta_L = 2.8$ inch as the initial output (to simplify the accounting we use 2.8 rather than 0.028). We set the relative width of the first input membership functions center at 250 based on the range between similar inputs. For instance, two similar input tuples are $x_1 = 12,250$ psi and $x_2 = 3750$ psi, and $x_1 = 12,100$ psi and $x_2 = 3750$ psi. The relative width of the second input membership function is set at 150 based on the information provided by the experimentalist. For the weighting factor we choose a value of 2.1 and set the test factor equal to 0.45. MLFE uses this to develop the remainder of the membership functions (Figs. 7.12 and 7.13) and the rule-base shown in Table 7.3. Note that the actual output is moved two decimal places to the left because of the simplification to the output mentioned above.

For example,

$$\text{If } A_1 \text{ and } B_1 \text{ then } f(x) = 2.80 \times 10^{-2}$$

$$\text{If } A_2 \text{ and } B_2 \text{ then } f(x) = 3.9218 \times 10^{-2}$$

$$\text{If } A_3 \text{ and } B_3 \text{ then } f(x) = 3.9218 \times 10^{-2}$$

We now have the information needed to employ the RLS algorithm to improve our fuzzy model.

Before we move on to specifying the rule parameters needed by the RLS algorithm we point out that the experimentalist, in this example, observed that the minor principal stress fluctuated by as much as 200 psi and the spread for this input developed by MLFE is as high as 1309 psi. This is a direct result of the weighting factor, which forces the adjacent membership functions to approach one another by increasing the spread. For instance, the spreads of the membership functions with $c_1 = 12,250$ psi and $c_1 = 12,000$ psi increase. Since the input membership functions are significantly apart from one another, a low value for the weighting factor forces the spread of the membership functions to be greater. Both of the inputs (x_1 and x_2) are based on the one weighting factor and we can improve the spreads by designating one

TABLE 7.3
Rule-base

Input parameters				Output parameter $f(x)$ (inch)
c_1(psi)	c_2 (psi)	σ_1	σ_2	
$\underset{\sim}{A_1} = 12,600$	$\underset{\sim}{B_1} = 9000$	250	25	2.80E-2
$\underset{\sim}{A_2} = 12,250$	$\underset{\sim}{B_2} = 3750$	166.7	2500	3.9218E-2
$\underset{\sim}{A_3} = 11,500$	$\underset{\sim}{B_3} = 6500$	357.143	1190.476	2.9030E-2
11,250	9000	119.048	1190.476	2.2519E-2
11,000	11,000	119.048	950.2381	1.6330E-2
11,960	9000	138.095	950.2381	2.5046E-2
13,000	3750	190.476	1309.524	4.3712E-2
13,800	3750	380.952	1309.524	4.9102E-2
12,950	6500	23.810	1190.476	3.6065E-2
12,930	13,000	9.524	952.381	2.0239E-2
13,000	12,500	23.81	238.095	2.3660E-2
12,000	12,000	19.048	238.095	1.9561E-2
14,490	14,450	328.571	690.476	2.4149E-2
13,220	6500	104.762	1166.667	3.8163

TABLE 7.4
Rule-base parameters for RLS algorithm

Input parameters				Regression parameter $\hat{\theta}$
c_1 (psi)	c_2 (psi)	σ_1	σ_2	
12,600	9000	250	250	5
12,250	3750	166.7	250	5
11,500	6500	357.143	250	5
11,250	9000	119.048	250	5
11,000	11,000	119.048	250	5
11,960	9000	138.095	250	5
13,000	3750	190.476	250	5
13,800	3750	380.952	250	5
12,950	6500	23.810	250	5
12,930	13,000	9.524	250	5
13,000	12,500	23.81	250	5
12,000	12,000	19.048	250	5
14,490	14,450	328.571	250	5
13,220	6500	104.762	250	5

weighting factor for each input. For now we simply modify the spread of the second input to about 250 psi and continue with the RLS algorithm.

The rule-base used by the RLS algorithm is shown in Table 7.4; additionally, we use the data-tuples of Table 7.2 as the training set, and the initial $\hat{\theta}$ (a vector, in this case consisting of 14 values for the 14 rules used) with values equal to 5. We use nonweighted least squares regression (so $\lambda = 1$), set $\alpha = 2000$, and cycle through the training set 100 times to develop a fuzzy model.

TABLE 7.5

Resulting values included in vector $\hat{\theta}$ from RLS algorithm

$\hat{\theta}$	2.8884	3.9018	3.0122	2.2226	1.6423	2.5246	4.2712	4.9832	3.5991
$\hat{\theta}$	2.0232	2.4142	1.9561	2.4149	3.8163				

TABLE 7.6

Predicted output values for the testing set

Testing set			Predicted output δ_L (inch)
Major (x_1 psi)	**Minor (x_2 psi)**	**δ_L (inch)**	
12,911.1	12,927	2.0273E-2	2.0231E-2
11,092.4	10,966.4	1.59737E-2	1.6423E-2
14,545.8	14,487.1	2.40827E-2	2.4149E-2
13,012.1	12,963.8	2.02157E-2	2.0820E-2
12,150	3750	3.8150E-2	3.9020E-2
12,904	3744.8	4.28953E-2	4.3184E-2
14,000	3770.4	5.05537E-2	4.9384E-2
11,406	6520.3	2.83967E-2	3.0122E-2
12,100	6535.5	3.2120E-2	3.0122E-2
13,109	6525.3	3.69773E-2	3.8163E-2
11,017.6	9000.7	2.11967E-2	2.2226E-2
12,105.9	8975.1	2.57443E-2	2.5963E-2
12,700	900.0	2.8504E-2	2.8884E-2

Once the fuzzy model has been developed we use the testing data of Table 7.2 to verify that the model is working properly. The resulting values included in vector $\hat{\theta}$ are shown in Table 7.5 and the testing set with its predicted output is shown Table 7.6.

SUMMARY

This chapter has summarized six methods for use in developing fuzzy systems from input–output data. Of these six methods, the LFE, MLFE, and clustering methods can be used to develop fuzzy systems from such data. The remaining three methods, RLS, BLS, and the gradient methods, can be used to take fuzzy systems which have been developed by the first group of methods and refine them (or *tune* them) with additional training data.

Fuzzy systems are useful in the formulation and quantification of human observations. These observations eventually are manifested in terms of input–output data; these data can take the form of numbers, images, audio records, etc., but they all can be reduced to unit-based quantities on the real line. We might ask: "in what forms are the human observations manifested?" We could suggest that one form is linguistic, which would be a *conscious* form where a human could express directly their observations as knowledge expressed in the form of rules. Or, the form of knowledge could be *subconscious* in the sense that the observations could be measured or monitored by some device, but where the behavior cannot yet be reduced to rules by a human observer (an example of this would be the parallel parking of a car: a human can do it, but they might have difficulty describing what they do in a canonical rule form). The latter form of information, that from measurable observations, comprises the input–output data that are dealt with in this chapter. Once the rules of the system are derived from the input–output data, they can be combined with other rules, perhaps rules that come from a human, and together (or separately) they provide a contextual meaning to the underlying physics of the problem.

But, whether the rules come directly from a human or from methods such as those illustrated in this chapter, the resulting simulations using fuzzy systems theory are the same. A unique property of a fuzzy system is that the rules derived from the observational data provide *knowledge* of the system being simulated; this knowledge is in the form of linguistic rules which are understandable to a human in terms of the underlying physical system behavior. In contrast to this, other model-free methods such as neural networks can also be used in simulation, but the information gleaned about the number of layers, number of neurons in each layer, path weights between neurons, and other features of a neural network reveals to the human almost nothing about the physical process being simulated. This is the power and utility of a fuzzy system model.

REFERENCES

Erikson, R. (1999). "Mechanical engineering," *Am. Soc. Mech. Engs.*, vol. 121, no. 1, pp. 58–62.
Passino, K. and Yurkovich, S. (1998). *Fuzzy Control*, Addison Wesley Longman, Menlo Park, CA.

PROBLEMS

7.1. Earlier in this chapter the *recursive least squares* algorithm was demonstrated using training set Z of Table 7.1; however, only one cycle was performed using the input data tuples. Perform an additional cycle with the input data-tuples to determine if the predicted output changes.

7.2. Earlier in this chapter the *recursive least squares* algorithm was demonstrated using training set Z of Table 7.1.

(*a*) Modify the input membership functions centers to the following values and develop a fuzzy model for the Z of Table 7.1 using the RLS algorithm (perform two cycles). Note that the remaining rule-base parameters are the same as those used in the text, e.g.,

$$\sigma_j^i = 2 \quad \text{and} \quad \hat{\theta}(0) = \begin{bmatrix} 0.3647 \\ 8.1775 \end{bmatrix}$$

$$c_1^1 = 1 \quad c_1^2 = 3$$

$$c_2^1 = 3 \quad c_2^2 = 6$$

(*b*) In part (*a*) the input membership function centers were slightly different than those used in the text but the spreads were the same values as those used in the text. Now change the spreads to the following values, $\sigma_1^1 = 3, \sigma_1^2 = 2, \sigma_2^1 = 1, \sigma_2^2 = 2$, and the remaining values should be the same as in part (*a*). Develop a fuzzy model for the Z of Table 7.1 using the RLS algorithm (perform two cycles.)

7.3. Using the software provided on the publisher's website (see the preface), improve the output of the *recursive least squares* model presented in Example 7.1 by modifying the input membership function parameters of Table 7.4 using the *gradient method*.

7.4. Earlier in the chapter an example using the *gradient method* was presented using only one time step; provide the second and third time steps.

7.5. Using the *clustering* method develop a fuzzy model for the input–output data presented in Table 7.2.

7.6. In Example 7.1 an initial rule was specified for the application of the MLFE algorithm. Change these values for the input membership function centers to $x_1 = 12,140$ psi and $x_2 = 6510$ psi and the output membership function center to $d_L = 3.22360$ inch. Also incorporate two test factors, one for x_1 and another for x_2, equal to 2.1 and 5 respectively. Keeping all other values the same as in Example 7.1 develop a fuzzy model using MLFE. Will the predictability of the developed fuzzy model be improved by changing the test factor for x_2? If so, does increasing or decreasing the test factor improve the output MLFE?

7.7. Using the *batch least squares* algorithm improve the fuzzy model produced in Problem 7.4.

CHAPTER
8

FUZZY SYSTEMS
SIMULATION

As the complexity of a system increases, our ability to make precise and yet significant statements about its behavior diminishes until a threshold is reached beyond which precision and significance (or relevance) become almost mutually exclusive characteristics.

Lotfi Zadeh
Professor, Systems Engineering, 1973

The real world is complex; complexity in the world generally arises from uncertainty in the form of ambiguity. Problems featuring complexity and ambiguity have been addressed subconsciously by humans since they could think; these ubiquitous features pervade most social, technical, and economic problems faced by the human race. Why then are computers, which have been designed by humans after all, not capable of addressing complex and ambiguous issues? How can humans reason about real systems, when the complete description of a real system often requires more detailed data than a human could ever hope to recognize simultaneously and assimilate with understanding? The answer is that humans have the capacity to reason approximately, a capability that computers currently do not have. In reasoning about a complex system, humans reason approximately about its behavior, thereby maintaining only a generic understanding about the problem. Fortunately, this generality and ambiguity are sufficient for human comprehension of complex systems. As the quote above from Dr. Zadeh's *principle of incompatibility* suggests, complexity and ambiguity (imprecision) are correlated: "The closer one looks at a real-world problem, the fuzzier becomes its solution" [Zadeh, 1973].

As we learn more and more about a system, its complexity decreases and our understanding increases. As complexity decreases, the precision afforded by computational methods becomes more useful in modeling the system. For systems with little complexity and little uncertainty, closed-form mathematical expressions provide precise descriptions of the systems. For systems that are a little more complex, but for which significant data

Fuzzy Logic with Engineering Applications, Second Edition   T. J. Ross
© 2004 John Wiley & Sons, Ltd   ISBNs: 0-470-86074-X (HB); 0-470-86075-8 (PB)

exist, model-free methods, such as artificial neural networks, provide a powerful and robust means to reduce some uncertainty through learning, based on patterns in the available data; unfortunately this learning is very shallow. For very complex systems where few numerical data exist and where only ambiguous or imprecise information may be available, fuzzy reasoning provides a way to understand system behavior by allowing us to interpolate approximately between observed input and output situations. Finally, for the most complex problems there are required forms of learning due to induction, or combinations of deduction and induction, that are necessary for even a limited level of understanding. This text does not address the most complex forms of learning due to induction, but the deductive methods involved in fuzzy reasoning are addressed here in terms of fuzzy systems models.

In constructing a fuzzy system model, Klir and Yuan [1995] describe the relationship among three characteristics that can be thought to maximize a model's usefulness. These characteristics of any model are: complexity, credibility, and uncertainty. This relationship is only known in an abstract sense. Uncertainty, of course, plays a crucial role in any efforts to maximize a systems model; but this crucial role can only be considered in the context of the other two characteristics. For example, allowing more uncertainty in a model reduces complexity and increases credibility of the resulting model. In developing models of complex systems one needs to seek a balance of uncertainty and utility; a model that is extremely limited in terms of its robustness is one which cannot accommodate much uncertainty.

All models are mathematical abstractions of the real physical world. The more assumptions one needs to make to get the model into a form where known mathematical structures can be used to address the real problem, the more uncertainty has crept into the modeling process. To ignore this uncertainty is to ignore the real world, and our understanding of it. But, we can make the models robust and credible by addressing the fact that complexity and uncertainty are inextricably related; when one is high, the other tends to be high, just as described by the quote above, by Zadeh.

The illustrations in Fig. 8.1 and 8.2 provide some thoughts on the relationship of complexity and uncertainty. In Fig. 8.1 we see that the case of ignorance is a situation involving high levels of complexity and uncertainty. Ignorance is the case where we have no specific information and we have no ideas about the physics that might describe the behavior of a system; an example might be our attempt to understand the concept of *infinity*. When we have information about a problem, perhaps described by a random collection of data, we see that our complexity and uncertainty are reduced somewhat, but we still do not have good understanding about a problem; an example might be an attempt to invest in a new technology in the stock market. As uncertainty and complexity diminish further we get to the case of knowledge. In this instance we have enough information and enough learning about a problem that we can actually pose rules, or algorithms that describe the essential characteristics of a problem; an example here is the knowledge of a nonlinear system on the basis of a robust collection of deductive inferences. The final step in understanding I shall term wisdom. This is the case where uncertainty and complexity are at their lowest levels. This is because with wisdom we fully understand a problem in all its manifestations and possible configurations. An example of wisdom might be Newton's second law, where we have a mathematical algorithm that describes fully the relationship among mass, acceleration, and force.

In Fig. 8.2 this idea is illustrated in a more specific way. We might have a collection of data points that, together as a population of information, is meaningless without some

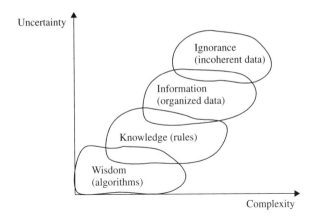

FIGURE 8.1
Complexity and uncertainty: relationships to ignorance, information, knowledge, and wisdom.

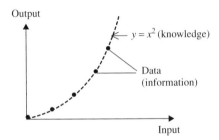

FIGURE 8.2
Specific example of information and knowledge in a simple one-dimensional relationship.

knowledge of how they are related. If we can find an algorithm that relates all of the data, and leaves no data point outside of its description, then we have a case of knowledge; we know the relationship between the inputs and outputs in a limited domain of applicability.

In the process of abstraction from the real world to a model, we need to match the model type with the character of the uncertainty exhibited in the problem. In situations where precision is available in the information fuzzy systems are less efficient than the more precise algorithms in providing us with the best understanding of the problem. On the other hand, fuzzy systems can focus on models characterized by imprecise or ambiguous information; such models are sometimes termed nonlinear models.

Virtually all physical processes in the real world are nonlinear or complex in some other way. It is our abstraction of the real world that leads us to the use of linear systems in modeling these processes. The linear systems are simple and understandable, and, in many situations, they provide acceptable simulations of the actual processes that we observe. Unfortunately, only the simplest of systems can be modeled with linear system theory and only a very small fraction of the nonlinear systems have verifiable solutions. The bulk of the physical processes that we must address are too complex to be reduced to algorithmic form – linear or nonlinear. Most observable processes have only a small amount of information available with which to develop an algorithmic understanding. The vast majority of information we have on most processes tends to be nonnumeric and nonalgorithmic. Most of the information is fuzzy and linguistic in form.

There is a quote from H. W. Brand [1961] that forms an appropriate introduction to matters in this chapter: "*There is no idea or proposition in the field, which cannot be put into mathematical language, although the utility of doing so can very well be doubted.*" We can always reduce a complicated process to simple mathematical form. And, for a while at least, we may feel comfortable that our model is a useful replicate of the process we seek to understand. However, reliance on simple linear, or nonlinear, models of the algorithmic form can lead to quite disastrous results, as many engineers have found in documented failures of the past.

A classic example in mechanics serves to illustrate the problems encountered in overlooking the simplest of assumptions. In most beginning textbooks in mechanics, Newton's second law is described by the following equation:

$$\sum F = m \cdot a \tag{8.1}$$

which states that the motion (acceleration) of a body under an imbalance of external forces acting on the body is equal to the sum of the forces ($\sum F$) divided by the body's mass (m). Specifically, the forces and acceleration of the body are vectors containing magnitude and direction. Unfortunately, Eq. (8.1) is not specifically Newton's second law. Newton hypothesized that the imbalance of forces was equivalent to the rate of change in the momentum ($m \cdot v$) of the body, i.e.,

$$\sum F = \frac{d(m \cdot v)}{dt} = m \cdot \frac{dv}{dt} + v \cdot \frac{dm}{dt} \tag{8.2}$$

where v is the velocity of the body and t is time. As one can see, Eqs. (8.1) and (8.2) are not equivalent unless the body's mass does not change with time. In many mechanics applications the mass does not change with time, but in other applications, such as in the flight of spacecraft or aircraft, where fuel consumption reduces total system mass, mass most certainly changes over time. It may be asserted that such an oversight has nothing to do with the fact that Newton's second law is not a valid algorithmic model, but rather it is a model that must be applied against an appropriate physical phenomenon. The point is this: algorithmic models are useful only when we understand and can observe all the underlying physics of the process. In the aircraft example, fuel consumption may not have been an observable phenomenon, and Eq. (8.1) could have been applied to the model. Most complex problems have only a few observables, and an understanding of all the pertinent physics is usually not available. As another example, Newton's first and second laws are not very useful in quantum mechanics applications.

If a process can be described algorithmically, we can describe the solution set for a given input set. If the process is not reducible to algorithmic form, perhaps the input–output features of the system are at least observable or measurable. This chapter deals with systems that cannot be simulated with conventional crisp or algorithmic approaches but that can be simulated because of the presence of other information – observed or linguistic – using fuzzy nonlinear simulation methods.

This chapter proposes to use fuzzy rule-based systems as suitable representations of simple and complex physical systems. For this purpose, a fuzzy rule-based system consists

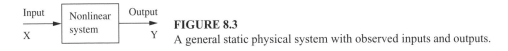

FIGURE 8.3
A general static physical system with observed inputs and outputs.

of (1) a set of rules that represent the engineer's understanding of the behavior of the system, (2) a set of input data observed going into the system, and (3) a set of output data coming from the system. The input and output data can be numerical, or they can be nonnumeric observations. Figure 8.3 shows a general static physical system, which could be a simple mapping from the input space to the output space, an industrial control system, a system identification problem, a pattern recognition process, or a decision-making process.

The system inputs and outputs can be vector quantities. Consider an n-input and m-output system. Let X be a Cartesian product of n universes X_i, for $i = 1, 2, \ldots, n$, i.e., $X = X_1 \times X_2 \times \cdots \times X_n$, and let Y be a Cartesian product of m universes Y_j for $j = 1, 2, \ldots, m$, i.e., $Y = Y_1 \times Y_2 \times \cdots \times Y_m$. The vector $\mathbf{X} = (X_1, X_2, \ldots, X_n)$ is the input vector to the system defined on real space R^n, and $\mathbf{Y} = (Y_1, Y_2, \ldots, Y_m)$ is the output vector of the system defined on real space R^m. The input data, rules, and output actions or consequences are generally fuzzy sets expressed by means of appropriate membership functions defined on an appropriate universe of discourse. The method of evaluation of rules is known as *approximate reasoning* or *interpolative reasoning* and is commonly represented by the composition of the fuzzy relations that are formed by the IF–THEN rules (see Chapter 5).

Three spaces are present in the general system posed in Fig. 8.3 [Ross, 1995]:

1. The space of possible conditions of the inputs to the system, which, in general, can be represented by a collection of fuzzy subsets $\underset{\sim}{A}^k$, for $k = 1, 2, \ldots$, which are fuzzy partitions of space X, expressed by means of membership functions

$$\mu_{\underset{\sim}{A}^k}(x) \quad \text{where } k = 1, 2, \ldots \tag{8.3}$$

2. The space of possible output consequences, based on some specific conditions of the inputs, which can be represented by a collection of fuzzy subsets $\underset{\sim}{B}^p$, for $p = 1, 2, \ldots$, which are fuzzy partitions of space Y, expressed by means of membership functions

$$\mu_{\underset{\sim}{B}^p}(y) \quad \text{where } p = 1, 2, \ldots \tag{8.4}$$

3. The space of possible mapping relations from the input space X onto the output space Y. The mapping relations are, in general, represented by fuzzy relations $\underset{\sim}{R}^q$, for $q = 1, 2, \ldots$, and expressed by means of membership functions

$$\mu_{\underset{\sim}{R}^q}(x, y) \quad \text{where } q = 1, 2, \ldots \tag{8.5}$$

A human perception of the system shown in Fig. 8.3 is based on experience and expertise, empirical observation, intuition, a knowledge of the physics of the system, or a set of subjective preferences and goals. The human observer usually puts this type of knowledge in the form of a set of unconditional as well as conditional propositions in natural language. Our understanding of complex systems is at a qualitative and declarative level,

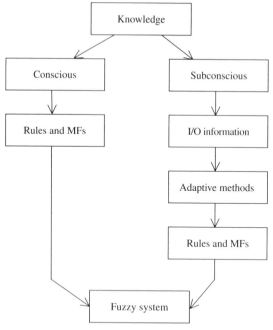

FIGURE 8.4
Two paths for knowledge to result in a fuzzy system.

based on vague linguistic terms; this is our so-called *fuzzy* level of understanding of the physical system. We have also seen in Chapters 6 and 7 that fuzzy rules and membership functions can be derived from a family of input–output data, but this form of the knowledge has the same utility whether it was derived by human understanding or observation, or developed in an automated fashion. Figure 8.4 shows this comparison. In the formulation of knowledge we can have two paths: a conscious path where the rules and membership functions (MFs) are derived intuitively by the human; and a subconscious path where we only have input–output (I/O) data or information and we use automated methods, such as those illustrated in Chapters 6 and 7, to derive the rules and MFs. The result of both paths is the construction of a fuzzy system, as shown in Fig. 8.4.

FUZZY RELATIONAL EQUATIONS

Consider a typical crisp nonlinear function relating elements of a single input variable, say x, to the elements of a single output variable, say y, as shown in Fig. 8.5. Notice in Fig. 8.5 that every x in the domain of the independent variable (each x') is "related" to a y (y') in the dependent variable (we call this relation a mapping in Chapter 12). The curve in Fig. 8.5 represents a transfer function, which, in generic terms, is a relation. In fact, any continuous-valued function, such as the curve in Fig. 8.5, can be discretized and reformulated as a matrix relation (see Chapter 12).

> **Example 8.1.** For the nonlinear function $y = x^2$, we can formulate a matrix relation to model the mapping imposed by the function. Discretize the independent variable x (the input variable)

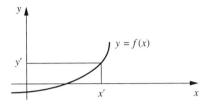

FIGURE 8.5
A crisp relation represented as a nonlinear function.

on the domain $x = -2, -1, 0, 1, 2$. We find that the mapping provides for the dependent variable y (the output variable) to take on the values $y = 0, 1, 4$. This mapping can be represented by a matrix relation, R, or

$$R = \begin{array}{c} \\ -2 \\ -1 \\ 0 \\ 1 \\ 2 \end{array} \begin{array}{ccc} 0 & 1 & 4 \\ \left[\begin{array}{ccc} 0 & 0 & 1 \\ 0 & 1 & 0 \\ 1 & 0 & 0 \\ 0 & 1 & 0 \\ 0 & 0 & 1 \end{array}\right] \end{array}$$

The elements in the crisp relation, R, are the indicator values as given later in Chapter 12 (by Eq. (12.2)).

We saw in Chapter 5 that a fuzzy relation can also represent a logical inference. The fuzzy implication IF $\underset{\sim}{A}$ THEN $\underset{\sim}{B}$ is known as the *generalized modus ponens* form of inference. There are numerous techniques for obtaining a fuzzy relation $\underset{\sim}{R}$ that will represent this inference in the form of a fuzzy relational equation given by

$$\underset{\sim}{B} = \underset{\sim}{A} \circ \underset{\sim}{R} \tag{8.6}$$

where $\circ$ represents a general method for composition of fuzzy relations. Equation (8.6) appeared previously in Chapter 5 as the generalized form of approximate reasoning, where Eqs. (5.4) and (5.5) provided two of the most common forms of determining the fuzzy relation $\underset{\sim}{R}$ from a single rule of the form IF $\underset{\sim}{A}$ THEN $\underset{\sim}{B}$.

NONLINEAR SIMULATION USING FUZZY SYSTEMS

Suppose our knowledge concerning a certain nonlinear process is not algorithmic, like the algorithm $y = x^2$ in Example 8.1, but rather is in some other more complex form. This more complex form could be data observations of measured inputs and measured outputs. Relations can be developed from these data that are analogous to a lookup table, and methods for this step have been given in Chapter 3. Alternatively, the complex form of the knowledge of a nonlinear process could be described with some linguistic rules of the form IF $\underset{\sim}{A}$ THEN $\underset{\sim}{B}$. For example, suppose we are monitoring a thermodynamic process involving an input heat, measured by temperature, and an output variable, pressure. We observe that when we use a "low" temperature, we get out of the process a "low" pressure; when we

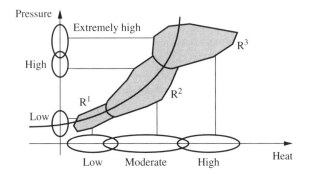

FIGURE 8.6
A fuzzy nonlinear relation matching patches in the input space to patches in the output space.

input a "moderate" temperature, we see a "high" pressure in the system; when we input "high" temperature into the thermodynamics of the system, the output pressure reaches an "extremely high" value; and so on. This process is shown in Fig. 8.6, where the inputs are now *not* points in the input universe (heat) and the output universe (pressure), but *patches* of the variables in each universe. These patches represent the fuzziness in describing the variables linguistically. Obviously, the mapping describing this relationship between heat and pressure is fuzzy. That is, patches from the input space map, or relate, to patches in the output space; and the relations R^1, R^2, and R^3 in Fig. 8.6 represent the fuzziness in this mapping. In general, all the patches, including those representing the relations, overlap because of the ambiguity in their definitions.

Each of the patches in the input space shown in Fig. 8.6 could represent a fuzzy set, say $\underset{\sim}{A}$, defined on the input variable, say x; each of the patches in the output space could be represented by a fuzzy set, say $\underset{\sim}{B}$, defined on the output variable, say y; and each of the patches lying on the general nonlinear function path could be represented by a fuzzy relation, say $\underset{\sim}{R}^k$, where $k = 1, 2, \ldots, r$ represents r possible linguistic relationships between input and output. Suppose we have a situation where a fuzzy input, say x, results in a series of fuzzy outputs, say y^k, depending on which fuzzy relation, $\underset{\sim}{R}^k$, is used to determine the mapping. Each of these relationships, as listed in Table 8.1, could be described by what is called a *fuzzy relational equation*, where y^k is the output of the system contributed by the kth rule, and whose membership function is given by $\mu_{y^k}(y)$. Both x and y^k $(k = 1, 2, \ldots, r)$ can be written as single-variable fuzzy relations of dimensions $1 \times n$ and $1 \times m$, respectively. The unary relations, in this case, are actually similarity relations between the elements of the fuzzy set and a most typical or prototype element, usually with membership value equal to unity.

The system of fuzzy relational equations given in Table 8.1 describes a general fuzzy nonlinear system. If the fuzzy system is described by a system of conjunctive rules, we could decompose the rules into a single aggregated fuzzy relational equation by making use of Eqs. (5.35)–(5.36) for each input, x, as follows:

$$y = (x \circ \underset{\sim}{R}^1) \text{ AND } (x \circ \underset{\sim}{R}^2) \text{ AND } \ldots \text{ AND } (x \circ \underset{\sim}{R}^r)$$

and equivalently,

$$y = x \circ (\underset{\sim}{R}^1 \text{ AND } \underset{\sim}{R}^2 \text{ AND } \ldots \text{ AND } \underset{\sim}{R}^r)$$

TABLE 8.1
System of fuzzy rela-
tional equations

$\underset{\sim}{R}^1$:	$y^1 = x \circ \underset{\sim}{R}^1$
$\underset{\sim}{R}^2$:	$y^2 = x \circ \underset{\sim}{R}^2$
$\vdots$	$\vdots$
$\underset{\sim}{R}^r$:	$y^r = x \circ \underset{\sim}{R}^r$

and finally

$$y = x \circ \underset{\sim}{R} \tag{8.7}$$

where $\underset{\sim}{R}$ is defined as

$$\underset{\sim}{R} = \underset{\sim}{R}^1 \cap \underset{\sim}{R}^2 \cap \cdots \cap \underset{\sim}{R}^r \tag{8.8}$$

The aggregated fuzzy relation $\underset{\sim}{R}$ in Eq. (8.8) is called the *fuzzy system transfer relation* for a single input, x. For the case of a system with n noninteractive fuzzy inputs (see Chapter 2), x_i, and a single output, y, described in Eq. (5.33), the fuzzy relational Eq. (8.7) can be written in the form

$$y = x_1 \circ x_2 \circ \cdots \circ x_n \circ \underset{\sim}{R} \tag{8.9}$$

If the fuzzy system is described by a system of disjunctive rules, we could decompose the rules into a single aggregated fuzzy relational equation by making use of Eqs. (5.37)–(5.38) as follows:

$$y = (x \circ \underset{\sim}{R}^1) \text{ OR } (x \circ \underset{\sim}{R}^2) \text{ OR } \ldots \text{ OR } (x \circ \underset{\sim}{R}^r)$$

and equivalently,

$$y = x \circ (\underset{\sim}{R}^1 \text{ OR } \underset{\sim}{R}^2 \text{ OR} \ldots \text{ OR } \underset{\sim}{R}^r)$$

and finally

$$y = x \circ \underset{\sim}{R} \tag{8.10}$$

where $\underset{\sim}{R}$ is defined as

$$\underset{\sim}{R} = \underset{\sim}{R}^1 \cup \underset{\sim}{R}^2 \cup \cdots \cup \underset{\sim}{R}^r \tag{8.11}$$

The aggregated fuzzy relation, i.e., $\underset{\sim}{R}$, again is called the *fuzzy system transfer relation*.

For the case of a system with n noninteractive (see Chapter 2) fuzzy inputs, x_i, and single output, y, described as in Eq. (5.33), the fuzzy relational Eq. (8.10) can be written in the same form as Eq. (8.9).

The fuzzy relation $\underset{\sim}{R}$ is very context-dependent and therefore has local properties with respect to the Cartesian space of the input and output universes. This fuzzy relation results from the Cartesian product of the fuzzy sets representing the inputs and outputs of the fuzzy nonlinear system. However, before the relation $\underset{\sim}{R}$ can be determined, one must consider the more fundamental question of how to *partition* the input and output spaces (universes of discourse) into meaningful fuzzy sets. Ross [1995] details methods in

TABLE 8.2
Canonical rule-based form of fuzzy
relational equations

$\underset{\sim}{R}^1$:	IF x is $\underset{\sim}{A}^1$, THEN y is $\underset{\sim}{B}^1$
$\underset{\sim}{R}^2$:	IF x is $\underset{\sim}{A}^2$, THEN y is $\underset{\sim}{B}^2$
$\vdots$	$\vdots$
$\underset{\sim}{R}^r$:	IF x is $\underset{\sim}{A}^r$, THEN y is $\underset{\sim}{B}^r$

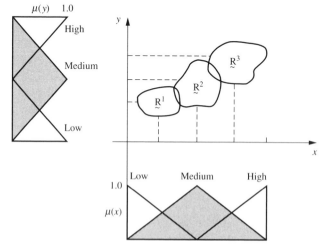

FIGURE 8.7
Fuzzy set inputs and fuzzy set outputs (the most general case).

partitioning that are due to human intuition, and Chapters 6 and 7 show how partitioning is a natural consequence of automated methods.

A general nonlinear system, such as that in Fig. 8.6, which is comprised of n inputs and m outputs, can be represented by fuzzy relational equations in the form expressed in Table 8.1. Each of the fuzzy relational equations, i.e., $\underset{\sim}{R}^r$, can also be expressed in canonical rule-based form, as shown in Table 8.2.

The rules in Table 8.2 could be connected logically by any of "and," "or," or "else" linguistic connectives; and the variables in Table 8.2, $\mathbf{x}$ and $\mathbf{y}$, are the input and output vectors, respectively, of the nonlinear system. Ross [1995] discusses in more details the various forms of nonlinear systems that can result from a rule-based approach, but this level of detail is not needed in conducting general nonlinear simulations. Only the most general form of a nonlinear system is considered here, shown in Fig. 8.7, where the inputs (x) and outputs (y) are considered as fuzzy sets, and where the input–output mappings (R) are considered as fuzzy relations.

FUZZY ASSOCIATIVE MEMORIES (FAMS)

Consider a fuzzy system with n noninteractive (see Chapter 2) inputs and a single output. Also assume that each input universe of discourse, i.e., $X_1, X_2, \ldots, X_n$, is partitioned into

k fuzzy partitions. Based on the canonical fuzzy model given in Table 8.2 for a nonlinear system, the total number of possible rules governing this system is given by

$$l = k^n \tag{8.12a}$$

$$l = (k + 1)^n \tag{8.12b}$$

where l is the maximum possible number of canonical rules. Equation (8.12b) is to be used if the partition "anything" is to be used, otherwise Eq. (8.12a) determines the number of possible rules. The actual number of rules, r, necessary to describe a fuzzy system is much less than l, i.e., $r \ll l$, because of the interpolative reasoning capability of the fuzzy model and because the fuzzy membership functions of the partitions overlap. If each of the n noninteractive inputs is partitioned into a different number of fuzzy partitions, say, X_1 is partitioned into k_1 partitions and X_2 is partitioned into k_2 partitions and so forth, then the maximum number of rules is given by

$$l = k_1 k_2 k_3 \cdots k_n \tag{8.13}$$

For a small number of inputs, e.g., $n = 1$ or $n = 2$, or $n = 3$, there exists a compact form of representing a fuzzy rule-based system. This form is illustrated for $n = 2$ in Fig. 8.8. In the figure there are seven partitions for input A (A_1 to A_7), five partitions for input B (B_1 to B_5), and four partitions for the output variable C (C_1 to C_4). This compact graphical form is called a fuzzy associative memory table, or FAM table. As can be seen from the FAM table, the rule-based system actually represents a general nonlinear mapping from the input space of the fuzzy system to the output space of the fuzzy system. In this mapping, the patches of the input space are being applied to the patches in the output space. Each rule or, equivalently, each fuzzy relation from input to the output represents a fuzzy point of data that characterizes the nonlinear mapping from the input to the output.

In the FAM table in Fig. 8.8 we see that the maximum number of rules for this situation, using Eq. (8.13), is $l = k_1 k_2 = 7(5) = 35$; but as seen in the figure, the actual number of rules is only $r = 21$.

We will now illustrate the ideas involved in simulation with three examples from various engineering disciplines.

Example 8.2. For the nonlinear function $y = 10 \sin x_1$, we will develop a fuzzy rule-based system using four simple fuzzy rules to approximate the output y. The universe of discourse for the input variable x_1 will be the interval $[-180, 180]$ in degrees, and the universe of discourse for the output variable y is the interval $[-10, 10]$.

Input B \ Input A	A_1	A_2	A_3	A_4	A_5	A_6	A_7
B_1	C_1		C_4	C_4		C_3	C_3
B_2		C_1				C_2	
B_3	C_4		C_1			C_2	C_2
B_4	C_3	C_3		C_1		C_1	C_2
B_5	C_3		C_4	C_4	C_1		C_3

FIGURE 8.8
FAM table for a two-input, single-output fuzzy rule-based system.

First, we will partition the input space x_1 into five simple partitions on the interval $[-180°, 180°]$, and we will partition the output space y on the interval $[-10, 10]$ into three membership functions, as shown in Figs. 8.9a and 8.9b, respectively. In these figures the abbreviations NB, NS, Z, PS, and PB refer to the linguistic variables "negative-big," "negative-small," "zero," "positive-small," and "positive-big," respectively.

Second, we develop four simple rules, listed in Table 8.3, that we think emulate the dynamics of the system (in this case the system is the nonlinear equation $y = 10 \sin x_1$ and we are observing the harmonics of this system) and that make use of the linguistic variables in Fig. 8.9. The FAM table for these rules is given in Table 8.4.

The FAM table of Table 8.4 is one-dimensional because there is only one input variable, x_1. As seen in Table 8.4, all rules listed in Table 8.3 are accommodated. Not all the four rules expressed in Table 8.3 are expressed in canonical form (some have disjunctive antecedents), but if they were transformed into canonical form, they would represent the five rules provided in the FAM table in Table 8.4.

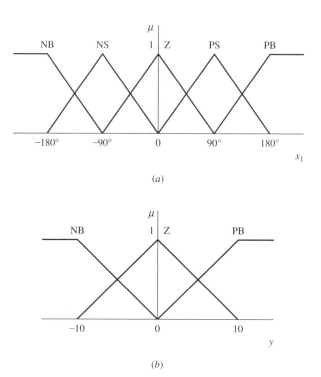

(a)

(b)

FIGURE 8.9
Fuzzy membership functions for the input and output spaces: (a) five partitions for the input variable, x_1; (b) three partitions for the output variable, y.

TABLE 8.3
Four simple rules for $y = 10 \sin x_1$

1	IF x_1 is Z or PB, THEN y is Z
2	IF x_1 is PS, THEN y is PB
3	IF x_1 is Z or NB, THEN y is Z
4	IF x_1 is NS, THEN y is NB

TABLE 8.4
FAM for the four simple rules in Table 8.3

x_i	NB	NS	Z	PS	PB
y	Z	NB	Z	PB	Z

In developing an approximate solution for the output y we select a few input points and employ a graphical inference method similar to that illustrated in Chapter 5. We will use the centroid method for defuzzification. Let us choose four crisp singletons as the input:

$$x_1 = \{-135°, -45°, 45°, 135°\}$$

For input $x_1 = -135°$, Rules 3 and 4 are fired, as shown in Figs. 8.10c and 8.10d. For input $x_1 = -45°$, Rules 1, 3, and 4 are fired. Figures 8.10a and 8.10b show the graphical inference for input $x_1 = -45°$ (which fires Rule 1), and for $x_1 = 45°$ (which fires Rule 2), respectively.

For input $x_1 = -45°$, Rules 3 and 4 are also fired, and we get results similar to those shown in Figs. 8.10c and 8.10d after defuzzification:

$$\text{Rule 3:} \quad y = 0$$

$$\text{Rule 4:} \quad y = -7$$

For $x_1 = 45°$, Rules 1, 2, and 3 are fired (see Fig. 8.10b for Rule 2), and we get the following results for Rules 1 and 3 after defuzzification:

$$\text{Rule 1:} \quad y = 0$$

$$\text{Rule 3:} \quad y = 0$$

For $x_1 = 135°$, Rules 1 and 2 are fired and we get, after defuzzification, results that are similar to those shown in Fig. 8.10b:

$$\text{Rule 1:} \quad y = 0$$

$$\text{Rule 2:} \quad y = 7$$

When we combine the results, we get an aggregated result summarized in Table 8.5 and shown graphically in Fig. 8.11. The y values in each column of Table 8.5 are the defuzzified results from various rules firing for each of the inputs, x_i. When we aggregate the rules using the union operator (disjunctive rules), the effect is to take the maximum value for y in each of the columns in Table 8.5 (i.e., maximum value irrespective of sign). The plot in Fig. 8.11 represents the maximum y for each of the x_i, and it represents a fairly accurate portrayal of the true solution, given only a crude discretization of four inputs and a simple simulation based on four rules. More rules would result in a closer fit to the true sine curve (see Problem 8.8).

Example 8.3. Suppose we want to model a serial transmission of a digital signal over a channel using RS232 format. Packets of information transmitted over the channel are ASCII characters composed of start and stop bits plus the appropriate binary pattern. If we wanted to know whether a valid bit was sent we could test the magnitude of the signal at the receiver using an absolute value function. For example, suppose we have the voltage (V) versus time trace shown in Fig. 8.12, a typical pattern. In this pattern the ranges for a valid mark and a valid space are as follows:

−12 to −3 V or +3 to +12 V	A valid mark (denoted by a one)
−3 to +3	A valid space (denoted by a zero)

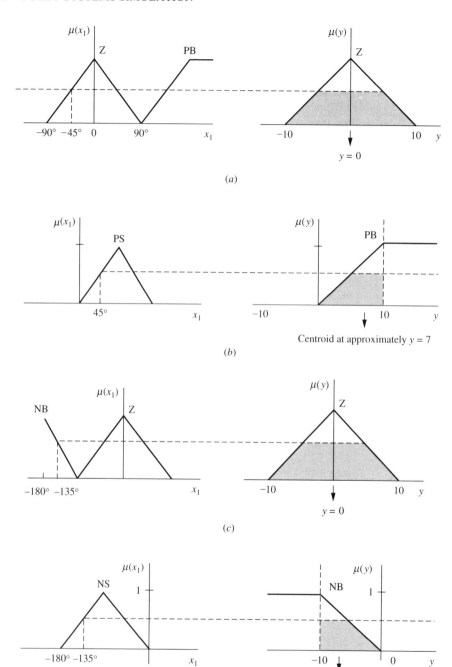

FIGURE 8.10

Graphical inference method showing membership propagation and defuzzification: (a) input $x_1 = -45°$ fires Rule 1; (b) input $x_1 = 45°$ fires Rule 2; (c) input $x_1 = -135°$ fires Rule 3; (d) input $x_1 = -135°$ fires Rule 4.

TABLE 8.5
Defuzzified results for simulation of $y = 10 \sin x_1$

x_1	$-135°$	$-45°$	$45°$	$135°$
y	0	0	0	0
	-7	0	0	7
		-7	7	

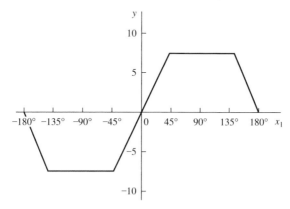

FIGURE 8.11
Simulation of nonlinear system $y = 10 \sin x_1$ using a four-rule fuzzy rule-base.

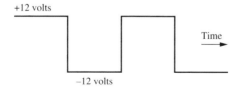

FIGURE 8.12
Typical pattern of voltage vs. time for a valid bit mark.

The absolute value function used to make this distinction is a nonlinear function, as shown in Fig. 8.13. To use this function on the scale of voltages $[-12, +12]$, we will attempt to simulate the nonlinear function $y = 12|x|$, where the range of x is $[-1, 1]$. First, we partition the input space, $x = [-1, 1]$, into five linguistic partitions as in Fig. 8.14. Next, we partition the output space. This task can usually be accomplished by mapping prototypes of input space to corresponding points in output space, if such information is available. Because we know the functional mapping (normally we would not know this for a real, complex, or nonlinear problem), the partitioning can be accomplished readily; we will use three equally spaced output partitions as shown in Fig. 8.15.

Since this function is simple and nonlinear, we can propose a few simple rules to simulate its behavior:

1. IF x = zero, THEN y = zero.
2. IF x = NS or PS, THEN y = PS.
3. IF x = NB or PB, THEN y = PB.

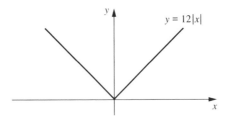

FIGURE 8.13
Nonlinear function $y = 12|x|$.

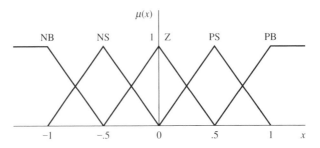

FIGURE 8.14
Partitions on the input space for $x = [-1, 1]$.

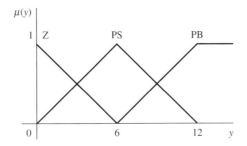

FIGURE 8.15
Output partitions on the range $y = [0, 12]$.

We can now conduct a graphical simulation of the nonlinear function expressed by these three rules. Let us assume that we have five input values, the crisp singletons $x = -0.6$, -0.3, 0, 0.3, and 0.6. The input $x = -0.6$ invokes (fires) Rules 2 and 3, as shown in Fig. 8.16. The defuzzified output, using the centroid method, for the truncated union of the two consequents is approximately 8. The input $x = -0.3$ invokes (fires) Rules 1 and 2, as shown in Fig. 8.17. The defuzzified output for the truncated union of the two consequents is approximately 5. The input $x = 0$ invokes (fires) Rule 1 only, as shown in Fig. 8.18. The defuzzified output for the truncated consequent ($y = Z$) is a centroidal value of 2. By symmetry it is easy to see that crisp inputs $x = 0.3$ and $x = 0.6$ result in defuzzified values for $y \approx 5$ and $y \approx 8$, respectively.

If we plot these simulated results and compare them to the exact relationship (which, again, we would not normally know), we get the graph in Fig. 8.19; the simulation, although approximate, is quite good.

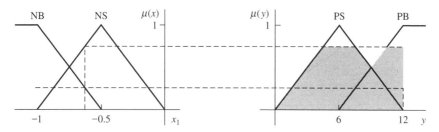

FIGURE 8.16
Graphical simulation for crisp input $x = -0.6$.

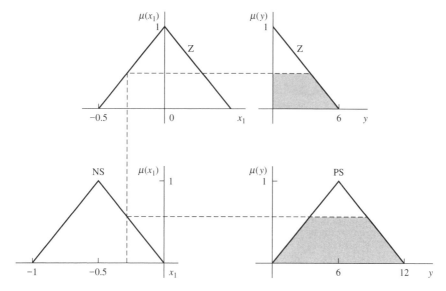

FIGURE 8.17
Graphical simulation for crisp input $x = -0.3$.

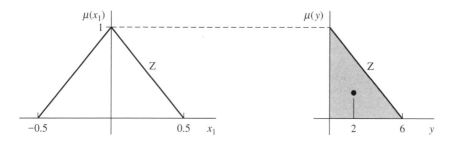

FIGURE 8.18
Graphical simulation for crisp input $x = 0$.

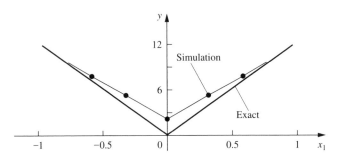

FIGURE 8.19
Simulated versus exact results for Example 8.3.

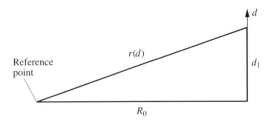

FIGURE 8.20
Schematic of aircraft SAR problem.

Example 8.4. When an aircraft forms a synthetic aperture radar (SAR) image, the pilot needs to calculate the range to a reference point, based on the position of the craft and the output of an inertial navigator, to within some fraction of a wavelength of the transmitted radar pulse. Assume that at position $d = 0$, the aircraft *knows* that the reference point is distance R_0 off the left broadside (angle $= 90°$) of the aircraft, and that the aircraft flies in a straight line; see Fig. 8.20. The question is: What is the range, $r(d)$, to the reference point when the aircraft is at the position d_1? The exact answer is $r(d) = (R_0^2 + d_1^2)^{1/2}$; however, the square root operation is nonlinear, cumbersome, and computationally slow to evaluate. In a typical computation this expression is expanded into a Taylor series. In this example, we wish to use a fuzzy rule-based approach instead.

If we normalize the range, i.e., let $d_1/R_0 = k_1 \cdot x_1$, then $r(x_1) = R_0(1 + k_1^2 x_1^2)^{1/2}$, where now x_1 is a scaled range and k_1 is simply a constant in the scaling process. For example, suppose we are interested in the range $|d_1/R_0| \le 0.2$; then $k_1 = 0.2$ and $|x_1| \le 1$. For this particular problem we will let $R_0 = 10,000 \text{ meters} = 10 \text{ kilometers (km)}$; then $r(x_1) = 10,000[1 + (0.04)x_1^2]^{1/2}$. Table 8.6 shows exact values of $r(x_1)$ for typical values of x_1.

Let $y = r(x_1)$ with x_1 partitioned as shown in Fig. 8.21, and let the output variable, y, be partitioned as shown in Fig. 8.22. In Fig. 8.22, the partitions S̰ and L̰ have symmetrical membership functions. We now pose three simple rules that relate the input and output variables:

$$\text{Rule 1: IF } x \in \underset{\sim}{Z}, \text{ THEN } y \in \underset{\sim}{S}.$$

$$\text{Rule 2: IF } x \in \underset{\sim}{PS} \text{ or } \underset{\sim}{NS}, \text{ THEN } y \in \underset{\sim}{M}.$$

$$\text{Rule 3: IF } x \in \underset{\sim}{PB} \text{ or } \underset{\sim}{NB}, \text{ THEN } y \in \underset{\sim}{L}.$$

If we conduct a graphical simulation like that in Example 8.2 we achieve the results shown in Fig. 8.23. In this figure the open circle denotes exact values and the cross denotes the

TABLE 8.6
Relationships for distance in SAR problem

x_1	$r(x_1)$
−1.0	10,198
−0.5	10,050
0.0	10,000
0.5	10,050
1.0	10,198

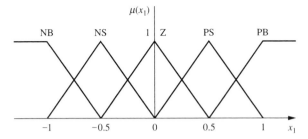

FIGURE 8.21
Partitioning for the input variable, x_1.

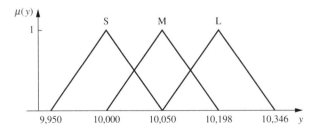

FIGURE 8.22
Partitioning for the output variable, y.

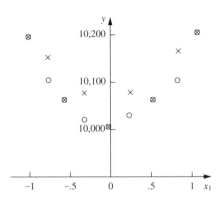

FIGURE 8.23
Exact and fuzzy values compared for SAR problem.

centroidal value of the fuzzy output as determined in the graphical simulation (in some cases the exact value and the fuzzy value coincide; this is represented by an open circle with a cross in it). The "approximate" curve follows the exact curve quite well. As a reminder, we would not normally know the exact values for a real problem whose algorithmic description was not known (this would be the case of knowledge, as described earlier in Fig. 8.2).

SUMMARY

A wide class of complex dynamic processes exists where the knowledge regarding the functional relationship between the input and output variables may be established on numerical or nonnumerical information. The numerical information is usually from a limited number of data points and the nonnumerical information is in the form of vague natural language protocols gathered from interviews with humans familiar with the input–output behavior or the real-time control of the system or process. Complexity in the system model arises as a result of many factors such as (1) high dimensionality, (2) too many interacting variables, and (3) unmodeled dynamics such as nonlinearities, time variations, external noise or disturbance, and system perturbations [Ross, 1995]. Hence, the information gathered on the system behavior is never complete, sharp, or comprehensive.

It has been shown that fuzzy systems theory is analogous to both a linear and an abstract algebra [Lucero, 2004]. The context in which fuzzy systems theory is analogous to linear algebra and to abstract algebra is that they are common for the concepts of mapping and domain. A mapping is intuitively a correspondence between two elements. But, when used with an aggregate of various mappings, the simple relations are weighted and the mapping is no longer intuitive. Stated simply, a fuzzy system is a mapping of a state. This state is defined on restricted domains. And the input variables are partitioned using a series of functions (membership functions) that transform the variable to a degree on the interval [0 1]. This degree is used to weigh the importance of a rule. More rules are defined and used, as the complexity of the system requires. The final output is a weighted value. The field of algebra encompasses a vast wealth of theories. In this field are the general disciplines of abstract algebra and linear algebra. Abstract algebra describes sets, relations, algebraic systems in general, and a linear algebra in part. Fuzzy systems do this abstraction as well, with sets which are isomorphic with linguistic knowledge. Linear algebra, as the computational kernel of this theory, contains the actual implementations, analogous to fuzzy compositions and implications. The foundation on which fuzzy systems theory is a universal approximator is based upon a fundamental theorem from real analysis, the Stone–Weierstrass theorem (see references and discussion in Chapter 1).

Fuzzy mathematics provides a range of mathematical tools that helps the analyst formalize ill-defined descriptions about complex systems into the form of linguistic rules and then eventually into mathematical equations, which can then be implemented on digital computers. These rules can be represented by fuzzy associative memories (FAMs). At the expense of relaxing some of the demands on the requirements for precision in some nonlinear systems, a great deal of simplification, ease of computation, speed, and efficiency are gained when using fuzzy models. The ill-defined nonlinear systems can be described with fuzzy relational equations. These relations are expressed in the form of various fuzzy composition operations, which are carried out on classes of membership functions defined on a number of overlapping partitions of the space of possible inputs (antecedents), possible mapping restrictions, and possible output (consequent) responses.

The membership functions used to describe linguistic knowledge are enormously subjective and context-dependent [Vadiee, 1993]. The input variables are assumed to be noninteractive, and the membership functions for them are assigned based on the degree of similarity of a corresponding prototypical element. Appropriate nonlinear transformations or sensory integration and fusion on input and/or output spaces are often used to reduce a complex process to a fuzzy system model. The net effect of this preprocessing on the input data is to decouple and linearize the system dynamics.

This chapter has dealt with the idea of fuzzy nonlinear simulation. The point made in this chapter is not that we can make crude approximations to well-known functions; after all, if we know a function, we certainly do not need fuzzy logic to approximate it. But there are many situations where we can only observe a complicated nonlinear process whose functional relationship we do not know, and whose behavior is known only in the form of linguistic knowledge, such as that expressed for the sine curve example in Table 8.3 or, for more general situations, as that expressed in Table 8.2. Then the power of fuzzy nonlinear simulation is manifested in modeling nonlinear systems whose behavior we can express in the form of input–output data-tuples, or in the form of linguistic rules of knowledge, and whose exact nonlinear specification we do not know. Fuzzy models to address such complex systems are being published in the literature at an accelerating pace; see, for example, Huang and Fan [1993] who address complex hazardous waste problems and Sugeno and Yasukawa [1993] who address problems ranging from a chemical process to a stock price trend model. The ability of fuzzy systems to analyze dynamical systems that are so complex that we do not have a mathematical model is the point made in this chapter. As we learn more about a system, the data eventually become robust enough to pose the model in analytic form; at that point we no longer need a fuzzy model.

REFERENCES

Brand, H. W. (1961). *The fecundity of mathematical methods in economic theory*, trans. Edwin Holstrom, D. Reidel, Dordrecht.

Huang, Y., and Fan, L. (1993). "A fuzzy logic-based approach to building efficient fuzzy rule-based expert systems," *Comput. Chem. Eng.*, vol. 17, no. 2, pp. 188–192.

Klir, G. and Yuan, B. (1995). *Fuzzy Sets and Fuzzy Logic: Theory and Application*, Prentice Hall, Upper Saddle River, NJ.

Lucero, J. (2004). "Fuzzy systems methods in structural engineering", PhD dissertation, Department of Civil Engineering, University of New Mexico, Albuquerque, NM.

Ross, T. (1995). *Fuzzy logic with engineering applications*, McGraw-Hill, New York.

Sugeno, M. (ed.). (1985). *Industrial applications of fuzzy control*, North-Holland, New York.

Sugeno, M. and Yasukawa, T. (1993). "A fuzzy-logic-based approach to qualitative modeling," *IEEE Trans. Fuzzy Syst.*, vol. 1, no. 1, pp. 7–31.

Vadiee, N. (1993). "Fuzzy rule-based expert systems – I and II," in M. Jamshidi, N. Vadiee, and T. Ross, (eds.), *Fuzzy logic and control: software and hardware applications*, Prentice Hall, Englewood Cliffs, NJ, chapters 4 and 5.

Zadeh, L. (1975). "The concept of a linguistic variable and its application to approximate reasoning – I," *Inf. Sci.*, vol. 8, pp. 199–249.

PROBLEMS

8.1. A video monitor's CRT has a nonlinear characteristic between the illuminance output and the voltage input. This nonlinear characteristic is $y = x^{2.2}$, where y is the illumination and x is the voltage. The CCD (Charge-Coupled Device) in a video camera has a linear light-in to voltage-out characteristic. To compensate for the nonlinear characteristic of the monitor, a "gamma correction" circuit is usually employed in a CCD camera. This nonlinear circuit has a transfer function of $y = x^{\text{gamma}}$, where the gamma factor is usually 0.45 (i.e., 1/2.2) to compensate for the 2.2 gamma characteristic of the monitor. The net result should be a linear response between the light incident on the CCD and the light produced by the monitor. Figure P8.1 shows the nonlinear gamma characteristic of a CCD camera (y_{actual}). Both the input, x, and the output, y, have a universe of discourse of $[0, 1]$.

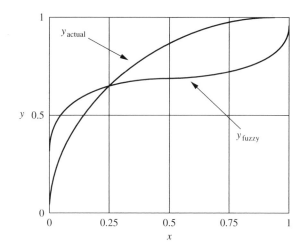

FIGURE P8.1

Partition the input variable, x, into three partitions, say small, S, medium, M, and big, B, and partition the output variable, y, into two partitions, say small, SM, and large, L. Using your own few simple rules for the nonlinear function $y = x^{0.45}$ and the crisp inputs $x = 0, 0.25, 0.5, 0.75, 1.0$, determine whether your results produce a solution roughly similar to y_{fuzzy} in Fig. P8.1 (which was developed with another fuzzy model [Ross, 1995]). Comment on the form of your solution and why it does or does not conform to the actual result.

8.2. A very widely used component in electrical engineering is the diode. The voltage–current relation is extremely nonlinear and is modeled by the expression

$$V_f = V_t \ln(I_f/I_s)$$

where V_f = forward voltage developed across the diode
  V_t = terminal voltage (~ 0.026 V)
  I_s = saturation current of a given diode (assume $\sim 10^{-12}$ A)
  I_f = forward current flowing through the diode

The resulting exact voltage–current curve is shown in Fig. P8.2 (rotated 90°). For this highly nonlinear function discuss the following:

(a) How would you go about partitioning the input space (I_f) and the output space (V_f)?

(b) Propose three to five simple rules to simulate the nonlinearity.

8.3. One of the difficulties with the Gaussian probability distribution is that it has no closed-form integral. Integration of this function must be conducted numerically. Because of this difficulty, approximations to the Gaussian distribution have been developed over the years. One of these approximations is the expression shown in Fig. P8.3a.

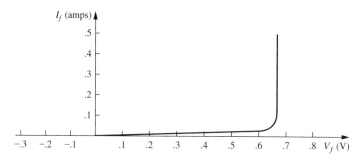

FIGURE P8.2

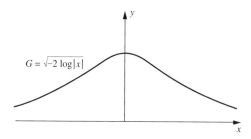

$$G = \sqrt{-2 \log|x|}$$

FIGURE P8.3a

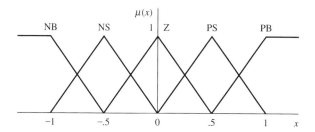

FIGURE P8.3b

This expression provides a reasonably good approximation to the Gaussian except for values of x near zero; as can be seen, the function G has a singularity at $x = 0$. Table P8.3 shows the exact values for this approximate function, G, and Fig. P8.3a shows the function.

If one uses the partitioning for the input variable, x, as shown in Fig. P8.3b, the discrete membership values for each of the quantities x shown in Table P8.3 for the following three fuzzy inputs,

1. $x_1 = $ NB or PB
2. $x_2 = $ Z or PS
3. $x_3 = $ Z or NS

TABLE P8.3

x	-1	-34	-12	-14	0.01	14	12	34	1
G	0	0.5	0.776	1.10	2	1.10	0.776	0.5	0

would be

$$x_1 = [1, 0.5, 0, 0, 0, 0, 0, 0.5, 1]$$

$$x_2 = [0, 0, 0, 0.5, 1, 0.5, 1, 0.5, 0]$$

$$x_3 = [0, 0.5, 1, 0.5, 1, 0.5, 0, 0, 0]$$

The membership functions for G for the first five elements in the table (the function is symmetric) corresponding to the three fuzzy inputs are

$$G_1 = \left\{ \frac{1}{0} + \frac{0.5}{0.5} + \frac{0}{0.776} + \frac{0}{1.10} + \frac{0}{2} \right\} = [1, 0.5, 0, 0, 0]$$

$$G_2 = [0, 0, 0, 0.5, 1]$$

$$G_3 = [0, 0.5, 1, 0.5, 1]$$

(a) Develop fuzzy relations (these matrices all will be of size 9×5) between the three fuzzy inputs and outputs using a Cartesian product operation.

(b) Find the overall fuzzy relation by taking the union of the three relations found in part (a).

(c) If the matrix relation in part (b) is replaced by a continuous surface, composition with crisp singleton inputs for x results in the following table of results for the output G. Verify some of these results.

x	G
-1	0.17
$-3/4$	0.88
$-1/2$	1.20
$-1/4$	1.09
0	1.20
1/4	1.09
1/2	1.20
3/4	0.88
1	0.17

8.4. A constant force, F, acts on a body with mass, m, moving on a smooth surface at velocity, v. The effective power of this force will be $EP = F(v) \cos \theta$ (Fig. P8.4a). Using the partitioning for the input variable, θ, as shown in Fig. P8.4b, and the partitioning for the output variable, EP, as shown in Fig. P8.4c, and the following three simple rules:

1. IF $\underset{\sim}{Z}$ THEN $\underset{\sim}{ME}$ (most efficient)
2. IF $\underset{\sim}{NS}$ or $\underset{\sim}{PS}$ THEN $\underset{\sim}{NE}$ (not efficient)
3. IF $\underset{\sim}{PB}$ or $\underset{\sim}{NB}$ THEN $\underset{\sim}{NME}$ (negative most efficient such as braking)

conduct a graphical simulation and plot the results on a graph of EP vs. θ. Show the associated exact solution on this same graph.

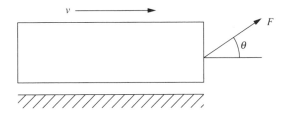

FIGURE P8.4a

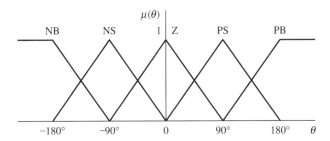

FIGURE P8.4b

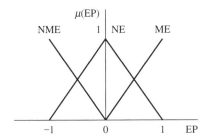

FIGURE P8.4c

8.5. Psycho-acoustic research has shown that *white noise* has different effects on people's moods, depending on the average pitch of the tones that make up the noise. Very high and very low pitches make people nervous, whereas midrange noise has a calming effect. The annoyance level of white noise can be approximated by a function of the square of the deviance of the average pitch of the noise from the central pitch of the human hearing range, approximately 10 kHz. As shown in Fig. P8.5a, the human annoyance level can be modeled by the nonlinear function $y = x^2$, where $x =$ deviance (in kHz) from 10 kHz. The range of x is $[-10, 10]$; outside that range pitches are not audible to humans.

The partitions for the input variable, x, are the five partitions on the range $[-10, 10]$ kHz, as shown in Fig. P8.5b, and the partitions for the output space for $y = x^2$ are shown in Fig. P8.5c. Using the following three simple rules,

1. IF $x = Z$, THEN $y = N$
2. IF $x = NS$ or PS, THEN $y = S$
3. IF $x = NB$ or PB, THEN $y = V$

show how a similar plot of fuzzy results as shown in Fig. 8.5d is determined.

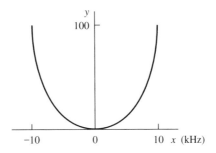

FIGURE P8.5a

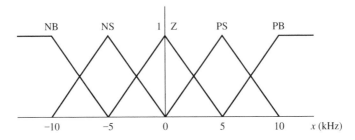

FIGURE P8.5b

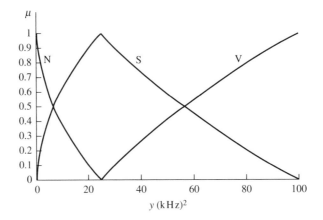

FIGURE P8.5c

8.6. Let us consider the case of a series motor under the influence of a load and a constant voltage source, as shown in Fig. P8.6a. A series motor should always be operated with some load, otherwise the speed of the motor will become excessively high, resulting in damage to the motor. The speed of the motor, N, in rpm, is inversely related to the armature current, I_a, in amps, by the expression $N = k/I_a$, where k is the flux. For this problem, we will estimate the flux parameter based on a motor speed of 1500 rpm at an armature current of 5 amps; hence, $k = 5(1500) = 7500$ rpm-amps. Suppose we consider the armature current to vary in the range $I_a = [-\infty, +\infty]$, and

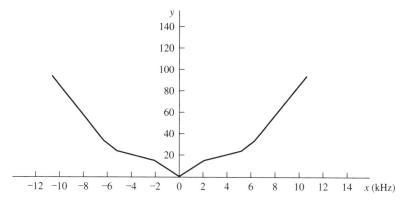

FIGURE P8.5d

we partition this universe of discourse as shown in Fig. 8.6b (note that the extremes at $-\infty$ and $+\infty$ are contained in the partitions NB and PB, respectively). Suppose we also partition the output variable, N, as shown in Fig. P8.6c. Using the input and output partitioning provided in Figs. P8.6b and P8.6c and the following five rules, conduct a graphical numerical simulation for the crisp inputs $I_a = -8, -2, 3, 9$ A. Plot this response on a graph of N vs. I_a.

IF I_a is Z, THEN N is HSC or HSAC

IF I_a is PS, THEN N is HSC

IF I_a is NS, THEN N is HSAC

IF I_a is PB, THEN N is MSC

IF I_a is NB, THEN N is MSAC

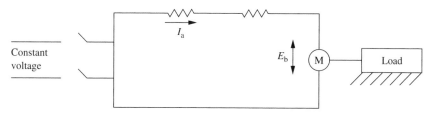

FIGURE P8.6a

8.7. In the field of image processing a *limiter* function is used to enhance an image when background lighting is too high. The limiter function is shown in Fig. P8.7a.

(a) Using the following rules, construct three matrix relations using the input (see Fig. P8.7b) and output (see Fig. P8.7c) partitions:

Rule 1: IF $x = Z$, THEN $y = S$

Rule 2: IF $x = PB$, THEN $y = PM$

Rule 3: IF $x = NB$, THEN $y = NM$

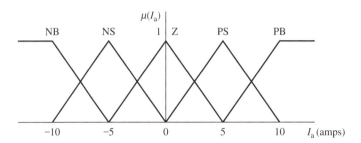

FIGURE P8.6*b*

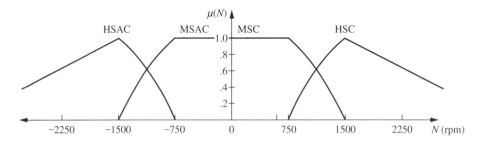

FIGURE P8.6*c*

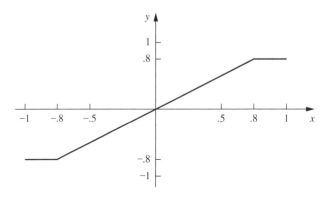

FIGURE P8.7*a*

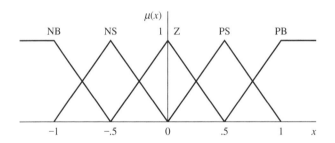

FIGURE P8.7*b*

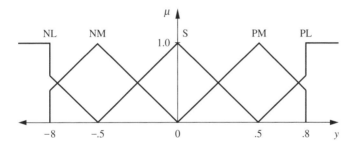

FIGURE P8.7c

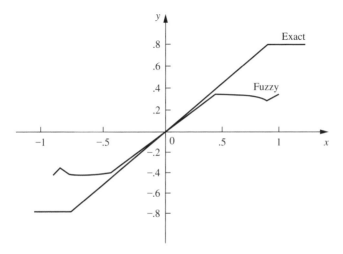

FIGURE P8.7d

(b) For crisp input values $x = -1, -0.8, -0.6, -0.4, -0.2$, and 0, use graphical techniques or max−min composition and centroidal defuzzification to determine the associated fuzzy outputs. Because of symmetry, values for $0 \le x \le 1$ are equal to $|x|$ for $-1 \le x \le 0$. Verify that these results follow Fig. P8.7d.

8.8. Do the example problem on the sine curve, Example 8.2, using (a) six rules and (b) eight rules. Does your result look more, or less, like a sine curve than the result in Example 8.2?

CHAPTER
9

RULE-BASE
REDUCTION
METHODS

In spite of the insurmountable computational limits, we continue to pursue the many problems that possess the characteristics of organized complexity. These problems are too important for our well being to give up on them. The main challenge in pursuing these problems narrows down fundamentally to one question: how can we deal with these problems if no computational power alone is sufficient?

George Klir
Professor of Systems Science, SUNY Binghamton, 1995

The quote, above, addresses two main concerns in addressing large, complex problems. First, organized complexity is a phrase describing problems that are neither linear with a small number of variables nor random with a large number of variables; they are typical in life, cognitive, social, environmental sciences, and medicine. These problems involve nonlinear systems with large numbers of components and rich interactions which are usually non-random and non-deterministic [Klir and Yuan, 1995]. Second, the matter of computational power was addressed by Hans Bremermann [1962] when he constructed a computational limit based on quantum theory: ''no data processing system, whether artificial or living, can process more than 2×10^{47} bits per second per gram of its mass.'' In other words, some problems involving organized complexity cannot be solved with an algorithmic approach because they exceed the physical bounds of speed and mass-storage. How can these problems be solved?

Such problems can be addressed by posing them as systems that are models of some aspect of reality. Klir and Yuan [1995] claim that such systems models contain three key characteristics: complexity, credibility, and uncertainty. Two of these three have been addressed in the previous chapters. Uncertainty is presumed to play a key role in any attempts to make a model useful. That is, the allowance for more uncertainty tends to

Fuzzy Logic with Engineering Applications, Second Edition   T. J. Ross
© 2004 John Wiley & Sons, Ltd   ISBNs: 0-470-86074-X (HB); 0-470-86075-8 (PB)

reduce complexity and increase credibility; Chapter 1 discusses how a model can be more robust by accommodating some levels of uncertainty. In a sense, then, fuzzy logic models provide for this robustness because they assess in a computational way the uncertainty associated with linguistic, or human, information expressed in terms of rules. However, even here there is a limit. A robust and credible fuzzy system would be represented by a very large rule-base, and as this rule-base gets more robust, and hence larger, the pressure on computational resources increases. This chapter addresses two methods to reduce the size of typical fuzzy rule-bases.

FUZZY SYSTEMS THEORY AND RULE REDUCTION

Fuzzy systems theory provides a powerful tool for system simulation and for uncertainty quantification. However, as powerful as it is, it can be limited when the system under study is complex. The reason for this is that the fuzzy system, as expressed in terms of a rule-base, grows exceedingly large in terms of the number of rules. Most fuzzy rule-bases are implemented using a conjunctive relationship of the antecedents in the rules. This has been termed an intersection rule configuration (IRC) by Combs and Andrews [1998] because the inference process maps the intersection of antecedent fuzzy sets to output consequent fuzzy sets. This IRC is the general exhaustive search of solutions that utilizes every possible combination of rules in determining an outcome. Formally, this method of searching is described as a k-tuple relation. A k-tuple is an ordered collection of k objects each with l_i possibilities [Devore, 1995]. In the present situation k represents the input variables with l_i linguistic labels for each. The product of these possible labels for each input variable gives the number of possible combinations for such k-tuples as $l_i l_{i+1} \ldots l_k$. This IRC method is not efficient since it uses significant computational time and results in the following exponential relation:

$$R = l^n \quad \text{or} \quad R = l_i l_{i+1} \ldots \tag{9.1}$$

where R = the number of rules
l = the number of linguistic labels for each input variable (assumed a constant for each variable)
n = the number of input variables

Equation (9.1) represents a combinatorial explosion in rules. Conceptually, this rule formulation can be thought of as a hypercube relating n input variables to single-output consequences. In the literature this is termed a fuzzy associative mapping (see Chapter 8).

This chapter discusses two rule-reduction schemes. These methods attempt to ameliorate the combinatorial explosion of rules due to the traditional IRC approach. The benefits of these methods are to simplify simulations and to make efficient use of computation time.

NEW METHODS

Two relatively recent methods of rule reduction are described and compared here. These two methods operate on fundamentally different premises. The first, *singular value decomposition* (SVD), uses the concepts of linear algebra and coordinate transformation to produce a reduced mapping in a different coordinate system. The second method, the *Combs method for rapid inference* (called the Combs method), is a result of a logical Boolean set-theoretic proof that transforms a multi-input, single-output system to a series of single-input, single-output rules.

The advantages of each are different and they address different needs for model development. The SVD method allows the practitioner to choose the amount of reduction based on error analysis. The Combs method gives good scalability; by this we mean that as new antecedents are added the number of rules grows linearly not exponentially. Therefore, the Combs method allows the user quick simulation times with transparent rules. Transparent rules are single-input, single-output rules making rule-base relations transparent.

The simulations in this chapter use triangular membership functions with a sum/product inference scheme on zero-order Takagi–Sugeno output functions. In other words, the output sets are singleton values. Furthermore, the weighted average defuzzification method is used to calculate the final output value, Z, from $i = 1$ to R output subsets, z_i, weighted by input fuzzy set membership values, μ_j, for $j = 1$ to n, antecedents as

$$Z = \frac{\sum_{i=1}^{R} z_i \prod_{j=1}^{n} \mu_j}{\sum_{i=1}^{R} \prod_{j=1}^{n} \mu_j} \tag{9.2}$$

The t-norm and t-conorm aggregators of this system are, respectively, the operators product and addition (sum). These reside in the numerator of Eq. (9.2) and are shown in Eq. (9.3). The product operator aggregates rule antecedent sets linguistically with "and," an intersection of input sets. And the addition operator (sum) aggregates the individual rules themselves linguistically with "or," a union of rules:

$$\sum_i z_i \prod_j \mu_j = \text{sum/product inference} \tag{9.3}$$

Singular Value Decomposition

The basic idea of the singular value decomposition (SVD) on a rule-base is to perform a coordinate transformation of the original rule-base, Z. This transformation uses singular values to illuminate information regarding rule importance within the rule-base because the largest singular values show their associated rules as column vectors having the biggest influence on the aggregated output of the rule-base. This method can be used to condense information from a rule-base by eliminating redundant or weakly contributing rules. This process allows the user to select and use the most contributing (important) rule antecedents forming a new reduced rule-base, Z_r, shown schematically in Fig. 9.1.

The effectiveness of the SVD on reducing fuzzy systems that model functions has been well documented [Yam, 1997]. SVD is used for diagonalization and approximation of linear maps in linear algebra typically defined as

$$Z = U \, \Sigma \, V^{\mathsf{T}} \tag{9.4}$$

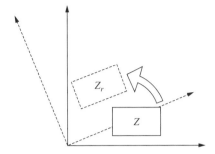

FIGURE 9.1
Coordinate transformation via the SVD method.

where U and V are orthogonal matrices and Σ is a diagonal matrix of singular values. Here the SVD is used to reduce fuzzy systems that approximate functions or other mappings. However, application of the SVD on the rule-base does not directly yield a fuzzy system with properly scaled membership functions. A few additional steps are necessary to supplement the SVD in order to produce properly scaled membership functions.

The foundation of a fuzzy system lies with the definitions of membership functions that relate the fuzzy sets that make up the system. The following three conditions define the membership functions and must be satisfied in order for an SVD application to a fuzzy system to be successful. First, the membership functions must have some degree of overlap such that $\sum \mu = 1.0$ for each x_i in the universe of discourse of the variable. Second, the membership functions must map membership values on the unit interval. And third, each membership function should have a prototypical value. Once these requirements are met, the fuzzy system will have resulted in a reduced system as desired. To meet these requirements, three sequential steps of linear algebra will be applied to the decomposed system of matrices.

The reduced-rule system is developed around the column vectors of the orthogonal matrices U and V from $Z = U\Sigma V^T$. Each matrix contains the fuzzy sets of an input variable. More specifically, each column vector represents individual fuzzy sets or labels for that input variable. As such, conditioning and conforming will take place on each column.

Overlapping membership functions

This conditioning will be the first step in the process that establishes a compact mapping of the input space to the output space. One reason why overlapping membership functions, where $\sum_{x_i} \mu = 1.0, \forall x_i$ in X, is important is because this allows for a good interpolation of the input values. In other words, the entire input space is accommodated. A proof showing how a series of column vectors representing membership functions sum to unity is given in Lucero [2004].

The orthogonal matrix, U, is separated according to the desired number of retained singular values, r, into two matrices, the reduced matrix U_r and the discarded matrix U_d:

$$U = [U_r | U_d] \tag{9.5}$$

where $d = n - r$ for n total singular values.

The reduced matrix, U_r, represents the input columns we wish to retain in our simulation. And the discarded matrix represents the collection of discarded columns of the orthogonal matrix U from the initial SVD. The objective in this separation process is to form a reduced matrix that includes the retained columns plus one column representing the discarded columns, condensed into one. Shown here is a brief justification that the matrix product of two matrices, one partitioned as reduced and discarded column vectors and the other partitioned as row sums of their transpose, gives this desired condition of row sums equaling one, and hence overlapping membership functions

$$[U_r|U_d]\begin{bmatrix} \text{sum}(U_r^T) \\ \text{sum}(U_d^T) \end{bmatrix} = [1]_{nx1} \tag{9.6}$$

It is important to note that the discarded set as a whole represents the last remaining column supplementing the retained columns. Completing this first step is another matrix product to get the first conditioned matrix, U_1:

$$U_1 = [U_r|U_d^* \text{sum}(U_d^T)]^* \mathcal{C} \tag{9.7}$$

where

$$\mathcal{C} = \text{diag}\begin{bmatrix} \text{sum}(U_r^T) \\ 1 \end{bmatrix} \tag{9.8}$$

More explicitly, the matrix is transformed as

$$U_1 = \begin{bmatrix} \vec{u}1_r & \vec{u}2_r \\ \vdots & \vdots \end{bmatrix} \begin{bmatrix} \begin{bmatrix} \vec{u}3_d & \vec{u}4_d & \vec{u}5_d \\ \vdots & \vdots & \vdots \end{bmatrix} * \begin{bmatrix} \text{sum}(\vec{u}3_d) \\ \text{sum}(\vec{u}4_d) \\ \text{sum}(\vec{u}5_d) \end{bmatrix} \end{bmatrix}$$

$$* \begin{bmatrix} \text{sum}(\vec{u}1_r) & 0 & 0 \\ 0 & \text{sum}(\vec{u}2_r) & 0 \\ 0 & 0 & 1 \end{bmatrix} \tag{9.9}$$

However, if $\text{sum}(U_d^T) = [0]_{nx1}$ then the last column can be omitted since there will be no information to condense from the discarded columns. As a result, Eq. (9.8) simply becomes $\mathcal{C} = \text{diag}[\text{sum}(U_r^T)]$. Again, we need a conditioned matrix, U_1, in this process to give us overlapping membership functions.

Non-negative membership values

The next step of the procedure is required because of the constraint that membership values must range from 0 to 1.0; hence, the values must also be ''non-negative.''
A matrix is formed by replacing the minimum element of U_1 with δ, by

$$\delta = \begin{cases} 1, & \text{if min } U_{1_{m,n}} \geq -1 \\ \dfrac{1}{|\text{min } U_{1_{m,n}}|}, & \text{otherwise} \end{cases} \tag{9.10}$$

A doubly stochastic matrix $Ø$ with the same dimension as U_1 can then be found,

$$Ø = \frac{1}{(n + \delta)} \begin{bmatrix} (1 + \delta) & 1 & \cdots & 1 \\ 1 & (1 + \delta) & \cdots & 1 \\ \vdots & \vdots & \ddots & \vdots \\ 1 & 1 & \cdots & (1 + \delta) \end{bmatrix} \tag{9.11}$$

where n is the number of columns of U_1. The doubly stochastic matrix is needed to shift the membership values from U_1 to the unit interval.

The matrix product $U_1 Ø$ generates the matrix U_2, which is now conditioned for both overlapping with non-negative elements.

$$U_2 = U_1 * Ø \tag{9.12}$$

Prototypical value

Here, the third step in the SVD process creates prototypical values for the fuzzy sets. A prototypical value is the value in the universe of a specific fuzzy set that has the highest membership value, usually a value of unity. This allows for ease of interpretation, especially for linguistic variables where the fuzzy set can be defined according to the input value that gives the maximum membership. In other words, having a prototypical value establishes a center value on the membership function that also helps to define the "nearness" of prototypical values of adjacent fuzzy sets. The *convex hull*, a concept necessary to achieve this third condition, is next defined.

Definition

Consider points on a two-dimensional plane; a convex hull (Fig. 9.2) is the smallest convex set (Fig. 9.3) that includes all of the points. This is the tightest polygon containing all the points. In other words, imagine a set of pegs. Stringing a rope around the extreme pegs such that all the pegs are enclosed forms a convex hull. However, stringing the rope around virtual pegs outside this set may give a convex set, but not a convex hull since it does not represent the smallest set.

Notice that a line drawn between any two points in either Figs. 9.2 or 9.3 does not exit the area created by the convex set. The convex hull becomes useful to rescale the columns of U_2, thus producing valid membership functions.

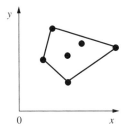

FIGURE 9.2
Convex hull.

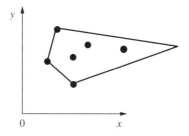

FIGURE 9.3
Convex set.

Let U_2 be defined as $U_2 = [\vec{u}_1 \vec{u}_2 \ldots \vec{u}_n]$ consisting of n column vectors and $i = 1, 2, \ldots, m$ row vectors. For $m > n$, U_2 has rank equal to n and is therefore of dimension n. As such, each row vector $U_{2_i} = [\vec{u}_{i,1} \vec{u}_{i,2} \ldots \vec{u}_{i,n}]$ represents a point in n-dimensional space. In other words, each row is a coordinate in n-dimensional space. The convex hull now becomes the $n - 1$ dimensional space onto which U_2 is projected, where the vertices represent the new prototypical values [Yam, 1997]. These vertices comprise a new matrix, $\not{E}$,

$$\not{E} = \begin{bmatrix} U_{2_j}(*,:) \\ U_{2_{j+1}}(*,:) \\ U_{2_n}(*,:) \end{bmatrix}^{-1} \qquad j = 1, 2, \ldots, n \tag{9.13}$$

The product of this matrix with the conditioned matrix of the second step, U_2, becomes the final matrix U_3, of the conditioned membership functions,

$$U_3 = U_2 * \not{E} \tag{9.14}$$

By rescaling the columns of U_2 as is accomplished in Eq. (9.14), this final operation allows for interpolation between the input column vectors and the new reduced rule-base.

In a similar fashion, steps 1, 2, and 3 are conducted on the orthogonal matrix V. After the conditioning of both matrices is complete the reduced rule-base is developed.

Reduced matrix of rule consequent values

In effect three matrix products are conducted on a reduced orthogonal matrix, i.e., $U_3 = U_r \cdot \mathcal{C} \cdot \not{D} \cdot \not{E}$. Since these products are not made on the original system, Z, in $Z \cong U_r \Sigma_r V_r^T$, a left-sided triple inverse matrix product must be made on Σ_r, the diagonal matrix of singular values to account for the conditioning of U_r. Likewise, a right-sided triple inverse matrix product must be made to Σ_r accounting for those matrix products made to V_r^T. Doing these multiplications leaves the system $Z \cong U_r \Sigma_r V_r^T$ unchanged since all matrix products cancel. A matrix multiplied by its inverse returns the identity matrix. This results in the following:

$$Z_r = \not{E}_U^{-1} \not{D}_U^{-1} \mathcal{C}_U^{-1} \Sigma_r \mathcal{C}_V^{T^{-1}} \not{D}_V^{T^{-1}} \not{E}_V^{T^{-1}} \tag{9.15}$$

which now becomes our new reduced matrix of rule consequent values. And thus our SVD procedure is complete providing a reduced rule-base approximation to the original rule-base.

Having obtained the reduced matrix of rule consequent values along with the conditioned column vectors, the system can then be implemented as the new reduced rule-base fuzzy system. However, sometimes the constraints from conditioning are only closely met, but are still effective as seen in Simon [2000] since the system is in a new coordinate system. Moreover, this new coordinate system might have negative values as a result of the numerical approximation of the convex hull and its inverse. Still, since the input sets are orthogonal, the interpolation will still yield a good approximation to the original system. In other words, some intuitive interpretation might be lost, but the procedure remains sound mathematically.

It is important to emphasize that keeping the entire set of singular values will give back the original rule-based system; but, retaining the most influential singular values and the corresponding column vectors will keep the essential features of the system (mapping). The most influential rules are those which are associated with the largest singular values of the initial decomposition. The singular values are positioned along the diagonal in descending order of impact from greatest to least. The largest (first) value will give the most important rule, then decreasing in importance to the least contributing rule to the system. Since this method reduces an already developed rule-base or matrix of function samplings, an error analysis can be conducted on the difference between the original system and the reduced system. If a user-specified error tolerance is not met, then one can include more singular values in the approximation resulting in more reduced rules to the system.

This method is analogous to a grid point function sampling. What this means is that the rule-base or function samplings must be taken from an evenly spaced grid. In fuzzy systems theory, the grid is the partition of variables into labels. So the labels (membership functions) must be evenly spaced.

Summary of operations

Given a rule-base or matrix of sampled function values, Z, the rules generalize to

Rule: If $(\underset{\sim}{A}(x_1)$ and $\underset{\sim}{B}(x_2))$ then Z

where $\underset{\sim}{A}$ and $\underset{\sim}{B}$ are fuzzy sets for the input values x_1 and x_2. Z, the rule consequent matrix, is decomposed using SVD. This becomes

$$Z = U \Sigma\, V^T \tag{9.16}$$

After choosing the most important or most contributing r singular values, Z gets approximated as

$$Z \approx \overline{Z} = U_r \Sigma_r V_r^T \tag{9.17}$$

Following the procedures described previously, the matrices U_r and V_r are conditioned so that U_r becomes A_r and V_r becomes B_r and Σ_r is updated to become Z_r. This yields the new approximation,

$$Z \approx \overline{Z} = A_r Z_r B_r^T \tag{9.18}$$

Now the columns of matrices A_r and B_r are the new fuzzy sets, membership functions, or labels for input values x_1 and x_2. And Z_r is the reduced-rule consequent matrix of this new system, i.e.,

New Rule: If $(A_r(x_1)$ and $B_r(x_2))$ then Z_r

Combs Method

Combs and Andrews [1998] discovered a classical logic equivalent to the traditional conjunctive fuzzy system rule-base, a disjunctive relation that inspired a new approach to fuzzy systems modeling. In other words, a traditional system that connects multi-antecedent subsets using an intersection operator, which Combs and Andrews called an intersection rule configuration or IRC, has been transformed to a system of single antecedent rules that use a union operator, which Combs and Andrews termed a union rule configuration or URC, as given in the proof table of Fig. 9.4.

$$\text{IRC} \longrightarrow [(p \cap q) \Rightarrow r]\widetilde{\Leftrightarrow}[(p \Rightarrow r) \cup (q \Rightarrow r)] \leftarrow \text{URC} \qquad (9.19)$$

In Eq. (9.19), p and q are the antecedents, r is the consequent, $\cap$ is intersection, $\cup$ is union, $\Rightarrow$ represents implication, and $\widetilde{\Leftrightarrow}$ represents equivalence but not necessarily equality for fuzzy systems. This means that for certain conditions equality between the IRC and URC can be achieved directly. These cases for which equality holds are defined as *additively separable* systems in Weinschenk, et al. [2003]. Additively separable systems are systems that satisfy the following condition:

$$F = \vec{x}_1 \oplus \vec{x}_2 \oplus \cdots \oplus \vec{x}_n, \qquad \oplus \text{ is the outer sum} \qquad (9.20)$$

where $\vec{x}_i$ are input vectors; in other words, the set of labels for each variable i. The outer sum of input vectors is analogous to the outer product; however, the sum operation is used in place of the product operator. When this is not the case, i.e., when the system of rules is not additively separable, the URC must be augmented with additional rules. The Combs method uses a series of single-input to single-output (SISO) rules to model the problem space with one powerful advantage: the avoidance of an explosion in the rule-base as the number of input variables increases. This method turns an exponential increase in the number of rules into a linear increase with the number of input variables.

The benefits of the Combs method are the following: the solution obtained is equivalent to the IRC in many cases, it is simple to construct, and it is fast computationally.

In the IRC approach it is the intersection of the input values that is related to the output. This intersection is achieved with an "and" operation. In the URC, accumulation of SISO rules is achieved with an "or" operation. An important but subtle feature of the "or" operation is that it is not an "exclusive or" (see Chapter 5) but an "inclusive or." This is important to satisfy the additively separable condition just described where each rule

$(p$ and $q)$ then r	IRC
not (p and q) or r	Classical implication
not p or not q or r	De Morgan's principle
not p or not q or (r or r)	Idempotency
(not p or r) or (not q or r)	Commutativity
$(p$ then $r)$ or $(q$ then $r)$	Classical implication, URC

FIGURE 9.4
Proof table of the IRC–URC equivalence.

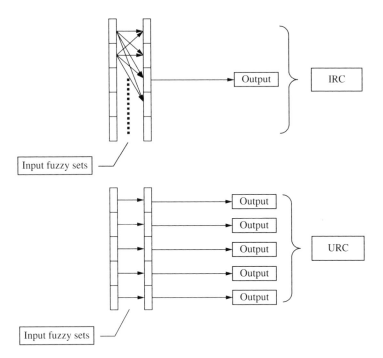

FIGURE 9.5
Schematic of IRC and URC input to output relations.

may contribute to the outer sum of input vectors. This accumulation of SISO rules gives the problem space representation, and is shown schematically in Fig. 9.5.

Specifically, an IRC rule such as

Rule: If (A and B) then Z

now becomes structured to the form

Rule: If (A then Z) or If (B then Z)

Unlike the IRC, which is described by a hypercube, relating n input variables to single-output consequences (also termed a fuzzy associative mapping), this method can use a union rule matrix (URM) or a network of SISO relations.

This matrix (URM) has as its cells the SISO rules. Product inference occurs in these individual cells as in Eq. (9.3) where only one antecedent membership value μ is multiplied by an output subset, z. Hence, each column in the URM is a series of such products as rules. Aggregating these rules is achieved by the union aggregator as in Eq. (9.3), an algebraic sum. An accumulator array accounts for this union of the rules as a summation of SISO products, μ^*z. This array is the last row of the matrix in Fig. 9.6. This is basically an accounting mechanism that results in a similar centroidal defuzzification as previously described in Eq. (9.2).

The Combs method generalizes to a system of double summations, as described in Eq. (9.21) [Weinschenk et al., 2003]. It becomes a double summation because the entries

Input A	If A is Low then z_1	If A is Medium then z_2	If A is High then z_3
Input B	If B is Low then z_1	If B is Medium then z_2	If B is High then z_3
Accumulator	$\sum \mu * z_1$	$\sum \mu * z_2$	$\sum \mu * z_3$

FIGURE 9.6
Union rule matrix (URM) showing the accounting mechanism for rules.

Input 1, A(x_1)	$\mu_{1,1} * z_{1,1}$	$\mu_{1,2} * z_{1,2}$	$\mu_{1,3} * z_{1,3}$
Input 2, B(x_2)	$\mu_{2,1} * z_{2,1}$	$\mu_{2,2} * z_{2,2}$	$\mu_{2,3} * z_{2,3}$
Accumulator	$\sum\limits_{i=1}^{2} \mu_{i,1} z_{i,1}$	$\sum\limits_{i=1}^{2} \mu_{i,2} z_{i,2}$	$\sum\limits_{i=1}^{2} \mu_{i,3} z_{i,3}$

FIGURE 9.7
A typical URM with cells replaced by membership values and output values according to Eq. (9.21).

of the accumulator array are summed. For each output subset there corresponds a sum of $i = 1$ to P(variables) as SISO rules. Consequently these sums are themselves summed over $j = 1$ to N_i fuzzy labels:

$$Z_{\text{URC}} = \frac{\sum\limits_{j=1}^{N_i} \sum\limits_{i=1}^{P} \mu_{i,j}(x_i) z_{i,j}}{\sum\limits_{j=1}^{N_i} \sum\limits_{i=1}^{P} \mu_{i,j}(x_i)} \qquad (9.21)$$

where x_i is the input crisp value to fuzzy variable i, $\mu_{i,j}$ is the membership value of x_i in fuzzy set j for variable i, and $z_{i,j}$ is the zero-order Takagi–Sugeno output function for fuzzy set j in variable i.

Figure 9.7 demonstrates this in the general URM form of Fig. 9.6 and the system is updated according to Eq. (9.21). Each cell in Fig. 9.7 represents a rule. The double sum takes place when summing the final row of Fig. 9.7 which has been defined in Combs and Andrews [1998] as the accumulator array in the URM format. It is important to note that not all input antecedents will have the same number of sets. This is what is meant by the summation from $j = 1$ to N_i where N_i is the total number of labels for the ith variable.

SVD AND COMBS METHOD EXAMPLES

Example 9.1. Fuzzy rule-bases will be developed to simulate the additively separable function

$$f(x, y) = x + y \qquad (9.22)$$

shown in Fig. 9.8. This function is additively separable conforming to Eq. (9.20) where the solution surface represents an outer sum of the vectors x and y in the range $[-5.0, 5.0]$. The first rule-base simulation will be a traditional IRC rule-base. The simulations to follow this will

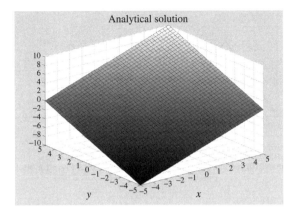

FIGURE 9.8
Solution surface for Eq. (9.22).

be reduced rule-bases using the SVD and the Combs method for rapid inference. This trivial linear example will highlight the essential features of each method.

IRC rule-base simulation

The IRC system begins by partitioning the input space into a representative number of fuzzy sets. Here five sets are chosen. This is an arbitrary selection and is based only on the user's judgment. Figures 9.9 and 9.10 show these five triangular membership functions representing the linguistic labels, i.e., fuzzy sets for the two input variables, x and y.

Figure 9.11 is the FAM table listing the 25 rules that will be used to simulate the function. The values in the cells represent the rule consequent prototypical values and are chosen based on the output of Eq. (9.22) and can be regarded as observations. Since the output sets are singleton values and not fuzzy sets, numbers are used to show the direct relationship to the analytical function values. The input fuzzy sets account for the interpolation between these output values.

Stepping along the x and y axes, the input x and y values map to the solution surface of Fig. 9.12.

SVD simulation

Now the number of rules shown in the previous IRC table, Fig. 9.11, is reduced using the SVD. First the original FAM is represented as matrix Z along orthogonal vectors x and y.

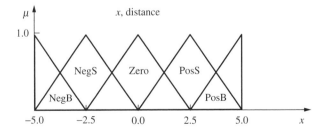

FIGURE 9.9
Fuzzy sets for the variable x represented by triangular membership functions.

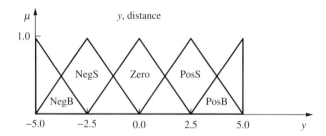

FIGURE 9.10
Fuzzy sets for the variable y represented by triangular membership functions.

	25 fuzzy rules	y				
		NegB	NegS	Zero	PosS	PosB
x	NegB	−10.0	−7.5	−5.0	−2.5	0.0
	NegS	−7.5	−5.0	−2.5	0.0	2.5
	Zero	−5.0	−2.5	0.0	2.5	5.0
	PosS	−2.5	0.0	2.5	5.0	7.5
	PosB	0.0	2.5	5.0	7.5	10.0

FIGURE 9.11
FAM table for the IRC simulation.

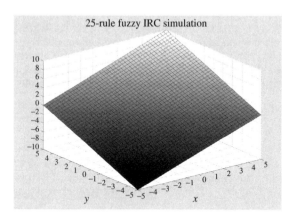

FIGURE 9.12
Fuzzy IRC simulation using 25 rules.

Here, like the previous IRC case, the rule consequent values are stored in each cell of matrix Z. Additionally, the fuzzy sets for x and y have been replaced with their prototypical values to emphasize the requirement of an evenly spaced grid as seen in Fig. 9.13. Therefore, it is seen that this system degenerates to a system of sampled points along the x and y axes.

The matrix Z is decomposed according to the SVD procedure previously described (see Eq. (9.4)). The initial decomposition results in three matrices. Matrices U and V are orthogonal and represent the input variables. Matrix Σ is the diagonal matrix of singular values. These singular values are used to relate the importance of the input sets yielding the most important

$$Z = f(x, y)$$

		y			
		−5	−2.5	0	2.5	5.0
	−5	−10	−7.5	−5	−2.5	0
	−2.5	−7.5	−5	−2.5	0	2.5
x	0	−5	−2.5	0	2.5	5
	2.5	−2.5	0	2.5	5	7.5
	5	0	2.5	5	7.5	10

FIGURE 9.13
Output of Eq. (9.22) to be used for SVD simulation.

rules of a reduced set to approximate the rule-based system. The singular values and the U and V matrices are in a new coordinate space. All the subsequent calculations occur in this space. For this simple example, i.e., for the data in Fig. 9.13, we get

$$Z = U\Sigma V^{\mathrm{T}} = \begin{bmatrix} -0.7303 & 0.2582 & -0.4620 & -0.3644 & -0.2319 \\ -0.5477 & 0.0 & 0.8085 & 0.1624 & -0.1411 \\ -0.3651 & -0.2582 & -0.1260 & 0.1537 & 0.8721 \\ -0.1826 & -0.5164 & -0.3255 & 0.6629 & -0.3932 \\ 0.0 & -0.7746 & 0.1050 & -0.6146 & -0.1059 \end{bmatrix}$$

$$* \begin{bmatrix} 17.6777 & 0 & 0 & 0 & 0 \\ 0 & 17.6777 & 0 & 0 & 0 \\ 0 & 0 & 0 & 0 & 0 \\ 0 & 0 & 0 & 0 & 0 \\ 0 & 0 & 0 & 0 & 0 \end{bmatrix} * \begin{bmatrix} 0.7746 & 0.0 & -0.0047 & 0.5588 & 0.2961 \\ 0.5164 & -0.1826 & 0.1155 & -0.8092 & 0.1782 \\ 0.2582 & -0.3651 & 0.2545 & 0.0958 & -0.8520 \\ 0.0 & -0.5477 & -0.8365 & 0.0008 & -0.0150 \\ -0.2582 & 0.7303 & 0.4712 & 0.1538 & 0.3927 \end{bmatrix}^{\mathrm{T}}$$

Before the matrices are conditioned to become a reduced fuzzy system, singular values are selected to simulate the system. This example yields only two singular values because this system is a rank 2 system. Only two independent vectors are necessary to define the space. However, the simulation continues as if there were more singular values from which to select, and the procedure uses the two largest singular values. The column vectors for the fuzzy system must now be conditioned. This means that the matrices, soon to contain membership values of fuzzy sets, must overlap, must lie on the unit interval, and must have a prototypical value (a single maximum membership value). We begin with operations on the matrix, U.

Overlapping membership functions: matrix U

Two submatrices representing the retained, U_r, and discarded U_d, columns of the U matrix are developed from the first two columns of U, and the last three columns of U, respectively:

$$U_r = \begin{bmatrix} -0.7303 & 0.2582 \\ -0.5477 & 0.0 \\ -0.3651 & -0.2582 \\ -0.1826 & -0.5164 \\ 0.0 & -0.7746 \end{bmatrix}, \quad U_d = \begin{bmatrix} -0.4620 & -0.3644 & -0.2319 \\ 0.8085 & 0.1624 & -0.1411 \\ -0.1260 & 0.1537 & 0.8721 \\ -0.3255 & 0.6629 & -0.3932 \\ 0.1050 & -0.6146 & -0.1059 \end{bmatrix}$$

The discarded columns, U_d, are then condensed into one column to augment U_r forming the first matrix term in the product of Eq. (9.7) as

$$[U_r \,|\, U_d * \mathrm{sum}(U_d^{\mathrm{T}})] = \begin{bmatrix} -0.7303 & 0.2582 \\ -0.5477 & 0.0 \\ -0.3651 & -0.2582 \\ -0.1826 & -0.5164 \\ 0.0 & -0.7746 \end{bmatrix} \begin{bmatrix} -0.4620 & -0.3644 & -0.2319 \\ 0.8085 & 0.1624 & -0.1411 \\ -0.1260 & 0.1537 & 0.8721 \\ -0.3255 & 0.6629 & -0.3932 \\ 0.1050 & -0.6146 & -0.1059 \end{bmatrix} * \begin{bmatrix} 0.0 \\ 0.0 \\ 0.0 \end{bmatrix}$$

Next, the column sums of U_r and one additional unity value form the diagonal matrix $\mathcal{C}_U$ as (see Eq. (9.8))

$$\mathcal{C}_U = \begin{bmatrix} -1.8257 & 0 & 0 \\ 0 & -1.291 & 0 \\ 0 & 0 & 1 \end{bmatrix} = \begin{bmatrix} \text{sum}(U_r^T) & \\ & 1 \end{bmatrix}$$

where the "1" is the required additional column to account for the discarded U columns.

Now applying Eq. (9.7), the product of $[U_r \,|\, U_d * \text{sum}(U_d^T)]$ and the diagonal matrix of column sums, $\mathcal{C}_U$, forms the matrix U_1 satisfying the special overlapping condition:

$$U_1 = \begin{bmatrix} -0.7303 & 0.2582 & 0.0 \\ -0.5477 & 0.0 & 0.0 \\ -0.3651 & -0.2582 & 0.0 \\ -0.1826 & -0.5164 & 0.0 \\ 0.0 & -0.7746 & 0.0 \end{bmatrix} * \begin{bmatrix} -1.8257 & 0 & 0 \\ 0 & -1.291 & 0 \\ 0 & 0 & 1 \end{bmatrix}$$

$$U_1 = \begin{bmatrix} 1.3333 & -0.3333 & 0.0 \\ 0.9999 & 0.0 & 0.0 \\ 0.6666 & 0.3333 & 0.0 \\ 0.3333 & 0.6666 & 0.0 \\ 0.0 & 1.0 & 0.0 \end{bmatrix} = \begin{bmatrix} 1.3333 & -0.3333 \\ 0.9999 & 0.0 \\ 0.6666 & 0.3333 \\ 0.3333 & 0.6666 \\ 0.0 & 1.0 \end{bmatrix}$$

The third column gives no useful information and is removed to provide the same U_1 that would have resulted using the smaller matrix $\mathcal{C}_U = \text{diag}[\text{sum}(U_r^T)]$ originally developed in Eq. (9.8).

Non-negative membership values: matrix U

This next operation relies on determining the minimum element in U_1 and assigning a variable, δ, a value 1 or $1/\min(U_1)$. We have $\min(U_1) = -0.3333$, therefore $\delta = 1.0$ in accordance with Eq. (9.10). We can now form a doubly stochastic matrix using Eq. (9.11),

$$\not{D}_U = \frac{1}{(n+\delta)} \begin{bmatrix} (1+\delta) & 1 & \cdots & 1 \\ 1 & (1+\delta) & \cdots & 1 \\ \vdots & \vdots & \ddots & \vdots \\ 1 & 1 & \cdots & (1+\delta) \end{bmatrix} = \frac{1}{(2+1)} \begin{bmatrix} 1+1 & 1 \\ 1 & 1+1 \end{bmatrix}$$

$$= \begin{bmatrix} 0.6667 & 0.3333 \\ 0.3333 & 0.6667 \end{bmatrix}$$

By Eq. (9.12), the product $U_1 \not{D}_U$ gives U_2, the matrix satisfying both the overlapping and non-negative conditions:

$$U_2 = U_1 * \not{D}_U = \begin{bmatrix} 1.3333 & -0.3333 \\ 0.9999 & 0.0 \\ 0.6666 & 0.3333 \\ 0.3333 & 0.6666 \\ 0.0 & 1.0 \end{bmatrix} * \begin{bmatrix} 0.6667 & 0.3333 \\ 0.3333 & 0.6667 \end{bmatrix} = \begin{bmatrix} 0.7778 & 0.2222 \\ 0.6666 & 0.3333 \\ 0.5555 & 0.4444 \\ 0.4444 & 0.5555 \\ 0.3333 & 0.6667 \end{bmatrix}$$

Prototypical membership value: matrix U

We now conduct the final step of conditioning the matrix U for membership functions. The convex hull of U_2, i.e., the extreme data points represented as rows in U_2, is used to construct the rescaling matrix, $\mathcal{E}_U$. In this case these minimum and maximum points occur at the first and fifth rows of U_2. The inverse of these two rows generates $\mathcal{E}_U$ according to Eq. (9.13):

$$\mathcal{E}_U = \begin{bmatrix} U_2(1,:) \\ U_2(5,:) \end{bmatrix}^{-1} = \begin{bmatrix} 0.7778 & 0.2222 \\ 0.3333 & 0.6667 \end{bmatrix}^{-1} = \begin{bmatrix} 1.4999 & -0.4999 \\ -0.7498 & 1.7498 \end{bmatrix}$$

Applying Eq. (9.14), the product of $\mathcal{E}_U$ with U_2 produces the final matrix U_3 that satisfies all the necessary requirements of overlapping, non-negative values and prototypical values:

$$U_3 = U_2 * \mathcal{E}_U = \begin{bmatrix} 0.7778 & 0.2222 \\ 0.6666 & 0.3333 \\ 0.5555 & 0.4444 \\ 0.4444 & 0.5555 \\ 0.3333 & 0.6667 \end{bmatrix} * \begin{bmatrix} 1.4999 & -0.4999 \\ -0.7498 & 1.7498 \end{bmatrix}$$

$$= \begin{bmatrix} 1.0000 & 0.0000 \\ 0.7500 & 0.2500 \\ 0.5000 & 0.5000 \\ 0.2500 & 0.7500 \\ 0.0000 & 1.0000 \end{bmatrix}$$

The columns of U_3 become the membership functions for the input x in the reduced rule-base system, Fig. 9.14.

Overlapping membership functions: matrix V

As with matrix U, two submatrices representing the retained, V_r, and discarded, V_d, columns of the V matrix are developed:

$$V_r = \begin{bmatrix} 0.7746 & 0.0 \\ 0.5164 & -0.1826 \\ 0.2582 & -0.3651 \\ 0.0 & -0.5477 \\ -0.2582 & -0.7303 \end{bmatrix}, V_d = \begin{bmatrix} -0.0048 & 0.5588 & 0.2961 \\ 0.1156 & -0.8092 & 0.1782 \\ 0.2545 & 0.0958 & -0.8521 \\ -0.8365 & 0.0008 & -0.0151 \\ 0.4712 & 0.1538 & 0.3928 \end{bmatrix}$$

The discarded columns V_d are condensed into one column to augment V_r forming the first matrix term in the product of Eq. (9.7) as

$$[V_r | V_d * \mathrm{sum}(V_d^T)] = \begin{bmatrix} 0.7746 & 0.0 \\ 0.5164 & -0.1826 \\ 0.2582 & -0.3651 \\ 0.0 & -0.5477 \\ -0.2582 & -0.7303 \end{bmatrix} \begin{bmatrix} -0.0048 & 0.5588 & 0.2961 \\ 0.1156 & -0.8092 & 0.1782 \\ 0.2545 & 0.0958 & -0.8521 \\ -0.8365 & 0.0008 & -0.0151 \\ 0.4712 & 0.1538 & 0.3928 \end{bmatrix} * \begin{bmatrix} 0.0 \\ 0.0 \\ 0.0 \end{bmatrix}$$

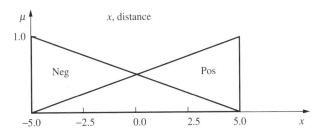

FIGURE 9.14
Plotted columns of U_3 representing two right triangle membership functions.

Using Eq. (9.8) the diagonal matrix $\mathcal{C}_V$ is formed from V as

$$\mathcal{C}_V = \begin{bmatrix} 1.2910 & 0 & 0 \\ 0 & -1.8257 & 0 \\ 0 & 0 & 1 \end{bmatrix} = \begin{bmatrix} \mathrm{sum}(V_r^T) & \\ & 1 \end{bmatrix}$$

where, again, the "1" is the required additional column to account for the discarded V columns. The product of $[V_r \mid V_d * \mathrm{sum}(V_d^T)]$ and the diagonal matrix of column sums, $\mathcal{C}_V$, forms the matrix V_1 by Eq. (9.7); this satisfies the special overlapping condition:

$$V_1 = \begin{bmatrix} 0.7746 & 0.0 & 0.0 \\ 0.5164 & -0.1826 & 0.0 \\ 0.2582 & -0.3651 & 0.0 \\ 0.0 & -0.5477 & 0.0 \\ -0.2582 & 0.7303 & 0.0 \end{bmatrix} * \begin{bmatrix} 1.2910 & 0 & 0 \\ 0 & -1.8257 & 0 \\ 0 & 0 & 1 \end{bmatrix}$$

$$V_1 = \begin{bmatrix} 1.0 & 0.0 & 0.0 \\ 0.6666 & 0.3333 & 0.0 \\ 0.3333 & 0.6666 & 0.0 \\ 0.0 & 1.0 & 0.0 \\ -0.3333 & 1.3333 & 0.0 \end{bmatrix} = \begin{bmatrix} 1.0 & 0.0 \\ 0.6666 & 0.3333 \\ 0.3333 & 0.6666 \\ 0.0 & 1.0 \\ -0.3333 & 1.3333 \end{bmatrix}$$

Once more, as with matrix U, the third column gives no useful information and is removed to provide the same V_1 that would have resulted using the smaller matrix $\mathcal{C}_V = \mathrm{diag}[\mathrm{sum}(V_r^T)]$ as in Eq. 9.8.

Non-negative membership values: matrix V

The minimum element in V_1 is $\min(V_1) = -0.3333$. Therefore, $\delta = 1.0$ by Eq. (9.10) and is used to form the doubly stochastic matrix (see Eq. 9.11)

$$\mathcal{D}_V = \frac{1}{(n+\delta)} \begin{bmatrix} (1+\delta) & 1 & \cdots & 1 \\ 1 & (1+\delta) & \cdots & 1 \\ \vdots & \vdots & \ddots & \vdots \\ 1 & 1 & \cdots & (1+\delta) \end{bmatrix} = \frac{1}{(2+1)} \begin{bmatrix} 1+1 & 1 \\ 1 & 1+1 \end{bmatrix} = \begin{bmatrix} 0.6667 & 0.3333 \\ 0.3333 & 0.6667 \end{bmatrix}$$

From Eq. (9.12) V_2 becomes

$$V_2 = V_1 * \mathcal{D}_V = \begin{bmatrix} 1.0 & 0.0 \\ 0.6666 & 0.3333 \\ 0.3333 & 0.6666 \\ 0.0 & 1.0 \\ -0.3333 & 1.3333 \end{bmatrix} * \begin{bmatrix} 0.6667 & 0.3333 \\ 0.3333 & 0.6667 \end{bmatrix} = \begin{bmatrix} 0.6667 & 0.3333 \\ 0.5555 & 0.4444 \\ 0.4444 & 0.5555 \\ 0.3333 & 0.6667 \\ 0.2222 & 0.7778 \end{bmatrix}$$

Prototypical membership value: matrix V

The final step of conditioning matrix V, following the same step as for matrix U, uses the convex hull of V_2. By Eq. (9.13), the inverse of the maximum and minimum rows of V_2 generates $\mathcal{E}_V$:

$$\mathcal{E}_V = \begin{bmatrix} V_2(1,:)^{(r)} \\ V_2(5,:)^{(r)} \end{bmatrix}^{-1} = \begin{bmatrix} 0.6667 & 0.3333 \\ 0.2222 & 0.7778 \end{bmatrix}^{-1} = \begin{bmatrix} 1.7498 & -0.7498 \\ -0.4999 & 1.4999 \end{bmatrix}$$

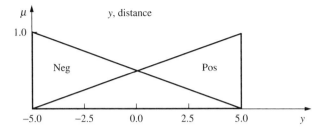

FIGURE 9.15
Plotted columns of V_3 representing two right triangle membership functions.

The final matrix V_3, satisfying the membership function constraints, is produced using Eq. (9.14):

$$V_3 = V_2 * E_V = \begin{bmatrix} 0.6667 & 0.3333 \\ 0.5555 & 0.4444 \\ 0.4444 & 0.5555 \\ 0.3333 & 0.6667 \\ 0.2222 & 0.7778 \end{bmatrix} * \begin{bmatrix} 1.7498 & -0.7498 \\ -0.4999 & 1.4999 \end{bmatrix}$$

$$= \begin{bmatrix} 1.0000 & 0.0000 \\ 0.7500 & 0.2500 \\ 0.5000 & 0.5000 \\ 0.2500 & 0.7500 \\ 0.0000 & 1.0000 \end{bmatrix}$$

The columns of V_3 become the membership functions for the input y of the reduced rule-base system, as seen in Fig. 9.15.

Reduced matrix of rule consequent values

Inverse matrix transformations of the previous operations help to form the reduced matrix of the rule consequent values, Z_r. This matrix will then be considered a FAM table for the new reduced system. The original orthogonal matrices U and V were conditioned by means of three matrix transformations $\mathcal{C}_{U,V}$, $\mathcal{D}_{U,V}$, and $\mathcal{E}_{U,V}$. Now, the inverse of each of these matrices is multiplied with Σ_r (the reduced diagonal matrix of singular values) on both the left side and right side resulting in a multiple matrix product, as described Eq. (9.15):

$$Z_r = E_U^{-1} D_U^{-1} C_U^{-1} \Sigma_r C_V^{T^{-1}} D_V^{T^{-1}} E_V^{T^{-1}}$$

The following substitution shows the process on Σ_r from the initial approximation, using unconditioned matrices, to one with conditioned matrices:

$$Z \cong U_r \Sigma_r V_r^T$$

$$Z \cong [[[U_r * \mathcal{C}_U] * \mathcal{D}_U] * \mathcal{E}_U] \Sigma_r [[[V_r * \mathcal{C}_V] * \mathcal{D}_V] * \mathcal{E}_V]^T$$

$$Z \cong [[[U_r * \mathcal{C}_U] * \mathcal{D}_U] * \mathcal{E}_U] \Sigma_r [\mathcal{E}_V^T * [\mathcal{D}_V^T * [\mathcal{C}_V^T * V_r^T]]]$$

The approximation of matrix Z remains unchanged by "undoing" all the matrix products by multiplying the conditioning matrices with their inverses. For example, $\mathcal{E}_U \mathcal{E}_U^{-1} = I$. Doing this gives the following system equality: $Z \cong U_r \Sigma_r V_r^T = U_r * I * \Sigma_r * I * V_r^T$. Now expanding this, including all the matrix products with their inverses, gives

$$Z \cong [[[U_r * \mathcal{C}_U] * \cancel{D}_U] * E_U]\{E_U^{-1}\cancel{D}_U^{-1}\mathcal{C}_U^{-1}\Sigma_r\mathcal{C}_V^{T^{-1}}\cancel{D}_V^{T^{-1}}E_V^{T^{-1}}\}[E_V^T * [\cancel{D}_V^T * [\mathcal{C}_V^T * V_r^T]]]$$

where $Z_r = \{E_U^{-1}\cancel{D}_U^{-1}\mathcal{C}_U^{-1}\Sigma_r\mathcal{C}_V^{T^{-1}}\cancel{D}_V^{T^{-1}}E_V^{T^{-1}}\}$ corresponds to the final reduced matrix of consequent output values.

Therefore, calculating the inverses from the previous matrices yields the following set:

$$\mathcal{C}_U^{-1} = \begin{bmatrix} -0.5477 & 0 \\ 0 & -0.7746 \end{bmatrix} \quad \mathcal{C}_V^{-1} = \begin{bmatrix} 0.7746 & 0 \\ 0 & -0.5477 \end{bmatrix}$$

$$\cancel{D}_U^{-1} = \begin{bmatrix} 1.9997 & -0.9997 \\ -0.9997 & 1.9997 \end{bmatrix} \quad \cancel{D}_V^{-1} = \begin{bmatrix} 1.9997 & -0.9997 \\ -0.9997 & 1.9997 \end{bmatrix}$$

$$E_U^{-1} = \begin{bmatrix} 0.7778 & 0.2222 \\ 0.3333 & 0.6667 \end{bmatrix} \quad E_V^{-1} = \begin{bmatrix} 0.6667 & 0.2222 \\ 0.3333 & 0.7778 \end{bmatrix}$$

This set of inverses, when multiplied with the matrix of retained singular values,

$$\Sigma_r = \begin{bmatrix} 17.6777 & 0 \\ 0 & 17.6777 \end{bmatrix}$$

gives the following matrix product:

$$Z_r = \begin{bmatrix} 0.7778 & 0.2222 \\ 0.3333 & 0.6667 \end{bmatrix}\begin{bmatrix} 1.9997 & -0.9997 \\ -0.9997 & 1.9997 \end{bmatrix}\begin{bmatrix} -0.5477 & 0 \\ 0 & -0.7746 \end{bmatrix}\begin{bmatrix} 17.6777 & 0 \\ 0 & 17.6777 \end{bmatrix}$$

$$* \begin{bmatrix} 0.7746 & 0 \\ 0 & -0.5477 \end{bmatrix}\begin{bmatrix} 1.9997 & -0.9997 \\ -0.9997 & 1.9997 \end{bmatrix}\begin{bmatrix} 0.6667 & 0.2222 \\ 0.3333 & 0.7778 \end{bmatrix}$$

Hence

$$Z_r = \begin{bmatrix} -10.0 & 0 \\ 0 & 10.0 \end{bmatrix}$$

becomes the reduced matrix of consequent values. In other words, the FAM table result for the reduced rule-base system is now determined, and is shown in Fig. 9.16.

A simulation using this reduced four-rule system is shown in Fig. 9.17; the results correlate exactly to the analytical and IRC cases.

Combs method for rapid inference

Here a somewhat more direct approach simulates the system using the same membership functions as in the 25-rule IRC approach as shown in Fig. 9.11. This is an additively separable system. Therefore the simulation is expected to be exact to that of the IRC. The URM as

Z_r		y	
		Neg	Pos
x	Neg	−10.0	0.0
	Pos	0.0	10.0

FIGURE 9.16
Reduced FAM table due to the SVD method.

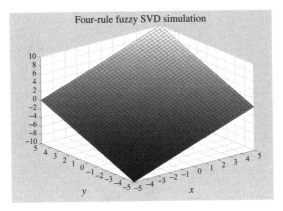

FIGURE 9.17
SVD simulation using four rules.

previously described maps the system as an outer sum. In other words, when x is -5.0 or "NegB" (Negative Big) the function is -10.0. And similarly, when y is -5.0 or "NegB" the function is -10.0. In progressing this way we get the results shown in the table in Fig. 9.18. Each row represents all the labels associated with that specific input variable. Moreover, there are two input variables in this example. Therefore, the URM consists of two rows plus one accumulator row. Here again, the output values are singletons using the defined crisp output values of Eq. (9.22). This method of rule-base development consists of 10 rules, the total number of input cells shown in Fig. 9.18.

Figure 9.18 gets transformed to the URM of Fig. 9.19 in accordance with Eq. (9.21). This shows how the computations of the system are implemented.

The results of this simulation do in fact correlate exactly to the IRC case, as can be seen in Fig. 9.20, where the solution surface simulates Eq. (9.22).

Error analysis of the methods

This section gives a brief comparison of the three methods in this example using absolute and relative errors. Since the operations are matrix computations on the rule consequent tables

$X(x)$, distance	If x is NegB then -10.0	If x is NegS then -5.0	If x is Zero then 0.0	If x is PosS then 5.0	If x is PosB then 10.0
$Y(y)$, distance	If y is NegB then -10.0	If y is NegS then -5.0	If y is Zero then 0.0	If y is PosS then 5.0	If y is PosB then 10.0
Accumulator	$\sum_{i=1}^{2} \mu_{i,1} z_{i,1}$	$\sum_{i=1}^{2} \mu_{i,2} z_{i,2}$	$\sum_{i=1}^{2} \mu_{i,3} z_{i,3}$	$\sum_{i=1}^{2} \mu_{i,4} z_{i,4}$	$\sum_{i=1}^{2} \mu_{i,5} z_{i,5}$

FIGURE 9.18
URM showing 10 rules, linguistic input labels, and output consequent values.

Input 1, $X(x)$	$\mu_{1,1} * -10.0$	$\mu_{1,2} * -5.0$	$\mu_{1,3} * 0.0$	$\mu_{1,4} * 5.0$	$\mu_{1,5} * 10.0$
Input 2, $Y(y)$	$\mu_{2,1} * -10.0$	$\mu_{2,2} * -5.0$	$\mu_{2,3} * 0.0$	$\mu_{2,4} * 5.0$	$\mu_{2,5} * 10.0$
Accumulator	$\sum_{i=1}^{2} \mu_{i,1} z_{i,1}$	$\sum_{i=1}^{2} \mu_{i,2} z_{i,2}$	$\sum_{i=1}^{2} \mu_{i,3} z_{i,3}$	$\sum_{i=1}^{2} \mu_{i,4} z_{i,4}$	$\sum_{i=1}^{2} \mu_{i,5} z_{i,5}$

FIGURE 9.19
URM showing 10 rules, membership values, and output consequent values for Eq. (9.21).

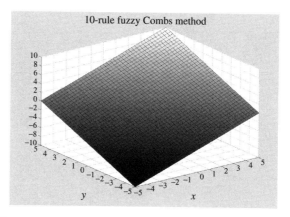

FIGURE 9.20
Combs Method simulation using 10 rules.

(FAM tables) the use of matrix norms is appropriate to estimate the size of error attributed to each method. The three matrix norms used are: the one norm, the Frobenius norm and the infinity norm.

Given a matrix A, the following equations define the three norms [Meyer, 2000]:

One norm:

$$\|A\|_1 = \max_j \sum_i |a_{ij}|$$
(9.23)

Frobenius norm:

$$\|A\|_F = \left(\sum_{ij} |a_{ij}|^2 \right)^{1/2}$$
(9.24)

Infinity norm:

$$\|A\|_\infty = \max_i \sum_j |a_{ij}|$$
(9.25)

We can use these norm measures to calculate the total error based on the difference between the exact solution matrix A and the approximate solution matrix Â, for example, using an absolute error metric,

Absolute error:

$$\|\hat{A} - A\|_p \quad p = 1, \text{ F, or } \infty \text{ (for the one, Frobenius, and infinity norms,}$$
(9.26)
$$\text{respectively)}$$

A more explicit indicator of an approximation is given by its relative error. The error can be relative to the approximate solution or to the exact solution. Here both absolute and relative errors complete our analysis.

Relative error 1 with respect to the exact solution A:

$$\frac{\|\hat{A} - A\|_p}{\|A\|_p} \quad p = 1, \text{ F, or } \infty$$
(9.27)

Relative error 2 with respect to the approximate solution $\hat{A}$:

$$\frac{\|\hat{A} - A\|_p}{\|\hat{A}\|_p} \qquad p = 1, \text{ F, or } \infty \tag{9.28}$$

All error measures for this problem are on the order of 10^{-14} or smaller since the function we are approximating is a simple linear surface. The small error ($<10^{-14}$) confirms that additively separable IRC fuzzy systems are equivalent to the Combs method. In addition to providing good simulation results the reduction methods also prove to be very effective. The IRC system is reduced from 25 rules to 10 rules using the Combs method. This is a 60% reduction. Similarly, the SVD reduced the 25-rule system by 84%. The benefit of these reduction methods would become even more apparent when applied to large 10,000 rule-base systems. Either a 60% or an 84% reduction in rules on such large systems would result in direct savings of time.

Example 9.2. To make the simulation a little more complex, we now apply the rule-reduction methods to simulate the function

$$f(x, y) = \sqrt{x} + y^2 \tag{9.29}$$

as shown in Fig. 9.21. Again, this function also represents an *additively separable* system, because the system can be broken into two vectors one, v_1 representing $\sqrt{x}$ and the other, v_2 representing y^2. The outer sum, Eq. (9.20), of these vectors provides the rule-base for simulation.

IRC rule-base simulation

As in Example 9.1, five partitions are used for the two input variables, x and y. They again consist of triangular membership functions representing the linguistic labels. However, the ranges of each variable change according to the function results plotted in Fig. 9.21, $x \rightarrow [0, 5.0]$ and $y \rightarrow [0, 1.0]$. (See Figs. 9.22–9.23.)

Like Example 9.1, an FAM table, Fig. 9.24, showing crisp rule consequent values maps the input fuzzy sets, x and y, to the output of Eq. (9.29).

This results in the solution surface of Fig. 9.25 by simply fitting a surface through the points given in Fig. 9.24.

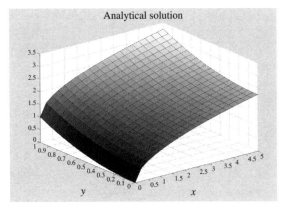

FIGURE 9.21
Solution surface for Eq. (9.29).

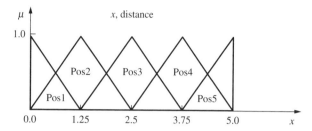

FIGURE 9.22
Fuzzy sets for the variable x represented by triangular membership functions.

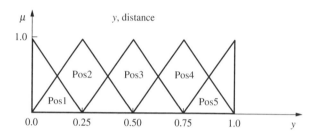

FIGURE 9.23
Fuzzy sets for the variable y represented by triangular membership functions.

25 fuzzy rules		y			
	Pos1	Pos2	Pos3	Pos4	Pos5
Pos1	0.0	0.0625	0.2500	0.4900	1.0000
Pos2	1.1180	1.1805	1.3680	1.6080	2.1180
Pos3	1.5811	1.6436	1.8311	2.0711	2.5811
Pos4	1.9365	1.9990	2.1865	2.4265	2.9365
Pos5	2.2361	2.2986	2.4861	2.7261	3.2361

FIGURE 9.24
FAM table for IRC simulation.

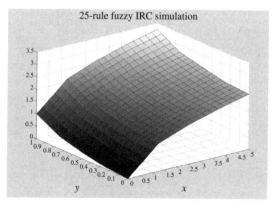

FIGURE 9.25
Fuzzy IRC simulation using 25 rules.

$Z = f(x, y)$		y			
		0	0.25	0.5	0.75	1.0
	0	0.0	0.0625	0.2500	0.4900	1.0000
x	1.25	1.1180	1.1805	1.3680	1.6080	2.1180
	2.5	1.5811	1.6436	1.8311	2.0711	2.5811
	3.75	1.9365	1.9990	2.1865	2.4265	2.9365
	5.0	2.2361	2.2986	2.4861	2.7261	3.2361

FIGURE 9.26
Output of Eq. (9.29) to be used for SVD simulation.

SVD simulation

The IRC FAM table, Fig. 9.26, is used as the matrix, Z, on which the SVD will be processed; refer to Example 9.1.

The matrix Z is decomposed as in Example 9.1 initially into three matrices, two orthogonal and one diagonal, U, V, and Σ, respectively. Again, the singular values of Σ are used to indicate the importance of the input sets to approximating a function or rule-base:

$$Z = U\Sigma V^T = \begin{bmatrix} -0.0967 & -0.8993 & 0.0738 & -0.0146 & -0.4198 \\ -0.3517 & -0.3128 & 0.2079 & -0.3126 & 0.7984 \\ -0.4573 & -0.0698 & -0.3168 & 0.8102 & 0.1709 \\ -0.5383 & 0.1166 & -0.6345 & -0.4955 & -0.2201 \\ -0.6067 & 0.2738 & 0.6696 & 0.0125 & -0.3294 \end{bmatrix} * \begin{bmatrix} 9.6575 & 0 & 0 & 0 & 0 \\ 0 & 0.7334 & 0 & 0 & 0 \\ 0 & 0 & 0 & 0 & 0 \\ 0 & 0 & 0 & 0 & 0 \\ 0 & 0 & 0 & 0 & 0 \end{bmatrix}$$

$$* \begin{bmatrix} -0.3640 & 0.5152 & -0.0883 & 0.1984 & -0.7449 \\ -0.3773 & 0.4392 & 0.1123 & 0.5241 & 0.6144 \\ -0.4171 & 0.2113 & -0.5536 & -0.6445 & 0.2439 \\ -0.4680 & -0.0805 & 0.7810 & -0.4047 & -0.0273 \\ -0.5763 & -0.7004 & -0.2513 & 0.3267 & -0.0860 \end{bmatrix}^T$$

Just as in Example 9.1, there are only two singular values. The reason for this is that the original system, Z, is rank 2, requiring only two singular values in its decomposition. Even though this makes for a simple problem, the same procedures will be highlighted that can be generalized to more complex problems, i.e., systems with more than two singular values. Therefore, the two singular values are used as the rule-base reduction continues. Matrix U is the subject of the first step of operations. The procedures in this example follow the same development as those in Example 9.1. Therefore, only the step-by-step calculations will be presented.

Overlapping membership functions: matrix U

Two submatrices representing the retained and discarded columns of the U partition are

$$U_r = \begin{bmatrix} -0.0967 & -0.8993 \\ -0.3517 & -0.3128 \\ -0.4573 & -0.0698 \\ -0.5383 & 0.1166 \\ -0.6067 & 0.2738 \end{bmatrix}, \quad U_d = \begin{bmatrix} 0.0738 & -0.0146 & -0.4198 \\ 0.2079 & -0.3126 & 0.7984 \\ -0.3168 & 0.8102 & 0.1709 \\ -0.6345 & -0.4955 & -0.2201 \\ 0.6696 & 0.0125 & -0.3294 \end{bmatrix}$$

The column sums of U_r and one additional unity value form the diagonal matrix $\mathcal{C}_U$, Eq. (9.8):

$$\mathcal{C}_U = \begin{bmatrix} -2.0507 & 0 & 0 \\ 0 & -0.8915 & 0 \\ 0 & 0 & 1 \end{bmatrix} = \begin{bmatrix} sum(U_r^T) & \\ & 1 \end{bmatrix}$$

From Eq. (9.7), we get

$$U_1 = \begin{bmatrix} -0.0967 & -0.8993 & 0.0 \\ -0.3517 & -0.3128 & 0.0 \\ -0.4573 & -0.0698 & 0.0 \\ -0.5383 & 0.1166 & 0.0 \\ -0.6067 & 0.2738 & 0.0 \end{bmatrix} * \begin{bmatrix} -2.0507 & 0 & 0 \\ 0 & -0.8915 & 0 \\ 0 & 0 & 1 \end{bmatrix}$$

$$U_1 = \begin{bmatrix} 0.1983 & 0.8017 & 0.0 \\ 0.7212 & 0.2788 & 0.0 \\ 0.9378 & 0.0622 & 0.0 \\ 1.1039 & -0.1039 & 0.0 \\ 1.2441 & -0.2441 & 0.0 \end{bmatrix} = \begin{bmatrix} 0.1983 & 0.8017 \\ 0.7212 & 0.2788 \\ 0.9378 & 0.0622 \\ 1.1039 & -0.1039 \\ 1.2441 & -0.2441 \end{bmatrix}$$

Non-negative membership values: matrix U

We assign δ equal to either 1 or $1/\min(U_1)$ according to Eq. (9.10) depending on the minimum element in U_1. This value is found to be $\min(U_1) = -0.2441$, therefore $\delta = 1.0$. Now Eq. (9.11) forms the doubly stochastic matrix, $\cancel{D}_U$,

$$\cancel{D}_U = \frac{1}{(n+\delta)} \begin{bmatrix} (1+\delta) & 1 & \cdots & 1 \\ 1 & (1+\delta) & \cdots & 1 \\ \vdots & \vdots & \ddots & \vdots \\ 1 & 1 & \cdots & (1+\delta) \end{bmatrix} = \frac{1}{(2+1)} \begin{bmatrix} 1+1 & 1 \\ 1 & 1+1 \end{bmatrix} = \begin{bmatrix} 0.6667 & 0.3333 \\ 0.3333 & 0.6667 \end{bmatrix}$$

From Eq. (9.12) we get

$$U_2 = U_1 * \cancel{D}_U = \begin{bmatrix} 0.1983 & 0.8017 \\ 0.7212 & 0.2788 \\ 0.9378 & 0.0622 \\ 1.1039 & -0.1039 \\ 1.2441 & -0.2441 \end{bmatrix} * \begin{bmatrix} 0.6667 & 0.3333 \\ 0.3333 & 0.6667 \end{bmatrix} = \begin{bmatrix} 0.3994 & 0.6006 \\ 0.5737 & 0.4263 \\ 0.6459 & 0.3541 \\ 0.7013 & 0.2987 \\ 0.7480 & 0.2520 \end{bmatrix}$$

Prototypical membership value: matrix U

The final step of conditioning matrix U uses Eq. (9.13) to form matrix $\cancel{E}_U$, the matrix used to assign prototypical membership values to the fuzzy sets:

$$\cancel{E}_U = \begin{bmatrix} U_2(1,:) \\ U_2(5,:) \end{bmatrix}^{-1} = \begin{bmatrix} 0.3994 & 0.6006 \\ 0.7480 & 0.2520 \end{bmatrix}^{-1} = \begin{bmatrix} -0.7229 & 1.7229 \\ 2.1457 & -1.1457 \end{bmatrix}$$

From Eq. (9.14) we get

$$U_3 = U_2 * \cancel{E}_U = \begin{bmatrix} 0.3994 & 0.6006 \\ 0.5737 & 0.4263 \\ 0.6459 & 0.3541 \\ 0.7013 & 0.2978 \\ 0.7480 & 0.2520 \end{bmatrix} * \begin{bmatrix} -0.7229 & 1.7229 \\ 2.1457 & -1.1457 \end{bmatrix}$$

$$= \begin{bmatrix} 1.0000 & 0.0000 \\ 0.5000 & 0.5000 \\ 0.2929 & 0.7071 \\ 0.1340 & 0.8660 \\ 0.0000 & 1.0000 \end{bmatrix}$$

The columns of U_3 become the membership functions for the input x, as seen in Fig. 9.27.

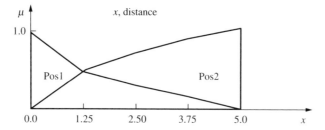

FIGURE 9.27
Plotted columns of U_3 representing two odd-shaped membership functions.

Overlapping membership functions: matrix V

Now matrix V is conditioned; two submatrices representing the retained and discarded columns of the V partition are

$$
V_r = \begin{bmatrix} -0.3640 & 0.5152 \\ -0.3773 & 0.4392 \\ -0.4171 & 0.2113 \\ -0.4680 & -0.0805 \\ -0.5763 & -0.7004 \end{bmatrix}, \; V_d = \begin{bmatrix} -0.0883 & 0.1984 & -0.7449 \\ 0.1123 & 0.5241 & 0.6144 \\ -0.5536 & -0.6445 & 0.2439 \\ 0.7810 & -0.4047 & -0.0273 \\ -0.2513 & 0.3267 & -0.0860 \end{bmatrix}
$$

The column sums of V_r and one additional unity value form the diagonal matrix $\mathcal{C}_V$, Eq. (9.8):

$$
\mathcal{C}_V = \begin{bmatrix} -2.2027 & 0 & 0 \\ 0 & 0.3848 & 0 \\ 0 & 0 & 1 \end{bmatrix} = \begin{bmatrix} \text{sum}(V_r^T) & \\ & 1 \end{bmatrix}
$$

From Eq. (9.7) we get

$$
V_1 = \begin{bmatrix} -0.3640 & 0.5152 & 0.0 \\ -0.3773 & 0.4392 & 0.0 \\ -0.4171 & 0.2113 & 0.0 \\ -0.4680 & -0.0805 & 0.0 \\ -0.5763 & -0.7004 & 0.0 \end{bmatrix} * \begin{bmatrix} -2.2027 & 0 & 0 \\ 0 & 0.3848 & 0 \\ 0 & 0 & 1 \end{bmatrix}
$$

$$
V_1 = \begin{bmatrix} 0.8018 & 0.1982 & 0.0 \\ 0.8310 & 0.1690 & 0.0 \\ 0.9187 & 0.0813 & 0.0 \\ 1.0309 & -0.0309 & 0.0 \\ 1.2694 & -0.2694 & 0.0 \end{bmatrix} = \begin{bmatrix} 0.8018 & 0.1982 \\ 0.8310 & 0.1690 \\ 0.9187 & 0.0813 \\ 1.0309 & -0.0309 \\ 1.2694 & -0.2694 \end{bmatrix}
$$

Non-negative membership values: matrix V

We find the minimum element in V_1 and substitute either 1 or $1/\min(V_1)$ for δ: $\min(V_1) = -0.2694$, therefore $\delta = 1.0$. Equation (9.11) forms the doubly stochastic matrix $\mathcal{D}_V$,

$$
\mathcal{D}_V = \frac{1}{(n+\delta)} \begin{bmatrix} (1+\delta) & 1 & \cdots & 1 \\ 1 & (1+\delta) & \cdots & 1 \\ \vdots & \vdots & \ddots & \vdots \\ 1 & 1 & \cdots & (1+\delta) \end{bmatrix} = \frac{1}{(2+1)} \begin{bmatrix} 1+1 & 1 \\ 1 & 1+1 \end{bmatrix}
$$

$$
= \begin{bmatrix} 0.6667 & 0.3333 \\ 0.3333 & 0.6667 \end{bmatrix}
$$

From Eq. (9.12) we determine V_2,

$$V_2 = V_1 * \cancel{D}_V = \begin{bmatrix} 0.8018 & 0.1982 \\ 0.0.8310 & 0.1690 \\ 0.9187 & 0.0813 \\ 1.0309 & -0.0309 \\ 1.2694 & -0.2694 \end{bmatrix} * \begin{bmatrix} 0.6667 & 0.3333 \\ 0.3333 & 0.6667 \end{bmatrix} = \begin{bmatrix} 0.6006 & 0.3994 \\ 0.6104 & 0.3896 \\ 0.6396 & 0.3604 \\ 0.6770 & 0.3230 \\ 0.7565 & 0.2435 \end{bmatrix}$$

Prototypical membership value: matrix V

Equation (9.13) is used for the final step of conditioning matrix V to form matrix $\cancel{E}_V$. Again, this matrix is used to assign prototypical membership values to the fuzzy sets:

$$\cancel{E}_V = \begin{bmatrix} V_2(1, :) \\ V_2(5, :) \end{bmatrix}^{-1} = \begin{bmatrix} 0.6006 & 0.3994 \\ 0.7565 & 0.2435 \end{bmatrix}^{-1} = \begin{bmatrix} -1.5618 & 2.5618 \\ 4.8522 & -3.8522 \end{bmatrix}$$

From Eq. (9.14) we get V_3,

$$V_3 = V_2 * \cancel{E}_V = \begin{bmatrix} 0.6006 & 0.3994 \\ 0.6104 & 0.3896 \\ 0.6396 & 0.3604 \\ 0.6770 & 0.3230 \\ 0.7565 & 0.2435 \end{bmatrix} * \begin{bmatrix} -1.5618 & 2.5618 \\ 4.8522 & -3.8522 \end{bmatrix}$$

$$= \begin{bmatrix} 1.0000 & 0.0000 \\ 0.9375 & 0.0625 \\ 0.7500 & 0.2500 \\ 0.5100 & 0.4900 \\ 0.0000 & 1.0000 \end{bmatrix}$$

The columns of V_3 become the membership functions for the input y, as seen in Fig. 9.28.

Reduced matrix of rule consequent values

The new reduced matrix of rule consequent values, Z_r, is now developed using Eq. (9.15) and the inverses of the matrices just produced. Hence, Z_r becomes the FAM table for this simulation; therefore, the inverses of the previous matrices yield the following set:

$$\cancel{C}_U^{-1} = \begin{bmatrix} -0.4876 & 0 \\ 0 & -1.1217 \end{bmatrix} \quad \cancel{C}_V^{-1} = \begin{bmatrix} -0.4540 & 0 \\ 0 & 2.5988 \end{bmatrix}$$

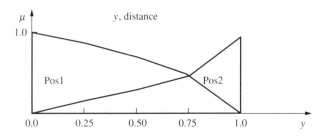

FIGURE 9.28
Plotted columns of V_3 representing two odd-shaped membership functions.

$$\not{D}_U^{-1} = \begin{bmatrix} 1.9997 & -0.9997 \\ -0.9997 & 1.9997 \end{bmatrix} \quad \not{D}_V^{-1} = \begin{bmatrix} 1.9997 & -0.9997 \\ -0.9997 & 1.9997 \end{bmatrix}$$

$$\not{E}_U^{-1} = \begin{bmatrix} 0.3994 & 0.6006 \\ 0.7480 & 0.2520 \end{bmatrix} \quad \not{E}_V^{-1} = \begin{bmatrix} 0.6006 & 0.7480 \\ 0.3994 & 0.2520 \end{bmatrix}$$

This set of matrices, multiplied with the matrix of retained singular values,

$$\Sigma_r = \begin{bmatrix} 9.6575 & 0 \\ 0 & 0.7334 \end{bmatrix}$$

gives the following matrix product using Eq. (9.15):

$$Z_r = \begin{bmatrix} 0.3994 & 0.6006 \\ 0.7480 & 0.2520 \end{bmatrix} \begin{bmatrix} 1.9997 & -0.9997 \\ -0.9997 & 1.9997 \end{bmatrix} \begin{bmatrix} -0.4876 & 0 \\ 0 & -1.1217 \end{bmatrix} \begin{bmatrix} 9.6575 & 0 \\ 0 & 0.7334 \end{bmatrix}$$

$$* \begin{bmatrix} -0.4540 & 0 \\ 0 & 2.5988 \end{bmatrix} \begin{bmatrix} 1.9997 & -0.9997 \\ -0.9997 & 1.9997 \end{bmatrix} \begin{bmatrix} 0.6006 & 0.7480 \\ 0.3994 & 0.2520 \end{bmatrix} = \begin{bmatrix} 0.0 & 1.0 \\ 2.2 & 3.2 \end{bmatrix}$$

The FAM table results are shown in Fig. 9.29.
The surface of Fig. 9.30 represents the reduced rule-base simulation of Eq. (9.29).

Combs method for rapid inference

This system, although still additively separable, is different enough from the first example to require a somewhat special approach to its simulation using the Combs method. Again, the

		\multicolumn{2}{c}{y}	
Z_r		Pos1	Pos2
x	Pos1	0.0	1.0
	Pos2	2.2	3.2

FIGURE 9.29
FAM table due to the SVD method.

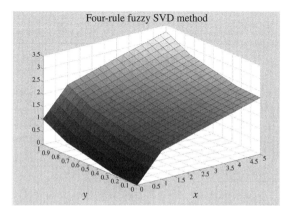

FIGURE 9.30
SVD simulation using four rules.

same membership functions as in the 25-rule IRC approach are used. However, instead of mapping the bilinear system as an outer sum, vectors corresponding to each nonlinear term now map the system as an outer sum. Weinschenk et al. [2003] derive the equality between IRC and URC for additively separable systems and state, "the consequent centers of mass of the URC system are scaled versions of the IRC projection vector elements." Hence, the output vectors of Eq. (9.20) (seen in Fig. 9.24 as the first row and first column) are multiplied by 2 and subsequently become the output values to the Combs method.

Using the generalized convention as previously described results in showing the input variables, x and y, each with five fuzzy sets as seen in Fig. 9.31. Also shown in this figure are the rule consequent values in each cell, as determined by using a weighting value of 2.

Figure 9.31 gets transformed like Example 9.1 to the URM of Fig. 9.32 in accordance with Eq. (9.21). This again shows how the computations of the system are implemented.

The results of this simulation are seen in Fig. 9.33 showing a reasonably good approximation to the solution surface of Eq. (9.29).

$X(x)$, distance	If x is Pos1 then 0.0	If x is Pos2 then 2.2	If x is Pos3 then 3.2	If x is Pos4 then 3.8	If x is Pos5 then 4.5
$Y(y)$, distance	If y is Pos1 then 0.0	If y is Pos2 then 0.12	If y is Pos3 then 0.5	If y is Pos4 then 1.12	If y is Pos5 then 2.0
Accumulator	$\sum_{i=1}^{2} \mu_{i,1} z_{i,1}$	$\sum_{i=1}^{2} \mu_{i,2} z_{i,2}$	$\sum_{i=1}^{2} \mu_{i,3} z_{i,3}$	$\sum_{i=1}^{2} \mu_{i,4} z_{i,4}$	$\sum_{i=1}^{2} \mu_{i,5} z_{i,5}$

FIGURE 9.31
URM showing 10 rules, linguistic input labels, and output consequent values.

Input 1, $X(x)$	$\mu_{1,1} * 0$	$\mu_{1,2} * 2.2$	$\mu_{1,3} * 3.2$	$\mu_{1,4} * 3.8$	$\mu_{1,5} * 4.5$
Input 2, $Y(y)$	$\mu_{2,1} * 0$	$\mu_{2,2} * 0.12$	$\mu_{2,3} * 0.5$	$\mu_{2,4} * 1.12$	$\mu_{2,5} * 2.0$
Accumulator	$\sum_{i=1}^{2} \mu_{i,1} z_{i,1}$	$\sum_{i=1}^{2} \mu_{i,2} z_{i,2}$	$\sum_{i=1}^{2} \mu_{i,3} z_{i,3}$	$\sum_{i=1}^{2} \mu_{i,4} z_{i,4}$	$\sum_{i=1}^{2} \mu_{i,5} z_{i,5}$

FIGURE 9.32
URM showing 10 rules, membership values, and output consequent values.

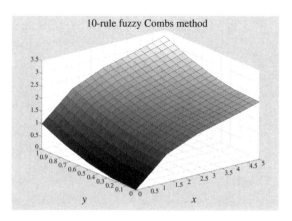

FIGURE 9.33
Combs method simulation using 10 rules.

Measures		IRC	SVD	Combs
One norm	*Absolute error*	5.9669	4.6802	5.7050
	Relative error1	0.1102	0.0864	0.1053
	Relative error2	0.1105	0.0865	0.1044
Frobenius norm	*Absolute error*	2.1612	1.6277	2.0357
	Relative error1	0.0533	0.0401	0.0502
	Relative error2	0.0543	0.0408	0.0508
Infinity norm	*Absolute error*	2.4691	2.0840	1.1910
	Relative error1	0.0476	0.0402	0.0229
	Relative error2	0.0485	0.0407	0.0234

FIGURE 9.34
Error comparison for all three fuzzy methods.

Error analysis of the methods

The error analysis here follows Eqs. (9.23)–(9.28), which give the relationships that define the error measures in the simulation. Again, the three matrix norms that are used include the one norm, the Frobenius norm, and the infinity norm. These norms are used on absolute and relative errors by means of Eqs. (9.26)–(9.28).

The absolute error has the same units of Eq. (9.29). However, since no units have been defined, consider the units to be displacement along the z axis. Relative error is a dimensionless value. Figure 9.34 gives the comparison results of this simulation showing values much greater than 0, unlike Example 9.1. The higher complexity in the analytical function is the reason for this, i.e., the sum of two nonlinear terms. The same reduction percentages are achieved as in Example 9.1: 60% for Combs and 84% for the SVD. This discussion concentrates on the accuracy of each method. The absolute error is the accumulation of differences between the simulation and the analytical solution for each point over the entire surface. Figure 9.34 shows that the IRC yields the largest error value for all the norms. Furthermore, the Combs method gives the second-largest value for the one norm and the Frobenius norm and the SVD is second for the infinity norm. What does this mean? The one norm gives the largest absolute column sum, and the infinity norm gives the largest absolute row sum. These measures give a sense of error in either the y direction (column) or the x direction (row). Therefore, the SVD gives the best approximation according to the one norm in the y direction, and the Combs method gives the best approximation in the x direction. Now the Frobenius norm is somewhat different. This norm takes all entries of the difference $\hat{A} - A$ as a single vector and finds its length. Hence, Fig. 9.34 reveals that the SVD gives the shortest length of errors, or the best approximation in terms of the Frobenius norm. This same reasoning extends to relative errors, and we see the same trends.

These results prove that not only can an IRC system be reduced, but also it can be improved. An IRC system maps the input space to the output space. But, based on the many influences that affect the system behavior, an IRC may not be as accurate as desired. Many times "fine tuning" the system is necessary. Fine tuning refers to the use of a different number of fuzzy sets, redefining membership functions ranges, prototypes, etc. The reduction methods in this example have effectively used either a different number of fuzzy sets or a different configuration to achieve improved accuracy. Although this may not be the case for every simulation, it is important to note that these refinements are possible with these new tools. In addition to the two new rule-reduction tools discussed in this chapter, there have been others published in the literature [Jamshidi, 1994].

SUMMARY

Building a robust yet economical fuzzy system can be a challenge. This can, however, be achieved by means of the two rule-base reduction methods presented in this chapter. Yam

[1997] explains how to expand the SVD to systems with more than two input dimensions. Similarly, Combs and Andrews [1998] describe systems with more than two input variables. Furthermore, these methods are founded on algebraic principles that allow for automatic implementation, unlike previous methods. Jamshidi [1994] describes three rule-reduction methods that rely heavily on developer acumen. First, the *hierarchical method* prioritizes the rules into input–output levels. This removes some of the combinatorial effect, but requires a good understanding about the system to be simulated. Second, the *sensory-fusion method* combines input algebraically. This allows fewer fuzzy sets as the number of inputs is combined. However, again a good understanding is necessary to determine the appropriate input to combine. Third, a hybrid of the first two is made as the *hierarchical with sensory-fusion method*. This method follows the same form as the first two but is effective only if combined with the best features of each method. Clearly, these methods, although sometimes effective, are not well-suited for automatic programming.

REFERENCES

Bremermann, H. J. (1962). "Quantum-theoretical Limitations of Data Processing," *Abstracts of Short Communications, International Congress of Mathematics, Stockholm.*

Combs, W. E. and Andrews, J. E. (1998). "Combinatorial Rule Explosion Eliminated by a Fuzzy Rule Configuration," IEEE Trans. Fuzzy Syst., vol. 6, no. 1, pp. 1–11.

Devore, J. L. (1995). *Probability and Statistics for Engineering and the Sciences*, 4th ed., Duxbury Press, Belmont, CA.

Jamshidi, M. (1994). *Large-Scale Systems: Modeling, Control and Fuzzy Logic*, Prentice Hall, Englewood Cliffs, NJ.

Klir, G. and Yuan, B. (1995). *Fuzzy Sets and Fuzzy Logic: Theory and applications*, Prentice Hall, Englewood Cliffs, NJ.

Lucero, J. (2004). "Fuzzy Systems Methods in Structural Engineering," PhD dissertation, University of New Mexico, Dept. of Civil Engineering, Albuquerque, NM.

Meyer, C. D. (2000). *Matrix Analysis and Applied Linear Algebra*, Society for Industrial and Applied Mathematics, Philadelphia, PA.

Simon, D. (2000). "Design and Rule Base Reduction of a Fuzzy Filter for the Estimation of Motor Currents," *Int. J. Approximate Reasoning*, vol. 25, pp. 145–167.

Weinschenk, J. J., Combs, W. E., and Marks II, R. J. (2003). "Avoidance of Rule Explosion by Mapping Fuzzy Systems to a Union Rule Configuration," *The IEEE International Conference on Fuzzy Systems*, pp. 43–48.

Yam, Y. (1997). "Fuzzy Approximation Via Grid Point Sampling and Singular Value Decomposition," *IEEE Trans. Syst., Man, Cybern. – Part B: Cybern.*, vol. 27, no. 6, pp. 933–951.

PROBLEMS

Singular Value Decomposition

9.1. This problem will demonstrate the progression of the singular value approximation using the MATLAB command "svd" in finding a singular value decomposition. A singular value decomposition diagonalizes and decomposes a matrix A into three matrices, one diagonal and two orthogonal.

(a) Given

$$A = \begin{bmatrix} 4 & 0 & 0 & 0 \\ 0 & 4 & 0 & 0 \\ 0 & 0 & 4 & 0 \\ 0 & 0 & 0 & 4 \end{bmatrix}$$

find its singular value decomposition using the MATLAB command "[U, S, V] = svd(A)."

(b) A matrix approximation is a summation of the products of all the singular values with their corresponding column vectors from matrices U and V as

$$A = u_1 s_1 v_1^T + \cdots + u_n s_n v_n^T$$

Perform a rank 1, 2, 3, and 4 approximation of the matrix in part (a).

9.2. Repeat Problem 9.1a and b on the following matrix:

$$A = \begin{bmatrix} 9 & 3 & 2 & 7 \\ 8 & 5 & 7 & 3 \\ 9 & 2 & 1 & 3 \\ 3 & 2 & 2 & 7 \end{bmatrix}$$

What are the two largest singular values?

9.3. Plot the set A and its convex hull

$$A = \begin{bmatrix} 1 & 2 \\ 1 & 3 \\ 4 & 3 \\ 4 & 5 \\ 7 & 2 \\ 5 & 2.5 \end{bmatrix}$$

9.4. Using the matrix

$$A = \begin{bmatrix} 0.3 & 0.2 & 0.4 & 0.1 \\ -0.4 & 0.5 & 0.1 & 0.8 \\ 1.2 & -0.4 & -0.3 & 0.5 \\ 0.15 & 0.7 & 0.35 & -0.2 \end{bmatrix}$$

show its corresponding doubly stochastic matrix.

9.5. This problem demonstrates the use of the "orth" command in MATLAB and how with an orthogonal matrix one can prove the "sum–normal" condition.

(a) Using the "orth" command in MATLAB verify that the orthonormal basis to matrix A is U.

$$A = \begin{bmatrix} 1 & 2 & 3 \\ 4 & 5 & 6 \\ 7 & 8 & 0 \end{bmatrix}, \quad U = \begin{bmatrix} -0.230357\ldots & -0.396071\ldots & -0.888855\ldots \\ -0.607283\ldots & -0.655207\ldots & 0.449343\ldots \\ -0.760356\ldots & 0.643296\ldots & -0.089595\ldots \end{bmatrix}$$

(b) Prove the sum–normal condition using matrix U.

9.6. For a given type of building with natural frequency ω, a vulnerability surface on damage has been developed based on analyst opinion and past measurements where damage ranges from 1 (slight) to 6 (collapse). The following table results:

		Maximum displacement, in.				
		1	2	3	4	5
Peak ground	0.1	1	3	2	2	1
acceleration,	0.2	2	3	2	2	1
in./s²	0.3	4	4	3	2	2
	0.4	5	6	5	4	3
	0.5	4	5	4	3	2

(a) Calculate U, S, and V using the MATLAB command, "[U S V] = svd(A)."

(b) Condition U and V according to the three steps outlined in this chapter.

Note: Peak ground acceleration corresponds to U and maximum displacement corresponds to V.

- Keep two singular values

$$\text{convex hull, } U = \begin{bmatrix} U_2(1,:) \\ U_2(3,:) \\ U_2(4,:) \end{bmatrix}$$

-

$$\text{convex hull, } V = \begin{bmatrix} V_2(1,:) \\ V_2(2,:) \\ V_2(5,:) \end{bmatrix}$$

(c) Form the matrix of rule consequent values, Z_r.

Combs Method for Rapid Inference

9.7. Using a truth table prove the set-theoretic equivalence of the IRC with the URC.

9.8. Input measurements are taken for fuzzy sets $\underset{\sim}{A}$, $\underset{\sim}{B}$, and $\underset{\sim}{C}$. They yield the following membership values:

$A(x)_{Low} = 0.25$	$B(y)_{Low} = 0.63$	$C(z)_{Low} = 0.18$
$A(x)_{Medium} = 0.4$	$B(y)_{Medium} = 0.37$	$C(z)_{Medium} = 0.04$
$A(x)_{High} = 0.9$	$B(y)_{High} = 0.1$	$C(z)_{High} = 0.0$

$$\text{Output Low} = 2$$
$$\text{Output Medium} = 5$$
$$\text{Output High} = 10$$

Fill in the following union rule matrix and calculate the final output, Z_{URC}.

	Low	Medium	High
Input A			
Input B			
Input C			
Accumulator array			

9.9. Take the outer sum of the following vectors:

$$A = \begin{bmatrix} 1.2 \\ 7.6 \\ 9.4 \\ 3.3 \end{bmatrix} \quad B = \begin{bmatrix} 5.5 \\ 3.9 \\ 2.0 \\ 6.8 \end{bmatrix}$$

9.10. Repeat Problem 9.8, except now A, B, and C are mapped to the following output sets:

Output $A_{Low} = 2$ Output $B_{Low} = 3$ Output $C_{Low} = 1$

Output $A_{Medium} = 5$ Output $B_{Medium} = 7$ Output $C_{Medium} = 2$

Output $A_{High} = 10$ Output $B_{High} = 8$ Output $C_{High} = 3$

CHAPTER
10

DECISION MAKING WITH FUZZY INFORMATION

To be, or not to be: that is the question:
Whether 'tis nobler in the mind to suffer
The slings and arrows of outrageous fortune,
Or to take arms against a sea of troubles,
And by opposing end them.

William Shakespeare
Hamlet, Act III, Scene I, 1602

The passage above represents a classic decision situation for humans. It is expressed in natural language – the form of information most used by humans and most ignored in computer-assisted decision making. But, as suggested many other times in this text, this is the nature of the problem engineers face every day: how do we embed natural fuzziness into our otherwise crisp engineering paradigms? Shakespeare would undoubtedly rejoice to learn that his question now has a whole range of possibilities available between the extremes of existence that he originally suggested. Ultimately, the decisions may be binary, as originally posed by Shakespeare in this passage from *Hamlet*, but there certainly should be no restrictions on the usefulness of fuzzy information in the *process* of making a decision or of coming to some consensus.

Decision making is a most important scientific, social, and economic endeavor. To be able to make consistent and correct choices is the essence of any decision process imbued with uncertainty. Most issues in life, as trivial as we might consider them, involve decision processes of one form or another. From the moment we wake in the morning to the time we place our bodies at rest at the day's conclusion we make many, many decisions. What should

Fuzzy Logic with Engineering Applications, Second Edition   T. J. Ross
© 2004 John Wiley & Sons, Ltd   ISBNs: 0-470-86074-X (HB); 0-470-86075-8 (PB)

we wear for the day; should we take an umbrella; what should we eat for breakfast, for lunch, for dinner; should we stop by the gas station on the way to work; what route should we take to work; should we attend that seminar at work; should we write the memorandum to our colleagues before we make the reservations for our next trip out of town; should we go to the store on our way home; should we take the kids to that new museum before, or after, dinner; should we watch the evening news before retiring; and so on and so forth?

We must keep in mind when dealing with decision making under uncertainty that *there is a distinct difference between a good decision and a good outcome*! In any decision process we weigh the information about an issue or outcome and choose among two or more alternatives for subsequent action. The information affecting the issue is likely incomplete or uncertain; hence, the outcomes are uncertain, irrespective of the decision made or the alternative chosen. We can make a good decision, and the outcome can be adverse. Alternatively, we can make a bad decision, and the outcome can be advantageous. Such are the vagaries of uncertain events. But in the long run, if we consistently make good decisions, advantageous situations will occur more frequently than bad ones.

To illustrate this notion, consider the choice of whether to take an umbrella on a cloudy, dark morning. As a simple binary matter, the outcomes can be rain or no rain. We have two alternatives: take an umbrella, or do not. The information we consider in making this decision could be as unsophisticated as our own feelings about the weather on similar days in the past or as sophisticated as a large-scale meteorological analysis from the national weather service. Whatever the source of information, it will be associated with some degree of uncertainty. Suppose we decide to take the umbrella after weighing all the information, and it does not rain. Did we make a bad decision? Perhaps not. Eight times out of 10 in circumstances just like this one, it probably rained. This particular occasion may have been one of the two out of 10 situations when it did not.

Despite our formal training in this area and despite our common sense about how clear this notion of uncertainty is, we see it violated every day in the business world. A manager makes a good decision, but the outcome is bad and the manager gets fired. A doctor uses the best established procedures in a medical operation and the patient dies; then the doctor gets sued for malpractice. A boy refuses to accept an unsolicited ride home with a distant neighbor on an inclement day, gets soaking wet on the walk home, ruins his shoes, and is reprimanded by his parent for not accepting the ride. A teenager decides to drive on the highway after consuming too many drinks and arrives home safely without incident. In all of these situations the outcomes have nothing to do with the quality of the decisions or with the process itself. The best we can do is to make consistently rational decisions every time we are faced with a choice with the knowledge that in the long run the ''goods'' will outweigh the ''bads.''

The problem in making decisions under uncertainty is that the bulk of the information we have about the possible outcomes, about the value of new information, about the way the conditions change with time (dynamic), about the utility of each outcome–action pair, and about our preferences for each action is typically vague, ambiguous, and otherwise fuzzy. In some situations the information may be robust enough so that we can characterize it with probability theory.

In making informed and rational decisions we must remember that individuals are essentially risk averse. When the consequences of an action might result in serious injury, death, economic ruin, or some other dreaded event, humans do not make decisions consistent with rational utility theory [Maes and Faber, 2004]. In fact, studies in cognitive psychology show that rationality is a rather weak hypothesis in decision making, easily refuted and therefore not always useful as an axiomatic explanation of the theory of decision making. Human risk preference in the face of high uncertainty is not easily modeled by its rational methods. In a narrow context of decision making, rational behavior is defined in terms of decision making which maximizes expected utility [von Neumann and Morgenstern, 1944]. Of course, this utility is a function of personal preferences of the decision maker. While much of the literature addresses decision making in the face of economic and financial risks, engineers are primarily concerned with two types of decisions [Maes and Faber, 2004]: (1) operational decisions, where for certain available resources an optimal action is sought to avoid a specific set of hazards; and (2) strategic decisions, which involve decisions regarding one's level of preparedness or anticipation of events in the future. Difficulties in human preference reversal, in using incomplete information, in bias towards one's own experience, and in using epistemic uncertainty (e.g., ambiguity, vagueness, fuzziness) are among the various issues cited by Maes and Faber [2004] as reasons why the independence axiom of an expected utility analysis (used in rational decision making) is violated by human behavior. While we do not address these matters in this text, it is nonetheless important to keep them in mind when using any of the methods developed here.

This chapter presents a few paradigms for making decisions within a fuzzy environment. Issues such as personal preferences, multiple objectives, nontransitive reasoning, and group consensus are presented. The chapter concludes with a rather lengthy development of an area known loosely as fuzzy Bayesian decision making, so named because it involves the introduction of fuzzy information, fuzzy outcomes, and fuzzy actions into the classical probabilistic method of Bayesian decision making. In developing this we are able to compare the value and differences of incorporating both fuzzy and random information into the same representational framework. Acceptance of the fuzzy approach is therefore eased by its natural accommodation within a classical, and historically popular, decision-making approach. Moreover, a recent book chapter [Ross et al., 2003] shows how the likelihood function in Bayes's rule has similar properties to a fuzzy membership function. This is not to suggest that Bayesian decision making is accepted universally; Maes and Faber [2004] highlight some problems with Bayesian updating of probabilities and utilities that are currently being debated in the literature.

FUZZY SYNTHETIC EVALUATION

The term *synthetic* is used here to connote the process of evaluation whereby several individual elements and components of an evaluation are synthesized into an aggregate form; the whole is a *synthesis* of the parts. The key here is that the various elements can be numeric or nonnumeric, and the process of fuzzy synthesis is naturally accommodated using synthetic evaluation. In reality, an evaluation of an object, especially an ill-defined one, is often vague and ambiguous. The evaluation is usually described in natural language terms, since a numerical evaluation is often too complex, too unacceptable, and too ephemeral (transient). For example, when grading a written examination, the professor might evaluate it from such perspectives as style, grammar, creativity, and so forth. The final grade on the paper might be linguistic instead of numeric, e.g., excellent, very good, good, fair, poor, and unsatisfactory. After grading many exams the professor might develop a relation by which a membership is assigned to the relations between the different perspectives, such as style and grammar, and the linguistic grades, such as fair and excellent. A fuzzy relation, $\underset{\sim}{R}$, such as the following one, might result that summarizes the professor's relationship between pairs of grading factors such as *creativity* and grade evaluations such as *very good*:

$$
\underset{\sim}{R} = \begin{array}{c} \text{Creativity} \\ \text{Grammar} \\ \text{Style} \\ \vdots \end{array}
\begin{array}{ccccc} \text{Excellent} & \text{Very good} & \text{Good} & \text{Fair} & \text{Poor} \\
\left[\begin{array}{ccccc} 0.2 & 0.4 & 0.3 & 0.1 & 0 \\ 0 & 0.2 & 0.5 & 0.3 & 0 \\ 0.1 & 0.6 & 0.3 & 0 & 0 \\ \vdots & \vdots & \vdots & \vdots & \vdots \end{array}\right] \end{array}
$$

The professor now wants to assign a grade to each paper. To formalize this approach, let X be a universe of factors and Y be a universe of evaluations, so

$$X = \{x_1, x_2, \ldots, x_n\} \quad \text{and} \quad Y = \{y_1, y_2, \ldots, y_m\}$$

Let $\underset{\sim}{R} = [r_{ij}]$ be a fuzzy relation, such as the foregoing grading example, where $i = 1, 2, \ldots, n$ and $j = 1, 2, \ldots, m$. Suppose we introduce a specific paper into the evaluation process on which the professor has given a set of "scores" (w_i) for each of the n grading factors, and we ensure, for convention, that the sum of the scores is unity. Each of these scores is actually a membership value for each of the factors, x_i, and they can be arranged in a fuzzy vector, $\underset{\sim}{w}$. So we have

$$\underset{\sim}{w} = \{w_1, w_2, \ldots, w_n\} \quad \text{where} \quad \sum_i w_i = 1 \tag{10.1a}$$

The process of determining a grade for a specific paper is equivalent to the process of determining a membership value for the paper in each of the evaluation categories, y_i. This process is implemented through the composition operation

$$\underset{\sim}{e} = \underset{\sim}{w} \circ \underset{\sim}{R} \tag{10.1b}$$

where $\underset{\sim}{e}$ is a fuzzy vector containing the membership values for the paper in each of the y_i evaluation categories.

Example 10.1. Suppose we want to measure the value of a microprocessor to a potential client. In conducting this evaluation, the client suggests that certain criteria are important. They can include performance (MIPS), cost ($), availability (AV), and software (SW). Performance is measured by millions of instructions per second, or MIPS; a minimum requirement is 10 MIPS. Cost is the cost of the microprocessor, and a cost requirement of "not to exceed" $500 has been set. Availability relates to how much time after the placement of an order the microprocessor vendor can deliver the part; a maximum of eight weeks has been set. Software represents the availability of operating systems, languages, compilers, and tools to be used with this microprocessor. Suppose further that the client is only able to specify a subjective criterion of having "sufficient" software.

A particular microprocessor (CPU) has been introduced into the market. It is measured against these criteria and given ratings categorized as excellent (e), superior (s), adequate (a), and inferior (i). "Excellent" means that the microprocessor is the best available with respect to the particular criterion. "Superior" means that microprocessor is among the best with respect to this criterion. "Adequate" means that, although not superior, the microprocessor can meet the minimum acceptable requirements for this criterion. "Inferior" means that the microprocessor cannot meet the requirements for the particular criterion. Suppose the microprocessor just introduced has been assigned the following relation based on the consensus of the design team:

$$
R = \begin{array}{c} \\ \text{MIPS} \\ \$ \\ \text{AV} \\ \text{SW} \end{array}
\begin{array}{cccc}
e & s & a & i \\
\left[\begin{array}{cccc}
0.1 & 0.3 & 0.4 & 0.2 \\
0 & 0.1 & 0.8 & 0.1 \\
0.1 & 0.6 & 0.2 & 0.1 \\
0.1 & 0.4 & 0.3 & 0.2
\end{array}\right]
\end{array}
$$

This relation could have been derived from data using similarity methods such as those discussed in Chapter 3.

If the evaluation team applies a scoring factor of 0.4 for performance, 0.3 for cost, 0.2 for availability, and 0.1 for software, which together form the factor vector, $\underset{\sim}{w}$, then the composition, $\underset{\sim}{e} = \underset{\sim}{w} \circ R = \{0.1, 0.3, 0.4, 0.2\}$, results in an evaluation vector that has its highest membership in the category "adequate."

It is important to point out in concluding this section that the relations expressed in this section are not constrained in that their row sums should equal unity. The examples given show the row sums equaling unity, a matter of convenience for illustration. However, since the entries in the synthetic evaluation matrix relations are membership values showing the degree of relation between the factors and the evaluations, these values can take on any number between 0 and 1. Hence, row sums could be larger, or smaller, than unity.

FUZZY ORDERING

Decisions are sometimes made on the basis of rank, or ordinal ranking: which issue is best, which is second best, and so forth. For issues or actions that are deterministic, such as $y_1 = 5$, $y_2 = 2$, $y_1 \geq y_2$, there is usually no ambiguity in the ranking; we might call this *crisp ordering*. In situations where the issues or actions are associated with uncertainty, either random or fuzzy, rank ordering may be ambiguous. This ambiguity, or uncertainty, can be demonstrated for both random and fuzzy variables. First, let us assume that the uncertainty in rank is random; we can use probability density functions (pdf) to illustrate the random case. Suppose we have one random variable, x_1, whose uncertainty is characterized by a Gaussian pdf with a mean of μ_1 and a standard deviation of σ_1, and another random variable, x_2, also Gaussian with a mean of μ_2 and standard deviation of σ_2. Suppose further that $\sigma_1 > \sigma_2$ and $\mu_1 > \mu_2$. If we plot the pdfs for these two random variables in Fig. 10.1, we see that the question of which variable is greater is not clear.

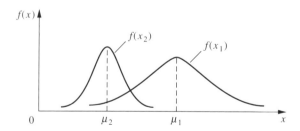

FIGURE 10.1
Density functions for two Gaussian random variables.

As an example of this uncertain ranking, suppose x_1 is the height of Italians and x_2 is the height of Swedes. Because this uncertainty is of the random kind, we cannot answer the question "Are Swedes taller than Italians?" unless we are dealing with two specific individuals, one each from Sweden and Italy, or we are simply assessing μ_1, average-height Swedes, and μ_2, average-height Italians. But we can ask the question, "How frequently are Swedes taller than Italians?" We can assess this frequency as the probability that one random variable is greater than another, i.e., $P(x_1 \geq x_2)$, with

$$P(x_1 \geq x_2) = \int_{-\infty}^{\infty} F_{x_2}(x_1) \, dx_1 \qquad (10.2a)$$

where F is a cumulative distribution function. Hence, with random variables we can quantify the uncertainty in ordering with a convolution integral, Eq. (10.2a).

Second, let us assume that the uncertainty in rank arises because of ambiguity. For example, suppose we are trying to rank people's preferences in colors. In this case the ranking is very subjective and not reducible to the elegant form available for some random variables, such as that given in Eq. (10.2a). For fuzzy variables we are also able to quantify the uncertainty in ordering, but in this case we must do so with the notion of membership.

A third type of ranking involves the notion of imprecision [Dubois and Prade, 1980]. To develop this, suppose we have two fuzzy numbers, $\underset{\sim}{I}$ and $\underset{\sim}{J}$. We can use tools provided in Chapter 12 on the extension principle to calculate the truth value of the assertion that fuzzy number $\underset{\sim}{I}$ is greater that fuzzy number $\underset{\sim}{J}$ (a fuzzy number was defined in Chapter 4):

$$T(\underset{\sim}{I} \geq \underset{\sim}{J}) = \sup_{x \geq y} \ \min(\mu_{\underset{\sim}{I}}(x), \mu_{\underset{\sim}{J}}(y)) \qquad (10.2b)$$

Figure 10.2 shows the membership functions for two fuzzy numbers $\underset{\sim}{I}$ and $\underset{\sim}{J}$. Equation (10.2b) is an extension of the inequality $x \geq y$ according to the extension principle. It represents the degree of possibility in the sense that if a specific pair (x, y) exists such that $x \geq y$ and $\mu_{\underset{\sim}{I}}(x) = \mu_{\underset{\sim}{J}}(y)$, then $T(\underset{\sim}{I} \geq \underset{\sim}{J}) = 1$. Since the fuzzy numbers $\underset{\sim}{I}$ and $\underset{\sim}{J}$ are convex, it can be seen from Fig. 10.2 that

$$T(\underset{\sim}{I} \geq \underset{\sim}{J}) = 1 \text{ if and only if } I \geq J \qquad (10.3a)$$

$$T(\underset{\sim}{J} \geq \underset{\sim}{I}) = \text{height}(\underset{\sim}{I} \cap \underset{\sim}{J}) = \mu_{\underset{\sim}{I}}(d) = \mu_{\underset{\sim}{J}}(d) \qquad (10.3b)$$

where d is the location of the highest intersection point of the two fuzzy numbers. The operation height $(\underset{\sim}{I} \cap \underset{\sim}{J})$ in Eq. (10.3b) is a good separation metric for two fuzzy numbers;

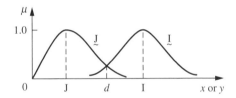

FIGURE 10.2
Two fuzzy numbers as fuzzy sets on the real line.

that is, the closer this metric is to unity, the more difficult it is to distinguish which of the two fuzzy numbers is largest. On the other hand, as this metric approaches zero, the easier is the distinction about rank order (which is largest). Unfortunately, the metric given in Eq. (10.3a) is not too useful as a ranking metric, because $T(I \geq J) = 1$ when I is slightly greater and when I is much greater than J. If we know that I and J are crisp numbers I and J, the truth value becomes $T(I \geq J) = 1$ for $I \geq J$ and $T(I \geq J) = 0$ for $I < J$.

The definitions expressed in Eqs. (10.2b) and (10.3) for two fuzzy numbers can be extended to the more general case of many fuzzy sets. Suppose we have k fuzzy sets $\underset{\sim}{I_1}, \underset{\sim}{I_2}, \ldots, \underset{\sim}{I_k}$. Then the truth value of a specified ordinal ranking is given by

$$T(\underset{\sim}{I} \geq \underset{\sim}{I_1}, \underset{\sim}{I_2}, \ldots, \underset{\sim}{I_k}) = T(\underset{\sim}{I} \geq \underset{\sim}{I_1}) \quad \text{and} \quad T(\underset{\sim}{I} \geq \underset{\sim}{I_2}) \quad \text{and} \ldots \text{and} \quad T(\underset{\sim}{I} \geq \underset{\sim}{I_k}) \qquad (10.4)$$

Example 10.2. Suppose we have three fuzzy sets, as described using Zadeh's notation:

$$\underset{\sim}{I_1} = \left\{ \frac{1}{3} + \frac{0.8}{7} \right\} \quad \text{and} \quad \underset{\sim}{I_2} = \left\{ \frac{0.7}{4} + \frac{1.0}{6} \right\} \quad \text{and} \quad \underset{\sim}{I_3} = \left\{ \frac{0.8}{2} + \frac{1}{4} + \frac{0.5}{8} \right\}$$

We can assess the truth value of the inequality, $\underset{\sim}{I_1} \geq \underset{\sim}{I_2}$, as follows:

$$T(\underset{\sim}{I_1} \geq \underset{\sim}{I_2}) = \max_{x_1 \geq x_2} \{\min(\mu_{\underset{\sim}{I_1}}(x_1), \mu_{\underset{\sim}{I_2}}(x_2))\}$$

$$= \max\{\min(\mu_{\underset{\sim}{I_1}}(7), \mu_{\underset{\sim}{I_2}}(4)), \min(\mu_{\underset{\sim}{I_1}}(7), \mu_{\underset{\sim}{I_2}}(6))\}$$

$$= \max\{\min(0.8, 0.7), \min(0.8, 1.0)\}$$

$$= 0.8$$

Similarly,

$$T(\underset{\sim}{I_1} \geq \underset{\sim}{I_3}) = 0.8 \quad T(\underset{\sim}{I_2} \geq \underset{\sim}{I_1}) = 1.0$$

$$T(\underset{\sim}{I_2} \geq \underset{\sim}{I_3}) = 1.0 \quad T(\underset{\sim}{I_3} \geq \underset{\sim}{I_1}) = 1.0$$

$$T(\underset{\sim}{I_3} \geq \underset{\sim}{I_2}) = 0.7$$

Then

$$T(\underset{\sim}{I_1} \geq \underset{\sim}{I_2}, \underset{\sim}{I_3}) = 0.8$$

$$T(\underset{\sim}{I_2} \geq \underset{\sim}{I_1}, \underset{\sim}{I_3}) = 1.0$$

$$T(\underset{\sim}{I_3} \geq \underset{\sim}{I_1}, \underset{\sim}{I_2}) = 0.7$$

The last three truth values in this example compared one fuzzy set to two others. This calculation is different from pairwise comparisons. To do the former, one makes use of the minimum function, as prescribed by Eq. (10.4). For example,

$$T(\underset{\sim}{I_1} \geq \underset{\sim}{I_2}, \underset{\sim}{I_3}) = \min\{(\underset{\sim}{I_1} \geq \underset{\sim}{I_2}), (\underset{\sim}{I_1} \geq \underset{\sim}{I_3})\}$$

Equation (10.4) can be used similarly to obtain the other multiple comparisons. Based on the foregoing ordering, the overall ordering for the three fuzzy sets would be $\underset{\sim}{I_2}$ first, $\underset{\sim}{I_1}$ second, and $\underset{\sim}{I_3}$ last.

Another procedure to compare two fuzzy numbers and find the larger or smaller of them is given by Klir and Yuan [1995]. In this case, we make use of what is known as the extended MIN and MAX operations; these are operations on fuzzy numbers, whereas the standard min and max are operations on real numbers. The reader is referred to the literature for more amplification on this.

NONTRANSITIVE RANKING

When we compare objects that are fuzzy, ambiguous, or vague, we may well encounter a situation where there is a contradiction in the classical notions of ordinal ranking and transitivity in the ranking. For example, suppose we are ordering on the preference of colors. When comparing red to blue, we prefer red; when comparing blue to yellow, we prefer blue; but when comparing red and yellow we might prefer yellow. In this case transitivity of sets representing preference in colors (red > blue and blue > yellow does *not* yield red > yellow) is not maintained.

To accommodate this form of nontransitive ranking (which can be quite normal for noncardinal-type concepts), we introduce a special notion of relativity [Shimura, 1973]. Let x and y be variables defined on universe X. We define a pairwise function

$$f_y(x) \text{ as the membership value of } x \text{ with respect to } y$$

and we define another pairwise function

$$f_x(y) \text{ as the membership value of } y \text{ with respect to } x$$

Then the relativity function given by

$$f(x \mid y) = \frac{f_y(x)}{\max[f_y(x), f_x(y)]} \tag{10.5}$$

is a measurement of the membership value of choosing x over y. The relativity function $f(x \mid y)$ can be thought of as the membership of preferring variable x over variable y. Note that the function in Eq. (10.5) uses arithmetic division.

To develop the general case of Eq. (10.5) for many variables, define variables $x_1, x_2, \ldots, x_i, x_{i+1}, \ldots, x_n$ all defined on universe X, and let these variables be collected in a set A, i.e., $A = \{x_1, x_2, \ldots, x_{i-1}, x_i, x_{i+1}, \ldots, x_n\}$. We then define a set identical to set A except this new set will be missing one element, x_i, and this set will be termed A'. The relativity function then becomes

$$f(x_i \mid A') = f(x_i \mid \{x_1, x_2, \ldots, x_{i-1}, x_{i+1}, \ldots, x_n\})$$
$$= \min\{f(x_i \mid x_1), f(x_i \mid x_2), \ldots, f(x_i \mid x_{i-1}), f(x_i \mid x_{i+1}), \ldots, f(x_i \mid x_n)\} \tag{10.6}$$

which is the fuzzy measurement of choosing x_i over all elements in the set A'. The expression in Eq. (10.6) involves the logical intersection of several variables; hence the minimum function is used. Since the relativity function of one variable with respect to itself is identity, i.e.,

$$f(x_i \mid x_i) = 1 \tag{10.7}$$

then

$$f(x_i \mid A') = f(x_i \mid A) \tag{10.8}$$

We can now form a matrix of relativity values, $f(x_i \mid x_j)$, where $i, j = 1, 2, \ldots, n$, and where x_i and x_j are defined on a universe X. This matrix will be square and of order n, and will be termed the C matrix (C for comparison). The C matrix can be used to rank many different fuzzy sets.

To determine the overall ranking, we need to find the smallest value in each of the rows of the C matrix; that is,

$$C_i' = \min f(x_i \mid X), \quad i = 1, 2, \ldots, n \tag{10.9}$$

where C_i' is the membership ranking value for the ith variable. We use the minimum function because this value will have the lowest weight for ranking purposes; further, the maximum function often returns a value of 1 and there would be ambivalence in the ranking process (i.e., ties will result).

Example 10.3. In manufacturing, we often try to compare the capabilities of various microprocessors for their appropriateness to certain applications. For instance, suppose we are trying to select from among four microprocessors the one that is best suited for image processing applications. Since many factors can affect this decision, including performance, cost, availability, software, and others, coming up with a crisp mathematical model for all these attributes is complicated. Another consideration is that it is much easier to compare these microprocessors subjectively in pairs rather than all four at one time.

Suppose the design team is polled to determine which of the four microprocessors, labeled x_1, x_2, x_3, and x_4, is the most preferred when considered as a group rather than when considered as pairs. First, pairwise membership functions are determined. These represent the subjective measurement of the appropriateness of each microprocessor when compared only to one another. The following pairwise functions are determined:

$$
\begin{array}{llll}
f_{x_1}(x_1) = 1 & f_{x_1}(x_2) = 0.5 & f_{x_1}(x_3) = 0.3 & f_{x_1}(x_4) = 0.2 \\
f_{x_2}(x_1) = 0.7 & f_{x_2}(x_2) = 1 & f_{x_2}(x_3) = 0.8 & f_{x_2}(x_4) = 0.9 \\
f_{x_3}(x_1) = 0.5 & f_{x_3}(x_2) = 0.3 & f_{x_3}(x_3) = 1 & f_{x_3}(x_4) = 0.7 \\
f_{x_4}(x_1) = 0.3 & f_{x_4}(x_2) = 0.1 & f_{x_4}(x_3) = 0.3 & f_{x_4}(x_4) = 1
\end{array}
$$

For example, microprocessor x_2 has membership 0.5 with respect to microprocessor x_1. Note that if these values were arranged into a matrix, it would not be symmetric. These membership values do not express similarity or relation; they represent membership values of ordering when considered in a particular order. If we now employ Eq. (10.5) to calculate all of the relativity values, the matrix shown below expresses these calculations; this is the so-called comparison, or C, matrix. For example,

$$f(x_2 \mid x_1) = \frac{f_{x_1}(x_2)}{\max[f_{x_1}(x_2), f_{x_2}(x_1)]} = \frac{0.5}{\max[0.5, 0.7]} = 0.71$$

$$
\begin{array}{c}
\begin{array}{ccccc}
\quad & x_1 & x_2 & x_3 & x_4 & \quad \min = f(x_i \mid X) \\
\end{array} \\
C = \begin{array}{c} x_1 \\ x_2 \\ x_3 \\ x_4 \end{array}
\left[
\begin{array}{cccc}
1 & 1 & 1 & 1 \\
0.71 & 1 & 0.38 & 0.11 \\
0.6 & 1 & 1 & 0.43 \\
0.67 & 1 & 1 & 1
\end{array}
\right]
\begin{array}{c}
1 \\
0.11 \\
0.43 \\
0.67
\end{array}
\end{array}
$$

The extra column to the right of the foregoing C matrix is the minimum value for each of the rows, i.e., for C_i', $i = 1, 2, 3, 4$, in Eq. (10.9). For this example problem, the order from best to worst is x_1, x_4, x_3, and x_2. This ranking is much more easily attained with this fuzzy approach than it would have been with some other method where the attributes of each microprocessor are assigned a value measurement and these values are somehow combined. This fuzzy method also contains the subjectiveness inherent in comparing one microprocessor to another. If all four were considered at once, a person's own bias might skew the value assessments to favor a particular microprocessor. By using pairwise comparisons, each microprocessor is compared individually against its peers, which should allow a more fair and less biased comparison.

PREFERENCE AND CONSENSUS

The goal of group decision making typically is to arrive at a consensus concerning a desired action or alternative from among those considered in the decision process. In this context, consensus is usually taken to mean a unanimous agreement by all those in the group concerning their choice. Despite the simplicity in defining consensus, it is another matter altogether to quantify this notion. Most traditional mathematical developments of consensus have used individual preference ranking as their primary feature. In these developments, the individual preferences of those in the decision group are collected to form a group metric whose properties are used to produce a scalar measure of "degree of consensus." However, the underlying axiomatic structure of many of these classical approaches is based on classical set theory. The argument given in this text is that the crisp set approach is too restrictive for the variables relevant to a group decision process. The information in the previous section showed individual preference to be a fuzzy relation.

There can be numerous outcomes of decision groups in developing consensus about a universe, X, of n possible alternatives, i.e., $X = \{x_1, x_2, \ldots, x_n\}$. To start the development, we define a *reciprocal* relation as a fuzzy relation, $\underset{\sim}{R}$, of order n, whose individual elements r_{ij} have the following properties [Bezdek et al., 1978]:

$$r_{ii} = 0 \quad \text{for } 1 \leq i \leq n \tag{10.10}$$

$$r_{ij} + r_{ji} = 1 \quad \text{for } i \neq j \tag{10.11}$$

This reciprocal relation, $\underset{\sim}{R}$, Eqs. (10.10)–(10.11), can be interpreted as follows: r_{ij} is the preference accorded to x_i relative to x_j. Thus, $r_{ij} = 1$ (hence, $r_{ji} = 0$) implies that alternative i is *definitely* preferred to alternative j; this is the crisp case in preference. At the other extreme, we have maximal fuzziness, where $r_{ij} = r_{ji} = 0.5$, and there is equal preference, pairwise. A definite choice of alternative i to all others is manifested in $\underset{\sim}{R}$ as the ith row being all ones (except $r_{ii} = 0$), or the ith column being all zeros.

Two common measures of preference are defined here as *average fuzziness* in $\underset{\sim}{R}$, Eq. (10.12), and *average certainty* in $\underset{\sim}{R}$, Eq. (10.13):

$$F(\underset{\sim}{R}) = \frac{\text{tr}(\underset{\sim}{R}^2)}{n(n-1)/2} \tag{10.12}$$

$$C(\underset{\sim}{R}) = \frac{\text{tr}(\underset{\sim}{R}\underset{\sim}{R}^{\text{T}})}{n(n-1)/2} \tag{10.13}$$

In Eqs. (10.12)–(10.13), tr() and ()^{T} denote the trace and transpose, respectively, and matrix multiplication is the algebraic kind. Recall that the trace of a matrix is simply the algebraic sum of the diagonal elements, i.e.,

$$\text{tr}(\underset{\sim}{R}) = \sum_{i=1}^{n} r_{ii}$$

The measure, $F(\underset{\sim}{R})$, averages the joint preferences in $\underset{\sim}{R}$ over all distinct pairs in the Cartesian space, $X \times X$. Each term maximizes the measure when $r_{ij} = r_{ji} = 0.5$ and minimizes the measure when $r_{ij} = 1$ and $r_{ji} = 0$; consequently, $F(\underset{\sim}{R})$ is proportional to the fuzziness or uncertainty (also, confusion) about pairwise rankings exhibited by the fuzzy preference relation, $\underset{\sim}{R}$. Conversely, the measure, $C(\underset{\sim}{R})$, averages the individual dominance (assertiveness) of each distinct pair of rankings in the sense that each term maximizes the measure when $r_{ij} = 1$ and $r_{ji} = 0$ and minimizes the measure when $r_{ij} = r_{ji} = 0.5$; hence, $C(\underset{\sim}{R})$ is proportional to the overall certainty in $\underset{\sim}{R}$. The two measures are dependent; they are both on the interval $[0, 1]$; and it can be shown [Bezdek et al., 1978] that

$$F(\underset{\sim}{R}) + C(\underset{\sim}{R}) = 1 \tag{10.14}$$

It can further be shown that C is a minimum and F is a maximum at $r_{ij} = r_{ji} = 0.5$, and that C is a maximum and F is a minimum at $r_{ij} = 1, r_{ji} = 0$. Also, at the state of maximum fuzziness ($r_{ij} = r_{ji} = 0.5$), we get $F(\underset{\sim}{R}) = C(\underset{\sim}{R}) = \frac{1}{2}$; and at the state of no uncertainty ($r_{ij} = 1, r_{ji} = 0$) we get $F(\underset{\sim}{R}) = 0$ and $C(\underset{\sim}{R}) = 1$. Moreover, the ranges for these two measures are $0 \leq F(\underset{\sim}{R}) \leq \frac{1}{2}$ and $\frac{1}{2} \leq C(\underset{\sim}{R}) \leq 1$.

Measures of preference can be useful in determining consensus. There are different forms of consensus. We have discussed the antithesis of consensus: complete ambivalence, or the maximally fuzzy case where all alternatives are rated equally; call this type of consensus M_1. For M_1 we have a matrix $\underset{\sim}{R}$ where all nondiagonal elements are equal to $\frac{1}{2}$. We have also discussed the converse of M_1, which is the nonfuzzy (crisp) preference where every pair of alternatives is definitely ranked; call this case M_2. In M_2 all nondiagonal elements in $\underset{\sim}{R}$ are equal to 1 or 0; however, there may not be a clear consensus. Consider following the reciprocity relation, M_2;

$$M_2 = \begin{bmatrix} 0 & 1 & 0 & 1 \\ 0 & 0 & 1 & 0 \\ 1 & 0 & 0 & 1 \\ 0 & 1 & 0 & 0 \end{bmatrix}$$

Here, the clear pairwise choices are these: alternative 1 over alternative 2, alternative 1 over alternative 4, alternative 2 over alternative 3, and alternative 3 over alternative 4. However, we do not have consensus because alternative 3 is preferred over alternative 1 and alternative 4 is preferred over alternative 2! So for relation M_1 we cannot have consensus and for relation M_2 we may not have consensus.

Three types of consensus, however, arise from considerations of the matrix $\underset{\sim}{R}$. The first type, known as Type I consensus, M_1^*, is a consensus in which there is one clear choice, say alternative i (the ith column is all zeros), and the remaining $(n - 1)$ alternatives all have equal secondary preference (i.e., $r_{kj} = \frac{1}{2}$, where $k \neq j$), as shown in the following example, where alternative 2 has a clear consensus:

$$M_1^* = \begin{bmatrix} 0 & 0 & 0.5 & 0.5 \\ 1 & 0 & 1 & 1 \\ 0.5 & 0 & 0 & 0.5 \\ 0.5 & 0 & 0.5 & 0 \end{bmatrix}$$

In the second type of consensus, called a Type II consensus, M_2^*, there is one clear choice, say alternative i (the ith column is all zeros), but the remaining $(n - 1)$ alternatives all have definite secondary preference (i.e., $r_{kj} = 1$, where $k \neq i$), as shown in this example:

$$M_2^* = \begin{bmatrix} 0 & 0 & 1 & 0 \\ 1 & 0 & 1 & 1 \\ 0 & 0 & 0 & 1 \\ 1 & 0 & 0 & 0 \end{bmatrix}$$

where alternative 2 has a clear consensus, but where there is no clear ordering after the first choice because alternative 1 is preferred to alternative 3, 3 to 4, but alternative 4 is preferred to alternative 1. There can be clear ordering after the first choice in Type II consensus matrices, but it is not a requirement.

Finally, the third type of consensus, called a Type *fuzzy* consensus, M_f^*, occurs where there is a unanimous decision for the most preferred choice, say alternative i again, but the remaining $(n - 1)$ alternatives have infinitely many fuzzy secondary preferences. The matrix shown here has a clear choice for alternative 2, but the other secondary preferences are fuzzy to various degrees:

$$M_f^* = \begin{bmatrix} 0 & 0 & 0.5 & 0.6 \\ 1 & 0 & 1 & 1 \\ 0.5 & 0 & 0 & 0.3 \\ 0.4 & 0 & 0.7 & 0 \end{bmatrix}$$

Mathematically, relations M_1 and M_2 are logical opposites, as are consensus relations M_1^* and M_2^* [Bezdek et al., 1978]. It is interesting to discuss the cardinality of these various preference and consensus relations. In this case, the cardinality of a relation is the number of possible combinations of that matrix type. It is obvious that there is only one possible matrix for M_1. The cardinality of all the preference or consensus relations discussed here is given in Eq. (10.15), where the symbol | | denotes cardinality of the particular relation:

$$|M_1| = 1$$

$$|M_2| = 2^{n(n-1)/2}$$

$$|M_1^*| = n \qquad \text{(Type I)} \qquad (10.15)$$

$$|M_2^*| = (2^{(n^2-3n+2)/2})(n) \quad \text{(Type II)}$$

$$|M_f^*| = \infty \qquad \text{(Type } \textit{fuzzy}\text{)}$$

So, for the examples previously illustrated for $n = 4$ alternatives, there are 64 (2^6) possible forms of the M_2 preference matrix, there are only four Type I (M_1^*) consensus matrices, and there are 32 ($2^3 \cdot 4$) possible Type II (M_2^*) consensus matrices.

From the *degree of preference* measures given in Eqs. (10.12)–(10.13), we can construct a *distance to consensus* metric, defined as

$$m(\underset{\sim}{R}) = 1 - (2C(\underset{\sim}{R}) - 1)^{1/2} \qquad (10.16)$$

where $m(\underset{\sim}{R}) = 1$ for an M_1 preference relation
$m(\underset{\sim}{R}) = 0$ for an M_2 preference relation

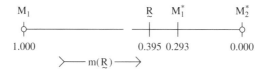

FIGURE 10.3
Illustration of *distance to consensus* [Bezdek et al., 1978].

$m(\underset{\sim}{R}) = 1 - (2/n)^{1/2}$ for a Type I (M_1^*)consensus relation

$m(\underset{\sim}{R}) = 0$ for a Type II (M_2^*) consensus relation

We can think of this metric, $m(\underset{\sim}{R})$, as being a distance between the points M_1 (1.0) and M_2 (0.0) in *n*-dimensional space. We see that $m(M_1^*) = m(M_2^*)$ for the case where we have only two ($n = 2$) alternatives. For the more general case, where $n > 2$, the distance between Type I and Type II consensus increases with n as it becomes increasingly difficult to develop a consensus choice and simultaneously rank the remaining pairs of alternatives.

> **Example 10.4.** Suppose a reciprocal fuzzy relation, $\underset{\sim}{R}$, is developed by a small group of people in their initial assessments for pairwise preferences for a decision process involving four alternatives, $n = 4$, as shown here:
>
> $$\underset{\sim}{R} = \begin{bmatrix} 0 & 1 & 0.5 & 0.2 \\ 0 & 0 & 0.3 & 0.9 \\ 0.5 & 0.7 & 0 & 0.6 \\ 0.8 & 0.1 & 0.4 & 0 \end{bmatrix}$$
>
> Notice that this matrix carries none of the properties of a consensus type; that is, the group does not reach consensus on their first attempt at ranking the alternatives. However, the group can assess their "degree of consensus" and they can measure how "far" they are from consensus prior to subsequent discussions in the decision process. So, for example, alternative 1 is *definitely* preferred to alternative 2, alternative 1 is rated *equal* to alternative 3, and so forth. For this matrix, $C(\underset{\sim}{R}) = 0.683$ (Eq. (10.13)), $m(\underset{\sim}{R}) = 0.395$, and $m(M_1^*) = 1 - (2/n)^{1/2} = 0.293$ (Eq. (10.16)). For their first attempt at ranking the four alternatives the group have a degree of consensus of 0.683 (recall a value of 0.5 is completely ambivalent (uncertain) and a value of 1.0 is completely certain). Moreover, the group are $1 - 0.395 = 0.605$, or 60.5% of the way from complete ambivalence (M_1) toward a Type II consensus, or they are $0.605/(1 - 0.293) = 85.5\%$ of the way toward a Type I consensus. These ideas are shown graphically in Fig. 10.3. The value of the *distance to consensus*, $m(\underset{\sim}{R})$, is its use in quantifying the dynamic evolution of a group as the group refines their preferences and moves closer to a Type I or Type II or Type *fuzzy* consensus. It should be noted that the vast majority of group preference situations eventually develop into Type *fuzzy* consensus; Types I and II are typically only useful as boundary conditions.

MULTIOBJECTIVE DECISION MAKING

Many simple decision processes are based on a single objective, such as minimizing cost, maximizing profit, minimizing run time, and so forth. Often, however, decisions must be made in an environment where more than one objective constrains the problem, and the relative value of each of these objectives is different. For example, suppose we are designing a new computer, and we want simultaneously to minimize cost, maximize CPU, maximize random access memory (RAM), and maximize reliability. Moreover, suppose cost is the most important of our objectives and the other three (CPU, RAM, reliability) carry lesser but equal weight when compared with cost. Two primary issues in multiobjective decision making are to acquire meaningful information regarding the satisfaction of the objectives by the various choices (alternatives) and to rank or weight the relative importance of each of the objectives. The approach illustrated in this section defines a decision calculus that requires only *ordinal* information on the ranking of preferences and importance weights [Yager, 1981].

The typical multiobjective decision problem involves the selection of one alternative, a_i, from a universe of alternatives A given a collection, or set, say {O}, of criteria or objectives that are important to the decision maker. We want to evaluate how well each alternative, or choice, satisfies each objective, and we wish to combine the weighted objectives into an overall decision function in some plausible way. This decision function essentially represents a mapping of the alternatives in A to an ordinal set of ranks. This process naturally requires subjective information from the decision authority concerning the importance of each objective. Ordinal orderings of this importance are usually the easiest to obtain. Numerical values, ratios, and intervals expressing the importance of each objective are difficult to extract and, if attempted and then subsequently altered, can often lead to results inconsistent with the intuition of the decision maker.

To develop this calculus we require some definitions. Define a universe of n alternatives, $A = \{a_1, a_2, \ldots, a_n\}$, and a set of r objectives, $O = \{O_1, O_2, \ldots, O_r\}$. Let O_i indicate the ith objective. Then the degree of membership of alternative a in O_i, denoted $\mu_{O_i}(a)$, is the degree to which alternative a satisfies the criteria specified for this objective. We seek a decision function that simultaneously satisfies all of the decision objectives; hence, the decision function, D, is given by the intersection of all the objective sets,

$$D = O_1 \cap O_2 \cap \cdots \cap O_r \tag{10.17}$$

Therefore, the grade of membership that the decision function, D, has for each alternative a is given by

$$\mu_D(a) = \min[\mu_{O_1}(a), \mu_{O_2}(a), \ldots, \mu_{O_r}(a)] \tag{10.18}$$

The optimum decision, a^*, will then be the alternative that satisfies

$$\mu_D(a^*) = \max_{a \in A}(\mu_D(a)) \tag{10.19}$$

We now define a set of preferences, {P}, which we will constrain to being linear and ordinal. Elements of this preference set can be linguistic values such as none, low, medium, high, absolute, or perfect; or they could be values on the interval [0, 1]; or they could be values on any other linearly ordered scale, e.g., [−1, 1], [1, 10], etc. These preferences will be attached to each of the objectives to quantify the decision maker's feelings about the influence that each objective should have on the chosen alternative. Let the parameter, b_i, be contained on the set of preferences, {P}, where $i = 1, 2, \ldots, r$. Hence, we have for each objective a measure of how important it is to the decision maker for a given decision.

The decision function, D, now takes on a more general form when each objective is associated with a weight expressing its importance to the decision maker. This function is represented as the intersection of r-tuples, denoted as a decision measure, $M(O_i, b_i)$, involving objectives and preferences,

$$D = M(O_1, b_1) \cap M(O_2, b_2) \cap \cdots \cap M(O_r, b_r) \tag{10.20}$$

A key question is what operation should relate each objective, O_i, and its importance, b_i, that preserves the linear ordering required of the preference set, and at the same time relates the two quantities in a logical way where negation is also accommodated. It turns out that the classical implication operator satisfies all of these requirements. Hence, the decision measure for a particular alternative, a, can be replaced with a classical implication of the form

$$M(O_i(a), b_i) = b_i \longrightarrow O_i(a) = \overline{b_i} \vee O_i(a) \tag{10.21}$$

Justification of the implication as an appropriate measure can be developed using an intuitive argument [Yager, 1981]. The statement "b_i implies O_i" indicates a unique relationship between a preference and its associated objective. Whereas various objectives can have the same preference weighting in a cardinal sense, they will be unique in an ordinal sense even though the equality situation $b_i = b_j$ for $i \neq j$ can exist for some objectives. Ordering will be preserved because $b_i \geq b_j$ will contain the equality case as a subset. Therefore, a reasonable decision model will be the joint intersection of r decision measures,

$$D = \bigcap_{i=1}^{r} (\overline{b_i} \cup O_i) \tag{10.22}$$

and the optimum solution, a^*, is the alternative that maximizes D. If we define

$$C_i = \overline{b_i} \cup O_i \qquad \text{hence} \qquad \mu_{C_i}(a) = \max[\mu_{\overline{b_i}}(a), \mu_{O_i}(a)] \tag{10.23}$$

then the optimum solution, expressed in membership form, is given by

$$\mu_D(a^*) = \max_{a \in A}[\min\{\mu_{C_1}(a), \mu_{C_2}(a), \ldots, \mu_{C_r}(a)\}] \tag{10.24}$$

This model is intuitive in the following manner. As the ith objective becomes more important in the final decision, b_i increases, causing $\overline{b_i}$ to decrease, which in turn causes $C_i(a)$ to decrease, thereby increasing the likelihood that $C_i(a) = O_i(a)$, where now $O_i(a)$ will be the value of the decision function, D, representing alternative a (see Eq. (10.22)). As we repeat this process for other alternatives, a, Eq. (10.24) reveals that the largest value $O_i(a)$ for other alternatives will eventually result in the choice of the optimum solution, a^*. This is exactly how we would want the process to work.

Yager [1981] gives a good explanation of the value of this approach. For a particular objective, the negation of its importance (preference) acts as a barrier such that all ratings of alternatives below that barrier become equal to the value of that barrier. Here, we disregard all distinctions less than the barrier while keeping distinctions above this barrier. This process is similar to the grading practice of academics who lump all students whose class averages fall below 60% into the F category while keeping distinctions of A, B, C, and D for students above this percentile. However, in the decision model developed here this barrier varies, depending upon the preference (importance) of the objective to the decision maker. The more important is the objective, the lower is the barrier, and thus the more levels of distinction there are. As an objective becomes less important the distinction barrier increases, which lessens the penalty to the objective. In the limit, if the objective becomes totally unimportant, then the barrier is raised to its highest level and all alternatives are given the same weight and no distinction is made. Conversely, if the objective becomes the most important, all distinctions remain. In sum, the more important an objective is in the decision process, the more significant its effect on the decision function, D.

A special procedure [Yager, 1981] should be followed in the event of a numerical tie between two or more alternatives. If two alternatives, x and y, are tied, their respective decision values are equal, i.e., $D(x) = D(y) = \max_{a \in A}[D(a)]$, where $a = x = y$. Since $D(a) = \min_i[C_i(a)]$ there exists some alternative k such that $C_k(x) = D(x)$ and some alternative g such that $C_g(y) = D(y)$. Let

$$\hat{D}(x) = \min_{i \neq k}[C_i(x)] \quad \text{and} \quad \hat{D}(y) = \min_{i \neq g}[C_i(y)] \tag{10.25}$$

Then, we compare $\hat{D}(x)$ and $\hat{D}(y)$ and if, for example, $\hat{D}(x) > \hat{D}(y)$ we select x as our optimum alternative. However, if a tie still persists, i.e., if $\hat{D}(x) = \hat{D}(y)$, then there exist some other alternatives j and h such that $\hat{D}(x) = C_j(x) = \hat{D}(y) = C_h(y)$. Then we formulate

$$\hat{\hat{D}}(x) = \min_{i \neq k, j}[C_i(x)] \quad \text{and} \quad \hat{\hat{D}}(y) = \min_{i \neq g, h}[C_i(y)] \tag{10.26}$$

and compare $\hat{\hat{D}}(x)$ and $\hat{\hat{D}}(y)$. The tie-breaking procedure continues in this manner until an unambiguous optimum alternative emerges or all of the alternatives have been exhausted. In the latter case where a tie still results some other tie-breaking procedure, such as a refinement in the preference scales, can be used.

Example 10.5. A geotechnical engineer on a construction project must prevent a large mass of soil from sliding into a building site during construction and must retain this mass of soil indefinitely after construction to maintain stability of the area around a new facility to be constructed on the site [Adams, 1994]. The engineer therefore must decide which type of retaining wall design to select for the project. Among the many alternative designs available, the engineer reduces the list of candidate retaining wall designs to three: (1) a mechanically stabilized embankment (MSE) wall, (2) a mass concrete spread wall (Conc), and (3) a gabion (Gab) wall. The owner of the facility (the decision maker) has defined four objectives that impact the decision: (1) the cost of the wall (Cost), (2) the maintainability (Main) of the wall, (3) whether the design is a standard one (SD), and (4) the environmental (Env) impact of the wall. Moreover, the owner also decides to rank the preferences for these objectives on the unit interval. Hence, the engineer sets up the problem as follows:

$$A = \{\text{MSE, Conc, Gab}\} = \{a_1, a_2, a_3\}$$

$$O = \{\text{Cost, Main, SD, Env}\} = \{O_1, O_2, O_3, O_4\}$$

$$P = \{b_1, b_2, b_3, b_4\} \longrightarrow [0, 1]$$

From previous experience with various wall designs, the engineer first rates the retaining walls with respect to the objectives, given here. These ratings are fuzzy sets expressed in Zadeh's notation.

$$\underset{\sim}{Q_1} = \left\{ \frac{0.4}{\text{MSE}} + \frac{1}{\text{Conc}} + \frac{0.1}{\text{Gab}} \right\}$$

$$\underset{\sim}{O_2} = \left\{ \frac{0.7}{\text{MSE}} + \frac{0.8}{\text{Conc}} + \frac{0.4}{\text{Gab}} \right\}$$

$$\underset{\sim}{O_3} = \left\{ \frac{0.2}{\text{MSE}} + \frac{0.4}{\text{Conc}} + \frac{1}{\text{Gab}} \right\}$$

$$\underset{\sim}{O_4} = \left\{ \frac{1}{\text{MSE}} + \frac{0.5}{\text{Conc}} + \frac{0.5}{\text{Gab}} \right\}$$

These membership functions for each of the alternatives are shown graphically in Fig. 10.4.

The engineer wishes to investigate two decision scenarios. Each scenario propagates a different set of preferences from the owner, who wishes to determine the sensitivity of the optimum solutions to the preference ratings. In the first scenario, the owner lists the preferences for each of the four objectives, as shown in Fig. 10.5. From these preference values, the following calculations result:

$$b_1 = 0.8 \quad b_2 = 0.9 \quad b_3 = 0.7 \quad b_4 = 0.5$$
$$\bar{b}_1 = 0.2 \quad \bar{b}_2 = 0.1 \quad \bar{b}_3 = 0.3 \quad \bar{b}_4 = 0.5$$

$$D(a_1) = D(\text{MSE}) = (\bar{b}_1 \cup O_1) \cap (\bar{b}_2 \cup O_2) \cap (\bar{b}_3 \cup O_3) \cap (\bar{b}_4 \cup O_4)$$
$$= (0.2 \vee 0.4) \wedge (0.1 \vee 0.7) \wedge (0.3 \vee 0.2) \wedge (0.5 \vee 1)$$
$$= 0.4 \wedge 0.7 \wedge 0.3 \wedge 1 = 0.3$$

$$D(a_2) = D(\text{Conc}) = (0.2 \vee 1) \wedge (0.1 \vee 0.8) \wedge (0.3 \vee 0.4) \wedge (0.5 \vee 0.5)$$
$$= 1 \wedge 0.8 \wedge 0.4 \wedge 0.5 = 0.4$$

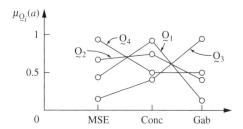

FIGURE 10.4
Membership for each alternative with respect to the objectives.

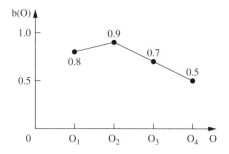

FIGURE 10.5
Preferences in the first scenario.

$$D(a_3) = D(\text{Gab}) = (0.2 \vee 0.1) \wedge (0.1 \vee 0.4) \wedge (0.3 \vee 1) \wedge (0.5 \vee 0.5)$$

$$= 0.2 \wedge 0.4 \wedge 1 \wedge 0.5 = 0.2$$

$$D^* = \max\{D(a_1), D(a_2), D(a_3)\} = \max\{0.3, 0.4, 0.2\} = 0.4$$

Thus, the engineer chose the second alternative, a_2, a concrete (Conc) wall as the retaining design under preference scenario 1.

Now, in the second scenario the engineer was given a different set of preferences by the owner, as shown in Fig. 10.6. From the preference values in Fig. 10.6, the following calculations result:

$$b_1 = 0.5 \quad b_2 = 0.7 \quad b_3 = 0.8 \quad b_4 = 0.7$$
$$\bar{b}_1 = 0.5 \quad \bar{b}_2 = 0.3 \quad \bar{b}_3 = 0.2 \quad \bar{b}_4 = 0.3$$

$$D(a_1) = D(\text{MSE}) = (\bar{b}_1 \cup O_1) \cap (\bar{b}_2 \cup O_2) \cap (\bar{b}_3 \cup O_3) \cap (\bar{b}_4 \cup O_4)$$

$$= (0.5 \vee 0.4) \wedge (0.3 \vee 0.7) \wedge (0.2 \vee 0.2) \wedge (0.3 \vee 1)$$

$$= 0.5 \wedge 0.7 \wedge 0.2 \wedge 1 = 0.2$$

$$D(a_2) = D(\text{Conc}) = (0.5 \vee 1) \wedge (0.3 \vee 0.8) \wedge (0.2 \vee 0.4) \wedge (0.3 \vee 0.5)$$

$$= 1 \wedge 0.8 \wedge 0.4 \wedge 0.5 = 0.4$$

$$D(a_3) = D(\text{Gab}) = (0.5 \vee 0.1) \wedge (0.3 \vee 0.4) \wedge (0.2 \vee 1) \wedge (0.3 \vee 0.5)$$

$$= 0.5 \wedge 0.4 \wedge 1 \wedge 0.5 = 0.4$$

Therefore, $D^* = \max\{D(a1), D(a2), D(a3)\} = \max\{0.2, 0.4, 0.4\} = 0.4$. But there is a tie between alternative a_2 and a_3. To resolve this tie, the engineer implements Eq. (10.25). The engineer looks closely at $D(a_2)$ and $D(a_3)$ and notes that the decision value of 0.4 for $D(a_2)$ came from the third term (i.e., $C_3(a_2)$; hence $k = 3$ in Eq. (10.25)), and that the decision value of 0.4 for $D(a_3)$ came from the second term (i.e., $C_2(a_3)$; hence $g = 2$ in Eq. (10.25)). Then the calculations proceed again between the tied choices a_2 and a_3:

$$\hat{D}(a_2) = \hat{D}(\text{Conc}) = (0.5 \vee 1) \wedge (0.3 \vee 0.8) \wedge (0.3 \vee 0.5)$$

$$= 1 \wedge 0.8 \wedge 0.5 = 0.5$$

$$\hat{D}(a_3) = \hat{D}(\text{Gab}) = (0.5 \vee 0.1) \wedge (0.2 \vee 1) \wedge (0.3 \vee 0.5)$$

$$= 0.5 \wedge 1 \wedge 0.5 = 0.5$$

Then $D^* = \max\{\hat{D}(a_2), \hat{D}(a_3)\} = \max\{0.5, 0.5\} = 0.5$, and there is still a tie between alternative a_2 and a_3. To resolve this second tie, the engineer implements Eq. (10.26). The engineer

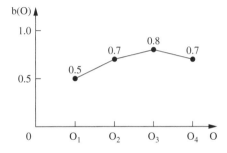

FIGURE 10.6
Preferences in the second scenario.

looks closely at $\hat{D}(a_2)$ and $\hat{D}(a_3)$ and notes that the decision value of 0.5 for $\hat{D}(a_2)$ came from the third term (i.e., $C_3(a_2)$; hence $j = 3$ in Eq. (10.26)), and that the decision value of 0.5 for $\hat{D}(a_3)$ came from the first term *and* the third term (i.e., $C_1(a_3) = C_3(a_3)$; hence $h = 1$ and $h = 3$ in Eq. (10.26)). Then the calculations proceed again between the tied choices a_2 and a_3:

$$\hat{\hat{D}}(a_2) = \hat{\hat{D}}(\text{Conc}) = (0.5 \vee 1) \wedge (0.3 \vee 0.8) = 0.8$$

$$\hat{\hat{D}}(a_3) = \hat{\hat{D}}(\text{Gab}) = (0.2 \vee 1) = 1$$

From these results, $D^* = \max\{\hat{\hat{D}}(a_2), \hat{\hat{D}}(a_3)\} = 1$; hence the tie is finally broken and the engineer chooses retaining wall a_3, a gabion wall, for the design under preference scenario 2.

FUZZY BAYESIAN DECISION METHOD

Classical statistical decision making involves the notion that the uncertainty in the future can be characterized probabilistically, as discussed in the introduction to this chapter. When we want to make a decision among various alternatives, our choice is predicated on information about the future, which is normally discretized into various "states of nature." If we knew with certainty the future states of nature, we would not need an analytic method to assess the likelihood of a given outcome. Unfortunately we do not know what the future will entail so we have devised methods to make the best choices given an uncertain environment. Classical Bayesian decision methods presume that future states of nature can be characterized as probability events. For example, consider the condition of "cloudiness" in tomorrow's weather by discretizing the state space into three levels and assessing each level probabilistically: the chance of a very cloudy day is 0.5, a partly cloudy day is 0.2, and a sunny (no clouds) day is 0.3. By convention the probabilities sum to unity. The problem with the Bayesian scheme here is that the events are vague and ambiguous. How many clouds does it take to transition between very cloudy and cloudy? If there is one small cloud in the sky, does this mean it is not sunny? This is the classic sorites paradox discussed in Chapter 5.

The following material first presents Bayesian decision making and then starts to consider ambiguity in the value of new information, in the states of nature, and in the alternatives in the decision process [see Terano et al., 1992]. Examples will illustrate these points.

First we shall consider the formation of probabilistic decision analysis. Let $S = \{s_1, s_2, \ldots, s_n\}$ be a set of possible states of nature; and the probabilities that these states will occur are listed in a vector,

$$\mathbf{P} = \{p(s_1), p(s_2), \ldots, p(s_n)\} \qquad \text{where} \qquad \sum_{i=1}^{n} p(s_i) = 1 \qquad (10.27)$$

The probabilities expressed in Eq. (10.27) are called "prior probabilities" in Bayesian jargon because they express prior knowledge about the true states of nature. Assume that the decision maker can choose among m alternatives, $A = \{a_1, a_2, \ldots, a_m\}$, and for a given alternative a_j we assign a utility value, u_{ji}, if the future state of nature turns out to be state s_i. These utility values should be determined by the decision maker since they express value, or cost, for each alternative-state pair, i.e., for each $a_j - s_i$ combination. The utility

TABLE 10.1
Utility matrix

States s_i Action a_j	s_1	s_2	$\cdots$	s_n
a_1	u_{11}	u_{12}	$\cdots$	u_{1n}
$\vdots$	$\vdots$	$\vdots$		$\vdots$
a_m	u_{m1}	u_{m2}	$\cdots$	u_{mn}

values are usually arranged in a matrix of the form shown in Table 10.1. The expected utility associated with the jth alternative would be

$$E(u_j) = \sum_{i=1}^{n} u_{ji} p(s_i) \tag{10.28}$$

The most common decision criterion is the *maximum* expected utility among all the alternatives, i.e.,

$$E(u^*) = \max_j E(u_j) \tag{10.29}$$

which leads to the selection of alternative a_k if $u^* = E(u_k)$.

Example 10.6. Suppose you are a geological engineer who has been asked by the chief executive officer (CEO) of a large oil firm to help make a decision about whether to drill for natural gas in a particular geographic region of northwestern New Mexico. You determine for your first attempt at the decision process that there are only two states of nature regarding the existence of natural gas in the region:

$$s_1 = \text{there is natural gas}$$

$$s_2 = \text{there is no natural gas}$$

and you are able to find from previous drilling information that the prior probabilities for each of these states is

$$p(s_1) = 0.5$$

$$p(s_2) = 0.5$$

Note these probabilities sum to unity. You suggest that there are two alternatives in this decision:

$$a_1 = \text{drill for gas}$$

$$a_2 = \bar{a}_1 = \text{do not drill for gas}$$

The decision maker (the CEO) helps you assemble a utility matrix to get the process started. The CEO tells you that the best situation for the firm is to decide to drill for gas, and subsequently find that gas, indeed, was in the geologic formation. The CEO assesses this value (u_{11}) as $+5$ in nondimensional units; in this case the CEO would have gambled (drilling costs big money) and won. Moreover, the CEO feels that the worst possible situation would be to drill for gas, and subsequently find that there was no gas in the area. Since this would cost time and money, the CEO determines that the value for this would be $u_{12} = -10$ units; the CEO would have gambled and lost – big. The other two utilities are assessed by the decision maker

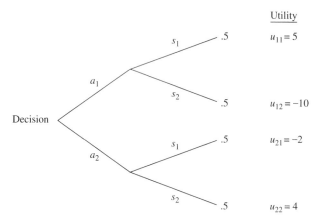

FIGURE 10.7
Decision tree for the two-alternative, two-state problem of Example 10.6.

in nondimensional units as $u_{21} = -2$ and $u_{22} = 4$. Hence, the utility matrix for this situation is given by

$$U = \begin{bmatrix} 5 & -10 \\ -2 & 4 \end{bmatrix}$$

Figure 10.7 shows the decision tree for this problem, of which the two initial branches correspond to the two possible alternatives, and the second layer of branches corresponds to the two possible states of nature. Superposed on the tree branches are the prior probabilities. The expected utility, in nondimensional units, for each alternative a_1 and a_2 is, from Eq. (10.28),

$$E(u_1) = (0.5)(5) + (0.5)(-10) = -2.5$$
$$E(u_2) = (0.5)(-2) + (0.5)(4) = 1.0$$

and we see that the maximum utility, using Eq. (10.29), is 1.0, which comes from alternative a_2; hence, on the basis of prior information only (prior probabilities) the CEO decides not to drill for natural gas (alternative a_2).

In many decision situations an intermediate issue arises: Should you get more information about the true states of nature prior to deciding? Suppose some new information regarding the true states of nature S is available from r experiments or other observations and is collected in a data vector, $\mathbf{X} = \{x_1, x_2, \ldots, x_r\}$. This information can be used in the Bayesian approach to update the prior probabilities, $p(s_i)$, in the following manner. First, the new information is expressed in the form of conditional probabilities, where the probability of each piece of data, x_k, where $k = 1, 2, \ldots, r$, is assessed according to whether the true state of nature, s_i, is known (not uncertain); these probabilities are presumptions of the future because they are equivalent to the following statement: *Given that we know that the true state of nature is s_i, the probability that the piece of new information x_k confirms that the true state is s_i is $p(x_k \mid s_i)$*. In the literature these conditional probabilities, denoted $p(x_k \mid s_i)$, are also called likelihood values. The likelihood values are then used as weights on the previous information, the prior probabilities $p(s_i)$, to find updated probabilities, known as posterior probabilities, denoted $p(s_i \mid x_k)$. The posterior probabilities are equivalent to this statement: *Given that the piece of new information x_k is true, the probability that the true* *state of nature is s_i is $p(s_i \mid x_k)$*. These updated probabilities are determined by Bayes's rule,

$$p(s_i \mid x_k) = \frac{p(x_k \mid s_i)}{p(x_k)} p(s_i) \tag{10.30}$$

where the term in the denominator of Eq. (10.30), $p(x_k)$, is the marginal probability of the data x_k and is determined using the total probability theorem

$$p(x_k) = \sum_{i=1}^{n} p(x_k \mid s_i) \cdot p(s_i) \tag{10.31}$$

Now the expected utility for the jth alternative, given the data x_k, is determined from the posterior probabilities (instead of the priors),

$$E(u_j \mid x_k) = \sum_{i=1}^{n} u_{ji} p(s_i \mid x_k) \tag{10.32}$$

and the maximum expected utility, given the new data x_k, is now given by

$$E(u^* \mid x_k) = \max_{j} E(u_j \mid x_k) \tag{10.33}$$

To determine the unconditional maximum expected utility we need to weight each of the r conditional expected utilities given by Eq. (10.33) by the respective marginal probabilities for each datum x_k, i.e., by $p(x_k)$, given in Eq. (10.34) as

$$E(u_x^*) = \sum_{k=1}^{r} E(u^* \mid x_k) \cdot p(x_k) \tag{10.34}$$

We can now introduce a new notion in the decision-making process, called the *value* of information, $V(x)$. In the case we have just introduced where there is some uncertainty about the new information, $X = \{x_1, x_2, \dots, x_r\}$, we call the information *imperfect* information. The value of this imperfect information, $V(x)$, can be assessed by taking the difference between the maximum expected utility without any new information, Eq. (10.29), and the maximum expected utility with the new information, Eq. (10.34), i.e.,

$$V(x) = E(u_x^*) - E(u^*) \tag{10.35}$$

We now introduce yet another notion in this process, called *perfect* information. This exercise is an attempt to develop a boundary condition for our problem, one that is altruistic in nature, i.e., can never be achieved in reality but nonetheless is quite useful in a mathematical sense to give us some scale on which to assess the value of imperfect information. If information is considered to be perfect (i.e., can predict the future states of nature precisely), we can say that the conditional probabilities are free of dissonance. That is, each new piece of information, or data, predicts one and only one state of nature; hence there is no ambivalence about what state is predicted by the data. However, if there is more than one piece of information the probabilities for a particular state of nature have to be shared by all the data. Mathematically, perfect information is represented by posterior probabilities of 0 or 1, i.e.,

$$p(s_i \mid x_k) = \begin{cases} 1 \\ 0 \end{cases} \tag{10.36}$$

We call this perfect information x_p. For perfect information, the maximum expected utility becomes (see Example 10.7)

$$E(u^*_{x_p}) = \sum_{k=1}^{r} E(u^*_{x_p} \mid x_k) p(x_k) \tag{10.37}$$

and the value of perfect information becomes

$$V(x_p) = E(u^*_{x_p}) - E(u^*) \tag{10.38}$$

Example 10.7. We continue with our gas exploration problem, Example 10.6. We had two states of nature – gas, s_1, and no gas, s_2 – and two alternatives – drill, a_1, and no drill, a_2. The prior probabilities were uniform,

$$p(s_1) = 0.5$$
$$p(s_2) = 0.5$$

Now, let us suppose the CEO of the natural gas company wants to reconsider the utility values. The CEO provides the utility matrix of Table 10.2 in the same form as Table 10.1. Further, the CEO has asked you to collect new information by taking eight geological boring samples from the region being considered for drilling. You have a natural gas expert examine the results of these eight tests, and get the expert's opinions about the conditional probabilities in the form of a matrix, given in Table 10.3. Moreover, you ask the natural gas expert for an assessment about how the conditional probabilities might change if they were perfect tests capable of providing perfect information. The expert gives you the matrix shown in Table 10.4.

TABLE 10.2
Utility matrix for natural gas example

u_{ji}	s_1	s_2
a_1	4	-2
a_2	-1	2

TABLE 10.3
Conditional probabilities for *imperfect* information

	x_1	x_2	x_3	x_4	x_5	x_6	x_7	x_8	
$p(x_k \mid s_1)$	0	0.05	0.1	0.1	0.2	0.4	0.1	0.05	$\sum$ row = 1
$p(x_k \mid s_2)$	0.05	0.1	0.4	0.2	0.1	0.1	0.05	0	$\sum$ row = 1

TABLE 10.4
Conditional probabilities for *perfect* information

	x_1	x_2	x_3	x_4	x_5	x_6	x_7	x_8	
$p(x_k \mid s_1)$	0	0	0	0	0.2	0.5	0.2	0.1	$\sum \text{row} = 1$
$p(x_k \mid s_2)$	0.1	0.2	0.5	0.2	0	0	0	0	$\sum \text{row} = 1$

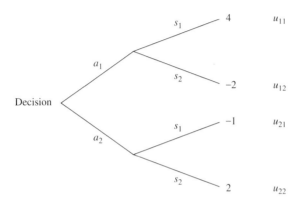

FIGURE 10.8
Decision tree showing utility values.

As the engineer assisting the CEO, you now conduct a decision analysis. Since the CEO changed the utility values, you have to recalculate the expected utility of making the decision on the basis of just the prior probabilities, before any new information is acquired. The decision tree for this situation is shown in Figure 10.8.

The expected utilities and maximum expected utility, based just on prior probabilities, are

$$E(a_1) = (4)(0.5) + (-2)(0.5) = 1.0$$

$$E(a_2) = (-1)(0.5) + (2)(0.5) = 0.5$$

$$E(u^*) = 1 \qquad \text{hence, you choose alternative } a_1, \text{ drill for natural gas.}$$

You are now ready to assess the changes in this decision process by considering additional information, both imperfect and perfect. Table 10.5 summarizes your calculations for the new prior probabilities, $p(s_1 \mid x_k)$ and $p(s_2 \mid x_k)$, the marginal probabilities for the new information, $p(x_k)$, the expected conditional utilities, $E(u^* \mid x_k)$, and the expected alternatives, $a_j \mid x_k$.

Typical calculations for the values in Table 10.5 are provided here. For the marginal probabilities for the new *imperfect* information, use Eq (10.31), the conditional probabilities from Table 10.3, and the prior probabilities:

$$p(x_1) = (0)(0.5) + (0.05)(0.5) = 0.025$$

$$p(x_4) = (0.1)(0.5) + (0.2)(0.5) = 0.15$$

The posterior probabilities are calculated with the use of Eq. (10.30), using the conditional probabilities from Table 10.3, the prior probabilities, and the marginal probabilities, $p(x_k)$,

TABLE 10.5
Posterior probabilities based on imperfect information

	x_1	x_2	x_3	x_4	x_5	x_6	x_7	x_8
$p(s_1 \mid x_k)$	0	$\dfrac{1}{3}$	$\dfrac{1}{5}$	$\dfrac{1}{3}$	$\dfrac{2}{3}$	$\dfrac{4}{5}$	$\dfrac{2}{3}$	1
$p(s_2 \mid x_k)$	1	$\dfrac{2}{3}$	$\dfrac{4}{5}$	$\dfrac{2}{3}$	$\dfrac{1}{3}$	$\dfrac{1}{5}$	$\dfrac{1}{3}$	0
$p(x_k)$	0.025	0.075	0.25	0.15	0.15	0.25	0.075	0.025
$E(u^* \mid x_k)$	2	1	$\dfrac{7}{5}$	1	2	$\dfrac{14}{5}$	2	4
$a_j \mid x_k$	a_2	a_2	a_2	a_2	a_1	a_1	a_1	a_1

just determined and summarized in Table 10.5 (third row); for example,

$$p(s_1 \mid x_2) = \frac{0.05(0.5)}{0.075} = \frac{1}{3} \qquad p(s_2 \mid x_2) = \frac{0.1(0.5)}{0.075} = \frac{2}{3} \quad \cdots$$

$$p(s_1 \mid x_6) = \frac{0.4(0.5)}{0.25} = \frac{4}{5} \qquad p(s_2 \mid x_6) = \frac{0.1(0.5)}{0.25} = \frac{1}{5} \quad \cdots$$

The conditional expected utilities, $E(u^* \mid x_k)$, are calculated using first Eq. (10.32), then Eq. (10.33); for example,

$$E(u_1 \mid x_3) = (\tfrac{1}{5})(4) + (\tfrac{4}{5})(-2) = -\tfrac{4}{5} \qquad E(u_2 \mid x_3) = (\tfrac{1}{5})(-1) + (\tfrac{4}{5})(2) = \tfrac{7}{5}$$

Hence $E(u^* \mid x_3) = \max(-\tfrac{4}{5}, \tfrac{7}{5}) = \tfrac{7}{5}$ (choose alternative a_2):

$$E(u_1 \mid x_8) = (1)(4) + (0)(-2) = 4 \qquad E(u_2 \mid x_8) = (1)(-1) + (0)(2) = -1$$

Hence $E(u^* \mid x_8) = \max(4, -1) = 4$ (choose alternative a_1).
    Now use Eq. (10.34) to calculate the overall unconditional expected utility for *imperfect* information, which is actually the sum of pairwise products of the values in the third and fourth rows of Table 10.5, e.g.,

$$E(u_x^*) = (0.025)(2) + (0.075)(1) + \cdots + (0.025)(4) = 1.875$$

and the value of the new *imperfect* information, using Eq. (10.35), is

$$V(x) = E(u_x^*) - E(u^*) = 1.875 - 1 = 0.875$$

To decide what alternative to choose, notice in Table 10.5 that the total utility favoring a_1 is 10.8 $(2 + \tfrac{14}{5} + 2 + 4)$ and the total utility favoring a_2 is 5.4 $(2 + 1 + \tfrac{7}{5} + 1)$. Hence, the CEO chooses alternative a_1, to drill for gas. In effect, the new information has not changed the CEO's mind about drilling.
    You begin the process of assessing the changes due to the consideration of the hypothetical *perfect* information. Table 10.6 summarizes your calculations for the new prior probabilities, $p(s_1 \mid x_k)$ and $p(s_2 \mid x_k)$, the marginal probabilities for the perfect information, $p(x_k)$, the expected conditional utilities, $E(u_{x_p}^* \mid x_k)$, and the expected alternatives, $a_j \mid x_k$.

TABLE 10.6
Posterior probabilities based on perfect information

	x_1	x_2	x_3	x_4	x_5	x_6	x_7	x_8
$p(s_1 \mid x_k)$	0	0	0	0	1	1	1	1
$p(s_2 \mid x_k)$	1	1	1	1	0	0	0	0
$p(x_k)$	0.05	0.1	0.25	0.1	0.1	0.25	0.1	0.05
$E(u^* \mid x_k)$	2	2	2	2	4	4	4	4
$a_j \mid x_k$	a_2	a_2	a_2	a_2	a_1	a_1	a_1	a_1

These are calculated in the same way as those in Table 10.5, except you make use of the *perfect* conditional probabilities of Table 10.4.

Equation (10.37) is used to calculate the overall unconditional expected utility for *perfect* information, which is actually the sum of pairwise products of the values in the third and fourth rows of Table 10.6, e.g.,

$$E(u^*_{x_p}) = (0.05)(2) + (0.1)(2) + \cdots + (0.05)(4) = 3.0$$

and the value of the new *perfect* information, using Eq. (10.38), is

$$V(x_p) = E(u^*_{x_p}) - E(u^*) = 3 - 1 = 2.0$$

Alternative a_1 is still the choice here. We note that the hypothetical information has a value of 2 and the imperfect information has a value of less than half of this, 0.875. This difference can be used to assess the value of the imperfect information compared to both no information (1) and perfect information (3).

We now discuss the fact that the new information might be inherently fuzzy [Okuda et al., 1974, 1978]. Suppose the new information, $X = \{x_1, x_2, \ldots, x_r\}$, is a universe of discourse in the units appropriate for the new information. Then we can define fuzzy events, $\underset{\sim}{M}$, on this information, such as "good" information, "moderate" information, and "poor" information. The fuzzy event will have membership function $\mu_{\underset{\sim}{M}}(x_k)$, $k = 1, 2, \ldots, r$. We can now define the idea of a "probability of a fuzzy event," i.e., the probability of $\underset{\sim}{M}$, as

$$P(\underset{\sim}{M}) = \sum_{k=1}^{r} \mu_{\underset{\sim}{M}}(x_k) p(x_k) \tag{10.39}$$

We note in Eq. (10.39) that if the fuzzy event is, in fact, crisp, i.e., $\underset{\sim}{M} = M$, then the probability reduces to

$$P(M) = \sum_{x_k \in M} p(x_k) \qquad \mu_M = \begin{cases} 1, & x_k \in M \\ 0, & \text{otherwise} \end{cases} \tag{10.40}$$

where Eq. (10.40) describes the probability of a crisp event simply as the sum of the marginal probabilities of those data points, x_k, that are defined to be in the event, M. Based on this, the posterior probability of s_i, given fuzzy information $\underset{\sim}{M}$, is

$$P(s_i \mid \underset{\sim}{M}) = \frac{\sum_{k=1}^{r} p(x_k \mid s_i) \mu_{\underset{\sim}{M}}(x_k) p(s_i)}{P(\underset{\sim}{M})} = \frac{P(\underset{\sim}{M} \mid s_i) p(s_i)}{P(\underset{\sim}{M})} \tag{10.41}$$

where

$$p(\underset{\sim}{M} \mid s_i) = \sum_{k=1}^{r} p(x_k \mid s_i)\mu_{\underset{\sim}{M}}(x_k) \tag{10.42}$$

We now define the collection of all the fuzzy events describing fuzzy information as an *orthogonal* fuzzy information system, $\Phi = \{\underset{\sim}{M_1}, \underset{\sim}{M_2}, \dots, \underset{\sim}{M_g}\}$ where by *orthogonal* we mean that the sum of the membership values for each fuzzy event, $\underset{\sim}{M_i}$, for every data point in the information universe, x_k, equals unity [Tanaka et al., 1976]. That is,

$$\sum_{t=1}^{g} \mu_{\underset{\sim}{M_t}}(x_k) = 1 \qquad \text{for all } x_k \in X \tag{10.43}$$

If the fuzzy events on the new information universe are *orthogonal*, we can extend the Bayesian approach to consider fuzzy information. The fuzzy equivalents of Eqs. (10.32), (10.33), and (10.34) become, for a fuzzy event $\underset{\sim}{M_t}$,

$$E(u_j \mid \underset{\sim}{M_t}) = \sum_{i=1}^{n} u_{ji} \cdot p(s_i \mid \underset{\sim}{M_t}) \tag{10.44}$$

$$E(u^* \mid \underset{\sim}{M_t}) = \max_{j} E(u_j \mid \underset{\sim}{M_t}) \tag{10.45}$$

$$E(u_{\Phi}^*) = \sum_{t=1}^{g} E(u^* \mid \underset{\sim}{M_t}) \cdot p(\underset{\sim}{M_t}) \tag{10.46}$$

Now the value of fuzzy information can be determined in an analogous manner as

$$V(\Phi) = E(u_{\Phi}^*) - E(u^*) \tag{10.47}$$

Example 10.8   [Continuation of Example 10.7]. Suppose the eight data samples are from overlapping, ill-defined parcels within the drilling property. We define an orthogonal fuzzy information system, Φ,

$$\Phi = \{\underset{\sim}{M_1}, \underset{\sim}{M_2}, \underset{\sim}{M_3}\} = \{\text{fuzzy parcel 1, fuzzy parcel 2, fuzzy parcel 3}\}$$

with membership functions in Table 10.7. The fourth row of Table 10.7 repeats the marginal probabilities for each data, x_k, from Table 10.5. As can be seen in Table 10.7 the sum of the membership values in each column (the first three rows) equals unity, as required for *orthogonal* fuzzy sets.

As before, we use Eq. (10.39) to determine the marginal probabilities for each fuzzy event,

$$p(\underset{\sim}{M_1}) = 0.225 \qquad p(\underset{\sim}{M_2}) = 0.55 \qquad p(\underset{\sim}{M_3}) = 0.225$$

and Eq. (10.42) to determine the fuzzy conditional probabilities,

$$p(\underset{\sim}{M_1} \mid s_1) = 0.1 \qquad p(\underset{\sim}{M_2} \mid s_1) = 0.55 \qquad p(\underset{\sim}{M_3} \mid s_1) = 0.35$$
$$p(\underset{\sim}{M_1} \mid s_2) = 0.35 \qquad p(\underset{\sim}{M_2} \mid s_2) = 0.55 \qquad p(\underset{\sim}{M_3} \mid s_2) = 0.1$$

TABLE 10.7
Orthogonal membership functions for orthogonal fuzzy events

	x_1	x_2	x_3	x_4	x_5	x_6	x_7	x_8
$\mu_{M_1}(x_k)$	1	1	0.5	0	0	0	0	0
$\mu_{M_2}(x_k)$	0	0	0.5	1	1	0.5	0	0
$\mu_{M_3}(x_k)$	0	0	0	0	0	0.5	1	1
$p(x_k)$	0.025	0.075	0.25	0.15	0.15	0.25	0.075	0.025

and Eq. (10.41) to determine the fuzzy posterior probabilities,

$$p(s_1 \mid M_1) = 0.222 \qquad p(s_1 \mid M_2) = 0.5 \qquad p(s_1 \mid M_3) = 0.778$$
$$p(s_2 \mid M_1) = 0.778 \qquad p(s_2 \mid M_2) = 0.5 \qquad p(s_2 \mid M_3) = 0.222$$

Now the conditional fuzzy expected utilities can be determined using Eq. (10.44),

$$M_1 : \quad E(u_1 \mid M_1) = (4)(0.222) + (-2)(0.778) = -0.668$$
$$E(u_2 \mid M_1) = (-1)(0.222) + (2)(0.778) = 1.334$$
$$M_2 : \quad E(u_1 \mid M_2) = (4)(0.5) + (-2)(0.5) = 1.0$$
$$E(u_2 \mid M_2) = (-1)(0.5) + (2)(0.5) = 0.5$$
$$M_3 : \quad E(u_1 \mid M_3) = (4)(0.778) + (-2)(0.222) = 2.668$$
$$E(u_2 \mid M_3) = (-1)(0.778) + (2)(0.222) = -0.334$$

and the maximum expected utility from Eq. (10.46), using each of the foregoing three maximum conditional probabilities,

$$E(u_\Phi^*) = (0.225)(1.334) + (0.55)(1) + (0.225)(2.668) = 1.45$$

and the value of the fuzzy information from Eq. (10.47),

$$V(\Phi) = 1.45 - 1 = 0.45$$

Here we see that the value of the fuzzy information is less than the value of the perfect information (2.0), and less than the value of the imperfect information (0.875). However, it may turn out that fuzzy information is *far less costly* (remember, precision costs) than either the imperfect or perfect (hypothetical) information. Although not developed in this text, this analysis could be extended to consider *cost* of information.

DECISION MAKING UNDER FUZZY STATES AND FUZZY ACTIONS

The Bayesian method can be further extended to include the possibility that the states of nature are fuzzy and the decision makers' alternatives are also fuzzy [Tanaka et al., 1976]. For example, suppose your company wants to expand and you are considering three fuzzy alternatives in terms of the size of a new facility:

$$A_1 = \text{small-scale project}$$
$$A_2 = \text{middle-scale project}$$
$$A_3 = \text{large-scale project}$$

Just as all fuzzy sets are defined on some universe of discourse, continuous or discrete, the fuzzy alternatives (actions) would also be defined on a universe of discourse, say values of square footage of floor space on a continuous scale of areas in some appropriate units of area. Moreover, suppose further that the economic climate in the future is very fuzzy and you pose the following three possible fuzzy states of nature ($\underset{\sim}{F}_s$, $s = 1, 2, 3$):

$$\underset{\sim}{F}_1 = \text{low rate of economic growth}$$

$$\underset{\sim}{F}_2 = \text{medium rate of economic growth}$$

$$\underset{\sim}{F}_3 = \text{high rate of economic growth}$$

all of which are defined on a universe of numerical rates of economic growth, say S, where $S = \{s_1, s_2, \dots, s_n\}$ is a discrete universe of economic growth rates (e.g., $-4\%, -3\%, \dots, 0\%, 1\%, 2\%, \dots$). The fuzzy states $\underset{\sim}{F}_s$ will be required to be orthogonal fuzzy sets, in order for us to continue to use the Bayesian framework. This orthogonal condition on the fuzzy states will be the same constraint as illustrated in Eq. (10.43), i.e.,

$$\sum_{s=1}^{3} \mu_{\underset{\sim}{F}_s}(s_i) = 1 \qquad i = 1, 2, \dots, n \tag{10.48}$$

Further, as we need utility values to express the relationship between crisp alternative-state pairs, we still need a utility matrix to express the value of all the fuzzy alternative-state pairings. Such a matrix will have the form shown in Table 10.8.

Proceeding as before, but now with fuzzy states of nature, the expected utility of fuzzy alternative $\underset{\sim}{A}_j$ is

$$E(u_j) = \sum_{s=1}^{3} u_{js} p(\underset{\sim}{F}_s) \tag{10.49}$$

where

$$p(\underset{\sim}{F}_s) = \sum_{i=1}^{n} \mu_{\underset{\sim}{F}_s}(s_i) p(s_i) \tag{10.50}$$

and the maximum utility is

$$E(u^*) = \max_{j} E(u_j) \tag{10.51}$$

TABLE 10.8
Utility values for fuzzy states and fuzzy alternatives

	$\underset{\sim}{F}_1$	$\underset{\sim}{F}_2$	$\underset{\sim}{F}_3$
$\underset{\sim}{A}_1$	u_{11}	u_{12}	u_{13}
$\underset{\sim}{A}_2$	u_{21}	u_{22}	u_{23}
$\underset{\sim}{A}_3$	u_{31}	u_{32}	u_{33}

We can have crisp or fuzzy information on a universe of information $X = \{x_1, x_2, \ldots, x_r\}$, e.g., rate of increase of gross national product. Our fuzzy information will again reside on a collection of orthogonal fuzzy sets on X, $\Phi = \{\underset{\sim}{M}_1, \underset{\sim}{M}_2, \ldots, \underset{\sim}{M}_g\}$, that are defined on X. We can now derive the posterior probabilities of fuzzy states $\underset{\sim}{F}_s$, given probabilistic information (Eq. (10.52a)), x_r, and fuzzy information $\underset{\sim}{M}_t$ (Eq. (10.52b)) as follows:

$$p(\underset{\sim}{F}_s \mid x_k) = \frac{\displaystyle\sum_{i=1}^{n} \mu_{\underset{\sim}{F}_s}(s_i) p(x_k \mid s_i) p(s_i)}{p(x_k)} \tag{10.52a}$$

$$p(\underset{\sim}{F}_s \mid \underset{\sim}{M}_t) = \frac{\displaystyle\sum_{i=1}^{n}\sum_{k=1}^{r} \mu_{\underset{\sim}{F}_s}(s_i) \mu_{\underset{\sim}{M}_t}(x_k) p(x_k \mid s_i) p(s_i)}{\displaystyle\sum_{k=1}^{r} \mu_{\underset{\sim}{M}_t}(x_k) p(x_k)} \tag{10.52b}$$

Similarly, the expected utility given the probabilistic (Eq. (10.53a)) and fuzzy (Eq. (10.53b)) information is then

$$E(u_j \mid x_k) = \sum_{s=1}^{3} u_{js} p(\underset{\sim}{F}_s \mid x_k) \tag{10.53a}$$

$$E(u_j \mid \underset{\sim}{M}_t) = \sum_{s=1}^{3} u_{js} p(\underset{\sim}{F}_s \mid \underset{\sim}{M}_t) \tag{10.53b}$$

where the maximum conditional expected utility for probabilistic (Eq. (10.54a)) and fuzzy (Eq. (10.54b)) information is

$$E(u^*_{x_k}) = \max_j E(u_j \mid x_k) \tag{10.54a}$$

$$E(u^*_{\underset{\sim}{M}_t}) = \max_j E(u_j \mid \underset{\sim}{M}_t) \tag{10.54b}$$

Finally, the unconditional expected utilities for fuzzy states and probabilistic information, (Eq. (10.55a)), or fuzzy information, (Eq. (10.55b)), will be

$$E(u^*_x) = \sum_{k=1}^{r} E(u^*_{x_k}) p(x_k) \tag{10.55a}$$

$$E(u^*_{\Phi}) = \sum_{t=1}^{g} E(u^*_{\underset{\sim}{M}_t}) p(\underset{\sim}{M}_t) \tag{10.55b}$$

The expected utility given in Eqs. (10.55) now enables us to compute the value of the fuzzy information, within the context of fuzzy states of nature, for probabilistic information (Eq. (10.35)) and fuzzy information (Eq. (10.56)):

$$V(\Phi) = E(u^*_{\Phi}) - E(u^*) \tag{10.56}$$

If the new fuzzy information is hypothetically *perfect* (since this would represent the ability to predict a fuzzy state F_s without dissonance, it is admittedly an untenable boundary condition on the analysis), denoted Φ_p, then we can compute the maximum expected utility of fuzzy *perfect* information using Eq. (10.57). The expected utility of ith alternative A_i for fuzzy perfect information on state F_s becomes (from Table 10.8)

$$u(A_i | F_s) = u(A_i, F_s) \tag{10.57}$$

Therefore, the optimum fuzzy action, $A_{F_s}^*$, is defined by

$$u(A_{F_s}^* | F_s) = \max_i u(A_i, F_s) \tag{10.58}$$

Hence, the total expected utility for fuzzy perfect information is

$$E(u_{\Phi_p}^*) = \sum_{j=1}^{3} u(A_{F_s}^* | F_s) p(F_s) \tag{10.59}$$

where $p(F_s)$ are the prior probabilities of the fuzzy states of nature given by Eq. (10.50). The result of Eq. (10.55b) for the fuzzy *perfect* case would be denoted $E(u_{\Phi_p}^*)$, and the value of the fuzzy *perfect* information would be

$$V(\Phi_p) = E(u_{\Phi_p}^*) - E(u^*) \tag{10.60}$$

Tanaka et al. [1976] have proved that the various values of information conform to the following inequality expression:

$$V(\Phi_p) \geq V(x_p) \geq V(x) \geq V(\Phi) \geq 0 \tag{10.61}$$

The inequalities in Eq. (10.61) are consistent with our intuition. The ordering, $V(x) \geq V(\Phi)$, is due to the fact that information Φ is characterized by fuzziness *and* randomness. The ordering, $V(x_p) \geq V(x)$, is true because x_p is better information than x; it is *perfect*. The ordering, $V(\Phi_p) \geq V(x_p)$, is created by the fact that the uncertainty expressed by the probability $P(F_i)$ still remains, even if we know the true state, s_i; hence, our interest is not in the crisp states of nature, S, but rather in the fuzzy states, F, which are defined on S.

To illustrate the development of Eqs. (10.48)–(10.61) in expanding a decision problem to consider fuzzy information, fuzzy states, and fuzzy actions in the Bayesian decision framework, the following example in computer engineering is provided.

Example 10.9. One of the decisions your project team faces with each new computer product is what type of printed circuit board (PCB) will be required for the unit. Depending on the density of tracks (metal interconnect traces on the PCB that act like wire to connect components together), which is related to the density of the components, we may use a single-layer PCB, a double-layer PCB, a four-layer PCB, or a six-layer PCB. A PCB layer is a two-dimensional plane of interconnecting tracks. The number of layers on a PCB is the number of parallel interconnection layers in the PCB. The greater the density of the interconnections in the design, the greater the number of layers required to fit the design onto a PCB of given size. One measure of board track density is the number of nodes required in the design. A node is created at a location in the circuit where two or more lines (wires, tracks) meet. The decision process will comprise the following steps.

1. *Define the fuzzy states of nature.* The density of the PCB is defined as three fuzzy sets on the singleton states $S = (s_1, s_2, s_3, s_4, s_5) = (s_i)$, $i = 1, 2, \ldots, 5$, where i defines the states in terms of a percentage of our most dense (in terms of components and interconnections) PCB. So, your team defines $s_1 = 20\%$, $s_2 = 40\%$, $s_3 = 60\%$, $s_4 = 80\%$, and $s_5 = 100\%$ of the density of the densest PCB; these are singletons on the universe of relative densities. Further, you define the following three fuzzy states, which are defined on the universe of relative density states S:

$$\underset{\sim}{F_1} = \text{low-density PCB}$$

$$\underset{\sim}{F_2} = \text{medium-density PCB}$$

$$\underset{\sim}{F_3} = \text{high-density PCB}$$

2. *Define fuzzy alternatives.* Your decision alternative will represent the type of the PCB we decide to use, as follows (these actions are admittedly not very fuzzy, but in general they can be):

$$\underset{\sim}{A_1} = \text{use a 2-layer PCB for the new design}$$

$$\underset{\sim}{A_2} = \text{use a 4-layer PCB for the new design}$$

$$\underset{\sim}{A_3} = \text{use a 6-layer PCB for the new design}$$

3. *Define new data samples (information).* The universe $X = (x_1, x_2, \ldots, x_5)$ represents the "measured number of nodes in the PCB schematic;" that is, the additional information is the measured number of nodes of the schematic that can be calculated by a *schematic capture system.* You propose the following discrete values for number of nodes:

$$x_1 = 100 \text{ nodes}$$

$$x_2 = 200 \text{ nodes}$$

$$x_3 = 300 \text{ nodes}$$

$$x_4 = 400 \text{ nodes}$$

$$x_5 = 500 \text{ nodes}$$

4. *Define orthogonal fuzzy information system.* You determine that the ambiguity in defining the density of nodes can be characterized by three linguistic information sets as $(\underset{\sim}{M_1}, \underset{\sim}{M_2}, \underset{\sim}{M_3})$, where

$$\underset{\sim}{M_1} = \text{low number of nodes on PCB [generally } < 300 \text{ nodes]}$$

$$\underset{\sim}{M_2} = \text{average (medium) number of nodes on PCB [about 300 nodes]}$$

$$\underset{\sim}{M_3} = \text{high number of nodes on PCB [generally } > 300 \text{ nodes]}$$

5. *Define the prior probabilities.* The prior probabilities of the singleton densities (states) are as follows:

$$p(s_1) = 0.2$$

$$p(s_2) = 0.3$$

$$p(s_3) = 0.3$$

$$p(s_4) = 0.1$$

$$p(s_5) = 0.1$$

The preceding numbers indicate that moderately dense boards are the most probable, followed by low-density boards, and high- to very high density boards are the least probable.

6. *Identify the utility values.* You propose the nondimensional utility values shown in Table 10.9 to represent the fuzzy alternative-fuzzy state relationships. The highest utility in Table 10.9 is achieved by the selection of a six-layer PCB for a high-density PCB, since the board layout is achievable. The same high-utility level of 10 is also achieved by selecting the two-layer PCB in conjunction with the low-density PCB, since a two-layer PCB is cheaper than a four- or six-layer PCB. The lowest utility is achieved by the selection of a two-layer PCB for a high-density PCB; since the layout cannot be done, it will not fit. The second to lowest utility is achieved when a six-layer PCB is chosen, but the design is of low density, so you are wasting money.

7. *Define membership values for each orthogonal fuzzy state.* The fuzzy sets in Table 10.10 satisfy the orthogonality condition, for the sum of each column equals 1, $\sum$ column $= \sum_s \mu_{F_s}(s_i) = 1$.

8. *Define membership values for each orthogonal fuzzy set on the fuzzy information system.* In Table 10.11, $\sum$ column $= \sum_t \mu_{M_t}(x_i) = 1$; hence the fuzzy sets are orthogonal.

9. *Define the conditional probabilities (likelihood values) for the uncertain information.* Table 10.12 shows the conditional probabilities for uncertain (probabilistic) information; note the sum of elements in each row equals unity.

10. *Define the conditional probabilities (likelihood values) for the probabilistic perfect information.* Table 10.13 shows the conditional probabilities for probabilistic perfect information; note that the sum of elements in each row equals unity and that each column only has one entry (i.e., no dissonance).

TABLE 10.9
Utilities for fuzzy states and alternatives

	$\underset{\sim}{F}_1$	$\underset{\sim}{F}_2$	$\underset{\sim}{F}_3$
$\underset{\sim}{A}_1$	10	3	0
$\underset{\sim}{A}_2$	4	9	6
$\underset{\sim}{A}_3$	1	7	10

TABLE 10.10
Orthogonal fuzzy sets for fuzzy states

	s_1	s_2	s_3	s_4	s_5
$\underset{\sim}{F}_1$	1	0.5	0	0	0
$\underset{\sim}{F}_2$	0	0.5	1	0.5	0
$\underset{\sim}{F}_3$	0	0	0	0.5	1

TABLE 10.11
Orthogonal fuzzy sets for fuzzy information

	x_1	x_2	x_3	x_4	x_5
$\underset{\sim}{M_1}$	1	0.4	0	0	0
$\underset{\sim}{M_2}$	0	0.6	1	0.6	0
$\underset{\sim}{M_3}$	0	0	0	0.4	1

TABLE 10.12
Conditional probabilities $p(x_k|s_i)$ for uncertain information

	x_1	x_2	x_3	x_4	x_5
$p(x_k \mid s_1)$	0.44	0.35	0.17	0.04	0
$p(x_k \mid s_2)$	0.26	0.32	0.26	0.13	0.03
$p(x_k \mid s_3)$	0.12	0.23	0.30	0.23	0.12
$p(x_k \mid s_4)$	0.03	0.13	0.26	0.32	0.26
$p(x_k \mid s_5)$	0	0.04	0.17	0.35	0.44

TABLE 10.13
Conditional probabilities $p(x_k \mid s_i)$ for fuzzy perfect
information

	x_1	x_2	x_3	x_4	x_5
$p(x_k \mid s_1)$	1	0	0	0	0
$p(x_k \mid s_2)$	0	1	0	0	0
$p(x_k \mid s_3)$	0	0	1	0	0
$p(x_k \mid s_4)$	0	0	0	1	0
$p(x_k \mid s_5)$	0	0	0	0	1

You are now ready to compute the values of information for this decision process involving fuzzy states, fuzzy alternatives, and fuzzy information.

Case 1. Crisp states and actions

(i) **Utility and optimum decision given no information.** Before initiating the no-information case, we must define the nondimensional utility values for this nonfuzzy state situation. Note that the utility values given in Table 10.14 compare the fuzzy alternatives to the singleton states (s_i), as opposed to the fuzzy states for which the utility is defined in Table 10.9. The expected values for this case are determined by using Eq. (10.28), e.g.,

$$E(u_1) = (10)(0.2) + (8)(0.3) + \cdots + (2)(0.1)$$

$$= 6.4$$

Similarly, $E(u_2) = 6.3$, and $E(u_3) = 4.4$. Hence, the optimum decision, given no information and with crisp (singleton) states, is alternative 1, $\underset{\sim}{A_1}$, i.e., $E(u_1) = 6.4$.

TABLE 10.14
Utility values for crisp states

	s_1	s_2	s_3	s_4	s_5
$\underset{\sim}{A}_1$	10	8	6	2	0
$\underset{\sim}{A}_2$	4	6	9	6	4
$\underset{\sim}{A}_3$	1	2	6	8	10

TABLE 10.15
Computed values for uncertain case (nonfuzzy states)

	x_1	x_2	x_3	x_4	x_5
$p(x_k)$	0.205	0.252	0.245	0.183	0.115
$p(s_1 \mid x_k)$	0.429	0.278	0.139	0.044	0.0
$p(s_2 \mid x_k)$	0.380	0.381	0.318	0.213	0.078
$p(s_3 \mid x_k)$	0.176	0.274	0.367	0.377	0.313
$p(s_4 \mid x_k)$	0.015	0.052	0.106	0.175	0.226
$p(s_5 \mid x_k)$	0.0	0.016	0.069	0.191	0.383
$E(u^* \mid x_k)$	8.42	7.47	6.68	6.66	7.67
$a_j \mid a_k$	1	1	2	2	3

(*ii*) **Utility and optimal decision given uncertain and perfect information.**

(*a*) *Probabilistic (uncertain) information.* Table 10.15 summarizes the values of the marginal probability $p(x_k)$, the posterior probabilities, and the maximum expected values for the uncertain case. The values in Table 10.15 have been calculated as in the preceding computation. For example, the marginal probabilities are calculated using Eq. (10.31):

$$p(x_1) = (0.44)(0.2) + (0.26)(0.3) + (0.12)(0.3) + (0.03)(0.1) + (0)(0.1)$$

$$= 0.205$$

$$p(x_3) = (0.17)(0.2) + (0.26)(0.3) + (0.3)(0.3) + (0.26)(0.1) + (0.17)(0.1)$$

$$= 0.245$$

The posterior probabilities are calculated using Eq. (10.30), the conditional probabilities, and the prior probabilities; for example,

$$p(s_1 \mid x_3) = \frac{(0.17)(0.2)}{(0.245)} = 0.139$$

$$p(s_3 \mid x_2) = \frac{(0.23)(0.3)}{(0.245)} = 0.274$$

$$p(s_5 \mid x_4) = \frac{(0.35)(0.1)}{(0.183)} = 0.191$$

The conditional expected utilities are calculated using Eqs. (10.32)–(10.33); for example, for datum x_1,

$$E(u_1|x_1) = (0.429)(10) + (0.380)(8) + (0.176)(6) + (0.015)(2) + (0)(0)$$

$$= 8.42$$

$$E(u_2|x_1) = (0.429)(4) + (0.380)(6) + (0.176)(9) + (0.015)(6) + (0)(4)$$

$$= 5.67$$

$$E(u_3|x_1) = (0.429)(1) + (0.380)(2) + (0.176)(6) + (0.015)(8) + (0)(10)$$

$$= 2.36$$

Therefore, the optimum decision for datum x_1, given uncertain information with crisp states, is

$$E(u^*|x_1) = \max(8.42, 5.67, 2.36) = 8.42 \text{ (choose action } \underset{\sim}{A}_1)$$

Now, using Eq. (10.34) to calculate the overall (for all data x_i) unconditional expected utility for the uncertain information, we get

$$E(u_x^*) = (8.42)(0.205) + (7.47)(0.252) + (6.68)(0.245) + \cdots + (7.67)(0.115)$$

$$= 7.37$$

The value of the uncertain information, using Eq. (10.35), is

$$V(x) = 7.37 - 6.4 = 0.97$$

(b) *Probabilistic perfect information.* Using the same utility values as before, and conditional probabilities as defined in Table 10.12, the marginal probabilities, posterior probabilities, and the expected values are shown in Table 10.16. The unconditional expected utility for probabilistic perfect information is given as

$$E(u_{x_p}^*) = (10)(0.2) + (8)(0.3) + \cdots + (10)(0.1)$$

$$= 8.9$$

and the value of the probabilistic perfect information from Eq. (10.35) is

$$V(x_p) = 8.9 - 6.4 = 2.5$$

Case 2. Fuzzy states and actions

(i) **Utility and optimum decision given no information.** The utility values for this case are shown in Table 10.9. We calculate the prior probabilities for the fuzzy states using Eq. (10.50). For example,

$$p(\underset{\sim}{F}_1) = (1)(0.2) + (0.5)(0.3) + (0)(0.3) + (0)(0.1) + (0)(0.1)$$

$$= 0.35$$

TABLE 10.16
Computed quantities for perfect information and crisp states

	x_1	x_2	x_3	x_4	x_5
$p(x_k)$	0.20	0.30	0.30	0.10	0.10
$p(s_1 \mid x_k)$	1.0	0.0	0.0	0.0	0.0
$p(s_2 \mid x_k)$	0.0	1.0	0.0	0.0	0.0
$p(s_3 \mid x_k)$	0.0	0.0	1.0	0.0	0.0
$p(s_4 \mid x_k)$	0.0	0.0	0.0	1.0	0.0
$p(s_5 \mid x_k)$	0.0	0.0	0.0	0.0	1.0
$E(u^* \mid x_k)$	10.0	8.0	9.0	8.0	10.0
$a_j \mid a_k$	1	1	2	3	3

Similarly, $p(\underset{\sim}{F}_2) = 0.5$, and $p(\underset{\sim}{F}_3) = 0.15$. Therefore, the expected utility is given by Eq. (10.49) as

$$E(u_j) = \begin{bmatrix} 5 \\ 6.8 \\ 5.35 \end{bmatrix}$$

The optimum expected utility of the fuzzy alternatives (actions) for the case of no information using Eq. (10.51) is

$$E(u^*) = 6.8$$

so alternative $\underset{\sim}{A}_2$ is the optimum choice.

(ii) **Utility and optimum decision given uncertain and perfect information**.

(a) *Probabilistic (uncertain) information.* Table 10.17 lists the posterior probabilities as determined by Eq. (10.52a). For example,

$$p(\underset{\sim}{F}_1 \mid x_1) = \frac{(1)(0.44)(0.2) + (0.5)(0.26)(0.3)}{0.205} = 0.620$$

The other values are calculated in a similar manner and are shown in Table 10.17. The expected utility values for each of the x_k are now calculated using Eq. (10.53a), and these values are given in Table 10.18. The optimum expected utilities for each

TABLE 10.17
Posterior probabilities for probabilistic information with fuzzy states

	$\underset{\sim}{F}_1$	$\underset{\sim}{F}_2$	$\underset{\sim}{F}_3$
x_1	0.620	0.373	0.007
x_2	0.468	0.49	0.042
x_3	0.298	0.58	0.122
x_4	0.15	0.571	0.279
x_5	0.039	0.465	0.496

TABLE 10.18
Expected utilities for fuzzy alternatives
with probabilistic information

	A_1	A_2	A_3
x_1	7.315	5.880	3.305
x_2	6.153	6.534	4.315
x_3	4.718	7.143	5.58
x_4	3.216	7.413	6.934
x_5	1.787	7.317	8.252

alternative are found by using Eq. (10.54a),

$$E(u^*_{x_k}) = \max_j E(u_j|x_k) = \{7.315, 6.534, 7.143, 7.413, 8.252\}$$

where the optimum choice associated with this value is obviously alternative A_3.
Finally, the expected utility, given by Eq. (10.55), is calculated to be

$$E(u^*_\Phi) = \sum_{k=1}^r E(u^*_{x_k})p(x_k)$$

$$= (7.315)(0.205) + (6.534)(0.252) + (7.143)(0.245)$$

$$+ (7.413)(0.183) + (8.252)(0.115) = 7.202$$

The value of the probabilistic uncertain information for fuzzy states is

$$V(x) = 7.202 - 6.8 = 0.402$$

(b) *Probabilistic perfect information.* Table 10.19 lists the posterior probabilities as
determined by Eq. (10.52a). For example,

$$p(F_1|x_1) = [(1)(1)(0.2) + (0.5)(0)(0.3)]/(0.2) = 1.0$$

The other values are calculated in a similar manner and are shown in Table 10.19.

TABLE 10.19
Posterior probabilities for probabilistic *perfect* information with
fuzzy states

	F_1	F_2	F_3
x_1	1.0	0.0	0.0
x_2	0.5	0.5	0.0
x_3	0.0	1.0	0.0
x_4	0.0	0.5	0.5
x_5	0.0	0.0	1.0

The expected utility values for each of the x_k are now calculated using Eq. (10.53a), and these values are given in Table 10.20. The optimum expected utilities for each alternative are found by using Eq. (10.54a),

$$E(u^*_{x_k}) = \max_j E(u_j | x_k) = \{10.0, 6.5, 9.0, 8.5, 10.0\}$$

where it is not clear which alternative is the optimum choice (i.e., there is a tie between alternatives 1 and 3). Finally, the expected utility, given by Eq. (10.37), is calculated to be

$$E(u^*_{x_p}) = \sum_{k=1}^{r} E(u^*_{x_p} \mid x_k)p(x_k)$$

$$= (10.0)(0.2) + (6.5)(0.3) + (9.0)(0.3) + (8.5)(0.1) + (10.0)(0.1)$$

$$= 8.5$$

The value of the probabilistic *perfect* information for fuzzy states is

$$V(x_p) = 8.5 - 6.8 = 1.7$$

(c) *Fuzzy information.* For the hypothetical fuzzy information, Table 10.21 summarizes the results of the calculations using Eq. (10.52b). An example calculation is shown here:

$$p(\underset{\sim}{F}_1 | \underset{\sim}{M}_1) = [(1)(1)(0.44)(0.2) + (1)(0.4)(0.35)(0.2) + (0.5)(1)(0.26)(0.3)$$

$$+ (0.5)(0.4)(0.32)(0.3)] \div [(1)(0.205) + (0.4)(0.252)] = 0.57$$

Similarly, Table 10.22 summarizes the calculations of the expected utilities using Eq. (10.53b).
Now, using Eq. (10.54b), we find the optimum expected utility for each of the fuzzy states is

$$E(u^*_{\underset{\sim}{M}_t}) = \max_j E(u_j | \underset{\sim}{M}_t) = \{6.932, 7.019, 7.740\}$$

TABLE 10.20
Expected utilities for fuzzy alternatives with probabilistic *perfect* information

	$\underset{\sim}{A}_1$	$\underset{\sim}{A}_2$	$\underset{\sim}{A}_3$
x_1	10.0	4.0	1.0
x_2	6.5	6.5	4.0
x_3	3.0	9.0	7.0
x_4	1.5	7.5	8.5
x_5	0.0	6.0	10.0

TABLE 10.21
Posterior probabilities for fuzzy information with fuzzy states

	$\underset{\sim}{M_1}$	$\underset{\sim}{M_2}$	$\underset{\sim}{M_3}$
$\underset{\sim}{F_1}$	0.570	0.317	0.082
$\underset{\sim}{F_2}$	0.412	0.551	0.506
$\underset{\sim}{F_3}$	0.019	0.132	0.411

TABLE 10.22
Posterior probabilities for fuzzy alternatives with fuzzy information

	$\underset{\sim}{M_1}$	$\underset{\sim}{M_2}$	$\underset{\sim}{M_3}$
$\underset{\sim}{A_1}$	6.932	4.821	2.343
$\underset{\sim}{A_2}$	6.096	7.019	7.354
$\underset{\sim}{A_3}$	3.638	5.496	7.740

where the optimum choice is again $\underset{\sim}{A_3}$. The marginal probabilities of the fuzzy
information sets are calculated using Eq. (10.39); for example, using the marginal
probabilities from Table 10.15 and the fuzzy information from Table 10.11, we find

$$p(\underset{\sim}{M_1}) = (1.0)(0.205) + (0.4)(0.252) = 0.306$$

and, along with the other two marginal probabilities, we get

$$p(\underset{\sim}{M_t}) = \begin{bmatrix} 0.306 \\ 0.506 \\ 0.188 \end{bmatrix}$$

The unconditional expected utility using Eq. (10.55b) is

$$E(u_\Phi^*) = \sum_{t=1}^{g} E(u_{\underset{\sim}{M_t}}^*) p(\underset{\sim}{M_t}) = 7.128$$

and the value of the perfect information for fuzzy states is $V(\Phi) = 7.128 - 6.8 =$
0.328.

TABLE 10.23
Expected utilities for fuzzy alternatives with fuzzy perfect information

	$\underset{\sim}{F_1}$	$\underset{\sim}{F_2}$	$\underset{\sim}{F_3}$
$\underset{\sim}{A_1}$	10.0	3.0	0.0
$\underset{\sim}{A_2}$	4.0	9.0	6.0
$\underset{\sim}{A_3}$	1.0	7.0	10.0

(d) *Fuzzy perfect information.* Table 10.23 summarizes the calculations of the expected utilities using Eq. (10.57). Note in Table 10.23 that the expected utilities are the same as the utilities in Table 10.9; this identity arises because the information is presumed perfect, and the conditional probability matrix, $p(\underset{\sim}{F}_s | \underset{\sim}{M}_t)$, in Eq. (10.53$b$) is the identity matrix.

Now, using Eq. (10.58), we find the optimum expected utility for each of the fuzzy states is

$$u(A^*_{\underset{\sim}{F}_s} | \underset{\sim}{F}_s) = \max_i u(\underset{\sim}{A}_i, \underset{\sim}{F}_s) = \{10.0, 9.0, 10.0\}$$

Finally, using the previously determined prior probabilities of the fuzzy states, $p(\underset{\sim}{F}_s)$ (see section (i)), we see the unconditional expected utility using Eq. (10.59) is

$$E(u^*_{\Phi_p}) = \sum_{j=1}^{3} u(A^*_{\underset{\sim}{F}_s} | \underset{\sim}{F}_s) p(\underset{\sim}{F}_s) = 10(0.35) + 9(0.5) + 10(0.15) = 9.5$$

where the value of the fuzzy perfect information for fuzzy states is

$$V(\Phi_p) = 9.5 - 6.8 = 2.7$$

Example summary

A typical decision problem is to decide on a basic policy in a fuzzy environment. This basic policy can be thought of as a fuzzy action. The attributes of such a problem are that there are many states, feasible policy alternatives, and available information. Usually, the utilities for all the states and all the alternatives cannot be formulated because of insufficient data, because of the high cost of obtaining this information, and because of time constraints. On the other hand, a decision maker in top management is generally not concerned with the detail of each element in the decision problem. Mostly, top managers want to decide roughly what alternatives to select as indicators of policy directions. Hence, an approach that can be based on fuzzy states and fuzzy alternatives and that can accommodate fuzzy information is a very powerful tool for making preliminary policy decisions.

The expected utilities and the value of information for the five cases, i.e., for no information, probabilistic (uncertain) information, probabilistic perfect information, fuzzy probabilistic (uncertain) information, and fuzzy perfect information, are summarized in Table 10.24. We can see from this table that the ordering of values of information is in accordance with that described in Eq. (10.61), i.e., $V(\Phi_p) \geq V(x_p) \geq V(x) \geq V(\Phi) \geq 0$. The probabilistic perfect information $(V(x_p) = 1.70)$ has a value much higher than the probabilistic information $(V(x) = 0.40)$. The decision maker needs to ascertain the cost of the hypothetical perfect information when compared with the cost of the uncertain information, the latter being more realistic. On the other hand, there is little difference between the value of fuzzy probabilistic information $(V(\Phi) = 0.33)$ and that of probabilistic information $(V(x) = 0.40)$. This result suggests that the fuzzy probabilistic information is sufficiently valuable compared with the probabilistic information for this problem, because fuzzy information generally costs far less than probabilistic (uncertain) information. Finally, the fact that the fuzzy perfect information $(V(\Phi_p) = 2.70)$ holds more value than the probabilistic perfect information $(V(x_p) = 1.70)$ confirms that our interest is more in the fuzzy states than the crisp states. When utility values, prior probabilities, conditional probabilities, and orthogonal membership values change for any of these scenarios, the elements in Table 10.24 will change and the conclusions derived from them will portray a different situation. The power of this approach is its ability to measure on an ordinal basis the value of the approximate information used in a decision-making problem. When the value of approximate (fuzzy) information approaches that of either probabilistic or perfect information, there is the potential for significant cost savings without reducing the quality of the decision itself.

TABLE 10.24
Summary of expected utility and value of information for fuzzy states and actions for the example

Information	Expected utility	Value of information
No information	6.8	–
Probabilistic information, $V(x)$	7.20	0.40
Perfect information, $V(x_p)$	8.5	1.7
Fuzzy probabilistic information, $V(\Phi)$	7.13	0.33
Fuzzy perfect information, $V(\Phi_p)$	9.5	2.7

SUMMARY

The literature is rich with references in the area of fuzzy decision making. This chapter has only presented a few rudimentary ideas in the hope of interesting the readers to continue their learning in this important area. One of the decision metrics in this chapter represents a philosophical approach where an existing crisp theory – Bayesian decision making – is reinterpreted to accept both fuzzy and random uncertainty. It is important to note that there have been significant challenges to the maximum expected utility theory on which Bayesian decision making is founded. Three violations of the independence axiom of this theory (the Allias paradox, the Bergen paradox, and sensitivity to tail affects) and one difficulty in representing epistemic uncertainty as a probabilistic belief (the Ellsburg paradox) have been reported in the literature [Maes and Faber, 2004]. One key problem in Bayesian decision making is that the updating (updating the priors to become posteriors) is not always applied correctly. Psychometric studies have shown [Tversky and Kahneman, 1974] that too little weight is given to prior information and too much importance is given to new data (likelihood function). Recent information tends to take precedence over long-accumulated prior knowledge.

Theoretical developments are expanding the field of fuzzy decision making; for example multiobjective situations represent an interesting class of problems that plague optimization in decision making [Sakawa, 1993] as do multiattribute decision problems [Baas and Kwakernaak, 1977]. This philosophical approach has been extended further where fuzzy utilities have been addressed with fuzzy states [Jain, 1976], and where fuzzy utilities are determined in the presence of probabilistic states [Jain, 1978; Watson et al., 1979]. Häage [1978] extended the Bayesian scheme to include possibility distributions (see Chapter 15 for definition of possibility) for the consequences of the decision actions. The other metrics in this chapter extend some specific problems to deal with issues like fuzzy preference relations, fuzzy objective functions, fuzzy ordering, fuzzy consensus, etc. In all of these, there is a compelling need to incorporate fuzziness in human decision making, as originally proposed by Bellman and Zadeh [1970]. In most decision situations the goals, constraints, and consequences of the proposed alternatives are not known with precision. Much of this imprecision is not measurable, and not random. The imprecision can be due to vague, ambiguous, or fuzzy information. Methods to address this form of imprecision are necessary to deal with many of the uncertainties we deal with in humanistic systems.

REFERENCES

Adams, T. (1994). ''Retaining structure selection with unequal fuzzy project-level objectives,'' *J. Intell. Fuzzy Syst.*, vol. 2, no. 3, pp. 251–266.

Baas, S. and Kwakernaak, H. (1977). ''Rating and ranking of multiple-aspect alternatives using fuzzy sets,'' *Automatica*, vol. 13, pp. 47–58.

Bellman, R. and Zadeh, L. (1970). ''Decision making in a fuzzy environment,'' *Manage. Sci.*, vol. 17, pp. 141–164.

Bezdek, J., Spillman, B. and Spillman, R. (1978). ''A fuzzy relation space for group decision theory,'' *Fuzzy Sets Syst.*, vol. 1, pp. 255–268.

Dubois, D. and Prade, H. (1980). *Fuzzy sets and systems: Theory and applications*, Academic Press, New York.

Häage, C. (1978). ''Possibility and cost in decision analysis,'' *Fuzzy Sets Syst.*, vol. 1, no. 2, pp. 81–86.

Jain, R. (1976). ''Decision-making in the presence of fuzzy variables,'' *IEEE Trans. Syst., Man, Cybern.*, vol. 6, no. 10, pp. 698–703.

Jain, R. (1978). ''Decision-making in the presence of fuzziness and uncertainty,'' *Proc. IEEE Conf. Decision Control, New Orleans*, pp. 1318–1323.

Klir, G. and Yuan, B. (1995). *Fuzzy sets and fuzzy logic*, Prentice Hall, Upper Saddle River, NJ.

Maes, M. and Faber, M. (2004). ''Issues in utility modeling and rational decision making,'' *Proc. 11th IFIP WG 7.5 Reliability and Optimization of Structural Systems*, M. Maes and L. Huyse (eds.) Balkema Publishers, London, pp. 95–104.

Okuda, T., Tanaka, H. and Asai, K. (1974). ''Decision making and information in fuzzy events,'' *Bull. Univ. Osaka Prefect., Ser. A*, vol. 23, no. 2, pp. 193–202.

Okuda, T., Tanaka, H., and Asai, K. (1978). ''A formulation of fuzzy decision problems with fuzzy information, using probability measures of fuzzy events,'' *Inf. Control*, vol. 38, no. 2, pp. 135–147.

Ross, T., Booker, J., and Parkinson, W. (2003). *Fuzzy Logic and Probability Applications: Bridging the Gap*, Society for Industrial and Applied Mathematics, Philadelphia, PA.

Sakawa, M. (1993). *Fuzzy sets and interactive multiobjective optimization*, Plenum Press, New York.

Shimura, M. (1973). ''Fuzzy sets concept in rank-ordering objects,'' *J. Math. Anal. Appl.*, vol. 43, pp. 717–733.

Tanaka, H., Okuda, T., and Asai, K. (1976). ''A formulation of fuzzy decision problems and its application to an investment problem,'' *Kybernetes*, vol. 5, pp. 25–30.

Terano, T., Asai, K., and Sugeno, M. (1992). *Fuzzy system theory and its applications*, Academic Press, San Diego, CA.

Tversky, A and Kahneman, D. (1974). ''Judgement under uncertainty: Heuristics and biases,'' *Science*, vol. 185, pp. 1124–1131.

Von Neumann, J. and Morgenstern, O. (1944). *Theory of Games and Economical Behavior*, Princeton University Press, Princeton, NJ.

Watson, S., Weiss, J., and Donnell, M. (1979). ''Fuzzy decision analysis,'' *IEEE Trans. Syst., Man, Cybern.*, vol. 9, no. 1, pp. 1–9.

Yager, R. (1981). ''A new methodology for ordinal multiobjective decisions based on fuzzy sets,'' *Decis. Sci.*, vol. 12, pp. 589–600.

PROBLEMS

Ordering and Synthetic Evaluation

10.1. For Example 10.2 change the first fuzzy set $\underset{\sim}{I}_1$ to $\left\{ \frac{1}{3} + \frac{0.7}{5} + \frac{0.4}{9} \right\}$ and recalculate the same quantities as those in Example 10.2.

10.2. Company Z makes chemical additives that are ultimately used for engine oil lubricants. Components such as surfactants, detergents, rust inhibitors, etc., go into the finished engine oil before it is sold to the public. Suppose that company Z makes a product D739.2 that is a detergent additive. You are asked to determine if a particular batch of D739.2 is good enough to be sold to an oil company, which will then make the final product. The detergent is evaluated on the following parameters: actual color of the material, consistency, base number (BN, measure of detergent capacity), and flash point (FP, ignition temperature of the material). After making several hundred batches of the detergent additive D739.2, the following relation matrix is obtained:

$$
\underset{\sim}{R} = \begin{array}{c} \text{color} \\ \text{consistency} \\ \text{BN} \\ \text{FP} \end{array}
\begin{array}{ccc}
\text{excellent} & \begin{array}{c}\text{very}\\\text{good}\end{array} & \text{fair} \\
\left[\begin{array}{ccc}
0.3 & 0.4 & 0.3 \\
0.1 & 0.5 & 0.4 \\
0.5 & 0.4 & 0.1 \\
0.4 & 0.3 & 0.3
\end{array}\right]
\end{array}
$$

The weight factor for the detergent is $\underset{\sim}{a} = \{0.1, 0.25, 0.4, 0.25\}$. Evaluate the quality of the detergent.

10.3. In making a decision to purchase an aircraft, airline management will consider the qualities of the plane's performance with respect to the competition. The Boeing 737 is the best-selling plane in aviation history and continues to outsell its more modern competitor, the A320, manufactured by the Airbus consortium. The four factors to be considered are these: range, payload, operating costs, and reliability. The criteria will be a comparison of the 737 with respect to the A320: superior (sup.), equivalent (eq.), and deficient (def.).

$$
\underset{\sim}{R} = \begin{array}{c} \text{range} \\ \text{payload} \\ \text{cost} \\ \text{reliability} \end{array}
\begin{array}{ccc}
\text{sup.} & \text{eq.} & \text{def.} \\
\left[\begin{array}{ccc}
0 & 0.7 & 0.3 \\
0.1 & 0.8 & 0.1 \\
0.1 & 0.5 & 0.4 \\
0.7 & 0.2 & 0.1
\end{array}\right]
\end{array}
$$

Given a typical airline's weighting factor of the four factors as $\underset{\sim}{a} = \{0.15, 0.15, 0.3, 0.4\}$, evaluate the performance of the 737 with respect to the A320.

10.4. A power supply needs to be chosen to go along with an embedded system. Four categories of evaluation criteria are important. The first is the physical size of the power supply. The second is the efficiency of the power supply. The third is the "ripple" voltage of the output of the power supply. This is a measure of how clean the power provided is. The fourth criterion is the peak current provided by the power supply. The following matrix defines the type of power supply required for the embedded system application:

$$
\underset{\sim}{R} = \begin{array}{c} \text{physical size} \\ \text{efficiency} \\ \text{ripple voltage} \\ \text{peak current} \end{array}
\begin{array}{ccccc}
\text{VG} & \text{G} & \text{F} & \text{B} & \text{VB} \\
\left[\begin{array}{ccccc}
0.2 & 0.7 & 0.1 & 0 & 0 \\
0.1 & 0.2 & 0.4 & 0.2 & 0.1 \\
0.3 & 0.4 & 0.2 & 0.1 & 0 \\
0 & 0.2 & 0.4 & 0.3 & 0.1
\end{array}\right]
\end{array}
$$

From this matrix, one can see that for the embedded system in mind, the power supply's physical size is very important as well as its ripple voltage. Of lesser importance is its efficiency, and lesser yet, is its peak current. So a small power supply with clean output voltage is needed. It needs to be somewhat efficient and is not required to provide very much "inrush current" or peak current. Evaluate a power supply with the following characteristics:

$$\underset{\sim}{\text{Power supply}} = \begin{bmatrix} 0.5 \\ 0.1 \\ 0.2 \\ 0.2 \end{bmatrix} \begin{matrix} \text{physical size} \\ \text{efficiency} \\ \text{ripple voltage} \\ \text{peak current} \end{matrix}$$

Nontransitive Ranking

10.5. An aircraft control system is a totally *nonlinear system* when the final approach and landing of an aircraft are considered. It involves maneuvering flight in an appropriate course to the airport and then along the optimum glide path trajectory to the runway. We know that this path is usually provided by an instrument landing system, which transmits two radio signals to the aircraft as a navigational aid. These orthogonal radio beams are known as the localizer and the glide slope and are transmitted from the ends of the runway in order to provide the approaching aircraft with the correct trajectory for landing. The pilot executing such a landing must monitor cockpit instruments that display the position of the aircraft relative to the desired flight path and make appropriate corrections to the controls. Presume that four positions are available to the pilot and that four corrections P_1, P_2, P_3, and P_4 from the actual position P are required to put the aircraft on the correct course. The pairwise comparisons for the four positions are as follows:

$$\begin{matrix} f_{P_1}(P_1) = 1 & f_{P_1}(P_2) = 0.5 & f_{P_1}(P_3) = 0.6 & f_{P_1}(P_4) = 0.8 \\ f_{P_2}(P_1) = 0.3 & f_{P_2}(P_2) = 1 & f_{P_2}(P_3) = 0.4 & f_{P_2}(P_4) = 0.3 \\ f_{P_3}(P_1) = 0.6 & f_{P_3}(P_2) = 0.4 & f_{P_3}(P_3) = 1 & f_{P_3}(P_4) = 0.6 \\ f_{P_4}(P_1) = 0 & f_{P_4}(P_2) = 0.3 & f_{P_4}(P_3) = 0.6 & f_{P_4}(P_4) = 1 \end{matrix}$$

Now, from these values, compute the comparison matrix, and determine the overall ranking.

10.6. When designing a radar system for imaging purposes, we frequently need to set priorities in accomplishing certain features. Some features that need to be traded off against each other are these:

1. The ability to penetrate foliage and even the ground to some depth.
2. The resolution of the resulting radar image.
3. The size of the antenna required for the radar system.
4. The amount of power required to operate at a given frequency.

It is useful to determine the order of importance of these features in selecting an operating frequency for the radar. Let x_1 represent penetration; x_2, resolution; x_3, antenna size; and x_4, power. A crisp ordering will have trouble resolving the importance of penetration compared to resolution, resolution compared to antenna size, and antenna size compared to penetration. These are entities that can only be compared in a very subjective manner, ideal for fuzzy techniques and difficult for crisp techniques.

Let $f_{x_i}(x_j)$ be the relative importance of feature x_j with respect to x_i. The comparisons $f_{x_i}(x_j)$ are subjectively assigned as follows:

		x_j			
		x_1	x_2	x_3	x_4
	x_1	1	0.6	0.5	0.9
	x_2	0.5	1	0.7	0.8
x_i					
	x_3	0.9	0.8	1	0.5
	x_4	0.3	0.2	0.3	1

Develop a comparison matrix and determine the overall ranking of the importance of each feature.

10.7. In tracking soil particles, a tracked particle can be occluded by other objects. To find out which one is the tracked particle, one can choose or pick a particle that is a certain distance from the tracked particle. Suppose there are four particles in the region of interest where the tracked particle is. Furthermore, let x_1, x_2, x_3, and x_4 resemble the tracked particle with fuzzy measurement 0.3, 0.4, 0.6, 0.7, respectively, when they alone are considered. Note $f_{x_j}(x_i)$ means how close x_i is to the tracked particle with respect to x_j.

$$
\begin{aligned}
&f_{x_1}(x_1) = 1 \quad &f_{x_1}(x_2) = 0.6 \quad &f_{x_1}(x_3) = 0.4 \quad &f_{x_1}(x_4) = 0.3 \\
&f_{x_2}(x_1) = 0.7 \quad &f_{x_2}(x_2) = 1 \quad &f_{x_2}(x_3) = 0.1 \quad &f_{x_2}(x_4) = 0.4 \\
&f_{x_3}(x_1) = 0.2 \quad &f_{x_3}(x_2) = 0.4 \quad &f_{x_3}(x_3) = 1 \quad &f_{x_3}(x_4) = 0.3 \\
&f_{x_4}(x_1) = 0.5 \quad &f_{x_4}(x_2) = 0.3 \quad &f_{x_4}(x_3) = 0.4 \quad &f_{x_4}(x_4) = 1
\end{aligned}
$$

Develop a comparison matrix, and determine which particle is closest to the tracked particle.

10.8. Suppose a wine manufacturer was interested in introducing a new wine to the market. A very good but somewhat expensive Chenin Blanc was already available to consumers and was very profitable to the company. To enter the lower-priced market of wine consumers, the wine manufacturer decided to make a less expensive wine that *tasted* similar to the very profitable Chenin Blanc already sold. After much market research and production knowledge, the manufacturer settled on four possible wines to introduce into the market. The fuzzy criteria of evaluation is *taste*, and we would like to know which wine tastes the most like the expensive Chenin Blanc. Define the following subjective estimations: universe $X = \{x_1, x_2, x_3, x_4\}$. A panel of wine tasters tasted each of the wines x_1, x_2, x_3, and x_4 and made the following estimations:

$$
\begin{aligned}
&f_{x_1}(x_1) = 1 \quad &f_{x_1}(x_2) = 0.4 \quad &f_{x_1}(x_3) = 0.8 \quad &f_{x_1}(x_4) = 0.5 \\
&f_{x_2}(x_1) = 0.2 \quad &f_{x_2}(x_2) = 1 \quad &f_{x_2}(x_3) = 0.7 \quad &f_{x_2}(x_4) = 0.4 \\
&f_{x_3}(x_1) = 0.3 \quad &f_{x_3}(x_2) = 0.2 \quad &f_{x_3}(x_3) = 1 \quad &f_{x_3}(x_4) = 0.5 \\
&f_{x_4}(x_1) = 0.7 \quad &f_{x_4}(x_2) = 0.5 \quad &f_{x_4}(x_3) = 0.8 \quad &f_{x_4}(x_4) = 1
\end{aligned}
$$

Develop a comparison matrix, and determine which of the four wines tastes most like the expensive Chenin Blanc.

Fuzzy Preference and Consensus

10.9. The Environmental Protection Agency (EPA) is faced with the challenge of cleaning up contaminated groundwaters at many sites around the country. In order to ensure an efficient cleanup process, it is crucial to select a firm that offers the best remediation technology at a reasonable cost. The EPA is deciding among four environmental firms. The professional engineers at the EPA compared the four firms and created a consensus matrix, shown here:

$$
\underset{\sim}{R} =
\begin{bmatrix}
0 & 0.5 & 0.7 & 0.4 \\
0.5 & 0 & 0.9 & 0.2 \\
0.3 & 0.1 & 0 & 0.3 \\
0.6 & 0.8 & 0.7 & 0
\end{bmatrix}
$$

Compute the distance to Type *fuzzy* consensus.

10.10. A manufacturing company is planning to purchase a lathe and is assessing the proposals from four lathe manufacturers. The company has developed a reciprocal relation for the four manufacturers based on the speed of delivery of the lathes and the cost. The relation is

$$\underset{\sim}{R} = \begin{bmatrix} 0 & 0.1 & 0.7 & 0.2 \\ 0.9 & 0 & 0.6 & 1 \\ 0.3 & 0.4 & 0 & 0.5 \\ 0.8 & 0 & 0.5 & 0 \end{bmatrix}$$

Calculate the degree of preference measures, and the distance to Type I, Type II, and Type *fuzzy* consensus. Explain the differences between the distances to the three consensuses.

10.11. Four methods for determining control are being considered for a navigation project. These methods include: Global Positioning System (GPS), Inertial Navigational System (INS), Surveying (S), and Astronomical Observation (ASTR). For the development of a network of control points, which will improve the accuracies of various applications that can use these points as a control point for their own networks, experts were asked to determine a reciprocal relation for these four methods. Generally, GPS is preferred to INS because of its superior long-term stability of measurements and it is not necessary to travel from one network point to another. GPS is somewhat less preferred than S because it cannot match the accuracy of S, which is important for a control network. GPS is much preferred over ASTR because it requires much less time and the skill requirement for ASTR is significant. INS is generally about the same as S; INS is preferred over ASTR for the same reasons as GPS. While S is somewhat less preferred than ASTR because both are slow and labor intensive, ASTR can generally offer better results and the equipment is not as expensive.

Using the following reciprocal relation determine the average fuzziness, average certainty, and the distance to consensus for Type I, II, and *fuzzy* consensus.

$$\underset{\sim}{R} = \begin{array}{c} \\ \text{GPS} \\ \text{INS} \\ \text{S} \\ \text{AST} \end{array} \begin{array}{c} \text{GPS} \quad \text{INS} \quad \text{S} \quad \text{AST} \\ \begin{bmatrix} 0 & 0.6 & 0.4 & 0.8 \\ 0.4 & 0 & 0.5 & 0.7 \\ 0.6 & 0.5 & 0 & 0.4 \\ 0.2 & 0.3 & 0.6 & 0 \end{bmatrix} \end{array}$$

10.12. A chemical plant reactor has yields lower than expected because the reactor is getting old. There are four feasible alternatives to solve the problem:

$A_1 = $ Buy a new reactor and replace the old one.

$A_2 = $ Buy a used reactor and replace the old one.

$A_3 = $ Add a new smaller unit at the end of the reactor to complete the reaction

to the expected yields.

$A_4 = $ Do major repair to the old reactor.

Each alternative has its own advantages and disadvantages according to cost, maintainability, and physical space available in the plant. The engineers involved in selecting one of the options have created a relation to show their consensus:

$$\underset{\sim}{R} = \begin{array}{c} \\ A_1 \\ A_2 \\ A_3 \\ A_4 \end{array} \begin{array}{c} A_1 \quad A_2 \quad A_3 \quad A_4 \\ \begin{bmatrix} 0 & 0.8 & 0.5 & 0.3 \\ 0.2 & 0 & 0.7 & 0.4 \\ 0.5 & 0.3 & 0 & 0.1 \\ 0.7 & 0.6 & 0.9 & 0 \end{bmatrix} \end{array}$$

Find the average fuzziness, average certainty, distance to consensus, and distances to consensus for a Type I, Type II, and Type fuzzy.

10.13. An automotive manufacturing company is buying vision sensors for its assembly verification to its client. The engineering team in the company has done some pairwise comparisons among four types of sensors (S); the results of these comparisons are given by the consensus relation matrix below:

$$\underset{\sim}{R} = \begin{array}{c} \\ S_1 \\ S_2 \\ S_3 \\ S_4 \end{array} \begin{array}{cccc} S_1 & S_2 & S_3 & S_4 \\ \left[\begin{array}{cccc} 0 & 0.6 & 0.4 & 0.8 \\ 0.4 & 0 & 0.5 & 0.7 \\ 0.6 & 0.5 & 0 & 0.4 \\ 0.2 & 0.3 & 0.6 & 0 \end{array}\right] \end{array}$$

Calculate the degree of preference measures, and the distance to Type I, Type II, and Type fuzzy consensus.

Multiobjective Decision Making

10.14. A carcinogen, trichloroethylene (TCE), has been detected in soil and groundwater at levels higher than the EPA maximum contaminant levels (MCLs). There is an immediate need to remediate soil and groundwater. Three remediation alternatives – (1) pump and treat with air stripping (PTA), (2) pump and treat with photooxidation (PTP), and (3) bioremediation of soil with pump and treat and air stripping (BPTA) – are investigated.

The objectives are these: cost (O_1), effectiveness (O_2, capacity to reduce the contaminant concentration), duration (O_3), and speed of implementation (O_4). The ranking of the alternatives on each objective are given as follows:

$$\underset{\sim}{O_1} = \left\{ \frac{0.7}{\text{PTA}} + \frac{0.9}{\text{PTP}} + \frac{0.3}{\text{BPTA}} \right\}$$

$$\underset{\sim}{O_2} = \left\{ \frac{0.4}{\text{PTA}} + \frac{0.6}{\text{PTP}} + \frac{0.8}{\text{BPTA}} \right\}$$

$$\underset{\sim}{O_3} = \left\{ \frac{0.7}{\text{PTA}} + \frac{0.3}{\text{PTP}} + \frac{0.6}{\text{BPTA}} \right\}$$

$$\underset{\sim}{O_4} = \left\{ \frac{0.8}{\text{PTA}} + \frac{0.5}{\text{PTP}} + \frac{0.5}{\text{BPTA}} \right\}$$

The preferences for each objective are P = {0.6, 0.8, 0.7, 0.5}. Determine the optimum choice of a remediation alternative.

10.15. Evaluate three different approaches to controlling conditions of an aluminum smelting cell (with respect to voltage across the cell and alumina concentration in the bath). The control approaches are

$$a_1 = \text{AGT: aggressive control tuning (very reactive)}$$

$$a_2 = \text{MOD: moderate control tuning (mildly reactive)}$$

$$a_3 = \text{MAN: essentially manual operation (very little computer control)}$$

There are several objectives to consider:

$$\underset{\sim}{O_1} : \text{minimum power consumption (power/lb of aluminum produced)}$$

$$\underset{\sim}{O_2} : \text{overall operating stability}$$

$$\underset{\sim}{O_3} : \text{minimum environmental impact}$$

The control approaches are rated as follows:

$$Q_1 = \left\{ \frac{0.7}{\text{AGT}} + \frac{0.6}{\text{MOD}} + \frac{0.3}{\text{MAN}} \right\}$$

$$Q_2 = \left\{ \frac{0.45}{\text{AGT}} + \frac{0.8}{\text{MOD}} + \frac{0.6}{\text{MAN}} \right\}$$

$$Q_3 = \left\{ \frac{0.5}{\text{AGT}} + \frac{0.62}{\text{MOD}} + \frac{0.4}{\text{MAN}} \right\}$$

The preferences are given by $b_1 = 0.8$, $b_2 = 0.5$, and $b_3 = 0.6$. What is the best choice of control?

10.16. In the tertiary treatment process for wastewater, the disinfection process is an important procedure that focuses on the destruction of disease-causing organisms. There are a lot of disinfection technologies available; of these three popular methods for disinfecting are to use: chlorine (Cl), ozone (Oz), or UV radiation (UV). A new wastewater treatment plant is to be built and the designers are having difficulty selecting a disinfecting method and thus elect to use a multiobjective decision approach. It is concluded that the selection of a disinfection method should be based on: efficiency and performance (EP), availability of large quantities of the disinfectants and reasonable prices (Av), maintenance and operation (MO), and environmental impact (Ev). The sets of alternatives (A), objectives (O), and preferences (P) are shown below. Using the ratings given for each objective and the preference specified by the facility owner, make a decision on which disinfection technology to use.

$$A = \{\text{Cl, Oz, UV}\} = \{a_1, a_2, a_3\}$$

$$O = \{\text{EP, Av, MO, Ev}\} = \{Q_1, Q_2, Q_3, Q_4\}$$

$$P = \{b_1, b_2, b_3, b_4\} = \{0.8, 0.9, 0.6, 0.5\}$$

Objectives:

$$Q_1 = \left\{ \frac{0.8}{a_1}, \frac{0.9}{a_2}, \frac{0.7}{a_3} \right\}, \quad Q_2 = \left\{ \frac{0.9}{a_1}, \frac{0.4}{a_2}, \frac{0.5}{a_3} \right\}, \quad Q_3 = \left\{ \frac{0.8}{a_1}, \frac{0.7}{a_2}, \frac{0.7}{a_3} \right\},$$

$$Q_4 = \left\{ \frac{0.5}{a_1}, \frac{0.8}{a_2}, \frac{0.9}{a_3} \right\}$$

10.17. For environmental modeling, remote sensing data play an important role in the data acquisition. Researchers must decide which type of sensor data best meet their preferences. Among the many alternative sensors available, the list of candidates has been reduced to three: LANTSAT 7 (LS7), GOES (GS), and TERRA (TA). The researchers have defined four objectives that impact their decision: (1) cost of the data (COST), (2) time to deliver data (TIME), (3) resolution of the data collected (RES), and (4) time for the sensor to return to the same spot cycle (CT). There was some disagreement as to how to define the importance of each objective in the preference set so the researchers decided to define two sets of preferences, P_1 and P_2.

$$\text{Alternatives:} \quad A = \{\text{LS7, GS, TA}\}$$

$$\text{Objectives:} \quad O = \{\text{COST, TIME, RES, CT}\}$$

$$\text{Preferences :} \quad P_1 = \{b_1, b_2, b_3, b_4\} = \{0.8, 0.4, 0.8, 0.7\}$$

$$P_2 = \{b_1, b_2, b_3, b_4\} = \{0.4, 0.6, 0.4, 0.5\}$$

The degree of membership of each alternative in the objectives is as follows:

$$\underset{\sim}{Q}_1 = \left\{ \frac{0.2}{LS7}, \frac{0.8}{GS}, \frac{0.4}{TA} \right\}, \quad \underset{\sim}{Q}_2 = \left\{ \frac{0.6}{LS7}, \frac{1}{GS}, \frac{0.2}{TA} \right\}, \quad \underset{\sim}{Q}_3 = \left\{ \frac{1}{LS7}, \frac{0.4}{GS}, \frac{0.8}{TA} \right\},$$

$$\underset{\sim}{Q}_4 = \left\{ \frac{0.8}{LS7}, \frac{0.7}{GS}, \frac{0.2}{TA} \right\}$$

Find the decision for each preference.

10.18. In the city of Calgary, Alberta, subdivisions constructed before 1970 were not required to retain overland storm-water flow on a site during major storm events to the level that has been accepted under current design criteria. In order to properly mitigate flooding and property damage in older subdivisions prone to flooding they are being upgraded based on technical feasibility and public acceptance of the works. Presently a subdivision is being considered for an upgrade of its storm-water sewer. It has been determined that there are two different methods to achieve the mitigation, either larger storm sewers have to be installed through the affected neighborhoods (pipe network) or storm-water retention facilities (pond) have to be built close enough to the neighborhood to reduce the flood threat. The mitigation alternatives (A) and the considered impacts or objectives (O) are described below.

$$\text{Alternatives:} \quad A = \{\text{pipe, pond}\}$$

Objectives: Additional land required (O_1), Cost (O_2), Flood damage (O_3), Public acceptance (O_4), and Environmental constraints (O_5):

$$O = \{\underset{\sim}{Q}_1, \underset{\sim}{Q}_2, \underset{\sim}{Q}_3, \underset{\sim}{Q}_4, \underset{\sim}{Q}_5\}$$

Based on previous experience with other subdivisions the city design engineer has determined the following ratings for this subdivision:

$$\underset{\sim}{Q}_1 = \left\{ \frac{0.8}{\text{pipe}}, \frac{0.6}{\text{pond}} \right\}, \quad \underset{\sim}{Q}_2 = \left\{ \frac{0.8}{\text{pipe}}, \frac{0.4}{\text{pond}} \right\}, \quad \underset{\sim}{Q}_3 = \left\{ \frac{0.6}{\text{pipe}}, \frac{0.8}{\text{pond}} \right\},$$

$$\underset{\sim}{Q}_4 = \left\{ \frac{0.4}{\text{pipe}}, \frac{0.9}{\text{pond}} \right\}, \quad \underset{\sim}{Q}_5 = \left\{ \frac{0.8}{\text{pipe}}, \frac{0.5}{\text{pond}} \right\}$$

The city council has given the administration the following preference values for each objective. Using the above objectives and preferences determine which system to use for this subdivision:

$$P = \{b_1, b_2, b_3, b_4, b_5\} = \{0.6, 0.4, 0.6, 0.7, 0.6\}$$

Bayesian Decision Making

10.19. A company produces printed circuit boards as a subcomponent for a system that is integrated (with other subcomponents) by another company. The system integration company cannot give precise information on how many PC boards it needs other than "approximately 10,000." It may require more or less than this number. The PC board manufacturer has three courses of action from which to choose: (1) build somewhat less than 10,000 PC boards, $\underset{\sim}{A}_1$; (2) build approximately 10,000 PC boards, $\underset{\sim}{A}_2$; and (3) build somewhat more than 10,000 PC boards, $\underset{\sim}{A}_3$.

The systems integration company will need the PC boards to meet the demand for its final product. The following are the three fuzzy states of nature:

1. Low demand, $\underset{\sim}{D}_1$

2. Medium demand, D_2
3. High demand, D_3

The utility function is given in this table:

	D_1	D_2	D_3
A_1	3	2	−1
A_2	−1	4	2
A_3	−5	2	5

There are six discrete states of nature, s_1-s_6, on which the fuzzy states are defined. The membership functions for the fuzzy states and the prior probabilities $p(s_i)$ of the discrete states are shown in the following table:

	s_1	s_2	s_3	s_4	s_5	s_6
μ_{D_1}	1.0	0.7	0.1	0.0	0.0	0.0
μ_{D_2}	0.0	0.3	0.9	0.9	0.3	0.0
μ_{D_3}	0.0	0	0.0	0.1	0.7	1.0
$p(s_i)$	0.2	0.1	0.4	0.1	0.1	0.1

The demand for the system integrator's product is related to the growth of refineries, as the final product is used in refineries. The new samples of refinery growth information are x; and M_i are the fuzzy sets on this information, defined as

1. Low growth, M_1
2. Medium growth, M_2
3. High growth, M_3

	x_1	x_2	x_3	x_4	x_5	x_6
μ_{M_1}	1.0	0.7	0.2	0.0	0.0	0.0
μ_{M_2}	0.0	0.3	0.8	0.8	0.3	0.0
μ_{M_3}	0.0	0.0	0.0	0.2	0.7	1.0

The likelihood values for the probabilistic uncertain information for the data samples are shown here:

	x_1	x_2	x_3	x_4	x_5	x_6
s_1	0.1	0.1	0.5	0.1	0.1	0.1
s_2	0.0	0.0	0.1	0.4	0.4	0.1
s_3	0.1	0.2	0.4	0.2	0.1	0.0
s_4	0.5	0.1	0.0	0.0	0.2	0.2
s_5	0.0	0.0	0.0	0.1	0.3	0.6
s_6	0.1	0.7	0.2	0.0	0.0	0.0

The likelihood values for the probabilistic perfect information for the data samples are shown next:

	x_1	x_2	x_3	x_4	x_5	x_6
s_1	0.0	0.0	1.0	0.0	0.0	0.0
s_2	0.0	0.0	0.0	1.0	0.0	0.0
s_3	0.0	0.0	0.0	0.0	1.0	0.0
s_4	1.0	0.0	0.0	0.0	0.0	0.0
s_5	0.0	0.0	0.0	0.0	0.0	1.0
s_6	0.0	1.0	0.0	0.0	0.0	0.0

For the information just presented, compare the following for perfect and imperfect information:

(a) Posterior probabilities of fuzzy state 2 ($\underset{\sim}{D}_2$) given the fuzzy information 3 ($\underset{\sim}{M}_3$).

(b) Conditional expected utility for action 1 ($\underset{\sim}{A}_1$) and fuzzy information 2 ($\underset{\sim}{M}_2$).

10.20. In a particular region a water authority must decide whether to build dikes to prevent flooding in case of excess rainfall. Three fuzzy courses of action may be considered:

1. Build a permanent dike ($\underset{\sim}{A}_1$).
2. Build a temporary dike ($\underset{\sim}{A}_2$).
3. Do not build a dike ($\underset{\sim}{A}_3$).

The sets $\underset{\sim}{A}_1$, $\underset{\sim}{A}_2$, and $\underset{\sim}{A}_3$ are fuzzy sets depending on the type and size of the dike to be built. The utility from each of these investments depends on the rainfall in the region. The crisp states of nature, $S = \{s_1, s_2, s_3, s_4, s_5\}$, are the amount of total rainfall in millimeters in the region. The utility for each of the alternatives has been developed for three levels of rainfall, (1) low ($\underset{\sim}{F}_1$), (2) medium ($\underset{\sim}{F}_2$), and (3) heavy ($\underset{\sim}{F}_3$), which are defined by fuzzy sets on S. The utility matrix may be given as follows:

u_{ij}	$\underset{\sim}{F}_1$	$\underset{\sim}{F}_2$	$\underset{\sim}{F}_3$
$\underset{\sim}{A}_1$	−2	4	10
$\underset{\sim}{A}_2$	1	8	−10
$\underset{\sim}{A}_3$	10	−5	−20

The membership functions of $\underset{\sim}{F}_1$, $\underset{\sim}{F}_2$, $\underset{\sim}{F}_3$, and the prior probabilities are given here:

	s_1	s_2	s_3	s_4	s_5
$\mu_{\underset{\sim}{F}_1}(s_i)$	1	0.4	0.05	0	0
$\mu_{\underset{\sim}{F}_2}(s_i)$	0	0.6	0.85	0.15	0
$\mu_{\underset{\sim}{F}_3}(s_i)$	0	0	0.1	0.85	1
$p(s_i)$	0.1	0.2	0.2	0.35	0.15

Let $X = \{x_1, x_2, x_3, x_4\}$ be the set of amount of rainfall in the next year. This represents the new information. The conditional probabilities $p(x_j|s_i)$ for probabilistic uncertain information are as given below:

	x_1	x_2	x_3	x_4
s_1	0.7	0.2	0.1	0.0
s_2	0.1	0.7	0.2	0.0
s_3	0.1	0.2	0.7	0.0
s_4	0.0	0.1	0.2	0.7
s_5	0.0	0.0	0.3	0.7

Consider a fuzzy information system,

$$M = \{M_1, M_2, M_3\}$$

where M_1 = rainfall is less than approximately 35 mm
M_2 = rainfall is equal to approximately 35 mm
M_3 = rainfall is greater than approximately 35 mm

The membership functions for the new fuzzy information that satisfy the orthogonality condition are given here:

	x_1	x_2	x_3	x_4
$\mu_{M_1}(x_i)$	1.0	0.3	0.1	0.0
$\mu_{M_2}(x_i)$	0.0	0.7	0.8	0.1
$\mu_{M_3}(x_i)$	0.0	0.0	0.1	0.9

Determine the following:
(a) Posterior probabilities for fuzzy state F_2 and fuzzy information M_1, and for fuzzy state F_3 and fuzzy information M_3.
(b) Conditional expected utility of building a permanent dike (A_1) when fuzzy information M_3 is given.

10.21. Your design team needs to determine what level of technology to incorporate in a new product. As is usually the case, current technology is least expensive whereas the most advanced or leading-edge technology is the most expensive. A given technology usually comes down in price with time. The decision cycle of your project is several years. The team must decide what level of technology to incorporate in the product based on the future expected cost. If the technology is still expensive by the time the product goes to the market, the product will not sell. If you do not incorporate the latest affordable technology, your product may not be so advanced as that of the competition and therefore sales may be poor. Consider the following:

Actual discrete states of nature:
s_1: Cost is low
s_2: Cost is moderate
s_3: Cost is high
Fuzzy actions:
A_1: Use current/well-established technology
A_2: Use newer/leading-edge/advanced technology
Fuzzy states on fuzzy information system, μ:
M_1: Cost is approximately the cost of implementing with current technology
M_2: Cost is approximately 2 times the cost of the current technology
M_3: Cost is approximately 10 times the cost of current technology
Let $X = \{x_1, x_2, x_3, x_4, x_5\}$ be the set of rates of increase in usage of advanced technology in the next term. Then we have the following:

Fuzzy states of nature:
F_1: Low cost
F_2: Medium cost
F_3: High cost

Prior probabilities:

$$p(s_i) = \begin{bmatrix} 0.25 \\ 0.5 \\ 0.25 \end{bmatrix} \begin{matrix} s_1 \\ s_2 \\ s_3 \end{matrix}$$

Utility matrix:

$$u = \begin{matrix} & s_1 & s_2 & s_3 \\ & \begin{bmatrix} -8 & -5 & 0 \\ 10 & -5 & -10 \end{bmatrix} & \begin{matrix} A_1 \\ A_2 \end{matrix} \end{matrix}$$

Membership values for each orthogonal fuzzy state on the actual state system:

$$\mu_F = \begin{matrix} & s_1 & s_2 & s_3 \\ & \begin{bmatrix} 0.8 & 0.1 & 0 \\ 0.2 & 0.8 & 0.2 \\ 0 & 0.1 & 0.8 \end{bmatrix} & \begin{matrix} F_1 \\ F_2 \\ F_3 \end{matrix} \end{matrix}$$

Membership values for each orthogonal fuzzy set on the fuzzy information system:

$$\mu_M = \begin{matrix} & x_1 & x_2 & x_3 & x_4 & x_5 \\ & \begin{bmatrix} 1 & 0.5 & 0 & 0 & 0 \\ 0 & 0.5 & 1 & 0.5 & 0 \\ 0 & 0 & 0 & 0.5 & 1 \end{bmatrix} & \begin{matrix} M_1 \\ M_2 \\ M_3 \end{matrix} \end{matrix}$$

Utility matrix for fuzzy information:

$$u = \begin{matrix} & F_1 & F_2 & F_3 \\ & \begin{bmatrix} -5 & 0 & 5 \\ 10 & 2 & -10 \end{bmatrix} & \begin{matrix} A_1 \\ A_2 \end{matrix} \end{matrix}$$

Likelihood values for probabilistic (uncertain) information for the data samples:

$$p(x_i|s_k) = \begin{matrix} & x_1 & x_2 & x_3 & x_4 & x_5 \\ & \begin{bmatrix} 0.1 & 0.25 & 0.15 & 0.35 & 0.15 \\ 0.3 & 0.05 & 0.1 & 0.1 & 0.45 \\ 0.2 & 0.4 & 0.35 & 0 & 0.05 \end{bmatrix} & \begin{matrix} s_1 \\ s_2 \\ s_3 \end{matrix} \end{matrix}$$

Likelihood values for probabilistic perfect information for the data samples:

$$p(x_i|s_k) = \begin{matrix} & x_1 & x_2 & x_3 & x_4 & x_5 \\ & \begin{bmatrix} 0 & 0 & 0 & 1 & 0 \\ 0.4 & 0 & 0 & 0 & 0.6 \\ 0 & 0.55 & 0.45 & 0 & 0 \end{bmatrix} & \begin{matrix} s_1 \\ s_2 \\ s_3 \end{matrix} \end{matrix}$$

(a) Determine the value of information for the fuzzy states and fuzzy actions for uncertain probabilistic information.

(b) Determine the value of information for the fuzzy states and fuzzy actions for perfect probabilistic information.

CHAPTER
11

FUZZY CLASSIFICATION AND PATTERN RECOGNITION

PART I CLASSIFICATION

From causes which appear similar, we expect similar effects. This is the sum total of all our experimental conclusions.

David Hume
Scottish philosopher, Enquiry Concerning Human Understanding 1748

There is structure in nature. Much of this structure is known to us and is quite beautiful. Consider the natural sphericity of rain drops and bubbles; why do balloons take this shape? How about the elegant beauty of crystals, rhombic solids with rectangular, pentagonal, or hexagonal cross sections? Why do these naturally beautiful, geometric shapes exist? What causes the natural repetition of the mounds of sand dunes? Some phenomena we cannot see directly: for example, the elliptical shape of the magnetic field around the earth; or we can see only when certain atmospheric conditions exist, such as the beautiful and circular appearance of a rainbow or the repetitive patterns of the aurora borealis in the night sky near the North Pole. Some patterns, such as the helical appearance of DNA or the cylindrical shape of some bacteria, have only appeared to us since the advent of extremely powerful electron microscopes. Consider the geometry and colorful patterns of a butterfly's wings; why do these patterns exist in our physical world? The answers to some of these questions are still unknown; many others have been discovered through increased understanding of physics, chemistry, and biology.

Just as there is structure in nature, we believe there is an underlying structure in most of the phenomena we wish to understand. Examples abound in image recognition, molecular biology applications such as protein folding and 3D molecular structure, oil exploration, cancer detection, and many others. For fields dealing with diagnosis we often seek to find structure in the data obtained from observation. Our observations can be visual, audio, or any of a variety of sensor-based electronic or optical signals. Finding the structure in data is the essence of classification. As the quotation at the beginning of this chapter suggests, our experimental observations lead us to develop relationships between the inputs and outputs of an experiment. As we are able to conduct more experiments we see the relationships forming some recognizable, or classifiable, structure. By finding structure, we are classifying the data according to similar patterns, attributes, features, and other characteristics. The general area is known as *classification*.

In classification, also termed *clustering,* the most important issue is deciding what criteria to classify against. For example, suppose we want to classify people. In describing people we will look at their height, weight, gender, religion, education, appearance, and so on. Many of these features are numerical quantities such as height and weight; other features are simply linguistic descriptors and these can be quite non-numeric. We can easily classify people according to gender, or one feature. For this classification the criterion is simple: female or male. We might want to classify people into three size categories: small, medium, and large. For this classification we might only need two of the features describing people: height and weight. Here, the classification criterion might be some algebraic combination of height and weight. Suppose we want to classify people according to whether we would want them as *neighbors*. Here the number of features to be used in the classification is not at all clear, and we might also have trouble developing a criterion for this classification. Nevertheless, a criterion for classification must be prepared before we can segregate the data into definable classes. As is often the case in classification studies, the number and kind of features and the type of classification criteria are choices that are continually changed as the data are manipulated; and this iteration continues until we think we have a grouping of the data that seems plausible from a structural and physical perspective.

This chapter summarizes only two popular methods of classification. The first is classification using equivalent relations [Zadeh, 1971; Bezdek and Harris, 1978]. This approach makes use of certain special properties of equivalent relations and the concept of defuzzification known as lambda-cuts on the relations. The second method of classification is a very popular method known as *fuzzy c-means*, so named because of its close analog in the crisp world, *hard c-means* [Bezdek, 1981]. This method uses concepts in n-dimensional Euclidean space to determine the geometric *closeness* of data points by assigning them to various clusters or classes and then determining the distance between the clusters.

CLASSIFICATION BY EQUIVALENCE RELATIONS

Crisp Relations

Define a set, $[x_i] = \{x_j \mid (x_i, x_j) \in R\}$, as the equivalent class of x_i on a universe of data points, X. This class is contained in a special relation, R, known as an equivalence relation (see Chapter 3). This class is a set of all elements related to x_i that have the following properties [Bezdek, 1974]:

1. $x_i \in [x_i]$ therefore $(x_i, x_i) \in R$
2. $[x_i] \neq [x_j] \Rightarrow [x_i] \cap [x_j] = \emptyset$
3. $\bigcup_{x \in X} [x] = X$

The first property is that of reflexivity (see Chapter 3), the second property indicates that equivalent classes do not overlap, and the third property simply expresses that the union of all equivalent classes exhausts the universe. Hence, the equivalence relation R can divide the universe X into mutually exclusive equivalent classes, i.e.,

$$X \mid R = \{[x] \mid x \in X\} \tag{11.1}$$

where X|R is called the quotient set. The quotient set of X relative to R, denoted X|R, is the set whose elements are the equivalence classes of X under the equivalence relation R. The cardinality of X|R (i.e., the number of distinct equivalence classes of X under R) is called the rank of the matrix R.

> **Example 11.1** **[Ross, 1995].** Define a universe of integers $X = \{1, 2, 3, \ldots, 10\}$ and define R as the crisp relation for "the identical remainder after dividing each element of the universe by 3." We have

$$R = \begin{array}{c} \\ 1 \\ 2 \\ 3 \\ 4 \\ 5 \\ 6 \\ 7 \\ 8 \\ 9 \\ 10 \end{array} \begin{array}{c} \begin{array}{cccccccccc} 1 & 2 & 3 & 4 & 5 & 6 & 7 & 8 & 9 & 10 \end{array} \\ \left[\begin{array}{cccccccccc} 1 & 0 & 0 & 1 & 0 & 0 & 1 & 0 & 0 & 1 \\ 0 & 1 & 0 & 0 & 1 & 0 & 0 & 1 & 0 & 0 \\ 0 & 0 & 1 & 0 & 0 & 1 & 0 & 0 & 1 & 0 \\ 1 & 0 & 0 & 1 & 0 & 0 & 1 & 0 & 0 & 1 \\ 0 & 1 & 0 & 0 & 1 & 0 & 0 & 1 & 0 & 0 \\ 0 & 0 & 1 & 0 & 0 & 1 & 0 & 0 & 1 & 0 \\ 1 & 0 & 0 & 1 & 0 & 0 & 1 & 0 & 0 & 1 \\ 0 & 1 & 0 & 0 & 1 & 0 & 0 & 1 & 0 & 0 \\ 0 & 0 & 1 & 0 & 0 & 1 & 0 & 0 & 1 & 0 \\ 1 & 0 & 0 & 1 & 0 & 0 & 1 & 0 & 0 & 1 \end{array} \right] \end{array}$$

We note that this relation is reflexive, it is symmetric, and, as can be determined by inspection (see Chapter 3), it is also transitive; hence the matrix is an equivalence relation. We can group the elements of the universe into classes as follows:

$$[1] = [4] = [7] = [10] = \{1, 4, 7, 10\} \quad \text{with remainder} = 1$$
$$[2] = [5] = [8] = \{2, 5, 8\} \quad \text{with remainder} = 2$$
$$[3] = [6] = [9] = \{3, 6, 9\} \quad \text{with remainder} = 0$$

Then we can show that the classes do not overlap, i.e., they are mutually exclusive:

$$[1] \cap [2] = \emptyset \text{ and } [2] \cap [3] = \emptyset$$

and that the union of all the classes exhausts (comprises) the universe.

$$\bigcup [x] = X$$

The quotient set is then determined to have three classes,

$$X \mid R = \{(1, 4, 7, 10), (2, 5, 8), (3, 6, 9)\}$$

Not all relations are equivalent, but if a relation is at least a tolerance relation (i.e., it exhibits properties of reflexivity and symmetry) then it can be converted to an equivalent relation through max–min compositions with itself.

Example 11.2. Suppose you have a collection (universe) of five data points,

$$X = \{x_1, x_2, x_3, x_4, x_5\}$$

and these data points show similarity to one another according to the following tolerance relation, which is reflexive and symmetric:

$$R_1 = \begin{bmatrix} 1 & 1 & 0 & 0 & 0 \\ 1 & 1 & 0 & 0 & 1 \\ 0 & 0 & 1 & 0 & 0 \\ 0 & 0 & 0 & 1 & 0 \\ 0 & 1 & 0 & 0 & 1 \end{bmatrix}$$

We see that this tolerance relation is not transitive from the expression

$$(x_1, x_2) \in R_1, \quad (x_2, x_5) \in R_1 \quad \text{but} \quad (x_1, x_5) \in R_1$$

As indicated in Chapter 3, any tolerance relation can be reformed into an equivalence relation through at most $n - 1$ compositions with itself. In this case one composition of R_1 with itself results in an equivalence relation,

$$R_1 \circ R_1 = \begin{bmatrix} 1 & 1 & 0 & 0 & 1 \\ 1 & 1 & 0 & 0 & 1 \\ 0 & 0 & 1 & 0 & 0 \\ 0 & 0 & 0 & 1 & 0 \\ 1 & 1 & 0 & 0 & 1 \end{bmatrix} = R$$

As one can see in the relation, R, there are three classes. The first, second and fifth columns are identical and the fourth and fifth columns are each unique. The data points can then be classified into three groups or classes, as delineated below:

$$[x_1] = [x_2] = [x_5] = \{x_1, x_2, x_5\} \quad [x_3] = \{x_3\} \quad [x_4] = \{x_4\}$$

Fuzzy Relations

As already illustrated, crisp equivalent relations can be used to divide the universe X into mutually exclusive classes. In the case of fuzzy relations, for all fuzzy equivalent relations, their λ-cuts are equivalent ordinary relations. Hence, to classify data points in the universe using fuzzy relations, we need to find the associated fuzzy equivalent relation.

Example 11.3. Example 3.11 had a tolerance relation, say R_t, describing five data points, that was formed into a fuzzy equivalence relation, R, by composition; this process is repeated here for this classification example.

$$R_t = \begin{bmatrix} 1 & 0.8 & 0 & 0.1 & 0.2 \\ 0.8 & 1 & 0.4 & 0 & 0.9 \\ 0 & 0.4 & 1 & 0 & 0 \\ 0.1 & 0 & 0 & 1 & 0.5 \\ 0.2 & 0.9 & 0 & 0.5 & 1 \end{bmatrix} \longrightarrow R = \begin{bmatrix} 1 & 0.8 & 0.4 & 0.5 & 0.8 \\ 0.8 & 1 & 0.4 & 0.5 & 0.9 \\ 0.4 & 0.4 & 1 & 0.4 & 0.4 \\ 0.5 & 0.5 & 0.4 & 1 & 0.5 \\ 0.8 & 0.9 & 0.4 & 0.5 & 1 \end{bmatrix}$$

TABLE 11.1
Classification of five data points according
to λ-cut level

λ-cut level	Classification
1.0	$[x_1][x_2][x_3][x_4][x_5]$
0.9	$[x_1]\{x_2.x_5\}[x_3][x_4]$
0.8	$\{x_1.x_2.x_5\}[x_3][x_4]$
0.5	$\{x_1.x_2.x_4.x_5\}[x_3]$
0.4	$\{x_1.x_2.x_3.x_4.x_5\}$

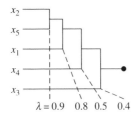

FIGURE 11.1
Classification diagram for Example 11.3.

By taking λ-cuts of fuzzy equivalent relation R at values of $\lambda = 1, 0.9, 0.8, 0.5,$ and 0.4, we get the following:

$$R_1 = \begin{bmatrix} 1 & & & & 0 \\ & 1 & & & \\ & & 1 & & \\ & & & 1 & \\ 0 & & & & 1 \end{bmatrix} \quad R_{0.9} = \begin{bmatrix} 1 & 0 & 0 & 0 & 0 \\ 0 & 1 & 0 & 0 & 1 \\ 0 & 0 & 1 & 0 & 0 \\ 0 & 0 & 0 & 1 & 0 \\ 0 & 1 & 0 & 0 & 1 \end{bmatrix} \quad R_{0.8} = \begin{bmatrix} 1 & 1 & 0 & 0 & 1 \\ 1 & 1 & 0 & 0 & 1 \\ 0 & 0 & 1 & 0 & 0 \\ 0 & 0 & 0 & 1 & 0 \\ 1 & 1 & 0 & 0 & 1 \end{bmatrix}$$

$$R_{0.5} = \begin{bmatrix} 1 & 1 & 0 & 1 & 1 \\ 1 & 1 & 0 & 1 & 1 \\ 0 & 0 & 1 & 0 & 0 \\ 1 & 1 & 0 & 1 & 1 \\ 1 & 1 & 0 & 1 & 1 \end{bmatrix} \quad R_{0.4} = \begin{bmatrix} 1 & 1 & 1 & 1 & 1 \\ 1 & 1 & 1 & 1 & 1 \\ 1 & 1 & 1 & 1 & 1 \\ 1 & 1 & 1 & 1 & 1 \\ 1 & 1 & 1 & 1 & 1 \end{bmatrix}$$

where we can see that the clustering of the five data points according to the λ-cut level is as shown in Table 11.1.

We can express the classification scenario described in Table 11.1 with a systematic classification diagram, as shown in Fig. 11.1. In the figure you can see that the higher the value of λ, the finer is the classification. That is, as λ gets larger the tendency of classification tends to approach the trivial case where each data point is assigned to its own class.

Another example in fuzzy classification considers grouping photographs of family members together according to visual similarity in attempting to determine genetics of the family tree when considering only facial image.

Example 11.4 [Tamura et al., 1971]. Three families exist that have a total of 16 people, all of whom are related by blood. Each person has their photo taken, and the 16 photos are mixed. A person not familiar with the members of the three families is asked to view the photographs to grade their resemblance to one another. In conducting this study the person assigns the similarity relation matrix, r_{ij} as shown in Table 11.2. The matrix developed by the person is a tolerance fuzzy relation, but it does not have properties of equivalence, i.e.,

TABLE 11.2
Similarity relation matrix. r_{ij}

	1	2	3	4	5	6	7	8	9	10	11	12	13	14	15	16
1	1.0															
2	0.0	1.0														
3	0.0	0.0	1.0													
4	0.0	0.0	0.4	1.0												
5	0.0	0.8	0.0	0.0	1.0											
6	0.5	0.0	0.2	0.2	0.0	1.0										
7	0.0	0.8	0.0	0.0	0.4	0.0	1.0									
8	0.4	0.2	0.2	0.5	0.0	0.8	0.0	1.0								
9	0.0	0.4	0.0	0.8	0.4	0.2	0.4	0.0	1.0							
10	0.0	0.0	0.2	0.2	0.0	0.0	0.2	0.0	0.2	1.0						
11	0.0	0.5	0.2	0.2	0.0	0.0	0.8	0.0	0.4	0.2	1.0					
12	0.0	0.0	0.2	0.8	0.0	0.0	0.0	0.0	0.4	0.8	0.0	1.0				
13	0.8	0.0	0.2	0.4	0.0	0.4	0.0	0.4	0.0	0.0	0.0	0.0	1.0			
14	0.0	0.8	0.0	0.2	0.4	0.0	0.8	0.0	0.2	0.2	0.6	0.0	0.0	1.0		
15	0.0	0.0	0.4	0.8	0.0	0.2	0.0	0.0	0.2	0.0	0.0	0.2	0.2	0.0	1.0	
16	0.6	0.0	0.0	0.2	0.2	0.8	0.0	0.4	0.0	0.0	0.0	0.0	0.4	0.2	0.4	1.0

$$r_{ij} = 1 \quad \text{for } i = j$$

$$r_{ij} = r_{ji}$$

$$r_{ij} \geq \lambda_1 \quad \text{and} \quad r_{jk} \geq \lambda_2 \quad \text{but} \quad r_{ik} < \min(\lambda_1, \lambda_2), \text{ i.e., transitivity does not hold}$$

For example,

$$r_{16} = 0.5, \qquad r_{68} = 0.8, \qquad \text{but} \qquad r_{18} = 0.4 < 0.5$$

By composition the equivalence relation shown in Table 11.3 is obtained.
When we take a λ-cut of this fuzzy equivalent relation at $\lambda = 0.6$, we get the defuzzified relation shown in Table 11.4.
Four distinct classes are identified:

$$\{1, 6, 8, 13, 16\}, \qquad \{2, 5, 7, 11, 14\}, \qquad \{3\}, \qquad \{4, 9, 10, 12, 15\}$$

From this clustering it seems that only photograph number 3 cannot be identified with any of the three families. Perhaps a lower value of λ might assign photograph 3 to one of the other three classes. The other three clusters are all correct in that the members identified in each class are, in fact, the members of the correct families as described in Tamura et al. [1971].

Classification using equivalence relations can also be employed to segregate data that are originally developed as a similarity relation using some of the similarity methods developed at the end of Chapter 3. The following problem is an example of this, involving earthquake damage assessment. It was first introduced in Chapter 3 as Example 3.12.

Example 11.5. Five regions have suffered damage from a recent earthquake (see Example 3.12). The buildings in each region are characterized according to three damage levels: no damage, medium damage, and serious damage. The percentage of buildings for a given region in each of the damage levels is given in Table 11.5.

TABLE 11.3
Equivalence relation

	1	2	3	4	5	6	7	8	9	10	11	12	13	14	15	16
1	1.0															
2	0.4	1.0														
3	0.4	0.4	1.0													
4	0.5	0.4	0.4	1.0												
5	0.4	0.8	0.4	0.4	1.0											
6	0.6	0.4	0.4	0.5	0.4	1.0										
7	0.4	0.8	0.4	0.4	0.8	0.4	1.0									
8	0.6	0.4	0.4	0.5	0.4	0.8	0.4	1.0								
9	0.5	0.4	0.4	0.8	0.4	0.5	0.4	0.5	1.0							
10	0.5	0.4	0.4	0.8	0.4	0.5	0.4	0.5	0.8	1.0						
11	0.4	0.8	0.4	0.4	0.8	0.4	0.8	0.4	0.4	0.4	1.0					
12	0.5	0.4	0.4	0.8	0.4	0.5	0.4	0.5	0.8	0.8	0.4	1.0				
13	0.8	0.4	0.4	0.5	0.4	0.6	0.4	0.6	0.5	0.5	0.4	0.5	1.0			
14	0.4	0.8	0.4	0.4	0.8	0.4	0.8	0.4	0.4	0.4	0.8	0.4	0.4	1.0		
15	0.5	0.4	0.4	0.8	0.4	0.5	0.4	0.5	0.8	0.8	0.4	0.8	0.5	0.4	1.0	
16	0.6	0.4	0.4	0.5	0.4	0.8	0.4	0.8	0.5	0.5	0.4	0.5	0.6	0.4	0.5	1.0

TABLE 11.4
Defuzzified relation

	1	2	3	4	5	6	7	8	9	10	11	12	13	14	15	16
1	1															
2	0	1														
3	0	0	1													
4	0	0	0	1												
5	0	1	0	0	1											
6	1	0	0	0	0	1										
7	0	1	0	0	1	0	1									
8	1	0	0	0	0	1	0	1								
9	0	0	0	1	0	0	0	0	1							
10	0	0	0	1	0	0	0	0	1	1						
11	0	1	0	0	1	0	1	0	0	0	1					
12	0	0	0	1	0	0	0	0	1	1	0	1				
13	1	0	0	0	0	1	0	1	0	0	0	0	1			
14	0	1	0	0	1	0	1	0	0	0	1	0	0	1		
15	0	0	0	1	0	0	0	0	1	1	0	1	0	0	1	
16	1	0	0	0	0	1	0	1	0	0	0	0	1	0	0	1

TABLE 11.5
Proportion of buildings damaged. in three levels by region

	Regions				
	x_1	x_2	x_3	x_4	x_5
x_{i1} – Ratio with no damage	0.3	0.2	0.1	0.7	0.4
x_{i2} – Ratio with medium damage	0.6	0.4	0.6	0.2	0.6
x_{i3} – Ratio with serious damage	0.1	0.4	0.3	0.1	0

TABLE 11.6
Classification of earthquake damage
by region for $\lambda = 0.934$

Regions	Mercalli intensity
$\{x_4\}$	VII
$\{x_1, x_5\}$	VIII
$\{x_2, x_3\}$	IX

Using the cosine amplitude approach, described in Chapter 3, we obtain the following tolerance relation, $\underset{\sim}{R}_1$:

$$\underset{\sim}{R}_1 = \begin{bmatrix} 1 & & & & \\ 0.836 & 1 & & \text{sym} & \\ 0.914 & 0.934 & 1 & & \\ 0.682 & 0.6 & 0.441 & 1 & \\ 0.982 & 0.74 & 0.818 & 0.774 & 1 \end{bmatrix}$$

Three max–min compositions produce a fuzzy equivalence relation,

$$\underset{\sim}{R} = \underset{\sim}{R}_1^3 = \begin{bmatrix} 1 & & & & \\ 0.914 & 1 & & \text{sym} & \\ 0.914 & 0.934 & 1 & & \\ 0.774 & 0.774 & 0.774 & 1 & \\ 0.982 & 0.914 & 0.914 & 0.774 & 1 \end{bmatrix}$$

Now, if we take λ-cuts at two different values of λ, say $\lambda = 0.914$ and $\lambda = 0.934$, the following defuzzified crisp equivalence relations and their associated classes are derived:

$$\lambda = 0.914: \quad R_\lambda = \begin{bmatrix} 1 & 1 & 1 & 0 & 1 \\ 1 & 1 & 1 & 0 & 1 \\ 1 & 1 & 1 & 0 & 1 \\ 0 & 0 & 0 & 1 & 0 \\ 1 & 1 & 1 & 1 & 1 \end{bmatrix}$$

$$\{x_1, \quad x_2, \quad x_3, \quad x_5\}, \quad \{x_4\}$$

$$\lambda = 0.934: \quad R_\lambda = \begin{bmatrix} 1 & 0 & 0 & 0 & 1 \\ 0 & 1 & 1 & 0 & 0 \\ 0 & 1 & 1 & 0 & 0 \\ 0 & 0 & 0 & 1 & 0 \\ 1 & 0 & 0 & 0 & 1 \end{bmatrix}$$

$$\{x_1, \quad x_5\}, \quad \{x_2, \quad x_3\}, \quad \{x_4\}$$

Hence, if we wanted to classify the earthquake damage for purposes of insurance payout into, say, two intensities on the modified Mercalli scale (the Mercalli scale is a measure of an earthquake's strength in terms of average damage the earthquake causes in structures in a given region), then regions 1, 2, 3, and 5 belong to a larger Mercalli intensity and region 4 belongs to a smaller Mercalli intensity (see $\lambda = 0.914$). But if we wanted to have a finer division for, say, three Mercalli scales, we could assign the regions shown in Table 11.6.

CLUSTER ANALYSIS

Clustering refers to identifying the number of subclasses of c clusters in a data universe X comprised of n data samples, and partitioning X into c clusters ($2 \leq c < n$). Note that

$c = 1$ denotes rejection of the hypothesis that there are clusters in the data, whereas $c = n$ constitutes the trivial case where each sample is in a "cluster" by itself. There are two kinds of c-partitions of data: hard (or crisp) and soft (or fuzzy). For numerical data one assumes that the members of each cluster bear more mathematical similarity to each other than to members of other clusters. Two important issues to consider in this regard are how to measure the similarity between pairs of observations and how to evaluate the partitions once they are formed.

One of the simplest similarity measures is distance between pairs of feature vectors in the feature space. If one can determine a suitable distance measure and compute the distance between all pairs of observations, then one may expect that the distance between points in the same cluster will be considerably less than the distance between points in different clusters. Several circumstances, however, mitigate the general utility of this approach, such as the combination of values of incompatible features, as would be the case, for example, when different features have significantly different scales. The clustering method described in this chapter defines "optimum" partitions through a global criterion function that measures the extent to which candidate partitions optimize a weighted sum of squared errors between data points and *cluster centers* in feature space. Many other clustering algorithms have been proposed for distinguishing substructure in high-dimensional data [Bezdek et al., 1986]. It is emphasized here that the method of clustering must be closely matched with the particular data under study for successful interpretation of substructure in the data.

CLUSTER VALIDITY

In many cases, the number c of clusters in the data is known. In other cases, however, it may be reasonable to expect cluster substructure at more than one value of c. In this situation it is necessary to identify the value of c that gives the most plausible number of clusters in the data for the analysis at hand. This problem is known as cluster validity [see Duda and Hart, 1973; or Bezdek, 1981]. If the data used are labeled, there is a unique and absolute measure of cluster validity: the c that is given. For unlabeled data, no absolute measure of clustering validity exists. Although the importance of these differences is not known, it is clear that the features nominated should be sensitive to the phenomena of interest and not to other variations that might not matter to the applications at hand.

c-MEANS CLUSTERING

Bezdek [1981] developed an extremely powerful classification method to accommodate fuzzy data. It is an extension of a method known as c-means, or hard c-means, when employed in a crisp classification sense. To introduce this method, we define a sample set of n data samples that we wish to classify:

$$X = \{\mathbf{x}_1, \mathbf{x}_2, \mathbf{x}_3, \ldots, \mathbf{x}_n\} \tag{11.2}$$

Each data sample, $\mathbf{x}_i$, is defined by m features, i.e.,

$$\mathbf{x}_i = \{x_{i1}, x_{i2}, x_{i3}, \ldots x_{im}\} \tag{11.3}$$

where each $\mathbf{x}_i$ in the universe X is an m-dimensional vector of m elements or m features. Since the m features all can have different units, in general, we have to normalize each of

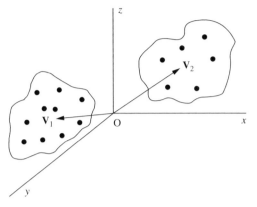

FIGURE 11.2
Cluster idea with hard *c*-means.

the features to a unified scale before classification. In a geometric sense, each $\mathbf{x}_i$ is a *point* in *m*-dimensional feature space, and the universe of the data sample, X, is a *point set* with *n* elements in the sample space.

Bezdek [1981] suggested using an objective function approach for clustering the data into hyperspherical clusters. This idea for hard clustering is shown in three-dimensional feature space in Fig. 11.2. In this figure, each cluster of data is shown as a hyperspherical shape with a hypothetical geometric cluster center. The objective function is developed so as to do two things simultaneously: first, minimize the Euclidean distance between each data point in a cluster and its cluster center (a calculated point), and second, maximize the Euclidean distance between cluster centers.

HARD *c*-MEANS (HCM)

HCM is used to classify data in a crisp sense. By this we mean that each data point will be assigned to one, and only one, data cluster. In this sense these clusters are also called *partitions* – that is, partitions of the data. Define a family of sets $\{A_i, i = 1, 2, \ldots, c\}$ as a hard *c*-partition of X, where the following set-theoretic forms apply to these partitions:

$$\bigcup_{i=1}^{c} A_i = X \tag{11.4}$$

$$A_i \cap A_j = \emptyset \qquad \text{all } i \neq j \tag{11.5}$$

$$\emptyset \subset A_i \subset X \qquad \text{all } i \tag{11.6}$$

again, where $X = \{\mathbf{x}_1, \mathbf{x}_2, \mathbf{x}_3, \ldots, \mathbf{x}_n\}$ is a finite set space comprised of the universe of data samples, and *c* is the number of classes, or partitions, or clusters, into which we want to classify the data. We note the obvious,

$$2 \leq c < n \tag{11.7}$$

where $c = n$ classes just places each data sample into its own class, and $c = 1$ places all data samples into the same class; neither case requires any effort in classification, and both are intrinsically uninteresting. Equation (11.4) expresses the fact that the set of all classes exhausts the universe of data samples. Equation (11.5) indicates that none of the classes overlap in the sense that a data sample can belong to more than one class. Equation (11.6) simply expresses that a class cannot be empty and it cannot contain all the data samples.

Suppose we have the case where $c = 2$. Equations (11.4) and (11.5) are then manifested in the following set expressions:

$$A_2 = \overline{A}_1 \qquad A_1 \cup \overline{A}_1 = X \qquad \text{and} \qquad A_1 \cap \overline{A}_1 = \emptyset$$

These set expressions are equivalent to the excluded middle axioms (Eqs. (2.12)).

The function-theoretic expressions associated with Eqs. (11.4), (11.5), and (11.6) are these

$$\bigvee_{i=1}^{c} \chi_{A_i}(\mathbf{x}_k) = 1 \qquad \text{for all } k \tag{11.8}$$

$$\chi_{A_i}(x_k) \wedge \chi_{A_j}(\mathbf{x}_k) = 0 \qquad \text{for all } k \tag{11.9}$$

$$0 < \sum_{k=1}^{n} \chi_{A_i}(\mathbf{x}_k) < n \qquad \text{for all } i \tag{11.10}$$

where the characteristic function $\chi_{A_i}(\mathbf{x}_k)$ is defined once again as

$$\chi_{A_i}(x_k) = \begin{cases} 1, & \mathbf{x}_k \in A_i \\ 0, & \mathbf{x}_k \notin A_i \end{cases} \tag{11.11}$$

Equations (11.8) and (11.9) explain that any sample $\mathbf{x}_k$ can only and definitely belong to one of the c classes. Equation (11.10) implies that no class is empty and no class is the whole set X (i.e., the universe).

For simplicity in notation, our membership assignment of the jth data point in the ith cluster, or class, is defined to be $\chi_{ij} \equiv \chi_{A_i}(\mathbf{x}_j)$. Now define a matrix U comprised of elements χ_{ij} $(i = 1, 2, \ldots, c; j = 1, 2, \ldots, n)$; hence, U is a matrix with c rows and n columns. Then we define a hard c-partition space for X as the following matrix set:

$$M_c = \left\{ U \mid \chi_{ij} \in \{0, 1\}, \sum_{i=1}^{c} \chi_{ik} = 1, 0 < \sum_{k=1}^{n} \chi_{ik} < n \right\} \tag{11.12}$$

Any matrix $U \in M_c$ is a hard c-partition. The cardinality of any hard c-partition, M_c, is

$$\eta_{M_c} = \left(\frac{1}{c!} \right) \left[\sum_{i=1}^{c} \binom{c}{i} (-1)^{c-i} \cdot i^n \right] \tag{11.13}$$

where the expression $\binom{c}{i}$ is the binomial coefficient of c things taken i at a time.

Example 11.6. Suppose we have five data points in a universe, $X = \{x_1, x_2, x_3, x_4, x_5\}$. Also, suppose we want to cluster these five points into two classes. For this case we have $n = 5$ and

$c = 2$. The cardinality, using Eq. (11.13), of this hard c-partition is given by

$$\eta_{M_c} = \tfrac{1}{2}[2(-1) + 2^5] = 15$$

Some of the 15 possible hard 2-partitions are listed here:

$$\begin{bmatrix} 1 & 1 & 1 & 1 & 0 \\ 0 & 0 & 0 & 0 & 1 \end{bmatrix} \quad \begin{bmatrix} 1 & 1 & 1 & 0 & 0 \\ 0 & 0 & 0 & 1 & 1 \end{bmatrix} \quad \begin{bmatrix} 1 & 1 & 0 & 0 & 0 \\ 0 & 0 & 1 & 1 & 1 \end{bmatrix} \quad \begin{bmatrix} 1 & 0 & 0 & 0 & 0 \\ 0 & 1 & 1 & 1 & 1 \end{bmatrix}$$

$$\begin{bmatrix} 1 & 0 & 1 & 0 & 0 \\ 0 & 1 & 0 & 1 & 1 \end{bmatrix} \quad \begin{bmatrix} 1 & 0 & 0 & 1 & 0 \\ 0 & 1 & 1 & 0 & 1 \end{bmatrix} \qquad \begin{bmatrix} 1 & 0 & 0 & 0 & 1 \\ 0 & 1 & 1 & 1 & 0 \end{bmatrix}$$

and so on.

Notice that these two matrices,

$$\begin{bmatrix} 1 & 1 & 1 & 1 & 0 \\ 0 & 0 & 0 & 0 & 1 \end{bmatrix} \quad \text{and} \quad \begin{bmatrix} 0 & 0 & 0 & 0 & 1 \\ 1 & 1 & 1 & 1 & 0 \end{bmatrix}$$

are not different-clustering 2-partitions. In fact, they are the same 2-partitions irrespective of an arbitrary row-swap. If we label the first row of the first U matrix class c_1 and we label the second row class c_2, we would get the same classification for the second U matrix by simply relabeling each row: the first row is c_2 and the second row is c_1. The cardinality measure given in Eq. (11.13) gives the number of *unique* c-partitions for n data points.

An interesting question now arises: Of all the possible c-partitions for n data samples, how can we select the most reasonable c-partition for the partition space M_c? For instance, in the example just provided, which of the 15 possible hard 2-partitions for five data points and two classes is the best? The answer to this question is provided by the objective function (or classification criteria) to be used to classify or cluster the data. The one proposed for the hard c-means algorithm is known as a within-class sum of squared errors approach using a Euclidean norm to characterize distance. This algorithm is denoted $J(U, \mathbf{v})$, where U is the partition matrix, and the parameter, $\mathbf{v}$, is a vector of cluster centers. This objective function is given by

$$J(U, \mathbf{v}) = \sum_{k=1}^{n} \sum_{i=1}^{c} \chi_{ik}(d_{ik})^2 \tag{11.14}$$

where d_{ik} is a Euclidean distance measure (in m-dimensional feature space, R^m) between the kth data sample $\mathbf{x}_k$ and ith cluster center $\mathbf{v}_i$, given by

$$d_{ik} = d(\mathbf{x}_k - \mathbf{v}_i) = \|\mathbf{x}_k - \mathbf{v}_i\| = \left[\sum_{j=1}^{m} (x_{kj} - v_{ij})^2 \right]^{1/2} \tag{11.15}$$

Since each data sample requires m coordinates to describe its location in R^m-space, each cluster center also requires m coordinates to describe its location in this same space. Therefore, the ith cluster center is a vector of length m,

$$\mathbf{v}_i = \{v_{i1}, v_{i2}, \ldots, v_{im}\}$$

where the jth coordinate is calculated by

$$v_{ij} = \frac{\sum\limits_{k=1}^{n} \chi_{ik} \cdot x_{kj}}{\sum\limits_{k=1}^{n} \chi_{ik}} \tag{11.16}$$

We seek the optimum partition, U*, to be the partition that produces the minimum value for the function, J. That is,

$$J(U^*, \mathbf{v}^*) = \min_{U \in M_c} J(U, \mathbf{v}) \tag{11.17}$$

Finding the optimum partition matrix, U*, is exceedingly difficult for practical problems because $M_c \to \infty$ for even modest-sized problems. For example, for the case where $n = 25$ and $c = 10$, the cardinality approaches an extremely large number, i.e., $M_c \to 10^{18}$! Obviously, a search for optimality by exhaustion is *not* computationally feasible for problems of reasonable interest. Fortunately, very useful and effective alternative search algorithms have been devised [Bezdek, 1981].

One such search algorithm is known as *iterative optimization*. Basically, this method is like many other iterative methods in that we start with an initial guess at the U matrix. From this assumed matrix, input values for the number of classes, and iteration tolerance (the accuracy we demand in the solution), we calculate the centers of the clusters (classes). From these cluster, or class, centers we recalculate the membership values that each data point has in the cluster. We compare these values with the assumed values and continue this process until the changes from cycle to cycle are within our prescribed tolerance level.

The step-by-step procedures in this iterative optimization method are provided here [Bezdek, 1981]:

1. Fix c ($2 \leq c < n$) and initialize the U matrix:

$$U^{(0)} \in M_c$$

Then do $r = 0, 1, 2, \ldots$.
2. Calculate the c center vectors:

$$\{\mathbf{v}_i^{(r)} \text{ with } U^{(r)}\}$$

3. Update $U^{(r)}$; calculate the updated characteristic functions (for all i, k):

$$\chi_{ik}^{(r+1)} = \begin{cases} 1, & d_{ik}^{(r)} = \min\{d_{jk}^{(r)}\} \text{ for all } j \in c \\ 0, & \text{otherwise} \end{cases} \tag{11.18}$$

4. If

$$\|U^{(r+1)} - U^{(r)}\| \leq \varepsilon \text{ (tolerance level)} \tag{11.19}$$

stop; otherwise set $r = r + 1$ and return to step 2.

In step 4, the notation $\| \; \|$ is any matrix norm such as the Euclidean norm.

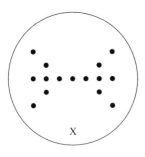

X

FIGURE 11.3
Butterfly classification problem [Bezdek, 1981].

Example 11.7   [Bezdek, 1981]. A good illustration of the iterative optimization method is provided with the "butterfly problem" shown in Fig. 11.3. In this problem we have 15 data points and one of them is on a vertical line of symmetry (the point in the middle of the data cluster). Suppose we want to cluster our data into two classes. We can see that the points to the left of the line of symmetry should be in one class and the points to the right of the line of symmetry should be in the other class. The problem lies in assigning the point on the line of symmetry to a class. To which class should this point belong? Whichever class the algorithm assigns this point to, there will be a good argument that it should be a member of the other class. Alternatively, the argument may revolve around the fact that the choice of two classes is a poor one for this problem. Three classes might be the best choice, but the physics underlying the data might be binary and two classes may be the only option.

In conducting the iterative optimization approach we have to assume an initial U matrix. This matrix will have two rows (two classes, $c = 2$) and 15 columns (15 data points, $n = 15$). It is important to understand that the classes may be unlabeled in this process. That is, we can look at the structure of the data without the need for the assignment of labels to the classes. This is often the case when one is first looking at a group of data. After several iterations with the data, and as we become more and more knowledgeable about the data, we can then assign labels to the classes. We start the solution with the assumption that the point in the middle (i.e., the eighth column) is assigned to the class represented by the bottom row of the initial U matrix, $U^{(0)}$:

$$U^{(0)} = \begin{bmatrix} 1 & 1 & 1 & 1 & 1 & 1 & 1 & 0 & 0 & 0 & 0 & 0 & 0 & 0 & 0 \\ 0 & 0 & 0 & 0 & 0 & 0 & 0 & 1 & 1 & 1 & 1 & 1 & 1 & 1 & 1 \end{bmatrix}$$

After four iterations [Bezdek, 1981] this method converges to within a tolerance level of $\varepsilon = 0.01$, as

$$U^{(4)} = \begin{bmatrix} 1 & 1 & 1 & 1 & 1 & 1 & 1 & 0 & 0 & 0 & 0 & 0 & 0 & 0 & 0 \\ 0 & 0 & 0 & 0 & 0 & 0 & 0 & 1 & 1 & 1 & 1 & 1 & 1 & 1 & 1 \end{bmatrix}$$

We note that the point on the line of symmetry (i.e., the eighth column) is still assigned to the class represented by the second row of the U matrix. The elements in the U matrix indicate membership of that data point in the first or second class. For example, the point on the line of symmetry has full membership in the second class and no membership in the first class; yet it is plain to see from Fig. 11.3 that physically it should probably share membership with each class. This is not possible with crisp classification; membership is binary – a point is either a member of a class or not.

The following example illustrates again the crisp classification method. The process will be instructive because of its similarity to the subsequent algorithm to be developed for the fuzzy classification method.

Example 11.8. In a chemical engineering process involving an automobile's catalytic converter (which converts carbon monoxide to carbon dioxide) we have a relationship between

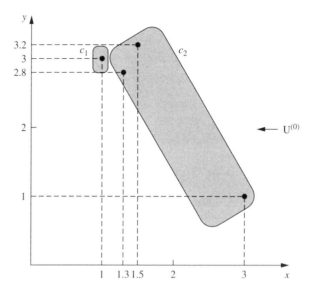

FIGURE 11.4
Four data points in two-dimensional feature space.

the conversion efficiency of the catalytic converter and the *inverse of the temperature* of the catalyst. Two classes of data are known from the reaction efficiency. Points of high conversion efficiency and high temperature are indicators of a nonpolluting system (class c_1), and points of low conversion efficiency and low temperature are indicative of a polluting system (class c_2). Suppose you measure the conversion efficiency and temperature (T) of four different catalytic converters and attempt to characterize them as polluting or nonpolluting. The four data points ($n = 4$) are shown in Fig. 11.4, where the y axis is conversion efficiency and the x axis is the inverse of the temperature (in a conversion process like this the exact solution takes the form of $\ln(1/T)$). The data are described by two features ($m = 2$), and have the following coordinates in 2D space:

$$\mathbf{x}_1 = \{1, 3\}$$
$$\mathbf{x}_2 = \{1.5, 3.2\}$$
$$\mathbf{x}_3 = \{1.3, 2.8\}$$
$$\mathbf{x}_4 = \{3, 1\}$$

We desire to classify these data points into two classes ($c = 2$). It is sometimes useful to calculate the cardinality of the possible number of crisp partitions for this system, i.e., to find η_{M_c} using Eq. (11.13); thus,

$$\eta_{M_c} = \left(\frac{1}{c!}\right)\left[\sum \binom{c}{i}(-1)^{c-i} i^n\right] = \frac{1}{2!}\left[\binom{2}{1}(-1)^1(1)^4 + \binom{2}{2}(-1)^0(2)^4\right]$$

$$= \frac{1}{2}[-2 + 16] = 7$$

which says that there are seven unique ways (irrespective of row-swaps) to classify the four points into two clusters. Let us begin the iterative optimization algorithm with an initial guess of the crisp partition, U, by assuming $\mathbf{x}_1$ to be in class 1 and $\mathbf{x}_2, \mathbf{x}_3, \mathbf{x}_4$ to be in class 2, as shown in Fig. 11.4, i.e.,

$$U^{(0)} = \begin{bmatrix} 1 & 0 & 0 & 0 \\ 0 & 1 & 1 & 1 \end{bmatrix}$$

Now, from the initial $U^{(0)}$ (which is one of the seven possible crisp partitions) we seek the optimum partition U^*, i.e.,

$$U^{(0)} \longrightarrow U^{(1)} \longrightarrow U^{(2)} \longrightarrow \cdots \longrightarrow U^*$$

Of course, optimality is defined in terms of the desired tolerance or convergence level, ε. In general, for class 1 we calculate the coordinates of the cluster center,

$$\begin{aligned} v_{1j} &= \frac{\chi_{11}x_{1j} + \chi_{12}x_{2j} + \chi_{13}x_{3j} + \chi_{14}x_{4j}}{\chi_{11} + \chi_{12} + \chi_{13} + \chi_{14}} \\ &= \frac{(1)x_{1j} + (0)x_{2j} + (0)x_{3j} + (0)x_{4j}}{1 + 0 + 0 + 0} \end{aligned}$$

and

$$\mathbf{v}_i = \{v_{i1}, v_{i2}, \dots, v_{im}\}$$

In this case $m = 2$, which means we deal with two coordinates for each data point. Therefore,

$$\mathbf{v}_i = \{v_{i1}, v_{i2}\}$$

where   for $c = 1$ (which is class 1), $\mathbf{v}_1 = \{v_{11}, v_{12}\}$
         for $c = 2$ (which is class 2), $\mathbf{v}_2 = \{v_{21}, v_{22}\}$

Therefore, using the expression for v_{ij} for $c = 1$, and $j = 1$ and 2, respectively,

$$\left. \begin{aligned} v_{11} &= \frac{1(1)}{1} = 1 & \longrightarrow \ x \text{ coordinate} \\ v_{12} &= \frac{1(3)}{1} = 3 & \longrightarrow \ y \text{ coordinate} \end{aligned} \right\} \Rightarrow \mathbf{v}_1 = \{1, 3\}$$

which just happens to be the coordinates of point x_1, since this is the only point in the class for the assumed initial partition, $U^{(0)}$. For $c = 2$ or class 2, we get cluster center coordinates

$$v_{2j} = \frac{(0)x_{1j} + (1)x_{2j} + (1)x_{3j} + (1)x_{4j}}{0 + 1 + 1 + 1} = \frac{x_{2j} + x_{3j} + x_{4j}}{3}$$

Hence, for $c = 2$ and $j = 1$ and 2, respectively,

$$\left. \begin{aligned} v_{21} &= \frac{1(1.5) + 1(1.3) + 1(3)}{3} = 1.93 & \longrightarrow \ x \text{ coordinate} \\ v_{22} &= \frac{1(3.2) + 1(2.8) + 1(1)}{3} = 2.33 & \longrightarrow \ y \text{ coordinate} \end{aligned} \right\} \Rightarrow \mathbf{v}_2 = \{1.93, 2.33\}$$

Now, we compute the values for d_{ik}, or the distances from the sample $\mathbf{x}_k$ (a data set) to the center, $\mathbf{v}_i$, of the ith class. Using Eq. (11.15),

$$d_{ik} = \left[\sum_{j=1}^{m} (x_{kj} - v_{ij})^2 \right]^{1/2}$$

we get, for example, for $c = 1 : d_{1k} = [(x_{k1} - v_{11})^2 + (x_{k2} - v_{12})^2]^{1/2}$. Therefore, for each data set $k = 1$ to 4, we compute the values of d_{ik} as follows: for cluster 1,

$$d_{11} = \sqrt{(1-1)^2 + (3-3)^2} = 0.0$$

$$d_{12} = \sqrt{(1.5-1)^2 + (3.2-3)^2} = 0.54$$

$$d_{13} = \sqrt{(1.3-1)^2 + (2.8-3)^2} = 0.36$$

$$d_{14} = \sqrt{(3-1)^2 + (1-3)^2} = 2.83$$

and for cluster 2,

$$d_{21} = \sqrt{(1-1.93)^2 + (3-2.33)^2} = 1.14$$

$$d_{22} = \sqrt{(1.5-1.93)^2 + (3.2-2.33)^2} = 0.97$$

$$d_{23} = \sqrt{(1.3-1.93)^2 + (2.8-2.33)^2} = 0.78$$

$$d_{24} = \sqrt{(3-1.93)^2 + (1-2.33)^2} = 1.70$$

Now, we update the partition to $U^{(1)}$ for each data point (for $(c-1)$ clusters) using Eq. (11.18). Hence, for class 1 we compare d_{ik} against the minimum of $\{d_{ik}, d_{2k}\}$:

For $k = 1$,

$$d_{11} = 0.0, \quad \min(d_{11}, d_{21}) = \min(0, 1.14) = 0.0; \quad \text{thus } \chi_{11} = 1$$

For $k = 2$,

$$d_{12} = 0.54, \quad \min(d_{12}, d_{22}) = \min(0.54, 0.97) = 0.54; \quad \text{thus } \chi_{12} = 1$$

For $k = 3$,

$$d_{13} = 0.36, \quad \min(d_{13}, d_{23}) = \min(0.36, 0.78) = 0.36; \quad \text{thus } \chi_{13} = 1$$

For $k = 4$,

$$d_{14} = 2.83, \quad \min(d_{14}, d_{24}) = \min(2.83, 1.70) = 1.70; \quad \text{thus } \chi_{14} = 0$$

Therefore, the updated partition is

$$U^{(1)} = \begin{bmatrix} 1 & 1 & 1 & 0 \\ 0 & 0 & 0 & 1 \end{bmatrix}$$

Since the partitions $U^{(0)}$ and $U^{(1)}$ are different, we repeat the same procedure based on the new setup of two classes. For $c = 1$, the center coordinates are

$$v_{1j} \quad \text{or} \quad v_j = \frac{x_{1j} + x_{2j} + x_{3j}}{1+1+1+0}, \quad \text{since} \quad \chi_{14} = 0$$

$$\left. \begin{aligned} v_{11} &= \frac{x_{11} + x_{21} + x_{31}}{3} = \frac{1 + 1.5 + 1.3}{3} = 1.26 \\ v_{12} &= \frac{x_{12} + x_{22} + x_{32}}{3} = \frac{3 + 3.2 + 2.8}{3} = 3.0 \end{aligned} \right\} v_1 = \{1.26, 3.0\}$$

and for $c = 2$, the center coordinates are

$$\mathbf{v}_{2j} \quad \text{or} \quad \mathbf{v}_j = \frac{x_{4j}}{0 + 0 + 0 + 1}, \quad \text{since } \chi_{21} = \chi_{22} = \chi_{23} = 0$$

$$\left. \begin{array}{l} \mathbf{v}_{21} = \frac{3}{1} = 3 \\ \mathbf{v}_{22} = \frac{1}{1} = 1 \end{array} \right\} \mathbf{v}_2 = \{3, 1\}$$

Now, we calculate distance measures again:

$d_{11} = \sqrt{(1 - 1.26)^2 + (3 - 3)^2} = 0.26$ $d_{21} = \sqrt{(1 - 3)^2 + (3 - 1)^2} = 2.83$

$d_{12} = \sqrt{(1.5 - 1.26)^2 + (3.2 - 3)^2} = 0.31$ $d_{22} = \sqrt{(1.5 - 3)^2 + (3.2 - 1)^2} = 2.66$

$d_{13} = \sqrt{(1.3 - 1.26)^2 + (2.8 - 3)^2} = 0.20$ $d_{23} = \sqrt{(1.3 - 3)^2 + (2.8 - 1)^2} = 2.47$

$d_{14} = \sqrt{(3 - 1.26)^2 + (1 - 3)^2} = 2.65$ $d_{24} = \sqrt{(3 - 3)^2 + (1 - 1)^2} = 0.0$

and again update the partition $U^{(1)}$ to $U^{(2)}$:

For $k = 1$,

$$d_{11} = 0.26, \quad \min(d_{11}, d_{21}) = \min(0.26, 2.83) = 0.26; \quad \text{thus } \chi_{11} = 1$$

For $k = 2$,

$$d_{12} = 0.31, \quad \min(d_{12}, d_{22}) = \min(0.31, 2.66) = 0.31; \quad \text{thus } \chi_{12} = 1$$

For $k = 3$,

$$d_{13} = 0.20, \quad \min(d_{13}, d_{23}) = \min(0.20, 2.47) = 0.20; \quad \text{thus } \chi_{13} = 1$$

For $k = 4$,

$$d_{14} = 2.65, \quad \min(d_{14}, d_{24}) = \min(2.65, 0.0) = 0.0; \quad \text{thus } \chi_{14} = 0$$

Because the partitions $U^{(1)}$ and $U^{(2)}$ are identical, we could say the iterative process has converged; therefore, the optimum hard partition (crisp) is

$$U^{(*)} = \begin{bmatrix} 1 & 1 & 1 & 0 \\ 0 & 0 & 0 & 1 \end{bmatrix}$$

This optimum partition tells us that for this catalytic converter example, the data points $\mathbf{x}_1$, $\mathbf{x}_2$, and $\mathbf{x}_3$ are more similar in the 2D feature space, and different from data point $\mathbf{x}_4$. We could say that points $\mathbf{x}_1$, $\mathbf{x}_2$, and $\mathbf{x}_3$ are more indicative of a nonpolluting converter than is data point $\mathbf{x}_4$.

FUZZY *c*-MEANS (FCM)

Let us consider whether the butterfly example in Fig. 11.3 could be improved with the use of fuzzy set methods. To develop these methods in classification, we define a family of fuzzy sets $\{\underset{\sim}{A}_i, i = 1, 2, \ldots, c\}$ as a fuzzy c-partition on a universe of data points, X. Because fuzzy sets allow for degrees of membership we can extend the crisp classification idea into a fuzzy classification notion. Then we can assign membership to the various data points in each fuzzy set (fuzzy class, fuzzy cluster). Hence, a single point can have partial membership in more than one class. It will be useful to describe the membership value that the kth data point has in the ith class with the following notation:

$$\mu_{ik} = \mu_{\underset{\sim}{A}_i}(x_k) \in [0, 1]$$

with the restriction (as with crisp classification) that the sum of all membership values for a single data point in all of the classes has to be unity:

$$\sum_{i=1}^{c} \mu_{ik} = 1 \qquad \text{for all } k = 1, 2, \ldots, n \qquad (11.20)$$

As before in crisp classification, there can be no empty classes and there can be no class that contains all the data points. This qualification is manifested in the following expression:

$$0 < \sum_{k=1}^{n} \mu_{ik} < n \qquad (11.21)$$

Because each data point can have partial membership in more than one class, the restriction of Eq. (11.9) is not present in the fuzzy classification case, i.e.,

$$\mu_{ik} \wedge \mu_{jk} \neq 0 \qquad (11.22)$$

The provisions of Eqs. (11.8) and (11.10) still hold for the fuzzy case, however,

$$\bigvee_{i=1}^{c} \mu_{\underset{\sim}{A}_i}(x_k) = 1 \qquad \text{for all } k \qquad (11.23)$$

$$0 < \sum_{k=1}^{n} \mu_{\underset{\sim}{A}_i}(x_k) < n \qquad \text{for all } i \qquad (11.24)$$

Before, in the case of $c = 2$, the classification problem reduced to that of the excluded middle axioms for crisp classification. Since we now allow partial membership, the case of $c = 2$ does not follow the restrictions of the excluded middle axioms, i.e., for two classes $\underset{\sim}{A}_i$ and $\underset{\sim}{A}_j$,

$$\underset{\sim}{A}_i \cap \underset{\sim}{A}_j \neq \emptyset \qquad (11.25)$$

$$\emptyset \subset \underset{\sim}{A}_i \subset X \qquad (11.26)$$

We can now define a family of fuzzy partition matrices, M_{fc}, for the classification involving c classes and n data points,

$$M_{fc} = \left\{ \underset{\sim}{U} \mid \mu_{ik} \in [0, 1]; \sum_{i=1}^{c} \mu_{ik} = 1; 0 < \sum_{k=1}^{n} \mu_{ik} < n \right\} \qquad (11.27)$$

where $i = 1, 2, \ldots, c$ and $k = 1, 2, \ldots, n$.

Any $U \in M_{fc}$ is a fuzzy c-partition, and it follows from the overlapping character of the classes and the infinite number of membership values possible for describing class membership that the cardinality of M_{fc} is also infinity, i.e., $\eta_{M_{fc}} = \infty$.

Example 11.9 **[Similar to Bezdek, 1981].** Suppose you are a fruit geneticist interested in genetic relationships among fruits. In particular, you know that a tangelo is a cross between a grapefruit and a tangerine. You describe the fruit with such features as color, weight, sphericity, sugar content, skin texture, and so on. Hence, your feature space could be highly dimensional. Suppose you have three fruits (three data points):

$$X = \{x_1 = \text{grapefruit}, x_2 = \text{tangelo}, x_3 = \text{tangerine}\}$$

These data points are described by m features, as discussed. You want to class the three fruits into two classes to determine the genetic assignment for the three fruits. In the crisp case, the classification matrix can take one of three forms, i.e., the cardinality for this case where $n = 3$ and $c = 2$ is $\eta_{M_c} = 3$ (see Eq. (11.13)). Suppose you arrange your U matrix as follows:

$$U = \begin{matrix} & x_1 & x_2 & x_3 \\ c_1 \\ c_2 \end{matrix} \begin{bmatrix} 1 & 0 & 0 \\ 0 & 1 & 1 \end{bmatrix}$$

The three possible partitions of the matrix are

$$\begin{bmatrix} 1 & 0 & 0 \\ 0 & 1 & 1 \end{bmatrix} \quad \begin{bmatrix} 1 & 1 & 0 \\ 0 & 0 & 1 \end{bmatrix} \quad \begin{bmatrix} 1 & 0 & 1 \\ 0 & 1 & 0 \end{bmatrix}$$

Notice in the first partition that we are left with the uncomfortable segregation of the grapefruit in one class and the tangelo and the tangerine in the other; the tangelo shares nothing in common with the grapefruit! In the second partition, the grapefruit and the tangelo are in a class, suggesting that they share nothing in common with the tangerine! Finally, the third partition is the most genetically discomforting of all, because here the tangelo is in a class by itself, sharing nothing in common with its progenitors! One of these three partitions will be the final partition when any algorithm is used. The question is, which one is best? Intuitively the answer is none, but in crisp classification we have to use one of these.

In the fuzzy case this segregation and genetic absurdity are not a problem. We can have the most intuitive situation where the tangelo shares membership with both classes with the parents. For example, the following partition might be a typical outcome for the fruit genetics problem:

$$U = \begin{matrix} & x_1 & x_2 & x_3 \\ 1 \\ 2 \end{matrix} \begin{bmatrix} 0.91 & 0.58 & 0.13 \\ 0.09 & 0.42 & 0.87 \end{bmatrix}$$

In this case, Eq. (11.24) shows that the sum of each row is a number between 0 and n, or

$$0 < \sum_k \mu_{1k} = 1.62 < 3$$

$$0 < \sum_k \mu_{2k} = 1.38 < 3$$

and for Eq. (11.22) there is overlap among the classes for each data point,

$$\mu_{11} \wedge \mu_{21} = \min(0.91, 0.09) = 0.09 \neq 0$$

$$\mu_{12} \wedge \mu_{22} = \min(0.58, 0.42) = 0.42 \neq 0$$

$$\mu_{13} \wedge \mu_{23} = \min(0.13, 0.87) = 0.13 \neq 0$$

Fuzzy c-Means Algorithm

To describe a method to determine the fuzzy c-partition matrix $\underset{\sim}{U}$ for grouping a collection of n data sets into c classes, we define an objective function J_m for a fuzzy c-partition,

$$J_m(\underset{\sim}{U}, \mathbf{v}) = \sum_{k=1}^{n} \sum_{i=1}^{c} (\mu_{ik})^{m'} (d_{ik})^2 \tag{11.28}$$

where

$$d_{ik} = d(\mathbf{x}_k - \mathbf{v}_i) = \left[\sum_{j=1}^{m} (x_{kj} - v_{ij})^2 \right]^{1/2} \tag{11.29}$$

and where μ_{ik} is the membership of the kth data point in the ith class.

As with crisp classification, the function J_m can have a large number of values, the smallest one associated with the *best* clustering. Because of the large number of possible values, now infinite because of the infinite cardinality of fuzzy sets, we seek to find the best possible, or optimum, solution without resorting to an exhaustive, or expensive, search. The distance measure, d_{ik} in Eq. (11.29), is again a Euclidean distance between the ith cluster center and the kth data set (data point in m-space). A new parameter is introduced in Eq. (11.28) called a weighting parameter, m' [Bezdek, 1981]. This value has a range $m' \in [1, \infty)$. This parameter controls the amount of fuzziness in the classification process and is discussed shortly. Also, as before, $\mathbf{v}_i$ is the ith cluster center, which is described by m features (m coordinates) and can be arranged in vector form as before, $\mathbf{v}_i = \{v_{i1}, v_{i2}, \ldots, v_{im}\}$.

Each of the cluster coordinates for each class can be calculated in a manner similar to the calculation in the crisp case (see Eq. (11.16)),

$$v_{ij} = \frac{\displaystyle\sum_{k=1}^{n} \mu_{ik}^{m'} \cdot x_{kj}}{\displaystyle\sum_{k=1}^{n} \mu_{ik}^{m'}} \tag{11.30}$$

where j is a variable on the feature space, i.e., $j = 1, 2, \ldots, m$.

As in the crisp case the optimum fuzzy c-partition will be the smallest of the partitions described in Eq. (11.28), i.e.,

$$J_m^*(\underset{\sim}{U}^*, \mathbf{v}^*) = \min_{M_{fc}} J(\underset{\sim}{U}, \mathbf{v}) \tag{11.31}$$

As with many optimization processes (see Chapter 14), the solution to Eq. (11.31) cannot be guaranteed to be a global optimum, i.e., the best of the best. What we seek is the best solution available within a prespecified level of accuracy. An effective algorithm for fuzzy classification, called iterative optimization, was proposed by Bezdek [1981]. The steps in this algorithm are as follows:

1. Fix c ($2 \leq c < n$) and select a value for parameter m'. Initialize the partition matrix, $\underset{\sim}{U}^{(0)}$. Each step in this algorithm will be labeled r, where $r = 0, 1, 2, \ldots$.
2. Calculate the c centers $\{\mathbf{v}_i^{(r)}\}$ for each step.

3. Update the partition matrix for the rth step, $\underset{\sim}{U}^{(r)}$ as follows:

$$\mu_{ik}^{(r+1)} = \left[\sum_{j=1}^{c} \left(\frac{d_{ik}^{(r)}}{d_{jk}^{(r)}} \right)^{2/(m'-1)} \right]^{-1} \qquad \text{for } I_k = \emptyset \qquad (11.32a)$$

or

$$\mu_{ik}^{(r+1)} = 0 \qquad \text{for all classes } i \text{ where } i \in \underset{\sim}{I}_k \qquad (11.32b)$$

where

$$I_k = \{i \mid 2 \leq c < n; d_{ik}^{(r)} = 0\} \qquad (11.33)$$

and

$$\underset{\sim}{I}_k = \{1, 2, \ldots, c\} - I_k \qquad (11.34)$$

and

$$\sum_{i \in I_k} \mu_{ik}^{(r+1)} = 1 \qquad (11.35)$$

4. If $\|\underset{\sim}{U}^{(r+1)} - \underset{\sim}{U}^{(r)}\| \leq \varepsilon_L$, stop; otherwise set $r = r + 1$ and return to step 2.

In step 4 we compare a matrix norm $\| \ \|$ of two successive fuzzy partitions to a prescribed level of accuracy, ε_L, to determine whether the solution is good enough. In step 3 there is a considerable amount of logic involved in Eqs. (11.32)–(11.35). Equation (11.32a) is straightforward enough, except when the variable d_{jk} is zero. Since this variable is in the denominator of a fraction, the operation is undefined mathematically, and computer calculations are abruptly halted. So the parameters I_k and $\tilde{I}_k$ comprise a bookkeeping system to handle situations when some of the distance measures, d_{ij}, are zero, or extremely small in a computational sense. If a zero value is detected, Eq. (11.32b) sets the membership for that partition value to be zero. Equations (11.33) and (11.34) describe the bookkeeping parameters I_k and $\underset{\sim}{I}_k$, respectively, for each of the classes. Equation (11.35) simply says that all the nonzero partition elements in each column of the fuzzy classification partition, $\underset{\sim}{U}$, sum to unity. The following example serves to illustrate Eqs. (11.32)–(11.35).

Example 11.10. Suppose we have calculated the following distance measures for one step in our iterative algorithm for a classification problem involving three classes and five data points. The values in Table 11.7 are simple numbers for ease of illustration. The bookkeeping parameters I_k and $\underset{\sim}{I}_k$, where in this example $k = 1, 2, 3, 4, 5$, are given next, as illustration of

TABLE 11.7
Distance measures for hypothetical example
($c = 3. n = 5$)

$d_{11} = 1$	$d_{21} = 2$	$d_{31} = 3$
$d_{12} = 0$	$d_{22} = 0.5$	$d_{32} = 1$
$d_{13} = 1$	$d_{23} = 0$	$d_{33} = 0$
$d_{14} = 3$	$d_{24} = 1$	$d_{34} = 1$
$d_{15} = 0$	$d_{25} = 4$	$d_{35} = 0$

the use of Eqs. (11.33) and (11.34):

$$I_1 = \emptyset \quad \underset{\sim}{I}_1 = \{1, 2, 3\} - \emptyset = \{1, 2, 3\}$$

$$I_2 = \{1\} \quad \underset{\sim}{I}_2 = \{1, 2, 3\} - \{1\} = \{2, 3\}$$

$$I_3 = \{2, 3\} \quad \underset{\sim}{I}_3 = \{1, 2, 3\} - \{2, 3\} = \{1\}$$

$$I_4 = \emptyset \quad \underset{\sim}{I}_4 = \{1, 2, 3\} - \emptyset = \{1, 2, 3\}$$

$$I_5 = \{1, 3\} \quad \underset{\sim}{I}_5 = \{1, 2, 3\} - \{1, 3\} = \{2\}$$

Now, Eqs. (11.32) and (11.35) are illustrated:

For data point 1: $\quad \mu_{11}, \mu_{21}, \mu_{31} \neq 0 \quad$ and $\quad \mu_{11} + \mu_{21} + \mu_{31} = 1$

For data point 2: $\quad \mu_{12} = 0 \quad$ and $\quad \mu_{22}, \mu_{32} \neq 0 \quad$ and $\quad \mu_{22} + \mu_{23} = 1$

For data point 3: $\quad \mu_{13} = 1 \quad$ and $\quad \mu_{23} = \mu_{33} = 0$

For data point 4: $\quad \mu_{14}, \mu_{24}, \mu_{34} \neq 0 \quad$ and $\quad \mu_{14} + \mu_{24} + \mu_{34} = 1$

For data point 5: $\quad \mu_{25} = 1 \quad$ and $\quad \mu_{15} = \mu_{35} = 0$

The algorithm given in Eq. (11.28) is a *least squares* function, where the parameter n is the number of data sets and c is the number of classes (partitions) into which one is trying to classify the data sets. The squared distance, d_{ik}^2, is then weighted by a measure, $(u_{ik})^{m'}$, of the membership of x_k in the ith cluster. The value of J_m is then a measure of the sum of all the *weighted* squared errors; this value is then minimized with respect to two constraint functions. First, J_m is minimized with respect to the squared errors within each cluster, i.e., for each specific value of c. Simultaneously, the distance between cluster centers is maximized, i.e., max $\|\mathbf{v}_i - \mathbf{v}_j\|, i \neq j$.

As indicated, the range for the membership exponent is $m' \in [1, \infty)$. For the case $m' = 1$, the distance norm is Euclidean and the FCM algorithm approaches a hard c-means algorithm, i.e., only zeros and ones come out of the clustering. Conversely, as $m' \to \infty$, the value of the function $J_m \to 0$. This result seems intuitive, because the membership values are numbers less than or equal to 1, and large powers of fractions less than 1 approach 0. In general, the larger m' is, the fuzzier are the membership assignments of the clustering; conversely, as $m' \to 1$, the clustering values become hard, i.e., 0 or 1. The exponent m' thus controls the extent of membership sharing between fuzzy clusters. If all other algorithmic parameters are fixed, then increasing m' will result in decreasing J_m. No theoretical optimum choice of m' has emerged in the literature. However, the bulk of the literature seems to report values in the range 1.25 to 2. Convergence of the algorithm tends to be slower as the value of m' increases.

The algorithm described here can be remarkably accurate and robust in the sense that poor guesses for the initial partition matrix, $\underset{\sim}{U}^{(0)}$, can be overcome quickly, as illustrated in the next example.

Example 11.11. Continuing with the chemical engineering example on a catalytic converter shown in Fig. 11.4, we can see that a visual display of these points in 2D feature space ($m = 2$) makes it easy for the human to cluster the data into two convenient classes based on the proximity of the points to one another. The fuzzy classification method generally converges quite rapidly, even when the initial guess for the fuzzy partition is quite poor, in a classification sense. The fuzzy iterative optimization method for this case would proceed as follows.

Using U* from the previous example as the initial fuzzy partition, $\underset{\sim}{U}^{(0)}$, and assuming a weighting factor of $m' = 2$ and a criterion for convergence of $\varepsilon_L = 0.01$, i.e.,

$$\max_{i,k} |\mu_{ik}^{(r+1)} - \mu_{ik}^{(r)}| \le 0.01$$

we want to determine the optimum fuzzy 2-partition $\underset{\sim}{U}^*$. To begin, the initial fuzzy partition is

$$\underset{\sim}{U}^{(0)} = \begin{bmatrix} 1 & 1 & 1 & 0 \\ 0 & 0 & 0 & 1 \end{bmatrix}$$

Next is the calculation of the initial cluster centers using Eq. (11.30), where $m' = 2$:

$$v_{ij} = \frac{\sum_{k=1}^{n} (\mu_{ik})^2 \cdot x_{kj}}{\sum_{k=1}^{n} (\mu_{ik})^2}$$

where for $c = 1$,

$$v_{1j} = \frac{\mu_1^2 x_{1j} + \mu_2^2 x_{2j} + \mu_3^2 x_{3j} + \mu_4^2 x_{4j}}{\mu_1^2 + \mu_2^2 + \mu_3^2 + \mu_4^2}$$

$$= \frac{(1)x_{1j} + (1)x_{2j} + (1)x_{3j} + (0)x_{4j}}{1 + 1 + 1 + 0} = \frac{x_{1j} + x_{2j} + x_{3j}}{1^2 + 1^2 + 1^2 + 0}$$

$$\left. \begin{array}{l} v_{11} = \dfrac{1 + 1.5 + 1.3}{3} = 1.26 \\[4mm] v_{12} = \dfrac{3 + 3.2 + 2.8}{3} = 3.0 \end{array} \right\} \quad \mathbf{v}_1 = \{1.26, 3.0\}$$

and for $c = 2$,

$$\mathbf{v}_{2j} \quad \text{or} \quad \mathbf{v}_j = \frac{x_{4j}}{0 + 0 + 0 + 1}, \quad \text{since } x_{21} = x_{22} = x_{23} = 0$$

$$\left. \begin{array}{l} \mathbf{v}_{21} = \frac{3}{1} = 3 \\[3mm] \mathbf{v}_{22} = \frac{1}{1} = 1 \end{array} \right\} \quad \mathbf{v}_2 = \{3, 1\}$$

Now the distance measures (distances of each data point from each cluster center) are found using Eq. (11.29):

$$d_{11} = \sqrt{(1 - 1.26)^2 + (3 - 3)^2} = 0.26 \qquad d_{21} = \sqrt{(1 - 3)^2 + (3 - 1)^2} = 2.82$$
$$d_{12} = \sqrt{(1.5 - 1.26)^2 + (3.2 - 3)^2} = 0.31 \qquad d_{22} = \sqrt{(1.5 - 3)^2 + (3.2 - 1)^2} = 2.66$$
$$d_{13} = \sqrt{(1.3 - 1.26)^2 + (2.8 - 3)^2} = 0.20 \qquad d_{23} = \sqrt{(1.3 - 3)^2 + (2.8 - 1)^2} = 2.47$$
$$d_{14} = \sqrt{(3 - 1.26)^2 + (1 - 3)^2} = 2.65 \qquad d_{24} = \sqrt{(3 - 3)^2 + (1 - 1)^2} = 0.0$$

With the distance measures, we can now update $\underset{\sim}{U}$ using Eqs. (11.33)–(11.35) (for $m' = 2$), i.e.,

$$\mu_{ik}^{(r+1)} = \left[\sum_{j=1}^{c} \left(\frac{d_{ik}^{(r)}}{d_{jk}^{(r)}} \right)^2 \right]^{-1}$$

and we get,

$$\mu_{11} = \left[\sum_{j=1}^{c}\left(\frac{d_{11}}{d_{j1}}\right)^2\right]^{-1} = \left[\left(\frac{d_{11}}{d_{11}}\right)^2 + \left(\frac{d_{11}}{d_{21}}\right)^2\right]^{-1} = \left[\left(\frac{0.26}{0.26}\right)^2 + \left(\frac{0.26}{2.82}\right)^2\right]^{-1} = 0.991$$

$$\mu_{12} = \left[\left(\frac{d_{12}}{d_{12}}\right)^2 + \left(\frac{d_{12}}{d_{22}}\right)^2\right]^{-1} = \left[1 + \left(\frac{0.31}{2.66}\right)^2\right]^{-1} = 0.986$$

$$\mu_{13} = \left[\left(\frac{d_{13}}{d_{13}}\right)^2 + \left(\frac{d_{13}}{d_{23}}\right)^2\right]^{-1} = \left[1 + \left(\frac{0.20}{2.47}\right)^2\right]^{-1} = 0.993$$

$$\mu_{14} = \left[\left(\frac{d_{14}}{d_{14}}\right)^2 + \left(\frac{d_{14}}{d_{24}}\right)^2\right]^{-1} = \left[1 + \left(\frac{2.65}{0}\right)^2\right]^{-1} \longrightarrow 0.0, \text{ for } I_4 \neq \emptyset$$

Using Eq. (11.20) for the other partition values, μ_{2j}, for $j = 1, 2, 3, 4$, the new membership functions form an updated fuzzy partition given by

$$\underset{\sim}{U}^{(1)} = \begin{bmatrix} 0.991 & 0.986 & 0.993 & 0 \\ 0.009 & 0.014 & 0.007 & 1 \end{bmatrix}$$

To determine whether we have achieved convergence, we choose a matrix norm $\|\;\|$ such as the maximum absolute value of pairwise comparisons of each of the values in $\underset{\sim}{U}^{(0)}$ and $\underset{\sim}{U}^{(1)}$, e.g.,

$$\max_{i,k} |\mu_{ik}^{(1)} - \mu_{ik}^{(0)}| = 0.0134 > 0.01$$

This result suggests our convergence criteria have not yet been satisfied, so we need another iteration of the method.

For the next iteration we proceed by again calculating cluster centers, but now from values from the latest fuzzy partition, $\underset{\sim}{U}^{(1)}$; for $c = 1$,

$$v_{1j} = \frac{(0.991)^2 x_{1j} + (0.986)^2 x_{2j} + (0.993)^2 x_{3j} + (0)x_{4j}}{0.991^2 + 0.986^2 + 0.993^2 + 0}$$

$$\left.\begin{array}{l} v_{11} = \dfrac{0.98(1) + 0.97(1.5) + 0.99(1.3)}{2.94} = \dfrac{3.719}{2.94} \approx 1.26 \\[3mm] v_{12} = \dfrac{0.98(3) + 0.97(3.2) + 0.99(2.8)}{2.94} = \dfrac{8.816}{2.94} \approx 3.0 \end{array}\right\} \; \mathbf{v}_1 = \{1.26, 3.0\}$$

and for $c = 2$,

$$v_{2j} = \frac{(0.009)^2 x_{1j} + (0.014)^2 x_{2j} + (0.007)^2 x_{3j} + (1)^2 x_{4j}}{0.009^2 + 0.014^2 + 0.007^2 + 1^2}$$

$$\left.\begin{array}{l} v_{21} = \dfrac{0.009^2(1) + 0.014^2(1.5) + 0.007^2(1.3) + 1(3)}{1.000} \approx 3.0 \\[3mm] v_{22} = \dfrac{0.009^2(3) + 0.014^2(3.2) + 0.007^2(2.8) + 1(1)}{1.000} \approx 1.0 \end{array}\right\} \; \mathbf{v}_2 = \{3.0, 1.0\}$$

We see that these two cluster centers are identical to those from the first step, at least to within the stated accuracy of (0.01); hence the final partition matrix will be unchanged, to an accuracy of two digits, from that obtained in the previous iteration. As suggested earlier, convergence

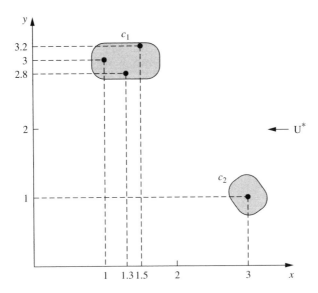

FIGURE 11.5
Converged fuzzy partition for catalytic converter example.

is rapid, at least for this example. The final partition, $\underset{\sim}{U}^{(2)}$, results in a classification shown in Fig. 11.5.

CLASSIFICATION METRIC

In most studies involving fuzzy pattern classification, the data used in the classification process typically come from electrically active transducer readings [Bezdek et al., 1986]. When a fuzzy clustering is accomplished, a question remains concerning the uncertainty of the clustering in terms of the features used. That is, our interest should lie with the extent to which pairs of fuzzy classes of $\underset{\sim}{U}$ overlap; a true classification with no uncertainty would contain classes with no overlap. The question then is: How fuzzy is a fuzzy c-partition? Suppose we compare two memberships for a given data set, x_k, pairwise, using the minimum function, i.e.,

$$\min\{u_i(x_k), u_j(x_k)\} > 0 \tag{11.36}$$

This comparison would indicate that membership of x_k is *shared* by u_i and u_j, whereas the minimum of these two values reflects the minimum amount of *unshared* membership x_k can claim in either u_i or u_j. Hence, a fuzziness measure based on functions $\min\{u_i(x_k), u_j(x_k)\}$ would constitute a point-by-point assessment – not of overlap, but of "anti-overlap" [Bezdek, 1974]. A more useful measure of fuzziness in this context, which has values directly dependent on the relative overlap between nonempty fuzzy class intersections, can be found with the algebraic product of u_i and u_j, or the form $u_i u_j(x_k)$. An interesting interpretation of the fuzzy clustering results is to compute the fuzzy partition coefficient,

$$F_c(\underset{\sim}{U}) = \frac{\mathrm{tr}(\underset{\sim}{U} * \underset{\sim}{U}^T)}{n} \tag{11.37}$$

where $\underset{\sim}{U}$ is the fuzzy partition matrix being segregated into c classes (partitions), n is the number of data sets, and the operation "$*$" is standard matrix multiplication. The product $\underset{\sim}{U} * \underset{\sim}{U}^T$ is a matrix of size $c \times c$. The partition coefficient, $F_c(\underset{\sim}{U})$, has some special properties [Bezdek, 1974]: $F_c(\underset{\sim}{U}) = 1$ if the partitioning in $\underset{\sim}{U}$ is crisp (comprised of zeros and ones); $F_c(\underset{\sim}{U}) = 1/c$ if all the values $u_i = 1/c$ (complete ambiguity); and in general $1/c \leq F_c(\underset{\sim}{U}) \leq 1$. The diagonal entries of $\underset{\sim}{U} * \underset{\sim}{U}^T$ are proportional to the amount of *unshared* membership of the data sets in the fuzzy clusters, whereas the off-diagonal elements of $\underset{\sim}{U} * \underset{\sim}{U}^T$ represent the amount of membership *shared* between pairs of fuzzy clusters of $\underset{\sim}{U}$. If the off-diagonal elements of $\underset{\sim}{U} * \underset{\sim}{U}^T$ are zero, then the partitioning (clustering) is crisp. As the partition coefficient approaches a value of unity, the fuzziness in overlap in classes is minimized. Hence, as $F_c(\underset{\sim}{U})$ increases, the decomposition of the data sets into the classes chosen is more successful.

> **Example 11.12   [Ross et al., 1993].** Forced response dynamics of a simple mechanical two-degrees-of-freedom (2-DOF) oscillating system, as shown in Fig. 11.6, are conducted. In the figure the parameters m_1, k_1, and c_1, and m_2, k_2, and c_2 are the mass, stiffness, and damping coefficients for the two masses, respectively, and the base of the system is assumed fixed to ground. The associated frequency and damping ratios for the two modes of free vibration for this system are summarized in Table 11.8. The 2-DOF system is excited with a force actuator on the larger of the two masses (see Fig. 11.6), and a displacement sensor on this same mass collects data on displacement vs. time.
>
> The response is computed for the displacement sensor in the form of frequency response functions (FRF). The derivatives of these FRF were computed for a specific exciting frequency of 1.0 rad/s with respect to the six modal mass and stiffness matrix elements (denoted in Table 11.9 as $x_1, x_2, \ldots, x_6$). Amplitude derivatives and the covariance matrix entries of these parameters are given in Table 11.9.
>
> A simple FCM classification approach was conducted on the feature data ($m = 2$) given in Table 11.9 and the fuzzy partitions shown in Table 11.10 resulted for a 2-class case ($c = 2$) and a 3-class case ($c = 3$).
>
> The resulting values of $F_c(\underset{\sim}{U})$ from Eq. (11.37) for the two clustering cases are listed in Table 11.11. Of course, the result for $c = 3$ is intuitively obvious from inspection of

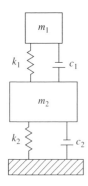

FIGURE 11.6
Mechanical system with two degrees of freedom.

TABLE 11.8
Modal response parameters for 2-DOF example

Mode	Frequency (rad/s)	Damping ratio
1	0.98617	0.01
2	1.17214	0.01

TABLE 11.9
FRF sensitivity-parameter uncertainty data sets

Data set	FRF derivative	Variance
x_1	3.5951	0.2370
x_2	1.7842	0.2906
x_3	0.1018	0.3187
x_4	3.3964	0.2763
x_5	1.7620	0.2985
x_6	0.1021	0.4142

TABLE 11.10
Clustering results for a simple 2-DOF problem

	Data pairs ($c = 2$)					
	x_1	x_2	x_3	x_4	x_5	x_6
Class 1	0.000	0.973	0.998	0.000	0.976	0.998
Class 2	1.000	0.027	0.002	1.000	0.024	0.002
	Data pairs ($c = 3$)					
Class 1	0	0	1	0	0	1
Class 2	0	1	0	0	1	0
Class 3	1	0	0	1	0	0

TABLE 11.11
Partitioning coefficient
for two different classes

c	$F_c(\underset{\sim}{U})$
2	0.982
3	1.000

Table 11.10, which is crisp. However, such obviousness is quickly lost when one deals with problems characterized by a large database.

HARDENING THE FUZZY c-PARTITION

There are two popular methods, among many others, to defuzzify fuzzy partitions, $\underset{\sim}{U}$, i.e., for hardening the fuzzy classification matrix. This defuzzification may be required in the ultimate assignment of data to a particular class. These two methods are called the *maximum membership method* and the *nearest center classifier*.

In the max membership method, the largest element in each column of the $\underset{\sim}{U}$ matrix is assigned a membership of unity and all other elements in each column are assigned a membership value of zero. In mathematical terms if the largest membership in the kth column is μ_{ik}, then x_k belongs to class i, i.e., if

$$\mu_{ik} = \max_{j \in c}\{\mu_{ik}\} \text{ then } \mu_{ik} = 1; \qquad \mu_{jk} = 0 \text{ for all } j \neq i \qquad (11.38)$$

for $i = 2, \ldots, c$ and $k = 1, 2, \ldots, n$.

In the nearest center classifier, each of the data points is assigned to the class that it is closest to; that is, the minimum Euclidean distance from a given data point and the c cluster centers dictates the class assignment of that point. In mathematical terms, if

$$d_{ik} = \min_{j \in c}\{d_{jk}\} = \min_{j \in c} \|\mathbf{x}_k - \mathbf{v}_j\|$$

then

$$\mu_{ik} = 1$$
$$\mu_{jk} = 0 \qquad \text{for all } j \neq i \qquad (11.39)$$

Example 11.13. If we take the partition matrix, $\underset{\sim}{U}$, developed on the catalytic converter in Example 11.8 as shown in Fig. 11.5, and harden it using the methods in Eqs. (11.38) and (11.39), we get the following:

$$\underset{\sim}{U} = \begin{bmatrix} 0.991 & 0.986 & 0.993 & 0 \\ 0.009 & 0.014 & 0.007 & 1 \end{bmatrix}$$

Max membership method:

$$U^{\text{Hard}} = \begin{bmatrix} 1 & 1 & 1 & 0 \\ 0 & 0 & 0 & 1 \end{bmatrix}$$

Nearest center classifier: If we take the distance measures from the catalytic converter problem, i.e.,

$$d_{12} = 0.26 \qquad d_{21} = 2.82$$
$$d_{12} = 0.31 \qquad d_{21} = 2.66$$
$$d_{12} = 0.20 \qquad d_{21} = 2.47$$
$$d_{12} = 2.65 \qquad d_{21} = 0$$

and arrange these values in a 2×4 matrix, such as

$$d_{ij} = \begin{bmatrix} 0.26 & 0.31 & 0.20 & 2.65 \\ 2.82 & 2.66 & 2.47 & 0 \end{bmatrix}$$

then the minimum value (distance) in each column is set to unity, and all other values (distances) in that column are set to zero. This process results in the following hard c-partition:

$$U^{\text{Hard}} = \begin{bmatrix} 1 & 1 & 1 & 0 \\ 0 & 0 & 0 & 1 \end{bmatrix}$$

which, for this example, happens to be the same partition that is derived using the max membership hardening method.

SIMILARITY RELATIONS FROM CLUSTERING

The classification idea can be recast in the form of a similarity relation that is also a tolerance relation. This idea represents another way to look at the structure in data, by comparing the data points to one another, pairwise, in a similarity analysis. In classification we seek to segregate data into clusters where points in each cluster are as "similar" to one another as possible and where clusters are dissimilar to one another. This notion of similarity, then, is central to classification. The use of a fuzzy similarity relation can be useful in the classification process [Bezdek and Harris, 1978].

A fuzzy relation R can be constructed from the fuzzy partition U as follows:

$$R = \left(U^{\mathrm{T}} \left(\sum \wedge \right) U \right) = [r_{kj}] \tag{11.40}$$

$$r_{kj} = \sum_{i=1}^{c} \mu_{ik} \wedge \mu_{ij} \tag{11.41}$$

where the symbol $(\sum \wedge)$ denotes "sum of mins."

Example 11.14. We take the fuzzy partition U from the fruit genetics example (Example 11.9) and perform the mixed algebraic and set operations as provided in Eqs. (11.40) and (11.41). So for

$$U^{\mathrm{T}} = \begin{bmatrix} 0.91 & 0.09 \\ 0.58 & 0.42 \\ 0.13 & 0.87 \end{bmatrix} \quad \text{and} \quad \tilde{U} = \begin{bmatrix} 0.91 & 0.58 & 0.13 \\ 0.09 & 0.42 & 0.87 \end{bmatrix}$$

we get

$$r_{11} = \min(0.91, 0.91) + \min(0.09, 0.09) = 1$$
$$r_{12} = \min(0.91, 0.58) + \min(0.09, 0.42) = 0.67$$
$$r_{13} = \min(0.91, 0.13) + \min(0.09, 0.87) = 0.22$$
$$r_{23} = \min(0.58, 0.13) + \min(0.42, 0.87) = 0.55$$

and so forth, and the following fuzzy similarity relation results:

$$R = \begin{bmatrix} 1 & 0.67 & 0.22 \\ 0.67 & 1 & 0.55 \\ 0.22 & 0.55 & 1 \end{bmatrix}$$

The fuzzy similarity relation R provides similar information about clustering as does the original fuzzy partition, U. The fuzzy classification partition groups the data according to class type; the fuzzy relation shows the pairwise similarity of the data without regard to class type. Data that have strong similarity, or high membership values in R, should tend to have high membership in the same class in U. Although the two measures are based on the same data (i.e., the features describing each data point), their information content is slightly different.

PART II   PATTERN RECOGNITION

There is no idea or proposition in the field, which cannot be put into mathematical language, although the utility of doing so can very well be doubted.

H. W. Brand
Mathematician, 1961

Pattern recognition can be defined as a process of identifying structure in data by comparisons to known structure; the known structure is developed through methods of classification [Bezdek, 1981] as illustrated in Part I. In the statistical approach to numerical pattern recognition, which is treated thoroughly by Fukunaga [1972], each input observation is represented as a multidimensional data vector (feature vector) where each component is called a feature. The purpose of the pattern recognition system is to assign each input to one of c possible pattern classes (or data clusters). Presumably, different input observations should be assigned to the same class if they have similar features and to different classes if they have dissimilar features. Statistical pattern recognition systems rest on mathematical models; it is crucial that the measure of mathematical similarity used to match feature vectors with classes assesses a property shared by physically similar components of the process generating the data.

The data used to design a pattern recognition system are usually divided into two categories: design (or training) data and test data, much like the categorization used in neural networks. Design data are used to establish the algorithmic parameters of the pattern recognition system. The design samples may be labeled (the class to which each observation belongs is known) or unlabeled (the class to which each data sample belongs is unknown). Test data are labeled samples used to test the overall performance of the pattern recognition system.

In the descriptions that follow, the following notation is used:

$$X = \{x_1, x_2, \ldots, x_n\} = \text{the universe of data samples}$$

$n = $ number of data samples in universe

$p = $ number of original (nominated) features

$x_k \in R^p$; kth data sample in X, in p-dimensional feature space

$x_{kj} \in R$; jth measured feature of x_k

$s = $ number of selected or extracted features

$c = $ number of clusters or classes

There are many similarities between classification and pattern recognition. The information provided in Fig. 11.7 summarizes the distinction between the two made in this textbook. Basically, classification establishes (or seeks to determine) the structure in data, whereas pattern recognition attempts to take new data and assign them to one of the classes defined in the classification process. Simply stated, classification *defines* the patterns and pattern recognition *assigns* data to a class; hence, the processes of define and assign are a coupled pair in the process described in Fig. 11.7. In both the classification process and the pattern recognition process there are necessary feedback loops: the first loop in classification is required when one is seeking a better segmentation of the data (i.e., better class distinctions), and the second loop is required when pattern matching fails (i.e., no useful assignment can be made).

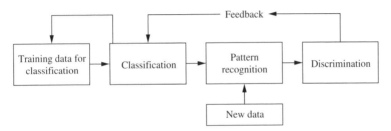

FIGURE 11.7
Difference between classification and pattern recognition.

FEATURE ANALYSIS

Feature analysis refers to methods for conditioning the raw data so that the information that is most relevant for classification and interpretation (recognition) is enhanced and represented by a minimal number of features. Feature analysis consists of three components: nomination, selection, and extraction. Feature nomination (FN) refers to the process of proposing the original p features; it is usually done by workers close to the physical process and may be heavily influenced by physical constraints, e.g., what can be measured by a particular sensor. For example, the nominated features can correspond to simple characteristics of various sensors that are represented by digitization of the sensor records. Feature selection (FS) refers to choosing the "best" subset of s features ($s < p$) from the original p features. Feature extraction (FE) describes the process of transforming the original p-dimensional feature space into an s-dimensional space in some manner that "best" preserves or enhances the information available in the original p-space. This is usually accomplished mathematically by means of some linear combination of the initial measurements.

Another method of feature extraction that lies closer to the expertise of the engineer is heuristic nomination and/or extraction. In other words, the process being examined may suggest choices for analytic features, e.g., in sensor records of measured pressures, the slopes (of rise and decay), areas (impulse or energy) during rise and decay, or even transformations (Fourier or Laplace) of the pressure waveform. Implicit in both FS and FE is a means for evaluating feature sets chosen by a particular procedure. The usual benchmark of feature quality is the empirical error rate achieved by a classifier on labeled test data. A second method of assessing feature quality is to refer algorithmic interpretations of the data to domain experts: do the computed results make sense? This latter test is less esoteric than the mathematical criteria, but very important.

PARTITIONS OF THE FEATURE SPACE

Partitioning the feature space into c regions, one for each subclass in the data, is usually in the domain of classifier design. More specifically, crisp classifiers partition R^p (or R^s) into disjoint subsets, whereas fuzzy classifiers assign fuzzy label vectors to each vector in feature space. The ideal classifier never errs; since this is usually impossible, one seeks designs that minimize the expected probability of error or, if some mistakes are more costly than others, minimize the average cost of errors, or both. An in-depth analysis of classifier design is available in Duda and Hart [1973]. Because many sources of data, such as those in image processing, do not lend themselves readily to classifier design, the discussion presented here does not consider this aspect of pattern classification. Part I of this chapter has discussed the nature and importance of cluster analysis and cluster validity in terms of feature selection.

SINGLE-SAMPLE IDENTIFICATION

A typical problem in pattern recognition is to collect data from a physical process and classify them into known patterns. The known patterns typically are represented as class structures, where each class structure is described by a number of features. For simplicity in presentation, the material that follows represents classes or patterns characterized by one feature; hence, the representation can be considered one-dimensional.

Suppose we have several typical patterns stored in our knowledge base (i.e., the computer), and we are given a new data sample that has not yet been classified. We want to determine which pattern the sample most closely resembles. Express the typical patterns as fuzzy sets $A_1, A_2, \ldots, A_m$. Now suppose we are given a new data sample, which is characterized by the crisp singleton, x_0. Using the simple criterion of maximum membership, the typical pattern that the data sample most closely resembles is found by the following expression:

$$\mu_{A_i}(x_0) = \max\{\mu_{A_1}(x_0), \mu_{A_2}(x_0), \ldots, \mu_{A_m}(x_0)\} \qquad (11.42)$$

where x_0 belongs to the fuzzy set A_i, which is the set indication for the set with the highest membership at point x_0. Figure 11.8 shows the idea expressed by Eq. (11.42), where clearly the new data sample defined by the singleton expressed by x_0 most closely resembles the pattern described by fuzzy set A_2.

> **Example 11.15 [Ross, 1995].** We can illustrate the single data sample example using the problem of identifying a triangle, as described in Chapter 6. Suppose the single data sample is described by a data triplet, where the three coordinates are the angles of a specific triangle, e.g., the triangle as shown in Fig. 6.2, $x_0 = \{A = 85°, B = 50°, C = 45°\}$. Recall from Chapter 6 that there were five known patterns stored: isosceles, right, right and isosceles, equilateral, and all other triangles. If we take this single triangle and determine its membership in each of the

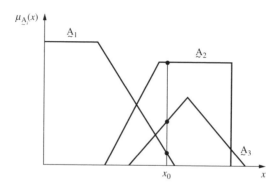

FIGURE 11.8
Single data sample using max member-ship criteria.

known patterns, we get the following results (as we did before in Chapter 6):

$$\mu_{\underset{\sim}{1}}(85, 50, 45) = 1 - \tfrac{5}{60} = 0.916$$

$$\mu_{\underset{\sim}{R}}(85, 50, 45) = 1 - \tfrac{5}{90} = 0.94$$

$$\mu_{\underset{\sim}{IR}}(85, 50, 45) = 1 - \max\left[\tfrac{5}{60}, \tfrac{5}{90}\right] = 0.916$$

$$\mu_{\underset{\sim}{E}}(85, 50, 45) = 1 - \tfrac{1}{180}(40) = 0.78$$

$$\mu_{\underset{\sim}{T}}(85, 50, 45) = \tfrac{1}{180} \min[3.35, 3(5), \tfrac{2}{5}, 40] = 0.05$$

Using the criterion of maximum membership, we see from these values that x_0 most closely resembles the right triangle pattern, $\underset{\sim}{R}$.

Now let us extend the paradigm to consider the case where the new data sample is not crisp, but rather a fuzzy set itself. Suppose we have m typical patterns represented as fuzzy sets $\underset{\sim}{A_i}$ on X ($i = 1, 2, \ldots, m$) and a new piece of data, perhaps consisting of a group of observations, is represented by a fuzzy set $\underset{\sim}{B}$ on X. The task now is to find which $\underset{\sim}{A_i}$ the sample $\underset{\sim}{B}$ most closely matches. To address this issue, we develop the notion of *fuzzy vectors*.

There are some interesting features and operations on fuzzy vectors that will become quite useful in the discipline of fuzzy pattern recognition [Dong, 1986]. Formally, a vector $\underset{\sim}{a} = (a_1, a_2, \ldots, a_n)$ is called a fuzzy vector if for any element we have $0 \leq a_i \leq 1$ for $i = 1, 2, \ldots, n$. Similarly, the transpose of the fuzzy vector $\underset{\sim}{a}$, denoted $\underset{\sim}{a}^T$, is a column vector if $\underset{\sim}{a}$ is a row vector, i.e.,

$$\underset{\sim}{a}^T = \begin{bmatrix} a_1 \\ a_2 \\ \vdots \\ a_n \end{bmatrix}$$

Let us define $\underset{\sim}{a}$ and $\underset{\sim}{b}$ as fuzzy vectors of length n, and define

$$\underset{\sim}{a} \bullet \underset{\sim}{b}^T = \overset{n}{\underset{i=1}{\vee}} (a_i \wedge b_i) \tag{11.43}$$

as the fuzzy *inner product* of $\underset{\sim}{a}$ and $\underset{\sim}{b}$, and

$$\underset{\sim}{a} \oplus \underset{\sim}{b}^T = \overset{n}{\underset{i=1}{\wedge}} (a_i \vee b_i) \tag{11.44}$$

as the fuzzy *outer product* of $\underset{\sim}{a}$ and $\underset{\sim}{b}$.

Example 11.16. We have two fuzzy vectors of length 4 as defined here, and want to find the inner product and the outer product for these two fuzzy vectors:

$$\underset{\sim}{a} = (0.3, 0.7, 1, 0.4) \quad \underset{\sim}{b} = (0.5, 0.9, 0.3, 0.1)$$

$$\underset{\sim}{a} \bullet \underset{\sim}{b}^T = (0.3, 0.7, 1.0, 0.4) \begin{pmatrix} 0.5 \\ 0.9 \\ 0.3 \\ 0.1 \end{pmatrix}$$

$$= (0.3 \wedge 0.5) \vee (0.7 \wedge 0.9) \vee (1 \wedge 0.3) \vee (0.4 \wedge 0.1) = 0.3 \vee 0.7 \vee 0.3 \vee 0.1 = 0.7$$

$$\underset{\sim}{a} \oplus \underset{\sim}{b}^{\mathrm{T}} = (0.3 \vee 0.5) \wedge (0.7 \vee 0.9) \wedge (1 \vee 0.3) \wedge (0.4 \vee 0.1) = 0.5 \wedge 0.9 \wedge 1 \wedge 0.4 = 0.4$$

The symbol $\oplus$ has also been used in the literature to describe the Boolean outer product. In this context we will use this symbol to refer to the outer product of two fuzzy vectors. An interesting feature of these products is found in comparing them to standard algebraic operations on vectors in physics. Whereas the inner and outer products on fuzzy vectors result in scalar quantities, only the algebraic inner product on vectors in physics produces a scalar; the outer product on two vectors in physics produces another vector, whose direction is orthogonal to the plane containing the original two vectors.

We now define the complement of the fuzzy vector, or fuzzy complement vector, as

$$\bar{\underset{\sim}{a}} = (1 - a_1, 1 - a_2, \ldots, 1 - a_n) = (\bar{a}_1, \bar{a}_2, \ldots, \bar{a}_n) \tag{11.45}$$

It should be obvious that since $\bar{\underset{\sim}{a}}$ is subject to the constraint $0 \le \bar{a}_i \le 1$ for $i = 1, 2, \ldots, n$, the fuzzy complement vector is also another fuzzy vector. Moreover, we define the largest element $\hat{a}$ in the fuzzy vector $\underset{\sim}{a}$ as its *upper bound*, i.e.,

$$\hat{a} = \max_i(a_i) \tag{11.46}$$

and the smallest element $\underset{\sim}{a}$ in the fuzzy vector $\underset{\sim}{a}$ as its *lower bound*, i.e.,

$$\underset{\sim}{a} = \min_i(a_i) \tag{11.47}$$

Some properties of fuzzy vectors that will become quite useful in the area of pattern recognition will be summarized here. For two fuzzy vectors, $\underset{\sim}{a}$ and $\underset{\sim}{b}$, both of length n, the following properties hold:

$$\overline{\underset{\sim}{a} \bullet \underset{\sim}{b}^{\mathrm{T}}} = \bar{\underset{\sim}{a}} \oplus \bar{\underset{\sim}{b}}^{\mathrm{T}} \quad \text{and alternatively} \quad \overline{\underset{\sim}{a} \oplus \underset{\sim}{b}^{\mathrm{T}}} = \bar{\underset{\sim}{a}} \bullet \bar{\underset{\sim}{b}}^{\mathrm{T}} \tag{11.48}$$

$$\underset{\sim}{a} \bullet \underset{\sim}{b}^{\mathrm{T}} \le (\hat{a} \wedge \hat{b}) \quad \text{and alternatively} \quad \underset{\sim}{a} \oplus \underset{\sim}{b}^{\mathrm{T}} \ge (\underset{\sim}{a} \vee \underset{\sim}{b}) \tag{11.49}$$

$$\underset{\sim}{a} \bullet \underset{\sim}{a}^{\mathrm{T}} = \hat{a} \quad \text{and} \quad \underset{\sim}{a} \oplus \underset{\sim}{a}^{\mathrm{T}} \ge \underset{\sim}{a} \tag{11.50}$$

$$\text{For } \underset{\sim}{a} \subseteq \underset{\sim}{b} \text{ then } \underset{\sim}{a} \bullet \underset{\sim}{b}^{\mathrm{T}} = \hat{a}; \quad \text{and for } \underset{\sim}{b} \subseteq \underset{\sim}{a} \text{ then } \underset{\sim}{a} \oplus \underset{\sim}{b}^{\mathrm{T}} = \underset{\sim}{a} \tag{11.51}$$

$$\underset{\sim}{a} \bullet \bar{\underset{\sim}{a}} \le \tfrac{1}{2} \quad \text{and} \quad \underset{\sim}{a} \oplus \bar{\underset{\sim}{a}} \ge \tfrac{1}{2} \tag{11.52}$$

From the fuzzy vector properties given in Eqs. (11.48)–(11.52), one can show (see Problems 11.15 and 11.16) that when two separate fuzzy vectors are identical, i.e., $\underset{\sim}{a} = \underset{\sim}{b}$, the inner product $\underset{\sim}{a} \bullet \underset{\sim}{b}^{\mathrm{T}}$ reaches a maximum while the outer product $\underset{\sim}{a} \oplus \underset{\sim}{b}^{\mathrm{T}}$ reaches a minimum. This is extremely powerful when used in any problem requiring a metric of similarity between two vectors. If two vectors are identical, the inner product metric will yield a maximum value, and if the two vectors are completely dissimilar the inner product will yield a minimum value. This chapter makes use of the inverse duality between the inner product and the outer product for fuzzy vectors and fuzzy sets in developing an algorithm for pattern recognition. These two norms, the inner product and the outer product, can be used simultaneously in pattern recognition studies because they measure *closeness* or *similarity*.

We can extend fuzzy vectors to the case of fuzzy sets. Whereas vectors are defined on a finite countable universe, sets can be used to address infinite-valued universes (see example below using Gaussian membership functions). Let $P^*(X)$ be a group of fuzzy sets with $\underset{\sim}{A}_i \neq \emptyset$, and $\underset{\sim}{A}_i \neq X$. Now we define two fuzzy sets from this family of sets, i.e., $\underset{\sim}{A}, \underset{\sim}{B} \in P^*(X)$; then either of the expressions (Eqs. (11.53) and (11.54))

$$(\underset{\sim}{A}, \underset{\sim}{B})_1 = (\underset{\sim}{A} \bullet \underset{\sim}{B}) \wedge (\overline{\underset{\sim}{A} \oplus \underset{\sim}{B}}) \tag{11.53}$$

$$(\underset{\sim}{A}, \underset{\sim}{B})_2 = \tfrac{1}{2}[(\underset{\sim}{A} \bullet \underset{\sim}{b}) + (\overline{\underset{\sim}{A} \oplus \underset{\sim}{B}})] \tag{11.54}$$

describes two metrics to assess the degree of similarity of the two sets $\underset{\sim}{A}$ and $\underset{\sim}{B}$:

$$(\underset{\sim}{A}, \underset{\sim}{B}) = (\underset{\sim}{A}, \underset{\sim}{B})_1 \quad \text{or} \quad (\underset{\sim}{A}, \underset{\sim}{B}) = (\underset{\sim}{A}, \underset{\sim}{B})_2 \tag{11.55}$$

In particular, when either of the values of $(\underset{\sim}{A}, \underset{\sim}{B})$ from Eq. (11.55) approaches 1, then the two fuzzy sets $\underset{\sim}{A}$ and $\underset{\sim}{B}$ are "more closely similar"; when either of the values $(\underset{\sim}{A}, \underset{\sim}{B})$ from Eq. (11.55) approaches a value of 0, the two fuzzy sets are "more far apart" (dissimilar). The metric in Eq. (11.53) uses a minimum property to describe similarity, and the expression in Eq. (11.54) uses an arithmetic metric to describe similarity. It can be shown (see Problem 11.19) that the first metric (Eq. (11.53)) always gives a value that is less than the value obtained from the second metric (Eq. (11.54)). Both of these metrics represent a concept that has been called the *approaching degree* [Wang, 1983].

Example 11.17. Suppose we have a universe of five discrete elements, $X = \{x_1, x_2, x_3, x_4, x_5\}$, and we define two fuzzy sets, $\underset{\sim}{A}$ and $\underset{\sim}{B}$, on this universe. Note that the two fuzzy sets are special: They are actually crisp sets and both are complements of one another:

$$\underset{\sim}{A} = \left\{ \frac{1}{x_1} + \frac{1}{x_2} + \frac{0}{x_3} + \frac{0}{x_4} + \frac{0}{x_5} \right\}$$

$$\underset{\sim}{B} = \overline{\underset{\sim}{A}} = \left\{ \frac{0}{x_1} + \frac{0}{x_2} + \frac{1}{x_3} + \frac{1}{x_4} + \frac{1}{x_5} \right\}$$

If we calculate the quantities expressed by Eqs. (11.53)–(11.55), we obtain the following values:

$$\underset{\sim}{A} \bullet \underset{\sim}{B} = 0 \quad \underset{\sim}{A} \oplus \underset{\sim}{B} = 1 \quad (\underset{\sim}{A}, \underset{\sim}{B})_1 = (\underset{\sim}{A}, \underset{\sim}{B})_2 = 0$$

The conclusion is that a crisp set and its complement are completely dissimilar.

The value of the approaching degree in the previous example should be intuitive. Since each set is the crisp complement of the other, they should be considered distinctly different patterns, i.e., there is no overlap. The inner product being zero and the outer product being unity confirm this mathematically. Conversely, if we assume fuzzy set $\underset{\sim}{B}$ to be identical to $\underset{\sim}{A}$, i.e., $\underset{\sim}{B} = \underset{\sim}{A}$, then we would find that the inner product equals 1 and the outer product equals 0 and the approaching degree (Eqs. (11.55)) would equal unity. The reader is asked to confirm this (see Problem 11.18). This proof simply reinforces the notion that a set is most similar to itself.

Example 11.18   [Ross, 1995]. Suppose we have a one-dimensional universe on the real line, $X = [-\infty, \infty]$; and we define two fuzzy sets having normal, Gaussian membership functions, $\underset{\sim}{A}, \underset{\sim}{B}$, which are defined mathematically as

$$\mu_{\underset{\sim}{A}}(x) = \exp\left[\frac{-(x-a)^2}{\sigma_a^2}\right] \qquad \mu_{\underset{\sim}{B}}(x) = \exp\left[\frac{-(x-b)^2}{\sigma_b^2}\right]$$

and shown graphically in Fig. 11.9. It can be shown that the inner product of the two fuzzy sets is equal to

$$\underset{\sim}{A} \bullet \underset{\sim}{B} = \exp\left[\frac{-(a-b)^2}{(\sigma_a + \sigma_b)^2}\right] = \mu_{\underset{\sim}{A}}(x_0) = \mu_{\underset{\sim}{B}}(x_0)$$

where $x_0 = \dfrac{\sigma_a \cdot b + \sigma_b \cdot a}{\sigma_a + \sigma_b}$

and that the outer product is calculated to be $\underset{\sim}{A} \oplus \underset{\sim}{B} = 0$. Hence, the values of Eqs. (11.53) and (11.54) are

$$(\underset{\sim}{A}, \underset{\sim}{B})_1 = \exp\left[\frac{-(a-b)^2}{(\sigma_a + \sigma_b)^2}\right] \wedge 1 \qquad \text{and} \qquad (\underset{\sim}{A}, \underset{\sim}{B})_2 = \frac{1}{2}\left\{\exp\left[\frac{-(a-b)^2}{(\sigma_a + \sigma_b)^2}\right] + 1\right\}$$

The preceding material has presented some examples in which a new data sample is compared to a single known pattern. In the usual pattern recognition problem we are interested in comparing a data sample to a number of known patterns. Suppose we have a collection of m patterns, each represented by a fuzzy set, $\underset{\sim}{A}_i$, where $i = 1, 2, \ldots, m$, and a sample pattern $\underset{\sim}{B}$, all defined on universe X. Then the question is: Which known pattern $\underset{\sim}{A}_i$ does data sample $\underset{\sim}{B}$ most closely resemble? A useful metric that has appeared in the literature is to compare the data sample to each of the known patterns in a pairwise fashion, determine the approaching degree value for each of these pairwise comparisons, then select the pair with the largest approaching degree value as the one governing the pattern recognition process. The known pattern that is involved in the maximum approaching degree value is then the pattern the data sample most closely resembles in a maximal sense. This concept has been termed the *maximum approaching degree* [Wang, 1983]. Equation (11.56) shows this concept for m known patterns:

$$(\underset{\sim}{B}, \underset{\sim}{A}_i) = \max\{(\underset{\sim}{B}, \underset{\sim}{A}_1), (\underset{\sim}{B}, \underset{\sim}{A}_2), \ldots, (\underset{\sim}{B}, \underset{\sim}{A}_m)\} \tag{11.56}$$

Example 11.19   [Ross, 1995]. Suppose you are an earthquake engineering consultant hired by the state of California to assess earthquake damage in a region just hit by a large earthquake.

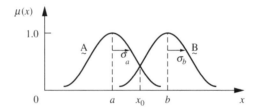

FIGURE 11.9
Two Gaussian membership functions.

Your assessment of damage will be very important to residents of the area because insurance companies will base their claim payouts on your assessment. You must be as impartial as possible. From previous historical records you determine that the six categories of the modified Mercalli intensity (I) scale (VI) to (XI) are most appropriate for the range of damage to the buildings in this region. These damage patterns can all be represented by Gaussian membership functions, $\underset{\sim}{A}_i$, $i = 1, 2, \ldots, 6$, of the following form:

$$\mu_{\underset{\sim}{A}_i}(x) = \exp\left(\frac{-(x - a_i)^2}{\sigma_a^2}\right)$$

where parameters a_i and σ_i define the shape of each membership function. Your historical database provides the information shown in Table 11.12 for the parameters for the six regions.

You determine via inspection that the pattern of damage to buildings in a given location is represented by a fuzzy set $\underset{\sim}{B}$, with the following characteristics:

$$\mu_{\underset{\sim}{B}}(x) = \exp\left(\frac{-(x - b)^2}{\sigma_b^2}\right); \qquad b = 41; \qquad \sigma_b = 10$$

The system you now have is shown graphically in Fig. 11.10. You then conduct the following calculations, using the similarity metric from Eq. (11.54) to determine the maximum approaching degree:

$$(\underset{\sim}{B}, \underset{\sim}{A}_1) = \tfrac{1}{2}(0.0004 + 1) \approx 0.5 \quad (\underset{\sim}{B}, \underset{\sim}{A}_4) = 0.98$$

$$(\underset{\sim}{B}, \underset{\sim}{A}_2) = 0.67 \qquad\qquad\quad (\underset{\sim}{B}, \underset{\sim}{A}_5) = 0.65$$

$$(\underset{\sim}{B}, \underset{\sim}{A}_3) = 0.97 \qquad\qquad\quad (\underset{\sim}{B}, \underset{\sim}{A}_6) = 0.5$$

TABLE 11.12
Parameters for Gaussian membership functions

	$\underset{\sim}{A}_1$, VI	$\underset{\sim}{A}_2$, VII	$\underset{\sim}{A}_3$, VIII	$\underset{\sim}{A}_4$, IX	$\underset{\sim}{A}_5$, X	$\underset{\sim}{A}_6$, XI
a_i	5	20	35	49	71	92
σ_{a_i}	3	10	13	26	18	4

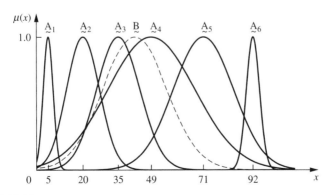

FIGURE 11.10
Six known patterns and a new fuzzy set data sample.

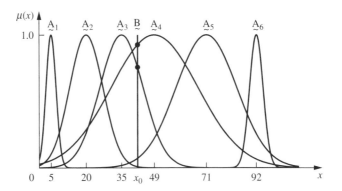

FIGURE 11.11
Six known patterns and a new singleton data sample.

From this list we see that Mercalli intensity IX (A_4) most closely resembles the damaged area because of the maximum membership value of 0.98.

Suppose you assume the membership function of the damaged region to be a simple singleton with the following characteristics:

$$\mu_B(41) = 1 \quad \text{and} \quad \mu_B(x \neq 41) = 0 \quad x_0 = 41$$

as shown in Fig. 11.11. This example reduces to the single data sample problem posed earlier, i.e.,

$$(B, A_i) = \mu_{A_i}(x_0) \wedge 1 = \mu_{A_i}(x_0)$$

Your calculations, again using Eq. (11.54), produce the following results:

$$\mu_{A_1}(41) \approx 0 \qquad \mu_{A_4}(41) = .91$$

$$\mu_{A_2}(41) = 0.01 \qquad \mu_{A_5}(41) = 0.06$$

$$\mu_{A_3}(41) = 0.81 \qquad \mu_{A_6}(41) \approx 0$$

Again, Mercalli scale IX (A_4) would be chosen on the basis of maximum membership (0.91). If we were to make the selection without regard to the shapes of the membership values, as shown in Fig. 11.11, but instead only considered the mean value of each region, we would be inclined erroneously to select region VIII because its mean value of 35 is closer to the singleton at 41 than it is to the mean value of region IX, i.e., to 49.

MULTIFEATURE PATTERN RECOGNITION

In the material covered so far in this chapter we have considered only one-dimensional pattern recognition; that is, the patterns here have been constructed based only on a single feature, such as Mercalli earthquake intensity. Suppose the preceding example on earthquake damage also considered, in addition to earthquake intensity, the importance of the particular building (schools versus industrial plants), seismicity of the region, previous history of damaging quakes, and so forth. How could we address the consideration of many features in the pattern recognition process? The literature develops many answers to this question, but this text summarizes three popular and easy approaches: (1) nearest neighbor classifier, (2) nearest center classifier, and (3) weighted approaching degree. The first two methods are restricted to the recognition of crisp singleton data samples.

In the *nearest neighbor classifier* we can consider m features for each data sample. So each sample $(\underset{\sim}{x}_i)$ is a vector of features,

$$\underset{\sim}{x}_i = \{x_{i1}, x_{i2}, x_{i3}, \dots, x_{im}\} \tag{11.57}$$

Now suppose we have n data samples in a universe, or $X = \{x_1, x_2, x_3, \dots, x_n\}$. Using a conventional fuzzy classification approach, we can cluster the samples into c-fuzzy partitions, then get c-hard partitions from these by using the equivalent relations idea or by "hardening" the soft partition $\underset{\sim}{U}$, both of which were described in Part I of this chapter. This would result in hard classes with the following properties:

$$X = \bigcup_{i=1}^{c} A_i \qquad A_i \cap A_j = \emptyset \qquad i \neq j$$

Now if we have a new singleton data sample, say $\mathbf{x}$, then the nearest neighbor classifier is given by the following distance measure, d:

$$d(\mathbf{x}, \mathbf{x}_i) = \min_{1 \leq k \leq n} \{d(\mathbf{x}, \mathbf{x}_k)\} \tag{11.58}$$

for each of the n data samples where $\mathbf{x}_i \in A_j$. That is, points $\mathbf{x}$ and $\mathbf{x}_i$ are nearest neighbors, and hence both would belong to the same class.

In another method for singleton recognition, the nearest center classifier method works as follows. We again start with n known data samples, $\mathbf{X} = \{x_1, x_2, x_3, \dots, x_n\}$, and each data sample is m-dimensional (characterized by m features). We then cluster these samples into c-classes using a fuzzy classification method such as the fuzzy c-means approach described in Part I of this chapter. These fuzzy classes each have a class center, so

$$V = \{v_1, v_2, v_3, \dots, v_c\}$$

is a vector of the c class centers. If we have a new singleton data sample, say $\mathbf{x}$, the nearest center classifier is then given by

$$d(\mathbf{x}, \mathbf{v}_i) = \min_{1 \leq k \leq c} \{d(\mathbf{x}, \mathbf{v}_k)\} \tag{11.59}$$

and now the data singleton, $\mathbf{x}$, is classified as belonging to fuzzy partition, $\underset{\sim}{A}_i$.

In the third method for addressing multifeature pattern recognition for a sample with several (m) fuzzy features, we will use the approaching degree concept again to compare the new data pattern with some known data patterns. Define a new data sample characterized by m features as a collection of noninteractive fuzzy sets, $\underset{\sim}{B} = \{\underset{\sim}{B}_1, \underset{\sim}{B}_2, \dots, \underset{\sim}{B}_m\}$. Because the new data sample is characterized by m features, each of the known patterns, $\underset{\sim}{A}_i$, is also described by m features. Hence, each known pattern in m-dimensional space is a fuzzy class (pattern) given by $\underset{\sim}{A}_i = \{\underset{\sim}{A}_{i1}, \underset{\sim}{A}_{i2}, \dots, \underset{\sim}{A}_{im}\}$, where $i = 1, 2, \dots, c$ describes c-classes (c-patterns). Since some of the features may be more important than others in the pattern recognition process, we introduce normalized weighting factors w_j, where

$$\sum_{j=1}^{m} w_j = 1 \tag{11.60}$$

Then either Eq. (11.53) or (11.54) in the approaching degree concept is modified for each of the known c-patterns ($i = 1, 2, \ldots, c$) by

$$(\underset{\sim}{B}, \underset{\sim}{A}_i) = \sum_{j=1}^{m} w_j (\underset{\sim}{B}_j, \underset{\sim}{A}_{ij}) \tag{11.61}$$

As before in the maximum approaching degree, sample $\underset{\sim}{B}$ is closest to pattern $\underset{\sim}{A}_j$ when

$$(\underset{\sim}{B}, \underset{\sim}{A}_j) = \max_{1 \le i \le c} \{(\underset{\sim}{B}, \underset{\sim}{A}_i)\} \tag{11.62}$$

Note that when the collection of fuzzy sets $\underset{\sim}{B} = \{\underset{\sim}{B}_1, \underset{\sim}{B}_2, \ldots, \underset{\sim}{B}_m\}$ reduces to a collection of crisp singletons, i.e., $\underset{\sim}{B} = \{\mathbf{x}_1, \mathbf{x}_2, \ldots, \mathbf{x}_m\}$, then Eq. (11.61) reduces to

$$\mu_{\underset{\sim}{A}_i}(x) = \sum_{j=1}^{m} w_j \cdot \mu_{\underset{\sim}{A}_{ij}}(\mathbf{x}_j) \tag{11.63}$$

As before in the maximum approaching degree, sample singleton, $\mathbf{x}$, is closest to pattern $\underset{\sim}{A}_j$ when Eq. (11.62) reduces to

$$\mu_{\underset{\sim}{A}}(\mathbf{x}) = \max_{1 \le i \le c} \{\mu_{\underset{\sim}{A}_i}(\mathbf{x})\} \tag{11.64}$$

Example 11.20. An example of multifeature pattern recognition is given where $m = 2$; the patterns can be illustrated in 3D images. Suppose we have a new pattern, $\underset{\sim}{B}$, that we wish to recognize by comparing it to other known patterns. This new pattern is characterized by two features; hence, it can be represented by a vector of its two noninteractive projections, $\underset{\sim}{B}_1$ and $\underset{\sim}{B}_2$. That is,

$$\underset{\sim}{B} = \left\{\underset{\sim}{B}_1, \underset{\sim}{B}_2\right\}$$

where the noninteractive patterns $\underset{\sim}{B}_1$ and $\underset{\sim}{B}_2$ are defined on their respective universes of discourse, X_1 and X_2. The two projections together (using Eqs. (2.35)) produce a 3D pattern in the shape of a pyramid, as shown in Fig. 11.12.

Further suppose that we have two ($c = 2$) patterns to which we wish to compare our new pattern; call them patterns $\underset{\sim}{A}_1$ and $\underset{\sim}{A}_2$. Each of these two known patterns could also

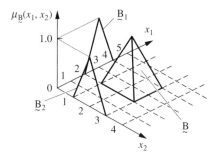

FIGURE 11.12
New pattern B and its noninteractive projections, $\underset{\sim}{B}_1$ and $\underset{\sim}{B}_2$.

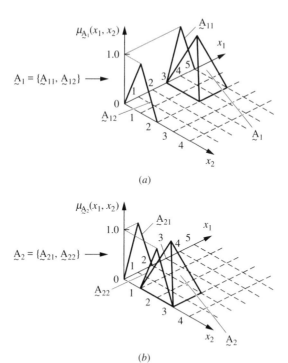

FIGURE 11.13

Multifeature pattern recognition: (*a*) known pattern A_1 and its noninteractive projections, A_{11} and A_{12}; (*b*) known pattern A_2 and its noninteractive projections, A_{21} and A_{22}.

be represented by their respective noninteractive projections, as shown in Figs. 11.13*a* and 11.13*b*, where the projections of each known pattern are also defined on X_1 and X_2.

The last step in this process is to assign weights to the various known patterns. Let us assume that $w_1 = 0.3$ and $w_2 = 0.7$, since $0.3 + 0.7 = 1$ by Eq. (11.60). We compare the new pattern with the two known patterns using Eq. (11.61),

$$\left(B, A_1 \right) = w_1 \left(B_1, A_{11} \right) + w_2 \left(B_2, A_{12} \right)$$

$$\left(B, A_2 \right) = w_1 \left(B_1, A_{21} \right) + w_2 \left(B_2, A_{22} \right)$$

where each of the operations in the preceding expressions is determined using the method of the approaching degree as described in Eqs. (11.53) or (11.54) for the zth pattern, i.e.,

$$\left(B, A_z \right) = \left[\left(B \bullet A_z \right) \wedge \overline{\left(B \oplus A_z \right)} \right] \quad \text{or} \quad \left(B, A_z \right) = \tfrac{1}{2} \left[\left(B \bullet A_z \right) + \left(B \oplus A_z \right) \right]$$

Then we assign the new pattern to the known pattern most closely resembling the new pattern using Eq. (11.62), i.e.,

$$\left(B, A_z \right) = \max \left\{ \left(B, A_1 \right), \left(B, A_2 \right) \right\}$$

The remainder of this example is left as an exercise for the reader.

Although it is not possible to sketch the membership functions for problems dealing with three or more features, the procedures outlined for multifeature pattern recognition work just as they did with the previous example. The following example in chemical engineering illustrates the multidimensional issues of Eqs. (11.60) to (11.64).

Example 11.21. A certain industrial production process can be characterized by three features: (1) pressure, (2) temperature, and (3) flow rate. Combinations of these features are used to indicate the current mode (pattern) of operation of the production process. Typical linguistic values for each feature for each mode of operation are defined by the fuzzy sets given in Table 11.13. The pattern recognition task is described as follows: the system *reads* sensor indicators of each feature (pressure, temperature, flow rate), manifested as crisp read-out values; it then determines the current mode of operation (i.e., it attempts to recognize a pattern of operation), and then the results are logged.

The four modes (patterns) of operation, and their associated linguistic values for each feature, are as follows:

1. *Autoclaving*: Here the pressure is high, temperature is high, and the flow rate is zero.
2. *Annealing*: Here the pressure is high, temperature is low, and the flow rate is zero.
3. *Sintering*: Here the pressure is low, temperature is zero, and the flow rate is low.
4. *Transport*: Here the pressure is zero, temperature is zero, and the flow rate is high.

This linguistic information is summarized in Table 11.13.

The features of pressure, temperature, and flow rate are expressed in the engineering units of kPa (kilopascals), °C (degrees Celsius), and gph (gallons per hour), respectively. Membership functions for these three features are shown in Figs. 11.14–11.16.

Now, suppose the system reads from a group of sensors a set of crisp readings (pressure = 5 kPa, temperature = 150°C, flow = 5 gph). We want to assign (recognize) this

TABLE 11.13
Relationships between operation mode and feature values

Mode (pattern)	Pressure	Temperature	Flow rate
Autoclaving	High	High	Zero
Annealing	High	Low	Zero
Sintering	Low	Zero	Low
Transport	Zero	Zero	High

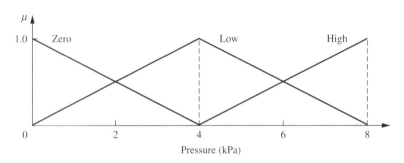

FIGURE 11.14
Membership functions for pressure.

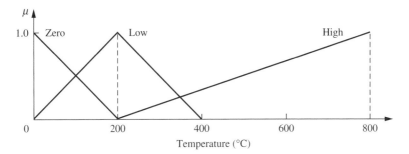

FIGURE 11.15
Membership functions for temperature.

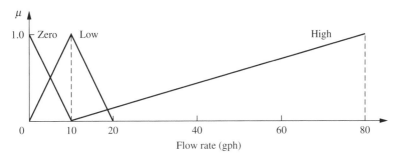

FIGURE 11.16
Membership functions for flow rate.

group of sensor readings to one of our four patterns (modes of operation). To begin, we need to assign weights to each of the features using Eq. (11.60). Since there is an explosion hazard associated with the production pressure value (5 kPa), we will weight it more heavily than the other two features:

$$w_{\text{pressure}} = 0.5$$

$$w_{\text{temperature}} = 0.25$$

$$w_{\text{flow}} = 0.25$$

Now we will use Eqs. (11.63) and (11.64) to employ the approaching degree to find which mode of operation is indicated by the above crisp values (5 kPa, 150°C, 5 gph). Using Eq. (11.63) and these two expressions,

$$X = \{5 \text{ kPa}, 150°\text{C}, 5 \text{ gph}\}$$

$$W = \{0.5, 0.25, 0.25\}$$

we find that

$$\mu_{\text{autoclaving}}(x) = (0.5) \cdot (0.25) + (0.25) \cdot (0) + (0.25) \cdot (0.5) = 0.25$$

$$\mu_{\text{annealing}}(x) = (0.5) \cdot (0.25) + (0.25) \cdot (0.75) + (0.25) \cdot (0.5) = 0.4375$$

$$\mu_{\text{sintering}}(x) = (0.5) \cdot (0.75) + (0.25) \cdot (0.25) + (0.25) \cdot (0.5) = 0.5625$$

$$\mu_{\text{transport}}(x) = (0.5) \cdot (0) + (0.25) \cdot (0.25) + (0.25) \cdot (0) = 0.0625$$

The crisp set, X = {5 kPa, 150°C, 5 gph}, most closely matches the values of pressure, temperature, and flow associated with *sintering*. Therefore, we write the production mode "sintering" in the logbook as the current production mode indicated by crisp readings from our three sensors.

Now suppose for this industrial process we use the same patterns (autoclaving, annealing, etc.). Suppose now that the readings or information on pressure, temperature, and flow are fuzzy sets rather than crisp singletons, i.e., B̰ = {B̰$_{pressure}$, B̰$_{temperature}$, B̰$_{flow}$}.

These fuzzy sets are defined in Figs. 11.17–11.19. Given these fuzzy definitions for our new pattern B̰, we use Eq. (11.62) to find which pattern is best matched by the new values B̰.

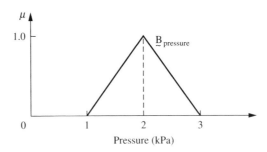

FIGURE 11.17
Fuzzy sensor reading for pressure.

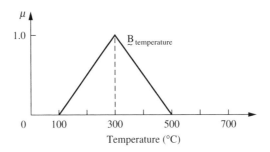

FIGURE 11.18
Fuzzy sensor reading for temperature.

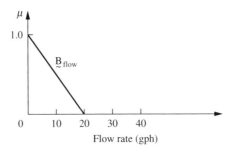

FIGURE 11.19
Fuzzy sensor reading for flow rate.

For the approaching degree between our new pattern's pressure feature and the stored autoclaving pattern's pressure feature we get

$$\underset{\sim}{B}_{pressure} \bullet autoclaving\ pressure = 0(max\ of\ mins)$$

$$\underset{\sim}{B}_{pressure} \oplus autoclaving\ pressure = 0(min\ of\ maxes)$$

as summarized in Fig. 11.20.

For the approaching degree between our new pattern's temperature feature and the stored autoclaving pattern's temperature feature we get

$$\underset{\sim}{B}_{temperature} \bullet autoclaving\ temperature = max[(0 \wedge 0), (0 \wedge 0.5), (0.166 \wedge 1),$$

$$(0.33 \wedge 0.5), (0 \wedge 0.5)] = 0.33$$

$$\underset{\sim}{B}_{temperature} \oplus autoclaving\ temperature = min[(0 \vee 0), (0 \vee 0.5), (0.166 \vee 1),$$

$$(0.33 \vee 0.5), (0 \vee 0.5)] = 0$$

as summarized in Fig. 11.21.

For the approaching degree between our new pattern's flow rate feature and the stored autoclaving pattern's flow rate feature we get

$$\underset{\sim}{B}_{flow} \bullet autoclaving\ flow = max[(1 \wedge 1), (0 \wedge 0.5), (0 \wedge 0)] = 1.0$$

$$\underset{\sim}{B}_{flow} \oplus autoclaving\ flow = min[(1 \vee 1), (0 \vee 0.5), (0 \vee 0)] = 0$$

as summarized in Fig. 11.22.

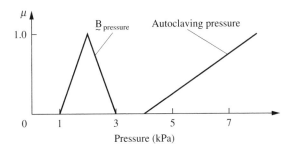

FIGURE 11.20
Pressure comparisons for autoclaving pattern.

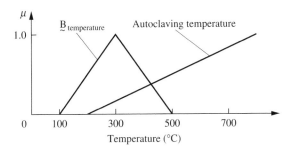

FIGURE 11.21
Temperature comparisons for autoclaving pattern.

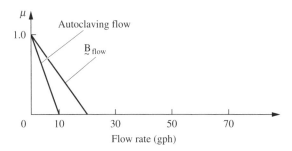

FIGURE 11.22
Flow rate comparisons for autoclaving pattern.

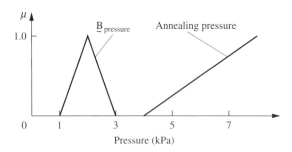

FIGURE 11.23
Pressure comparisons for annealing pattern.

Now the use of Eqs. (11.54) and (11.61) enables us to calculate the approaching degree value between the new sensor pattern and the autoclaving pattern:

$$(\underset{\sim}{B}, \text{autoclaving}) = 0.5 \left[\tfrac{1}{2}(0+1)\right] + 0.25 \left[\tfrac{1}{2}(0.33 + 1.0)\right] + 0.25 \left[\tfrac{1}{2}(1+1)\right]$$

$$= (0.5)(0.5) + (0.25)(0.66) + (0.25)(1) = 0.665$$

For the next possible pattern, annealing, we again use Eq. (11.62) to determine the approaching degree between the new sensor pressure and the annealing pressure. Because they are disjoint,

$$\underset{\sim}{B}_{\text{pressure}} \bullet \text{annealing pressure} = 0$$

$$\underset{\sim}{B}_{\text{pressure}} \oplus \text{annealing pressure} = 0$$

as summarized in Fig. 11.23.

The approaching degree between the new sensor temperature and the annealing temperature is given by

$$\underset{\sim}{B}_{\text{temperature}} \bullet \text{annealing temperature} = 0.75$$

by inspection of max of mins, and

$$\underset{\sim}{B}_{\text{temperature}} \oplus \text{annealing temperature}$$

$$= \min[(0 \vee 0), (0 \vee 0.5), (0.5 \vee 1), (1 \vee 0.5), (0.5 \vee 0), (0 \vee 0)] = 0$$

as summarized in Fig. 11.24.

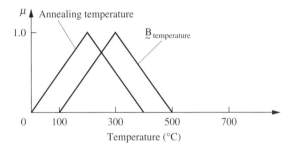

FIGURE 11.24
Temperature comparisons for annealing pattern.

The approaching degree between the new sensor flow rate and the annealing flow rate is given by

$$\underset{\sim}{B}_{flow} \bullet \text{annealing flow} = 1$$

$$\underset{\sim}{B}_{flow} \oplus \text{annealing flow} = 0$$

(identical to $\underset{\sim}{B}_{flow}$ and autoclaving flow).

Again using Eqs. (11.54) and (11.61), we get the approaching degree value for the annealing pattern,

$$(B, \text{annealing}) = 0.5[\tfrac{1}{2}(0+1)] + 0.25[\tfrac{1}{2}(0.75+1)] + 0.25[\tfrac{1}{2}(1+1)]$$

$$= (0.5)(0.5) + (0.25)(0.87) + (0.25)(1) = 0.7175$$

Now moving to the next pattern, sintering, we again use Eq. (11.62) for each of the features. The first is pressure:

$$\underset{\sim}{B}_{pressure} \bullet \text{sintering pressure} \approx 0.6$$

by inspection of max (mins), and

$$\underset{\sim}{B}_{pressure} \oplus \text{sintering pressure}$$

$$= \min[(0 \vee 0), (0 \vee 0.25), (1 \vee 0.5), (0 \vee 0.75), (0 \vee 1), \ldots] = 0$$

as summarized in Fig. 11.25.

Next is temperature:

$$\underset{\sim}{B}_{temperature} \bullet \text{sintering temperature} = 0.25$$

by inspection of max (mins), and

$$\underset{\sim}{B}_{temperature} \oplus \text{sintering temperature} = \min[(0 \vee 1), (0 \vee 0.5), (0.5 \vee 0),$$

$$(1 \vee 0), (0.5 \vee 0), (0 \vee 0)] = 0$$

as summarized in Fig. 11.26.

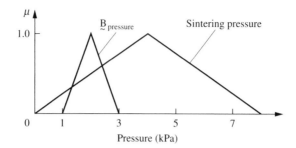

FIGURE 11.25
Pressure comparisons for sintering pattern.

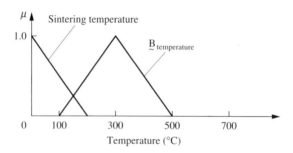

FIGURE 11.26
Temperature comparisons for sintering pattern.

Finally we consider the flow rate:

$$\underset{\sim}{B}_{flow} \bullet \text{sintering flow} = 0.7$$

by inspection of max (mins), and

$$\underset{\sim}{B}_{flow} \oplus \text{sintering flow} = \min[(0 \vee 1), (1 \vee 0), (0 \vee 0)] = 0$$

as summarized in Fig. 11.27.

Using Eqs. (11.54) and (11.61) with the metric from Eq. (11.54), we get

$$(\underset{\sim}{B}, \text{sintering}) = 0.5[\tfrac{1}{2}(0.6 + 1)] + 0.25[\tfrac{1}{2}(0.25 + 1)] + 0.25[\tfrac{1}{2}(0.7 + 1)]$$

$$= (0.5)(0.8) + (0.25)(0.625) + (0.25)(0.85) = 0.7687$$

Finally we consider the last pattern, the transport mode of operation. Using Eq. (11.62) for each feature, we begin first with pressure:

$$\underset{\sim}{B}_{pressure} \bullet \text{transport pressure} \approx 0.7$$

by inspection of max (mins), and

$$\underset{\sim}{B}_{pressure} \oplus \text{transport pressure} = \min[(0 \vee 1), (0 \vee 0.75), (1 \vee 0.5), (0.25 \vee 0), (0 \vee 0)] = 0$$

as summarized in Fig. 11.28.

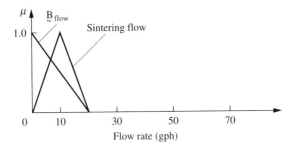

FIGURE 11.27
Flow rate comparisons for sintering pattern.

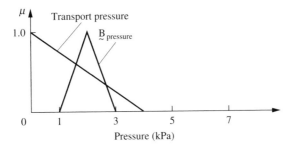

FIGURE 11.28
Pressure comparisons for transport pattern.

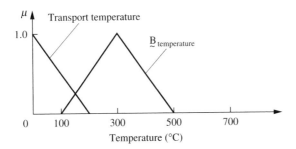

FIGURE 11.29
Temperature comparisons for transport pattern.

Then, moving to temperature:

$$B_{temperature} \bullet \text{transport temperature} = 0.25$$
$$B_{temperature} \oplus \text{transport temperature} = 0$$

as summarized in Fig. 11.29. And last, moving to flow rate:

$$B_{flow} \bullet \text{transport flow} = 0.1$$
$$B_{flow} \oplus \text{transport flow} = 0$$

as summarized in Fig. 11.30.

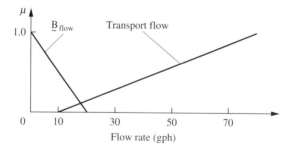

FIGURE 11.30
Flow rate comparisons for transport pattern.

To conclude the calculations using the approaching degree on the last pattern of transport, Eqs. (11.54) and (11.61) are used to determine

$$(\underset{\sim}{B}, transport) = 0.5 \left[\tfrac{1}{2}(0.7+1)\right] + 0.25 \left[\tfrac{1}{2}(0.25+1)\right] + 0.25 \left[\tfrac{1}{2}(0.1+1)\right]$$

$$= (0.5)(0.85) + (0.25)(0.625) + (0.25)(0.55) = 0.7188$$

Summarizing the results for the four possible patterns, we have

$$(\underset{\sim}{B}, \text{ autoclaving}) = 0.665$$

$$(\underset{\sim}{B}, \text{ annealing}) = 0.7175$$

$$(\underset{\sim}{B}, \text{ sintering}) = 0.7687 \qquad \text{(max is here)}$$

$$(\underset{\sim}{B}, \text{ transport}) = 0.7188$$

The fuzzy readings of pressure, temperature, and flow collectively match most closely, in an approaching degree sense, the *sintering* pattern. We therefore write the production mode "sintering" in our log.

IMAGE PROCESSING

An image (having various shades of gray) is represented mathematically by a spatial brightness function $f(m, n)$ where (m, n) denotes the spatial coordinate of a point in the (flat) image. The value of $f(m, n)$, $0 < f(m, n) < \infty$, is proportional to the brightness value or gray level of the image at the point (m, n). For computer processing, the continuous function $f(m, n)$ has been discretized both in spatial coordinates and in brightness. Such an approximated image X(digitized) can be considered as an $M \times N$ array,

$$X = f(m, n) = \begin{bmatrix} x_{11} & x_{12} & \cdots & x_{1n} & \cdots & x_{1N} \\ x_{21} & x_{22} & \cdots & x_{2n} & \cdots & x_{2N} \\ x_{31} & x_{32} & \cdots & x_{3n} & \cdots & x_{3N} \\ \vdots & \vdots & \cdots & \vdots & \cdots & \vdots \\ x_{M1} & x_{M2} & \cdots & x_{Mn} & \cdots & x_{MN} \end{bmatrix} \qquad (11.65)$$

whose row and column indices identify a point (m, n) in the image, and the corresponding matrix element value $x_{mn}[\sim f(m, n)]$ denotes the gray level at that point.

The right side of Eq. (11.65) represents what is called a digital image. Each element of the matrix, which is a discrete quantity, is referred to as an image element, picture element, pixel, or pel, with the last two names commonly used as abbreviations of picture element. From now on the terms *image* and *pixels* will be used to denote a digital image and its elements, respectively. For the purpose of processing, this image along with the coordinates of its pixels is stored in the computer in the form of an $M \times N$ array of numbers.

The methods so far developed for image processing may be categorized into two broad classes, namely, frequency domain methods and spatial domain methods. The techniques in the first category depend on modifying the Fourier transform of an image by transforming pixel intensity to pixel frequency, whereas in spatial domain methods the direct manipulation of the pixel is adopted. Some fairly simple and yet powerful processing approaches are formulated in the spatial domain.

In frequency domain methods, processing is done with various kinds of frequency filters. For example, low frequencies are associated with uniformly gray areas, and high frequencies are associated with regions where there are abrupt changes in pixel brightness. In the spatial domain methods pixel intensities can be modified independently of pixel location, or they can be modified according to their neighboring pixel intensities. Examples of these methods include (1) contrast stretching, where the range of pixel intensities is enlarged to accommodate a larger range of values, (2) image smoothing, where "salt and pepper" noise is removed from the image, and (3) image sharpening, which involves edge or contour detection and extraction.

Although many of the crisp methods for image processing have a good physical and theoretical basis, they often correlate poorly with the recognition of an image judged by a human because the human visual system does not process the image in a point-by-point fashion. When pattern recognition becomes indeterminate because the underlying variability is vague and imprecise, fuzzy methods can be very useful. A good example of this is the recognition of human speech. Speech carries information regarding the message and the speaker's sex, age, health, and mind; hence it is to a large extent fuzzy in nature. Similarly, an image carries significant fuzzy information. With this in mind we could consider an image as an array of fuzzy singletons, each with a value of membership function denoting the degree of brightness, or "grayness."

Before one is able to conduct meaningful pattern recognition exercises with images, one may need to preprocess the image to achieve the best image possible for the recognition process. The original image might be polluted with considerable noise, which would make the recognition process difficult. Processing, reducing, or eliminating this noise will be a useful step in the process. An image can be thought of as an ordered array of pixels, each characterized by gray tone. These levels might vary from a state of no brightness, or completely black, to a state of complete brightness, or totally white. Gray tone levels in between these two extremes would get increasingly lighter as we go from black to white. Various preprocessing techniques such as contrast enhancement, filtering, edge detection, ambiguity measure, segmentation, and others are described in the literature [Pal and Majumder, 1986]. For this chapter we will introduce only contrast enhancement using fuzzy procedures.

An image X of $M \times N$ dimensions can be considered as an array of fuzzy singletons, each with a value of membership denoting the degree of brightness level $p, p = 0, 1, 2, \ldots, P - 1$ (e.g., a range of densities from $p = 0$ to $p = 255$), or some relative pixel density. Using the notation of fuzzy sets, we can then write Eq. (11.65) as

$$
X = \begin{bmatrix}
\mu_{11}/x_{11} & \mu_{12}/x_{12} & \cdots & \mu_{1n}/x_{1n} & \cdots & \mu_{1N}/x_{1N} \\
\mu_{21}/x_{21} & \mu_{22}/x_{22} & \cdots & \mu_{2n}/x_{2n} & \cdots & \mu_{2N}/x_{2N} \\
\mu_{31}/x_{31} & \mu_{32}/x_{32} & \cdots & \mu_{3n}/x_{3n} & \cdots & \mu_{3N}/x_{3N} \\
\vdots & \vdots & \cdots & \vdots & \cdots & \vdots \\
\mu_{M1}/x_{M1} & \mu_{M2}/x_{M2} & \cdots & \mu_{Mn}/x_{Mn} & \cdots & \mu_{MN}/x_{MN}
\end{bmatrix}
\tag{11.66}
$$

where $m = 1, 2, \ldots, M$ and $n = 1, 2, \ldots, N$, and where μ_{mn}/x_{mn} $(0 \le \mu_{mn} \le 1)$ represents the grade of possessing some property μ_{mn} by the (m, n)th pixel x_{mn}. This fuzzy property μ_{mn} may be defined in a number of ways with respect to any brightness level (pixel density) depending on the problem.

Contrast within an image is the measure of difference between the gray levels in an image. The greater the contrast, the greater is the distinction between gray levels in the image. Images of high contrast have either all black or all white regions; there is very little gray in the image. Low-contrast images have lots of similar gray levels in the image, and very few black or white regions. High-contrast images can be thought of as crisp, and low-contrast ones as completely fuzzy. Images with good gradation of grays between black and white are usually the best images for purposes of recognition by humans. Heretofore, computers have worked best with images that have had high contrast, although algorithms based on fuzzy sets have been successful with both.

The object of contrast enhancement is to process a given image so that the result is more suitable than the original for a specific application in pattern recognition. As with all image processing techniques we have to be especially careful that the processed image is not distinctly different from the original image, making the identification process worthless. The technique used here makes use of modifications to the brightness membership value in stretching or contracting the contrast of an image. Many contrast enhancement methods work as shown in Fig. 11.31, where the procedure involves a primary enhancement of an image, denoted by E_1 in the figure, followed by a smoothing algorithm, denoted by S, and a subsequent final enhancement, step E_2. The fuzzy operator defined in Eq. (5.31), called intensification, is often used as a tool to accomplish the primary and final enhancement phases shown in Fig. 11.31.

The function of the smoothing portion of this method (the S block in Fig. 11.31) is to blur (make more fuzzy) the image, and this increased blurriness then requires the use of the final enhancement step, E_2. Smoothing is based on the property that adjacent image points (points that are close spatially) tend to possess nearly equal gray levels. Generally, smoothing algorithms distribute a portion of the intensity of one pixel in the image to adjacent pixels. This distribution is greatest for pixels nearest to the pixel being smoothed, and it decreases for pixels farther from the pixel being smoothed.

The contrast intensification operator, Eq. (5.31), on a fuzzy set $\underset{\sim}{A}$ generates another fuzzy set, $\underset{\sim}{A}' = \text{INT}(\underset{\sim}{A})$, in which the fuzziness is reduced by increasing the values of

FIGURE 11.31
Diagram of the enhancement model [adapted from Pal and Majumder, 1986].

$\mu_{\underset{\sim}{A}}(x)$ that are greater than 0.5 and by decreasing the values of $\mu_{\underset{\sim}{A}}(x)$ that are less than 0.5 [Pal and King, 1980]. If we define this transformation T_1, we can define T_1 for the membership values of brightness for an image as

$$T_1(\mu_{mn}) = T_1'(\mu_{mn}) = 2\mu_{mn}^2 \quad 0 \le \mu_{mn} \le 0.5$$
$$= T_1''(\mu_{mn}) = 1 - 2(1 - \mu_{mn})^2 \quad 0.5 \le \mu_{mn} \le 1 \tag{11.67}$$

In general, each μ_{mn} in X may be modified to μ_{mn}' to enhance the image X in the property domain by a transformation function, T_r, where

$$\mu_{mn}' = T_r(\mu_{mn}) = T_r'(\mu_{mn}) \quad 0 \le \mu_{mn} \le 0.5$$
$$= T_r''(\mu_{mn}) \quad 0.5 \le \mu_{mn} \le 1 \tag{11.68}$$

and $T_1(\mu_{mn})$ represents the operator INT as defined in Eq. (5.31). The transformation T_r is defined as successive applications of T_1 by the recursive relation,

$$T_r(\mu_{mn}) = T_1\{T_{r-1}(\mu_{mn})\} \quad r = 1, 2, \ldots \tag{11.69}$$

The graphical effect of this recursive transformation for a typical membership function is shown in Fig. 11.32. As r (i.e., the number of successive applications of the INT function) increases, the slope of the curve gets steeper. As r approaches infinity, the shape approaches a crisp (binary) function. The parameter r allows the user to use an appropriate level of enhancement for domain-specific situations.

Example 11.22. We will demonstrate enhancement of the image shown in Fig. 11.33a. The dark square image of Fig. 11.33a has a lighter square box in it that is not very apparent because the shade of the background is very nearly the same as that of the lighter box itself. Table 11.14 shows the 256 gray-scale intensity values of pixels of the 10×10 pixel array of the image shown in Fig. 11.33a. If we take the intensity values from Table 11.15 and scale them on the

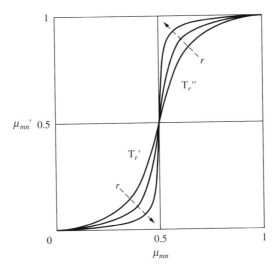

FIGURE 11.32
INT transformation function for contrast enhancement [Pal and Majumder, 1986].

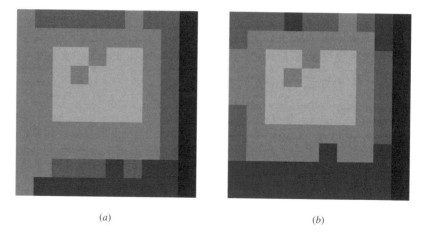

(a)   (b)

FIGURE 11.33
Lighter square inside smaller square: (a) original image; (b) image after one application of INT operator (Eq. (11.67)).

interval [0, 255], we get membership values in the density set *white* (low values are close to black, high values close to white). These values, of course, will be between 0 and 1 as membership values.

Using Eq. (11.67), we modify the pixel values to obtain the matrix shown in Table 11.16. The reader should notice that the intensity values above and below 0.5 have been suitably modified to increase the contrast between the intensities. The enhanced image is shown in Fig. 11.33b. Results of successive enhancements of the image by using Eq. (11.67) repeatedly are shown in Figs. 11.33c–11.33h.

A useful smoothing algorithm is called defocusing. The (m, n)th smoothed pixel intensity is found [Pal and King, 1981] from

$$\mu'_{mn} = a_0 \mu_{mn} + a_1 \sum_{Q_1} \mu_{ij} + a_2 \sum_{Q_2} \mu_{ij} + \cdots + a_s \sum_{Q_s} \mu_{ij} \qquad (11.70)$$

where

$$a_0 + N_1 a_1 + N_2 a_2 + \cdots + N_s a_s = 1$$

$$1 > a_1 > a_2 \cdots a_s > 0$$

$$(i, j) \neq (m, n)$$

In Eq. (11.70), μ_{mn} represents the (m, n)th pixel intensity, expressed as a membership value; Q_1 denotes a set of N_1 coordinates (i, j) that are on or within a circle of radius R_1 centered at the point (m, n); Q_s denotes a set of N_s coordinates (i, j) that are on or within a circle of radius R_s centered at the (m, n)th point but that do not fall into Q_{s-1}; and so on. For example, Q = {$(m, n + 1), (m, n - 1), (m + 1, n), (m - 1, n)$} is a set of coordinates that are on or within a circle of unit radius from a point (m, n). Hence, in this smoothing algorithm, a part of the intensity of the (m, n)th pixel is being distributed to its neighbors. The amount of energy distributed to a neighboring point decreases as its distance from the (m, n)th pixel increases. The parameter a_0 represents the fraction retained by a pixel after

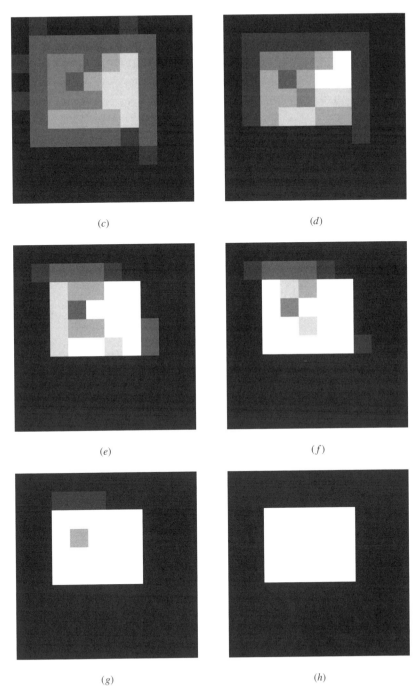

(c)   *(d)*

(e)   *(f)*

(g)   *(h)*

FIGURE 11.33

(*Continued*) Lighter square inside smaller square: (*c*)–(*h*) successive applications of INT operator on original image (*a*).

TABLE 11.14

Gray-scale intensity values of pixels in a 10×10 pixel array of the image shown in Fig. 11.33a

77	89	77	64	77	71	99	56	51	38
77	122	125	125	125	122	117	115	51	26
97	115	140	135	133	153	166	112	56	31
82	112	145	130	150	166	166	107	74	23
84	107	140	138	135	158	158	120	71	18
77	110	143	148	153	145	148	122	77	13
79	102	99	102	97	94	92	115	77	18
71	77	74	77	71	64	77	89	51	20
64	64	48	51	51	38	51	31	26	18
51	38	26	26	31	13	26	26	26	13

TABLE 11.15

Scaled matrix of the intensity values in Table 11.14

0.30	0.35	0.30	0.25	0.30	0.28	0.39	0.22	0.20	0.15
0.30	0.48	0.49	0.49	0.49	0.48	0.46	0.45	0.20	0.10
0.38	0.45	0.55	0.53	0.52	0.60	0.65	0.44	0.22	0.12
0.32	0.44	0.57	0.51	0.59	0.65	0.65	0.42	0.29	0.09
0.33	0.42	0.55	0.54	0.53	0.62	0.62	0.47	0.28	0.07
0.30	0.43	0.56	0.58	0.60	0.57	0.58	0.48	0.30	0.05
0.31	0.40	0.39	0.40	0.38	0.37	0.36	0.45	0.30	0.07
0.28	0.30	0.29	0.30	0.28	0.25	0.30	0.35	0.20	0.08
0.25	0.25	0.19	0.20	0.20	0.15	0.20	0.12	0.10	0.07
0.20	0.15	0.10	0.10	0.12	0.05	0.10	0.10	0.10	0.05

TABLE 11.16

Intensity matrix after applying the enhancement algorithm

0.18	0.24	0.18	0.12	0.18	0.16	0.30	0.10	0.08	0.05
0.18	0.46	0.48	0.48	0.48	0.46	0.42	0.40	0.08	0.02
0.29	0.40	0.60	0.56	0.54	0.68	0.75	0.39	0.10	0.03
0.20	0.39	0.63	0.52	0.66	0.75	0.75	0.35	0.17	0.02
0.22	0.35	0.60	0.58	0.56	0.71	0.71	0.44	0.16	0.01
0.18	0.37	0.61	0.65	0.68	0.63	0.65	0.46	0.18	0.01
0.19	0.32	0.30	0.32	0.29	0.27	0.26	0.40	0.18	0.01
0.16	0.18	0.17	0.18	0.16	0.12	0.18	0.24	0.08	0.01
0.12	0.12	0.07	0.08	0.08	0.05	0.08	0.03	0.02	0.01
0.01	0.01	0.01	0.01	0.01	0.01	0.01	0.01	0.01	0.01

distribution of part of its energy (intensity) to its neighbors. The set of coefficients a_i is important in the algorithm, and specific values are problem-dependent.

Example 11.23. The final enhanced image in Example 11.22 is used here with some random "salt and pepper" noise introduced into it. "Salt and pepper" noise is the occurrence of black and white pixels scattered randomly throughout the image. In Fig. 11.34, five pixels are shown to have intensity values different from what they should be (i.e., compared with the image shown in Fig. 11.33h, for example).

FIGURE 11.34
Image with five "salt and pepper" noise points.

FIGURE 11.35
The pixels required around the center pixel to use the smoothing algorithm for reducing "salt and pepper" noise.

TABLE 11.17
Scaled intensity values (black=0, white=1) for the image shown in Fig. 11.34

0	0	0	0	0	0	0	0	0	0
0	0	0	0	0	0	0	0	0	0
0	0	1	1	1	1	1	0	0	0
0	0	1	1	0	1	1	0	1	0
0	0	1	1	1	0	1	0	0	0
0	0	1	1	1	1	1	0	0	0
0	0	0	0	0	0	0	0	0	0
0	0	0	0	0	0	0	0	0	0
0	1	0	0	0	0	0	1	0	0
0	0	0	0	0	0	0	0	0	0

We use the image-smoothing algorithm presented in Eq. (11.70) to reduce the "salt and pepper" noise. Using $a_0 = a_2 = a_3 = a_4 = \cdots = 0$ as the values for the coefficients in Eq. (11.70) gives us the expression for the intensity (membership value) for a pixel as

$$\mu_{00} = \tfrac{1}{4}(\mu_{-10} + \mu_{10} + \mu_{01} + \mu_{0-1})$$

as shown in Fig. 11.35. The expression for μ_{00} does limit the pixels that can be smoothed. The pixels along the edges of the image cannot be smoothed, because all the intensity values around the pixel of interest would not be available. The user should thus be careful when programming for this algorithm.

To start the algorithm, we begin with the initial pixel values describing the image in Fig. 11.34. These initial values are presented in a normalized fashion as membership values in the set *white* ($\mu = 0$ means the pixel has no membership in *white*, and complete membership in the complement of white, or *black*) as seen in Table 11.17.

TABLE 11.18
Intensity matrix after smoothing the image once

0.00	0.00	0.00	0.00	0.00	0.00	0.00	0.00	0.00	0.00
0.00	0.00	0.25	0.31	0.33	0.33	0.33	0.08	0.02	0.00
0.00	0.25	0.62	0.73	0.52	0.71	0.51	0.15	0.29	0.00
0.00	0.31	0.73	0.62	0.78	0.62	0.53	0.42	0.18	0.00
0.00	0.33	0.77	0.85	0.66	0.82	0.59	0.25	0.11	0.00
0.00	0.33	0.52	0.59	0.56	0.60	0.30	0.14	0.06	0.00
0.00	0.08	0.15	0.19	0.19	0.20	0.12	0.07	0.03	0.00
0.00	0.27	0.11	0.07	0.07	0.07	0.05	0.28	0.08	0.00
0.00	0.07	0.04	0.03	0.02	0.02	0.27	0.14	0.05	0.00
0.00	0.00	0.00	0.00	0.00	0.00	0.00	0.00	0.00	0.00

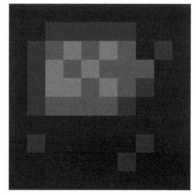

FIGURE 11.36
Image from Fig. 11.34 after one application of smoothing algorithm.

After one application of the smoothing algorithm, Eq. (11.70), the intensity matrix of Table 11.17 is modified as shown in Table 11.18, and the associated smoothed image is shown in Fig. 11.36.

SYNTACTIC RECOGNITION

In many recognition problems structural information plays an important role in describing the patterns. Some examples include image recognition, fingerprint recognition, chromosome analysis, character recognition, scene analysis, etc. In such cases, when the patterns are complex and the number of possible descriptions is very large, it is impractical to regard each description as defining a class; rather, description of the patterns in terms of small sets of simple subpatterns of primitives and grammatical rules derived from formal language theory becomes necessary [Fu, 1982].

The application of the concepts of formal language theory to structural pattern recognition problems can be illustrated as follows. Let us consider the simplest case in which there are two pattern classes, C_1 and C_2. Let us consider a set of the simplest subpatterns, in terms of which the patterns of both classes can be described completely. We call these the primitives. They can be identified with the terminals of formal language theory. Accordingly, we denote the set of primitives by V_T. Each pattern may then be looked upon as a string or sentence. Let us suppose we can find grammars G_1 and G_2 such that the sets of strings generated by them, $L(G_1)$ and $L(G_2)$, are exactly the same as those corresponding to pattern classes C_1 and C_2, respectively. Clearly, then, if a string corresponding to an unknown pattern is seen to be a member of $L(G_i)$, $i = 1, 2$, we can classify the pattern into C_i. Of course, if it is certain that the unknown string can only come from either of the two classes, then it is sufficient to have just one grammar corresponding to any one of the two classes, say, C_1. In this case, if a string is not from $L(G_1)$, it is automatically assumed to be from C_2. The procedure by which one determines whether a given string is syntactically correct with respect to a given grammar is called syntax analysis or parsing [Fu, 1982].

Example 11.24   [Pal and Majumder, 1986]. Let us consider a very simple problem in which we wish to distinguish "squares" from "rectangles." Obvious primitives for this problem are horizontal and vertical line segments of unit length, which we denote by a and b, respectively. Let us suppose for the sake of simplicity that the dimensions of the figures under consideration are integral multiples of the unit used. Then the two classes can be described as

$$C_{\text{squares}} = \{a^n b^n a^n b^n \mid n \geq 1\} \quad C_{\text{rectangles}} = \{a^m b^n a^m b^n \mid m, n \geq 1, m \neq n\}$$

One can easily see that C_{squares} is the same as $L(G)$, the language generated by a grammar (see Eq. 11.72)

$$G = (Y_N, V_T, P, S)$$

where $V_N = \{S, A, B, C\}$
$V_T = \{a, b\}$
$P = \{Sa, aAb, Sabab, aAbaaAbb, aAbaaBbb, BbCa, bCabbCaa, Cba\}$

where P is a collection of various concatenations making up squares and rectangles. Therefore, any pattern for which the corresponding string can be parsed by G is classified as a square. Otherwise, it is a rectangle.

The foregoing approach has a straightforward generalization for an m-class pattern recognition problem. Depending on whether or not the m classes exhaust the pattern space, we choose $m - 1$ or m grammars. In the first case, we classify a pattern into C_i if the corresponding string belongs to $L(G_i)$, $i = 1, 2, \ldots, m - 1$. Otherwise, the pattern is classified into C_m. In the second case, a pattern is identified as coming from C_i, $i = 1, 2, \ldots, m$, if the string corresponding to it can be parsed by G_i. If not, the pattern is reckoned to be noisy and is rejected.

The syntactic (structural) approach, which draws an analogy between the hierarchical structure of patterns and the syntax of languages, is a powerful method. After identifying each of the primitives within the pattern, the recognition process is accomplished by performing a syntax analysis of the "sentence" describing the given pattern. In the syntactic method, the ability to select and classify the simple pattern primitives and their relationships represented by the composition operations is the vital means by which a system is effective. Since the techniques of composition of primitives into patterns are usually governed by the formal language theory, the approach is often referred to as a *linguistic approach*. This learned description is then used to analyze and to produce syntactic descriptions of unknown patterns.

For the purposes of recognition by computer, the patterns are usually digitized (in time or in space) to carry out the previously mentioned preprocessing tasks. In the syntax analyzer the decision is made whether the pattern representation is syntactically correct. If the pattern is not syntactically correct, it is either rejected or analyzed on the basis of other grammars, which presumably represent other possible classes of patterns under consideration. For the purpose of recognition, the string of primitives of an input pattern is matched with those of the prototypes representing the classes. The class of the reference pattern having the best match is decided to be an appropriate category.

In practical situations, most patterns encountered are noisy or distorted. That is, the string corresponding to a noisy pattern may not be recognized by any of the pattern grammars, or ambiguity may occur in the sense that patterns belonging to the different classes may appear to be the same. In light of these observations, the foregoing approach may seem to have little practical importance. However, efforts have been made to incorporate features that can help in dealing with noisy patterns. The more noteworthy of them are the following [Fu, 1982]:

- The use of approximation
- The use of transformational grammars
- The use of similarity and error-correcting parsing
- The use of stochastic grammars
- The use of fuzzy grammars

The first approach proposes to reduce the effect of noise and distortion by approximation at the preprocessing and primitive extraction stage. The second approach attempts to define the relation between noisy patterns and their corresponding noise-free patterns by a transformational grammar. If it is possible to determine such a transformational grammar, the problem of recognizing noisy or distorted patterns can be transformed into one of recognizing noise-free patterns. The third approach defines distance measures between two strings and extends these to define distance measures between a string and a language.

In the stochastic approach, when ambiguity occurs, i.e., when two or more patterns have the same structural description or when a single (noisy) string is accepted by more than one pattern grammar, it means that languages describing different classes overlap. The incorporation of the element of probability into the pattern grammars gives a more realistic model of such situations and gives rise to the concept of stochastic languages.

One natural way of generating stochastic language from ordinary formal language is to randomize the productions of the corresponding grammars. This leads to the concept of stochastic phrase-structure grammars, which are the same as ordinary phrase-structure grammars except that every production rule has a probability associated with it. Also, if a pattern grammar is being heuristically constructed, then we can tackle the problem of ''unwanted'' strings (strings that do not represent patterns in the class) by assigning very small probabilities to such strings. Reviews of the syntactic methods and their applications are available in Fu [1982]. The last approach is addressed next.

Formal Grammar

The concept of a grammar was formalized by linguists with a view to finding a means of obtaining structural descriptions of sentences of a language that could not only recognize but also generate the language [Fu, 1982; Hopcroft and Ullman, 1969]. Although satisfactory formal grammars have not been obtained to date for describing the English language, the concept of a formal grammar can easily be explained with certain ideas borrowed from English grammar.

An *alphabet* or vocabulary is any finite set of symbols. A *sentence* (or string or word) over an alphabet is any *string* of finite length composed of symbols from the alphabet. If V is an alphabet, then V^* denotes the set of all sentences composed of symbols of V, including the empty sentence Λ. A *language* is any set of sentences over an alphabet. For example,

$$\text{If } V = \{a, b\}, \text{ then } V^* = \{\Lambda, a, b, ab, ba, \ldots\}$$

and a few examples of languages over V are

$$L_1 = \{a^n, n = 1, 2, \ldots \text{ denoted as written, or repeated, } n \text{ times}\}$$

$$L_2 = \{a^m b^n, m \neq n + 1, m, n = 1, 2, \ldots\}$$

$$L_3 = \{a, b, ab\}$$

and so forth [Pal and Majumder, 1986]. The formal prescription of a language theory is useful in syntactic recognition from the axiom, "If there exists a procedure (or algorithm) for recognizing a language, then there also exists a procedure for generating it."

Suppose we want to parse a simple English sentence, "The athlete jumped high." The rules that one applies to parsing can easily be described as follows:

$$\langle\text{sentence}\rangle \longrightarrow \langle\text{noun phrase}\rangle\langle\text{verb phrase}\rangle$$

$$\langle\text{noun phrase}\rangle \longrightarrow \langle\text{article}\rangle\langle\text{noun}\rangle$$

$$\langle\text{verb phrase}\rangle \longrightarrow \langle\text{verb}\rangle\langle\text{adverb}\rangle$$

$$\langle\text{article}\rangle \longrightarrow \text{The} \tag{11.71}$$

$$\langle\text{noun}\rangle \longrightarrow \text{athlete}$$

$$\langle\text{verb}\rangle \longrightarrow \text{jumped}$$

$$\langle\text{adverb}\rangle \longrightarrow \text{high}$$

where the symbol $\rightarrow$ denotes "can be written as." We can now define a formal phrase grammar, G, as a four-tuple [Pal and Majumder, 1986],

$$G = (V_N, V_T, P, S) \tag{11.72}$$

where V_N and V_T are the nonterminal and terminal vocabularies of G. Essentially, the nonterminal vocabularies are the phrases just illustrated and the terminal vocabularies are the alphabet of the language; in a sense V_T is the collection of singletons of the language (smallest elements) and V_N is the collections, or sets, containing the singletons. The symbols P and S denote the finite set of production rules of the type $\alpha \rightarrow \beta$ where α and β are strings over $V = V_N \cup V_T$, with α having at least one symbol of V_N and $S \in V_N$ is a starting symbol or a sentence (sentence or object to be recognized). In the preceding example,

$$V_N = \{\langle\text{sentence}\rangle, \langle\text{noun phrase}\rangle, \langle\text{verb phrase}\rangle, \langle\text{article}\rangle, \langle\text{noun}\rangle, \langle\text{verb}\rangle, \langle\text{adverb}\rangle\}$$

$$V_T = \{\text{the, athlete, jumped, high}\}$$

$$P = \text{the set of rules, Eqs. (11.71)}$$

$$S = \langle\text{sentence}\rangle$$

Fuzzy Grammar and Syntactic Recognition

The concept of a formal grammar is often found to be too rigid to handle real patterns, which are usually distorted or noisy yet still retain underlying structure. When the indeterminacy of the patterns is due to inherent vagueness rather than randomness, fuzzy language can be a better tool for describing the ill-defined structural information [see Pal and Majumder, 1986]. In this case, the generative power of a grammar is increased by introducing fuzziness either in the definition of primitives (labels of the fuzzy sets) or in the physical relations among primitives (fuzzified production rules), or in both of these. A fuzzy grammar produces a language that is a fuzzy set of strings with the membership value of each string denoting the degree of belonging of the string in that language. The grade of membership of an unknown pattern in a class described by the grammar is obtained using a max–min composition rule.

Let V_T^* denote the set of finite strings of alphabet V_T, including the null string, Λ. Then, a fuzzy language (FL) on V_T is defined as a fuzzy subset of V_T^*. Thus, FL is the fuzzy set

$$FL = \sum_{x \in V_T^*} \frac{\mu_{FL}(x)}{x} \qquad (11.73)$$

where $\mu_{FL}(x)$ is the grade of membership of the string x in FL. It is further assumed that all other strings in V_T^* have 0 membership in FL.

For two fuzzy languages FL_1 and FL_2 the operations of containment, equivalence, union, intersection, and complement follow the same definitions for the resulting membership as those delineated in Chapter 2.

Informally, a fuzzy grammar may be viewed as a set of rules for generating a fuzzy subset of V_T^*. A fuzzy grammar FG is a 6-tuple given by

$$FG = (V_N, V_T, P, S, J, \mu) \qquad (11.74)$$

where, in addition to the definitions given for Eq. (11.71), we have J: $\{r_i \mid i = 1, 2, \ldots n,$ and n = cardinality of P$\}$, i.e., the number of production rules, and μ is a mapping $\mu: J \rightarrow [0, 1]$, such that $\mu(r_i)$ denotes the membership in P of the rule labeled r_i. A fuzzy grammar generates a fuzzy language L(FG) as follows.

A string $x \in V_T^*$ is said to be in L(FG) if and only if it is derivable from S and its grade of membership $\mu_{L(FG)}(x)$ in L(FG) is greater than 0, where

$$\mu_{L(FG)}(x) = \max_{1 \leq k \leq m} \left[\min_{1 \leq k \leq l_k} \mu(r_i^k) \right] \qquad (11.75)$$

where m is the number of derivations that x has in FG; l_k is the length of the kth derivation chain, $k = 1(1)m$; and r_i^k is the label of the ith production used in the kth derivation chain, $i = 1, 2, \ldots, l_k$. Clearly if the production rule $\alpha \rightarrow \beta$ is visualized as a chain link of strength $\mu(r)$, r being the label of the rule, then the strength of a derivation chain is the strength of its weakest link, and hence

$$\mu_{L(FG)}(x) = \text{strength of the strongest derivation chain for S to } x, \text{ for all } x \in V_T^*$$

Example 11.25 [Pal and Majumder, 1986]. Suppose $FG_1 = (\{A, B, S\}, \{a, b\}, P, S, \{1, 2, 3, 4\}, \mu)$, where J, P, and μ are as follows:

1. $S \rightarrow AB$ with $\mu(1) = 0.8$
2. $S \rightarrow aSb$ with $\mu(2) = 0.2$
3. $A \rightarrow a$ with $\mu(3) = 1$
4. $B \rightarrow b$ with $\mu(4) = 1$

Then the fuzzy language must be $FL_1 = \{x \mid x = a^n b^n, n = 1, 2, \ldots\}$ with

$$\mu_{FL_1}(ab) = \begin{cases} 0.8, & \text{if } n = 1 \\ 0.2, & \text{if } n \geq 2 \end{cases}$$

Careful inspection of Rules 1 and 2 shows that Rule 2 can be repeated over and over again, generating, first, *ab* (Rule 1), second, *aabb* (Rule 2 using the new value for S), third, *aaabbb*, and so on, recursively, with increasing n.

These grammars, of course, can be used in a wide array of pattern recognition problems where the alphabets become line segments and the words, or vocabulary, become geometric shapes as illustrated in the following three examples. The following example from Pal and Majumder [1986], for a right triangle, illustrates this idea.

Example 11.26   **[Pal and Majumder, 1986].**   Suppose a fuzzy grammar is given by

$$FG_2 = (\{S, A, B\}, (a, b, c), P, S, J, \mu)$$

where J, P, and m are as follows:

$$r_1 : S \longrightarrow aA \quad \text{with} \quad \mu(r_1) = \mu_H(a)$$
$$r_2 : A \longrightarrow bB \quad \text{with} \quad \mu(r_2) = \mu_V(a)$$
$$r_3 : B \longrightarrow c \quad\;\; \text{with} \quad \mu(r_3) = \mu_{ob}(a)$$

with the primitives a, b, and c being *horizontal*, *vertical*, and *oblique* directed line segments, respectively, as seen in Fig. 11.37. The membership functions for these line segments are given

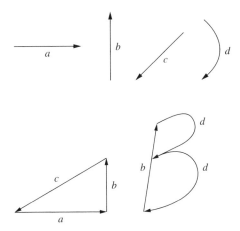

FIGURE 11.37
Primitive line segments and production of a triangle [Pal and Majumder, 1986].

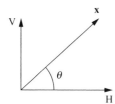

FIGURE 11.38
Membership functions for horizontal, vertical, and oblique lines.

here with reference to Fig. 11.38:

$$\mu_H(\theta) = 1 - |\tan\theta|, \qquad \mu_V(\theta) = 1 - \left|\frac{1}{\tan\theta}\right|, \quad \text{and} \quad \mu_{ob}(\theta) = 1 - \left|\frac{\theta - 45°}{45°}\right|$$

From the three rules, the only string generated is $x = abc$. This string is, of course, a right triangle, as seen in Fig. 11.38, and is formed from the specified sequence in the string abc. In this syntactic recognition, the primitives (line segments) will be concatenated in the "head-to-tail" style.

Another example for a geometric shape is given in Example 11.27 for a trapezoid.

Example 11.27. Consider a fuzzy grammar

$$FG = (\{S, A, B, C\}, \{a, b, c, d\}, P, S, J, \mu)$$

with

$$P : r_1 : S \longrightarrow a + A \quad \mu(r_1) = \mu_{PO}(a)$$
$$r_2 : A \longrightarrow b + B \quad \mu(r_2) = \mu_{FH}(b)$$
$$r_3 : B \longrightarrow c + C \quad \mu(r_3) = \mu_{NO}(c)$$
$$r_4 : C \longrightarrow \sim b \qquad \mu(r_4) = \mu_{IH}(\sim b)$$

The primitives $a, b, c,$ and $\sim b$ represent *positive oblique, forward horizontal, negative oblique,* and *inverse horizontal* directed line segments, respectively, as in Fig. 11.39.

The fuzzy membership functions for positive oblique (PO), forward horizontal (FH), negative oblique (NO), and inverse horizontal (IH) are as follows:

$$\mu_{PO} = \begin{cases} 1, & 0 \le \theta \le 90° \\ 0, & \text{elsewhere} \end{cases}$$

$$\mu_{NO} = \begin{cases} 1, & 90° \le \theta \le 180° \\ 0, & \text{elsewhere} \end{cases}$$

$$\mu_{FH} = \cos\theta$$

$$\mu_{IH} = -\cos(\theta + 90°)$$

where the angle θ is as defined in Fig. 11.40.

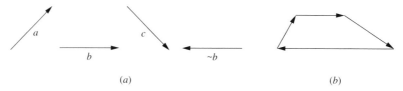

(a)   (b)

FIGURE 11.39
Directed line segments for Example 11.27: (*a*) primitives; (*b*) production of trapezoid.

Vertical | Oblique

θ

Horizontal

FIGURE 11.40
Geometric orientation for membership functions.

Making use of a head–tail type of concatenation, the foregoing fuzzy grammar FG produces a structure of a trapezoid with the membership function

$$\mu_{L(FG)}(x) = \min(\mu_{PO}(a), \mu_{FH}(b), \mu_{NO}(c), \mu_{IH}(\sim b))$$

where $x = a + b + c + (\sim b)$. The structure is shown in Fig. 11.39b. If membership values are $\mu_{PO}(a) = 1$, $\mu_{FH}(b) = 1$, $\mu_{NO}(c) = 1$, and $\mu_{IH}(\sim b) = 0.9$, then the membership value of the trapezoid is

$$\min(1, 1, 1, 0.9) = 0.9$$

This geometric idea in syntactic recognition can be extended to include other symbols, such as illustrated in the following example.

Example 11.28. For the syntactic pattern recognition of electrical symbols in an electrical design diagram such as a resistor, an inductor, and a capacitor, etc., several fuzzy grammars can be used for each different symbol. An inductor and a resistor can be assigned to the same fuzzy language. If the fuzzy grammar for this language is called FG_1, then FG_1 is defined as

$$FG_1 = (\{A, S\}, \{a_i\}, P, S, \{1, 2, 3\}, \mu)$$

where J, P, and μ are as follows:

1. $S \longrightarrow A$ with $\mu(1) = 0.15$
2. $S \longrightarrow a_i S$ with $\mu(2) = 0.85$
3. $A \longrightarrow a_i$ with $\mu(3) = 1$

and a_i is a primitive. Suppose a_1 ($i = 1$) represents the primitive (inductor symbol) ⨍⨍⨍⨍ and a_2 ($i = 2$) represents the primitive (resistor symbol) ∿∿∿. The preceding fuzzy grammar generates the fuzzy language as

$$L(FG_1) = \{x \mid x = a_i^n, n = 1, 2, 3, \ldots\}$$

Further, if the concatenation of a head–tail type is used for the generation of string x, then $x = a_i^n$ infers the pattern of an inductor when $i = 1$ or a resistor when $i = 2$. These ideas are shown in Fig. 11.41. The membership values for these two patterns can be expressed as

a_1 a_2

(a)

a_1^3 a_2^3

(b)

FIGURE 11.41
Directed line segments for Example 11.28: (a) primitives and (b) patterns produced by the primitives.

TABLE 11.19
The recognition of electrical elements of inductors and resistors

Number	a_1	a_2	Inference of the element
$n = 1$	0.15	0	◊◊◊◊
$n \geq 2$	0.85	0	
$n = 1$	0	0.15	∧∧∧
$n \geq 2$	0	0.85	

1. $i = 1$, which infers a pattern of an inductor

$$\mu_{L(FG_1)}(a_1^n) = \begin{cases} 0.15, & n = 1 \\ 0.85, & n \geq 2 \end{cases}$$

2. $i = 2$, which infers a pattern of a resistor

$$\mu_{L(FG_1)}(a_2^n) = \begin{cases} 0.15, & n = 1 \\ 0.85, & n \geq 2 \end{cases}$$

and these patterns and associated membership values are summarized in Table 11.19.
  Another fuzzy grammar FG_2 can be developed to recognize the pattern of a capacitor. If FG_2 is expressed in a general form,

$$FG_2 = (V_N, V_T, P, S, J, \mu)$$

then

$$V_N = \{S, R\}$$
$$V_T = \{a_3, a_4\} \quad \text{or} \quad V_T = \{a_4, a_5\}$$
$$P : S \longrightarrow L(a_3, a_4) \quad \text{or} \quad S \longrightarrow L(a_4, a_5)$$

where a_i $(i = 3, 4, 5)$ is a primitive in which a_3 represents the symbol ")", a_4 represents the symbol "|", and a_5 represents the symbol "(". $L(x, y)$ means "x is to the right of y." Therefore, the fuzzy language decided by FG_2 represents a capacitor in reality. A pattern S that meets FG_2 can be considered a capacitor. Besides, FG_2 belongs to a context-free grammar.
  If an unknown pattern S has the primitives of a_3 and a_4, and the membership values for a_3 and a_4 are given by

$$\mu_{L(FG_2)}(a_3) = 0.8 \quad \mu_{L(FG_2)}(a_4) = 1$$

then the membership of S representing a capacitor is

$$\mu_c(S) = \min(\mu_{L(FG_2)}(a_3), \mu_{L(FG_2)}(a_4)) = \min(0.8, 1) = 0.8$$

Similarly, if an unknown pattern S has the primitives of a_4 and a_5, and the membership values for a_4 and a_5 are

$$\mu_{L(FG_2)}(a_4) = 0.9 \quad \mu_{L(FG_2)}(a_5) = 0.8$$

then the membership for S as a capacitor is

$$\mu_c(S) = \min(\mu_{L(FG_2)}(a_4), \mu_{L(FG_2)}(a_5)) = \min(0.9, 0.8) = 0.8$$

By modification of the fuzzy grammar used for capacitors, FG_2 can be used to develop a fuzzy language for the electrical source (AC, DC) patterns of ⊖ and ⵏ. In those cases, V_T and P are changed to meet requirements of different patterns. In general, for the pattern recognition of an electrical element, the unknown pattern is first classified into a certain grammar, such as FG_1 or FG_2, then the recognition is carried out according to different primitives, production rules, and membership functions.

SUMMARY

The concept of a fuzzy set first arose in the study of problems related to pattern classification [Bellman et al., 1966]. Since the recognition and classification of patterns is integral to human perception, and since these perceptions are fuzzy, this study seems a likely beginning. This chapter has presented a simple idea in the area of classification involving equivalence relations and has dealt in depth with a particular form of classification using a popular clustering method: *fuzzy c-means*. The objective in clustering is to partition a given data set into homogeneous clusters; by homogeneous we mean that all points in the same cluster share *similar* attributes and they do not share similar attributes with points in other clusters. However, the separation of clusters and the meaning of *similar* are fuzzy notions and can be described as such. One of the first introductions to the clustering of data was in the area of fuzzy partitions [Ruspini, 1969, 1970, 1973*a*], where similarity was measured using membership values. In this case, the classification metric was a function involving a distance measure that was minimized. Ruspini [1973*b*] points out that a definite benefit of fuzzy clustering is that stray points (outliers) or points isolated between clusters (see Fig. 11.2) may be classified this way; they will have low membership values in the clusters from which they are isolated. In crisp classification methods these stray points need to belong to at least one of the clusters, and their membership in the cluster to which they are assigned is unity; their distance, or the extent of their isolation, cannot be measured by their membership. These notions of fuzzy classification described in this chapter provide for a point of departure in the recognition of known patterns, which is the subject of Part II of this chapter.

This chapter has introduced only the most elementary forms of fuzzy pattern recognition. A simple similarity metric called the *approaching degree* (the name is arbitrary; other pseudonyms are possible) is used to assess "closeness" between a known one-dimensional element and an unrecognized one-dimensional element. The idea involved in the approaching degree can be extended to higher-dimensional problems, as illustrated in this chapter, with the use of noninteractive membership functions. The areas of image processing and syntactic recognition are just briefly introduced to stimulate the reader into some exploratory thinking about the wealth of other possibilities in both these fields. The references to this chapter can be explored further to enrich the reader's background in these, and other pattern recognition, applications.

REFERENCES

Bellman, R., Kalaba, R., and Zadeh, L. (1966). "Abstraction and pattern classification," *J. Math. Anal. Appl.*, vol. 13, pp. 1–7.

Bezdek, J. (1974). "Numerical taxonomy with fuzzy sets," *J. Math. Biol.*, vol. 1, pp. 57–71.

Bezdek, J. (1981). *Pattern recognition with fuzzy objective function algorithms*, Plenum, New York.

Bezdek, J. and Harris, J. (1978). "Fuzzy partitions and relations: An axiomatic basis for clustering," *Fuzzy Sets Syst.*, vol. 1, pp. 111–127.

Bezdek, J., Grimball, N., Carson, J., and Ross, T. (1986). "Structural failure determination with fuzzy sets," *Civ. Eng. Syst.*, vol. 3, pp. 82–92.

Dong, W. (1986). "Applications of fuzzy sets theory in structural and earthquake engineering," PhD dissertation, Department of Civil Engineering, Stanford University, Stanford, CA.

Duda, R., and Hart, R. (1973). *Pattern classification and scene analysis*, John Wiley & Sons, New York.

Fu, K. S. (1982). *Syntactic pattern recognition and applications*, Prentice Hall, Englewood Cliffs, NJ.

Fukunaga, K. (1972). *Introduction to statistical pattern recognition*, Academic Press, New York.

Hopcroft, J. and Ullman, J. (1969). *Formal languages and their relation to automata*, Addison-Wesley, Reading, MA.

Pal, S. and King, R. (1980). "Image enhancement using fuzzy sets," *Electron. Lett.*, vol. 16, pp. 376–378.

Pal, S. and King, R. (1981). "Image enhancement using smoothing with fuzzy sets," *IEEE Trans. Syst., Man, Cybern.*, vol. SMC-11, pp. 494–501.

Pal, S. and Majumder, D. (1986). In *Fuzzy mathematical approach to pattern recognition*, John Wiley & Sons, New York.

Ross, T. (1995). *Fuzzy Logic with Engineering Applications*, McGraw-Hill, New York.

Ross, T., Hasselman, T., and Chrostowski, J. (1993). "Fuzzy set methods in assessing uncertainty in the modeling and control of space structures," *J. Intell. Fuzzy Syst.*, vol. 1, no. 2, pp. 135–155.

Ruspini, E. (1969). "A new approach to clustering," *Inf. Control*, vol. 15, pp. 22–32.

Ruspini, E. (1970). "Numerical methods for fuzzy clustering," *Inf. Sci.*, vol. 2, pp. 319–350.

Ruspini, E. (1973a). "New experimental results in fuzzy clustering," *Inf. Sci.*, vol. 6, pp. 273–284.

Ruspini, E. (1973b). "A fast method for probabilistic and fuzzy cluster analysis using association measures," *Proc. 6th Int. Conf. Syst. Sci., Hawaii*, pp. 56–58.

Tamura, S., Higuchi, S., and Tanaka, K. (1971). "Pattern classification based on fuzzy relations," *IEEE Trans. Syst., Man, Cybern.*, vol. 1, pp. 61–66.

Wang, P. (1983). "Approaching degree method," in *Fuzzy sets theory and its applications*, Science and Technology Press, Shanghai, PRC (in Chinese).

Zadeh, L. (1971). "Similarity relations and fuzzy orderings," *Inf. Sci.*, vol. 3, pp. 177–200.

PROBLEMS

Exercises for Equivalence Classification

11.1. A fuzzy tolerance relation, R, is reflexive and symmetric. Find the equivalence relation R_e and then classify it according to λ-cut levels = {0.9, 0.8, 0.5}.

$$R = \begin{bmatrix} 1 & 0.8 & 0 & 0.2 & 0.1 \\ 0.8 & 1 & 0.9 & 0 & 0.4 \\ 0 & 0.9 & 1 & 0 & 0.3 \\ 0.2 & 0 & 0 & 1 & 0.5 \\ 0.1 & 0.4 & 0.3 & 0.5 & 1 \end{bmatrix}$$

11.2. In a pattern recognition test, four unknown patterns need to be classified according to three known patterns (primitives) a, b, and c. The relationship between primitives and unknown patterns is in the following table:

	x_1	x_2	x_3	x_4
a	0.6	0.2	0.1	0.8
b	0.3	0.2	0.7	0.1
c	0.1	0.6	0.2	0.1

If a λ-cut level is 0.5, then into how many classes can these patterns be divided?

Hint: Use a max–min method (see Chapter 3) to first generate a fuzzy similarity relation $\underset{\sim}{R}$.

11.3. As a first step in automatic segmentation of magnetic resonance imaging (MRI) data regarding the head, it is necessary to determine the orientation of a data set to be segmented. The standard radiological orientations are sagittal, coronal, and horizontal. One way to classify the orientation of the new data would be to compare a slice of the new data to slices of known orientation. To do the classification we will use a simple metric obtained by overlaying slice images and obtaining an area of intersection, then normalizing these, based on the largest area of intersection. This metric will be our "degree of resemblance" for the equivalence relation. From data you have the following fuzzy relation:

$$
\begin{array}{c c}
& \begin{array}{cccc} \text{S} & \text{C} & \text{H} & \text{N} \end{array} \\
\begin{array}{c} \text{Sagittal} \\ \text{Coronal} \\ \text{Horizontal} \\ \text{New slice} \end{array} &
\begin{bmatrix}
1 & 0.6 & 0.4 & 0.7 \\
0.6 & 1 & 0.5 & 0.7 \\
0.4 & 0.5 & 1 & 0.5 \\
0.7 & 0.7 & 0.5 & 1
\end{bmatrix}
\end{array}
$$

(*a*) What kind of relation is this?

(*b*) Determine the equivalence relation and conduct a classification at λ-cut levels of 0.4, 0.6, and 0.7.

Exercises for Fuzzy *c*-Means

11.4. (*Note:* This problem will require a computerized form of the *c*-means algorithm.) Suppose we conduct a tensile strength test of four kinds of unidentified material. We know from other sources that the materials are from two different categories. From the yield stress, σ_y, and yield strain, Δ_y, data determine which materials are from the two different categories (see Fig. P11.4).

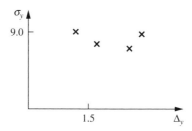

FIGURE P11.4

	m_1	m_2	m_3	m_4
σ_y	8.9	8.1	7.3	8.3
Δ_y	1.4	1.6	1.8	1.9

Determine which values for m' and ε_L would give the following results after 25 iterations, i.e.,

$$
\underset{\sim}{U}^{(25)} = \begin{bmatrix}
0.911 & 0.824 & 0.002 & 0.906 \\
0.089 & 0.176 & 0.998 & 0.094
\end{bmatrix}
$$

and final cluster centers of

$$\mathbf{v}_1 = \{8.458 \; 1.634\} \qquad \mathbf{v}_2 = \{7.346 \; 1.792\}$$

11.5. A problem in construction management is to allocate four different job sites to two different construction teams such that the time wasted in shuttling between the sites is minimized. Let the job sites be designated as x_i and combined to give a universe, $X = \{\mathbf{x}_1, \mathbf{x}_2, \mathbf{x}_3, \mathbf{x}_4\}$. If the head office, where the construction teams start every day, has coordinates $\{0, 0\}$, the following vectors give the locations of the four job sites:

$$\mathbf{x}_1 = \{4, 5\}$$

$$\mathbf{x}_2 = \{3, 4\}$$

$$\mathbf{x}_3 = \{8, 10\}$$

$$\mathbf{x}_4 = \{9, 12\}$$

Conduct a fuzzy c-means calculation to determine the optimum partition, $\underset{\sim}{U}^*$. Start with the following initial 2-partition:

$$\underset{\sim}{U}^{(0)} = \begin{Bmatrix} 1 & 1 & 0 & 0 \\ 0 & 0 & 1 & 1 \end{Bmatrix}$$

(Use $m' = 2.0$ and $\varepsilon_L \leq 0.01$.)

11.6. A radar image of a vehicle is a mapping of the bright (most reflective) parts of it. Suppose we have a radar image that we know contains two vehicles parked close together. The threshold on the instrument has been set such that the image contains seven bright dots. We wish to classify the dots as belonging to one or the other vehicle with a fuzzy membership before we conduct a recognition of the vehicle type. The seven bright dots are arranged in a matrix X, and we seek to find an optimum membership matrix $\underset{\sim}{U}^*$. The features defining each of the seven dots are given here:

2	9	9	5	8	5	6
7	3	5	6	8	10	4

Start the calculation with the following initial 2-partition:

$$\underset{\sim}{U}^0 = \begin{bmatrix} 0 & 0 & 0 & 0 & 0 & 0 & 1 \\ 1 & 1 & 1 & 1 & 1 & 1 & 0 \end{bmatrix}$$

Find the converged optimal 2-partition. (Use $m' = 2.0$ and $\varepsilon_L \leq 0.01$.)

11.7. In a magnetoencephalography (MEG) experiment, we attempt to partition the space of dipole model order versus reduced chi-square value for the dipole fit. This could be useful to an MEG researcher in determining any trends in his or her data-fitting procedures. Typical ranges for these parameters would be as follows:

$$\text{Dipole model order} = (1, 2, \dots, 6) = x_{1i}$$

$$\text{Reduced } \chi^2 \in (1, 3) = x_{2i}$$

Suppose we have three MEG data points, $\mathbf{x}_i = (x_{1i}, x_{2i})$, $i = 1, 2, 3$, to classify into two classes. The data are

$$\mathbf{x}_1 = (2, 1.5) \qquad \mathbf{x}_2 = (3, 2.2) \qquad \mathbf{x}_3 = (4, 2)$$

Find the optimum fuzzy 2-partition using the following initial partition:

$$\underset{\sim}{U}^{(0)} = \begin{bmatrix} 1 & 0 & 0 \\ 0 & 1 & 1 \end{bmatrix}$$

(Use $m' = 2.0$ and $\varepsilon_L \leq 0.01$.)

11.8. Suppose we want to sample a complex signal from a demodulator circuit and classify it into two sets, $A_\emptyset$ or A_1. The sample points are $x_1 = (-3, 1)$, $x_2 = (-2, 2)$, $x_3 = (-1, 1.5)$, and $x_4 = (1, 2)$ as shown in Fig. P11.8. If the first row of your initial 2-partition is $[1\ 0\ 0\ 0]$, find the fuzzy 2-partition after three iterations, i.e., find $\underset{\sim}{U}^{(3)}$. (Use $m' = 2.0$ and $\varepsilon_L \leq 0.01$.)

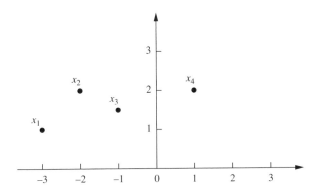

FIGURE P11.8

11.9. A small number of sequential operations can effectively limit the speedup of a parallel algorithm. Let f be the fraction of operations in a computation that must be performed sequentially, where $0 \leq f \leq 1$. According to Amdahl's law, the maximum speedup s achievable by a parallel computer with p processors is

$$s \leq \frac{1}{f + (1 - f)/p}$$

Suppose we have three data points, each described by two features: the fraction of sequential operations (f), and the maximum efficiency (s). These data points and their features are given in the following table:

	x_1	x_2	x_3
f	0.3	0.2	0.1
s	0.5	0.4	0.8

We want to classify these three data points into two classes ($c = 2$) according to the curves of Amdahl's law. Of the three possible hard partitions ($\eta_{M_c} = 3$), the one that seems graphically plausible is

$$\begin{bmatrix} 1 & 1 & 0 \\ 0 & 0 & 1 \end{bmatrix}$$

Using this partition as the initial guess, find the fuzzy 2-partition after two cycles, $\underset{\sim}{U}^{(2)}$. (Use $m' = 2.0$ and $\varepsilon_L \leq 0.01$.)

11.10. We want to classify the performance of three computer systems based on throughput (in mips) and response time (in seconds). The data points in our sample, $X = \{x_1, x_2, x_3\}$, are $x_1 = (50, 10)$, $x_2 = (40, 12)$, and $x_3 = (20, 5)$. Using the initial 2-partition,

$$\underset{\sim}{U}^{(0)} = \begin{bmatrix} 1 & 1 & 0 \\ 0 & 0 & 1 \end{bmatrix}$$

verify that the optimum fuzzy 2-partition after two cycles is

$$\underset{\sim}{U}^{(2)} = \begin{bmatrix} 0.974 & 0.9418 & 0 \\ 0.026 & 0.0582 & 1 \end{bmatrix}$$

(Use $m' = 2.0$ and $\varepsilon_L \leq 0.01$.)

Exercises for Classification Metric and Similarity

11.11. There are many different grades of naphtha, which is a mixture of hydrocarbons characterized by a boiling point range between $80°C$ and $250°C$. There are four types of naphtha ($n = 4$) that are characterized based on density, average molecular weight, and hydrogen-to-carbon (H/C) molar ratio ($m = 3$):

	Type I	Type II	Type III	Type IV
Density (g/m³)	0.679	0.7056	0.701	0.718
Avg molecular wt. (g)	85.5	93.0	91.0	98.3
H/C molar ratio	2.25	2.177	2.253	2.177

There are several studies that predict the products of the naphtha pyrolysis based on light and medium naphtha. It would be useful to classify the above four types into either light or medium classes ($c = 2$).

Using $m' = 2$ and $\varepsilon = 0.01$ conduct the following:
(a) Hard c-means.
(b) Fuzzy c-means.
(c) Find the classification metric.
(d) Find the similarity relation that results from the U-partition found in part (b).

11.12. In gas systems there are two basic properties of gas (temperature and pressure) that can be used to determine whether calculations for the system can be done using the ideal gas law or if a more robust property package is required. The ideal gas law applies to systems with high temperature (T) and low pressure (P). The drawback of employing a more robust approach is that the computational requirements increase tremendously.

P (atm)	T (K)
1	400
1.7	370
1	280
9	300
9.5	280

Classify the given set of data for P and T into two classes to determine which systems require a robust property package for thermodynamic calculations. Use $m' = 2$ and $\varepsilon = 0.01$ for the following:

(a) Hard c-means.

(b) Fuzzy c-means.

(c) Find the classification metric.

(d) Find the similarity relation using the U-partition from part (b).

11.13. The following data points describe the temperature, density and mono-ethanol-amine (MEA) weight fraction for an MEA/water solution. Classify the data into three classes using $m' = 2$ and $\varepsilon = 0.001$.

Temperature (°C)	Density (kg/m³)	Composition (weight fraction of MEA)
10	1001.96974	0.1
20	1000.798826	0.1
30	998.611997	0.1
10	1003.339776	0.15
20	1002.280297	0.15
30	1000.204904	0.15
10	1004.709813	0.2
20	1003.761769	0.2
30	1001.79781	0.2

Conduct the following:

(a) Hard c-means.

(b) Fuzzy c-means.

(c) Discuss the classification and how it relates to the original data and how the partitions from parts (a) and (b) are related.

(d) Find the classification metric.

11.14. The biomechanical department of a prominent university is conducting research in bone structure. One study involves developing a relationship between the wrist joint angle and the sarcomere length in the lower arm. In this study the following data were obtained:

Wrist joint angle (deg)	−75	−50	−25	0	25
Sarcomere length (μm)	3	3.25	3.5	2.75	3

(a) Classify these data, in one cycle, into two classes using the hard c-means method.

(b) Classify these data into two classes using the fuzzy c-means method; use $m' = 2$ and $\varepsilon = 0.01$ and conduct two cycles. What is the value of the accuracy at the end of two cycles?

(c) Find the classification metric.

(d) Find the similarity relation for the U-partition that results from part (b).

Exercises for Fuzzy Vectors

11.15. Show that when two separate fuzzy vectors are identical, i.e., $\underset{\sim}{a} = \underset{\sim}{b}$, the inner product $\underset{\sim}{a} \bullet \underset{\sim}{b}^T$ reaches a maximum as the outer product $\underset{\sim}{a} \oplus \underset{\sim}{b}^T$ reaches a minimum.

11.16. For two fuzzy vectors $\underset{\sim}{a}$ and $\underset{\sim}{b}$ and the particular case where $\hat{a} = \hat{b} = 1$ and $\underset{\sim}{a} = \underset{\sim}{b} = 0$, show that when $\underset{\sim}{a} = \underset{\sim}{b}$, then the inner product $\underset{\sim}{a} \bullet \underset{\sim}{b}^T = 1$ and the outer product $\underset{\sim}{a} \oplus \underset{\sim}{b}^T = 0$.

11.17. For two fuzzy vectors $\underset{\sim}{a}$ and $\underset{\sim}{b}$, prove the following expressions (transpose on the second vector in each operation is presumed):

(a) $\overline{\underset{\sim}{a} \bullet \underset{\sim}{b}} = \overline{\underset{\sim}{a}} \oplus \overline{\underset{\sim}{b}}$

(b) $\underset{\sim}{a} \bullet \underset{\sim}{a} \leq 0.5$

11.18. Prove the following:

(a) For any $\underset{\sim}{A} \in P^*(X)$, prove that $(\underset{\sim}{A}, \underset{\sim}{A})_{1 \text{ or } 2} = 1$.

(b) For any $\underset{\sim}{A}$ on X, prove that

$$(\underset{\sim}{A}, \overline{\underset{\sim}{A}})_1 \leq \tfrac{1}{2}$$

$$(\underset{\sim}{A}, \overline{\underset{\sim}{A}})_2 \leq \tfrac{1}{2}$$

11.19. Show that the metric in Eq. (11.53) always gives a value less than or equal to the metric in Eq. (11.54) for any pair of fuzzy sets.

Exercises for Multifeature Pattern Recognition

11.20. In signal processing the properties of an electrical signal can be important. The most sought-after properties of continuous time signals are their magnitude, phase, and frequency exponents. Three of these properties together determine one sinusoidal component of a signal where a sinusoid can be represented by the following voltage:

$$V(t) = A \sin(f_0 t - \theta)$$

where
A = magnitude (or amplitude) of the sinusoidal component
f_0 = fundamental frequency of the sinusoidal component
θ = phase of the sinusoidal component

With each of these properties representing a "feature" of the electrical signal, it is possible to model a fuzzy pattern recognition system to detect what type of sinusoidal components are present. Let us define the prototypical values for patterns of magnitude, frequency, and phase that we are interested in:

Components	Prototypical values
$0 \leq A \leq 12$ V	3 V, 6 V, 9 V, 12 V
$0 \leq f_0 \leq 80$ Hz	20 Hz, 40 Hz, 60 Hz, 80 Hz
$0 \leq \theta \leq 180°$	$45°, 90°, 135°, 180°$

Draw the resulting three-feature membership graphs.

Now let the input sinusoidal signal vector **B** comprise three crisp singletons, i.e., **B** = {5 V, 45 Hz, 45°}, with weights of 0.6, 0.2, and 0.2 assigned to each of the corresponding features. Determine which pattern vector **B** most closely resembles.

11.21. Using the same patterns as in Problem 11.20, but with a new input fuzzy pattern $\underset{\sim}{B}$ and features given by

$$\underset{\sim}{B}_{\text{voltage}} = \left\{ \frac{0}{1} + \frac{0.2}{2} + \frac{0.7}{3} + \frac{1.0}{5} + \frac{0}{6} \right\}$$

$$\underset{\sim}{B}_{\text{frequency}} = \left\{ \frac{0}{20} + \frac{0.5}{30} + \frac{1.0}{40} + \frac{0.4}{50} + \frac{0}{60} \right\}$$

$$\underset{\sim}{B}_{\text{phase}} = \left\{ \frac{0}{50} + \frac{0.3}{70} + \frac{0.7}{90} + \frac{1}{110} + \frac{0.7}{120} + \frac{0}{130} \right\}$$

determine the pattern that most closely matches the input pattern.

11.22. Transuranic waste will be stored at a southeastern New Mexico site known as WIPP. The site has underlying strata of rock salt, which is well-known for its healing and creeping properties. Healing is the tendency of a material to close fractures or other openings, and creep is the capacity of the material to deform under constant load. The radioactive wastes are stored in rooms excavated deep underground. Because of the creep of the ceiling, these rooms will eventually collapse, thus permanently sealing the wastes in place. The creep properties of salt depend on the depth, moisture content, and clay content of the salt at the location being considered. Rock salt from specified depths was studied through numerous tests conducted at various labs nationwide. These data comprise the known patterns. Hence, each pattern has three features. Now, the possibility of locating a room at a certain depth is being investigated.

We wish to determine the creep properties at some depth of salt with a certain clay and moisture content. Membership functions for each of the patterns are shown in Fig. P11.22.

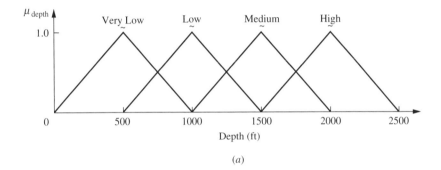

(a)

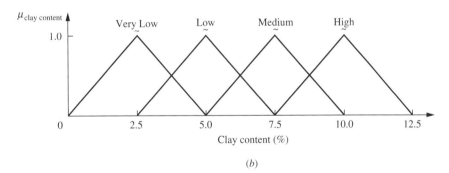

(b)

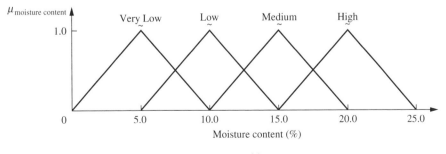

(c)

FIGURE P11.22

Find which known pattern the unknown pattern matches the best. The features for the unknown pattern are given by the crisp singletons

$$B = \{depth = 1750 \text{ ft}, \text{ clay content} = 6.13\%, \text{ moisture content} = 12.5\%\}$$

The weights given to the features are $W = \{0.5, 0.3, 0.2\}$.

11.23. Using the same known pattern as in Problem 11.22, and using fuzzy features for the new pattern, find which known pattern matches the new pattern most closely. Features for the new pattern are as follows:

$$\underset{\sim}{B}_{depth} = \left\{ \frac{0}{1700} + \frac{0.5}{1725} + \frac{1}{1750} + \frac{0.5}{1775} + \frac{0}{1800} \right\}$$

$$\underset{\sim}{B}_{clay \ content} = \left\{ \frac{0}{5.5} + \frac{0.5}{5.813} + \frac{1.0}{6.13} + \frac{0.5}{6.44} + \frac{0}{6.75} \right\}$$

$$\underset{\sim}{B}_{moisture \ content} = \left\{ \frac{0}{11.0} + \frac{0.5}{11.75} + \frac{1}{12.5} + \frac{0.5}{13.25} + \frac{0}{14.0} \right\}$$

11.24. A member of the police bomb squad has to be able to assess the type of bomb used in a terrorist activity in order to gain some knowledge that might lead to the capture of the culprit. The most commonly used explosive device is the pipe bomb. Pipe bombs can be made from a variety of explosives ranging from large-grain black powder and gunpowder to more sophisticated compounds, such as PETN or RDX. Identification of the explosive material used in the pipe bomb (after detonation) will tell a bomb squad investigator where the materials might have been purchased, the level of sophistication of the terrorist, and other important identifiers about the criminal.

Four basic types of energetic materials are used in making pipe bombs, each with its own distinctive pattern of post-mortem damage.

1. *Explosives.* Those that detonate at a velocity equal to the compressional sound speed of the explosive material itself. These materials are by far the most energetic (also the most difficult to acquire) and are characterized (after explosion) by very small pipe fragments, highly discolored (usually bluish in tint) fragments, and extreme collateral damage (especially close to the detonation).

2. *Propellants.* Usually formed from some compound based on nitrocellulose. These materials do not detonate, but burn very rapidly. Usually propellants are formed in special geometric shapes that allow their surface area to remain constant or increase as they burn, thus causing the burning rate to increase until the compound has been completely exhausted. The destructive force of propellant-based pipe bombs is somewhat less than that of true explosives and is characterized by medium fragment size, little discoloration, and moderate collateral damage.

3. *Large-grain black powder.* Has been around since about 600 BC, when the Chinese discovered the carbon–sulfur–potassium nitrate mixture. The size of black powder grains can vary tremendously, but the geometry is such that the powder always burns down (the burn rate always decreases once the entire surface of the mixture is burning). Although still very deadly, the damage from these types of pipe bombs is less than that of the other two. The residual fragment size is larger, and the discoloration of the fragments is slight.

4. *Gunpowder.* A subclass of black powder, usually considered to be homemade. It is characterized by very large fragments (one or two in number), almost no discoloration, and little collateral damage. Black powder is still very common among terrorists.

We can form a table of patterns for each feature:

	Features		
	Fragment size	**Fragment discoloration**	**Damage**
Explosives	Small (S)	High (H)	Extreme (E)
Propellants	Medium (M)	Medium (M)	Large (L)
Black powder	Large (L)	Low (L)	Medium (M)
Gunpowder	Very large (VL)	None (N)	Small (S)

Assume that the membership space for each feature can be partitioned similarly into four sections on a normalized abscissa, as shown in Fig. P11.24. The other two graphs (for discoloration and collateral damage) would look identical to the one in Fig. P11.24 (with different labels). The weights assigned to each feature are 0.5, 0.3, and 0.2, respectively. Now, say a new bombing has taken place and the aftereffects measured over three features are denoted as singletons, given as

$$B = \left\{ \text{fragment size} = \frac{1}{0.7}, \text{fragment color} = \frac{1}{0.6}, \text{damage} = \frac{1}{0.5} \right\}$$

Determine the composition of the bomb used in the bombing.

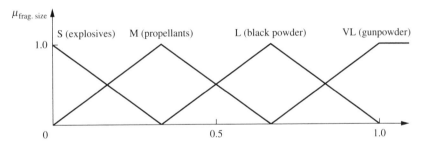

FIGURE P11.24

11.25. Using the information in Problem 11.24, but selecting a fuzzy input, perform a pattern recognition. The fuzzy input patterns in the form of triangular fuzzy numbers based on the three features are as follows:

$$B_{\text{frag. size}} = \left\{ \frac{0}{0.43} + \frac{1}{0.5} + \frac{0}{0.57} \right\}$$

$$B_{\text{frag. color}} = \left\{ \frac{0}{0.53} + \frac{1}{0.6} + \frac{0}{0.67} \right\}$$

$$B_{\text{damage}} = \left\{ \frac{0}{0.63} + \frac{1}{0.7} + \frac{0}{0.77} \right\}$$

11.26. We intend to recognize preliminary data coming off a satellite. Each of the five data packets has a unique packet header identifier, as follows:

$$A_1 = \text{satellite performance metrics}$$

$$A_2 = \text{ground positioning system}$$

$$A_3 = \text{IR sensor}$$

$$A_4 = \text{visible camera}$$

$$A_5 = \text{star mapper}$$

The three header values each set will look for are (1) signal type, (2) terminal number, and (3) data identifier. The weights assigned to each of the headers are 0.3, 0.3, and 0.4, respectively. Let us define the fuzzy pattern as

$$A_1 = \left\{ \frac{0.2}{x_1} + \frac{0.2}{x_2} + \frac{0.6}{x_3} \right\}$$

$$A_2 = \left\{ \frac{0.3}{x_1} + \frac{0.4}{x_2} + \frac{0.7}{x_3} \right\}$$

$$A_3 = \left\{ \frac{0.4}{x_1} + \frac{0.6}{x_2} + \frac{0.8}{x_3} \right\}$$

$$A_4 = \left\{ \frac{0.5}{x_1} + \frac{0.8}{x_2} + \frac{0.9}{x_3} \right\}$$

$$A_5 = \left\{ \frac{0.6}{x_1} + \frac{1.0}{x_2} + \frac{1.0}{x_3} \right\}$$

A data stream given by the crisp singleton

$$B = \left\{ \frac{1.0}{x_1} + \frac{1.0}{x_2} + \frac{1.0}{x_3} \right\}$$

is received. Determine which of the five different packets we are receiving at the present time.

11.27. Signals are investigated from the following four digital signal processing plants: $A_1 =$ least mean squares, $A_2 =$ root-mean square, $A_3 =$ Newton's method, and $A_4 =$ steepest descent method. The three ($m = 3$) important parameters that will be considered in each c-space are convergence rate, tracking, and stability. The weights assigned to each of the features are 0.4, 0.4, and 0.2, respectively. The data patterns corresponding to the features are membership triangles

$$A_1 = \{0.2, 0.3, 0.8\}$$

$$A_2 = \{0.4, 0.4, 0.6\}$$

$$A_3 = \{0.6, 0.2, 0.4\}$$

$$A_4 = \{0.8, 0.5, 0.2\}$$

The sample data set has the following vector pattern as a membership triangle:

$$B = \{0.5, 0.5, 0.5\}$$

Determine the pattern most closely represented by the sample data set.

11.28. Lube oils are classified by three features: color, viscosity, and flash point. Depending on the values of these features, the lube oil is classified as 100 neutral (100N), 150 neutral (150N), heavy solvent neutral (HSN), and 500 neutral (500N). Among the features, color is the most important, followed by viscosity, then flash point. The reason for this ordering is that it is easier to blend lube oils to obtain correct viscosity and flash point than it is to blend to obtain proper color. Any material not falling into one of these lube oil categories is downgraded to catalyst cracker feed (PGO), where it is converted to gasoline.

Fuzzy patterns for each of these features are shown in Fig. P11.28. The weights for these features are 0.5 for color, 0.3 for viscosity, and 0.2 for flash point. You receive a lab analysis for a sample described by the crisp singleton

$$B = \{color = 6.5, viscosity = 825\ m^2/s, flash\ point = 750°C\}$$

Under what category do you classify this sample?

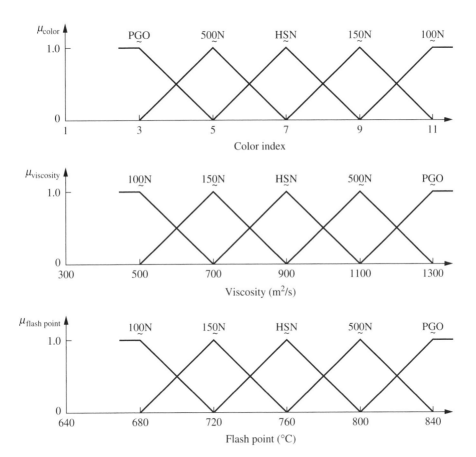

FIGURE P11.28

11.29. Over several years a satellite tracking facility has classified several objects on the universe of signal to noise ratio (SNR), total signal (TS), and radius (R). The fuzzy sets are shown here for four satellites:

$$\text{ASTEX} \quad \underset{\sim}{A}_1 = \left\{ \frac{0.1}{SNR} + \frac{0.15}{TS} + \frac{0.2}{R} \right\}$$

$$\text{DMSP} \quad \underset{\sim}{A}_2 = \left\{ \frac{0.2}{SNR} + \frac{0.2}{TS} + \frac{0.3}{R} \right\}$$

$$\text{SEASAT} \quad \underset{\sim}{A_3} = \left\{ \frac{0.5}{\text{SNR}} + \frac{0.7}{\text{TS}} + \frac{0.5}{\text{R}} \right\}$$

$$\text{MIR} \quad \underset{\sim}{A_4} = \left\{ \frac{0.9}{\text{SNR}} + \frac{0.9}{\text{TS}} + \frac{0.9}{\text{R}} \right\}$$

with weights $w_1 = 0.4$, $w_2 = 0.4$, $w_3 = 0.2$. One night an unknown object is tracked, and the following observation is made:

$$\underset{\sim}{B} = \left\{ \frac{0.3}{\text{SNR}} + \frac{0.3}{\text{TS}} + \frac{0.3}{\text{R}} \right\}$$

Which satellite does the object most closely resemble?

11.30. A set of patterns indicating the performance of an aluminum smelting cell is developed. The important features are bath temperature T ($^\circ$C), cell voltage V, and noise N (standard deviation of the cell resistance). The cell conditions (patterns) are described as follows:

$\underset{\sim}{A_1}$	Cell has a very small anode–cathode distance. Characterized by low temperature, low voltage, and high noise.
$\underset{\sim}{A_2}$	Cell is in good condition. Characterized by moderately low temperature, moderately low voltage, and low noise.
$\underset{\sim}{A_3}$	Cell has a very large anode–cathode distance. Characterized by high temperature, high voltage, and low noise.
$\underset{\sim}{A_4}$	Cell has deposits on bottom cathode. Characterized by moderately high temperature, high voltage, and high noise.

The fuzzy sets are represented by Gaussian membership functions:

$$\underset{\sim}{A_1} = \left\{ \exp\left[-\frac{(T-945)^2}{4^2}\right], \exp\left[-\frac{(V-4.2)^2}{(0.1)^2}\right], \exp\left[-\frac{(N-26)^2}{5^2}\right] \right\}$$

$$\underset{\sim}{A_2} = \left\{ \exp\left[-\frac{(T-950)^2}{4^2}\right], \exp\left[-\frac{(V-4.4)^2}{(0.1)^2}\right], \exp\left[-\frac{(N-6)^2}{2^2}\right] \right\}$$

$$\underset{\sim}{A_3} = \left\{ \exp\left[-\frac{(T-970)^2}{8^2}\right], \exp\left[-\frac{(V-4.8)^2}{(0.1)^2}\right], \exp\left[-\frac{(N-6)^2}{2^2}\right] \right\}$$

$$\underset{\sim}{A_4} = \left\{ \exp\left[-\frac{(T-965)^2}{6^2}\right], \exp\left[-\frac{(V-4.7)^2}{(0.1)^2}\right], \exp\left[-\frac{(N-20)^2}{5^2}\right] \right\}$$

To reflect the relative importance of the features, select $w_1 = 0.3$, $w_2 = 0.5$, and $w_3 = 0.2$. Now a new data sample (measurements from a smelting cell) yields temperature $= 953^\circ$C, voltage $= 4.5$ V, and noise $= 12$ (a data singleton). Classify the operating conditions of the cell.

11.31. Use the same patterns as in Problem 11.30. But now use a sample comprising fuzzy sets. This is appropriate because measurements such as temperature are subject to substantial error, and electrical signals fluctuate over time as disturbances affect the system. The new sample is represented as the following:

$$B = \left\{ \exp\left[-\frac{(T-957)^2}{3^2}\right], \exp\left[-\frac{(V-4.6)^2}{(0.2)^2}\right], \exp\left[-\frac{(N-16)^2}{3^2}\right] \right\}$$

Classify the operating conditions of the cell based on this information.

11.32. Skis are classified on the basis of three features: weight, performance stiffness, and response times in turns. There are four different types of skis: freestyle, giant slalom (GS), slalom, and all-around. These skis have Gaussian distributions on each of the features and the parameters for the distribution are given in following table. A Gaussian distribution has the form

$$\mu_{\underset{\sim}{A_{ij}}}(x) = \exp\left[-\left(\frac{x_j - a_{ij}}{\sigma^2_{a_{ij}}}\right)^2\right]$$

The fuzzy patterns are defined on a normalized scale as follows:

	Feature					
	Weight		Stiffness		Response time	
	a_i	σ	a_i	σ	a_i	σ
All-around	50	10	40	12	60	7
Slalom	40	3	90	5	75	10
Giant slalom	30	15	80	10	60	10
Freestyle	40	10	20	6	70	3

The weights given to the features are

$$w_{weight} = 0.3$$

$$w_{stiffness} = 0.4$$

$$w_{response} = 0.3$$

A new ski whose features are given on a normalized scale by a crisp singleton,

$$B = \{weight = 45, stiffness = 60, response\ time = 65\}$$

is introduced into the market. Determine what type of ski the new ski should be labeled.

11.33. In Problem 11.32 the new ski introduced into the market was given by a crisp singleton. However, given the uncertainty in measurements, it is more appropriate to define a ski by fuzzy parameters. For the same problem, and with the same weights assigned to each of the features, classify the new ski if it is given by a fuzzy set whose membership functions are given by a Gaussian distribution whose parameters (mean and standard deviation) are given in the following table:

	Feature					
	Weight		**Stiffness**		**Response time**	
	a_i	σ	a_i	σ	a_i	σ
$\underset{\sim}{B}$	45	10	60	12	65	20

Exercises for Syntactic Pattern Recognition

11.34. Generate a fuzzy grammar for the syntactic pattern recognition of an isosceles trapezoid.
11.35. Generate a fuzzy grammar for the syntactic pattern recognition of an equilateral triangle.
11.36. Continue Example 11.28 by developing fuzzy grammars for the pattern recognition of the two electric sources, symbols ⊖ and ⊩.

Exercises for Image Processing

11.37. The accompanying table shows the intensity values (for an 8-bit image) associated with an array of 25 pixels. Use the image enhancement algorithm on these intensity values to enhance the image. Do you recognize the pattern in the image?

111	105	140	107	110
110	132	111	120	105
140	105	105	115	154
137	135	145	150	145
140	118	115	109	148

11.38. The following table shows the intensity values (for an 8-bit image) associated with an array of 25 pixels. Use the image-softening algorithm on these intensity values to remove the "salt and pepper" noise (shown as shaded pixels) from the image of the alphabetic character M.

220	30	10	15	250
205	230	0	239	230
225	20	225	20	220
217	255	30	10	215
220	25	15	255	235

FUZZY ARITHMETIC
AND THE EXTENSION
PRINCIPLE

*Said the Mock Turtle with a sigh, "I only took the regular course." "What was that?" inquired
Alice. "Reeling and Writing, of course, to begin with," the Mock Turtle replied; "and the
different branches of Arithmetic – Ambition, Distraction, Uglification, and Derision."*

Lewis Carroll
Alice in Wonderland, 1865

As Lewis Carroll so cleverly implied as early as 1865 (he was, by the way, a brilliant
mathematician), there possibly could be other elements of arithmetic: consider those of
ambition, distraction, uglification, and derision. Certainly fuzzy logic has been described in
worse terms by many people over the last four decades! Perhaps Mr. Carroll had a presage
of fuzzy set theory exactly 100 years before Dr. Zadeh; perhaps, possibly.

In this chapter we see that standard arithmetic and algebraic operations, which are
based after all on the foundations of classical set theory, can be extended to fuzzy arithmetic
and fuzzy algebraic operations. This extension is accomplished with Zadeh's extension
principle [Zadeh, 1975]. Fuzzy numbers, briefly described in Chapter 4, are used here
because such numbers are the basis for fuzzy arithmetic. In this context the arithmetic
operations are not fuzzy; the numbers on which the operations are performed are fuzzy and,
hence, so too are the results of these operations. Conventional interval analysis is reviewed
as a prelude to some improvements and approximations to the extension principle, most
notably the fuzzy vertex method and its alternative forms.

EXTENSION PRINCIPLE

In engineering, mathematics, and the sciences, functions are ubiquitous elements in
modeling. Consider a simple relationship between one independent variable and one

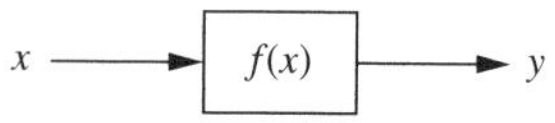

$$x \longrightarrow \boxed{f(x)} \longrightarrow y$$

FIGURE 12.1
A simple single-input, single-output mapping (function).

dependent variable as shown in Fig. 12.1. This relationship is a single-input, single-output process where the transfer function (the box in Fig. 12.1) represents the mapping provided by the general function f. In the typical case, f is of analytic form, e.g., $y = f(x)$, the input, x, is deterministic, and the resulting output, y, is also deterministic.

How can we extend this mapping to the case where the input, x, is a fuzzy variable or a fuzzy set, and where the function itself could be fuzzy? That is, how can we determine the fuzziness in the output, y, based on either a fuzzy input or a fuzzy function or both (mapping)? An extension principle developed by Zadeh [1975] and later elaborated by Yager [1986] enables us to extend the domain of a function on fuzzy sets.

The material of the next several sections introduces the extension principle by first reviewing theoretical issues of classical (crisp) transforms, mappings, and relations. The theoretical material then moves to the case where the input is fuzzy but the function itself is crisp, then to the case where the input and the function both are fuzzy. Simple examples to illustrate the ideas are provided. The next section serves as a more practical guide to the implementation of the extension principle, with several numerical examples. The extension principle is a very powerful idea that, in many situations, provides the capabilities of a "fuzzy calculator."

Crisp Functions, Mapping, and Relations

Functions (also called transforms), such as the logarithmic function, $y = \log(x)$, or the linear function $y = ax + b$, are mappings from one universe, X, to another universe, Y. Symbolically, this mapping (function, f) is sometimes denoted $f : X \to Y$. Other terminology calls the mapping $y = f(x)$ the *image of x under f*, and the inverse mapping, $x = f^{-1}(y)$, is termed the *original image of y*. A mapping can also be expressed by a relation R (as described in Chapter 3), on the Cartesian space X $\times$ Y. Such a relation (crisp) can be described symbolically as R $= \{(x, y) \mid y = f(x)\}$, with the characteristic function describing membership of specific x, y pairs to the relation R as

$$\chi_R(x, y) = \begin{cases} 1, & y = f(x) \\ 0, & y \neq f(x) \end{cases} \qquad (12.1)$$

Now, since we can define transform functions, or mappings, for specific elements of one universe (x) to specific elements of another universe (y), we can also do the same thing for collections of elements in X mapped to collections of elements in Y. Such collections have been referred to in this text as sets. Presumably, then, all possible sets in the power set of X can be mapped in some fashion (there may be null mapping for many of the combinations) to the sets in the power set of Y, i.e., $f : P(X) \to P(Y)$. For a set A defined on universe X, its image, set B on the universe Y, is found from the mapping, B $= f(A) = \{y \mid$ for all $x \in A, y = f(x)\}$, where B will be defined by its characteristic value

$$\chi_B(y) = \chi_{f(A)}(y) = \bigvee_{y = f(x)} \chi_A(x) \qquad (12.2)$$

Example 12.1. Suppose we have a crisp set A = {0, 1}, or, using Zadeh's notation,

$$A = \left\{ \frac{0}{-2} + \frac{0}{-1} + \frac{1}{0} + \frac{1}{1} + \frac{0}{2} \right\}$$

defined on the universe X = {−2, −1, 0, 1, 2} and a simple mapping $y = |4x| + 2$. We wish to find the resulting crisp set B on an output universe Y using the extension principle. From the mapping we can see that the universe Y will be Y = {2, 6, 10}. The mapping described by Eq. (12.2) will yield the following calculations for the membership values of each of the elements in universe Y:

$$\chi_B(2) = \vee \{\chi_A(0)\} = 1$$

$$\chi_B(6) = \vee \{\chi_A(-1), \chi_A(1)\} = \vee \{0, 1\} = 1$$

$$\chi_B(10) = \vee \{\chi_A(-2), \chi_A(2)\} = \vee \{0, 0\} = 0$$

Notice there is only one way to get the element 2 in the universe Y, but there are two ways to get the elements 6 and 10 in Y. Written in Zadeh's notation this mapping results in the output

$$B = \left\{ \frac{1}{2} + \frac{1}{6} + \frac{0}{10} \right\}$$

or, alternatively, B = {2, 6}.

Suppose we want to find the image B on universe Y using a relation that expresses the mapping. This transform can be accomplished by using the composition operation described in Chapter 3 for finite universe relations, where the mapping $y = f(x)$ is a general relation. Again, for X = {−2, −1, 0, 1, 2} and a generalized universe Y = {0, 1, ..., 9, 10}, the crisp relation describing this mapping ($y = |4x| + 2$) is

$$R = \begin{array}{c} \\ -2 \\ -1 \\ 0 \\ 1 \\ 2 \end{array} \begin{array}{cccccccccccc} 0 & 1 & 2 & 3 & 4 & 5 & 6 & 7 & 8 & 9 & 10 \\ \left[\begin{array}{ccccccccccc} 0 & 0 & 0 & 0 & 0 & 0 & 0 & 0 & 0 & 0 & 1 \\ 0 & 0 & 0 & 0 & 0 & 0 & 1 & 0 & 0 & 0 & 0 \\ 0 & 0 & 1 & 0 & 0 & 0 & 0 & 0 & 0 & 0 & 0 \\ 0 & 0 & 0 & 0 & 0 & 0 & 1 & 0 & 0 & 0 & 0 \\ 0 & 0 & 0 & 0 & 0 & 0 & 0 & 0 & 0 & 0 & 1 \end{array}\right] \end{array}$$

The image B can be found through composition (since X and Y are finite): that is, B = A∘R (we note here that any set, say A, can be regarded as a one-dimensional relation), where, again using Zadeh's notation,

$$A = \left\{ \frac{0}{-2} + \frac{0}{-1} + \frac{1}{0} + \frac{1}{1} + \frac{0}{2} \right\}$$

and B is found by means of Eq. (3.9) to be

$$\chi_B(y) = \bigvee_{x \in X} (\chi_A(x) \wedge \chi_R(x, y)) = \begin{cases} 1, & \text{for } y = 2, 6 \\ 0, & \text{otherwise} \end{cases}$$

or in Zadeh's notation on Y,

$$B = \left\{ \frac{0}{0} + \frac{0}{1} + \frac{1}{2} + \frac{0}{3} + \frac{0}{4} + \frac{0}{5} + \frac{1}{6} + \frac{0}{7} + \frac{0}{8} + \frac{0}{9} + \frac{0}{10} \right\}$$

Functions of Fuzzy Sets – Extension Principle

Again we start with two universes of discourse, X and Y, and a functional transform (mapping) of the form $y = f(x)$. Now suppose that we have a collection of elements in universe x that form a fuzzy set $\underset{\sim}{A}$. What is the image of fuzzy set $\underset{\sim}{A}$ on X under the mapping f? This image will also be fuzzy, say we denote it fuzzy set $\underset{\sim}{B}$; and it will be found through the same mapping, i.e., $\underset{\sim}{B} = f(\underset{\sim}{A})$.

The membership functions describing $\underset{\sim}{A}$ and $\underset{\sim}{B}$ will now be defined on the universe of a unit interval $[0, 1]$, and for the fuzzy case Eq. (12.2) becomes

$$\mu_{\underset{\sim}{B}}(y) = \bigvee_{f(x)=y} \mu_{\underset{\sim}{A}}(x) \tag{12.3}$$

A convenient shorthand for many fuzzy calculations that utilize matrix relations involves the *fuzzy vector*. Basically, a fuzzy vector is a vector containing fuzzy membership values. Suppose the fuzzy set $\underset{\sim}{A}$ is defined on n elements in X, for instance on $x_1, x_2, \ldots, x_n$, and fuzzy set $\underset{\sim}{B}$ is defined on m elements in Y, say on $y_1, y_2, \ldots, y_m$. The array of membership functions for each of the fuzzy sets $\underset{\sim}{A}$ and $\underset{\sim}{B}$ can then be reduced to fuzzy vectors by the following substitutions:

$$\underset{\sim}{a} = \{a_1, \ldots, a_n\} = \{\mu_{\underset{\sim}{A}}(x_1), \ldots, \mu_{\underset{\sim}{A}}(x_n)\} = \{\mu_{\underset{\sim}{A}}(x_i)\}, \text{ for } i = 1, 2, \ldots, n \tag{12.4}$$

$$\underset{\sim}{b} = \{b_1, \ldots, b_m\} = \{\mu_{\underset{\sim}{B}}(y_1), \ldots, \mu_{\underset{\sim}{B}}(y_m)\} = \{\mu_{\underset{\sim}{B}}(y_j)\}, \text{ for } j = 1, 2, \ldots, m \tag{12.5}$$

Now, the image of fuzzy set $\underset{\sim}{A}$ can be determined through the use of the composition operation, or $\underset{\sim}{B} = \underset{\sim}{A} \circ \underset{\sim}{R}$, or when using the fuzzy vector form, $\underset{\sim}{b} = \underset{\sim}{a} \circ \underset{\sim}{R}$ where $\underset{\sim}{R}$ is an $n \times m$ fuzzy relation matrix.

More generally, suppose our input universe comprises the Cartesian product of many universes. Then the mapping f is defined on the power sets of this Cartesian input space and the output space, or

$$f : P(X_1 \times X_2 \times \cdots \times X_n) \longrightarrow P(Y) \tag{12.6}$$

Let fuzzy sets $\underset{\sim}{A_1}, \underset{\sim}{A_2}, \ldots, \underset{\sim}{A_n}$ be defined on the universes $X_1, X_2, \ldots, X_n$. The mapping for these particular input sets can now be defined as $\underset{\sim}{B} = f(\underset{\sim}{A_1}, \underset{\sim}{A_2}, \ldots, \underset{\sim}{A_n})$, where the membership function of the image $\underset{\sim}{B}$ is given by

$$\mu_{\underset{\sim}{B}}(y) = \max_{y=f(x_1,x_2,\ldots,x_n)} \{\min[\mu_{\underset{\sim}{A_1}}(x_1), \mu_{\underset{\sim}{A_2}}(x_2), \ldots, \mu_{\underset{\sim}{A_n}}(x_n)]\} \tag{12.7}$$

In the literature Eq. (12.7) is generally called Zadeh's *extension principle*. Equation (12.7) is expressed for a discrete-valued function, f. If the function, f, is a continuous-valued expression, the max operator is replaced by the sup (supremum) operator (the supremum is the least upper bound).

Fuzzy Transform (Mapping)

The material presented in the preceding two sections is associated with the issue of "extending" fuzziness in an input set to an output set. In this case, the input is fuzzy, the output is fuzzy, but the transform (mapping) is crisp, or $f : \underset{\sim}{A} \to \underset{\sim}{B}$. What happens in a more restricted case where the input is a single element (a nonfuzzy singleton) and this single element maps to a fuzzy set in the output universe? In this case the transform, or mapping, is termed a fuzzy transform.

Formally, let a mapping exist from an element x in universe X ($x \in$ X) to a fuzzy set B in the power set of universe Y, P(Y). Such a mapping is called a fuzzy mapping, $\underset{\sim}{f}$, where the output is no longer a single element, y, but a fuzzy set $\underset{\sim}{B}$, i.e.,

$$\underset{\sim}{B} = \underset{\sim}{f}(x) \tag{12.8}$$

If X and Y are finite universes, the fuzzy mapping expressed in Eq. (12.8) can be described as a fuzzy relation, $\underset{\sim}{R}$, or, in matrix form,

$$\underset{\sim}{R} = \begin{array}{c} \\ x_1 \\ x_2 \\ x_i \\ x_n \end{array} \begin{array}{cccccc} y_1 & y_2 & \cdots & y_j & \cdots & y_m \\ \left[\begin{array}{cccccc} r_{11} & r_{12} & \cdots & r_{1j} & \cdots & r_{1m} \\ r_{21} & r_{22} & \cdots & r_{2j} & \cdots & r_{2m} \\ r_{i1} & r_{i2} & \cdots & r_{ij} & \cdots & r_{im} \\ r_{n1} & r_{n2} & \cdots & r_{nj} & \cdots & r_{nm} \end{array} \right] \end{array} \tag{12.9}$$

For a particular single element of the input universe, say x_i, its fuzzy image, $\underset{\sim}{B}_i = \underset{\sim}{f}(x_i)$, is given in a general symbolic form as

$$\mu_{B_i}(y_j) = r_{ij} \tag{12.10}$$

or, in fuzzy vector notation,

$$\underset{\sim}{b}_i = \{r_{i1}, r_{i2}, \ldots, r_{im}\} \tag{12.11}$$

Hence, the fuzzy image of the element x_i is given by the elements in the ith row of the fuzzy relation, $\underset{\sim}{R}$, defining the fuzzy mapping, Eq. (12.9).

Suppose we now further generalize the situation where a fuzzy input set, say $\underset{\sim}{A}$, maps to a fuzzy output through a fuzzy mapping, or

$$\underset{\sim}{B} = \underset{\sim}{f}(\underset{\sim}{A}) \tag{12.12}$$

The extension principle again can be used to find this fuzzy image, $\underset{\sim}{B}$, by the following expression:

$$\mu_B(y) = \bigvee_{x \in X} (\mu_A(x) \wedge \mu_R(x, y)) \tag{12.13}$$

The preceding expression is analogous to a fuzzy composition performed on fuzzy vectors, or $\underset{\sim}{b} = \underset{\sim}{a} \circ \underset{\sim}{R}$, or in vector form,

$$\underset{\sim}{b}_j = \max_i(\min(a_i, r_{ij})) \tag{12.14}$$

where $\underset{\sim}{b}_j$ is the jth element of the fuzzy image $\underset{\sim}{B}$.

Example 12.2. Suppose we have a fuzzy mapping, $\underset{\sim}{f}$, given by the following fuzzy relation, $\underset{\sim}{R}$:

$$\underset{\sim}{R} = \begin{array}{c} \\ \\ \\ \\ \\ \end{array} \begin{array}{cccccc} 1.4 & 1.5 & 1.6 & 1.7 & 1.8 & (m) \\ \left[\begin{array}{ccccc} 1 & 0.8 & 0.2 & 0.1 & 0 \\ 0.8 & 1 & 0.8 & 0.2 & 0.1 \\ 0.2 & 0.8 & 1 & 0.8 & 0.2 \\ 0.1 & 0.2 & 0.8 & 1 & 0.8 \\ 0 & 0.1 & 0.2 & 0.8 & 1 \end{array} \right] & \begin{array}{c} 40 \\ 50 \\ 60 \\ 70 \\ 80 \end{array} & \begin{array}{c} (kg) \\ \\ \\ \\ \\ \end{array} \end{array}$$

which represents a fuzzy mapping between the length and mass of test articles scheduled for flight in a space experiment. The mapping is fuzzy because of the complicated relationship between mass and the cost to send the mass into space, the constraints on length of the test articles fitted into the cargo section of the spacecraft, and the scientific value of the experiment. Suppose a particular experiment is being planned for flight, but specific mass requirements have not been determined. For planning purposes the mass (in kilograms) is presumed to be a fuzzy quantity described by the following membership function:

$$A = \left\{ \frac{0.8}{40} + \frac{1}{50} + \frac{0.6}{60} + \frac{0.2}{70} + \frac{0}{80} \right\} \text{ kg}$$

or as a fuzzy vector $\underset{\sim}{a} = \{0.8, 1, 0.6, 0.2, 0\}$ kg.

The fuzzy image $\underset{\sim}{B}$ can be found using the extension principle (or, equivalently, composition for this fuzzy mapping), $\underset{\sim}{b} = \underset{\sim}{a} \circ R$ (recall that a set is also a one-dimensional relation). This composition results in a fuzzy output vector describing the fuzziness in the length of the experimental object (in meters), to be used for planning purposes, or $\underset{\sim}{b} = \{0.8, 1, 0.8, 0.6, 0.2\}$ m.

Practical Considerations

Heretofore we have discussed features of fuzzy sets on certain universes of discourse. Suppose there is a mapping between elements, u, of one universe, U, onto elements, v, of another universe, V, through a function f. Let this mapping be described by $f : u \rightarrow v$. Define $\underset{\sim}{A}$ to be a fuzzy set on universe U; that is, $\underset{\sim}{A} \subset U$. This relation is described by the membership function

$$\underset{\sim}{A} = \left\{ \frac{\mu_1}{u_1} + \frac{\mu_2}{u_2} + \cdots + \frac{\mu_n}{u_n} \right\} \tag{12.15}$$

Then the extension principle, as manifested in Eq. (12.3), asserts that, for a function f that performs a one-to-one mapping (i.e., maps one element in universe U to one element in universe V), an obvious consequence of Eq. (12.3) is

$$f(\underset{\sim}{A}) = f\left(\frac{\mu_1}{u_1} + \frac{\mu_2}{u_2} + \cdots + \frac{\mu_n}{u_n} \right)$$
$$= \left\{ \frac{\mu_1}{f(u_1)} + \frac{\mu_2}{f(u_2)} + \cdots + \frac{\mu_n}{f(u_n)} \right\} \tag{12.16}$$

The mapping in Eq. (12.16) is said to be *one-to-one*.

Example 12.3. Let a fuzzy set $\underset{\sim}{A}$ be defined on the universe U = $\{1, 2, 3\}$. We wish to map elements of this fuzzy set to another universe, V, under the function

$$v = f(u) = 2u - 1$$

We see that the elements of V are V = $\{1, 3, 5\}$. Suppose the fuzzy set $\underset{\sim}{A}$ is given by

$$\underset{\sim}{A} = \left\{ \frac{0.6}{1} + \frac{1}{2} + \frac{0.8}{3} \right\}$$

Then the fuzzy membership function for $v = f(u) = 2u - 1$ would be

$$f(\underset{\sim}{A}) = \left\{ \frac{0.6}{1} + \frac{1}{3} + \frac{0.8}{5} \right\}$$

For cases where this functional mapping f maps products of elements from two universes, say U_1 and U_2, to another universe V, and we define $\underset{\sim}{A}$ as a fuzzy set on the Cartesian space $U_1 \times U_2$, then

$$f(\underset{\sim}{A}) = \left\{ \sum \frac{\min [\mu_1(i), \mu_2(j)]}{f(i, j)} \mid i \in U_1, j \in U_2 \right\} \tag{12.17}$$

where $\mu_1(i)$ and $\mu_2(j)$ are the separable membership projections of $\mu(i, j)$ from the Cartesian space $U_1 \times U_2$ when $\mu(i, j)$ cannot be determined. This projection involves the invocation of a condition known as *noninteraction* (see Chapter 2) between the separate universes. It is analogous to the assumption of independence employed in probability theory, which reduces a joint probability density function to the product of its separate marginal density functions. In the fuzzy noninteraction case we are doing a kind of intersection; hence, we use the minimum operator (some logics use operators other than the minimum operator) as opposed to the product operator used in probability theory.

Example 12.4. Suppose we have the integers 1 to 10 as the elements of two identical but different universes; let

$$U_1 = U_2 = \{1, 2, 3, \ldots, 10\}$$

Then define two fuzzy numbers $\underset{\sim}{A}$ and $\underset{\sim}{B}$ on universe U_1 and U_2, respectively:

$$\text{Define } \underset{\sim}{A} = \underset{\sim}{2} = \text{``approximately 2''} = \left\{ \frac{0.6}{1} + \frac{1}{2} + \frac{0.8}{3} \right\}$$

$$\text{Define } \underset{\sim}{B} = \underset{\sim}{6} = \text{``approximately 6''} = \left\{ \frac{0.8}{5} + \frac{1}{6} + \frac{0.7}{7} \right\}$$

The product of (''approximately 2'') × (''approximately 6'') should map to a fuzzy number ''approximately 12,'' which is a fuzzy set defined on a universe, say V, of integers, $V = \{5, 6, \ldots, 18, 21\}$, as determined by the extension principle, Eq. (12.7), or

$$\underset{\sim}{2} \times \underset{\sim}{6} = \left(\frac{0.6}{1} + \frac{1}{2} + \frac{0.8}{3} \right) \times \left(\frac{0.8}{5} + \frac{1}{6} + \frac{0.7}{7} \right)$$

$$= \left\{ \frac{\min(0.6, 0.8)}{5} + \frac{\min(0.6, 1)}{6} + \cdots + \frac{\min(0.8, 1)}{18} + \frac{\min(0.8, 0.7)}{21} \right\}$$

$$= \left\{ \frac{0.6}{5} + \frac{0.6}{6} + \frac{0.6}{7} + \frac{0.8}{10} + \frac{1}{12} + \frac{0.7}{14} + \frac{0.8}{15} + \frac{0.8}{18} + \frac{0.7}{21} \right\}$$

In this example each of the elements in the universe, V, is determined by a unique mapping of the input variables. For example, $1 \times 5 = 5$, $2 \times 6 = 12$, etc. Hence, the maximum operation expressed in Eq. (12.7) is not necessary. It should also be noted that the result of this arithmetic product is not convex, and hence does not appear to be a fuzzy number (i.e., normal and convex). However, the nonconvexity arises from numerical aberrations from the discretization of the two fuzzy numbers, $\underset{\sim}{2}$ and $\underset{\sim}{6}$, and not from any inherent problems in the extension principle. This issue is discussed at length later in this chapter in Example 12.14.

The complexity of the extension principle increases when we consider if more than one of the combinations of the input variables, U_1 and U_2, are mapped to the same variable in the output space, V, i.e., if the mapping is not one-to-one. In this case we take the maximum membership grades of the combinations mapping to the same output variable, or, for the following mapping, we get

$$\mu_{\underset{\sim}{A}}(u_1, u_2) = \max_{v=f(u_1,u_2)} [\min\{\mu_1(u_1), \mu_2(u_2)\}] \tag{12.18}$$

Example 12.5. We have two fuzzy sets $\underset{\sim}{A}$ and $\underset{\sim}{B}$, each defined on its own universe as follows:

$$\underset{\sim}{A} = \left\{ \frac{0.2}{1} + \frac{1}{2} + \frac{0.7}{4} \right\} \quad \text{and} \quad \underset{\sim}{B} = \left\{ \frac{0.5}{1} + \frac{1}{2} \right\}$$

We wish to determine the membership values for the algebraic product mapping

$$f(\underset{\sim}{A}, \underset{\sim}{B}) = \underset{\sim}{A} \times \underset{\sim}{B} \text{ (arithmetic product)}$$

$$= \left\{ \frac{\min(0.2, 0.5)}{1} + \frac{\max[\min(0.2, 1), \min(0.5, 1)]}{2} \right.$$

$$\left. + \frac{\max[\min(0.7, 0.5), \min(1, 1)]}{4} + \frac{\min(0.7, 1)}{8} \right\}$$

$$= \left\{ \frac{0.2}{1} + \frac{0.5}{2} + \frac{1}{4} + \frac{0.7}{8} \right\}$$

In this case, the mapping involves two ways to produce a 2 (1×2 and 2×1) and two ways to produce a 4 (4×1 and 2×2); hence the maximum operation expressed in Eq. (12.7) is necessary.

The extension principle can also be useful in propagating fuzziness through generalized relations that are discrete mappings of ordered pairs of elements from input universes to ordered pairs of elements in an output universe.

Example 12.6. We want to map ordered pairs from input universes $X_1 = \{a, b\}$ and $X_2 = \{1, 2, 3\}$ to an output universe, $Y = \{x, y, z\}$. The mapping is given by the crisp relation, R,

$$R = \begin{array}{c} a \\ b \end{array} \begin{bmatrix} \begin{array}{ccc} 1 & 2 & 3 \\ x & z & x \\ x & y & z \end{array} \end{bmatrix}$$

We note that this relation represents a mapping, and it does not contain membership values. We define a fuzzy set $\underset{\sim}{A}$ on universe X_1 and a fuzzy set $\underset{\sim}{B}$ on universe X_2 as

$$\underset{\sim}{A} = \left\{ \frac{0.6}{a} + \frac{1}{b} \right\} \quad \text{and} \quad \underset{\sim}{B} = \left\{ \frac{0.2}{1} + \frac{0.8}{2} + \frac{0.4}{3} \right\}$$

We wish to determine the membership function of the output, $\underset{\sim}{C} = f(\underset{\sim}{A}, \underset{\sim}{B})$, whose relational mapping, f, is described by R. This is accomplished with the extension principle, Eq. (12.7), as follows:

$$\mu_{\underset{\sim}{C}}(x) = \max[\min(0.2, 0.6), \min(0.2, 1), \min(0.4, 0.6)] = 0.4$$

$$\mu_{\underset{\sim}{C}}(y) = \max[\min(0.8, 1)] = 0.8$$

$$\mu_{\underset{\sim}{C}}(z) = \max[\min(0.8, 0.6), \min(0.4, 1)] = 0.6$$

Hence,

$$\underset{\sim}{C} = \left\{ \frac{0.4}{x} + \frac{0.8}{y} + \frac{0.6}{z} \right\}$$

The extension principle is also useful in mapping fuzzy inputs through continuous-valued functions. The process is the same as for a discrete-valued function, but the effort involved in the computations is more rigorous.

Example 12.7   [Wong and Ross, 1985]. Suppose we have a nonlinear system given by the harmonic function $\underset{\sim}{x} = \cos(\underset{\sim}{\omega}t)$, where the frequency of excitation, $\underset{\sim}{\omega}$, is a fuzzy variable described by the membership function shown in Fig. 12.2a. The output variable, $\underset{\sim}{x}$, will be fuzzy because of the fuzziness provided in the mapping from the input variable, $\underset{\sim}{\omega}$. This function

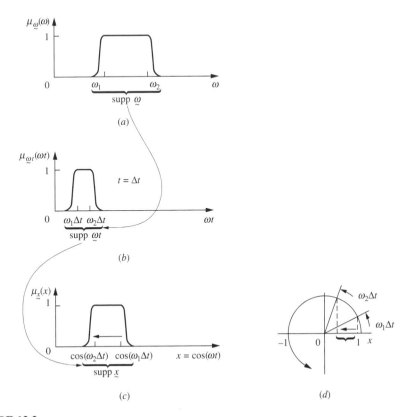

FIGURE 12.2
Extension principle applied to $\underset{\sim}{x} = \cos(\underset{\sim}{\omega}t)$, at $t = \Delta t$.

represents a one-to-one mapping in two stages, $\omega \rightarrow \omega t \rightarrow x$. The membership function of x will be determined through the use of the extension principle, which for this example will take on the following form:

$$\mu_x(x) = \bigvee_{x=\cos(\omega t)} [\mu_\omega(\omega)]$$

To show the development of this expression, we will take several time points, such as $t = 0, 1, \ldots$. For $t = 0$, all values of ω map into a single point in the ωt domain, i.e., $\omega t = 0$, and into a single point in the x universe, i.e., $x = 1$. Hence, the membership of x is simply a singleton at $x = 1$, i.e.,

$$\mu_x(x) = \begin{cases} 1, & \text{if } x = 1 \\ 0, & \text{otherwise} \end{cases}$$

For a nonzero but small t, say $t = \Delta t$, the support of ω, denoted in Fig. 12.2a as $supp\ \omega$, is mapped into a small but finite support of x, denoted in Fig. 12.2c as $supp\ x$ (the support of a fuzzy set was defined in Chapter 4 as the interval corresponding to a λ-cut of $\lambda = 0^+$). The membership value for each x in this interval is determined directly from the membership of ω in a one-to-one mapping. As can be seen in Fig. 12.2, as t increases, the support of x increases, and the fuzziness in the response spreads with time. Eventually, there will be a value of t when

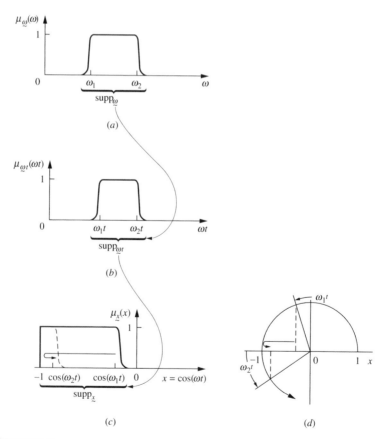

FIGURE 12.3
Extension principle applied to $x = \cos(\omega t)$ when t causes overlap in support of x.

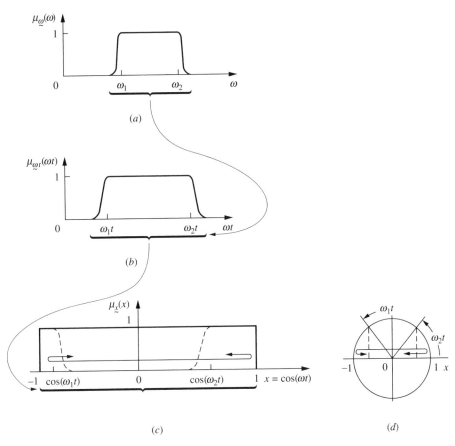

FIGURE 12.4
Extension principle applied to $x = \cos(\omega t)$ when t causes complete fuzziness.

the support of $\underset{\sim}{x}$ folds partly onto itself, i.e., we have multi-ω-to-single-x mapping. In this event, the maximum of all candidate membership values of $\underset{\sim}{\omega}$ is used as the membership value of x according to the extension principle, Eq. (12.3), as shown in Fig. 12.3(c).

When t is of such magnitude that the support of x occupies the interval $[-1, 1]$ completely, the largest support possible, the membership $\mu_{\underset{\sim}{x}}(x)$ will be unity for all x within this interval. This is the state of complete fuzziness, as illustrated in Fig. 12.4. In the equation $\underset{\sim}{x} = \cos(\underset{\sim}{\omega}t)$, the output can have any value in the interval $[-1, 1]$ with equal and complete membership. Once this state is reached the system remains there for all future time.

FUZZY ARITHMETIC

Chapter 4 defines a fuzzy number as being described by a normal, convex membership function on the real line; fuzzy numbers *usually* also have symmetric membership functions. In this chapter we wish to use the extension principle to perform algebraic operations on fuzzy numbers (as illustrated in previous examples in this chapter). We define a normal, convex fuzzy set on the real line to be a fuzzy number, and denote it $\underset{\sim}{I}$.

Let $\underset{\sim}{I}$ and $\underset{\sim}{J}$ be two fuzzy numbers, with $\underset{\sim}{I}$ defined on the real line in universe X and $\underset{\sim}{J}$ defined on the real line in universe Y, and let the symbol $*$ denote a general arithmetic operation, i.e., $* \equiv \{+, -, \times, \div\}$. An arithmetic operation (mapping) between these two number, denoted $\underset{\sim}{I} * \underset{\sim}{J}$, will be defined on universe Z, and can be accomplished using the extension principle, by

$$\mu_{\underset{\sim}{I}*\underset{\sim}{J}}(z) = \bigvee_{x*y=z} (\mu_{\underset{\sim}{I}}(x) \wedge \mu_{\underset{\sim}{J}}(y)) \tag{12.19}$$

Equation (12.19) results in another fuzzy set, the fuzzy number resulting from the arithmetic operation on fuzzy numbers $\underset{\sim}{I}$ and $\underset{\sim}{J}$.

Example 12.8. We want to perform a simple addition ($* \equiv +$) of two fuzzy numbers. Define a fuzzy *one* by the normal, convex membership function defined on the integers,

$$\underset{\sim}{1} = \left\{ \frac{0.2}{0} + \frac{1}{1} + \frac{0.2}{2} \right\}$$

Now, we want to add "fuzzy one" plus "fuzzy one," using the extension principle, Eq. (12.19), to get

$$\underset{\sim}{1} + \underset{\sim}{1} = \underset{\sim}{2} = \left(\frac{0.2}{0} + \frac{1}{1} + \frac{0.2}{2} \right) + \left(\frac{0.2}{0} + \frac{1}{1} + \frac{0.2}{2} \right)$$

$$= \left\{ \frac{\min(0.2, 0.2)}{0} + \frac{\max[\min(0.2, 1), \min(1, 0.2)]}{1} \right.$$

$$+ \frac{\max[\min(0.2, 0.2), \min(1, 1), \min(0.2, 0.2)]}{2}$$

$$\left. + \frac{\max[\min(1, 0.2), \min(0.2, 1)]}{3} + \frac{\min(0.2, 0.2)}{4} \right\}$$

$$= \left\{ \frac{0.2}{0} + \frac{0.2}{1} + \frac{1}{2} + \frac{0.2}{3} + \frac{0.2}{4} \right\}$$

Note that there are two ways to get the resulting membership value for a 1 ($0 + 1$ and $1 + 0$), three ways to get a 2 ($0 + 2$, $1 + 1$, $2 + 0$), and two ways to get a 3 ($1 + 2$ and $2 + 1$). These are accounted for in the implementation of the extension principle.

The support for a fuzzy number, say $\underset{\sim}{I}$ (see Chapter 4), is given by

$$supp\, \underset{\sim}{I} = \{x \mid \mu_{\underset{\sim}{I}}(x) > 0\} = I \tag{12.20}$$

which is an interval on the real line, denoted symbolically as I. Since applying Eq. (12.19) to arithmetic operations on fuzzy numbers results in a quantity that is also a fuzzy number, we can find the support of the fuzzy number resulting from the arithmetic operation, $\underset{\sim}{I} * \underset{\sim}{J}$, i.e.,

$$supp_{\underset{\sim}{I}*\underset{\sim}{J}}(z) = I * J \tag{12.21}$$

which is seen to be the arithmetic operation on the two individual supports (crisp intervals), I and J, for fuzzy numbers $\underset{\sim}{I}$ and $\underset{\sim}{J}$, respectively.

Chapter 4 revealed that the support of a fuzzy set is equal to its λ-cut value at $\lambda = 0^+$. In general, we can perform λ-cut operations on fuzzy numbers for any value of λ. A result we saw in Chapter 4 for set operations (Eq. (4.1)) is also valid for general arithmetic operations:

$$\left(\underset{\sim}{I} * \underset{\sim}{J} \right)_\lambda = I_\lambda * J_\lambda \tag{12.22}$$

Equation 12.22 shows that the λ-cut on a general arithmetic operation ($* \equiv \{+, -, \times, \div\}$) on two fuzzy numbers is equivalent to the arithmetic operation on the respective λ-cuts of the two fuzzy numbers. Both $(\underset{\sim}{I} * \underset{\sim}{J})_\lambda$ and $I_\lambda * J_\lambda$ are interval quantities; and manipulations of these quantities can make use of classical interval analysis, the subject of the next section.

INTERVAL ANALYSIS IN ARITHMETIC

As alluded to in Chapter 2, a fuzzy set can be thought of as a crisp set with ambiguous boundaries. In this sense, as Chapter 4 illustrated, a convex membership function defining a fuzzy set can be described by the intervals associated with different levels of λ-cuts. Let I_1 and I_2 be two interval numbers defined by ordered pairs of real numbers with lower and upper bounds:

$$I_1 = [a, b], \quad \text{where } a \leq b$$

$$I_2 = [c, d], \quad \text{where } c \leq d$$

When $a = b$ and $c = d$, these interval numbers degenerate to a scalar real number. We again define a general arithmetic property with the symbol $*$, where $* \equiv \{+, -, \times, \div\}$. Symbolically, the operation

$$I_1 * I_2 = [a, b] * [c, d] \tag{12.23}$$

represents another interval. This interval calculation depends on the magnitudes and signs of the elements a, b, c, and d. Table 12.1 shows the various combinations of set-theoretic intersection ($\cap$) and set-theoretic union ($\cup$) for the six possible combinations of these elements ($a < b$ and $c < d$ still hold). Based on the information in Table 12.1, the four arithmetic interval operations associated with Eq. (12.23) are given as follows:

$$[a, b] + [c, d] = [a + c, b + d] \tag{12.24}$$

$$[a, b] - [c, d] = [a - d, b - c] \tag{12.25}$$

$$[a, b] \cdot [c, d] = [\min(ac, ad, bc, bd), \max(ac, ad, bc, bd)] \tag{12.26}$$

$$[a, b] \div [c, d] = [a, b] \cdot \left[\frac{1}{d}, \frac{1}{c} \right] \text{ provided that } 0 \notin [c, d] \tag{12.27}$$

$$\alpha[a, b] = \begin{cases} [\alpha a, \alpha b] & \text{for } \alpha > 0 \\ [\alpha b, \alpha a] & \text{for } \alpha < 0 \end{cases} \tag{12.28}$$

where ac, ad, bc, and bd are arithmetic products and $1/d$ and $1/c$ are quotients.

The caveat applied to Eq. (12.27) is that the equivalence stated is not valid for the case when $c \leq 0$ and $d \geq 0$ (obviously the constraint $c < d$ still holds), i.e., zero cannot be contained within the interval $[c, d]$. Interval arithmetic follows properties of associativity

TABLE 12.1
Set operations on intervals

Cases	Intersection ($\cap$)	Union ($\cup$)
$a > d$	$\emptyset$	$[c, d] \cup [a, b]$
$c > b$	$\emptyset$	$[a, b] \cup [c, d]$
$a > c, b < d$	$[a, b]$	$[c, d]$
$c > a, d < b$	$[c, d]$	$[a, b]$
$a < c < b < d$	$[c, b]$	$[a, d]$
$c < a < d < b$	$[a, d]$	$[c, b]$

and commutativity for both summations and products, but it does not follow the property of distributivity. Rather, intervals do follow a special subclass of distributivity known as *subdistributivity*, i.e., for three intervals, I, J, and K,

$$I \cdot (J + K) \subset I \cdot J + I \cdot K \qquad (12.29)$$

The failure of distributivity to hold for intervals is due to the treatment of two occurrences of identical interval numbers (i.e., I) as two independent interval numbers [Dong and Shah, 1987].

Example 12.9.

$$-3 \cdot [1, 2] = [-6, -3]$$
$$[0, 1] - [0, 1] = [-1, 1]$$
$$[1, 3] \cdot [2, 4] = [\min(2, 4, 6, 12), \max(2, 4, 6, 12)] = [2, 12]$$
$$[1, 2] \div [1, 2] = [1, 2] \cdot [\tfrac{1}{2}, 1] = [\tfrac{1}{2}, 2]$$

Consider the following example of subdistributivity. For $I = [1, 2]$, $J = [2, 3]$, $K = [1, 4]$, then

$$I \cdot (J - K) = [1, 2] \cdot ([2, 3] - [1, 4]) = [1, 2] \cdot [-2, 2] = [-4, 4]$$
$$I \cdot J - I \cdot K = [1, 2] \cdot [2, 3] - [1, 2] \cdot [1, 4] = [2, 6] - [1, 8] = [-6, 5]$$

Now, $[-4, 4] \neq [-6, 5]$, but $[-4, 4] \subset [-6, 5]$.

Interval arithmetic can be thought of in the following way. When we add or multiply two crisp numbers, the result is a crisp singleton. When we add or multiply two intervals we are essentially performing these operations on the infinite number of combinations of pairs of crisp singletons from each of the two intervals; hence, in this sense, an interval is expected as the result. In the simplest case, when we multiply two intervals containing only positive real numbers, it is easy conceptually to see that the interval comprising the solution is found by taking the product of the two lowest values from each of the intervals to form the solution's lower bound, and by taking the product of the two highest values from each of the intervals to form the solution's upper bound. Even though we can see conceptually that an infinite number of combinations of products between these two intervals exist, we need only the endpoints of the intervals to find the endpoints of the solution.

APPROXIMATE METHODS OF EXTENSION

A serious disadvantage of the discretized form of the extension principle in propagating fuzziness for continuous-valued mappings is the irregular and erroneous membership functions determined for the output variable if the membership functions of the input variables are discretized for numerical convenience (this problem is demonstrated in Example 12.14). The reason for this anomaly is that the solution to the extension principle, as expressed in Eq. (12.7), is really a nonlinear programming problem for continuous-valued functions. It is well-known that, in any optimization process, discretization of any variables can lead to an erroneous optimum solution because portions of the solution space are omitted in the calculations. For example, try to plot a 10th-order curve with a series of equally spaced points; some local minimum and maximum points on the curve are going to be missed if the discretization is not small enough. Again, these problems do not arise because of any inherent problems in the extension principle itself; they arise when continuous-valued functions are discretized, then allowed to propagate from the input domain to the output domain using the extension principle.

Other methods have been proposed to ease the computational burden in implementing the extension principle for continuous-valued functions and mappings. Among the alternative methods proposed in the literature to avoid this disadvantage for continuous fuzzy variables are three approaches that are summarized here along with illustrative numerical examples. All of these approximate methods make use of intervals, at various λ-cut levels, in defining membership functions.

Vertex Method

A procedure known as the vertex method [Dong and Shah, 1987] greatly simplifies manipulations of the extension principle for continuous-valued fuzzy variables, such as fuzzy numbers defined on the real line. The method is based on a combination of the λ-cut concept and standard interval analysis. The vertex method can prevent abnormality in the output membership function due to application of the discretization technique on the fuzzy variables' domain, and it can prevent the widening of the resulting function value set due to multiple occurrences of variables in the functional expression by conventional interval analysis methods. The algorithm is very easy to implement and can be computationally efficient.

The algorithm works as follows. Any continuous membership function can be represented by a continuous sweep of λ-cut intervals from $\lambda = 0^+$ to $\lambda = 1$. Figure 12.5 shows a typical membership function with an interval associated with a specific value of λ. Suppose we have a single-input mapping given by $y = f(x)$ that is to be extended for fuzzy sets, or $B = f(A)$, and we want to decompose A into a series of λ-cut intervals, say I_λ.

When the function $f(x)$ is continuous and monotonic on $I_\lambda = [a, b]$, the interval representing B at a particular value of λ, say B_λ, can be obtained by

$$B_\lambda = f(I_\lambda) = [\min(f(a), f(b)), \max(f(a), f(b))] \qquad (12.30)$$

Equation (12.30) has reduced the interval analysis problem for a functional mapping to a simple procedure dealing only with the endpoints of the interval. When the mapping

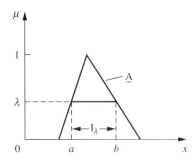

FIGURE 12.5
Interval corresponding to a λ-cut level on fuzzy set $\underset{\sim}{A}$.

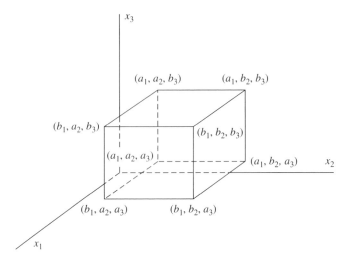

FIGURE 12.6
Three-dimensional Cartesian region involving intervals for three input variables, x_1, x_2, and x_3.

is given by n inputs, i.e., $y = f(x_1, x_2, x_3, \ldots, x_n)$, then the input space can be represented by an n-dimensional Cartesian region; a 3D Cartesian region is shown in Fig. 12.6. Each of the input variables can be described by an interval, say $I_{i\lambda}$, at a specific λ-cut, where

$$I_{i\lambda} = [a_i, b_i] \quad i = 1, 2, \ldots, n \tag{12.31}$$

As seen in Fig. 12.6, the endpoint pairs of each interval given in Eq. (12.31) intersect in the 3D space and form the vertices (corners) of the Cartesian space. The coordinates of these vertices are the values used in the vertex method when determining the output interval for each λ-cut. The number of vertices, N, is a quantity equal to $N = 2^n$, where n is the number of fuzzy input variables. When the mapping $y = f(x_1, x_2, x_3, \ldots, x_n)$ is continuous in the n-dimensional Cartesian region and when also there is no extreme point in this region (or along the boundaries), the value of the interval function for a particular

λ-cut can be obtained by

$$B_{\lambda} = f(I_{1\lambda}, I_{2\lambda}, I_{3\lambda}, \ldots, I_{n\lambda}) = \left[\min_{j}(f(c_j)), \max_{j}(f(c_j)) \right] \quad j = 1, 2, \ldots, N \quad (12.32)$$

where c_j is the coordinate of the jth vertex representing the n-dimensional Cartesian region.

The vertex method is accurate only when the conditions of continuity and no extreme point are satisfied. When extreme points of the function $y = f(x_1, x_2, x_3, \ldots, x_n)$ exist in the n-dimensional Cartesian region of the input parameters, the vertex method will miss certain parts of the interval that should be included in the output interval value, B_{λ}. Extreme points can be missed, for example, in certain mappings that are not one-to-one. If the extreme points can be identified, they are simply treated as additional vertices, E_k, in the Cartesian space and Eq. (12.32) becomes, because the continuity property still holds,

$$B_{\lambda} = \left[\min_{j,k}(f(c_j), f(E_k)), \max_{j,k}(f(c_j), f(E_k)) \right] \quad (12.33)$$

where $j = 1, 2, \ldots, N$ and $k = 1, 2, \ldots, m$ for m extreme points in the region.

Example 12.10. We wish to determine the fuzziness in the output of a simple nonlinear mapping given by the expression $y = f(x) = x(2 - x)$, seen in Fig. 12.7a, where the fuzzy input variable, x, has the membership function shown in Fig. 12.7b.

We shall solve this problem using the fuzzy vertex method at three λ-cut levels, for $\lambda = 0^+, 0.5, 1$. As seen in Fig. 12.7b, the intervals corresponding to these λ-cuts are $I_{0^+} = [0.5, 2], I_{.5} = [0.75, 1.5], I_1 = [1, 1]$ (a single point). Since the problem is one-dimensional, the vertices, c_j, are described by a single coordinate; there are $N = 2^1 = 2$ vertices ($j = 1, 2$). In addition, an extreme point does exist within the region of the membership function and is determined using a derivative of the function, $df(x)/dx = 2 - 2x = 0, x_0 = E_1 = 1$ (E_k, where $k = 1$). This extreme point is within each of the three λ-cut intervals, so will be involved in all the following calculations for B_{λ}:

$$I_{0^+} = [0.5, 2]$$

$$c_1 = 0.5, \qquad c_2 = 2, \qquad E_1 = 1$$

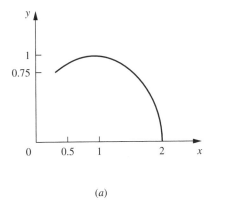

(a)

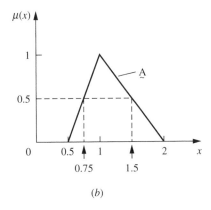

(b)

FIGURE 12.7
Nonlinear function and fuzzy input membership.

$$f(c_1) = 0.5(2 - 0.5) = 0.75, \qquad f(c_2) = 2(2 - 2) = 0,$$
$$f(E_1) = 1(2 - 1) = 1$$
$$B_{0^+} = [\min(0.75, 0, 1), \max(0.75, 0, 1)] = [0, 1]$$

$$I_{0.5} = [0.75, 1.5]$$
$$c_1 = 0.75, \quad c_2 = 1.5, \quad E_1 = 1$$
$$f(c_1) = 0.75(2 - 0.75) = 0.9375, f(c_2) = 1.5(2 - 1.5) = 0.75,$$
$$f(E_1) = 1(2 - 1) = 1$$
$$B_{0.5} = [\min(0.9375, 0.75, 1), \max(0.9375, 0.75, 1)] = [0.75, 1]$$

$$I_1 = [1, 1]$$
$$c_1 = 1, \quad c_2 = 1, \quad E_1 = 1$$
$$f(c_1) = f(c_2) = f(E_1) = 1(2 - 1) = 1$$
$$B_1 = [\min(1, 1, 1), \max(1, 1, 1)] = [1, 1] = 1$$

Figure 12.8 provides a plot of the intervals B_{0^+}, $B_{0.5}$, and B_1 to form the fuzzy output, y.

DSW Algorithm

The DSW algorithm [Dong, Shah, and Wong, 1985] also makes use of the λ-cut representation of fuzzy sets, but, unlike the vertex method, it uses the full λ-cut intervals in a standard interval analysis. The DSW algorithm consists of the following steps:

1. Select a λ value where $0 \le \lambda \le 1$.
2. Find the interval(s) in the input membership function(s) that correspond to this λ.
3. Using standard binary interval operations, compute the interval for the output membership function for the selected λ-cut level.
4. Repeat steps 1–3 for different values of λ to complete a λ-cut representation of the solution.

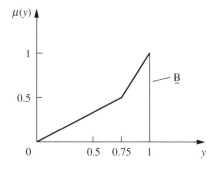

FIGURE 12.8
Fuzzy membership function for the output to $y = x(2 - x)$.

Example 12.11. Let us consider a nonlinear, 1D expression similar to the previous example, or $y = x(2 + x) = 2x + x^2$, where we again use the fuzzy input variable shown in Fig. 12.7b. The new function is shown in Fig. 12.9a, along with the fuzzy input in Fig. 12.9b. Again, if we decompose the membership function for the input into three λ-cut intervals, for $\lambda = 0^+$, 0.5, and 1, we get the intervals $I_{0^+} = [0.5, 2]$, $I_{0.5} = [0.75, 1.5]$, and $I_1 = [1, 1]$ (a single point). In terms of binary interval operations, the functional mapping on the intervals would take place as follows for each λ-cut level:

$$I_{0^+} = [0.5, 2]$$
$$B_{0^+} = 2[0.5, 2] + [0.5^2, 2^2] = [1, 4] + [0.25, 4] = [1.25, 8]$$

$$I_{0.5} = [0.75, 1.5]$$
$$B_{0.5} = 2[0.75, 1.5] + [0.75^2, 1.5^2] = [1.5, 3] + [0.5625, 2.25] = [2.0625, 5.25]$$

$$I_1 = [1, 1]$$
$$B_1 = 2[1, 1] + [1^2, 1^2] = [2, 2] + [1, 1] = [3, 3] = 3$$

Figure 12.10 provides a plot of the intervals B_{0^+}, $B_{0.5}$, and B_1 to form the fuzzy output, y.

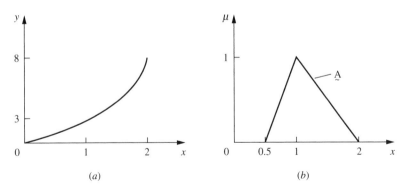

FIGURE 12.9
Nonlinear function and fuzzy input membership.

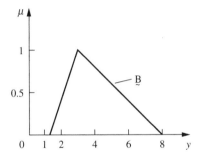

FIGURE 12.10
Fuzzy membership function for the output to $y = x(2 + x)$.

The previous example worked with a fuzzy input that was defined on the positive side of the real line; hence DSW operations were conducted on positive quantities. Suppose we want to conduct the same DSW operations, but on a fuzzy input that is defined on both the positive and negative side of the real line. The user of the DSW algorithm must be careful in this case. If the lower bound of an interval is negative and the upper bound is positive (i.e., if the interval contains zero) and if the function involves a square or an even-power operation, then the lower bound of the result should be zero. This feature of an interval analysis, like the DSW method, is demonstrated in the following example.

Example 12.12. Let us consider the nonlinear, 1D expression from Example 12.11, i.e., $y = x(2 + x) = 2x + x^2$, which is shown in Fig. 12.9a and is repeated in Fig. 12.11a. Suppose we change the domain of the input variable, x, to include negative numbers, as shown in Fig. 12.11b. Again, if we decompose the membership function for the input into three λ-cut intervals, for $\lambda = 0^+$, 0.5, and 1, we get the intervals $I_{0^+} = [-0.5, 1]$, $I_{0.5} = [-0.25, 0.5]$, and $I_1 = [0, 0]$ (a single point). In terms of binary interval operations, the functional mapping on the intervals would take place as follows for each λ-cut level:

$$I_{0^+} = [-0.5, 1]$$

$$B_{0^+} = 2[-0.5, 1] + [\mathbf{0}, 1^2] = [-1, 2] + [0, 1] = [-1, 3]$$

(*Note:* The boldface zero is taken as the minimum, since $(-0.5)^2 > 0$; because zero is contained in the interval $[-0.5, 1]$ the minimum of squares of any number in the interval will be zero.)

$$I_{0.5} = [-0.25, 0.5]$$

$$B_{0.5} = 2[-0.25, 0.5] + [\mathbf{0}, 0.5^2] = [-0.5, 1] + [0, 0.25] = [-0.5, 1.25]$$

$$I_1 = [0, 0]$$

$$B_1 = 2[0, 0] + [0, 0] = [0, 0]$$

Figure 12.12 is a plot of the intervals B_{0^+}, $B_{0.5}$, and B_1 that form the fuzzy output, y.

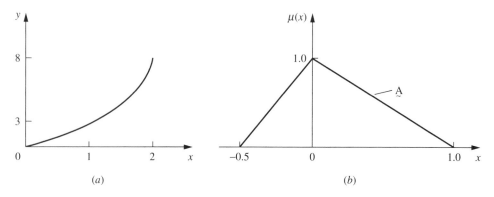

(a)   (b)

FIGURE 12.11
Nonlinear function and fuzzy input membership.

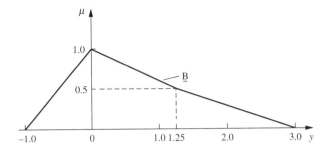

FIGURE 12.12
Fuzzy membership function for the output to $y = x(2 + x)$.

Restricted DSW Algorithm

This method, proposed by Givens and Tahani [1987], is a slight restriction of the original DSW algorithm. Suppose we have two interval numbers, $I = [a, b]$ and $J = [c, d]$. For the special case where neither of these intervals contains negative numbers, i.e., $a, b, c, d \geq 0$, and none of the calculations using these intervals involves subtraction, the definitions of interval multiplication, Eq. (12.26), and interval division, Eq. (12.27), can be simplified as follows:

$$I \cdot J = [a, b] \cdot [c, d] = [ac, bd] \tag{12.34}$$

$$I/J = [a, b] \div [c, d] = \left[\frac{a}{d}, \frac{b}{c}\right] \tag{12.35}$$

These definitions of interval multiplication and division require only one-fourth the number of multiplications (or divisions), and there is no need for the min or max operations, unlike the previous definitions, Eqs. (12.26, 12.27).

> **Example 12.13.** Let us consider the function in Example 12.12, $y = x(2 + x)$, and another nonlinear, 1D expression of the form $y = x/(2 + x)$, where we again use the fuzzy input variable shown in Fig. 12.7b in both functions. In interval calculations we can represent the scalar value 2 by the interval [2, 2]. The λ-cut interval calculations using the restricted DSW calculations are now as follows:
>
> $y = x(2 + x)$
>
> $I_{0^+} = [0.5, 2]$
> $B_{0^+} = [0.5, 2] \cdot \{[2, 2] + [0.5, 2]\} = [0.5, 2] \cdot [2.5, 4] = [1.25, 8]$
>
> $I_{0.5} = [0.75, 1.5]$
> $B_{0.5} = [0.75, 1.5] \cdot \{[2, 2] + [0.75, 1.5]\} = [0.75, 1.5] \cdot [2.75, 3.5] = [2.0625, 5.25]$
>
> $I_1 = [1, 1]$
> $B_1 = [1, 1] \cdot \{[2, 2] + [1, 1]\} = [1, 1] \cdot [3, 3] = 3$

Note that these three intervals for the output $\underset{\sim}{B}$ are identical to those in the previous example.

$y = x/(2 + x)$

$I_{0^+} = [0.5, 2]$

$B_{0^+} = [0.5, 2] \div \{[2, 2] + [0.5, 2]\} = [0.5, 2] \div [2.5, 4] = [0.125, 0.8]$

$I_{0.5} = [0.75, 1.5]$

$B_{0.5} = [0.75, 1.5] \div \{[2, 2] + [0.75, 1.5]\} = [0.75, 1.5] \div [2.75, 3.5]$
$\qquad = [0.2143, 0.5455]$

$I_1 = [1, 1]$

$B_1 = [1, 1] \div \{[2, 2] + [1, 1]\} = [1, 1] \div [3, 3] = 0.3333$

Comparisons

It will be useful at this point to compare the three methods discussed so far – the extension principle, the vertex method, and the DSW algorithm – by applying them to the same problem. This comparison will illustrate the problems faced with using the extension principle on discretized membership functions, as compared to the other two methods.

Example 12.14. We define fuzzy sets X and Y with the membership functions as shown in Fig. 12.13. We will use the following methods to compute $X * Y$ and to demonstrate the similarity of results:

- The extension principle
- The vertex method
- The DSW algorithm

Extension principle using discretized fuzzy sets. The fuzzy variables may be discretized at seven points as

$$X = \left\{ \frac{0}{1} + \frac{0.33}{2} + \frac{0.66}{3} + \frac{1.0}{4} + \frac{0.66}{5} + \frac{0.33}{6} + \frac{0}{7} \right\}$$

and

$$Y = \left\{ \frac{0}{2} + \frac{0.33}{3} + \frac{0.66}{4} + \frac{1.0}{5} + \frac{0.66}{6} + \frac{0.33}{7} + \frac{0}{8} \right\}$$

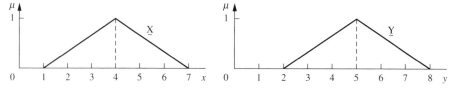

FIGURE 12.13
Fuzzy sets X and Y.

Their product would then give us

$$
\underset{\sim}{X} \times \underset{\sim}{Y} = \left\{ \frac{0}{2} + \frac{0}{3} + \frac{0}{4} + \frac{0}{5} + \frac{0.33}{6} + \frac{0}{7} + \frac{0.33}{8} + \frac{0.33}{9} + \frac{0.33}{10} + \frac{0.66}{12} \right.
$$

$$
+ \frac{0.33}{14} + \frac{0.66}{15} + \frac{0.33}{16} + \frac{0.66}{18} + \frac{1.0}{20} + \frac{0.33}{21} + \frac{0.66}{24} + \frac{0.66}{25} + \frac{0.33}{28}
$$

$$
\left. + \frac{0.66}{30} + \frac{0}{32} + \frac{0.33}{35} + \frac{0.33}{36} + \frac{0.0}{40} + \frac{0.33}{42} + \frac{0.0}{48} + \frac{0.0}{49} + \frac{0.0}{56} \right\}
$$

The result of the operation $\underset{\sim}{X} \times \underset{\sim}{Y}$ for a discretization level of seven points is plotted in Fig. 12.14a. Figures 12.14b, c, and d show the product function $\underset{\sim}{X} \times \underset{\sim}{Y}$ for greater discretization levels of the fuzzy variables $\underset{\sim}{X}$ and $\underset{\sim}{Y}$.

Vertex method. I_{0^+}: Support for X is the interval [1, 7] and support for Y is the interval [2, 8].

(a) $x = 1,$ $y = 2,$ $f(a) = 2$
(b) $x = 1,$ $y = 8,$ $f(b) = 8$
(c) $x = 7,$ $y = 2,$ $f(c) = 14$
(d) $x = 7,$ $y = 8,$ $f(d) = 56$

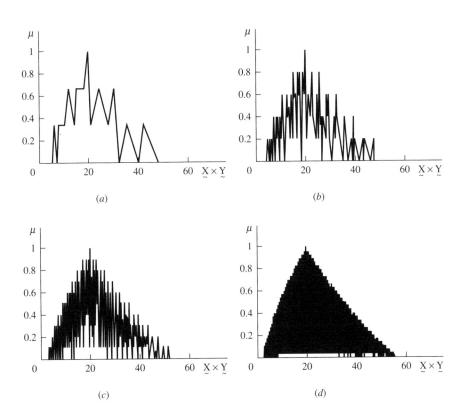

FIGURE 12.14
$\underset{\sim}{X} \times \underset{\sim}{Y}$ for increasing discretization of both X and Y (both variables are discretized for the same number of points): (a) 7 points; (b) 13 points; (c) 23 points; (d) 63 points.

Therefore, min = 2, max = 56, and $B_{0^+} = [2, 56]$.

$I_{0.33}$: X[2, 6], Y[3, 7].

$$\begin{array}{llll}
(a) \ x = 2, & y = 3, & f(a) = 6 \\
(b) \ x = 2, & y = 7, & f(b) = 14 \\
(c) \ x = 6, & y = 3, & f(c) = 18 \\
(d) \ x = 6, & y = 7, & f(d) = 42
\end{array}$$

Therefore, min = 6, max = 42, and $B_{0.33} = [6, 42]$.

$I_{0.66}$: X[3, 5], Y[4, 6].

$$\begin{array}{llll}
(a) \ x = 3, & y = 4, & f(a) = 12 \\
(b) \ x = 3, & y = 6, & f(b) = 18 \\
(c) \ x = 5, & y = 4, & f(c) = 20 \\
(d) \ x = 5, & y = 6, & f(d) = 30
\end{array}$$

Therefore, min = 12, max = 30, and $B_{0.66} = [12, 30]$.

$I_{1.0}$: X[4, 4], Y[5, 5].

$$\begin{array}{llll}
(a) \ x = 4, & y = 5, & f(a) = 20
\end{array}$$

Therefore, min = 20, max = 20, and $B_{1.0} = [20, 20]$.

The results of plotting the four λ-cut levels is shown in Fig. 12.15.

DSW method.

$$I_{0^+} : [1, 7] \bullet [2, 8] = [\min(2, 14, 8, 56), \max(2, 14, 8, 56)] = [2, 56]$$

$$I_{0.33} : [2, 6] \bullet [3, 7] = [\min(6, 18, 14, 42), \max(6, 18, 14, 42)] = [6, 42]$$

$$I_{0.66} : [3, 5] \bullet [4, 6] = [\min(12, 20, 18, 30), \max(12, 20, 18, 30)] = [12, 30]$$

$$I_1 : [4, 4] \bullet [5, 5] = [\min(20, 20, 20, 20), \max(20, 20, 20, 20)] = [20, 20]$$

The results of plotting the four λ-cut levels are shown in Fig. 12.16.

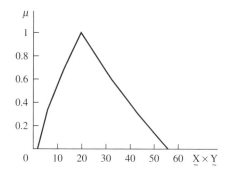

FIGURE 12.15
Output profile of $\underset{\sim}{X} \times \underset{\sim}{Y}$ determined using the vertex method.

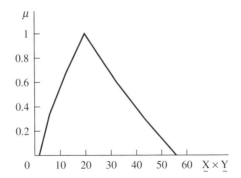

FIGURE 12.16
Output profile of $\underset{\sim}{X} \times \underset{\sim}{Y}$ determined using the DSW algorithm.

Comparing the results from the foregoing three methods, we see that the results are the same. The discretization method was performed for increasing levels of discretization. In each case the outer envelope of the curves due to the discretized method gave the correct results. It should be intuitively obvious that as the discretization is increased (the equation is exactly simulated) the resulting curve approaches the true values of membership functions. Also note that the discretization technique is computationally expensive for complex problems.

SUMMARY

The extension principle is one of the most basic ideas in fuzzy set theory. It provides a general method for extending crisp mathematical concepts to address fuzzy quantities, such as real algebra operations on fuzzy numbers. These operations are computationally effective generalizations of interval analysis. Several methods to convert extended fuzzy operations into efficient computational algorithms have been presented in this chapter. All of these approximations make use of the decomposition of a membership function into a series of λ-cut intervals. The employment of the extension principle on discretized fuzzy numbers can lead to counterintuitive results, unless sufficient resolution in the discretization is maintained. This statement is simply a caution to potential users of some of the simpler ideas in the extension principle. Although the set of real fuzzy numbers equipped with an extended addition or multiplication is no longer a group, many structural properties of the resulting fuzzy numbers are preserved in the process [Dubois and Prade, 1980].

REFERENCES

Dong, W. and Shah, H. (1987). ''Vertex method for computing functions of fuzzy variables,'' *Fuzzy Sets Syst.*, vol. 24, pp. 65–78.

Dong, W., Shah, H., and Wong, F. (1985). ''Fuzzy computations in risk and decision analysis,'' *Civ. Eng. Syst.*, vol. 2, pp. 201–208.

Dubois, D. and Prade, H. (1980). *Fuzzy sets and systems: Theory and applications*, Academic Press, New York.

Givens, J. and Tahani, H. (1987). "An improved method of performing fuzzy arithmetic for computer vision," *Proceedings of North American Information Processing Society (NAFIPS)*, Purdue University, West Lafayette, IN, pp. 275–280.

Wong, F. and Ross, T. (1985). "Treatment of uncertainties in structural dynamics models," in F. Deyi and L. Xihui (eds.), *Proceedings of the International Symposium on Fuzzy Mathematics in Earthquake Research*, Seismological Press, Beijing.

Yager, R. R. (1986). "A characterization of the extension principle," *Fuzzy Sets Syst.*, vol. 18, pp. 205–217.

Zadeh, L. (1975). "The concept of a linguistic variable and its application to approximate reasoning, Part I," *Inf. Sci.*, vol. 8, pp. 199–249.

PROBLEMS

12.1. Perform the following operations on intervals:
   (a) $[2, 3] + [3, 4]$
   (b) $[1, 2] \times [1, 3]$
   (c) $[4, 6] \div [1, 2]$
   (d) $[3, 5] - [4, 5]$

12.2. Given the following fuzzy numbers and using Zadeh's extension principle, calculate $\underset{\sim}{K} = \underset{\sim}{I} \cdot \underset{\sim}{J}$ and explain (or show) why $\underset{\sim}{6}$ is nonconvex:

$$\underset{\sim}{I} = \underset{\sim}{3} = \frac{0.2}{2} + \frac{1}{3} + \frac{0.1}{4}$$

$$\underset{\sim}{J} = \underset{\sim}{2} = \frac{0.1}{1} + \frac{1}{2} + \frac{0.3}{3}$$

12.3. This problem makes use of Zadeh's extension principle. You are given the fuzzy sets $\underset{\sim}{A}$ and $\underset{\sim}{B}$ on the real line as follows:

$\mu(x_i)$	0	1	2	3	4	5	6	7
$\underset{\sim}{A}$	0.0	0.1	0.6	0.8	0.9	0.7	0.1	0.0
$\underset{\sim}{B}$	0.0	1.0	0.7	0.5	0.2	0.1	0.0	0.0

If x and y are real numbers defined by sets $\underset{\sim}{A}$ and $\underset{\sim}{B}$, respectively, calculate the fuzzy set $\underset{\sim}{C}$ representing the real numbers z given by
   (a) $z = 3x - 2$
   (b) $z = 4x^2 + 3$
   (c) $z = x^2 + y^2$
   (d) $z = x - x$
   (e) $z = \min(x, y)$

12.4. For the function $y = x_1^2 + x_2^2 - 4x_1 + 4$ and the membership functions for fuzzy variables x_1 and x_2 shown in Fig. P12.4, find and plot the membership function for the fuzzy output variable, y, using
   (a) A discretized form of the extension principle
   (b) The vertex method
   (c) The DSW algorithm

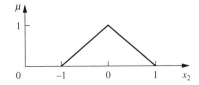

FIGURE P12.4

12.5. The voltage drop across an element in a series circuit is equal to the series current multiplied by the element's impedance. The current, $\underset{\sim}{I}$, impedance, $\underset{\sim}{R}$, and voltage, $\underset{\sim}{V}$, are presumed to be fuzzy variables. Membership functions for the current and impedance are as follows:

$$\underset{\sim}{I} = \left\{ \frac{0}{0} + \frac{0.8}{0.5} + \frac{1}{1} + \frac{0.8}{1.5} + \frac{0}{2} \right\}$$

$$\underset{\sim}{R} = \left\{ \frac{0.5}{500} + \frac{0.9}{750} + \frac{1}{1000} + \frac{0.9}{1250} + \frac{0.5}{1500} \right\}$$

Find the arithmetic product for $\underset{\sim}{V} = \underset{\sim}{I} \cdot \underset{\sim}{R}$ using the extension principle.

12.6. Determine equivalent resistance of the circuit shown in Fig. P12.6, where $\underset{\sim}{R_1}$ and $\underset{\sim}{R_2}$ are fuzzy sets describing the resistance of resistors R_1 and R_2, respectively, expressed in ohms. Since the resistors are in series they can be added arithmetically. Using the extension principle, find the equivalent resistance,

$$\underset{\sim}{R_{eq}} = \underset{\sim}{R_1} + \underset{\sim}{R_2}$$

The membership functions for the two resistors are

$$\underset{\sim}{R_1} = \left\{ \frac{0.5}{3} + \frac{0.8}{4} + \frac{0.6}{5} \right\} \quad \text{and} \quad \underset{\sim}{R_2} = \left\{ \frac{0.3}{8} + \frac{1.0}{9} + \frac{0.4}{10} \right\}$$

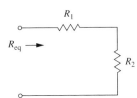

FIGURE P12.6

12.7. In Newtonian mechanics the equivalent force on a body in motion can be found by taking the product of its mass and acceleration; this is commonly referred to as Newton's second law. For an object in a particular state, suppose the acceleration under the present force is given by the fuzzy set

$$\underset{\sim}{A} = \left\{ \frac{0}{0} + \frac{0.2}{1} + \frac{0.7}{2} + \frac{1}{3} + \frac{0}{4} \right\}$$

and the mass is given by the fuzzy set

$$\underset{\sim}{M} = \left\{ \frac{0}{1} + \frac{0.5}{2} + \frac{1}{3} + \frac{0.5}{4} + \frac{0}{5} \right\}$$

Assume both sets are in nondimensionalized units.

(a) Find the fuzzy set representing the force on the object using the extension principle.

(b) Develop analogous continuous membership functions, and plot them, for the fuzzy acceleration and mass and solve for the fuzzy force using (i) the vertex method and (ii) the restricted DSW algorithm.

12.8. For fluids, the product of the pressure (P) and the volume (V) of the fluid is a constant for a given temperature, i.e.,

$$PV = \text{constant}$$

Assume that at a given temperature a fluid of fuzzy volume

$$V_1 = \left\{ \frac{0.0}{0.5} + \frac{0.5}{0.75} + \frac{1.0}{1.0} + \frac{0.5}{1.25} + \frac{0.0}{1.5} \right\}$$

is under a fuzzy pressure

$$P_1 = \left\{ \frac{0.0}{0.5} + \frac{0.5}{1.75} + \frac{1.0}{2.0} + \frac{0.5}{2.25} + \frac{0.0}{2.5} \right\}$$

(a) Using the extension principle, determine the pressure P_2 if the volume is reduced to

$$V_2 = \left\{ \frac{0.0}{0.4} + \frac{0.5}{0.45} + \frac{1.0}{0.5} + \frac{0.5}{0.55} + \frac{0.0}{0.6} \right\}$$

(b) Develop analogous continuous membership functions for the fuzzy pressure P_1 and volume V_1 and solve for the pressure P_2 using (i) the vertex method and (ii) the DSW algorithm. Plot the resulting membership function.

(c) Explain why $P_2 \cdot V_2$ would not be the same as $P_1 \cdot V_1$.

12.9. A circle is governed by the equation $x^2 + y^2 = 8^2$. Its fuzzy x coordinate is defined by the fuzzy set

$$x = \left\{ \frac{0}{0} + \frac{0.6}{2} + \frac{0.65}{3} + \frac{0.7}{4} + \frac{0.75}{5} + \frac{0.8}{6} \right\}$$

Find the fuzzy y coordinate, and plot its membership function for the equation of a circle.

(a) Use the DSW algorithm.

(b) Perform the same calculation using the restricted DSW algorithm.

(c) Comment on the nature of the results using a fuzzy x that is non-normal.

12.10. For the function $y = x_1^2 \cdot x_2 - 3x_2$, where the membership functions of x_1 and x_2 are given in Fig. P12.10, find and plot the fuzzy membership function for y using

(a) The vertex method

   (i) Ignoring any extreme points

   (ii) Including any extreme points

(b) The restricted DSW algorithm

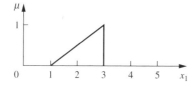 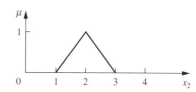

FIGURE P12.10

12.11. Define a fuzzy set $\underset{\sim}{X}$ with the membership function

$$\underset{\sim}{x} = \left\{ \frac{0.1}{1} + \frac{1}{2} + \frac{0.4}{3} \right\}$$

Using the extension principle, determine the membership function for $\underset{\sim}{z}$, written in two different forms, i.e., for

(a) $\underset{\sim}{z} = \underset{\sim}{x}^2$

(b) $\underset{\sim}{z} = \underset{\sim}{x} \cdot \underset{\sim}{x}$

For parts (a) and (b) use the direct extension principle, the vertex method, and the DSW method, and compare the three results.

(c) $\underset{\sim}{z} = \underset{\sim}{x}^2$ and $\underset{\sim}{z} = \underset{\sim}{x} \cdot \underset{\sim}{x}$ using the vertex method

(d) $\underset{\sim}{z} = \underset{\sim}{x}^2$ and $\underset{\sim}{z} = \underset{\sim}{x} \cdot \underset{\sim}{x}$ using the DSW algorithm

(e) Discuss your answers from the different forms and methods.

12.12. Now suppose $\underset{\sim}{x}$ has membership function

$$\underset{\sim}{x} = \left\{ \frac{0.1}{-3} + \frac{0.3}{-2} + \frac{0.7}{-1} + \frac{1}{0} + \frac{0.7}{1} + \frac{0.3}{2} + \frac{0.1}{3} \right\}$$

Repeat steps (a), (b), (c), and (d) of Problem 12.11 and (e) comment on any differences or similarities.

12.13. When taking hydrostatic measurements, hydrostatic pressure is given by

$$P = \rho g h$$

where ρ is density, g is acceleration due to gravity, and h is the height of the column of fluid. In a well-drilling environment, the density of the drilling mud has some uncertainty due to the inconsistent nature of the fluid. The well depth can also possess significant uncertainty due to stretching in the drill pipe used to measure the well depth. A membership function for density and depth is given in Fig. P12.13. Using the DSW algorithm determine the membership function for hydrostatic pressure, P.

12.14. A dissipated power, P, in a resistor can be described by $P = R \cdot I^2$, where R is the resistance, in ohms, and I is the current, measured in amps, passing through the resistor. Let a fuzzy set $\underset{\sim}{R}$ be defined on the universe $x_1 = \{10, 20, 30, 40, 50\}$ ohms and a fuzzy set $\underset{\sim}{I}$ be defined on the universe $x_2 = \{0, 1, 2, 3\}$ amps. We wish to map elements of these fuzzy sets to the dissipated power universe, y, under the relation $P = R \cdot I^2$. We have a medium resistance given by

$$\underset{\sim}{R} = \left\{ \frac{0.3}{10} + \frac{0.9}{20} + \frac{1}{30} + \frac{0.8}{40} + \frac{0.4}{50} \right\} = \text{``medium resistance''}$$

and a low current given by

$$\underset{\sim}{I} = \left\{ \frac{0.5}{0} + \frac{1}{1} + \frac{0.4}{2} + \frac{0.1}{3} \right\} = \text{``low current''}$$

Using the discretized form of the extension principle, determine the membership values for P.

12.15. An airport passenger terminal has two activities with specific time intervals: processing times (t_1) and waiting times (t_2). The universe of time is $X = \{10, 20, 30\}$ in minutes. For each of these two activities there is a membership function relating the level of service to the total time the passengers spend waiting in line: (1) tolerable service, or (2) good service. For this

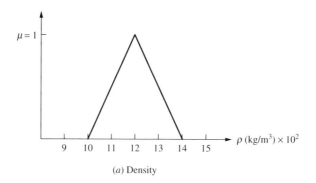

(a) Density

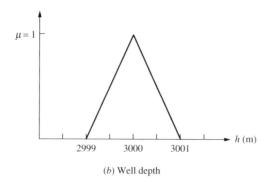

(b) Well depth

FIGURE P12.13
(a) Density, (b) well depth.

exercise, suppose each of the times is ''good,'' as given below:

$$t_1 = \left\{ \frac{1.0}{10} + \frac{0.8}{20} + \frac{0.5}{30} \right\}$$

$$t_2 = \left\{ \frac{1.0}{20} + \frac{0.6}{30} + \frac{0.3}{40} \right\}$$

Using a discretized form of the extension principle, find the membership function for the total time (processing time + waiting time), i.e., for the total time defined as $t = t_1 + t_2$.

12.16. Flue gas is used to heat a process stream using a counter-current heat exchanger. The process stream is intended to meet a required temperature of 190°C with an average heat capacity rate WCp_{ps}. The flue gas entering the heat exchanger has $WCp = 0.3$ kW/°C and $T = 1000$°C (Fig. P12.16).

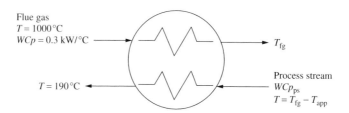

Flue gas
$T = 1000$°C
$WCp = 0.3$ kW/°C

T_{fg}

$T = 190$°C

Process stream
WCp_{ps}
$T = T_{fg} - T_{app}$

FIGURE P12.16

The outlet temperature of the gas (T_{fg}) is considered a discrete fuzzy set of values related to optimum operation conditions $(\mu(T_{fg}))$, and the process stream inlet temperature depends on both T_{fg} and the approach temperature (ΔT_{app}) that is also considered a discrete fuzzy set of values $(\mu(\Delta T_{app}))$:

$$\mu_{T_{fg}} = \left\{ \frac{0}{160°C} + \frac{0.5}{170°C} + \frac{0.75}{180°C} + \frac{0.9}{190°C} + \frac{1}{200°C} \right\}$$

$$\mu_{\Delta T_{app}} = \left\{ \frac{0}{20°C} + \frac{0.55}{30°C} + \frac{0.85}{40°C} + \frac{1}{50°C} + \frac{0}{60°C} \right\}$$

The following equation from the energy balance is needed:

$$WCp_{ps} = \frac{(0.3 \text{ kW}/°C)(1000°C - T_{fg})}{190°C - T} \quad \text{with } T = T_{fg} - \Delta T_{app}$$

Implementing the extension principle, find the fuzzy values of the heat capacity rate of the process stream for optimum operation.

CHAPTER
13

FUZZY
CONTROL
SYSTEMS

The decision to reject one paradigm is always simultaneously the decision to accept another, and the judgment leading to that decision involves the comparison of both paradigms with nature and with each other ... the search for assumptions (even for non-existent ones) can be an effective way to weaken the grip of a tradition upon the mind and to suggest the basis for a new one.

Thomas Kuhn
The Structure of Scientific Revolutions, 1962

Control applications are the kinds of problems for which fuzzy logic has had the greatest success and acclaim. Many of the consumer products that we use today involve fuzzy control. And, even though fuzzy control is now a standard within industry, the teaching of this subject on academic campuses is still far from being a *standard* offering. But, a paradigm shift is being realized in the area of fuzzy control, given its successes for some problems where classical control has not been effective or efficient. In the quote, above, such a paradigm shift can be explained. It was not long ago that fuzzy logic and fuzzy systems were the subject of ridicule and scorn in the scientific communities, but the control community moved quickly in accepting the *new paradigm* and its success is now manifested in the marketplace.

Control systems abound in our everyday life; perhaps we do not see them as such, perhaps because some of them are larger than what a single individual can deal with, but they are ubiquitous. For example, economic systems are large, global systems that can be controlled; ecosystems are large, amorphous, and long-term systems that can be controlled. Systems that can be controlled have three key features: inputs, outputs, and control parameters (or actions) which are used to perturb the system into some desirable state. The system is monitored in some fashion and left alone if the desired state is realized,

Fuzzy Logic with Engineering Applications, Second Edition T. J. Ross
© 2004 John Wiley & Sons, Ltd ISBNs: 0-470-86074-X (HB); 0-470-86075-8 (PB)

or perturbed with control actions until the desired state is reached. Usually, the control parameters (actions) are used to perturb the inputs to the system. For example, in the case of economic systems the inputs might be the balance of trade index, the federal budget deficit, and the consumer price index; outputs might be the inflation rate and the Dow Jones Industrial index; a control parameter might be the federal lending rate that gets adjusted occasionally by the US Federal Reserve Board. In the case of ecosystems the inputs could be the rate of urbanization, automobile traffic, and water use; the outputs could be reductions in green spaces, or habitat erosion; a control action could be federal laws and policy on pollution prevention. Other, everyday, control situations are evident in our daily lives. Traffic lights are control mechanisms: inputs are arrival rates of cars at an intersection and time of day, outputs are the length of the lines at the lights, and the control parameters are the length of the various light actions (green, yellow, green arrow, etc.). And, construction projects involve control scenarios. The inputs on these projects would include the weather, availability of materials, and labor; outputs could be the daily progress toward goals and the dates of key inspections; the control actions could include rewards for finishing on time or early, and penalties for finishing the project late. There are numerous texts which focus just on fuzzy control; a single chapter on this subject could not possibly address all the important topics in this field. References at the end of this chapter are provided for the interested reader. So, in this chapter, we choose to focus on only two types of control: physical system control and industrial process control.

A control system for a physical system is an arrangement of hardware components designed to alter, to regulate, or to command, through a *control action*, another physical system so that it exhibits certain desired characteristics or behavior. Physical control systems are typically of two types: open-loop control systems, in which the control action is independent of the physical system output, and closed-loop control systems (also known as *feedback control systems*), in which the control action depends on the physical system output. Examples of open-loop control systems are a toaster, in which the amount of heat is set by a human, and an automatic washing machine, in which the controls for water temperature, spin-cycle time, and so on are preset by the human. In both these cases the control actions are not a function of the output of the toaster or the washing machine. Examples of feedback control are a room temperature thermostat, which senses room temperature and activates a heating or cooling unit when a certain threshold temperature is reached, and an autopilot mechanism, which makes automatic course corrections to an aircraft when heading or altitude deviations from certain preset values are sensed by the instruments in the plane's cockpit.

In order to control any physical variable, we must first measure it. The system for measurement of the *controlled signal* is called a *sensor*. The physical system under control is called a *plant*. In a closed-loop control system, certain forcing signals of the system (the *inputs*) are determined by the responses of the system (the *outputs*). To obtain satisfactory responses and characteristics for the closed-loop control system, it is necessary to connect an additional system, known as a *compensator*, or a *controller*, into the loop. The general form of a closed-loop control system is illustrated in Fig. 13.1 [Phillips and Harbor, 1996].

Control systems are sometimes divided into two classes. If the object of the control system is to maintain a physical variable at some constant value in the presence of disturbances, the system is called a *regulatory* type of control, or a regulator. Sometimes

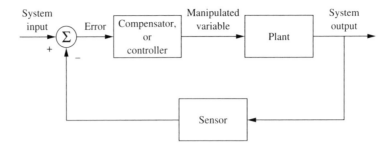

FIGURE 13.1
A closed-loop control system.

this type is also referred to as *disturbance-rejection*. The room temperature control and autopilot are examples of regulatory controllers. The second class of control systems are *set-point tracking* controllers. In this scheme of control, a physical variable is required to follow or track some desired time function. An example of this type of system is an automatic aircraft landing system (see Example 13.3), in which the aircraft follows a ''ramp'' to the desired touchdown point.

The control problem is stated as follows [Phillips and Harbor, 1996]. The output, or response, of the physical system under control (i.e., the plant) is adjusted as required by the *error signal.* The error signal is the difference between the actual response of the plant, as measured by the sensor system, and the desired response, as specified by a *reference input.* In the following section we describe a typical control system – a closed-loop (feedback) control system.

CONTROL SYSTEM DESIGN PROBLEM

The general problem of feedback control system design is defined as obtaining a generally nonlinear vector-valued function $\mathbf{h}(\ )$, defined for some time, t, as follows [Vadiee, 1993]:

$$\mathbf{u}(t) = \mathbf{h}[t, \mathbf{x}(t), \mathbf{r}(t)] \tag{13.1}$$

where $\mathbf{u}(t)$ is the control input to the plant or process, $\mathbf{r}(t)$ is the system reference (desired) input, and $\mathbf{x}(t)$ is the system state vector; the state vector might contain quantities like the system position, velocity, or acceleration. The feedback control law $\mathbf{h}$ is supposed to stabilize the feedback control system and result in a satisfactory performance.

In the case of a time-invariant system with a regulatory type of controller, where the reference input is a constant setpoint, the vast majority of controllers are based on one of the general models given in Eqs. (13.2) and (13.3); that is, either full state feedback or output feedback, as shown in the following:

$$\mathbf{u}(t) = \mathbf{h}[\mathbf{x}(t)] \tag{13.2}$$

$$\mathbf{u}(t) = \mathbf{h}\left[y(t), \dot{y}, \int y \, dt\right] \tag{13.3}$$

where $y()$ is the system output or response function. In the case of a simple single-input, single-output system and a regulatory type of controller, the function **h** takes one of the following forms:

$$\mathbf{u}(t) = K_P \cdot e(t) \tag{13.4}$$

for a proportional, or P, controller;

$$\mathbf{u}(t) = K_P \cdot e(t) + K_I \cdot \int e(t)\, dt \tag{13.5}$$

for a proportional-plus-integral, or PI, controller;

$$\mathbf{u}(t) = K_P \cdot e(t) + K_D \cdot \dot{e}(t) \tag{13.6}$$

for a proportional-plus-derivative, or PD, controller (see Example 13.1 for a PD controller);

$$\mathbf{u}(t) = K_P \cdot e(t) + K_I \cdot \int e(t)\, dt + K_D \cdot \dot{e}(t) \tag{13.7}$$

for a proportional-plus-derivative-plus-integral, or PID, controller, where $e(t)$, $\dot{e}(t)$, and $\int e(t)\, dt$ are the output error, error derivative, and error integral, respectively; and

$$\mathbf{u}(t) = -[k_1 \cdot x_1(t) + k_2 \cdot x_2(t) + \cdots + k_n \cdot x_n(t)] \tag{13.8}$$

for a full state-feedback controller.

The problem of control system design is defined as obtaining the generally nonlinear function $\mathbf{h}(\ )$ in the case of nonlinear systems; coefficients K_P, K_I, and K_D in the case of output-feedback systems; and coefficients $k_1, k_2, \ldots k_n$ in the case of a full state-feedback control policy for linear systems. The function $\mathbf{h}(\ )$ in Eqs. (13.2) and (13.3) describes a general nonlinear surface that is known as a control, or decision, surface, discussed in the next section.

Control (Decision) Surface

The concept of a control surface, or decision surface, is central in fuzzy control systems methodology [Ross, 1995]. In this section we define this very important concept. The function **h** as defined in Eqs. (13.1), (13.2), and (13.3) is, in general, defining P nonlinear hypersurfaces in an n-dimensional space. For the case of linear systems with output feedback or state feedback it generally is a hyperplane in an n-dimensional space. This surface is known as the control, or decision, surface. The control surface describes the dynamics of the controller and is generally a time-varying nonlinear surface. Owing to unmodeled dynamics present in the design of any controller, techniques should exist for adaptively tuning and modifying the control surface shape.

Fuzzy rule-based systems use a collection of fuzzy conditional statements derived from a knowledge base to approximate and construct the control surface [Mamdani and Gaines, 1981; Kiszka et al., 1985; Sugeno, 1985]. This paradigm of control system design is based on interpolative and approximate reasoning. Fuzzy rule-based controllers or system identifiers, are generally model-free paradigms. Fuzzy rule-based systems are universal nonlinear function approximators, and any nonlinear function (e.g., control surface) of n independent variables and one dependent variable can be approximated to any desired precision.

Alternatively, artificial neural networks are based on analogical learning and try to learn the nonlinear decision surface through adaptive and converging techniques, based on numerical data available from input–output measurements of the system variables and some performance criteria.

Assumptions in a Fuzzy Control System Design

A number of assumptions are implicit in a fuzzy control system design. Six basic assumptions are commonly made whenever a fuzzy rule-based control policy is selected.

1. The plant is observable and controllable: state, input, and output variables are usually available for observation and measurement or computation.
2. There exists a body of knowledge comprised of a set of linguistic rules, engineering common sense, intuition, or a set of input–output measurements data from which rules can be extracted (see Chapter 7).
3. A solution exists.
4. The control engineer is looking for a ''good enough'' solution, not necessarily the optimum one.
5. The controller will be designed within an acceptable range of precision.
6. The problems of stability and optimality are not addressed explicitly; such issues are still open problems in fuzzy controller design.

The following section discusses the procedure for obtaining the control surface, $\mathbf{h}(\ )$, from approximations based on a collection of fuzzy IF–THEN rules that describe the dynamics of the controller.

Simple Fuzzy Logic Controllers

First-generation (nonadaptive) simple fuzzy controllers can generally be depicted by a block diagram such as that shown in Fig. 13.2.

The knowledge-base module in Fig. 13.2 contains knowledge about all the input and output fuzzy partitions. It will include the term set and the corresponding membership functions defining the input variables to the fuzzy rule-base system and the output variables, or control actions, to the plant under control.

The steps in designing a simple fuzzy control system are as follows:

1. Identify the variables (inputs, states, and outputs) of the plant.
2. Partition the universe of discourse or the interval spanned by each variable into a number of fuzzy subsets, assigning each a linguistic label (subsets include all the elements in the universe).
3. Assign or determine a membership function for each fuzzy subset.

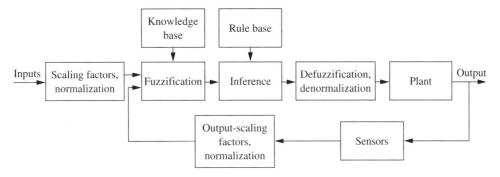

FIGURE 13.2
A simple fuzzy logic control system block diagram.

4. Assign the fuzzy relationships between the inputs' or states' fuzzy subsets on the one hand and the outputs' fuzzy subsets on the other hand, thus forming the rule-base.
5. Choose appropriate scaling factors for the input and output variables in order to normalize the variables to the [0, 1] or the [−1, 1] interval.
6. Fuzzify the inputs to the controller.
7. Use fuzzy approximate reasoning to infer the output contributed from each rule.
8. Aggregate the fuzzy outputs recommended by each rule.
9. Apply defuzzification to form a crisp output.

EXAMPLES OF FUZZY CONTROL SYSTEM DESIGN

Most control situations are more complex than we can deal with mathematically. In this situation fuzzy control can be developed, provided a body of knowledge about the control process exists, and formed into a number of fuzzy rules. For example, suppose an industrial process output is given in terms of the pressure. We can calculate the difference between the desired pressure and the output pressure, called the pressure error (e), and we can calculate the difference between the desired rate of change of the pressure, dp/dt, and the actual pressure rate, called the pressure error rate, ($\dot{e}$). Also, assume that knowledge can be expressed in the form of IF–THEN rules such as

> IF pressure error (e) is "positive big (PB)" or "positive medium (PM)" and
>
> IF pressure error rate ($\dot{e}$) is "negative small (NS),"
>
> THEN heat input change is "negative medium (NM)."

The linguistic variables defining the pressure error, "PB" and "PM," and the pressure error rate, "NS" and "NM," are fuzzy; but the measurements of both the pressure and pressure rate as well as the control value for the heat (the control variable) ultimately applied to the system are precise (crisp). The schematic in Fig. 13.3 shows this idea. An input to the industrial process (physical system) comes from the controller. The physical system responds with an output, which is sampled and measured by some device. If the measured output is a crisp quantity it can be fuzzified into a fuzzy set (see Chapter 4). This

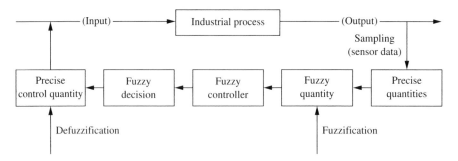

FIGURE 13.3
Typical closed-loop fuzzy control situation.

fuzzy output is then considered as the fuzzy input into a fuzzy controller, which consists of linguistic rules. The output of the fuzzy controller is then another series of fuzzy sets. Since most physical systems cannot interpret fuzzy commands (fuzzy sets), the fuzzy controller output must be converted into crisp quantities using defuzzification methods (again, see Chapter 4). These crisp (defuzzified) control-output values then become the input values to the physical system and the entire closed-loop cycle is repeated.

Example 13.1. For some industrial plants a human operator is sometimes more efficient than an automatic controller. These intuitive control strategies, which provide a possible method to handle qualitative information, may be modeled by a fuzzy controller. This example looks at a pressure process controlled by a fuzzy controller. The controller is formed by a number of fuzzy rules, such as: if pressure error is "positive big" or "positive medium," and if the rate of change in the pressure error is "negative small," then heat input change is "negative medium." This example is illustrated in four steps.

Step 1. Value assignment for the fuzzy input and output variables. We will let the error (e) be defined by eight linguistic variables, labeled $A_1, A_2, \ldots, A_8$, partitioned on the error space of $[-e_m, +e_m]$, and the error rate ($\dot{e}$, or de/dt) be defined by seven variables, labeled $B_1, B_2, \ldots, B_7$, partitioned on the error rate space of $[-\dot{e}_m, \dot{e}_m]$. We will normalize these ranges to the same interval $[-a, +a]$ by

$$e_1 = \left(\frac{a}{e_m} \right) \cdot e$$

$$\dot{e}_1 = \left(\frac{a}{\dot{e}_m} \right) \cdot \dot{e}$$

For the error, the eight fuzzy variables, A_i ($i = 1, 2, \ldots, 8$), will conform to the linguistic variables NB, NM, NS, N0, P0, PS, PM, PB. For the error rate, $\dot{e}$, the seven fuzzy variables, B_j ($j = 1, 2, \ldots, 7$), will conform to the linguistic variables NB, NM, NS, 0, PS, PM, PB. The membership functions for these quantities will be on the range $[-a, a]$, where $a = 6$, and are shown in Tables 13.1 and 13.2 (in the tables $x = e$ and $y = \dot{e}$).

The fuzzy output variable, the control quantity (z), will use seven fuzzy variables on the normalized universe, $z = \{-7, -6, -5, \ldots, +7\}$. The control variable will be described by fuzzy linguistic control quantities, C_k ($k = 1, 2, \ldots, 7$), which are partitioned on the control universe. Table 13.3 is the normalized control quantity, z, which is defined by seven linguistic variables.

TABLE 13.1*

Membership functions for error (e)

x		-6	-5	-4	-3	-2	-1	$0-$	$0+$	1	2	3	4	5	6
A_i															
A_8	PB	0	0	0	0	0	0	0	0	0	0	0.1	0.4	0.8	1
A_7	PM	0	0	0	0	0	0	0	0	0	0.2	0.7	1	0.7	0.2
A_6	PS	0	0	0	0	0	0	0	0.3	0.8	1	0.5	0.1	0	0
A_5	P0	0	0	0	0	0	0	0	1	0.6	0.1	0	0	0	0
A_4	N0	0	0	0	0	0.1	0.6	1	0	0	0	0	0	0	0
A_3	NS	0	0	0.1	0.5	1	0.8	0.3	0	0	0	0	0	0	0
A_2	NM	0.2	0.7	1	0.7	0.2	0	0	0	0	0	0	0	0	0
A_1	NB	1	0.8	0.4	0.1	0	0	0	0	0	0	0	0	0	0

* In the case of crisp control the membership values in the shaded boxes become unity and all other values become zero.

TABLE 13.2*

Membership functions for error rate (de/dt)

y		-6	-5	-4	-3	-2	-1	0	1	2	3	4	5	6
B_j														
B_7	PB	0	0	0	0	0	0	0	0	0	0.1	0.4	0.8	1
B_6	PM	0	0	0	0	0	0	0	0	0.2	0.7	1	0.7	0.2
B_5	PS	0	0	0	0	0	0	0	0.9	1	0.7	0.2	0	0
B_4	0	0	0	0	0	0	0.5	1	0.5	0	0	0	0	0
B_3	NS	0	0	0.2	0.7	1	0.9	0	0	0	0	0	0	0
B_2	NM	0.2	0.7	1	0.7	0.2	0	0	0	0	0	0	0	0
B_1	NB	1	0.8	0.4	0.1	0	0	0	0	0	0	0	0	0

* In the case of crisp control the membership values in the shaded boxes become unity and all other values become zero.

TABLE 13.3*

Membership functions for the control quantity (z)

z		-7	-6	-5	-4	-3	-2	-1	0	1	2	3	4	5	6	7
C_k																
C_1	PB	0	0	0	0	0	0	0	0	0	0	0	0.1	0.4	0.8	1
C_2	PM	0	0	0	0	0	0	0	0	0	0.2	0.7	1	0.7	0.2	0
C_3	PS	0	0	0	0	0	0	0	0.4	1	0.8	0.4	0.1	0	0	0
C_4	0	0	0	0	0	0	0	0.5	1	0.5	0	0	0	0	0	0
C_5	NS	0	0	0	0.1	0.4	0.8	1	0.4	0	0	0	0	0	0	0
C_6	NM	0	0.2	0.7	1	0.7	0.2	0	0	0	0	0	0	0	0	0
C_7	NB	1	0.8	0.4	0.1	0	0	0	0	0	0	0	0	0	0	0

* In the case of crisp control the membership values in the shaded boxes become unity and all other values become zero.

TABLE 13.4
Control rules (FAM table)

	A_i							
	NB	**NM**	**NS**	**N0**	**P0**	**PS**	**PM**	**PB**
B_j								
NB	PB	PM	NB	NB	NB	NB		
NM	PB	PM	NM	NM	NS	NM		
NS	PB	PM	NS	NS	NS	NS	NM	NB
0	PB	PM	PS	0	0	NS	NM	NB
PS	PB	PM	PS	PS	PS	PS	NM	NB
PM			PS	PS	PM	PM	NM	NB
PB			PB	PB	PB	PB	NM	NB

Step 2. Summary of control rules. According to human operator experience, control rules are of the form

$$\text{If } e \text{ is } \underset{\sim}{A}_1 \text{ and } \dot{e} \text{ is } \underset{\sim}{B}_1, \text{ then } z \text{ is } \underset{\sim}{C}_1.$$

$$\text{If } e \text{ is } \underset{\sim}{A}_1 \text{ and } \dot{e} \text{ is } \underset{\sim}{B}_2, \text{ then } z \text{ is } \underset{\sim}{C}_{12}.$$

$$\text{If } e \text{ is } \underset{\sim}{A}_i \text{ and } \dot{e} \text{ is } \underset{\sim}{B}_j, \text{ then } z \text{ is } \underset{\sim}{C}_k.$$

Each rule can be translated into a fuzzy relation, $\underset{\sim}{R}$. Using such an approach will result in linguistic variables, $\underset{\sim}{C}_k$, shown as control entries in Table 13.4.

Step 3. Conversion between fuzzy variables and precise quantities. From the output of the system we can use an instrument to measure the error (e) and calculate the error rate ($\dot{e}$), both of which are precise numbers. A standard defuzzification procedure to develop membership functions, such as the maximum membership principle (see Chapter 4), can be used to get the corresponding fuzzy quantities ($\underset{\sim}{A}_i$, $\underset{\sim}{B}_i$). Sending the $\underset{\sim}{A}$ and $\underset{\sim}{B}$ obtained from the output of the system to the fuzzy controller will yield a fuzzy action variable $\underset{\sim}{C}$ (control rules) as discussed in step 2. But before implementing the control, we have to enter the precise control quantity z into the system. We need another conversion from $\underset{\sim}{C}$ to z. This can be done by a maximum membership principle, or by a weighted average method (see Chapter 4).

Step 4. Development of control table. When the procedures in step 3 are used for all e and all $\dot{e}$, we obtain a control table as shown in Table 13.5. This table now contains precise numerical quantities for use by the industrial system hardware. If the values in Table 13.5 are plotted, they represent a control surface. Figure 13.4 is the control surface for this example, and Fig. 13.5 would be the control surface for this example if it had been conducted using only crisp sets and operations (for the crisp case, the values in Table 13.5 will be different). The volume under a control surface is proportional to the amount of energy expended by the controller. It can be shown that the fuzzy control surface (Fig. 13.4) will actually fit underneath the crisp control surface (Fig. 13.5), indicating that the fuzzy control expends less energy than the crisp control. Fuzzy control methods, such as this one, have been used for some industrial systems and have achieved significant efficiency [Mamdani, 1974; Pappas and Mamdani, 1976].

In the foregoing example, we did not conduct a simulation of a control process because we do not have a model for the controller. The development of the control surface is derived simply from the control rules and associated membership functions. After the control surface

TABLE 13.5
Control actions

						y							
	−6	−5	−4	−3	−2	−1	0	1	2	3	4	5	6
x													
−6	7	6	7	6	7	7	7	4	4	2	0	0	0
−5	6	6	6	6	6	6	6	4	4	2	0	0	0
−4	7	6	7	6	7	7	7	4	4	2	0	0	0
−3	6	6	6	6	6	6	6	3	2	0	−1	−1	−1
−2	4	4	4	5	4	4	4	1	0	0	−1	−1	−1
−1	4	4	4	5	4	4	1	0	0	0	−3	−2	−1
0+	4	4	4	5	1	1	0	−1	−1	−1	−4	−4	−4
0+	4	4	4	5	1	1	0	−1	−1	−1	−4	−4	−4
1	2	2	2	2	0	0	−1	−4	−4	−3	−4	−4	−4
2	1	1	1	−2	0	−3	−4	−4	−4	−3	−4	−4	−4
3	0	0	0	0	−3	−3	−6	−6	−6	−6	−6	−6	−6
4	0	0	0	−2	−4	−4	−7	−7	−7	−6	−7	−6	−7
5	0	0	0	−2	−4	−4	−6	−6	−6	−6	−6	−6	−6
6	0	0	0	−2	−4	−4	−7	−7	−7	−6	−7	−6	−7

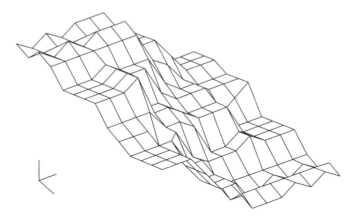

FIGURE 13.4
Control surface for fuzzy process control in Example 13.1.

is developed, a simulation can be conducted if a mathematical or linguistic (rule-based) model of the control process is available.

AIRCRAFT LANDING CONTROL PROBLEM

The following example shows the flexibility and reasonable accuracy of a typical application in fuzzy control.

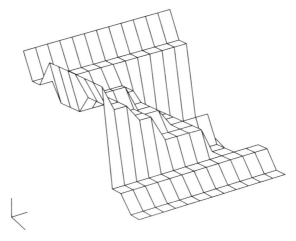

FIGURE 13.5
Control surface for crisp process control in Example 13.1.

Example 13.2. We will conduct a simulation of the final descent and landing approach of an aircraft. The desired profile is shown in Fig. 13.6. The desired downward velocity is proportional to the square of the height. Thus, at higher altitudes, a large downward velocity is desired. As the height (altitude) diminishes, the desired downward velocity gets smaller and smaller. In the limit, as the height becomes vanishingly small, the downward velocity also goes to zero. In this way, the aircraft will descend from altitude promptly but will touch down very gently to avoid damage.

The two state variables for this simulation will be the height above ground, h, and the vertical velocity of the aircraft, v (Fig. 13.7). The control output will be a force that, when applied to the aircraft, will alter its height, h, and velocity, v. The differential control equations are loosely derived as follows. See Fig. 13.8. Mass m moving with velocity v has momentum $p = mv$. If no external forces are applied, the mass will continue in the same direction at the same velocity, v. If a force f is applied over a time interval Δt, a change in velocity of $\Delta v = f \Delta t / m$ will result. If we let $\Delta t = 1.0$ (s) and $m = 1.0$ (lb s^2/ft), we obtain $\Delta v = f$ (lb), or the change in velocity is proportional to the applied force.

In difference notation we get

$$v_{i+1} = v_i + f_i$$
$$h_{i+1} = h_i + v_i \cdot \Delta t \tag{1}$$

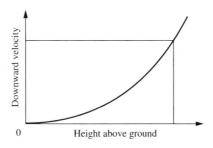

FIGURE 13.6
The desired profile of downward velocity vs. altitude.

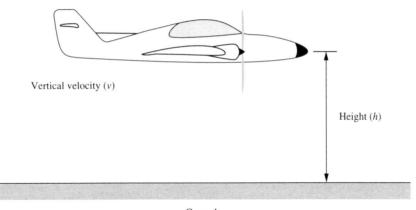

FIGURE 13.7
Aircraft landing control problem.

FIGURE 13.8

where v_{i+1} is the new velocity, v_i is the old velocity, h_{i+1} is the new height, and h_i is the old height. These two "control equations" define the new value of the state variables v and h in response to control input and the previous state variable values. Next, we construct membership functions for the height, h, the vertical velocity, v, and the control force, f:

     Step 1. Define membership functions for state variables as shown in Tables 13.6 and 13.7 and Figs. 13.9 and 13.10.

     Step 2. Define a membership function for the control output, as shown in Table 13.8 and Fig. 13.11.

     Step 3. Define the rules and summarize them in an FAM table (Table 13.9). The values in the FAM table, of course, are the control outputs.

TABLE 13.6
Membership values for height

	Height (ft)										
	0	**100**	**200**	**300**	**400**	**500**	**600**	**700**	**800**	**900**	**1000**
Large (L)	0	0	0	0	0	0	0.2	0.4	0.6	0.8	1
Medium (M)	0	0	0	0	0.2	0.4	0.6	0.8	1	0.8	0.6
Small (S)	0.4	0.6	0.8	1	0.8	0.6	0.4	0.2	0	0	0
Near zero (NZ)	1	0.8	0.6	0.4	0.2	0	0	0	0	0	0

TABLE 13.7
Membership values for velocity

	Vertical velocity (ft/s)													
	−30	−25	−20	−15	−10	−5	0	5	10	15	20	25	30
Up large (UL)	0	0	0	0	0	0	0	0	0	0.5	1	1	1
Up small (US)	0	0	0	0	0	0	0	0.5	1	0.5	0	0	0
Zero (Z)	0	0	0	0	0	0.5	1	0.5	0	0	0	0	0
Down small (DS)	0	0	0	0.5	1	0.5	0	0	0	0	0	0	0
Down large (DL)	1	1	1	0.5	0	0	0	0	0	0	0	0	0

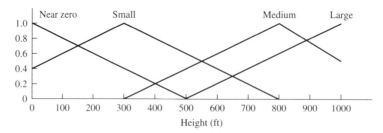

FIGURE 13.9
Height, h, partitioned.

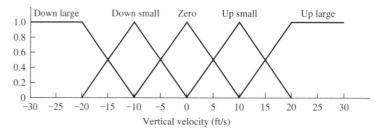

FIGURE 13.10
Velocity, v, partitioned.

TABLE 13.8
Membership values for control force

	Output force (lb)													
	−30	−25	−20	−15	−10	−5	0	5	10	15	20	25	30
Up large (UL)	0	0	0	0	0	0	0	0	0	0.5	1	1	1
Up small (US)	0	0	0	0	0	0	0	0.5	1	0.5	0	0	0
Zero (Z)	0	0	0	0	0	0.5	1	0.5	0	0	0	0	0
Down small (DS)	0	0	0	0.5	1	0.5	0	0	0	0	0	0	0
Down large (DL)	1	1	1	0.5	0	0	0	0	0	0	0	0	0

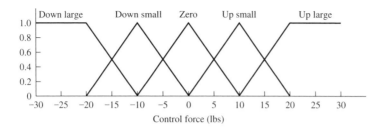

FIGURE 13.11
Control force, f, partitioned.

TABLE 13.9
FAM table

Height	Velocity				
	DL	**DS**	**Zero**	**US**	**UL**
L	Z	DS	DL	DL	DL
M	US	Z	DS	DL	DL
S	UL	US	Z	DS	DL
NZ	UL	UL	Z	DS	DS

Step 4. Define the initial conditions, and conduct a simulation for four cycles. Since the task at hand is to control the aircraft's vertical descent during approach and landing, we will start with the aircraft at an altitude of 1000 feet, with a downward velocity of -20 ft/s. We will use the following equations to update the state variables for each cycle:

$$v_{i+1} = v_i + f_i$$

$$h_{i+1} = h_i + v_i$$

Initial height, h_0: 1000 ft

Initial velocity, v_0: -20 ft/s

Control f_0: to be computed

Height h fires L at 1.0 and M at 0.6

Velocity v fires only DL at 1.0

Height		Velocity		Output
L (1.0)	AND	DL (1.0)	$\Rightarrow$	Z (1.0)
M (0.6)	AND	DL (1.0)	$\Rightarrow$	US (0.6)

We defuzzify using the centroid method and get $f_0 = 5.8$ lb. This is the output force computed from the initial conditions. The results for cycle 1 appear in Fig. 13.12.

Now, we compute new values of the state variables and the output for the next cycle,

$$h_1 = h_0 + v_0 = 1000 + (-20) = 980 \text{ ft}$$

$$v_1 = v_0 + f_0 = -20 + 5.8 = -14.2 \text{ ft/s}$$

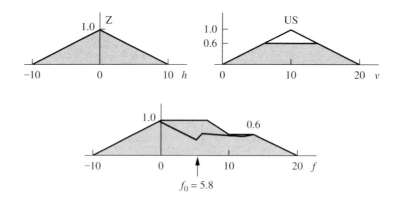

FIGURE 13.12
Truncated consequents and union of fuzzy consequent for cycle 1.

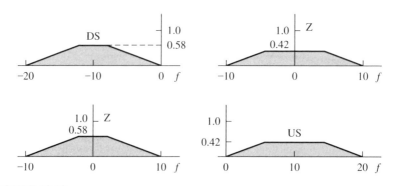

FIGURE 13.13
Truncated consequents for cycle 2.

Height $h_1 = 980$ ft fires L at 0.96 and M at 0.64

Velocity $v_1 = -14.2$ ft/s fires DS at 0.58 and DL at 0.42

Height		Velocity		Output
L (0.96)	AND	DS (0.58)	⇒	DS (0.58)
L (0.96)	AND	DL (0.42)	⇒	Z (0.42)
M (0.64)	AND	DS (0.58)	⇒	Z (0.58)
M (0.64)	AND	DL (0.42)	⇒	US (0.42)

We find the centroid to be $f_1 = -0.5$ lb. Results are shown in Fig. 13.13.
    We compute new values of the state variables and the output for the next cycle.

$$h_2 = h_1 + v_1 = 980 + (-14.2) = 965.8 \text{ ft}$$

$$v_2 = v_1 + f_1 = -14.2 + (-0.5) = -14.7 \text{ ft/s}$$

$$h_2 = 965.8 \text{ ft fires L at 0.93 and M at 0.67}$$

$$v_2 = -14.7 \text{ ft/s fires DL at 0.43 and DS at 0.57}$$

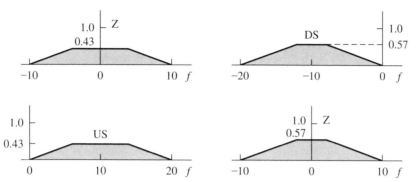

FIGURE 13.14
Truncated consequents for cycle 3.

Height		Velocity		Output
L (0.93)	AND	DL (0.43)	$\Rightarrow$	Z (0.43)
L (0.93)	AND	DS (0.57)	$\Rightarrow$	DS (0.57)
M (0.67)	AND	DL (0.43)	$\Rightarrow$	US (0.43)
M (0.67)	AND	DS (0.57)	$\Rightarrow$	Z (0.57)

We find the centroid for this cycle to be $f_2 = -0.4$ lb. Results appear in Fig. 13.14. Again, we compute new values of state variables and output:

$$h_3 = h_2 + v_2 = 965.8 + (-14.7) = 951.1 \text{ ft}$$

$$v_3 = v_2 + f_2 = -14.7 + (-0.4) = -15.1 \text{ ft/s}$$

and for one more cycle we get

$$h_3 = 951.1 \text{ ft fires L at } 0.9 \text{ and M at } 0.7$$

$$v_3 = -15.1 \text{ ft/s fires DS at } 0.49 \text{ and DL at } 0.51$$

Height		Velocity		Output
L (0.9)	AND	DS (0.49)	$\Rightarrow$	DS (0.49)
L (0.9)	AND	DL (0.51)	$\Rightarrow$	Z (0.51)
M (0.7)	AND	DS (0.49)	$\Rightarrow$	Z (0.49)
M (0.7)	AND	DL (0.51)	$\Rightarrow$	US (0.51)

The results are shown in Fig. 13.15, with a defuzzified centroid value of $f_3 = 0.3$ lb. Now, we compute the final values for the state variables to finish the simulation,

$$h_4 = h_3 + v_3 = 951.1 + (-15.1) = 936.0 \text{ ft}$$

$$v_4 = v_3 + f_3 = -15.1 + 0.3 = -14.8 \text{ ft/s}$$

The summary of the four-cycle simulation results is presented in Table 13.10. If we look at the downward velocity vs. altitude (height) in Table 13.10, we get a descent profile which appears to be a reasonable start at the desired parabolic curve shown in Fig. 13.6 at the beginning of the example.

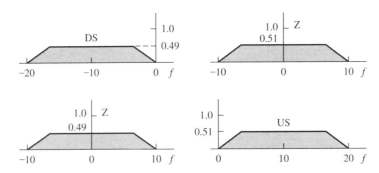

FIGURE 13.15
Truncated consequents for cycle 4.

TABLE 13.10
Summary of four-cycle simulation results

	Cycle 0	Cycle 1	Cycle 2	Cycle 3	Cycle 4
Height, ft	1000.0	980.0	965.8	951.1	936.0
Velocity, ft/s	−20	−14.2	−14.7	−15.1	−14.8
Control force	5.8	−0.5	−0.4	0.3	

FUZZY ENGINEERING PROCESS CONTROL [Parkinson, 2001]

Engineering process control, or the automatic control of physical processes, is a rather large complex field. We discuss first some simple concepts from classical process control in order to provide a background for fuzzy process control concepts. Since fuzzy process control systems can be very complex and diverse, we present only enough information here to provide an introduction to this very interesting topic. We first discuss the classical proportional–integral–derivative (PID) controller (see Eq. (13.7)), then some fuzzy logic controllers. Of the two types of control problems, setpoint-tracking and disturbance rejection, we will illustrate only the setpoint-tracking problem. Most industrial problems are single-input, single-output (SISO), or at least treated that way because multi-input, multi-output (MIMO) problems are normally significantly more difficult. Fuzzy MIMO problems will be discussed in this chapter because fuzzy controllers usually handle these problems quite well; of the many types, only feedback control systems will be illustrated.

Classical Feedback Control

The classical feedback control system can be described using a block flow diagram like the one shown in Fig. 13.16.

The first rectangular block in this figure represents the controller. The second rectangular block represents the system to be controlled, often called the *plant*. The block in the feedback loop is a converter. The *converter* converts the feedback signal to a signal useable by the *summer*, which is the circle at the far left-hand side of the diagram. The letter

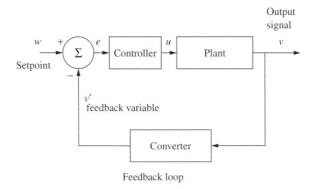

FIGURE 3.16
Standard block flow diagram for a control system.

w represents the setpoint value or the desired control point. This is the desired value of the variable that we are controlling. The letter v represents the output signal or the current value of the variable that we are controlling. The symbol v' represents the feedback variable, essentially the same signal as the output signal but converted to a form that is compatible with the setpoint value. The letter e represents the error, or the difference between the setpoint value and the feedback variable value. The letter u represents the control action supplied by the *controller* to the *plant*. A short example will clarify this explanation.

> **Example 13.3   Liquid-Level Control.** Consider the tank with liquid in it shown in Fig. 13.17. We want to design a controller that will either maintain that liquid level at a desired point, a disturbance rejection problem, or one that can be used to move the level set point from, say, 4 feet to 6 feet, the setpoint-tracking problem, or both. We can do either one or both, but for purposes of illustration, it is easier to confine our explanation to the setpoint-tracking problem. Suppose that the tank in Fig. 13.17 is 10 feet tall and the tank is empty. We want to fill the tank to a level of 5 feet, so we make the current setpoint, w, equal to 5. The idea is to fill the tank to the desired setpoint as quickly and smoothly as possible. We want to minimize the amount of overshoot, or the time that the tank has a level greater than the setpoint value before it finally settles down. The current level at any time, t, is designated as h. Liquid flows out of the tank through an open valve. This flow is designated by the letter q. Liquid flows into

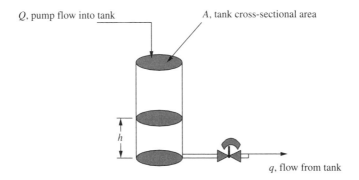

FIGURE 13.17
Tank with a liquid level that needs to be controlled.

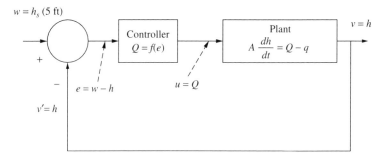

FIGURE 13.18
Block flow diagram for Example 13.3.

the tank by means of a pump. The pump flow, Q, can be regulated by the controller. The tank cross-sectional area is designated by the letter A. Equation (13.9) describes the mass balance for the liquid in the tank as a function of time:

$$A\frac{dh}{dt} = Q - q \tag{13.9}$$

Flow out of the tank, q, through the outlet pipe and the valve is described by

$$q = \Phi A_p \sqrt{2gh} \tag{13.10}$$

where Φ is a friction coefficient for flow through both the small exit pipe and the valve. It can be calculated with fair precision, or better, measured. The term A_p represents the cross-sectional area of the small exit pipe. The gravitational constant, g, is equal to 32.2 ft/s² in US engineering units. Figure 13.18 shows the block flow diagram for this example.

Classical PID Control

The PID control algorithm is described by

$$u = K_P e + K_I \int_0^T e\,dt + K_D\frac{de}{dt} \tag{13.11}$$

where the symbols K_P, K_I, and K_D are the proportional, integral, and derivative control constants, respectively (as in Eq. (13.7)). These constants are specific to the system in question. They are usually picked to optimize the controller performance and insure that the system remains stable for all possible control actions.

If we use a PID controller, which is linear, or any other linear controller with a linear plant, then the system is called a linear system. Linear systems have nice properties. The control engineer can use Laplace transforms to convert the linear equations in the blocks in Fig. 13.18 to the Laplace domain. The blocks can then be combined to form a single transfer function for the entire system. Most of the systems studied in the control systems literature are linear systems. In the real world, many systems are at least slightly nonlinear. However, often this fact is ignored or the system is linearized so that linear control systems theory can be used to solve the problem. There are several techniques for linearizing control systems. The most common is to expand the nonlinear function in a truncated Taylor's series. Equation (13.10)

TABLE 13.11
Approximate linearization

h (ft)	$\sqrt{h}$	Eq. (13.12)
10	3.162	3.354
9	3.0	3.130
8	2.828	2.907
7	2.646	2.683
6	2.449	2.460
5	2.236	2.236
4	2.0	2.012
3	1.732	1.789
2	1.414	1.565
1	1.0	1.342
0	0	1.118

shows that our plant in Example 13.3 is not linear. The truncated Taylor's series for linearizing Eq. (13.10) about a steady state value, in this case our setpoint, is given by

$$\sqrt{h} \approx \sqrt{h_s} + \frac{1}{2\sqrt{h_s}}(h - h_s) \tag{13.12}$$

If we choose $h_s = 5$ ft, we can linearize the radical term over some of the control range. Table 13.11 demonstrates how well this works.

The reader can see that the Taylor's approximation is not a bad one, at least until one gets near zero. The point of all this is that once the equation for the plant becomes linear, we can take Laplace transforms of the plant equation. We can also take Laplace transforms of the control equation, Eq. (13.11), which is already linear. The control engineer typically redefines the variables e and h in Eqs. (13.9), (13.10), and (13.12) as deviation variables. That is, variables that deviate about some steady state. This causes the constants and boundary conditions to go to zero and the equations become much easier to deal with. The Laplace transforms for the controller and the plant are combined with one for the feedback loop to form an algebraic transfer function, $G(s)$ (Fig. 13.19), in the Laplace domain. This transfer function form has some very nice properties from a control system point of view.

In Fig. 13.19 $G(s)$ is the transfer function for the entire block flow diagram including the feedback loop. The variables $w(s)$ and $v(s)$ are the setpoint and output variables, respectively, converted to the Laplace, s, domain. The control engineer can work with the system transfer function and determine the range in which the control constants, K_P, K_I, and K_D must fall in order to keep the system stable. Electrical engineers design controllers for a wide variety of systems. Many of these systems can become unstable as a result of a sudden change in the control action. An example might be an aircraft control system. Chemical engineers, on the other hand, usually design control systems only for chemical processes. Many of these systems are not as likely to become unstable from a sudden change in control action. Our liquid-level controller is an example. A sudden change in the control action, say a response to a leak in

FIGURE 13.19
Overall system block flow diagram or transfer function for the Laplace domain.

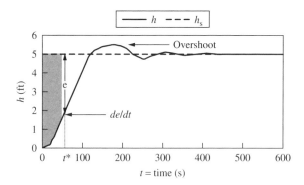

FIGURE 13.20
Time- level response for the tank-filling problem in Example 13.3.

the tank, is not likely to make the system become unstable, no matter how abrupt the change. Laplace transforms and the s domain are a very important part of classical control theory, but are not needed to illustrate our example.

The PID controller accounts for the error, the integral of the error, and the derivative of the error, in order to provide an adequate response to the error. Figure 13.20 shows a typical time-domain response curve for a PID controller, for a problem like the tank-filling problem of Example 13.3.

Figure 13.20 represents a typical response curve for a setpoint-tracking problem. At time t^* the tank level $h(t^*)$ is at the point on the response curve that is pointed to by the de/dt arrow. The error, e, is the distance between the setpoint level line, h_s, and the current tank level $h(t^*)$. Since in this case

$$de/dt = d(h_s - h(t))/dt = -dh/dt$$

the derivative term in the PID control equation is shown as the negative derivative of the response curve at time t^*. The shaded area between the h_s line and the h curve between time t_0 and t^* is an approximation to the integral term, in Eq. (13.11). One criterion for an optimal controller is to find control constants than minimize the error integral. That is, find K_P, K_I, and K_D such that $\int_{t_0}^{t_\infty} e\,dt$ is minimized. The term t_∞ is the time at which the controlled variable, tank-level in this case, actually reaches and stays at the setpoint. The proportional term in Eq. (13.11) drives the control action hard when there is a large error and slower when the error is small. The derivative term helps to home in the controller variable to the setpoint. It also reduces overshoot because of its response to the change in the sign and the rate of change of the error. Without the integral term, however, the controlled variable would never hold at the setpoint, because at the setpoint both e and de/dt are equal to 0. A pure PD controller would become an on–off controller when operating about the setpoint. The overshoot shown in Fig. 13.20 is something that control engineers normally try to minimize. It can really present a problem for a tank-filling exercise, especially if the level setpoint is near the top of an open tank.

Fuzzy Control

In the simplest form, a fuzzy control system connects input membership functions, functions representing the input to the controller, e, to output membership functions that represent the control action, u. A good example for the fuzzy control system is a controller that controls the liquid level in the tank shown in Fig. 13.17. This time we want to design a controller that will allow us to change the setpoint either up or down, and one that will correct itself in the case of overshoot. A simple fuzzy control system designed for our tank-level setpoint-tracking problem consists of three rules.

1. If the *Level Error* is *Positive* Then the *Change in Control Action* is *Positive.*
2. If the *Level Error* is *Zero* Then the *Change in Control Action* is *Zero.*
3. If the *Level Error* is *Negative* Then the *Change in Control Action* is *Negative.*

The input membership functions are shown in Fig. 13.21. The reader should also notice the "dead band" or "dead zone" in the membership function *Zero* between about ±3 inches. This is optional and is a feature commonly used with on–off controllers. It is easy to implement with a fuzzy controller and is useful if the control engineer wishes to minimize control response to small transient-level changes. This step can save wear and tear on equipment.

    The output membership functions for this controller are shown in Fig. 13.22. In this figure the defuzzified output value from the controller is a fractional value representing the

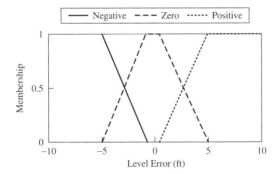

FIGURE 13.21
Input membership functions for the fuzzy tank-level controller.

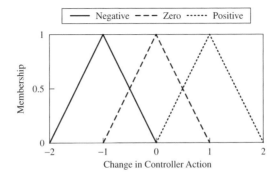

FIGURE 13.22
Output membership functions for the fuzzy liquid-level controller.

required pump output for the desired level change. It is defined by the following expression:

$$\textbf{\textit{Change in Controller Action}} = \Delta u = (Q_i - Q_{sp})/Range \qquad (13.13)$$

where the term Q_{sp} represents the pump output (gallons per minute) required to maintain the setpoint level. The term Q_i is the new pump output requested by the controller. If $\Delta u > 0$ then the *Range* is defined as $Q_{max} - Q_{sp}$, where Q_{max} is the maximum pump output. If $\Delta u < 0$ then the *Range* is defined as Q_{sp}. The term Q_{sp} must be calculated using a steady state mass balance for the tank or it must be estimated in some fashion. The steady state calculation requires only algebra. It requires only a knowledge of the parameter Φ in Eq. (13.10). This value can either be measured by experiment, or approximated quite closely from resistance coefficients found in any fluid mechanics text (e.g., Olsen [1961]). This is an engineering calculation that is quite different from, and usually easier to do than, calculations needed to compute K_P, K_I, and K_D for the PID controller.

The ranges of the fuzzy output sets *Positive* and *Negative* are +2.0 to 0.0 and −2.0 to 0.0, respectively. Since the *Change in Controller Action* is a fraction between either 0.0 and 1.0 or 0.0 and −1.0, it is clear that we will never obtain a control action outside of the range of −1.0 to 1.0. Our defuzzification technique will require that we include numbers up to 2.0 in the fuzzy set or membership function *Positive* and numbers down to −2.0 in the fuzzy set *Negative*. Even though numbers of this magnitude can never be generated by our fuzzy system, we can still include them in our fuzzy sets. The users can define their fuzzy sets however they wish. The fuzzy mathematics described in earlier chapters is capable of handling objects of this type. The user has to define the fuzzy sets so that they make sense for the particular problem. In our case we are going to use the centroid technique for defuzzification. We therefore need to extend our membership functions so that it is possible to obtain centroids of ±1.0. We need this capability in order for the control system to either turn the pump on "full blast" or turn it completely off.

We can describe our simple fuzzy controller as an approximation to an I or Integral controller. Our rules are of the form $\Delta u = f(e)$, where Δu is the *Control Action Change* for the sample time interval Δt. We can make the approximation that $\Delta u / \Delta t \approx du/dt \approx K_I e$ and $\int du = K_I \int e\, dt$ and that $u = u_0 + K_I \int_{t_0}^{T} e\, dt$ or that

$$u = K_I \int_0^T e\, dt$$

Example 13.4 (continuation of Example 13.3). Suppose that we decide to change our setpoint level from 5 feet to 8 feet in the tank described in Example 13.3. The error is defined as the setpoint level, 8 ft, minus the current level, 5ft, or +3 ft. The three rules are fired, producing the following results:

1. *Positive* error is 0.5.
2. *Zero* error is 0.5.
3. *Negative* error is 0.0,

The results are shown graphically in Fig. 13.23.

In this example an error of +3 ft intersects the membership function *Zero* at approximately 0.5 and the membership function *Positive* at approximately 0.5. We say that Rules 1

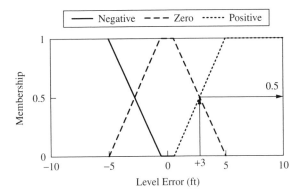

FIGURE 13.23
Resolution of input for Example 13.4.

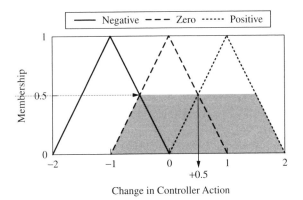

FIGURE 13.24
Resolution of output for Example 13.4.

and 2 were each fired with strength 0.5. The output membership functions corresponding to Rules 1 and 2 are each "clipped" at 0.5. See Fig. 13.24.

The centroid of the "clipped" membership functions, the shaded area in Fig. 13.24, is +0.5. This centroid becomes the term Δu in Eq. (13.13). Since Δu is greater than 0.0, Eq. (13.13) can be rewritten as

$$Q_i = (Q_{max} - Q_{sp})\Delta u + Q_{sp} \quad \text{or} \quad Q_i = 0.5(Q_{max} + Q_{sp}) \quad \text{since } \Delta u = 0.5 \quad (13.14)$$

This says that the new pump output, Q_i, should be adjusted to be halfway between the current or setpoint output and the maximum heater output. After an appropriate time interval, corresponding to a predetermined sample rate, the same procedure will be repeated until the setpoint level, 8 ft, is achieved. The setpoint-tracking response curve for this problem will probably look something like the one shown in Fig. 13.20. Hopefully, the overshoot will be reduced by the addition of the dead band in the input membership functions and a judicious choice of the sample interval time, Δt. In level control problems like this one, dead bands can be very useful, because the physical action of the liquid pouring from the pump outlet onto the liquid surface in the tank will cause the fluid in the tank to "slosh" around. A good sensor will pick up these level changes and overwork the controller. For the same reason classical

PI controllers are often used for level control problems like this, instead of PID controllers, because the fluid movement keeps the derivative portion very active.

Unfortunately, this simple fuzzy control system will not handle disturbance rejection problems very well. This is because of the method that we have chosen to solve this problem. The problem is the term Q_{sp}. This term is reasonably easy to measure or calculate, but it is no longer valid if there is a hole in the tank, or a plugged valve, which are probably the most likely causes of disturbances in this system. This is reasonably easy to fix with more rules, but the explanation is quite lengthy. The interested reader is referred to Parkinson [2001] or Ross et. al. [2002]. The PID controller will solve both the disturbance rejection problem and the setpoint-tracking problem, with one set of control constants. Often, however, PID control constants that are optimized for one type of solution are not very efficient for the other type.

Multi-input, Multi-output (MIMO) Control Systems

The classical multi-input, multi-output control system is much more complicated. The textbook approach assumes linear systems, uses a lot of linear algebra, and often the best that can come out of these models is a set of proportional-only controllers. Phillips and Harbor [1996] have a good readable chapter devoted to this approach. There are also entire textbooks and graduate-level control courses devoted to linear control systems for MIMO systems. A common industrial approach, at least in the chemical industry where systems tend to be highly nonlinear, is to use multiple PID controllers. Because of this "brute force" approach, these controllers tend to interact and "fight" one another. There are methods for decoupling multiple controllers, but it is a great deal of work. A good discussion for learning how to decouple multiple controllers is given in Ogunnaike and Ray [1994]. One of the advantages of fuzzy controllers is that it is reasonably easy to write good MIMO control systems for highly nonlinear MIMO problems. The next example illustrates this.

> **Example 13.5 The Three-tank MIMO Problem.** The three-tank system described here was a real experiment [Parkinson, 2001]. It is an extension of the single-tank system discussed in Examples 13.3 and 13.4. The tanks are smaller, however. The experimental apparatus consisted of three Lucite tanks, in series, each holding slightly less than 0.01 m^3 of liquid; the system is shown in Fig. 13.25. The tanks are numbered from left to right in this figure as: tank 1, tank 3, and tank 2. All three tanks are connected, with the third tank in the series, tank 2, draining to the system exit. Liquid is pumped into the first and the third tanks to maintain their levels. The levels in the first and third tanks control the level in the middle tank. The level in the middle tank affects the levels in the two end tanks.

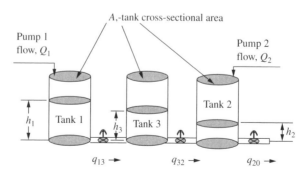

FIGURE 13.25
The experimental three-tank system.

The differential equations that describe this experimental system are Eqs. (13.15)–(13.17). The tank flows are described by Eqs. (13-18)–(13.20). The symbols used in these equations have the same meaning as those used in Examples 13.3 and 13.4, except they now have subscripts as described in Fig. 13.25.

$$A\frac{dh_1}{dt} = Q_1 - q_{13} \tag{13.15}$$

$$A\frac{dh_2}{dt} = Q_2 + q_{32} - q_{20} \tag{13.16}$$

$$A\frac{dh_3}{dt} = q_{13} - q_{32} \tag{13.17}$$

$$q_{13} = \Phi_1 A_p \, \text{Sign}(h_1 - h_3)\sqrt{2g|h_1 - h_3|} \tag{13.18}$$

$$q_{32} = \Phi_3 A_p \, \text{Sign}(h_3 - h_2)\sqrt{2g|h_3 - h_2|} \tag{13.19}$$

$$q_{20} = \Phi_2 A_p \sqrt{2gh_2} \tag{13.20}$$

Again we will only describe the setpoint-tracking portion of the controller and the results for the setpoint-tracking experiments. For the interested reader the description of the disturbance rejection portion of the controller and the results of the disturbance rejection experiments are given in Parkinson [2001].

The fuzzy rules for this setpoint-tracking module are given in Table 13.12. The rules are of the form

If $Error(i)$ is ... then Flow_Change(i) is ...

The term $Error(i)$ in Table 13.12 is defined as follows for both tanks 1 and 2:

$$Error(i) = \frac{w_i - h_i}{Rangeh(i)}, \quad \text{for } i = 1 \text{ or } 2 \tag{13.21}$$

If $w_i > h_i$ then $Rangeh(i)$ equals w_i
Else $Rangeh(i)$ equals $h_{i\,max} - w_i$, for $i = 1$ or 2
The variable h_i is the current level for tank i and w_i is the setpoint for tank i. The term $h_{i\,max}$ is the maximum level for tank i, or the level when the tank is full. The Flow_Change(i) variable is defined by

$$Flow\_Change(i) = \frac{q_i - q_{iss}}{Rangeq(i)} \tag{13.22}$$

If Flow_Change$(i) > 0$ then $Rangeq(i)$ equals q_{iss}
Otherwise, $Rangeq(i)$ equals $q_{i\,max} - q_{iss}$, for $i = 1$ or 2

TABLE 13.12
Fuzzy rules for setpoint tracking

Rule number	Error(i)	Flow_Change(i)
1	Error(1) = Negative	Flow_Change(1) = Negative
2	Error(1) = Zero	Flow_Change(1) = Zero
3	Error(1) = Positive	Flow_Change(1) = Positive
4	Error(2) = Negative	Flow_Change(2) = Negative
5	Error(2) = Zero	Flow_Change(2) = Zero
6	Error(2) = Positive	Flow_Change(2) = Positive

The term $q_{i\,max}$ is the maximum possible flow from pump i, 6.0 L/min for each pump. The variable q_i is the current pump flow, from pump i, and q_{iss} is the steady state, or setpoint, flow for pump i, computed from a mass balance calculation. The mass balance calculation is a simple algebraic calculation. It is based on Eqs. (13.15)–(13.20) with derivatives set to zero and the tank levels set at the setpoints. The calculation simplifies to three equations with three unknowns. The three unknowns obtained from the solution are the two steady state pump flows and one independent steady state tank level. One difficulty is determining the flow coefficients, Φ_i. These coefficients can be easily measured or calculated using textbook values; measurement of these values would be best. However, if the coefficients are calculated and the calculation is not correct, the disturbance rejection mode will be invoked automatically and the values will be determined by the control system.

The membership function universes for Error(i) and for Flow_Change(i) have been normalized from -1 to 1. This was done in order to construct generalized functions. In this type of control problem, we want one set of rules to apply to all tank-level setpoints. Generalizing these functions makes it possible to use only six rules to handle all tank levels and all tank-level changes. Figure 13.26 shows the input membership functions for the fuzzy controller used to solve this three-tank problem. Figure 13.27 shows the output membership functions for the fuzzy control system. These membership function ranges have been expanded to the limits of -2 and 2 so that the pumps can be turned on full blast if needed. These inputs and outputs are connected by the rules shown in Table 13.12.

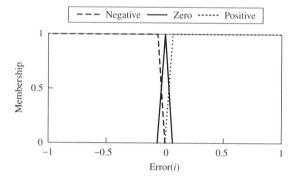

FIGURE 13.26
Membership functions for input Error(i) for $i =$ both 1 and 2 (pumps 1 and 2).

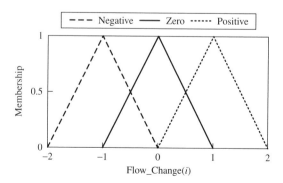

FIGURE 13.27
Membership functions for Flow_Change(i) for $i =$ both 1 and 2 (pumps 1 and 2).

Example 13.6   Three-tank problem (continued). For this example we select the setpoints, w_1 and w_2, for tanks 1 and 2 to be 0.40 and 0.20 m, respectively. These are the same setpoints that were used in the actual experiment [Parkinson, 2001]. We start with three empty tanks. The steady state flows required to maintain these levels are computed from the mass balance calculation to be $q_{1ss} = 1.99346$ L/min and $q_{2ss} = 2.17567$ L/min. The maximum pump flows are $q_{1\,max} = q_{2\,max} = 6.0$ L/min. For the example, we assume that the tank levels have nearly reached their setpoint values. The level in tank 1, h_1, is 0.38 m and the level in tank 2, h_2, is 0.19 m. The maximum tank levels are $h_{1\,max} = h_{2\,max} = 0.62$ m.

Error(1) and Error(2) are both calculated to be -0.05 from Eq. (13.21). In both cases the memberships from Fig. 13.26 are computed to be **Negative** $= 0.775$ and **Zero** $= 0.225$. These values cause Rules 1 and 4 to be fired with a weight of 0.775, and Rules 2 and 5 to be fired with a weight of 0.225. The output membership functions shown in Fig. 13.27 are truncated at these values. The defuzzification method used with the setpoint-tracking control for this controller is the correlation minimum encoding (CME) technique. This method computes the areas and centroids of the entire truncated triangles. That is, the negative and positive Flow_Change membership functions are extended to -2 and $+2$, respectively, in order to compute the centroids and the areas. This is one of many techniques used with fuzzy output membership functions when it is important to use the centroid of the membership function to designate complete "shut off" or "go full blast." The crisp, or defuzzified, value used in the control equations, and obtained from firing these rules, is defined by Eq. (13.22). Both Flow_Change(1) and Flow_Change(2) are computed to be -0.704. For both pumps, Flow_Change(i) is less than 0, so $Rangeq(i)$ is equal to $q_{i\,max} - q_{iss}$. $Rangeq(1)$ is then 4.00654 and $Rangeq(2)$ is equal to 3.82433. By manipulating Eq. (13.22) we obtain the following relationship:

$$q_i = q_{iss} - \text{Flow\_Change}(i) * Rangeq(i) \qquad (13.23)$$

From Eq. (13.23) the value of q_1 is computed as

$$q_1 = 1.99346 + (0.704)(4.00654) = 4.8141$$

and q_2 is

$$q_2 = 2.17567 + (0.704)(3.82433) = 4.8680$$

This setpoint tracking test case was a tank-filling problem. The setpoint level was changed from 0 m, an empty condition, to 0.4 m for tank 1, and 0.2 m for tank 2. The level for tank 3 cannot be set independently. For this test tank 3 found its own level at about 0.3 m. The run time was set to 7.5 min, or 450 s. The results for the fuzzy controller are shown in Fig. 13.28.

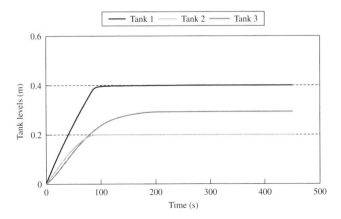

FIGURE 13.28
Tank levels versus time for the fuzzy controller.

Attention is called to the fact that there is no overshoot on any of the three tanks, even without a dead band in the input membership functions. Also, the filling time is very rapid. In the actual experiment, Parkinson [2001] details several tests for both setpoint tracking and disturbance rejection., and the fuzzy controller performed better than the classical controllers every time.

FUZZY STATISTICAL PROCESS CONTROL [PARKINSON AND ROSS, 2002]

There are two basic types of statistical process control (SPC) problems. One of the problem types deals with measurement data. An example would be the measurement of the diameter of a cylindrical part that is produced by a machining operation. The typical SPC method for dealing with measurement data is to use $\overline{X} - R$ charts [Shewhart, 1986]. These charts work well with one input variable and will be discussed in the next section. The second problem type deals with attribute data. In this case, instead of dealing with the actual measurement information, the process control person assigns an attribute like ''pass'' or ''fail'' to the item. A common SPC technique is to use a p-chart based on binary inputs and using the binomial distribution [Shewhart, 1986]. This technique will be discussed in the section on attribute data. These traditional techniques work very well as long as only a single input is required for the measurement data problems and as long as only binary input is required for the attribute data problems.

Fuzzy SPC is useful when multiple inputs are required for each of these two SPC problem types. Two separate studies were conducted to determine the usefulness of the fuzzy SPC technique [Parkinson and Ross, 2002]. These case studies, illustrated here as Examples 13.7 and 13.8, were directed at beryllium part manufacture. Although the numbers presented here are not the actual values used in the studies, they are useful in illustrating the fuzzy SPC technique. The fuzzy technique using measurement data was for beryllium exposure, but would apply equally well to quality control. The other study used attribute data for quality control of the manufacturing of beryllium parts. The beryllium manufacturing process is quite interesting because the manufacturing process is atypical; it is atypical because the process almost always involves small lots, and often a different part is processed each time. In both Examples 13.7 and 13.8 a computer model of the beryllium plant was used. The purpose of the studies was to compare fuzzy SPC techniques with traditional SPC techniques for these atypical cases. The exposure control problem in the beryllium plant is atypical because it must consider several variables. The traditional SPC procedure with multiple input data is to apply a least squares technique to regress the multiple data to one input and then use the $\overline{X} - R$ chart. The fuzzy technique, illustrated in Example 13.7, uses rules and membership functions to reduce the multiple variables to a single variable and then applies the $\overline{X} - R$ chart [Parkinson, 2001].

In the quality control study utilizing attribute data, illustrated in Example 13.8, instead of the usual binary pass–fail situation, we have multiple classifications such as firsts, seconds, recycle, and discard. The traditional SPC method of dealing with this problem is to use a generalized p-chart based upon the chi-square distribution [Shewhart, 1986]. The fuzzy approach to this multiple-input problem is to use fuzzy rules to combine the multiple variables and then use a fuzzy chart that is somewhat similar to the standard p-chart technique [Parkinson, 2001].

In both case studies, the fuzzy method proved slightly superior and much easier to use than the standard statistical techniques. The purpose of these illustrations is not to discuss the relative merits of the techniques, but to demonstrate the use of the fuzzy methods. Interested students are encouraged to study the various techniques and to make their own decisions about which technique to use in various applications.

Measurement Data – Traditional SPC

Suppose that the operation that produces the cylindrical parts mentioned above produces a thousand parts a day. If an 8 hour day were used, this would be 125 parts an hour. Suppose, further, that the process control engineer decides to measure the diameter of five parts every 2 hours for a week. The engineer then plots the results on an $\overline{X}$–R chart. Figure 13.29 is a simulated $\overline{X}$ chart for this situation. Figure 13.30 is the associated simulated R chart. The first point on Fig. 13.29, or set number 1, is the average of the five diameter measurements taken in the first sample. In this case the five measurements were 0.6, 0.59, 0.54, 0.57, and 0.58 inches. The average or $\overline{X}$ for set number 1 is 0.576 inches. The first point on Fig. 13.30, or set number 1, is the range of the five diameter measurements taken in the first sample, $0.6 - 0.54 = 0.06$. In this example four sets were sampled every day for 5 days for a total of 20 sets. The $\overline{X}$ average and range, R, of every set were determined and plotted in Fig. 13.29 and 13.30, respectively. The average of the 20 $\overline{X}$, or the grand average $(\overline{\overline{X}})$ is also plotted in Fig. 13.29. The average of the 20 ranges, $\overline{R}$, is plotted in Fig. 13.30. The upper and lower control limits shown in Figures 13.29 and 13.30 are computed using Table 13.13 and Eqs. (13.24)–(13.27),

$$\mathrm{UCL}_{\overline{X}} = \overline{\overline{X}} + A_2 \overline{R} \tag{13.24}$$

$$\mathrm{LCL}_{\overline{X}} = \overline{\overline{X}} - A_2 \overline{R} \tag{13.25}$$

$$\mathrm{UCL}_R = D_4 \overline{R} \tag{13.26}$$

$$\mathrm{LCL}_R = D_3 \overline{R} \tag{13.27}$$

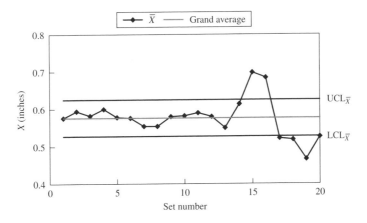

FIGURE 13.29
Simulated $\overline{X}$ chart for the measurement of cylinder diameters.

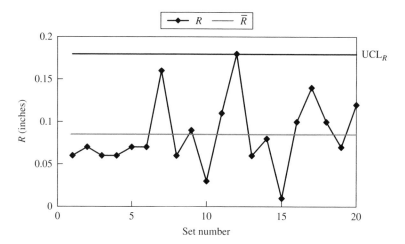

FIGURE 13.30
Simulated R chart for the measurement of cylinder diameters.

TABLE 13.13
Factors for determining control limits

n	A_2	D_3	D_4
2	1.880	–	3.268
3	1.023	–	2.574
4	0.729	–	2.282
5	0.577	–	2.114
6	0.483	–	2.004
7	0.419	0.076	1.924
8	0.373	0.136	1.864
9	0.337	0.184	1.816
10	0.308	0.223	1.777
11	0.285	0.256	1.744
12	0.266	0.283	1.717
13	0.249	0.307	1.693
14	0.235	0.328	1.672
15	0.223	0.347	1.653

where $\text{UCL}_{\overline{X}}$ and $\text{LCL}_{\overline{X}}$ are the upper and lower control limits, respectively, for the $\overline{X}$ chart and UCL_R and LCL_R are the upper and lower control limits, respectively, for the R chart [Shewhart, 1986]. The symbols A_2, D_3, and D_4 are listed in Table 13.13 for various sample sizes, n. In the case of our example, the sample set size, n, is 5. A_2 is 0.577. There is no D_3, or lower limit for the range for set sizes smaller than 7. D_4 is 2.004.

The upper and lower control limits are roughly three standard deviations away from the average line. This means normal or random deviations will fall between the upper and lower control limits 99 to 100% of the time. Values falling outside of these lines should nearly all be "special cause events" that need to be addressed. Values falling inside the control limits are normal events, due to normal random deviations in the process. The $\overline{X}$ chart, Fig. 13.29, shows process deviations over a period of time, the deviation between samples. The R chart, Fig. 13.30, shows deviations within the sample set. An example of a problem that might be detected with the R chart would be a bad sensor. In this case, Fig. 13.29 shows a process that is going out of control as time moves on. This could be caused from a cutting tool wearing out, a situation that needs to be corrected. Figure 13.30 shows one range point ''out of control.'' This is probably due to one bad measurement within the set. If this problem continued, it could indicate a bad set of calipers or other measuring device.

Our beryllium problem is a little more complex than the cylindrical parts example, but a similar technique can be applied. Exposure to beryllium particulate matter, especially very small particles, has long been a concern to the beryllium industry because of potential human health problems – inhaled beryllium is extremely toxic. The industrial exposure limit has been $2 \, \mu g/m^3$ per worker per 8 hour shift, but recently these limits have been reset to $0.2 \, \mu g/m^3$, or 10 times lower than the previous industrial standard. The beryllium manufacturing facility investigated in our study has a workload that changes from day to day. Also the type of work done each day can vary dramatically. This makes the average beryllium exposure vary widely from day to day. This in turn makes it very difficult to determine a degree of control with the standard statistical control charts.

Example 13.7   Measurement Data – Fuzzy SPC. The simulated plant has four workers and seven machines. Each worker wears a device that measures the amount of beryllium inhaled during his or her shift. The devices are analyzed in the laboratory and the results are reported the next day after the exposure has occurred. An $\overline{X} - R$ chart can be constructed with these data and presumably answer the questions of control and quality improvement. Any standard text on SPC will contain a thorough discussion on control limits for these charts; for example, see Wheeler and Chambers [1992] or Mamzic [1995]. Although such a chart can be useful, because of the widely fluctuating daily circumstances, these tests for controllability are not very meaningful.

There are four variables that have a large influence upon the daily beryllium exposure. They are the number of parts machined, the size of the part, the number of machine setups performed, and the type of machine cut (rough, medium, or fine). In our fuzzy model, a semantic description of these four variables and the beryllium exposure are combined to produce a semantic description of the type of day that each worker has had. The day type is then averaged and a distribution is found. These values are then used to produce fuzzy Shewhart-type $\overline{X}$ and R charts. These charts take into account the daily variability. They provide more realistic control limits than the traditional $\overline{X} - R$ charts.

The fuzzy system consists of five input variables or universes of discourse and one output variable. Each input universe has two membership functions and the output universe has five membership functions. The inputs and the outputs are related by 32 rules. The five input variables are:

1. Number of Parts – with a range of 0 to 10 and membership functions
   (*a*) Few and (*b*) Many.
2. Size of Parts – with a range of 0 to 135 and membership functions
   (*a*) Small and (*b*) Large.
3. Number of Setups – with a range of 0 to 130 and membership functions
   (*a*) Few and (*b*) Many.
4. Type of Cut – with a range of 1 to 5 and membership functions
   (*a*) Fine and (*b*) Rough.

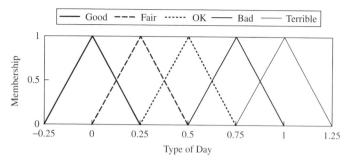

FIGURE 13.31
Output membership functions.

5. Beryllium Exposure – with a range of 0 to 0.4 and membership functions
(*a*) Low and (*b*) High.

   The output variable is:

1. The Type of Day – with range from 0 to 1 and membership functions
(*a*) Good, (*b*) Fair, (*c*) OK, (*d*) Bad, and (*e*) Terrible.

For each of the five input variables there are two membership functions represented in each case by two equal triangles. Figure 13.31 shows the five output membership functions. The rules are based on some simple ideas. For example, if all of the four mitigating input variables indicate that the beryllium exposure should be low, and it is low, then the Type of Day is OK. Likewise, if all four indicate that the exposure should be high, and it is high, then the Type of Day is also OK. If all four indicate that the exposure should be low, and it is high, then the Type of Day is Terrible. If all four indicate that the exposure should be high, and it is low, then the Type of Day is Good. Fair and Bad days fall in between the OK days and the Good and Terrible extremes. The form of the rules is:

> If (**Number of Parts**) is ... and If (**Size of Parts**) is
>
>   ... and If (**Number of Setups**) is ... and If (**Type of Cut**) is
>
>   ... and If (**Beryllium Exposure**) is ... Then (**The Type of Day**) is ... .

The Size of Parts is determined as the number of parts multiplied by the average diameter of each part, measured in centimeters. The Type of Cut is determined by a somewhat complicated formula based on a roughness factor for each part, the number of parts and the size of those parts, and the number of setups required for each worker each day. A fine cut has a roughness factor (rf) of 1, a medium cut has an rf equal to 3, and a rough cut has an rf equal to 5. The calculation for Type of Cut is a bit complicated but it provides a daily number between 1 and 5 (fine to rough) for each worker, which is meaningful. An example of the use of the fuzzy technique will follow a discussion of the plant simulation.

Plant simulation

The model has the following limitations or boundary conditions:

1. There are four machinists.
2. There are seven machines.

3. Machines 1 and 2 do rough cuts only.
4. Machines 3 and 4 do both rough cuts and medium cuts.
5. Machines 5, 6, and 7 do only fine cuts.
6. Machine 7 accepts only work from machines 3 and 4.
7. Machines 5 and 6 accept only work from machines 1 and 2.
8. Each machinist does all of the work on one order.
9. All machinists have an equally likely chance of being chosen to do an order.
10. There are 10 possible paths through the plant (at this point all are equally likely).

The simulation follows the algorithm below:

1. A random number generator determines how many orders will be processed on a given day (1 to 40).
2. Another random number generator picks a machinist.
3. A third random number generator picks a part size.
4. A fourth random number generator picks a path through the plant. For example, machine 1 to machine 3 to machine 7.
5. The machine and path decide the type of cut (rough, medium, or fine). Machines 1 and 2 are for rough cuts only, machines 5, 6, and 7 are for fine cuts only, and machines 3 and 4 do rough cuts if they are the first machines in the path and medium cuts if they are the second machines in the path.
6. A random number generator picks the number of setups for each machine on the path.
7. Another random generator picks the beryllium exposure for the operator at each step.

The above procedure is carried out for each part, each day. The entire procedure is repeated the following day, until the required number of days has passed. For this study the procedure was run for 30 days to generate some sample control charts. A description of the fuzzy control chart construction follows.

Establishing fuzzy membership values

We follow each step of the process for a specific machinist for a given day; this process is then extended to the work for the entire day for all machinists. From the simulation, on day 1, 13 part orders were placed. Machinist 2 processed four of these orders, machinist 1 processed three, machinist 3 processed four, and machinist 4 processed two orders. Machinist 2 will be used to demonstrate the fuzzy system.

The cumulative size of the four parts that machinist 2 processed on day 1 was calculated to be 64.59. The number of setups that he or she performed was 45. The numeric value for the type of cuts he or she performed on that day was 1.63. Finally, the machinist's beryllium exposure was 0.181 $\mu g/m^3$ for that 8 hour period.

Upon inserting the input values into the simple input membership functions described above, we obtain the following values. For Number of Parts = 4 the membership in Many is 0.4 and the membership in Few is 0.6. For Size of Parts = 64.59 the membership in Small is 0.52 and the membership in Large is 0.48. For Number of Setups = 45 the membership in Many is 0.35 and the membership in Few is 0.65. For Type of Cuts = 1.63 the membership in Rough is 0.16 and the membership in Fine is 0.84. The Beryllium Exposure is 0.181. The membership in High is 0.45 and the membership in Low is 0.55.

In this example a max – min Mamdani inference was used on the rules and the centroid method was used for defuzzification. For example, Rule 1 is fired with the following weights:

- Number of Parts – Few = 0.6
- Size of Parts – Small = 0.52

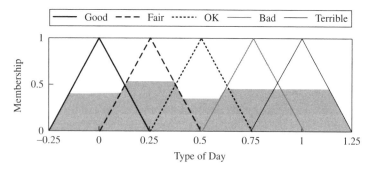

FIGURE 13.32
"Truncated" output membership functions for the example.

- Number of Setups – Few = 0.65
- Type of Cut – Fine = 0.84
- Beryllium Exposure – Low = 0.55

The consequent of Rule 1 is Fair and takes the minimum value, 0.52. Of the 32 rules that are all fired for this example, Fair is the consequent of 10 of them with membership values ranging from 0.16 to 0.52. The max–min rule assigns the maximum value of 0.52 to the consequent Fair. Similarly, the consequent Terrible appears five times with a maximum value of 0.45, OK appears twice with a maximum value of 0.35, Bad appears 10 times with a maximum value of 0.45, and Good appears five times with a maximum value of 0.4. The Mamdani inference process truncates the output membership functions at their maximum value (see Chapter 5). In this example the membership functions are truncated as follows: Good = 0.4, Fair = 0.52, OK = 0.35, Bad = 0.45, and Terrible = 0.45. The shaded area in Fig. 13.32 shows the results of the truncation in this example. The defuzzified value is the centroid (see Chapter 4) of the shaded area in Fig. 13.32, and is equal to 0.5036. So on day 1, machinist 2 had an OK Type of Day (0.5036 ≈ 0.5).

The next step is to provide an average and a distribution for the entire day based on the results from each machinist. The procedure outlined above can be followed for each machinist, for day 1. The Type of Day results for the other machinists were: machinist 1 = 0.4041, machinist 3 = 0.4264, and machinist 4 = 0.4088. The set average, or $\overline{X}$, for day 1 is 0.4357. This is the point for the first data set shown in Fig. 13.33, the fuzzy Type of Day $\overline{X}$ chart. For the same 30 day run, the daily average beryllium exposure and beryllium exposure ranges were also plotted in the form of $\overline{X} - R$ charts. The $\overline{X}$ chart is presented in Fig. 13.34. In Fig. 13.33, all of the important variables are taken into account and the control chart indicates that nothing is out of control. This is the result that we would expect from this simulation since it is based on random numbers, representing only the normal or "common-cause" variation. In Fig. 13.34, the traditional SPC technique, two points are above the upper control limit, representing an out-of-control situation. This represents two false alarms generated because all of the important variables are not factored into the solution of the problem. The corresponding R charts are not shown because, for this example, they did not add any information. Other simulations were run, in which the system was purposely perturbed. In all cases where the significant variables influenced the outcome the fuzzy SPC technique significantly outperformed the traditional SPC technique.

Attribute Data – Traditional SPC

The p chart is probably the most common test used with attribute data. Other common attribute data test charts are: np charts, c charts, and u charts. Wheeler and Chambers

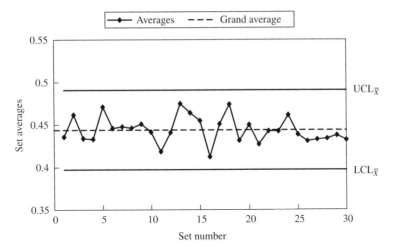

FIGURE 13.33
Fuzzy $\overline{X}$ chart for a ''normal'' 30 day beryllium plant run.

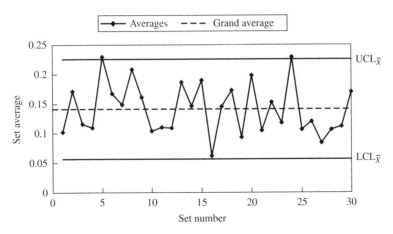

FIGURE 13.34
Beryllium exposure $\overline{X}$ chart for a ''normal'' 30 day beryllium plant run.

[1992] and Mamzic [1995] both give very good descriptions of these types of tests. We will use the p chart test as an illustration here. The p chart looks very much like the $\overline{X}$ chart. If we use the same example as we did for the cylindrical parts measurement data problem, we can construct a p chart. This time, suppose we accept only cylindrical parts with a measured diameter of 0.575 ± 0.020 inches; we will reject the rest. This creates the binary attribute system of ''accept'' and ''reject.'' In the first sample set, we have five measurements of 0.60, 0.59, 0.54, 0.57, and 0.58 inches respectively. We will reject the parts with diameters of 0.60 and 0.54 inches. The proportion rejected, p, is then 2/5 or 0.4. The p chart corresponding to the original example with this ''accept – reject'' criteria is shown in Fig. 13.35, where the p-value for set number 1 is 0.4. $\overline{p}$ is computed as the

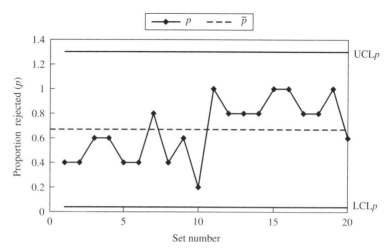

FIGURE 13.35
The p chart for the example problem.

average proportion rejected for the entire 20 sets. The upper and lower control limits can be computed as follows:

$$UCL_p = \overline{p} + 3\sqrt{\frac{\overline{p}(1 - \overline{p})}{\overline{n}}} \tag{13.28}$$

$$LCL_p = \overline{p} - 3\sqrt{\frac{\overline{p}(1 - \overline{p})}{\overline{n}}} \tag{13.29}$$

These limits are ± 3 standard deviations from the mean, based upon having a true binomial distribution to represent the data. The reader will probably notice that these limits are wider than those for the $\overline{X}$ chart. The reason for the tighter limits for the $\overline{X}$ chart is that the values for $\overline{X}$ have already been averaged, and the standard deviation must be again divided by $\sqrt{n}$. The p chart in Fig. 13.35 would have much tighter bounds if it had been developed totally with data that were ''in control''; the p chart would then tell us a different story. One advantage of the p chart is that the simple formulation for the control limits is based upon the convenient properties of the binomial distribution. Our questions is: What happens if there are more than two attributes and the binomial distribution no longer applies? Our beryllium quality control study addresses this question.

The quality control issue in our beryllium manufacturing plant is a multivalued attribute problem. The beryllium material is very expensive; therefore, care is taken not to waste it. Suppose the parts from this plant are required to satisfy two different applications: one application requires very stringent quality control, and the second application requires somewhat less quality control. An effort is made to rework parts that do not meet specification because of the cost of the material. Therefore parts produced from the plant can fall into one of five categories: Premium or ''Firsts,'' ''Seconds,'' ''Culls,'' ''Possible Rework to a First,'' and ''Possible Rework to a Second.'' In SPC terms, these are attribute data, normally analyzed with a p chart. Two papers by Raz and Wang [1990] presented some fuzzy solutions to multivalued or multinomial attribute problems. This work was criticized by Laviolette and Seaman [1992, 1994] and Laviolette et al. [1995]. But one problem with the work presented by Raz and Wang was that their example membership functions were not well-defined. There has to be some physical justification for assigning the membership functions. In our beryllium simulation we take special care in matching fuzzy membership functions to the physical world.

Our computer simulation is the same as that for Example 13.7 with the addition of the randomly generated flaws that represent the number of scratches, their length, and their depth. A 30 day simulation was accomplished in order to generate some sample control charts. This inspection technique is only one of several proposed, and the number and size of flaws in the material are strictly arbitrary.

Example 13.8   Attribute Data, Fuzzy SPC. Our fuzzy system is divided into two parts. A fuzzy rule-base is used to assign inspected parts to the proper category. This is the connection to the physical world that provides realistic membership functions for each category. The second part is the fuzzy approach to developing a multinomial p chart. The development of the p chart works like this:

- First we choose our sample set size to be the total daily plant production.
- Then we count the number of parts that fall into each category, Firsts, Seconds, etc., from the sample set.
- We use the fraction in each category to define a fuzzy sample set in terms of membership functions.
- We defuzzify the sample to get a single representative value for the sample set and use that to construct a p chart.
- Finally, we use the fuzzy p chart just like a binomial p chart to determine the state of our process.

All parts going through the beryllium plant are inspected twice. The first inspector measures part dimensions with a machine to see if they are within the desired tolerance limits. This machine is a precision instrument, so parts are assigned to categories in a crisp manner. There are three major categories, "Firsts," "Seconds," and "Culls." Suppose there is a substantial demand for the "Seconds," and a tight tolerance for the "Firsts." Enough "Seconds" are generated by this procedure to satisfy demand. The "Firsts" are assigned a score of 0.0, the "Seconds" a score of 0.5, and the "Culls" a score of 1.0. Due to the high cost of the beryllium, two more categories are added; they represent the possibility of reworking the part to qualify it to be either a "First" or a "Second." A second inspector looks at the surface finish of the part. This inspector visually checks for scratches, and records the number of scratches, the average length, and the average depth of any scratch. The numbers used in this example are not the true values for scratch sizes; they are representative values created by the plant simulator.

A fuzzy rule-based system determines how much to "downgrade" each part from the first inspection category, based on the number, depth, and length of the scratches. The fuzzy membership functions that describe the beryllium parts after the inspections are shown in Fig. 13.36.

The fuzzy system used to downgrade parts consists of three input variables and one output variable. The input variables are:

- Number of Scratches – with a range of 0–9.
- Length of Scratches – with a range of 0–2.4 cm.
- Depth of Scratches – with a range of 0–10 microns.

The output variable is the Amount of Downgrade.

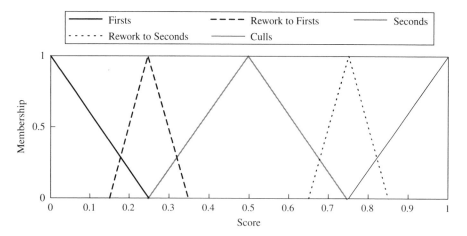

FIGURE 13.36
Fuzzy membership functions describing beryllium parts after inspections.

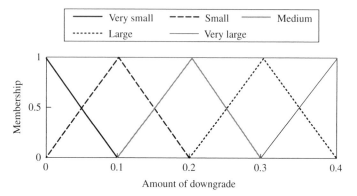

FIGURE 13.37
Output membership functions for rules to downgrade beryllium parts.

Each input variable has three membership functions and the output variable has five membership functions. The inputs and the outputs are related by 27 rules. The input membership functions are quite simple; each variable is represented by three triangles; two equal-sized right triangles, with an equilateral triangle in the middle. The area of the equilateral triangle is twice the area of one of the right triangles. The configuration is such that the sum of membership values for each variable adds to unity (the so-called orthogonal membership functions defined in Chapter 10). The output membership functions are a little bit more interesting and are shown in Fig. 13.37. The form of the rules is

If (Number of Scratches) is . . . and If (Depth of Scratches) is . . .

and If (Length of Scratches) is . . . Then the Amount of Downgrade is . . . .

The input variables are determined by measurements or estimates by the second inspector. The appropriate rules are fired using a max–min Mamdani procedure, and a centroid method is used to defuzzify the result. This defuzzified value is added to the score given by the first inspector producing a final score. The final score is used with Fig. 13.36 to place the part in its final category (Firsts, Rework to Firsts, Seconds, Rework to Seconds, Culls). The categorization is accomplished by projecting a vertical line onto the score chart, Fig. 13.36, at the point on the abscissa corresponding to the final score, and then picking the category with the highest membership value.

We used the information produced by our 30 day simulation of the plant operation to build a control chart for the beryllium manufacturing process. For each part, in each set, for each day, the categorization procedure described above is carried out. The next step is to provide a fuzzy representation for each day based on the categorization of each part for that day. This is done using the *extension principle* (see Chapter 12) and the concept of a triangular fuzzy number (TFN) [Kaufmann and Gupta, 1985]. A TFN can be completely described by the vector $[t_1, t_2, t_3]^T$. The values t_1, t_2, and t_3 are the x values of the $x - y$ pairs representing the corners of a triangle with the base resting on the x axis ($y = 0$) and the apex resting on the line $y = 1$. For example, the triangular membership function Seconds in Fig. 13.36 can be described as a TFN with $t_1 = 0.25$, $t_2 = 0.5$, and $t_3 = 0.75$, or $[0.25, 0.5, 0.75]^T$. The other four output membership functions are described as follows:

- Firsts $= [0.0, 0.0, 0.25]^T$
- Rework to Firsts $= [0.15, 0.25, 0.35]^T$
- Rework to Seconds $= [0.65, 0.75, 0.85]^T$
- Culls $= [0.75, 1.0, 1.0]^T$

A matrix, called the A matrix, can be constructed with columns comprised of the five-output membership function TFNs. For this example the A matrix is

$$A = \begin{bmatrix} 0.0 & 0.15 & 0.25 & 0.65 & 0.75 \\ 0.0 & 0.25 & 0.5 & 0.75 & 1.0 \\ 0.25 & 0.35 & 0.75 & 0.85 & 1.0 \end{bmatrix}$$

Next a five-element vector called **B** is constructed. The first element of the **B** vector is the fraction of the daily readings that were Firsts. The second element is the fraction of the readings that were Rework to Seconds, and so on. A **B** vector can be constructed for every day of the run. For example, for day 1, our simulation generated 13 parts and categorized them in the following manner:

- Number of parts categorized as Firsts $= 6$
- Number of parts categorized as Rework to Firsts $= 2$
- Number of parts categorized as Seconds $= 2$
- Number of parts categorized as Rework to Seconds $= 1$
- Number of parts categorized as Culls $= 2$

The following value is obtained for the **B** vector:

$$\mathbf{B} = [6/13, 2/13, 2/13, 1/13, 2/13]^T = [0.462, 0.154, 0.154, 0.077, 0.154]^T$$

The product AB is a TFN that represents the fuzzy distribution for the day. For day 1 of the 30 day run, the TFN AB $\approx [0.227, 0.327, 0.504]^T$. This is a triangular distribution that is approximately halfway between Rework to Firsts and Seconds. Figure 13.38 shows how day 1 is distributed on a ''Score chart'' like Fig. 13.36. The shaded area is the TFN, or fuzzy distribution for day 1. A different distribution is obtained every day. In order to construct a control chart, values for both a centerline and control limits must be determined. There are several metrics that can be used to represent the central tendency of a fuzzy set; the metric we used is the α-level (we called this a λ-level in Chapter 4) fuzzy midrange, as in Laviolette

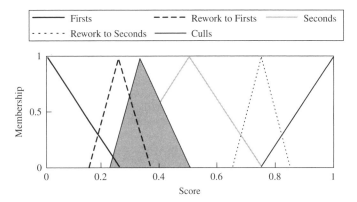

FIGURE 13.38
The fuzzy distribution for day 1, the shaded area, shown on the "Score chart."

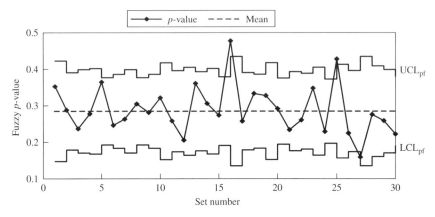

FIGURE 13.39
Fuzzy p chart using the α-level fuzzy midrange technique with $\alpha = 1/3$.

and Seaman [1994]. The α-level fuzzy midrange, for $\alpha = 1/3$, for day 1 is 0.353. This is the midrange of the shaded triangle in Fig. 13.38. The mean or the centerline for the 30 day run has a value of 0.285. These are the values shown in Fig. 13.39 for the set 1 p value and the mean. Figure 13.39 is the p chart for the 30 day simulation run.

The α-level fuzzy midrange is defined for a fuzzy set as the midpoint of the crisp interval that divides the set into two subsets. One subset contains all of the values that have a membership in the original set of greater than or equal to α. The other subset contains all of the values with memberships less than α. This interval is called the α-cut (or the λ-cut in Chapter 4). The central tendency of the TFN **AB** described above can be represented by Eq. (13.30).

$$R = \underline{\alpha}^{T} \mathbf{AB} \qquad (13.30)$$

where R is the α-level fuzzy midrange and $\underline{\alpha}$ is a vector defined by

$$\underline{\alpha} = \begin{bmatrix} \dfrac{1-\alpha}{2} & \alpha & \dfrac{1-\alpha}{2} \end{bmatrix}^{T} \qquad (13.31)$$

where α is the scalar value chosen for the α-cut. If we pick α equal to 1/3 then the vector $\underline{\alpha} = [1/3, 1/3, 1/3]$. For sufficiently large sample size, n, the vector $\mathbf{B}$ constitutes an observation from a multivariate normal distribution with a rank of $c - 1$, where c is the number of categories in the problem. The scalar R is then an observation from a univariate normal distribution with a mean $\mu = \underline{\alpha}^{\mathrm{T}} A \pi$ and a variance $\sigma^2 = \underline{\alpha}^{\mathrm{T}} A \Sigma A^{\mathrm{T}} \underline{\alpha}$ where π and Σ are the respective mean vector and covariance matrix of the set of $\mathbf{B}$ vectors. The covariance matrix Σ is defined by

$$\Sigma = [\sigma_{ij}] = \begin{cases} \pi_i(1 - \pi_i)/n, & i = j \\ -\pi_i\pi_j/n, & i \neq j \end{cases} \tag{13.32}$$

This is convenient because it provides upper and lower control limits, $\mathrm{UCL_{pf}}$ and $\mathrm{LCL_{pf}}$, for the fuzzy p chart:

$$\mathrm{UCL_{pf}} = \mu + z_{\mathrm{c}}\sigma \tag{13.33}$$

$$\mathrm{LCL_{pf}} = \mu - z_{\mathrm{c}}\sigma \tag{13.34}$$

The factor z_{c} is a function of the confidence level for the normalized Gaussian random variable. These factors are readily available; for example, see Williams [1991]. For this problem we picked $z_{\mathrm{c}} = 1.96$, for a 95% confidence level. The control limits for each sample will be a function of n, the sample size. This can be seen in Fig. 13.39 where the irregular shapes of the upper and lower control limits, $\mathrm{UCL_{pf}}$ and $\mathrm{LCL_{pf}}$, were computed using the technique described above. Figure 13.39 shows that two points are beyond the 95% confidence limit; these are points 16 and 25. Of 30 points, 28 are within the control limits; this represents 93.33% of the data, which is very close to 95% level that we would expect.

INDUSTRIAL APPLICATIONS

Two recent papers have provided an excellent review of the wealth of industrial products and consumer appliances that are bringing fuzzy logic applications to the marketplace. One paper describes fuzzy logic applications in a dozen household appliances [Quail and Adnan, 1992] and the other deals with a large suite of electronics components in the general area of image processing equipment [Takagi, 1992]. A conference in 1992 dealt entirely with industrial applications of fuzzy control [Yen et al., 1992].

Few of us could have foreseen the revolution that fuzzy set theory has already produced. Dr. Zadeh himself predicts that fuzzy logic will be part of every appliance when he says that we will "see appliances rated not on horsepower but on IQ" [Rogers and Hoshai, 1990]. In Japan, the revolution has been so strong that "fuzzy logic" has become a common advertising slogan [Reid, 1990]. Whereas the Eastern world equates the word *fuzzy* with a form of computer intelligence, the Western world still largely associates the word derisively within the context of "imprecise or approximate science."

The consumer generally purchases new appliances based on their ability to streamline housework and to use the consumer's available time more effectively. Fuzzy logic is being incorporated worldwide in appliances to accomplish these goals, primarily in the control mechanisms designed to make them work. Appliances with fuzzy logic controllers provide the consumer with optimum settings that more closely approximate human perceptions and reactions than those associated with standard control systems. Products with fuzzy logic monitor user-defined settings, then automatically set the equipment to function at the user's preferred level for a given task. For example, fuzzy logic is well-suited to making adjustments in temperature, speed, and other control conditions found in a wide variety of consumer products [Loe, 1991] and in image processing applications [Takagi, 1992]. Ross [1995] provided a good summary of several industrial applications using fuzzy control, including blood pressure control during anesthesia [Meier et al., 1992], autofocusing for a 35 mm camera [Shingu and Nishimori, 1989], image stabilization for video camcorders [Egusa et al., 1992], adaptive control of a home heating system [Altrock et al., 1993], and adaptive control of an automobile's throttle system [Cox, 1993]. The literature abounds in papers and books pertaining to fuzzy control systems [see, for example, Passino and Yurkovitz, 1998].

SUMMARY

New generations of fuzzy logic controllers are based on the integration of conventional and fuzzy controllers. Fuzzy clustering techniques have also been used to extract the linguistic IF–THEN rules from the numerical data. In general, the trend is toward the compilation and fusion of different forms of knowledge representation for the best possible identification and control of ill-defined complex systems. The two new paradigms – artificial neural networks and fuzzy systems – try to understand a real-world system starting from the very fundamental sources of knowledge, i.e., patient and careful observations, measurements, experience, and intuitive reasoning and judgments, rather than starting from a preconceived theory or mathematical model. Advanced fuzzy controllers use adaptation capabilities to tune the vertices or supports of the membership functions or to add or delete rules to optimize the performance and compensate for the effects of any internal or external perturbations. Learning fuzzy systems try to learn the membership functions or the rules. In addition, principles of genetic algorithms, for example, have been used to find the best string representing an optimum class of input or output symmetrical triangular membership functions (see Chapter 4).

It would take an entire book to thoroughly discuss the subject of classical control theory and there are many good ones available. The interested reader should see, for example, Phillips and Harbor [1996], Shinskey [1988], Ogunnaike and Ray [1994], or Murrill [1991]. For more information on the fuzzy logic control aspects the reader is referred to Ross et al. [2003], Passino and Yurkovich [1998], Wang [1997], and Parkinson [2001].

All the engineering process control examples are for setpoint-tracking control. For a problem involving the tougher problem of disturbance rejection, the reader is referred to Problem 13.10 at the end of the chapter.

For the illustrations of fuzzy statistical process control, the fuzzy ''type of day'' Shewhart-type $\overline{X} - R$ control chart has the potential to take into account task-dependent beryllium exposure for beryllium plant operations. Based upon the studies completed to this point, we believe these control charts will provide more realistic information than the standard single-variable $\overline{X} - R$ chart using only beryllium exposure information. Because of the ability to take into account task-dependency, ''the type of day'' chart can be used to determine the significance of plant improvements as well as initiate ''out-of-control'' alarms. This fuzzy technique should work well with many other task-dependent problems, which are characteristic of small-lot problems, as long as they are well-defined semantically. A least squares approach, which was studied but not presented here, will also work for this type of problem, but in many cases will not be as descriptive as the fuzzy approach. The least squares approach can produce problems if the data used to develop a control chart have many out-of-control points in them. This is because the technique squares the difference between the expected value and the measured value.

The fuzzy technique for dealing with multinomial attribute data works quite well if the problem is defined well. With our simulation, we have also compared the fuzzy technique with individual p charts which deal with multinomial attribute data. While not presented here, these comparisons have shown the fuzzy technique to be superior. We have also compared our fuzzy technique with the chi-square technique for multinomial data, and have discovered that the chi-square technique works nearly as well as the fuzzy technique, but has only one control limit. The single control limit is not a problem in the examples illustrated here because all efforts were directed toward keeping the process below the upper limit. If both upper and lower limits are important, additional work will be required to determine the meaning given by the chi-square chart.

The computer models of the beryllium plant operation described here were built from a semantic description of the process as was the fuzzy rule-base and membership functions. Consequently the correlation between the fuzzy model and the plant simulation was quite good. Both models come from the same description. It is important when developing a fuzzy model of a process that a lot of care is taken to listen to the experts and get the best model possible. If the domain expert is knowledgeable this task is usually not that difficult, but it may require several iterations to achieve validity. The fuzzy control chart will only be as good as the fuzzy rules and membership functions that comprise the system.

REFERENCES

Altrock, C., Arend, H., Krause, B., Steffens, C., and Behrens-Rommler, E. (1993). *Customer-adaptive fuzzy control of home heating system*, IEEE Press, Piscataway, NJ, pp. 115–119.

Cox, E. (1993). "Adaptive fuzzy systems," *IEEE Spectrum*, February, pp. 27–31.

Egusa, Y., Akahori, H., Morimura, A., and Wakami, N. (1992). "An electronic video camera image stabilizer operated on fuzzy theory," *IEEE international conference on fuzzy systems*, IEEE Press, San Diego, pp. 851–858.

Kaufmann, A. and Gupta, M. M. (1985). *Introduction to Fuzzy Arithmetic- Theory and Applications*. Van Nostrand Reinhold, New York.

Kiszka, J. B., Gupta, M. M., and Nikfrouk, P. N. (1985). "Some properties of expert control systems," in M. M. Gupta, A. Kendal, W. Bandler, and J. B. Kiszka (eds.), *Approximate reasoning in expert systems*, Amsterdam, Elsevier Science, pp. 283–306.

Laviolette, M. and Seaman, Jr., J. W. (1992). "Evaluating fuzzy representations of uncertainty," *Math. Sci.*, vol. 17, pp. 26–41.

Laviolette, M. and Seaman, Jr., J. W. (1994). "The efficacy of fuzzy representations of uncertainty," *IEEE Trans. Fuzzy Syst.*, vol. 2, pp. 4–15.

Laviolette, M., Seaman, Jr., J. W., Barrett, J. D., and Woodall, W. H. (1995). "A probabilistic and statistical view of fuzzy methods," *Technometrics*, vol. 37, no. 3, pp. 249–261.

Loe, S. (1991). "SGS-Thomson launches fuzzy-logic research push," *Electron. World News*, August 12, p. 1.

Mamdani, E. (1974). "Application of fuzzy algorithms for control of simple dynamic plant," *Proc. IEEE*, pp. 1585–1588.

Mamdani, E. H. and Gaines, R. R. (eds.) (1981). *Fuzzy reasoning and its applications*, Academic Press, London.

Mamzic, C. L. (ed.) (1995), *Statistical Process Control*, ISA Press, Research Triangle Park, NC.

Meier, R., Nieuwland, J., Zbinden, A., and Hacisalihzade, S. (1992). "Fuzzy logic control of blood pressure during anesthesia," *IEEE Control Syst.*, December, pp. 12–17.

Murrill, P. W. (1991). *Fundamentals of Process Control Theory*, 2nd ed., Instrument Society of America, Research Triangle Park, NC.

Ogunnaike, B. A. and Ray, W. H. (1994). *Process Dynamics, Modeling, and Control*, Oxford University Press, New York.

Olsen R. M. (1961). *Essentials of Engineering Fluid Mechanics*, International Textbook, Scranton, PA.

Pappas, C. and Mamdani, E. (1976). "A fuzzy logic controller for a traffic junction," Research report, Queen Mary College, London.

Parkinson, W. J. (2001). "Fuzzy and probabilistic techniques applied to the chemical process industries," PhD dissertation, Department of Electrical and Computer Engineering, University of New Mexico, Albuquerque, NM.

Parkinson, J. and Ross, T. (2002). "Control charts for statistical process control," in T. J. Ross, J. M. Booker, and W. J. Parkinson, (eds), *Fuzzy Logic and Probability Applications: Bridging the Gap*, Society for Industrial and Applied Mathematics, Philadelphia, PA.

Passino, K. and Yurkovich, S. (1998). *Fuzzy Control*, Addison-Wesley, Menlo Park, CA.

Phillips, C. L. and Harbor, R. D. (1996). *Feedback Control Systems*, 3rd ed., Prentice Hall, Englewood Cliffs, NJ.

Quail, S. and Adnan, S. (1992). "State of the art in household appliances using fuzzy logic," in J. Yen, R. Langari, and L. Zadeh (eds.), *Proceedings of the second international workshop – industrial fuzzy control and intelligent systems*, IEEE Press, College Station, TX, pp. 204–213.

Raz, T. and Wang, J. (1990). "Probabilistic and membership approaches in the construction of control charts for linguistic data," *Prod. Plann. Control*, vol. 1, pp. 147–157.

Reid, T. (1990). "The future of electronics looks 'fuzzy'; Japanese firms selling computer logic products," *Washington Post*, Financial Section, p. H–1.

Rogers, M. and Hoshai, Y. (1990). "The future looks 'fuzzy,' " *Newsweek*, vol. 115, no. 22.

Ross, T. (1995). *Fuzzy Logic with Engineering Applications*, McGraw-Hill, New York.

Ross, T. J., Booker, J. M., and Parkinson, W. J. (eds.) (2002). *Fuzzy Logic and Probability Applications: Bridging the Gap*, Society for Industrial and Applied Mathematics, Philadelphia, PA.

Shewhart, Walter A. (1986). *Statistical Method from the Viewpoint Economic of Quality Control*, Dover, New York.

Shingu, T. and Nishimori, E. (1989). "Fuzzy-based automatic focusing system for compact camera," *Proceedings of the third international fuzzy systems association congress*, Seattle, WA, pp. 436–439.

Shinskey, F. G. (1988). *Process Control Systems – Applications, Design, and Tuning*, 3rd ed., McGraw-Hill, New York.

Sugeno, M. (ed.) (1985). *Industrial application of fuzzy control*, North-Holland, New York.

Takagi, H. (1992). "Survey: Fuzzy logic applications to image processing equipment," in J. Yen, R. Langari, and L. Zadeh (eds.), *Proceedings of the second international workshop – industrial fuzzy control and intelligent systems*, IEEE Press, College Station, TX, pp. 1–9.

Vadiee, N. (1993). "Fuzzy rule-based expert systems – I," in M. Jamshidi, N. Vadiee, and T. Ross (eds.), *Fuzzy logic and control: software and hardware applications*, Prentice Hall, Englewood Cliffs, NJ, pp. 51–85.

Wang, J. and Raz, T. (1990). "On the construction of control charts using linguistic variables," *Int. J. Prod. Res.*, vol. 28, pp. 477–487.

Wang, L. X. (1997). *A course in fuzzy systems and control*, Prentice Hall, Upper Saddle River, NJ.

Wheeler, D. J. and Chambers, D. S. (1992). *Understanding Statistical Process Control*, 2nd ed., SPC Press, Knoxville, TN.

Williams, Richard H. (1991). *Electrical Engineering Probability*, West, St. Paul, MN.

Yen, J., Langari, R., and Zadeh, L. (eds.) (1992). *Proceedings of the second international workshop on industrial fuzzy control and intelligent systems*, IEEE Press, College Station, TX.

PROBLEMS

13.1. The interior temperature of an electrically heated oven is to be controlled by varying the heat input, u, to the jacket. The oven is shown in Fig. P13.1a. Let the heat capacities of the oven interior and of the jacket be c_1 and c_2, respectively. Let the interior and the exterior jacket surface areas be a_1 and a_2, respectively. Let the radiation coefficients of the interior and exterior jacket surfaces be r_1 and r_2, respectively. Assume that there is uniform and instantaneous distribution of temperature throughout, and the rate of loss of heat is proportional to area and the excess of temperature over that of the surroundings. If the external temperature is T_0, the jacket temperature is T_1, and the oven interior temperature is T_2, then we have

$$c_1 \dot{T}_1 = -a_2 r_2 (T_1 - T_0) - a_1 r_1 (T_1 - T_2) + u$$

$$c_2 \dot{T}_2 = a_1 r_1 (T_1 - T_2)$$

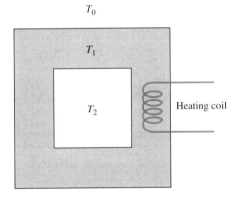

FIGURE P13.1a

Let the state variables be the excess of temperature over the exterior, i.e., $x_1 = T_1 - T_0$ and $x_2 = T_2 - T_0$. With these substituted into the preceding equations we find that they can be written as

$$\dot{x}_1 = -\frac{(a_2 r_2 + a_1 r_1)}{c_1} \cdot x_1 + \frac{a_1 r_1}{c_1} \cdot x_2 + \frac{1}{c_1} \cdot u$$

and

$$\dot{x}_2 = \frac{a_1 r_1}{c_2} \cdot x_1 - \frac{a_1 r_1}{c_2} \cdot x_2$$

Assuming that

$$\frac{a_2 r_2}{c_1} = \frac{a_1 r_1}{c_1} = \frac{a_1 r_1}{c_2} = \frac{1}{c_1} = 1$$

we have

$$\dot{x}_1(t) = -2x_1(t) + x_2(t) + u(t)$$

and

$$\dot{x}_2(t) = x_1(t) - x_2(t)$$

and

$$\dot{x}_i(t+1) = x_i(t) + \alpha \dot{x}_i(t), \quad i = 1, 2$$

Let $\alpha = 1/10$.

The membership functions for each of x_1, x_2, and u, each given on the same universe, are shown in Fig. P13.1b. For each of the variables, membership functions are taken to be low (L), medium (M), and high (H) temperatures.

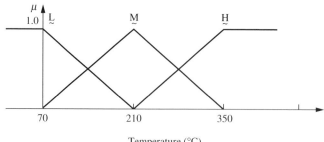

FIGURE P13.1b

Using the accompanying FAM table, conduct a graphical simulation of this control problem. The entries in the table are the control actions (u). Conduct at least four simulation cycles similar to Example 13.2. Use initial conditions of $x_1(0) = 80°$, and $x_2(0) = 85°$.

	Interior temperature excess, x_2		
Jacket temperature excess, x_1	**L**	**M**	**H**
L	H	M	L
M	H	–	L
H	H	M	L

13.2. Conduct a simulation of an automobile cruise control system. The input variables are speed and angle of inclination of the road, and the output variable is the throttle position. Let speed $= 0$ to 100 (mph), incline $= -10°$ to $+10°$, and throttle position $= 0$ to 10. The dynamics of the system are given by the following:

$$T = k_1 v + \theta k_2 + m\dot{v}$$

$$\dot{v} = v(n+1) - v(n)$$

$$T(n) = k_1 v(n) + \theta(n)k_2 + m(v_{n+1} - v_n)$$

$$v_{n+1} = \left(1 - \frac{k_1}{m}\right) v(n) + T(n) - \frac{k_2}{m}\theta(n)$$

$$v_{n+1} = k_a v(n) + [1 - k_b] \begin{bmatrix} T(n) \\ \theta(n) \end{bmatrix}$$

where T = throttle position
k_1 = viscous friction
v = speed
θ = angle of incline
$k_2 = mg \sin\theta$
$\dot{v}$ = acceleration
m = mass

$$k_a = 1 - \frac{k_1}{m} \quad \text{and} \quad k_b = \frac{k_2}{m}$$

Assign $k_1/m = k_2/m = 0.1$. The membership function for speed is determined by the cruise control setting, which we will assume to be 50 mph. The membership functions are shown in Fig. P13.2.

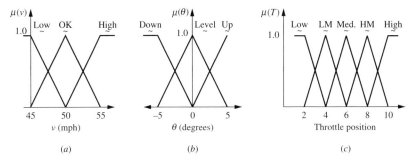

FIGURE P13.2

The FAM table is shown next:

Speed	Inclination of the road		
	Up	**Level**	**Down**
High	LM	LM	Low
OK	HM	Medium	LM
Low	High	HM	HM

Use initial conditions of speed = 52.5 mph and angle of incline = $-5°$. Conduct at least four simulation cycles.

13.3. A printer drum is driven by a brushless DC motor. The moment of inertia of the drum is $J = 0.00185$ kgm². The motor resistance is $R = 1.12\ \Omega$. The torque constant for the motor is $K_T = 0.0363$ Nm/A. The back EMF constant is $k = 0.0363$ V/(rad/s). The equation of the system is

$$J\ddot{\theta} = \frac{K_T(V - \dot{\theta}k)}{R}$$

where $\dfrac{(V - \dot{\theta}k)}{R}$ I = motor current
θ = rotational angle
V = motor control voltage

The state variables are $x_1 = \theta$ and $x_2 = \dot\theta$. Also

$$\ddot\theta = \frac{K_T}{JR}V - \frac{K_T k}{JR}\dot\theta$$

Now

$$x_2 = \dot x_1$$

Therefore,

$$\dot x_2 = \frac{K_T}{JR}V - \frac{K_T k}{JR}x_2$$

Substituting in the values of the constants, we find

$$\dot x_2 + 0.64 x_2 = 17.5\ V$$

The resulting difference equations will be

$$x_1(k+1) = x_2(k) + x_1(k)$$
$$x_2(k+1) = 17.5V(k) + 0.36x_2(k)$$

The motor can be controlled to run at constant speed or in the position mode. The membership functions for x_1, x_2, and V are shown in Fig. P13.3. The rule-based system is summarized in the following FAM table:

		x_2	
x_1	**Negative**	**Zero**	**Positive**
N	PB	P	N
Z	P	Z	N
P	Z	Z	NB

Using the initial conditions of $x_1 = 7.5°$ and $x_2 = -150$ rad/s and the difference equations, conduct at least four graphical simulation cycles.

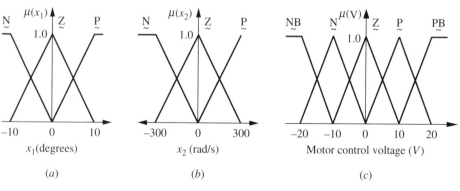

(a) (b) (c)

FIGURE P13.3

13.4. The basic mechanical system behind clocks that are enclosed in glass domes is the torsional pendulum. The general equation that describes the torsional pendulum is

$$J\frac{d^2\theta(t)}{dt^2} = \tau(t) - B\frac{d\theta(t)}{dt} - k\theta(t)$$

The moment of inertia of the pendulum bob is represented by J, the elasticity of the brass suspension strip is represented by k, and the friction between the bob and the air is represented by B. The controlling torque $\tau(t)$ is applied at the bob. When this device is used in clocks the actual torque is not applied at the bob but is applied through a complex mechanism at the main spring. The foregoing differential equation is the sum of the torques of the pendulum bob. The numerical values are $J = 1$ kgm^2, $k = 5$ Nm/rad, and $B = 2$ Nms/rad. The final differential equation with the foregoing constants incorporated is given by

$$\frac{d^2\theta(t)}{dt^2} + 2\frac{d\theta(t)}{dt} + 5\theta(t) = \tau(t)$$

The state variables are

$$x_1 = \theta(t) \text{ and } x_2(t) = \dot{\theta}(t)$$

$$\dot{x}_1 = \dot{\theta}(t) \text{ and } \dot{x}_2(t) = \ddot{\theta}(t)$$

Rewriting the differential equation using state variables, we have

$$\dot{x}_2(t) = \tau(t) - 2x_2(t) - 5x_1(t) \tag{P13.4.1}$$

$$\dot{x}_1(t) = x_2(t) \tag{P13.4.2}$$

Using these equations,

$$\dot{x}_1(t) = x_1(k+1) - x_1(k)$$

$$\dot{x}_2(t) = x_2(k+1) - x_2(k)$$

in Eqs. (P13.4.1) and (P13.4.2), and rewriting the equations in terms of θ and $\dot{\theta}$ in matrix form, we have

$$\begin{bmatrix} \theta(k+1) \\ \dot{\theta}(k+1) \end{bmatrix} = \begin{bmatrix} \theta(k) + \dot{\theta}(k) \\ -5\theta(k) - \dot{\theta}(k) \end{bmatrix} + \begin{bmatrix} 0 \\ \tau(k) \end{bmatrix}$$

The membership values for θ, $\dot{\theta}$, and τ are shown in Fig. P13.4. The rules for the control system are summarized in the accompanying FAM table.

	$\dot{\theta}$		
θ	**Positive**	**Zero**	**Negative**
P	NB	N	Z
Z	Z	Z	Z
N	Z	P	PB

The initial conditions are given as

$$\theta(0) = 0.7°$$

$$\dot{\theta}(0) = -0.2 \text{ rad/s}$$

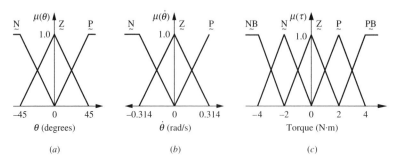

(a)   (b)   (c)

FIGURE P13.4

Conduct a graphical simulation for the control system.

13.5. On the electrical circuit shown in Fig. P13.5a it is desired to control the output current at inductor L_2 by using a variable voltage source, V. By using Kirchhoff's voltage law, the differential equation for loop 1 is given in terms of the state variables as

$$\frac{dL_1(t)}{dt} = -2L_1(t) + 2L_2(t) + 2V(t) \qquad \text{(P13.5.1)}$$

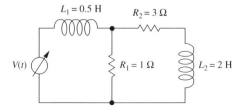

FIGURE P13.5a

and that for loop 2 is

$$\frac{dL_2(t)}{dt} = 0.5L_1(t) - 2L_2(t) \qquad \text{(P13.5.2)}$$

Converting the system of differential equations into a system of difference equations, we get

$$L_1(k + 1) = -L_1(k) + 2L_2(k) + 2V(k)$$

$$L_2(k + 1) = 0.5L_1(k) - L_2(k)$$

Rewriting the equations in matrix form, we get

$$\begin{bmatrix} L_1(k + 1) \\ L_2(k + 1) \end{bmatrix} = \begin{bmatrix} -L_1(k) + 2L_2(k) \\ 0.5L_1(k) - L_2(k) \end{bmatrix} + \begin{bmatrix} 2V(k) \\ 0 \end{bmatrix}$$

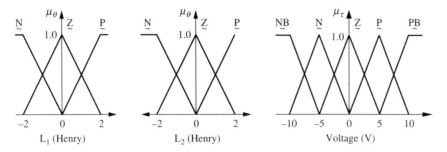

FIGURE P13.5*b*

The membership functions are given in Fig. P13.5*b* and the rules are presented in this table:

L_1	L_2 Negative	Zero	Positive
N	PB	P	Z
Z	Z	Z	Z
P	Z	N	NB

The initial conditions are $L_1(0) = 1$ H and $L_2(0) = -1$ H. Conduct a simulation of the system.

13.6 We have a cylindrical tank with cross-sectional area, A_c. Liquid flows in at a rate F_i and liquid flows out at a constant rate F_o. We want to control the tank liquid level h using a level controller to change the liquid level set height h_s. The available tank liquid height is H_T. The flow rate in the tank (F_i) is proportional to the percentage that the value is opened. We call this set flow into the tank F_{is}.

$$F_i - F_o = A_c \frac{dh}{dt}$$

$$\frac{dh}{dt} = \frac{F_i - F_o}{A_c}$$

The difference between the liquid-level setpoint and the actual tank liquid level is

$$e = h - h_s$$

and the percentage difference is

$$e = \frac{h - h_s}{h}$$

The percentage difference is used to govern the flow into the system with the following rules:

If $e = 0\%$ then $\Delta F_i = 0$

If $e > 10\%$ then $\Delta F_i = 4\%$

If $e < -10\%$ then $\Delta F_i = -4\%$

The percentage change in flow into the system (ΔF_i) is

$$\Delta F_i\% = \frac{F_{is} - F_i}{F_{is}}$$

The initial values are

$$F_{is} = 0.3 \text{ m}^3/\text{s}$$
$$F_{is} = 0.3 \text{ m}^3/\text{s}$$
$$H_T = 2 \text{ m}$$
$$A_c = 3 \text{ m}^2$$
$$h_s = 1 \text{ m}$$

At $t = 0$ the disturbance in the inlet flow is

$$F_i = 0.4 \text{ m}^3/\text{s}$$
$$F_o = 0.3 \text{ m}^3/\text{s}$$
$$e = 0$$

At $t = 0.5$ s $\Delta h = (0.4 - 0.3)/3 = 0.03$ m; thus $h = 1.03$ m and $e = (1.03 - 1)/1 = 3\%$.
  We now make use of the fuzzy controller. The single-input membership function is as shown in Fig. P13.6*a*:

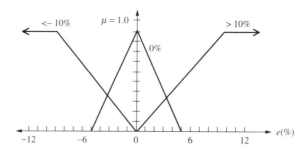

FIGURE P13.6*a*

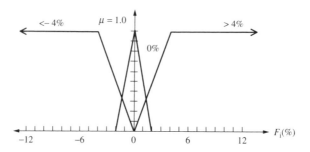

FIGURE P13.6*b*

while the output membership function is as shown in Fig. P13.6*b*:
Using a weighted average defuzzification, conduct a three-cycle simulation of this system.

13.7 The transport of toxic chemicals in water principally depends on two phenomena: advection and dispersion. In advection, the mathematical expression for time-variable diffusion is a partial differential equation accounting for concentration difference in space and time, which is derived from Fick's first law:

$$J = -DA \cdot \frac{\Delta C}{\Delta x}$$

$$V \cdot \frac{\Delta C}{\Delta t} = -DA \cdot \frac{\Delta C}{\Delta x} \quad \text{and} \quad V = A \cdot \Delta\alpha$$

$$\frac{\Delta C}{\Delta t} = -D \cdot \frac{\Delta C}{\Delta x \, \Delta x} \quad \Rightarrow \quad \frac{\partial C}{\partial t} = -D\frac{\partial^2 C}{\partial x^2}$$

where J = the mass flux rate due to molecular diffusion, mg/s
D = the molecular diffusion coefficient, cm^2/s
A = the area of the cross section, cm^2
$\dfrac{\Delta C}{\Delta t}$ = the concentration gradient in time, mg/cm^3/s
Δx = movement distance, cm

So if we want to control $\Delta C/\Delta t$, we can set a control with the following inputs:

$$W_1 = C \text{ (concentration, mg/cm}^3\text{cm)}$$

$$W_2 = \frac{\Delta C}{\Delta x} \text{ (concentration gradient in space, mg/cm}^3\text{s)}$$

and the output: $\Delta C/\Delta t = \alpha$. So

$$\frac{dW_1}{dx} = W_2 \text{ and } \frac{dW_2}{dx} = -\alpha \quad \text{(if } D = 1.0 \text{ cm}^2/\text{s)}$$

Therefore

$$W_1(k+1) = W_1(k) + W_2(k) \quad \text{and} \quad W_2(k+1) = W_2(k) - \alpha(k)$$

For this problem, we assume

$$0 \le W_1 \le 2000 \text{ mg/cm}^3$$

$$-400 \le W_2 \le 0 \text{ mg/cm}^3$$

$$0 \le \alpha \le 80 \text{ mg/cm}^3$$

(W_2 is negative because flow direction is from high concentration to low concentration).
   Step 1: Partition W_1 to zero (PZ), low (PL), high (PH) (Fig. P13.7a).

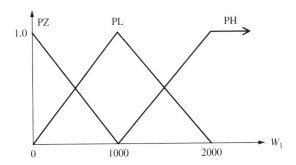

FIGURE P13.7a

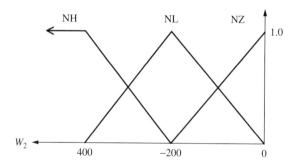

FIGURE P13.7b

Partition W_2 to zero (Z), low (NL), high (NH) (Fig. P13.7b)
Step 2: Partition α to low (L) and high (H) (Fig. P13.7c).

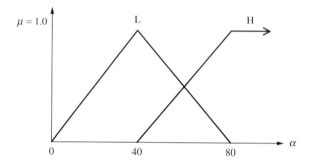

FIGURE P13.7c

Step 3: Construct rules based on experience as in the following FAM table:

		W_2	
W_1	NZ	NL	NH
PZ	L	L	L
PL	L	L	H
PH	L	H	H

Step 4: Initial conditions:

$$W_1(0) = 800 \text{ mg/cm}^3 \quad \text{and} \quad W_2(0) = -280 \text{ mg/cm}^3 \text{ cm}$$

Now conduct a two-cycle simulation using a centroidal defuzzification method.

13.8 GIS (Global Information System) is a powerful tool in environmental modeling. It integrates geographical information with data stored in databases. The main issue in using GIS is selecting the appropriate spatial resolution. If the spatial resolution selected is low, then the mapping tool cannot fully represent the true topography. On the other hand, if the spatial resolution selected is too high, then the database size will be larger than necessary, thus increasing storage requirement and processing speed.

Two parameters can be used in a fuzzy control system to govern the GIS. The first one is the digital elevation (DE) value. This value is the difference between the highest and lowest elevation in a certain geographical area. The second parameter is the area of coverage (AC). The update equation is defined as follows:

$$DE_{new} = \frac{SR^2}{AC} DE_{old} + DE_{old}$$

In the above equation SR represents spatial resolution (meters).

The first input is DE and can be either {small, medium, large} (in meters) as shown in Fig. P13.8a, the second input is AC and can be either {small, medium, large} (in meters^2) as shown in P13.8b, and SR is the output, which can either be {increase (I), decrease (D)} as seen in Fig. P13.8c and the FAM table.

	DE		
AC	**L**	**M**	**S**
L	D	I	I
M	I	D	I
S	D	D	D

Initial condition for DE is DE(0) = 2000 m.
Initial condition for AC is AC(0) = 5000 m^2.

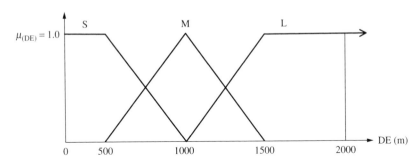

FIGURE P13.8a

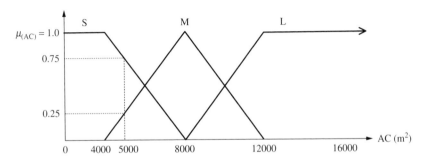

FIGURE P13.8b

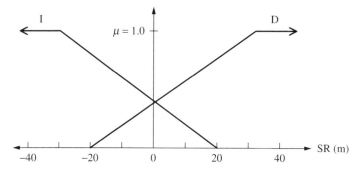

FIGURE P13.8*c*

Using a weighted average defuzzification, conduct a two-cycle simulation.

13.9. A businessman employs five people: one engineer to do his SPC work and four woodcarvers. The woodcarvers sit by the side of the road and carve figures of small animals for tourists. For the purposes of this problem each figure is equally hard to carve. The tourist picks the type of figure and the type of wood that the figure is to be carved from. There are several types of wood with ratings of 0 for very soft to 10 for very hard. The tourist pays a price for the carving based upon the number of flaws in the final product. The businessman wants to be able to keep track of the quality of the work, but knows that number of flaws alone is not a good metric. The number of flaws per worker per day is a function of the hardness of the wood and the number of carvings each worker has to do each day. The businessman decides to use the fuzzy "type of day" approach discussed in this chapter for his SPC work. He develops rules of the form

If the wood hardness is ... and the number of carvings is ... and the number of flaws is ...

Then the type of day is ....

The input membership functions are described by the following triangular fuzzy numbers:

Wood hardness: Soft (0.0, 0.0, 10.0); Hard (0.0, 10.0, 10.0)

Number of carvings: Small (0.0, 0.0, 5.0); Medium (0.0, 5.0, 10.0); Large (5.0, 10.0, 10.0)

Number of flaws: Small (0.0,0.0, 50.0); Medium (0.0, 50.0, 100.0); Large (50.0, 100.0, 100.0)

The output membership functions are the day types Good, Fair, OK, Bad, and Terrible, and are exactly the same as those shown in the body of the text.

There are 18 rules and they are given in Table P13.9a.

The businessman uses $\overline{X} - R$ charts to gain information about the quality of his product. For these charts he computes a type of day for each worker, each day, using his fuzzy rule-based system. He then uses the four type of day readings to compute his set average and set range. He does this for about 20 working days and then computes his grand average, average range, and control limits. Since he is paying his woodcarvers the minimum wage, there is quite a bit of turnover. For this reason he keeps his R charts to see if a statistical difference between workers develops. He also keeps the $\overline{X}$ charts to see if the average type of day is changing with any statistical significance over a period of time. Since the turnover

TABLE P13.9a
Woodcutters' rules

Rule number	Wood hardness	Number of carvings	Number of flaws	Type of day
1	Soft	Small	Small	Fair
2	Soft	Small	Medium	OK
3	Soft	Small	Large	Bad
4	Soft	Medium	Small	OK
5	Soft	Medium	Medium	OK
6	Soft	Medium	Large	Bad
7	Soft	Large	Small	Good
8	Soft	Large	Medium	Fair
9	Soft	Large	Large	OK
10	Hard	Small	Small	OK
11	Hard	Small	Medium	Bad
12	Hard	Small	Large	Terrible
13	Hard	Medium	Small	Fair
14	Hard	Medium	Medium	OK
15	Hard	Medium	Large	OK
16	Hard	Large	Small	Fair
17	Hard	Large	Medium	OK
18	Hard	Large	Large	Bad

rate is high, he does not know his carver's names. They are just called A, B, C, and D. One other thing that the businessman is looking for is: Has there been an out-of-control situation during the last control period? He makes his carvers work out of doors, because it attracts tourists. But the number of flaws in the carvings also influence the price of the carvings and his profit.

The 20 day period has ended and the SPC engineer has nearly completed the analysis. There was some bad weather during this period and the businessman wants to know if there was a statistically significant effect on the quality of work, or on the type of day. Unfortunately, his engineer left work early before the calculation was completed. Table P13.9b lists the partially completed work of the SPC engineer.

Day 20 was nearly completed but not quite. Worker A had a type of day of 0.45, worker B had a type of day of 0.45, and worker C had a type of day of 0.43. The type of day calculation was not finished for worker D. Worker D had the following statistics: the number of carvings was 5, the total number of flaws was 35, and the average wood hardness for these wood carvings for the day was 8.

Assume that you are the businessman. Finish the calculations by computing the following:

(a) The type of day for worker D.
(b) The set average and set range for day 20.
(c) The grand average, the average range, and all of the control limits for the 20 $\overline{X}$ and R values.
(d) Determine from the $\overline{X}$ chart if the system was ever "out of control".
(e) Is there anything in the R chart that would indicate a significant difference between the workers at any time?

TABLE P13.9*b*

Partially completed work

Day or set number	Worker average type of day ($\overline{X}$)	Range (R)
1	0.43	0.10
2	0.45	0.07
3	0.41	0.06
4	0.41	0.08
5	0.62	0.12
6	0.59	0.06
7	0.58	0.03
8	0.44	0.03
9	0.44	0.06
10	0.43	0.02
11	0.40	0.08
12	0.43	0.08
13	0.47	0.08
14	0.46	0.05
15	0.45	0.06
16	0.40	0.04
17	0.45	0.06
18	0.47	0.02
19	0.42	0.10
20	–	–

13.10. In this problem we have a hot liquid that is a product stream coming from a poorly mixed stirred tank reactor. The reaction is exothermic so that the fluid leaving the reactor can get very hot. This fluid is cooled with cooling water flowing through a counter-current heat exchanger, with the hot fluid on the shell side. The situation is depicted in Fig. P13.10. The hot fluid temperature needs to be maintained at about 110°F, because it is used in another process. There is some leeway. The process that is accepting the new hot fluid can easily handle a fluctuation of ± 5°F. Differences much larger than this start to become a problem. The cooling water comes from a cooling tower where the temperature of the cold water is maintained at 85°F. Since this temperature is constant, the only way of controlling the hot fluid temperature is to allow more or less cooling water to flow through the heat exchanger. The flow is controlled by opening and closing a control valve. The amount of flow through the valve is relative to the valve stem position. A stem position of 1 represents fully opened and a position of 0 equals fully closed. The system is controlled using an SISO fuzzy control system. The fuzzy rules for the system are of the form:

$$\text{If the } \Delta T \text{ is } \ldots \text{ Then the valve fractional change is } \ldots .$$

where ΔT is $T - T_s$, or the current hot fluid temperature minus the setpoint temperature.

The valve fractional change, f, is a fraction defined by the output membership functions of a Range. If the fraction, f, is greater than 0, the Range is defined as "full open" (1.0) minus the current valve position. If f is less than 0, the Range is defined as the current valve position. The control action described by the fuzzy controller is:

$$\text{New valve position} = \text{old valve position} + f * \text{Range}$$

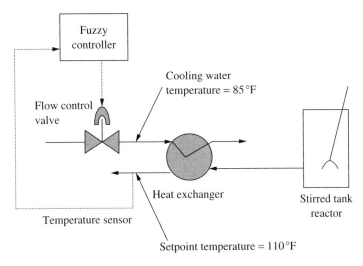

FIGURE P13.10
SISO fluid cooling problem.

TABLE P13.10
Fuzzy rules for hot fluid problem

Rule number	ΔT	Valve fractional change, f
1	Large positive	Large positive
2	Small positive	Small positive
3	Zero	Zero
4	Small negative	Small negative
5	Large negative	Large negative

The fuzzy rules are given in Table P13.10.

There are five input membership functions for ΔT. Since they are not all triangles, we will give the (x, y) coordinates rather than the TFNs:

- Large negative: $(-20.0, 1.0)\ (-15.0, 1.0)\ (-10.0, 0.0)$
- Small negative: $(-15.0, 0.0)\ (-10.0, 1.0)\ (-5.0, 0.0)$
- Zero: $(-10.0, 0.0)\ (-5.0, 1.0)\ (+5.0, 1.0)\ (+10.0, 0.0)$ – note the dead band
- Small positive: $(+5.0, 0.0)\ (+10.0, 1.0)\ (+15.0, 0.0)$
- Large positive: $(+10.0, 0.0)\ (+15.0, 1.0)\ (+20.0, 1.0)$

There are five output membership functions for valve fractional change. Since they are all triangles, we will give their TFNs:

- Large negative: $(-1.5, -1.0, -0.5)$
- Small negative: $(-1.0, -0.5, 0.0)$
- Zero: $(-0.5, 0.0, 0.5)$
- Small positive: $(0.0, 0.5, 1.0)$
- Large positive: $(0.5, 1.0, 1.5)$

Assuming that the current valve position is 0.6 calculate the following:

(a) If the hot fluid temperature suddenly increases to 113°F, what is the new valve position recommended by the fuzzy controller?

(b) If the hot fluid temperature suddenly rises to 122°F, what is the new valve position recommended by the fuzzy controller?

(c) If the hot fluid temperature suddenly drops to 98°F, what would be the new valve position recommended by the fuzzy controller?

CHAPTER
14

MISCELLANEOUS TOPICS

Knowing ignorance is strength, and ignoring knowledge is sickness.

Lao Tsu
Chinese philosopher, in Tao Te Ching, circa 600 BC

This chapter exposes the reader to a few of the additional application areas that have been extended with fuzzy logic. These few areas cannot cover the wealth of other applications, but they give to the reader an appreciation of the potential influence of fuzzy logic in almost any technology area. Addressed in this chapter are just four additional application areas: optimization, fuzzy cognitive mapping, system identification, and linear regression.

FUZZY OPTIMIZATION

Most technical fields, including all those in engineering, involve some form of optimization that is required in the process of design. Since design is an open-ended problem with many solutions, the quest is to find the "best" solution according to some criterion. In fact, almost any optimization process involves trade-offs between costs and benefits because finding optimum solutions is analogous to creating designs – there can be many solutions, but only a few might be optimum, or useful, particularly where there is a generally nonlinear relationship between performance and cost. Optimization, in its most general form, involves finding the most optimum solution from a family of reasonable solutions according to an optimization criterion. For all but a few trivial problems, finding the global optimum (the best optimum solution) can never be guaranteed. Hence, optimization in the last three decades has focused on methods to achieve the best solution per unit computational cost.

In cases where resources are unlimited and the problem can be described analytically and there are no constraints, solutions found by exhaustive search [Akai, 1994] can

Fuzzy Logic with Engineering Applications, Second Edition   T. J. Ross
© 2004 John Wiley & Sons, Ltd   ISBNs: 0-470-86074-X (HB); 0-470-86075-8 (PB)

guarantee global optimality. In effect, this global optimum is found by setting all the derivatives of the criterion function to zero, and the coordinates of the stationary point that satisfy the resulting simultaneous equations represent the solution. Unfortunately, even if a problem can be described analytically there are seldom situations with unlimited search resources. If the optimization problem also requires the simultaneous satisfaction of several constraints and the solution is known to exist on a boundary, then constraint boundary search methods such as Lagrangian multipliers are useful [deNeufville, 1990]. In situations where the optimum is not known to be located on a boundary, methods such as the steepest gradient, Newton–Raphson, and penalty function have been used [Akai, 1994], and some very promising methods have used genetic algorithms [Goldberg, 1989].

For functions with a single variable, search methods such as Golden section and Fibonacci are quite fast and accurate. For multivariate situations, search strategies such as parallel tangents and steepest gradients have been useful in some situations. But most of these classical methods of optimization [Vanderplaats, 1984] suffer from one or more disadvantages: the problem of finding higher order derivatives of a process, the issue of describing the problem as an analytic function, the problem of combinatorial explosion when dealing with many variables, the problem of slow convergence for small spatial or temporal step sizes, and the problem of overshoot for step sizes too large. In many situations, the precision of the optimization approach is greater than the original data describing the problem, so there is an *impedance mismatch* in terms of resolution between the required precision and the inherent precision of the problem itself.

In the typical scenario of an optimization problem, fast methods with poorer convergence behavior are used first to get the process near a solution point, such as a Newton method, then slower but more accurate methods, such as gradient schemes, are used to converge to a solution. Some current successful optimization approaches are now based on this hybrid idea: fast, approximate methods first, slower and more precise methods second. Fuzzy optimization methods have been proposed as the first steps in hybrid optimization schemes. One of these methods will be introduced here. More methods can be found in Sakawa [1993].

One-dimensional Optimization

Classical optimization for a one-dimensional (one independent variable) relationship can be formulated as follows. Suppose we wish to find the optimum solution, x^*, which maximizes the objective function $y = f(x)$, subject to the constraints

$$g_i(x) \leq 0, \qquad i = 1, m \tag{14.1}$$

Each of the constraint functions $g_i(x)$ can be aggregated as the intersection of all the constraints. If we let $C_i = \{x \mid g_i(x) \leq 0\}$, then

$$C = C_1 \cap C_2 \cap \cdots \cap C_m = \{x \mid g_1(x) \leq 0, \ g_2(x) \leq 0, \ldots, g_m(x) \leq 0\} \tag{14.2}$$

which is the feasible domain described by the constraints C_i. Thus, the solution is

$$f(x^*) = \max_{x \in C}\{f(x)\} \tag{14.3}$$

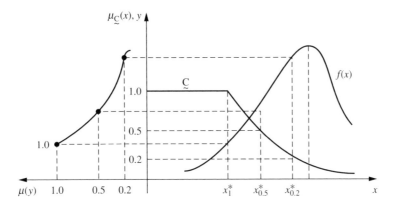

FIGURE 14.1
Function to be optimized, $f(x)$, and fuzzy constraint, $\underset{\sim}{C}$.

In a real environment, the constraints might not be so crisp, and we could have fuzzy feasible domains (see Fig. 14.1) such as "x could exceed x_0 a little bit." If we use λ-cuts on the fuzzy constraints $\underset{\sim}{C}$, fuzzy optimization is reduced to the classical case. Obviously, the optimum solution x^* is a function of the threshold level λ, as given in Eq. (14.4):

$$f(x_\lambda^*) = \max_{x \in C_\lambda}\{f(x)\} \tag{14.4}$$

Sometimes, the goal and the constraint are more or less contradictory, and some trade-off between them is appropriate. This can be done by converting the objective function $y = f(x)$ into a pseudogoal $\underset{\sim}{G}$ [Zadeh, 1972] with membership function

$$\mu_{\underset{\sim}{G}}(x) = \frac{f(x) - m}{M - m} \tag{14.5}$$

where $m = \inf_{x \in X} f(x)$
$\ M = \sup_{x \in X} f(x)$

Then the fuzzy solution set $\underset{\sim}{D}$ is defined by the intersection

$$\underset{\sim}{D} = \underset{\sim}{C} \cap \underset{\sim}{G} \tag{14.6}$$

membership is described by

$$\mu_{\underset{\sim}{D}}(x) = \min\{\mu_{\underset{\sim}{C}}(x), \mu_{\underset{\sim}{G}}(x)\} \tag{14.7}$$

and the optimum solution will be x^* with the condition

$$\mu_{\underset{\sim}{D}}(x^*) \geq \mu_{\underset{\sim}{D}}(x) \qquad \text{for all } x \in X \tag{14.8}$$

where $\mu_{\underset{\sim}{C}}(x)$, $\underset{\sim}{C}$, x should be substantially greater than x_0. Figure 14.2 shows this situation.

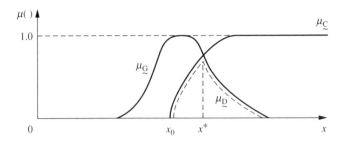

FIGURE 14.2
Membership functions for goal and constraint [Zadeh, 1972].

Example 14.1. Suppose we have a deterministic function given by

$$f(x) = xe^{(1-x/5)}$$

for the region $0 \le x \le 5$, and a fuzzy constraint given by

$$\mu_{\underset{\sim}{C}}(x) = \begin{cases} 1, & 0 \le x \le 1 \\ \dfrac{1}{1 + (x-1)^2}, & x > 1 \end{cases}$$

Both of these functions are illustrated in Fig. 14.3. We want to determine the solution set $\underset{\sim}{D}$ and the optimum solution x^*, i.e., find $f(x^*) = y^*$. In this case we have $M = \sup[f(x)] = 5$ and $m = \inf[f(x)] = 0$; hence Eq. (14.5) becomes

$$\mu_{\underset{\sim}{G}}(x) = \frac{f(x) - 0}{5 - 0} = \frac{x}{5} e^{(1-x/5)}$$

which is also shown in Fig. 14.3. The solution set membership function, using Eq. (14.7), then becomes

$$\mu_{\underset{\sim}{D}}(x) = \begin{cases} \dfrac{x}{5} e^{(1-x/5)}, & 0 \le x \le x^* \\ \dfrac{1}{1 + (x-1)^2}, & x > x^* \end{cases}$$

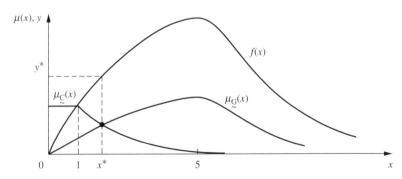

FIGURE 14.3
Problem domain for Example 14.1.

and the optimum solution x^*, using Eq. (14.8), is obtained by finding the intersection

$$\frac{x^*}{5}e^{(1-x^*/5)} = \frac{1}{1+(x-1)^2}$$

and is shown in Fig. 14.3.

When the goal and the constraint have unequal importance the solution set $\underset{\sim}{D}$ can be obtained by the convex combination, i.e.,

$$\mu_{\underset{\sim}{D}}(x) = \alpha\mu_{\underset{\sim}{C}}(x) + (1-\alpha)\mu_{\underset{\sim}{G}}(x) \tag{14.9}$$

The single-goal formulation expressed in Eq. (14.9) can be extended to the multiple-goal case as follows. Suppose we want to consider n possible goals and m possible constraints. Then the solution set, $\underset{\sim}{D}$, is obtained by the aggregate intersection, i.e., by

$$\underset{\sim}{D} = \left(\underset{i=1,m}{\cap} \underset{\sim}{C}_i\right) \cap \left(\underset{j=1,n}{\cap} \underset{\sim}{G}_j\right) \tag{14.10}$$

Example 14.2. A beam structure is supported at one end by a hinge and at the other end by a roller. A transverse concentrated load P is applied at the middle of the beam, as in Fig. 14.4. The maximum bending stress caused by P can be expressed by the equation $\sigma_b = Pl/w_z$, where w_z is a coefficient decided by the shape and size of a beam and l is the beam's length. The deflection at the centerline of the beam is $\delta = Pl^3/(48EI)$ where E and I are the beam's modulus of elasticity and cross-sectional moment of inertia, respectively. If $0 \le \delta \le 2$ mm, and $0 \le \sigma_b \le 60$ MPa, the constraint conditions are these: the span length of the beam,

$$l = \begin{cases} l_1, & 0 \le l_1 \le 100 \text{ m} \\ 200 - l_1, & 100 < l_1 \le 200 \text{ m} \end{cases}$$

and the deflection,

$$\delta = \begin{cases} 2 - \delta_1, & 0 \le \delta_1 \le 2 \text{ mm} \\ 0, & \delta_1 > 2 \text{ mm} \end{cases}$$

To find the minimum P for this two-constraint and two-goal problem (the goals are the stress, σ_b, and the deflection, δ), we first find the membership function for the two goals and two constraints.

1. The μ_{G_1} for bending stress σ_b is given as follows:

$$P(0) = 0, \; P(60 \text{ MPa}) = \frac{w_z 60}{l}, \; P(\sigma_b) = \frac{w_z\sigma_b}{l}; \text{ thus, } \mu_{G_1} = \frac{\sigma_b}{60}$$

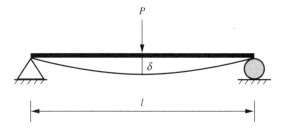

FIGURE 14.4
Simply supported beam with a transverse concentrated load.

To change the argument in μ_{G_1} into a unitless form, let $x = \sigma_b/60$, where $0 \le x \le 1$. Therefore, $\mu_{G_1}(x) = x$ when $0 \le x \le 1$.

2. The μ_{G_2} for deflection δ is as follows:

$$P(\delta) = \frac{48EI\delta}{l^3}, \; P(0) = 0, \; P(2) = \frac{48EI \times 2}{l^3}; \; \text{thus,} \; \mu_{G_2} = \frac{\delta}{2}$$

Let $x = \delta/2$, so that the argument of μ_{G_2} is unitless. Therefore, $\mu_{G_2} = x$, $0 \le x \le 1$.

3. Using Eq. (14.10), we combine $\mu_{G_1}(x)$ and $\mu_{G_2}(x)$ to find $\mu_G(x)$:

$$\mu_G(x) = \min(\mu_{G_1}(x), \mu_{G_2}(x)) = x \qquad 0 \le x \le 1$$

4. The fuzzy constraint function μ_{C_1} for the span is

$$\mu_{C_1}(x) = \begin{cases} 2x, & 0 \le x \le 0.5 \\ 2 - 2x, & 0.5 < x \le 1 \end{cases}$$

where $x = l_1/200$. Therefore, the constraint function will vary according to a unitless argument x.

5. The fuzzy constraint function μ_{C_2} for the deflection δ can be obtained in the same way as in point 4:

$$\mu_{C_2}(x) = \begin{cases} 1 - x, & 0 \le x \le 1 \\ 0, & x > 1 \end{cases}$$

where $x = \delta/2$.

6. The fuzzy constraint function $\mu_C(x)$ for the problem can be found by the combination of $\mu_{C_1}(x)$ and $\mu_{C_2}(x)$, using Eq. (14.10):

$$\mu_C(x) = \min(\mu_{C_1}(x), \mu_{C_2}(x))$$

and $\mu_C(x)$ is shown as the bold line in Fig. 14.5.

Now, the optimum solutions P can be found by using Eq. (14.10):

$$D = (G \cap C)$$

$$\mu_D(x) = \mu_C(x) \wedge \mu_G(x)$$

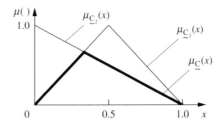

FIGURE 14.5
Minimum of two constraint functions.

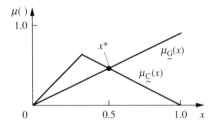

FIGURE 14.6
Graphical solution to minimization problem of Example 14.2.

The optimum value can be determined graphically, as seen in Fig. 14.6, to be $x^* = 0.5$. From this, we can obtain the optimum span length, $l = 100$ m, optimum deflection, $\delta = 1$ mm, and optimum bending stress, $\sigma_b = 30$ MPa. The minimum load P is

$$P = \min\left(\frac{\sigma_b w_z}{l}, \frac{48 E I \delta}{l^3}\right)$$

Suppose that the importance factor for the goal function $\mu_G(x)$ is 0.4. Then the solution for this same optimization problem can be determined using Eq. (14.9) as

$$\mu_D = 0.4\mu_G + 0.6\mu_C$$

where μ_C can be expressed by the function (see Fig. 14.5)

$$\mu_C(x) = \begin{cases} 2x, & 0 \le x \le \frac{1}{3} \\ 1 - x, & \frac{1}{3} < x \le 1 \end{cases}$$

Therefore,

$$0.6\mu_C(x) = \begin{cases} 1.2x, & 0 \le x \le \frac{1}{3} \\ 0.6 - 0.6x, & \frac{1}{3} < x \le 1 \end{cases}$$

and $0.4\mu_G(x) = 0.4x$. The membership function for the solution set, from Eq. (14.9), then is

$$\mu_D(x) = \begin{cases} 1.6x, & 0 \le x \le \frac{1}{3} \\ 0.6 - 0.2x, & \frac{1}{3} < x \le 1 \end{cases}$$

The optimum solution for this is $x^* = 0.33$, which is shown in Fig. 14.7.

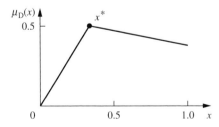

FIGURE 14.7
Solution of Example 14.2 considering an importance factor.

FUZZY COGNITIVE MAPPING

Cognitive maps (CMs) were introduced by Robert Axelrod [1976] as a formal means of modeling decision making in social and political systems. CMs are a type of directed graph that offers a means to model interrelationships or causalities among concepts; there are various forms of CMs, such as signed digraphs, weighted graphs, and functional graphs. The differences amongst these various forms can be found in Kardaras and Karakostas [1999]. CMs can also be used for strategic planning, prediction, explanation, and for engineering concept development. The use of simple binary relationships (i.e., increase and decrease) is done in a *conventional* (crisp) CM. All CMs offer a number of advantages that make them attractive as models for engineering planning and concept development. CMs have a clear way to visually represent causal relationships, they expand the range of complexity that can be managed, they allow users to rapidly compare their mental models with reality, they make evaluations easier, and they promote new ways of thinking about the issue being evaluated.

Concept variables and causal relations

CMs graphically describe a system in terms of two basic types of elements: concept variables and causal relations. Nodes represent concept variables, C_x, where $x = 1, \ldots, N$. A concept variable at the origin of an arrow is a cause variable, whereas a concept variable at the endpoint of an arrow is an effect variable. For example, for $C_h \rightarrow C_i$, C_h is the cause variable that impacts C_i, which is the effect variable. Figure 14.8 represents a simple CM, in which there are four concept variables (C_h represents utilization of waste steam for heat, C_i represents the amount of natural gas required to generate heat, C_k represents economic gain for the local economy, and C_j represents the market value of waste steam (which is dependent on the price of natural gas)).

Arrows represent the causal relations between concept variables, which can be positive or negative. For example, for $C_h \overset{-}{\rightarrow} C_i$, C_h has a negative causal relationship on C_i. Therefore, an increase in C_h results in a decrease in C_i.

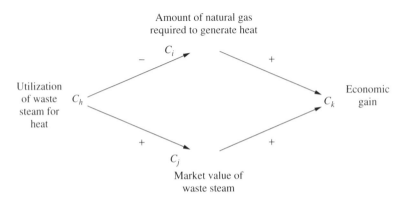

FIGURE 14.8
A conventional cognitive map for the utilization of waste steam.

Paths and cycles

A path between two concept variables, C_h and C_k, denoted by $P(h, k)$, is a sequence of all the nodes which are connected by arrows from the first node (C_h) to the last node (C_k) (Fig. 14.8) [Kosko, 1986]. A cycle is a path that has an arrow from the last point of the path to the first point.

Indirect effect

The indirect effect of a path from the cause variable C_h to the effect variable C_k, which is denoted by $I(h, k)$, is the product of the causal relationships that form the path from the cause variable to the effect variable [Axelrod, 1976]. If a path has an even number of negative arrows, then the indirect effect is positive. If the path has an odd number of negative arrows, then the indirect effect is negative. In Fig. 14.8 the indirect effect of cause variable C_h on the effect variable C_k through path $P(h, i, k)$ is negative; the indirect effect of the cause variable C_h on the effect variable C_k through path $P(h, j, k)$ is positive.

Total effect

The total effect of the cause variable C_h on the effect variable C_k, which is denoted by $T(h, k)$, is the union of all the indirect effects of all the paths from the cause variable to the effect variable [Axelrod, 1976]. If all the indirect effects are positive, the total effect is positive. If all the indirect effects are negative, so is the total effect. If some indirect effects are positive and some are negative, the sum is indeterminate [Kosko, 1986]. A large CM, that is one with a large number of concepts and paths, will therefore be dominated by the characteristic of being indeterminate. In Fig. 14.8 the total effect of cause variable C_h to effect variable C_k is the collection of the indirect effect of C_h to C_k through the paths $P(h, i, k)$ and $P(h, j, k)$. Since one indirect effects is positive and the other is negative in this case, this means that the total effect is indeterminate.

Indeterminacy

The character of a conventional CM being indeterminate can be resolved, but it comes at a computational and conceptual price. To do so, the CM must accommodate a numerical weighting scheme [Kosko, 1986]. If the causal edges are weighted with positive or negative real numbers, then the indirect effect is the product of each of the weights in a given path, and the total effect is the sum of the path products. This scheme of weighting the path relationships removes the problem of indeterminacy from the total effect calculation, but it also requires a finer causal discrimination. Such a fineness may not be available from the analysts or experts who formulate the CM. This finer discrimination between concepts in the CM would make knowledge acquisition a more onerous process–forced numbers from insufficient decision information, different numbers from different experts, or from the same expert on different days, and so on. However, causal relationships could be represented by linguistic quantities as opposed to numerical ones. Such is the context of a *fuzzy* CM (FCM).

Fuzzy Cognitive Maps

If one were to emphasize that the simple binary relationship of a CM needed to be extended to include various degrees of increase or decrease (small decrease, large increase, almost no increase, etc.), then a *fuzzy* cognitive map (FCM) is more appropriate. An FCM extends the idea of conventional CMs by allowing concepts to be represented linguistically with an associated fuzzy set, rather than requiring them to be precise. Extensions by Taber [1994] and Kosko [1992] allow fuzzy numbers or linguistic terms to be used to describe the degree of the relationship between concepts in the FCM. FCMs are analyzed either geometrically or numerically [Pelaez and Bowles, 1996]. A geometric analysis is used primarily for small FCMs, where it simply traces the increasing and decreasing effects along all paths from one concept to another. For larger FCMs, such as those illustrated later in this section, a numerical analysis is required, where the concepts are represented by a state vector and the relations between concepts are represented by a fuzzy relational matrix, called an *adjacency* matrix. This, along with a few other key features of FCMs that distinguish them from CMs, are mentioned below.

Adjacency matrix

A CM can be transformed using a matrix called an adjacency matrix [Kosko, 1986]. An adjacency matrix is a square matrix that denotes the effect that a cause variable (row) given in the CM has on the effect variable (column). Figure 14.9 is an adjacency matrix for the CM displayed in Fig. 14.8. In other words, the adjacency matrix for a CM with n nodes uses an $n \times n$ matrix in which an entry in the (i, j) position of the matrix denotes an arrow between nodes C_h and C_i. This arrow (as shown in Fig. 14.8) simply represents the "strength" of the effect between the two nodes (i.e., a "+1" represents that the effect is to increase, whereas a "−1" represents that the effect is to decrease).

Threshold function

Concept states are held within defined boundaries through the threshold function. The type of threshold function chosen determines the behavior of a CM. A bivalent threshold function requires concepts to have a value of 1 or 0, which is equivalent to "on" or "off":

$$f(x_i) = 0, \quad x_i \leq 0$$
$$f(x_i) = 1, \quad x_i > 0$$

The trivalent threshold function includes negative activation. Therefore, concepts have a value of 1, 0, or −1, which is equivalent to "positive effect", "no effect," and "negative effect", respectively:

$$f(x_i) = -1, \quad x_i \leq -0.5$$

$$E = \begin{array}{c} \\ C_h \\ C_i \\ C_k \\ C_j \end{array} \begin{array}{cccc} C_h & C_i & C_k & C_j \\ \left[\begin{array}{cccc} 0 & -1 & 0 & +1 \\ 0 & 0 & +1 & 0 \\ 0 & 0 & 0 & 0 \\ 0 & 0 & +1 & 0 \end{array}\right] \end{array}$$

FIGURE 14.9
The adjacency matrix for the cognitive map in Fig. 14.8.

$$f(x_i) = 0, \qquad -0.5 < x_i < 0.5$$

$$f(x_i) = 1, \quad x_i \geq 0.5$$

Concepts are multiplied by their connecting causal relation weights to give the total input to the effect concept. In cases where there are multiple paths connecting a concept, the sum of all the causal products is taken as the input [Tsadiras and Margaritis, 1996]:

$$x_i = \sum_{\substack{j=1 \\ j \neq i}}^{n} C_j w_{ji} \qquad (14.11)$$

where x_i = input
$\quad C_j$ = concept state
$\quad w_{ji}$ = weight of the causal relations

Feedback

For FCMs we can model dynamic systems that are cyclic, and therefore, feedback within a cycle is allowed. Each concept variable is given an initial value based on the belief of the expert(s) of the current state. The FCM is then free to interact until an equilibrium is reached [Kosko, 1997]. An equilibrium is defined to be the case when a new state vector is equal to a previous state vector.

Min–max inference approach

The min–max inference approach is a technique that can be used to evaluate the indirect and total effects of an FCM. The causal relations between concepts are often defined by linguistic variables, which are words that describe the strength of the relationship. The min–max inference approach can be utilized to evaluate these linguistic variables [Pelaez and Bowles, 1995]. The minimum value of the links in a path is considered to be the path strength. If more than one path exists between the cause variable and the effect variable, the maximum value of all the paths is considered to be the overall effect. In other words, the indirect effect amounts to specifying the weakest linguistic variable in a path, and the total effect amounts to specifying the strongest of the weakest paths.

Example 14.3. Figure 14.10 depicts an FCM with five concept variables (C_1 represents *utilization of waste steam for heat*, C_2 represents amount of *natural gas required* to produce heat, C_3 represents the resulting carbon dioxide (CO_2) *emissions* produced from the burning of a methane-based gas, C_4 represents *carbon credits* that would need to be purchased, and C_5 represents the *economic gain*). Carbon credits are credits a company would receive from reducing its CO_2 emissions below the required level stipulated by the government's Kyoto implementation plan. Those companies not meeting their required level may need to purchase credits from others. In the FCM the "effects" of the paths, P, are linguistic instead of simple binary quantities like a "+1" (increase) or a "−1" (decrease). However, the numerical quantities +1, 0, and −1 for a trivalent FCM are still used to convey the signs of the linguistic term. For example, a linguistic effect of "significant, +1" means that the effect is "significantly positive." A linguistic effect of "a lot, −1" means that the effect is "negatively a lot." The values of the paths can be in matters of degree such as "none," "some," "much," or "a lot."

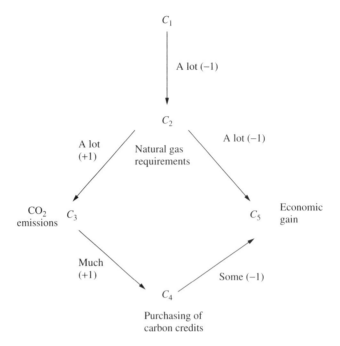

FIGURE 14.10
A fuzzy cognitive map involving waste steam and greenhouse gas emissions.

In this example, then, $P = \{$none $<$ some $<$ much $<$ a lot$\}$. These P values would be the linguistic values that would be contained within the adjacency matrix of the FCM:

$$
\begin{bmatrix}
0 & -1 & 0 & 0 & 0 \\
0 & 0 & +1 & 0 & -1 \\
0 & 0 & 0 & +1 & 0 \\
0 & 0 & 0 & 0 & -1 \\
0 & 0 & 0 & 0 & 0
\end{bmatrix}
$$

To implement the FCM we start by activating C_1 (i.e., we begin the process by assessing the impact of an increase in waste steam for a facility); this results in the initial state vector

$$[1, 0, 0, 0, 0]$$

This state vector is activating only concept C_1 in Fig. 14.10. Causal flow in the FCM was determined with repeated vector–matrix operations and thresholding [Pelaez and Bowles, 1995].

The new state is the old state multiplied by the adjacency matrix [Pelaez and Bowles, 1996]:

$$
[C_1 C_2 \ldots C_n]_{\text{new}} = [C_1 C_2 \ldots C_n]_{\text{old}} *
\begin{bmatrix}
C_{11} & \cdots & C_{1n} \\
\vdots & \ddots & \vdots \\
C_{n1} & \cdots & C_{nn}
\end{bmatrix}
\tag{14.12}
$$

The values of the state vector were thresholded to keep their values in the set $\{-1, 0, 1\}$, and the activated concept (in this case C_1) was reset to 1 after each matrix multiplication. Using the algorithm developed by Pelaez and Bowles [1995] we premultiply the trivalent adjacency matrix shown above by this initial state vector. At each iteration of this multiplication, the trivalent threshold function is invoked. This multiplication is continued until the output vector reaches a limiting state (i.e., it stabilizes). For this simple example, the resulting state vector stabilized after four iterations to the following form:

$$[1.0, -1.0, -1.0, -1.0, 1.0]$$

This stabilized output vector can be understood in the following sense. For an *increase* in the waste steam ($+1$), the natural gas requirements will *decrease* (-1), the CO_2 emissions will *decrease* (-1), and carbon credits also *decrease* (-1). Finally, there is an *increase* in economic gain ($+1$).

With a conventional CM we would get the following results. First, we see that path $I_1 = (1, 2, 5)$ has two negative causal relationships, and its indirect path effect would be positive (two negatives yield a positive). For path $I_2 = (1, 2, 3, 4, 5)$ we see that it has one negative causal relationship (between C_1 and C_2) and three positive relationships for the other three elements of the path; hence, the indirect effect of this path is negative (one negative and three positives yield a negative effect). Hence, in the conventional CM characterization of this simple example, the results would be *indeterminate* (one positive indirect effect and a negative indirect effect). For an FCM, we can accommodate linguistic characterizations of the elements, as discussed previously, and as seen in Fig. 14.10. Two unique paths that exist from the cause variable (C_1) to the effect variable (C_5) are $I_1 = (1, 2, 5)$ and $I_2 = (1, 2, 3, 4, 5)$. The indirect effects of C_1 on C_5, expressed in terms of the linguistic values of P, are [Kosko, 1992]

$$I_1(C_1, C_5) = \min\{e_{12}, e_{25}\} = \min\{\text{a lot, a lot}\}$$
$$= \text{a lot}$$
$$I_2(C_1, C_5) = \min\{e_{12}, e_{23}, e_{34}, e_{45}\} = \min\{\text{a lot, a lot, much, some}\}$$
$$= \text{some}$$

Therefore, the linguistic total effect is expressed as

$$T(C_1, C_5) = \max\{I_1(C_1, C_5), I_2(C_1, C_5)\}$$
$$= \max\{\text{a lot, some}\} = \text{a lot}$$

Applying these linguistic results to the stabilized vector above, i.e., [1.0, -1.0, -1.0, -1.0, 1.0], we come to the conclusion that an increase in waste steam for this facility results in "*a lot of increase*" in economic gain.

To conclude the example, fuzzy cognitive mapping does suffer in comparison with other methods in that there is a large degree of subjectivity. But fuzzy cognitive mapping does allow for varying degrees of magnitude or significance of relationships, which is a limitation of other crisp or standard methods. Therefore, much of the grayness in subjectivity is captured and accounted for, resulting in a more balanced assessment. Thus, with appropriate expert-based professional judgment (likely by a panel of experts in the field of the subject matter), FCM can be an effective assessment tool. Problems 14.12–14.14 at the end of this chapter are given as exercises to illustrate how various changes to the paths in this example result in different results and conclusions of the FCM approach.

SYSTEM IDENTIFICATION

Suppose we have a standard fuzzy relational equation of the form $B = A \circ R$. In the normal situation we have a fuzzy relation R from either rules or data, and the information contained in A is also known from data or is assumed. The determination of B usually is accomplished through some form of composition. Suppose, however, that we know B and R, and we are interested in finding A. There are many physical problems where this situation arises. Foremost among these is the field of *system identification*. We might have a system (modeled by R) which is subjected to an input which is unknown (represented by A), but we are able to observe or measure the output of the system (given by B). So, we want to find out what possible sets of the input could generate the observed output.

   If the relational equation were linear, we would find the inverse of the equation, i.e., $A = B \circ R^{-1}$.

   Unfortunately, a fuzzy relational equation is not linear, and the inverse cannot provide a unique solution, in general, or any solution in some situations. In fact, the inverse is difficult to find for most situations [Terano et al., 1992].

   A fuzzy relational equation can be expressed in expanded form by

$$(a_1 \wedge r_{11}) \vee (a_2 \wedge r_{12}) \vee \cdots \vee (a_n \wedge r_{1n}) = b_1$$

$$(a_1 \wedge r_{21}) \vee (a_2 \wedge r_{22}) \vee \cdots \vee (a_n \wedge r_{2n}) = b_2$$

$$\vdots$$

$$(a_1 \wedge r_{m1}) \vee (a_2 \wedge r_{m2}) \vee \cdots \vee (a_n \wedge r_{mn}) = b_m$$

$$(14.13)$$

where a_i, r_{ij}, and b_i are membership values for A, R, and B, respectively.

   To solve $A = \{a_1, a_2, \ldots, a_n\}$ given r_{ij} and b_j ($i = 1, 2, \ldots, n$ and $j = 1, 2, \ldots, m$), we can use a method reported by Tsukamoto and Terano [1977] that produces interval values for the solution. In this approach there are two standard definitions that first must be presented: a fuzzy equality and a fuzzy inequality. An *equality* is expressed as

$$\text{Equality} \qquad a \wedge r = b \qquad (14.14)$$

The inverse solution for a in the equality in Eq. (14.14), given r and b are known, is generally an interval number and is denoted by an operator $b \diamond r$,

$$b \diamond r = \begin{cases} b, & r > b \\ [b, 1], & r = b \\ \emptyset, & r < b \end{cases} \qquad (14.15)$$

An *inequality* is defined as

$$\text{Inequality} \qquad a \wedge r \leq b \qquad (14.16)$$

For the inequality (14.16) the inverse solution for a is also an interval number, denoted by the operator $b \hat{\diamond} r$, and is given by

$$b \hat{\diamond} r = \begin{cases} [0, b], & r > b \\ [0, 1], & r \leq b \end{cases} \qquad (14.17)$$

A typical row of the standard fuzzy relational equation system (see Eqs. (14.13)), say the first row, can be represented in a simpler form,

$$(a_1 \wedge r_1) \vee (a_2 \wedge r_2) \vee \cdots \vee (a_n \wedge r_n) = b \tag{14.18}$$

The expression in Eq. (14.18) can be subdivided into n *equalities* of the type

$$(a_1 \wedge r_1) = b, \qquad (a_2 \wedge r_2) = b, \qquad \ldots, \qquad (a_n \wedge r_n) = b \tag{14.19}$$

and n *inequalities* of the type

$$(a_1 \wedge r_1) \leq b, \qquad (a_2 \wedge r_2) \leq b, \qquad \ldots, \qquad (a_n \wedge r_n) \leq b \tag{14.20}$$

A solution, represented as an interval vector, to the fuzzy relational Eq. (14.18) exists if and only if there is at least one equality and no more than $(n-1)$ inequalities in the solution. That is, the inverse solution for a_i in Eq. (14.18), denoted W_i, is

$$a_i = W_1 \quad \text{or} \quad W_2 \quad \text{or} \quad \ldots \quad \text{or} \quad W_n \tag{14.21}$$

where

$$W_i = (b \,\hat{\diamond}\, r_1, \ldots, b \,\hat{\diamond}\, r_{i-1}, b \,\diamond\, r_i, b \,\hat{\diamond}\, r_{i+1}, \ldots, b \,\hat{\diamond}\, r_n) \tag{14.22}$$

Note that the ith term in Eq. (14.22) is an equality and the other terms are inequalities.

Example 14.4. Suppose we want to solve for a_i ($i = 1, 2, 3, 4$) in the single inverse fuzzy equation

$$(a_1 \wedge 0.7) \vee (a_2 \wedge 0.8) \vee (a_3 \wedge 0.6) \vee (a_4 \wedge 0.3) = 0.6$$

Making use of expressions (14.19)–(14.20), we subdivide the single fuzzy equation into $n = 4$ equalities:

$$\begin{aligned}
Y_{\text{eq}} &= \{b \diamond r_1, b \diamond r_2, \ldots, b \diamond r_n\} \\
&= \{0.6 \diamond 0.7, 0.6 \diamond 0.8, 0.6 \diamond 0.6, 0.6 \diamond 0.3\} \\
&= \{0.6, 0.6, [0.6, 1], \emptyset\} \quad \text{(four } equality \text{ values)}
\end{aligned}$$

and into $n = 4$ inequalities:

$$\begin{aligned}
Y_{\text{ineq}} &= \{b \,\hat{\diamond}\, r_1, b \,\hat{\diamond}\, r_2, \ldots, b \,\hat{\diamond}\, r_n\} \\
&= \{0.6 \,\hat{\diamond}\, 0.7, 0.6 \,\hat{\diamond}\, 0.8, 0.6 \,\hat{\diamond}\, 0.6, 0.6 \,\hat{\diamond}\, 0.3\} \\
&= \{[0, 0.6], [0, 0.6], [0, 1], [0, 1]\} \quad \text{(four } inequality \text{ values)}
\end{aligned}$$

Then Eq. (14.22) provides for the $n = 4$ potential solutions, W_i, where $i = 1, 2, 3, 4$:

$$W_1 = \{0.6, [0, 0.6], [0, 1], [0, 1]\} \quad \text{(position 1 is the first equality value)}$$
$$W_2 = \{[0, 0.6], 0.6, [0, 1], [0, 1]\} \quad \text{(position 2 is the second equality value)}$$
$$W_3 = \{[0, 0.6], [0, 0.6], [0.6, 1], [0, 1]\} \quad \text{(position 3 is the third equality value)}$$
$$W_4 = \{[0, 0.6], [0, 0.6], [0, 1], \emptyset\} = \emptyset \quad \text{(position 4 is the fourth equality value)}$$

where all values are the inequalities, except those equalities noted specifically. Equation (14.21) provides for the aggregated solution in interval form; note that W_4 does not contribute to the aggregated solution because it has a value of null. Hence,

$$a_i = W_1 \text{ or } W_2 \text{ or } W_3$$

Its maximum solution is $a_{max} = \{0.6, 0.6, 1, 1\}$. Its minimum solutions are

$$
\begin{aligned}
a_{min} &= \{0.6, 0, 0, 0\} &\quad \text{from } W_1 \\
&= \{0, 0.6, 0, 0\} &\quad \text{from } W_2 \\
&= \{0, 0, 0.6, 0\} &\quad \text{from } W_3
\end{aligned}
$$

Now suppose instead of a single equation we want to find the solution for an equation set; that is, a collection of m simultaneous equations of the form given in Eqs. (14.13). Then the solution set will consist of an m set of n equalities, expressed in an $m \times n$ matrix (denoted Y) and an m set of n inequalities, also expressed in an $m \times n$ matrix (denoted $\hat{Y}$):

$$
Y = \begin{bmatrix}
b_1 \diamond r_{11} & b_1 \diamond r_{12} & \cdots & b_1 \diamond r_{1n} \\
b_2 \diamond r_{21} & b_2 \diamond r_{22} & \cdots & b_2 \diamond r_{2n} \\
& \vdots & & \\
b_m \diamond r_{m1} & b_m \diamond r_{m2} & \cdots & b_m \diamond r_{mn}
\end{bmatrix}
\tag{14.23}
$$

$$
\hat{Y} = \begin{bmatrix}
b_1 \hat{\diamond} r_{11} & b_1 \hat{\diamond} r_{12} & \cdots & b_1 \hat{\diamond} r_{1n} \\
b_2 \hat{\diamond} r_{21} & b_2 \hat{\diamond} r_{22} & \cdots & b_2 \hat{\diamond} r_{2n} \\
& \vdots & & \\
b_m \hat{\diamond} r_{m1} & b_m \hat{\diamond} r_{m2} & \cdots & b_m \hat{\diamond} r_{mn}
\end{bmatrix}
\tag{14.24}
$$

Taking an element for each row from Y and replacing the corresponding element in $\hat{Y}$, we get an array solution for each element, ij, in the $m \times n$ equation matrix:

$$
W^*_{(i_1, i_2, \dots, i_m)} = \begin{bmatrix}
b_1 \hat{\diamond} r_{11} & \cdots & b_1 \diamond r_{1i} & \cdots & b_1 \hat{\diamond} r_{1n} \\
b_2 \hat{\diamond} r_{21} & \cdots & b_2 \diamond r_{2i} & \cdots & b_2 \hat{\diamond} r_{2n} \\
& & \vdots & & \\
b_m \hat{\diamond} r_{m1} & \cdots & b_m \diamond r_{mi} & \cdots & b_m \hat{\diamond} r_{mn}
\end{bmatrix} = (w^*_{ij})
\tag{14.25}
$$

where indices $i_1 = (1, 2, \dots, n)$, $i_2 = (1, 2, \dots, n)$, and $i_m = (1, 2, \dots, n)$ represent m arrays that are all of length n. Each array is a solution to one row of the original equations (i.e., of Eqs. (14.13)), and we need at most m arrays for the complete solution. Hence, we can have (n^m) solutions, including null solutions, of the type expressed by Eq. (14.25). To establish a sign convention we denote a particular solution, (w^*_{ij}), of the $m \times n$ array by identifying its location (ij) in the array. The complete solution incorporating all $m \times n$ possible solutions will be denoted

$$W_{(i_1, i_2, \dots, i_m)} = \{w_1, w_2, \dots, w_m\} \tag{14.26}$$

where

$$w_j = \bigcap_i w^*_{ij} \tag{14.27}$$

and where $i = 1, 2, \dots, n$ and $j = 1, 2, \dots, m$.

In the foregoing development, each possible solution $w_1, w_2, \ldots, w_m$ in Eq. (14.26) is found by taking the intersection of all the solutions in the jth column, i.e., Eq. (14.27), of the solution arrays described by Eq. (14.25).

Example 14.5. Suppose we have a system of three simultaneous equations as given here. In this example, $m = 3$ and $n = 3$. Hence, there is a potential for $3^3 = 27$ distinct solutions that need to be explored in order to develop the full solution. Our goal is to find interval values for the three unknown quantities, $\{a_1, a_2, a_3\}$, in the following inverse equations:

$$\{a_1, a_2, a_3\} \circ \begin{bmatrix} 0.3 & 0.5 & 0.2 \\ 0.2 & 0 & 0.4 \\ 0 & 0.6 & 0.1 \end{bmatrix} = [\,0.2 \quad 0.4 \quad 0.2\,]$$

To begin the solution process, we need to find the individual equality sets, Y, and inequality sets, $\hat{Y}$, using Eqs. (14.23) and (14.24), respectively. So for the first row of Y we operate b_1 on the first column of the r matrix, i.e., using Eq. (14.15),

$$\{b_1 \diamond r_{11}, b_1 \diamond r_{21}, b_1 \diamond r_{31}\} = \{0.2 \diamond 0.3, 0.2 \diamond 0.2, 0.2 \diamond 0\} = \{0.2, [0.2, 1], \varnothing\}$$

The second row of Y is found, by operating b_2 on the second column of the r matrix, to be

$$\{b_2 \diamond r_{12}, b_2 \diamond r_{22}, b_2 \diamond r_{32}\} = \{0.4 \diamond 0.5, 0.4 \diamond 0, 0.4 \diamond 0.6\} = \{0.4, \varnothing, 0.4\}$$

The third row of Y is found, by operating b_3 on the third column of the r matrix, to be

$$\{b_3 \diamond r_{13}, b_3 \diamond r_{23}, b_3 \diamond r_{33}\} = \{0.2 \diamond 0.2, 0.2 \diamond 0.4, 0.2 \diamond 0.1\} = \{[0.2, 1], 0.2, \varnothing\}$$

Therefore, we have

$$Y = \begin{bmatrix} 0.2 & [0.2, 1] & \varnothing \\ 0.4 & \varnothing & 0.4 \\ [0.2, 1] & 0.2 & \varnothing \end{bmatrix}$$

To calculate the inequality matrix, $\hat{Y}$, we have the same sequence of operations of elements in the b vector on columns in the r matrix, but we use the operator $\hat{\diamond}$. For the first row of Y we operate b_1 on the first column of the r matrix, i.e., using Eq. (14.17),

$$\{b_1 \hat{\diamond} r_{11}, b_1 \hat{\diamond} r_{21}, b_1 \hat{\diamond} r_{31}\} = \{0.2 \hat{\diamond} 0.3, 0.2 \hat{\diamond} 0.2, 0.2 \hat{\diamond} 0\} = \{[0, 0.2], [0, 1], [0, 1]\}$$

The second row of Y is found, by operating b_2 on the second column of the r matrix, to be

$$\{b_2 \hat{\diamond} r_{12}, b_2 \hat{\diamond} r_{22}, b_2 \hat{\diamond} r_{32}\} = \{0.4 \hat{\diamond} 0.5, 0.4 \hat{\diamond} 0, 0.4 \hat{\diamond} 0.6\} = \{[0, 0.4], [0, 1], [0, 0.4]\}$$

The third row of Y is found, by operating b_3 on the third column of the r matrix, to be

$$\{b_3 \hat{\diamond} r_{13}, b_3 \hat{\diamond} r_{23}, b_3 \hat{\diamond} r_{33}\} = \{0.2 \hat{\diamond} 0.2, 0.2 \hat{\diamond} 0.4, 0.2 \hat{\diamond} 0.1\} = \{[0, 1], [0, 0.2], [0, 1]\}$$

Therefore, we have

$$\hat{Y} = \begin{bmatrix} [0, 0.2] & [0, 1] & [0, 1] \\ [0, 0.4] & [0, 1] & [0, 0.4] \\ [0, 1] & [0, 0.2] & [0, 1] \end{bmatrix}$$

Now, using Eq. (14.25), we can construct the W_{ij}^* matrices. The first matrix is W_{111}^*, the subscripts denoting that the equality element for the first row of W_{111}^* comes from the *first* position in Y (the first subscript 1) in the first row; the equality element for the second row of W_{111}^* comes from the *first* position in Y (the second subscript 1) in the second row; the equality element for the third row of W_{111}^* comes from the *first* position in Y (the third subscript 1) in the third row. All other elements for W_{111}^* come from the same positions they are in for the matrix $\hat{Y}$. Hence, W_{111}^* looks like

$$W_{111}^* = \begin{bmatrix} 0.2 & [0, 1] & [0, 1] \\ 0.4 & [0, 1] & [0, 0.4] \\ [0.2, 1] & [0, 0.2] & [0, 1] \end{bmatrix}$$

Finally, using Eqs. (14.26)–(14.27), we take the intersection of the elements in each column of W_{111}^* to get

$$W_{111} = (\emptyset, [0, 0.2], [0, 0.4]) = \emptyset$$

Since there is a null element in W_{111}, we set the entire value equal to null.

Continuing in a similar fashion, the second matrix is W_{112}^*, with the subscripts denoting that the equality element for the first row of W_{112}^* comes from the *first* position in Y (the first subscript 1) in the first row; the equality element for the second row of W_{112}^* comes from the *first* position in Y (the second subscript 1) in the second row; the equality element for the third row of W_{112}^* comes from the *second* position in Y (the third subscript 2) in the third row. All other elements for W_{112}^* come from the same positions they are in for the matrix $\hat{Y}$. Hence, W_{112}^* looks like

$$W_{112}^* = \begin{bmatrix} 0.2 & [0, 1] & [0, 1] \\ 0.4 & [0, 1] & [0, 0.4] \\ [0, 1] & 0.2 & [0, 1] \end{bmatrix}$$

and using Eqs. (14.26)–(14.27) again, we get

$$W_{112} = (\emptyset, 0.2, [0, 0.4]) = \emptyset$$

because there is at least one null element in W_{112}^*.

Now, moving to other elements in w_{ij}^*, such as W_{131}^*, we get

$$W_{131}^* = \begin{bmatrix} 0.2 & [0, 1] & [0, 1] \\ [0, 0.4] & [0, 1] & 0.4 \\ [0.2, 1] & [0, 0.2] & [0, 1] \end{bmatrix}$$

and the resulting solution after performing intersections on the elements in each column is

$$W_{131} = (0.2, [0, 0.2], 0.4)$$

This process continues for the other 24 solutions, e.g., W_{211}, W_{232}, W_{333}, etc., out of the total of 27 (3^3).

Of all 27 possible solutions, only four are non-null (nonempty). These four are

$$W_{131} = (0.2, [0, 0.2], 0.4)$$
$$W_{132} = (0.2, 0.2, 0.4)$$
$$W_{231} = (0.2, 0.2, 0.4)$$
$$W_{232} = ([0, 0.2], 0.2, 0.4)$$

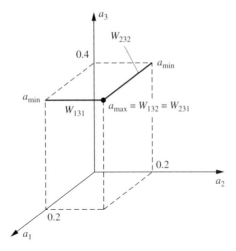

FIGURE 14.11
Solution for Example 14.5.

By inspection we can see that $W_{131} \supseteq W_{132}$ and $W_{232} \supseteq W_{231}$, and that $W_{231} = W_{132}$; hence, the solution can be expressed by the intervals W_{132} and W_{232}. This solution space in the three dimensions governed by the original coordinates a_1, a_2, and a_3 is shown in Fig. 14.11.

In the figure we see that the minimum solution is given by two points

$$a_{min,1} = \{a_1, a_2, a_3\} = \{0.2, 0, 0.4\} \quad \text{from } W_{131}$$

$$a_{min,2} = \{a_1, a_2, a_3\} = \{0, 0.2, 0.4\} \quad \text{from } W_{232}$$

and that the maximum solution is given by the single point

$$a_{max} = \{a_1, a_2, a_3\} = \{0.2, 0.2, 0.4\}$$

The entire solution in this three-dimensional example comprises the two edges that are darkened in Fig. 14.11.

It is perhaps clear from Examples 14.4 and 14.5 that when the cardinal numbers n and m are large, the number of analytical solutions becomes exponentially large; also, sometimes the final solution can be null. For both these cases other approaches, such as those in pattern recognition, might be more practical. In any case, fuzzy inverses are only approximate; they are not unique in general, even for linear equations.

FUZZY LINEAR REGRESSION

Regression analysis is used to model the relationship between dependent and independent variables. In regression analysis, the dependent variable, y, is a function of the independent variables; and the degree of contribution of each variable to the output is represented by coefficients on these variables. The model is empirically developed from data collected from observations and experiments. A crisp linear regression model is shown in Eq. (14.28),

$$y = f(x, a) = a_0 + a_1 x_1 + a_2 x_2 + \cdots + a_n x_n \tag{14.28}$$

In conventional regression techniques, the difference between the observed values and the values estimated from the model is assumed to be due to observational errors, and the difference is considered a random variable. Upper and lower bounds for the estimated value are established, and the probability that the estimated value will be within these two bounds represents the confidence of the estimate. In other words, conventional regression analysis is probabilistic. But in fuzzy regression, the difference between the observed and the estimated values is assumed to be due to the ambiguity inherently present in the system. The output for a specified input is assumed to be a range of possible values, i.e., the output can take on any of these possible values. Therefore, fuzzy regression is *possibilistic* in nature. Moreover, fuzzy regression analyses use fuzzy functions to represent the coefficients as opposed to crisp coefficients used in conventional regression analysis [Terano et al., 1992]. Equation (14.29) shows a typical fuzzy linear regression model,

$$\underset{\sim}{Y} = f(x, \underset{\sim}{A}) = \underset{\sim}{A}_0 + \underset{\sim}{A}_1 x_1 + \underset{\sim}{A}_2 x_2 + \cdots + \underset{\sim}{A}_n x_n \tag{14.29}$$

where $\underset{\sim}{A}_i$ is the ith fuzzy coefficient (usually a fuzzy number).

Fuzzy regression estimates a range of possible values that are represented by a possibility distribution (a more rigorous definition of possibilities is given in Chapter 15), termed here a membership function. Membership functions are formed by assigning a specific membership value (degree of belonging) to each of the estimated values (Fig. 14.12). Such membership functions are also defined for the coefficients of the independent variables. Triangular membership functions for the fuzzy coefficients, like those shown in Fig. 14.12, allow for the solution to be found via a linear programming formulation; other membership functions for the coefficients require alternative approaches [Kikuchi and Nanda, 1991].

The membership function $\mu_{\underset{\sim}{A}}$ for each of the coefficients is expressed as

$$\mu_{\underset{\sim}{A}_i}(a_i) = \begin{cases} 1 - \dfrac{\mid p_i - a_i \mid}{c_i}, & p_i - c_i \leq x_i \leq p_i + c_i \\ 0, & \text{otherwise} \end{cases} \tag{14.30}$$

The fuzzy function $\underset{\sim}{A}$ is a function of two parameters, p and c, known as the middle value and the spread, respectively. The spread denotes the fuzziness of the function. The figure shows the membership function for a fuzzy number "approximately p_i." A more detailed explanation of membership functions, fuzzy numbers, and operations on fuzzy numbers

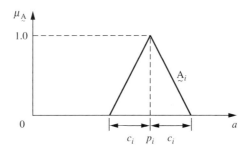

FIGURE 14.12
Membership function for the fuzzy coefficient $\underset{\sim}{A}$.

is given in Chapters 4 and 12. The fuzzy parameters $A = \{A_1, \ldots, A_n\}$ can be denoted in the vector form of $A = \{\mathbf{p}, \mathbf{c}\}$, where $\mathbf{p} = (p_0, \ldots, p_n)$ and $\mathbf{c} = (c_0, \ldots, c_n)$. Therefore, the output is a revised version of Eq. (14.29),

$$Y = (p_0, c_0) + (p_1, c_1)x_1 + (p_2, c_2)x_2 + \cdots + (p_n, c_n)x_n$$

The membership function for the output fuzzy parameter, Y, is given by

$$\mu_Y(y) = \begin{cases} \max_i(\min[\mu_A(a_i)]), & \{a \mid y = f(x, a)\} \neq \emptyset \\ 0, & \text{otherwise} \end{cases} \quad (14.31)$$

Substituting Eq. (14.30) into Eq. (14.31), we get [see Tanaka et al., 1982]

$$\mu_Y(y) = \begin{cases} 1 - \dfrac{\left| y - \sum_{i=1}^{n} p_i x_i \right|}{\sum_{i=1}^{n} c_i \mid x_i \mid}, & x_i \neq 0 \\ 1, & x_i = 0, \quad y = 0 \\ 0, & x_i = 0, \quad y \neq 0 \end{cases} \quad (14.32)$$

The foregoing equations are applied to m data sets that can be obtained from sampling. The output and the input data can be either fuzzy or nonfuzzy. Table 14.1 shows an example of the data sets for the nonfuzzy data. In the table, y_j is the output for the jth sample, and x_{ij} is the ith input variable for the jth sample.

The Case of Nonfuzzy Data

Tanaka et al. [1982] have determined the solution to the regression model by converting it to a linear programming problem (this is not the only approach, as is discussed shortly). For nonfuzzy data the objective of the regression model is to determine the optimum parameters A^* such that the fuzzy output set, which contains y_i, is associated with a membership value greater than h, i.e.,

$$\mu_{Y_j}(y_j) \geq h, \qquad j = 1, \ldots, m \quad (14.33)$$

The degree h is specified by the user; as h increases, the fuzziness of the output increases [Kikuchi and Nanda, 1991]. Figure 14.13 shows the membership function for the fuzzy

TABLE 14.1
An example of the data sets for nonfuzzy data

Sample number, j	Output, y_j	n inputs, x_{ij}
1	y_1	$x_{11}, x_{21}, \ldots, x_{n1}$
$\vdots$	$\vdots$	$\vdots$
m	y_m	$x_{1m}, x_{2m}, \ldots, x_{nm}$

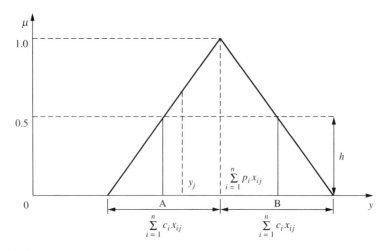

FIGURE 14.13
Fuzzy output function.

output. Equation (14.33) states that the fuzzy output should lie between A and B of Fig. 14.13. In the figure the middle value ($\sum_{i=1}^{n} p_i x_i$) and the spread ($\sum_{i=1}^{n} c_i |x_i|$) are obtained by considering Eq. (14.32), where h is specified by the user.

In regression we seek to find the fuzzy coefficients that minimize the spread of fuzzy output for all the data sets. Equation (14.34) shows the objective function that has to be minimized.

$$O = \min \left\{ mc_0 + \sum_{j=1}^{m} \sum_{i=0}^{n} c_i x_{ij} \right\} \tag{14.34}$$

where $x_0 j = 1$, for $j = 1, \ldots, m$. The objective function given in Eq. (14.34) is minimized, subject to two constraints. The constraints are obtained by substituting Eq. (14.32) into Eq. (14.33); they become

$$y_j \geq \sum_{i=1}^{n} p_i x_{ij} - (1 - h) \sum_{i=1}^{n} c_i x_{ij} \tag{14.35}$$

and

$$y_j \leq \sum_{i=1}^{n} p_i x_{ij} + (1 - h) \sum_{i=1}^{n} c_i x_{ij} \tag{14.36}$$

Since each data set produces two constraints, there is a total of $2m$ constraints for each data set.

The Case of Fuzzy Data

When human judgment or imprecise measurements are involved in determining the output, the output is seldom a crisp number. The output in such situations is best represented by a

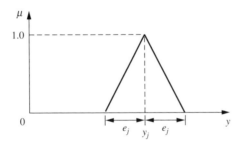

FIGURE 14.14
An example of fuzzy output.

fuzzy number as $\underset{\sim}{Y}_j = (y_j, e_j)$, where y_j is the middle value and e_j represents the ambiguity in the output, as seen in Fig. 14.14.

The membership function for the observed fuzzy output is given as

$$\mu_{\underset{\sim}{Y}_j}(y) = 1 - \frac{|y_j - y|}{e_j} \tag{14.37}$$

An estimate of this fuzzy output can be obtained from Eq. (14.32) as

$$\mu_{\underset{\sim}{Y}_j^*}(y) = 1 - \frac{\left|y_j - \sum\limits_{i=1}^{n} p_i x_{ij}\right|}{\sum\limits_{i=1}^{n} c_i |x_{ij}|} \qquad \text{for } j = 1, m \tag{14.38}$$

The degree of fitting of the estimated fuzzy output $\underset{\sim}{Y}_j^*$ to the given data $\underset{\sim}{Y}_j$ is determined by h_j, which maximizes h subject to $\underset{\sim}{Y}_j^h \subset \underset{\sim}{Y}_j^{h^*}$ where

$$\underset{\sim}{Y}_j^h = \left\{ y \mid \mu_{\underset{\sim}{Y}_j}(y) \geq h \right\}$$
$$\underset{\sim}{Y}_j^{h^*} = \left\{ y \mid \mu_{\underset{\sim}{Y}_j^*}(y) \geq h \right\} \tag{14.39}$$

Figure 14.15 illustrates these concepts. The objective of the fuzzy linear regression model is to determine fuzzy parameters $\underset{\sim}{A}^*$ that minimize the spread subject to the constraint that $h_j \geq H$ for all j, where H is chosen by the user as the degree of fitting of the fuzzy linear model. The jth fitting parameter, h_j, is computed from Fig. 14.15 as

$$h_j = 1 - \frac{\left|y_j - \sum\limits_{i=1}^{n} p_i x_{ij}\right|}{\sum\limits_{i=1}^{n} c_i |x_{ij}| - e_j} \tag{14.40}$$

In summary the objective function to be minimized is

$$O_f = \min \left\{ mc_0 + \sum\limits_{j=1}^{m} \sum\limits_{i=0}^{n} c_i x_{ij} \right\} \tag{14.41}$$

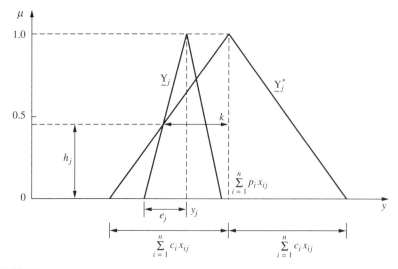

FIGURE 14.15
Degree of fitting of estimated fuzzy output to the given fuzzy output.

subject to the constraints

$$y_j \geq \sum_{i=1}^{n} p_i x_{ij} - (1 - H) \sum_{i=1}^{n} c_i x_{ij} + (1 - H) e_j \qquad (14.42)$$

and

$$y_j \leq \sum_{i=1}^{n} p_i x_{ij} + (1 - H) \sum_{i=1}^{n} c_i x_{ij} - (1 - H) e_j \qquad (14.43)$$

for each data set where $j = 1, \ldots, m$. In Eqs. (14.42) and (14.43) we note that $e_j = y_j - y$. The important equations used in the fuzzy linear regression model for both the nonfuzzy output (y_j) and the fuzzy output $(\underset{\sim}{Y}_j)$ cases, along with the equation numbers, are summarized in Table 14.2.

Example 14.6. The concepts of fuzzy regression are illustrated through a trivial one-dimensional problem. Here, we will only use two data points to illustrate the linear regression approach. The data set is shown in Table 14.3. For these two points the equation of the line that runs through them both is $y = 2.9762x + 0.5524$. We should get these values in our fuzzy analysis since, for two points, we do not have uncertainty in the regression analysis.

Let this data set be represented by a linear regression model, $\underset{\sim}{Y} = \underset{\sim}{A}_0 + \underset{\sim}{A}_1 x$, where the coefficients $\underset{\sim}{A}_0$ and $\underset{\sim}{A}_1$ are fuzzy numbers. For the data set given in the Table 14.3, $i = 1$ and $j = 2$. From Eq. (14.34) the objective function to be minimized is given by

$$O = \left\{ \sum_{j=1}^{m} \sum_{i=0}^{n} c_i x_{ij} \right\} = m c_0 + (x_{11} + x_{12}) c_1 = m c_0 + (0.52 + 1.36) c_1$$

$$= 2 c_0 + 1.88 c_1$$

TABLE 14.2
Summary of equations used in fuzzy linear regression model (Eq. (14.29)), $\underset{\sim}{Y} = f(x, \underset{\sim}{A}) = \underset{\sim}{A}_0 + \underset{\sim}{A}_1 x_1 + \underset{\sim}{A}_2 x_2 + \cdots + \underset{\sim}{A}_n x_n$

Membership functions		
$\underset{\sim}{A}$	$\mu_{\underset{\sim}{A}_i}(a_i) = \begin{cases} 1 - \dfrac{|p_i - a_i|}{c_i} & p_i - c_i \leq x_i \leq p_i + c_i \\ 0 & \text{otherwise} \end{cases}$	Eq. (14.30)
$\underset{\sim}{Y}$	$\mu_{\underset{\sim}{Y}}(y) = \begin{cases} 1 - \dfrac{\left| y - \sum\limits_{i=1}^{n} p_i x_i \right|}{\sum\limits_{i=1}^{n} c_i |x_i|} & x_i \neq 0 \\ 1 & x_i = 0, y = 0 \\ 0 & x_i = 0, y \neq 0 \end{cases}$	Eq. (14.32)

Solution

Nonfuzzy data

Objective function	$O = \min \left\{ mc_0 + \sum\limits_{j=1}^{m} \sum\limits_{i=0}^{n} c_i x_{ij} \right\}$	Eq. (14.34)
Constraints		
1	$y_j \geq \sum\limits_{i=1}^{n} p_i x_{ij} - (1-h) \sum\limits_{i=1}^{n} c_i x_{ij}$	Eq. (14.35)
2	$y_j \leq \sum\limits_{i=1}^{n} p_i x_{ij} + (1-h) \sum\limits_{i=1}^{n} c_i x_{ij}$	Eq. (14.36)

Fuzzy data

Objective function	$O_f = \min \left\{ mc_0 + \sum\limits_{j=1}^{m} \sum\limits_{i=0}^{n} c_i x_{ij} \right\}$	Eq. (14.41)
Constraints		
1	$y_j \geq \sum\limits_{i=1}^{n} p_i x_{ij} - (1-H) \sum\limits_{i=1}^{n} c_i x_{ij} + (1-H)e_j$	Eq. (14.42)
2	$y_j \leq \sum\limits_{i=1}^{n} p_i x_{ij} + (1-H) \sum\limits_{i=1}^{n} c_i x_{ij} - (1-H)e_j$	Eq. (14.43)

TABLE 14.3
Two data sets describing one-dimensional problem

y_i	x_{ij}
2.1	0.52
4.6	1.36

Since there are two data sets, the objective function has to be minimized subject to four constraints as shown here:

$$y_1 \geq p_0 + 0.52p_1 - (1-h)(c_0 + 0.52c_1)$$

$$y_1 \leq p_0 + 0.52p_1 + (1-h)(c_0 + 0.52c_1)$$

$$y_2 \geq p_0 + 1.36p_1 - (1-h)(c_0 + 1.36c_1)$$

$$y_2 \leq p_0 + 1.36p_1 + (1-h)(c_0 + 1.36c_1)$$

Substituting the values of y_j $(i = 1, 2)$, and setting $h = 0.5$, we get

$$2.1 \geq p_0 + 0.52p_1 - 0.5c_0 - 0.26c_1$$

$$2.1 \leq p_0 + 0.52p_1 + 0.5c_0 + 0.26c_1$$

$$4.6 \geq p_0 + 1.36p_1 - 0.5c_0 - 0.68c_1$$

$$4.6 \leq p_0 + 1.36p_1 + 0.5c_0 + 0.68c_1$$

The linear programming problem is now solved using the simplex method; a good explanation of this method is available in Hillier and Lieberman [1980]. Since the constraint equations are expressed by inequality relationships, basic variables (D_i) are introduced to convert the inequalities to equations (these variables are equated to the *slack* in the inequality). Also, since basic variables cannot be negative (this requirement arises due to the "less than" inequality), artificial variables are introduced to account for the negative sign. The basic variables are variables in the objective function that have a zero coefficient. The artificial variables are denoted by a bar on top of the letter, i.e., $\overline{D}_i$.

$$2.1 = p_0 + 0.52p_1 - 0.5c_0 - 0.26c_1 + D_1 \tag{1}$$

$$2.1 = p_0 + 0.52p_1 + 0.5c_0 + 0.26c_1 - D_2 + \overline{D}_3 \tag{2}$$

$$4.6 = p_0 + 1.36p_1 - 0.5c_0 - 0.68c_1 + D_4 \tag{3}$$

$$4.6 = p_0 + 1.36p_1 + 0.5c_0 + 0.68c_1 - D_5 + \overline{D}_6 \tag{4}$$

In the simplex method, an additional variable, denoted M, is used to weight the artificial variables (there are two, $\overline{D}_3$ and $\overline{D}_6$) in the objective function. The updated objective function is then

$$O = 2c_0 + 1.88c_1 + M\overline{D}_3 + M\overline{D}_6$$

where M is a very large number compared with the magnitudes of the numbers in the original data set; in this example M is set at 100. Therefore, the final objective function is

$$-O + 2c_0 + 1.88c_1 + 100\overline{D}_3 + 100\overline{D}_6 = 0 \tag{0}$$

The basic variables are D_1, $\overline{D}_3$, D_4, and $\overline{D}_6$. Since the stopping rule of the simplex method requires the basic variables to have a coefficient of 0, Eq. (0) is subtracted from M times the equations containing the artificial variables, Eqs. (2) and (4). This procedure works very much like a Gaussian elimination method for solving simultaneous equations. The calculations are shown in Table 14.4.

The values of c_0, c_1, p_0, and p_1 are determined so that $-O$ has the maximum value (or O has the minimum value). The calculations performed to determine c_0, c_1, p_0, and p_1 by the simplex method are shown in Tables 14.4 and 14.5. The simplex method is an iterative process in which the basic variable is replaced by the nonbasic variable that lies in the column of the highest negative coefficient in row (0) (the nonbasic variable in this column becomes an *entering* basic variable). This is the nonbasic variable that would increase the objective function at the fastest rate. For example, in the first block of Table 14.5 (blocks are separated by single horizontal lines), the value of -200 is the largest negative number in Eq. (0); therefore, p_0 is selected as the entering basic variable.

The basic variable that is replaced by the entering basic variable is called the *leaving* basic variable. The leaving basic variable is determined by dividing y (right-hand side) by the positive coefficients in the column containing the entering basic variable; the row that yields the lowest value contains the leaving basic variable. This is the basic variable that reaches zero first as the entering basic variable is increased. In this example, for block 1 the lowest value (2.1/1, 2.1/1, 4.6/1, 4.6/1) lies in rows (1) and (2); either one of the rows can be chosen for selecting the leaving basic variable.

The column that contains the entering basic variable is called the pivot column and the row containing the leaving basic variable is the pivot row. As the basic variables must have a coefficient of $+1$, the entire pivot row is divided by the pivot number (placed at the intersection of the pivot row and pivot column). The new basic variable is now eliminated from all the other equations by the following formula:

$$\text{New row} = \text{old row} - (\text{pivot column coefficient}) \times \text{new pivot row}$$

Therefore, row (0) of the second block is

$$
\begin{array}{rrrrrrrrrrrr}
& [-99 & -92.1 & -200 & -188 & 0 & 100 & 0 & 0 & 100 & 0 & -670] \\
-(-200) \times & [\ 0.5 & 0.26 & 1 & 0.52 & 0 & -1 & 1 & 0 & 0 & 0 & 2.1] \\
\text{New row} = & [\ 1 & -40.1 & 0 & -84 & 0 & -100 & 200 & 0 & 100 & 0 & -250]
\end{array}
$$

Similar calculations are conducted on all the other equations of the first block.

Row (1) (row (2) is the pivot row):

$$
\begin{array}{rrrrrrrrrrr}
& [-0.5 & -0.26 & 1 & 0.52 & 1 & 0 & 0 & 0 & 0 & 0 & 2.1] \\
-(1) \times & [\ 0.5 & 0.26 & 1 & 0.52 & 0 & -1 & 1 & 0 & 0 & 0 & 2.1] \\
\text{New row} = & [-1 & -0.52 & 0 & 0 & 1 & 1 & -1 & 0 & 0 & 0 & 0\ ]
\end{array}
$$

Row (3):

$$
\begin{array}{rrrrrrrrrrr}
& [-0.5 & -0.68 & 1 & 1.36 & 0 & 0 & 0 & 1 & 0 & 0 & 4.6] \\
-(1) \times & [0.5 & 0.26 & 1 & 0.52 & 0 & -1 & 1 & 0 & 0 & 0 & 2.1] \\
\text{New row} = & [-1 & -0.94 & 0 & 0.84 & 0 & 1 & -1 & 1 & 0 & 0 & 2.5]
\end{array}
$$

Row (4):

$$
\begin{array}{rrrrrrrrrrr}
& [0.5 & 0.68 & 1 & 1.36 & 0 & 0 & 0 & 0 & -1 & 1 & 4.6] \\
-(1) \times & [0.5 & 0.26 & 1 & 0.52 & 0 & -1 & 1 & 0 & 0 & 0 & 2.1] \\
\text{New row} = & [0 & 0.42 & 0 & 0.84 & 0 & 1 & -1 & 0 & -1 & 1 & 2.5]
\end{array}
$$

In Table 14.5 the leaving basic variable in each block of calculations is identified with an asterisk.

From the last column in the final block of Table 14.6, we see that $A_0 = (p_0, c_0) = (0.55, 0)$ and $A_1 = (p_1, c_1) = (2.97, 0)$. Substituting these values into $Y = A_0 + A_1 x$ yields

$$Y = (0.55, 0) + (2.97, 0) \times 0.52 = 2.09$$

which is essentially (to within computational error) the same as the actual y value (i.e., 2.1).

TABLE 14.4
First step in the simplex method

Equation	c_0	c_1	p_0	p_1	D_1	D_2	$\overline{D_3}$	D_4	D_5	$\overline{D_6}$	y
Eq. (0)	1	1.88	0	0	0	0	100	0	0	100	0
Eq. (2)*M	50	26	100	52	0	−100	100	0	0	0	210
Eq. (4)*M	50	68	100	136	0	0	0	0	−100	100	460
Eq. (0)−Eq. (2)−Eq. (4)	−99	−92.1	−200	−188	0	100	0	0	100	0	−670

TABLE 14.5
Calculations of the simplex method

Basic variable	Equation	O	c_0	c_1	p_0 (pivot column)	p_1	D_1	D_2	$\overline{D_3}$ (pivot row)	D_4	D_5	$\overline{D_6}$	y
Block 1													
O	(0)	−1	−99	−92.1	−200	188	0	100	0	0	100	0	−670
D_1	(1)	0	−0.5	−0.26	1	0.52	1	0	0	0	0	0	2.1
*$\overline{D_3}$	(2)	0	0.5	0.26	1	0.52	0	−1	1	0	0	0	2.1
D_4	(3)	0	−0.5	−0.68	1	1.36	0	0	0	1	0	0	4.6
$\overline{D_6}$	(4)	0	0.5	0.68	1	1.36	0	0	0	0	−1	1	4.6
Block 2													
O	(0)	−1	1	−40.1	0	−84	0	−100	200	0	100	0	−250
*D_1	(1)	0	−1	−0.52	0	0	1	1	−1	0	0	0	0
p_0	(2)	0	0.5	0.26	1	0.52	0	−1	1	0	0	0	2.1
D_4	(3)	0	−1	−0.94	0	0.84	0	1	−1	1	0	0	2.5
$\overline{D_6}$	(4)	0	0	0.42	0	0.84	0	1	−1	0	−1	1	2.5
Block 3													
O	(0)	−1	−99	−90.1	0	−84	100	0	100	0	100	0	−250
D_2	(1)	0	−1	−0.52	1	0	1	1	−1	0	0	0	0
p_0	(2)	0	−0.5	−0.26	0	0.52	1	0	0	0	0	0	2.1
D_4	(3)	0	0	−0.42	0	0.84	−1	0	0	1	0	0	2.5
*$\overline{D_6}$	(4)	0	1	0.94	0	0.84	−1	0	0	0	−1	1	2.5
Block 4													
O	(0)	−1	0	0.94	0	−0.84	1	0	100	0	1	99	−2.5
*D_2	(1)	0	0	0.42	0	0.84	0	1	−1	0	−1	1	2.5
p_0	(2)	0	0	0.21	1	0.94	0.5	0	0	0	−0.5	0.5	3.35
D_4	(3)	0	0	−0.42	0	0.84	−1	0	0	1	0	0	2.5
c_0	(4)	0	1	0.94	0	0.84	−1	0	0	0	−1	1	2.5

*Leaving variable

In Example 14.5 the fuzziness in the coefficients is 0; we should get the exact solution because we only had two points in the data set. For nontrivial data sets this outcome is not the case, and the coefficients turn out to be fuzzy sets as represented by a triangular membership function of the form expressed in Fig. 14.12. The following example illustrates this idea.

TABLE 14.6
Final block of simplex calculations

Basic variable	Equation	Coefficient of											
		O	c_0	c_1	p_0	p_1	D_1	D_2	$\overline{D}_3$	D_4	D_5	$\overline{D}_6$	y
O	(0)	-1	0	3.29	0	0	1	1	99	0	0	2	0
p_1	(1)	0	0	0.5	0	1	0	1.2	-1.2	0	-1.2	1.2	2.97
p_0	(2)	0	0	-0.26	1	0	0.5	-1.1	1.1	0	0.62	-0.44	0.55
D_4	(3)	0	0	-0.84	0	0	-1	-1	1	1	1	-1	0
c_0	(4)	0	1	0.52	0	0	-1	-1	1	0	0	0	0

TABLE 14.7
Five data samples

y	x_1	x_2
3.54	0.84	0.46
4.05	0.65	0.52
4.51	0.76	0.57
2.63	0.7	0.3
1.9	0.73	0.2

Example 14.7 **[Kikuchi and Nanda, 1991].** Consider the data set given in Table 14.7. In this case there are five data points; hence there will be $2 \times 5 = 10$ constraints.

The fuzzy linear regression equation, $\underset{\sim}{Y} = \underset{\sim}{A}_0 + \underset{\sim}{A}_1 x_1 + \underset{\sim}{A}_2 x_2$ is used to fit the data set. The objective function to be minimized is

$$O = 5c_0 + \sum_j x_{1j}c_1 + \sum_j x_{2j}c_2 + M\overline{D}_3 + M\overline{D}_6 + M\overline{D}_9 + M\overline{D}_{12} + M\overline{D}_{15}$$

$$O = 5c_0 + 3.68c_1 + 2.05c_2 + M\overline{D}_3 + M\overline{D}_6 + M\overline{D}_9 + M\overline{D}_{12} + M\overline{D}_{15}$$

Using an h value of 0.5 and Eqs. (14.35)–(14.36) for each of the m data points, we get the following constraint equations:

$$3.54 = p_0 + 0.84p_1 + 0.46p_2 - 0.5[c_0 + 0.84c_1 + 0.46c_2] + D_1 \tag{1}$$

$$3.54 = p_0 + 0.84p_1 + 0.46p_2 + 0.5[c_0 + 0.84c_1 + 0.46c_2] - D_2 + \overline{D}_3 \tag{2}$$

$$4.05 = p_0 + 0.65p_1 + 0.52p_2 - 0.5[c_0 + 0.65c_1 + 0.52c_2] + D_4 \tag{3}$$

$$4.05 = p_0 + 0.65p_1 + 0.52p_2 + 0.5[c_0 + 0.65c_1 + 0.52c_2] - D_5 + \overline{D}_6 \tag{4}$$

$$4.51 = p_0 + 0.76p_1 + 0.57p_2 - 0.5[c_0 + 0.76c_1 + 0.57c_2] + D_7 \tag{5}$$

$$4.51 = p_0 + 0.76p_1 + 0.57p_2 + 0.5[c_0 + 0.76c_1 + 0.57c_2] - D_8 + \overline{D}_9 \tag{6}$$

$$2.63 = p_0 + 0.7p_1 + 0.3p_2 - 0.5[c_0 + 0.7c_1 + 0.3c_2] + D_{10} \tag{7}$$

$$2.63 = p_0 + 0.7p_1 + 0.3p_2 + 0.5[c_0 + 0.7c_1 + 0.3c_2] - D_{11} + \overline{D}_{12} \tag{8}$$

$$1.9 = p_0 + 0.73p_1 + 0.2p_2 - 0.5[c_0 + 0.73c_1 + 0.2c_2] + D_{13} \tag{9}$$

$$1.9 = p_0 + 0.73p_1 + 0.2p_2 + 0.5[c_0 + 0.73c_1 + 0.2c_2] - D_{14} + \overline{D}_{15} \tag{10}$$

As explained in Example 14.6, the final objective function equation is obtained by considering the equations containing the artificial variables:

$$-1(O) - 1.5M(c_0) - 1.84M(c_1) - 1.025M(c_2) - 5M(p_0) - 3.68M(p_1)$$

$$-2.05M(p_2) + 0(D_1) + (1)M(D_2) + 0(\overline{D}_3) + 0(D_4) + (1)M(D_5) + 0(\overline{D}_6)$$

$$+0(D_7) + (1)M(D_8) + 0(\overline{D}_9) + 0(D_{10}) + (1)M(D_{11}) + 0(\overline{D}_{12}) + 0(D_{13}) \tag{0}$$

$$+(1)M(D_{14}) + 0(\overline{D}_{15}) - 16.63M(y) = 0$$

where the zeros in expression (0) are coefficients on the many slack variables that do not appear in the objective equation. The same solution procedure as in Example 14.6 is used to determine the parameters of the coefficients of the fuzzy linear regression model. The fuzzy coefficients computed [Kikuchi and Nanda, 1991] are

$$\underset{\sim}{A}_0 = (1.242, 0); \qquad \underset{\sim}{A}_1 = (0, 1.4); \qquad \underset{\sim}{A}_2 = (5.843, 0)$$

We see that coefficient $\underset{\sim}{A}_1$ is fuzzy because it has a nonzero spread. It can be determined that as h increases, the fuzziness of the output increases (see Problem 14.8 at the end of the chapter).

We conclude this section with a few comments on fuzzy regression. First, the fact that the estimated value for the output variable, y, is given as a fuzzy number represents a drawback when outlier points exist in the data set. The presence of an outlier point makes the spread of the estimate very large since the estimate must cover that point at least to the level of confidence, h. However, Kikuchi and Nanda [1991] have shown that, although the spread of the fuzzy numbers may become large in fuzzy regression, the prototypical values (modal values) of the estimates remain relatively stable. Second, if a high value of h is given, the spread of the estimate of the output, y, increases in order to satisfy the increased measure of goodness of fit. Third, each data point requires two constraint equations; see Eqs. (14.35)–(14.36) or (14.42)–(14.43). The first of each of these equation pairs represents the case when y_j lies in the interval to the left of the prototypical value and the latter represents the case when y_j lies in the interval to the right of the prototypical value. For m data sets this means $2m$ constraints; for large data sets the computational load associated with these equations can become a deterrent to this regression method.

There are, of course, alternatives to the solution of fuzzy regression equations using a linear programming approach. For example, Tanaka et al. [1982] solve the *dual problem* expressed by Eqs. (14.42)–(14.43). Here the number of equations is related to the number of independent variables, n, as opposed to the number of constraints, $2N$. This relation makes the computational burden significantly lower than with conventional linear programming methods for situations where $N \geq n$. In addition to the problem of $N \geq n$, when n changes the entire set of constraints has to be reformulated; this characteristic also limits the utility of the fuzzy regression method. Any linear programming formulation requires that all the unknown variables must be positive, as demonstrated in the examples here. Hence, a prior knowledge of the effect of each variable on the outcome is useful. If ambiguity in the sign of a variable is a feature of the problem, the unknown value must be presented as a linear combination of two positive numbers – yet another growth in the number of equations. To overcome these difficulties, some investigators are using other methods to solve fuzzy regression problems, such as artificial neural networks [Kikuchi and Nanda, 1991].

SUMMARY

This chapter summarizes fuzzy logic applications in the areas of optimization, cognitive mapping, system identification, and regression. This only begins to scratch the surface of the plethora of applications being developed in the rapidly expanding field of fuzzy logic. The reader is referred to the literature for many other applications projects, which are summarized in such works or collected bibliographies as Schmucker [1984], Klir and Folger [1988], McNeill and Freiberger [1993], Kosko [1993], Dubois and Prade [1980], Ross et al. [2003], and Cox [1994], or discussed in some of the active international research journals focusing on fuzzy applications such as *Intelligent and Fuzzy Systems*, *Fuzzy Sets and Systems*, and *IEEE Transactions on Fuzzy Systems*.

REFERENCES

Akai, T. (1994). *Applied numerical methods for engineers*, John Wiley & Sons, New York, chapter 10.

Axelrod, R. (1976). *Structure of Decision*, Princeton University Press, Princeton, NJ.

Cox, E. (1994). *The fuzzy systems handbook*, Academic Press Professional, Cambridge, MA.

deNeufville, R. (1990). *Applied systems analysis: engineering planning and technology management*, McGraw-Hill, New York.

Dubois, D. and Prade, H. (1980). *Fuzzy sets and systems: Theory and applications*, Academic Press, New York.

Fuzzy Sets and Systems. C. Negoita, L. Zadeh, and H. Zimmerman (eds.), Elsevier Science, Amsterdam.

Goldberg, D. (1989). *Genetic algorithms in search, optimization and machine learning*, Addison-Wesley, Reading, MA.

Hillier, F. and Lieberman, G. (1980). *Introduction to operations research*, Holden-Day, San Francisco.

IEEE Transactions on Fuzzy Systems. IEEE Press, New York.

Intelligent and Fuzzy Systems. R. Langari (ed.) IOS Press, Amsterdam.

Kardaras, D. and Karakostas, B. (1999). ''The use of fuzzy cognitive maps to simulate the information systems strategic planning process,'' *Inf. Software Technol.*, vol. 41, pp. 197–210.

Kikuchi, S. and Nanda, R. (1991). ''Fuzzy regression analysis using a neural network: Application to trip generation model,'' unpublished manuscript.

Klir, G. and Folger, T. (1988). *Fuzzy sets, uncertainty, and information*, Prentice Hall, Englewood Cliffs, NJ.

Kosko, B. (1986). ''Fuzzy cognitive maps,'' *Int. J. Man-Mach. Stud.*, vol. 24, pp. 65–75.

Kosko, B. (1992). *Neural Networks and Fuzzy Systems*, Prentice-Hall, Englewood Cliff, NJ.

Kosko, B. (1993). *Fuzzy thinking*, Hyperion Press, New York.

Kosko, B. (1997). *Fuzzy Engineering*, Prentice Hall, Upper Saddle River, NJ.

McNeill, D. and Freiberger, P. (1993). *Fuzzy logic*, Simon & Schuster, New York.

Pelaez, C. E. and Bowles, J. B. (1995). ''Applying fuzzy cognitive-maps knowledge-representation to failure modes effects analysis,'' *Proceedings Annual Reliability and Maintainability Symposium*, IEEE, New York.

Pelaez, C. E. and Bowles, J. B. (1996). ''Using fuzzy cognitive maps as a system model for failure modes and effects analysis,'' *Inf. Sci.*, vol. 88, pp. 177–199.

Ross, T., Booker, J., and Parkinson, J. (2003). *Fuzzy Logic and Probability Applications: Bridging the Gap*, Society for Industrial and Applied Mathematics, Philadelphia, PA.

Sakawa, M. (1993). *Fuzzy sets and interactive multiobjective optimization*, Plenum Press, New York.

Schmucker, K. (1984). *Fuzzy sets, natural language computations, and risk analysis*, Computer Science Press, Rockville, MD.

Taber, R. (1994). "Fuzzy cognitive maps model social systems," *AI Expert*, July, pp. 19–23.

Tanaka, H., Uejima, S., and Asai, K. (1982). "Linear regression analysis with fuzzy model," *IEEE Trans. Syst., Man, Cybern.*, vol. 12, no. 6, pp. 903–907.

Terano, T., Asai, K., and Sugeno, M. (1992). *Fuzzy system theory and its applications*, Academic Press, San Diego.

Tsadiras, A. K. and Margaritis, K. G. (1996). "Using certainty neurons in fuzzy cognitive maps," *Neural Network World*, vol. 4, pp. 719–728.

Tsukamoto, Y. and Terano, T. (1977). "Failure diagnosis by using fuzzy logic," *Proceedings of the IEEE Conference an Decision Control, New Orleans*. vol. 2, pp. 1390–1395.

Vanderplaats, G. (1984). *Numerical optimization techniques for engineering design with applications*, McGraw-Hill, New York.

Zadeh, L. (1972). "On fuzzy algorithms," Memo UCB/ERL M-325, University of California, Berkeley.

PROBLEMS

Fuzzy Optimization

14.1. The feedforward transfer function for a unit-feedback control system is $1/(s-1)$, as shown in Fig. P14.1. A unit step signal is input to the system. Determine the minimum error of the system response by using a fuzzy optimization method for the time period, $0 < t < 10$ s, and a fuzzy constraint given by

$$u_c(t) = \begin{cases} 1, & 0 \le t \le 1 \\ e^{1-t}, & t > 1 \end{cases}$$

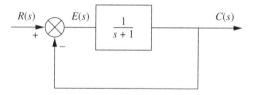

FIGURE P14.1

14.2. A beam structure is forced by an axial load P (Fig. P14.2). When P is increased to its *critical* value, the beam will buckle. Prove that the critical force P to cause buckling can be expressed by a function

$$P = \frac{n^2 \pi^2 E I}{L}$$

where EI = stiffness of the beam

L = span length of the beam

n = number of sine waves the beam shape takes when it buckles (assume it to be continuous)

If $0 \le n \le 2$, assume that n is constrained by the fuzzy member function

$$u_c(n) = \begin{cases} 1 - n, & 1 \le n \le 2 \\ 0, & n < 1 \end{cases}$$

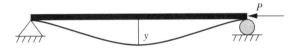

FIGURE P14.2

14.3. Suppose that the beam structure in Problem 14.2 also has a transverse load P applied at the middle of the beam. Then the maximum bending stress can be calculated by the equation $\sigma_b = Pl/4w_z$, where w_z, with units m^3, is a coefficient based on the shape and size of the cross section of the beam, and l is in meters. If $0 \le \sigma_b \le 60$ MPa, and the fuzzy constraint function for σ_b is

$$\mu_c(\sigma_b) = \begin{cases} \dfrac{1}{(x+1)^2}, & 0 \le x \le 1 \\ 0, & x > 1 \end{cases}$$

where $x = \dfrac{\sigma}{60 \text{MPa}}$

combine the conditions given in Problem 14.2 to find the optimum load P, in newtons.
Hint: This problem involves multiple constraints.

14.4. In the metallurgical industry, the working principle for a cold rolling mill is to extrude a steel strip through two rows of working rollers, as shown in Fig. P14.4. The size of the roller is very important. The stress between the roller and the strip can be expressed by the following function:

$$\sigma_H = 0.564 \sqrt{\dfrac{PE}{LR}}$$

where $E =$ Young's modulus (kN/cm^2)
$P =$ loading force (N)
$L =$ contact length between roll and strip (cm)
$R =$ radius of a roller (cm)

If $\sigma_H = 2.5$ kN/cm^2 and $10 < R < 20$, find the minimum R in which σ_H has a maximum value. The radius R has a fuzzy constraint of

$$u_c(R) = \begin{cases} 1, & 10 \le R \le 15 \\ \dfrac{20-R}{5}, & 15 < R \le 20 \end{cases}$$

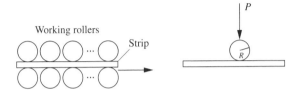

FIGURE P14.4

System Identification

14.5. In a fuzzy relation equation, $\underset{\sim}{A} \circ \underset{\sim}{R} = \underset{\sim}{B}$, r_i is known as $(0.5, 0.7, 0.9)$ and b_i is 0.6. Use the Tsukamoto method for an inverse fuzzy equation to find a_i $(i = 1, 2, 3)$.

14.6. A fuzzy relation has an expression given as

$$\{a_1, a_2\} \circ \begin{bmatrix} 0.4 & 0.6 \\ 0.8 & 0.1 \end{bmatrix} = [0.3 \quad 0.1]$$

Find a_i ($i = 1, 2$) by using the Tsukamoto method for inverse fuzzy relations.

14.7. A system having a single degree of freedom has the following ordinary differential governing equation:

$$a_1 \ddot{x} + a_2 \dot{x} + a_3 x = b$$

If b is the input signal, in which $b_1 = 0.5$ and $b_2 = 0.6$, and two sets of response data of x are (0.4, 0.6, 0.8) and (0.5, 0.7, 0.9), respectively, find the system coefficients a_i ($i = 1, 2, 3$).

Regression

14.8. Show that as the parameter h increases, the fuzziness of the output increases in the five-point regression problem (Example 14.7) [Kikuchi and Nanda, 1991]. Use the simplex method for values of $h = 0.2$ and 0.8.

14.9. Risk assessment is fast becoming the basis of many EPA guidelines that determine whether a site contaminated with hazardous substances needs to be remediated or not. Risk is defined as the likelihood of an adverse health impact to the public due to exposure to environmental hazards. Risk assessment consists of four parts: (1) hazard identification; (2) dose–response assessment – assessing the health response to a certain dose (concentration) of the chemical; (3) exposure assessment – assessing the duration and concentration of exposure; and (4) risk characterization – quantification and presentation of risks. Part (2) of the risk assessment process consists of exposing a controlled population of animals to various doses of the chemical and fitting a dose–response curve to the experimental data. The following table comprises the data derived from the tests:

Dose (mg/L)	Response
0	0
2	0.02
5	1.0
10	2.3

Assuming that the dose–response relationship can be expressed by a fuzzy linear regression model, $\underset{\sim}{Y} = \underset{\sim}{A_0} + \underset{\sim}{A_1} x$, where x represents the dose in mg/L, determine the fuzzy coefficients $\underset{\sim}{A_0}$ and $\underset{\sim}{A_1}$. Use an h value of 0.5.

14.10. In a survey on costs for the construction of new houses, the number of rooms (including bedrooms, kitchen, bathroom, and living rooms) in a house was compared with the material costs of the house. The following table gives the results of the survey:

Cost ($)	Number of rooms
10,000	2
25,000	5
100,000	8

Assuming an h of 0.5, determine the coefficients of the fuzzy one-dimensional linear regression model, $\underset{\sim}{Y} = \underset{\sim}{A_0} + \underset{\sim}{A_1} x$.

14.11. In fuzzy regression, the output y is a triangular fuzzy number with the spread e_j representing the error in measurement (Fig. P14.11).

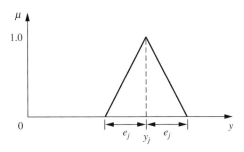

FIGURE P14.11

The accompanying table shows the fuzzy output and the corresponding crisp input:

y, e	x
2.1, 0.2	0.52
4.6, 0.35	1.36

Using Eqs. (14.41)–(14.43) for fuzzy output, and $h = 0.4$, determine the fuzzy coefficients for a simple fuzzy linear regression model, $\underset{\sim}{Y} = \underset{\sim}{A}_0 + \underset{\sim}{A}_1 x$.

Cognitive Mapping

14.12. For the information in Example 14.3 find the stabilized state vector corresponding to an initial state vector of [0, 1, 0, 0, 0].

14.13. For the information in Example 14.3 find the stabilized state vector corresponding to a fuzzy linguistic effect on the path from C_1 to C_2 that is "Much."

14.14. For the information in Example 14.3 find the stabilized state vector corresponding to an initial state vector of [0, 0, 1, 0, 0] and a fuzzy linguistic effect on the path from C_2 to C_3 that is "Some."

CHAPTER
15

MONOTONE MEASURES: BELIEF, PLAUSIBILITY, PROBABILITY, AND POSSIBILITY

... whenever you find yourself getting angry about a difference of opinion, be on your guard; you will probably find, on examination, that your belief is getting beyond what the evidence warrants.

Bertrand Russell
British philosopher and Nobel Laureate Unpopular Essays, 1923

A bag contains 2 counters, as to which nothing is known except that each is either black or white. Ascertain their colours without taking them out of the bag.

Louis Carroll
Author and mathematician Pillow Problems, 1895

Most of this text has dealt with the quantification of various forms of non-numeric uncertainty. Two prevalent forms of uncertainty are those arising from vagueness and from imprecision. How do vagueness and imprecision differ as forms of uncertainty? Often vagueness and imprecision are used synonymously, but they can differ in the following sense. Vagueness can be used to describe certain kinds of uncertainty associated with linguistic information or intuitive information. Examples of vague information are that the image quality is "good," or that the transparency of an optical element is "acceptable." Imprecision can be associated with quantitative or countable data as well as noncountable

Fuzzy Logic with Engineering Applications, Second Edition   T. J. Ross
© 2004 John Wiley & Sons, Ltd   ISBNs: 0-470-86074-X (HB); 0-470-86075-8 (PB)

data. As an example of the latter, one might say the length of a bridge span is "long." An example of countable imprecision would be to report the length to be 300 meters. If we take a measuring device and measure the length of the bridge 100 times we likely will come up with 100 different values; the differences in the numbers will no doubt be on the order of the precision of the measuring device. Measurements using a 10 meter chain will be less precise than those developed from a laser theodolite. If we plot the bridge lengths on some sort of probit paper and develop a Gaussian distribution to describe the length of this bridge, we could state the imprecision in probabilistic terms. In this case the length of the bridge is uncertain to some degree of precision that is quantified in the language of statistics. Since we are not able to make this measurement an infinite number of times, there is also uncertainty in the statistics describing the bridge length. Hence, imprecision can be used to quantify random variability in quantitative uncertainty and it can also be used to describe a lack of knowledge for descriptive entities (e.g., acceptable transparency, good image quality). Vagueness is usually related to nonmeasurable issues.

This chapter develops the relationships between probability theory and evidence theory; to a limited extent it also shows the relationship between a possibility theory, founded on crisp sets, and a fuzzy set theory. All of these theories are related under an umbrella theory termed monotone measures [see Klir and Smith, 2001] (which was termed fuzzy measure theory for a couple decades despite the confusion this generates when we try to distinguish other theories from fuzzy set theory); all of these theories have been used to characterize and model various forms of uncertainty. That they are all related mathematically is an especially crucial advantage in their use in quantifying the uncertainty spectrum because, as more information about a problem becomes available, the mathematical description of uncertainty can easily transform from one theory to the next in the characterization of the uncertainty. This chapter begins by developing monotone measures as an overarching framework for the other theories used to characterize various forms of uncertainty. The development continues with specific forms of monotone measures such as belief, plausibility, possibility, and probability. The chapter illustrates a new method in developing possibility distributions from empirical data, and it briefly describes a special kind of relationship between a possibility distribution and a fuzzy set. Examples are provided to illustrate the various theories.

MONOTONE MEASURES

A monotone measure describes the vagueness or imprecision in the assignment of an element a to two or more crisp sets. Figure 15.1 shows this idea. In the figure the universe of discourse comprises a collection of sets and subsets, or the power set. In a monotone measure what we are trying to describe is the vagueness or imprecision in assigning this point to any of the crisp sets on the power set. This notion is not random; the crisp sets have no uncertainty about them. The uncertainty is about the assignment. This uncertainty is usually associated with evidence to establish an assignment. The evidence can be completely lacking – the case of total ignorance – or the evidence can be complete – the case of a probability assignment. Hence, the difference between a monotone measure and a fuzzy set on a universe of elements is that, in the former, the imprecision is in the assignment of an element to one of two or more crisp sets, and in the latter the imprecision is in the prescription of the boundaries of a set.

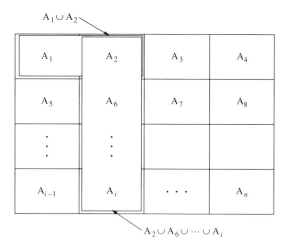

FIGURE 15.1
A monotone measure.

BELIEF AND PLAUSIBILITY

There are special forms of monotone measures. A form associated with preconceived notions is called a belief measure. A form associated with information that is possible, or plausible, is called a plausibility measure. Specific forms of belief measures and plausibility measures are known as certainty and possibility measures, respectively. The intersection of belief measures and plausibility measures (i.e., where belief equals plausibility) will be shown to be a probability. Monotone measures are defined by weaker axioms than probability theory, thus subsuming probability measures as specific forms of monotone measures.

Basically, a belief measure is a quantity, denoted bel(A), that expresses the degree of support, or evidence, for a collection of elements defined by one or more of the crisp sets existing on the power set of a universe. The plausibility measure of this collection A is defined as the "complement of the Belief of the complement of A," or

$$pl(A) = 1 - bel(\overline{A}) \tag{15.1}$$

Since belief measures are quantities that measure the degree of support for a collection of elements or crisp sets in a universe, it is entirely possible that the belief measure of some set A plus the belief measure of $\overline{A}$ will not be equal to unity (the total belief, or evidence, for all elements or sets on a universe is equal to 1, by convention). When this sum equals 1, we have the condition where the belief measure is a probability; that is, the evidence supporting set A can be described probabilistically. The difference between the sum of these two quantities (bel(A) + bel($\overline{A}$)) and 1 is called the ignorance, i.e., ignorance $= 1 - [bel(A) + bel(\overline{A})]$. When the ignorance equals 0 we have the case where the evidence can be described by probability measures.

Say we have evidence about a certain prospect in our universe of discourse, evidence of some set occurring or some set being realized, and we have no evidence (zero evidence) of the complement of that event. In probability theory we must assume, because of the excluded middle axioms, that if we know the probability of A then the probability of $\overline{A}$ is also known, because we have in all cases involving probability measures, $\text{prob}(A) + \text{prob}(\overline{A}) = 1$. This constraint of the excluded middle axioms is not a requirement in evidence theory. The probability of $\overline{A}$ also has to be supported with some sort of evidence. If there is no evidence (zero degree of support) for $\overline{A}$ then the degree of ignorance is large. This distinction between evidence theory and probability theory is important. It will also be shown that this is an important distinction between fuzzy set theory and probability theory (see Appendix A).

Monotone measures are very useful in quantifying uncertainty that is difficult to measure or that is linguistic in nature. For example, in assessing structural damage in buildings and bridges after an earthquake or hurricane, evidence theory has proven quite successful because what we have are nonquantitative estimates from experts; the information concerning damage is not about how many inches of displacement or microinches per inch of strain the structure might have undergone, but rather is about expert judgment concerning the suitability of the structure for habitation or its intended function. These kinds of judgments are not quantitative; they are qualitative.

The mathematical development for monotone measures follows [see, for example, Klir and Folger, 1988]. We begin by assigning a value of membership to each crisp set existing in the power set of a universe, signifying the degree of evidence or belief that a particular element from the universe, say x, belongs in any of the crisp sets on the power set. We will label this membership $g(A)$, where it is a mapping between the power set and the unit interval,

$$g : P(X) \longrightarrow [0, 1] \tag{15.2}$$

and where $P(X)$ is the power set of all crisp subsets on the universe, X (see Chapter 2). So, the membership value $g(A)$ represents the degree of available evidence of the belief that a given element x belongs to a crisp subset A.

The collection of these degrees of belief represents the *fuzziness* associated with several *crisp* alternatives. This type of uncertainty, which we call a monotone measure, is *different* from the uncertainty associated with the boundaries of a single set, which we call a fuzzy set. Monotone measures are defined for a *finite* universal set by at least three axioms, two of which are given here (a third axiom is required for an infinite universal set):

1. $g(\emptyset) = 0, \qquad g(X) = 1$

2. $g(A) \le g(B)$ for A, B $\in P(X), A \subseteq B$ \tag{15.3}

The first axiom represents the boundary conditions for the monotone measure, $g(A)$. It says that there is no evidence for the null set and there is complete (i.e., unity) membership for the universe. The second axiom represents monotonicity by simply stating that if one set A is completely contained in another set B then the evidence supporting B is at least as great as the evidence supporting the subset A.

A belief measure also represents a mapping from the crisp power set of a universe to the unit interval representing evidence, denoted

$$\text{bel} : P(X) \longrightarrow [0, 1] \tag{15.4}$$

Belief measures can be defined by adding a third axiom to those represented in Eq. (15.3), given by

$$\text{bel}(A_1 \cup A_2 \cup \cdots \cup A_n) \geq \sum_i \text{bel}(A_i) - \sum_{i<j} \text{bel}(A_i \cap A_j) + \cdots$$

$$+ (-1)^{n+1}\text{bel}(A_1 \cap A_2 \cap \cdots \cap A_n) \qquad (15.5)$$

where there are n crisp subsets on the universe X. For each crisp set $A \in P(X)$, bel(A) is the degree of belief in set A based on available evidence. When the sets A_i in Eq. (15.5) are pairwise disjoint, i.e., where $A_i \cap A_j = \emptyset$, then Eq. (15.5) becomes

$$\text{bel}(A_1 \cup A_2 \cup \cdots \cup A_n) \geq \text{bel}(A_1) + \text{bel}(A_2) + \cdots + \text{bel}(A_n) \qquad (15.6)$$

For the special case where $n = 2$, we have two disjoint sets A and $\overline{A}$, and Eq. (15.6) becomes

$$\text{bel}(A) + \text{bel}(\overline{A}) \leq 1 \qquad (15.7)$$

A plausibility measure is also a mapping on the unit interval characterizing the total evidence, i.e.,

$$\text{pl} : P(X) \longrightarrow [0, 1] \qquad (15.8)$$

Plausibility measures satisfy the basic axioms of monotone measures, Eq. (15.3), and one additional axiom (different from Eq. (15.5) for beliefs),

$$\text{pl}(A_1 \cap A_2 \cap \cdots \cap A_n) \leq \sum_i \text{pl}(A_i) - \sum_{i<j} \text{pl}(A_i \cup A_j) + \cdots$$

$$+ (-1)^{n+1}\text{pl}(A_1 \cup A_2 \cup \cdots \cup A_n) \qquad (15.9)$$

From Eq. (15.1) we have a mutually dual system between plausibility and belief [see Shafer, 1976],

$$\text{pl}(A) = 1 - \text{bel}(\overline{A})$$

$$\text{bel}(A) = 1 - \text{pl}(\overline{A}) \qquad (15.10)$$

For the specific case of $n = 2$, i.e., for two disjoint sets A and $\overline{A}$, Eq. (15.10) produces

$$\text{pl}(A) + \text{pl}(\overline{A}) \geq 1 \qquad (15.11)$$

By combining Eq. (15.7) and Eq. (15.10) it can be shown that

$$\text{pl}(A) \geq \text{bel}(A) \qquad (15.12)$$

Equation (15.12) simply states that for whatever evidence supports set A, its plausibility measure is always at least as great as its belief measure.

We now define another function on the crisp sets (A ∈ P(X)) of a universe, denoted $m(A)$, which can be used to express and determine both belief and plausibility measures. This measure is also a mapping from the power set to the unit interval,

$$m : P(X) \longrightarrow [0, 1] \tag{15.13}$$

This measure, called a *basic evidence assignment (bea)* has been termed a basic probability assignment (bpa) before in the literature, and has boundary conditions

$$m(\varnothing) = 0 \tag{15.14}$$

$$\sum_{A \in P(X)} m(A) = 1 \tag{15.15}$$

The measure $m(A)$ is the degree of belief that a specific element, x, of the universe X belongs to the set A, *but not to any specific subset of* A. In this way $m(A)$ differs from both beliefs and plausibility. It's important to remark here, to avoid confusion with probability theory, that there is a distinct difference between a *basic evidence assignment (bea)* and a probability density function (pdf). The former are defined on sets of the power set of a universe (i.e., on $A \in P(X)$), whereas the latter are defined on the singletons of the universe (i.e., on $x \in P(X)$). This difference will be reinforced through some examples in this chapter. To add to the jargon of the literature, the first boundary condition, Eq. (15.14), provides for a *normal bea*.

The bea is used to determine a belief measure by

$$\text{bel}(A) = \sum_{B \subseteq A} m(B) \tag{15.16}$$

In Eq. (15.16) note that $m(A)$ is the degree of evidence in set A *alone*, whereas bel(A) is the total evidence in set A *and* all subsets (B) of A. The measure $m(A)$ is used to determine a plausibility measure by

$$\text{pl}(A) = \sum_{B \cap A \neq \varnothing} m(B) \tag{15.17}$$

Equation (15.17) shows that the plausibility of an event A is the total evidence in set A plus the evidence in all sets of the universe that intersect with A (including those sets that are also subsets of A). Hence, the plausibility measure in set A contains all the evidence contained in a belief measure (bel(A)) plus the evidence in sets that intersect with set A. Hence, Eq. (15.12) is verified.

> **Example 15.1.** A certain class of short-range jet aircraft has had, for the shorter fuselage versions, a history of an oscillatory behavior described as *vertical bounce*. This is due to the in-flight flexing of the fuselage about two body-bending modes. Vertical bounce is most noticeable at the most forward and aft locations in the aircraft. An acceptable acceleration threshold of $\pm 0.1g$ has been set as the point at which aft lower-body vortex generators should be used to correct this behavior. In order to avoid the cost of instrumented flight tests, expert engineers often decide whether vertical bounce is present in the aircraft. Suppose an expert engineer is asked to assess the evidence in a particular plane for the following two conditions:
>
> 1. Are oscillations caused by other phenomena? (O)
> 2. Are oscillations characteristic of the vertical bounce? (B)

TABLE 15.1
Measures of evidence for aircraft bounce

Focal element, A_i	Expert		
	$m(A_i)$	$bel(A_i)$	$pl(A_i)$
Ø	0	0	0
O	0.4	0.4	0.8
B	0.2	0.2	0.6
O ∪ B	0.4	1	1

This universe is a simple one, consisting of the singleton elements O and B. The non-null (Eq. (15.14) reminds us that the null set contains no evidence, i.e., $m(\emptyset) = 0$) power set then consists simply of the two singletons and the union of these two, O ∪ B; including the null set there are $2^2 = 4$ elements in the power set. All the elements in the power set are called *focal elements*. Suppose the expert provides the measures of evidence shown in Table 15.1 for each of the focal elements (i.e., the expert gives $m(A_i)$, for $i = 1, \ldots, 4$). Note that the sum of the evidences in the $m(A)$ column equals unity, as required by Eq. (15.15). We now want to calculate the degrees of belief and plausibility for this evidence set. Using Eq. (15.16), we find

$$bel(O) = m(O) = 0.4 \quad \text{and} \quad bel(B) = m(B) = 0.2$$

as seen in Table 15.1. The singletons O and B have no other subsets in them. Using Eq. (15.16) we find

$$bel(O \cup B) = m(O) + m(B) + m(O \cup B) = 0.4 + 0.2 + 0.4 = 1$$

as seen in Table 15.1. Using Eq. (15.17), we find

$$pl(O) = m(O \cap O) + m(O \cap (O \cup B)) = 0.4 + 0.4 = 0.8$$

and

$$pl(B) = m(B \cap B) + m(B \cap (O \cup B)) = 0.2 + 0.4 = 0.6$$

since sets O and B both intersect with the set O ∪ B; and finally,

$$pl(O \cup B) = m((O \cup B) \cap O) + m((O \cup B) \cap B)$$
$$+ m((O \cup B) \cap (O \cup B)) = 0.4 + 0.2 + 0.4 = 1$$

since all sets in the power set intersect with (O ∪ B). These quantities are included in the fourth column of Table 15.1. Thus, the engineer *believes* the evidence supporting set O (other oscillations) is at least 0.4 and *possibly* as high as 0.8 (plausibility), and *believes* the evidence supporting set B (vertical bounce) is at least 0.2 and *possibly* as high as 0.6 (plausibility). Finally, the evidence supporting either of these sets (O ∪ B) is full, or complete (i.e., bel = pl = 1).

EVIDENCE THEORY

The material presented in the preceding section now sets the stage for a more complete assessment of evidence, called evidence theory [Shafer, 1976]. Suppose the evidence for certain monotone measures comes from more than one source, say two experts. Evidence obtained in the same context (e.g., for sets A_i on a universe X) from two independent sources (e.g., two experts) and expressed by two *beas* (e.g., m_1 and m_2) on some power set P(X) can be combined to obtain a joint *bea*, denoted m_{12}, using Dempster's rule of combined evidence [Dempster, 1967]. The procedure to combine evidence is given here in Eqs. (15.18) and (15.19):

$$m_{12}(A) = \frac{\sum_{B \cap C = A} m_1(B) \cdot m_2(C)}{1 - K} \quad \text{for A} \neq \emptyset \quad (15.18)$$

where the denominator is a normalizing factor such that

$$K = \sum_{B \cap C = \emptyset} m_1(B) \cdot m_2(C) \quad (15.19)$$

Dempster's rule of combination combines evidence in a manner analogous to the way in which joint probability density functions (pdfs) in probability theory are calculated from two independent marginal pdfs. We can define a *body of evidence*, then, as a pair (A, *m*) where A are sets with available evidence *m*(A).

Example 15.2. If a generator is to run untended, the external characteristic of the shunt machine may be very unsatisfactory, and that of a series even more so, since a source of constant potential difference supplying a varying load current is usually required. The situation is even less satisfactory if the load is supplied via a feeder with appreciable resistance, since this will introduce an additional drop in potential at the load end of the feeder. What is required is a generator with rising external characteristics, since this would counteract the effect of feeder resistance. Such a characteristic may be obtained from a compound generator. In a compound generator we can get variable induced electromotive force (emf) with increase of load current by arranging the field magnetomotive force (mmf). So, shunt and field windings are used in the generator as shown in Fig. 15.2. In this figure, R_a, R_c, R_f, and R_s are the armature, compound,

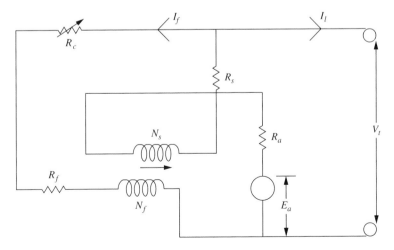

FIGURE 15.2
Electrical diagram of a compound generator.

field, and series resistance, respectively; I_f and I_l are the field and load current, respectively; E_a and V_t are the induced armature and terminal voltages, respectively; and N_f and N_s are the number of turns in the series and field windings, respectively. By arranging the field winding in different combinations (i.e., varying the difference between N_s and N_f), we can get different combinations of compound generators; in particular we can get (1) overcompounded (OC), (2) flatcompounded (FC), and (3) undercompounded (UC) generators. Each type of compounded generator has its own external and internal characteristics.

We can say that these three types comprise a universal set of generators, X. Let us consider two experts, E_1 an electrical engineer and E_2 a marketing manager, called to evaluate the efficiency and performance of a compound generator. We can come to some conclusion that, say for a particular outdoor lighting situation, the machine chosen by the two experts may be different for various reasons. On the one hand, the electrical engineer may think about performance issues like minimizing the error, maintaining constant voltage, and other electrical problems. On the other hand, the marketing manager may be concerned only with issues like minimizing the cost of running or minimizing maintenance and depreciation costs. In reality both experts may have valid reasons for the selection of a specific machine required for final installation.

Suppose that a company hires these two individuals to help it decide on a specific kind of generator to buy. Each expert is allowed to conduct tests or surveys to collect information (evidence) about the value of each of the three generators. The universe showing the individual sets of the power set is illustrated in Fig. 15.3. The focal elements of the universe in this figure are OC, FC, UC, OC ∪ FC, OC ∪ UC, FC ∪ UC, and OC ∪ FC ∪ UC (hereafter we ignore the null set in determining evidence since this set contains no evidence by definition, Eq. (15.14)), as listed in the first column of Table 15.2. Note that there are $2^3 - 1 = 7$ non-null elements. Suppose that the two experts, E_1 and E_2, give their information (evidence measures) about each focal element A_i, where $i = 1, 2, \ldots, 7$; that is, they provide $m_1(A_i)$ and $m_2(A_i)$, respectively (the second and fourth columns in Table 15.2). Note that the sum of the entries in the second and fourth columns equals unity, again guaranteeing Eq. (15.15).

Using this information, we can calculate the belief measures for each expert. For example,

$$\text{bel}_1(\text{OC} \cup \text{FC}) = \sum_{B \subseteq (\text{OC} \cup \text{FC})} m_1(B) = m_1(\text{OC} \cup \text{FC}) + m_1(\text{OC}) + m_1(\text{FC})$$

$$= 0.15 + 0 + 0.05 = 0.20$$

$$\text{bel}_2(\text{FC} \cup \text{UC}) = \sum_{B \subseteq (\text{FC} \cup \text{UC})} m_2(B) = m_2(\text{FC} \cup \text{UC}) + m_2(\text{FC}) + m_2(\text{UC})$$

$$= 0.20 + 0.15 + 0.05 = 0.40$$

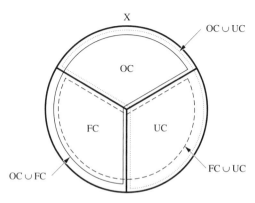

FIGURE 15.3
Universe X of compound generators.

TABLE 15.2
Focal elements and evidence for compound generators

Focal elements, A_i	Expert 1		Expert 2		Combined evidence	
	$m_1(A_i)$	$\text{bel}_1(A_i)$	$m_2(A_i)$	$\text{bel}_2(A_i)$	$m_{12}(A_i)$	$\text{bel}_{12}(A_i)$
OC	0	0	0	0	0.01	0.01
FC	0.05	0.05	0.15	0.15	0.21	0.21
UC	0.05	0.05	0.05	0.05	0.09	0.09
OC ∪ FC	0.15	0.20	0.05	0.20	0.12	0.34
OC ∪ UC	0.05	0.10	0.05	0.10	0.06	0.16
FC ∪ UC	0.10	0.20	0.20	0.40	0.20	0.50
OC ∪ FC ∪ UC	0.60	1	0.50	1	0.30	1

The remaining calculated belief values for the two experts are shown in the third and fifth columns of Table 15.2.

Using the Dempster rule of combination (Eqs. (15.18)–(15.19)), we can calculate the combined evidence (m_{12}) measures (calculations will be to two significant figures). First we must calculate the normalizing factor, K, using Eq. (15.19). In calculating this expression we need to sum the multiplicative measures of all those focal elements whose intersection is the null set, i.e., all those focal elements that are disjoint:

$$K = m_1(\text{FC})m_2(\text{OC}) + m_1(\text{FC})m_2(\text{UC}) + m_1(\text{FC})m_2(\text{OC} \cup \text{UC}) + m_1(\text{OC})m_2(\text{FC})$$

$$+ m_1(\text{OC})m_2(\text{UC}) + m_1(\text{OC})m_2(\text{FC} \cup \text{UC}) + m_1(\text{UC})m_2(\text{FC})$$

$$+ m_1(\text{UC})m_2(\text{OC}) + m_1(\text{UC})m_2(\text{FC} \cup \text{OC}) + m_1(\text{FC} \cup \text{OC})m_2(\text{UC})$$

$$+ m_1(\text{FC} \cup \text{UC})m_2(\text{OC}) + m_1(\text{OC} \cup \text{UC})m_2(\text{FC}) = 0.03$$

Hence $1 - K = 0.97$. So, for example, combined evidence on the set FC can be calculated using Eq. (15.18):

$$m_{12}(\text{FC}) = \{m_1(\text{FC})m_2(\text{FC}) + m_1(\text{FC})m_2(\text{OC} \cup \text{FC}) + m_1(\text{FC})m_2(\text{FC} \cup \text{UC})$$

$$+ m_1(\text{FC})m_2(\text{FC} \cup \text{UC} \cup \text{OC}) + m_1(\text{OC} \cup \text{FC})m_2(\text{FC})$$

$$+ m_1(\text{OC} \cup \text{FC})m_2(\text{FC} \cup \text{UC}) + m_1(\text{FC} \cup \text{UC})m_2(\text{FC})$$

$$+ m_1(\text{FC} \cup \text{UC})m_2(\text{OC} \cup \text{FC}) + m_1(\text{FC} \cup \text{UC} \cup \text{OC})m_2(\text{FC})\}/0.97$$

$$= \{0.05 \times 0.15 + 0.05 \times 0.05 + 0.05 \times 0.2 + 0.05 \times 0.5 + 0.15 \times 0.15$$

$$+ 0.15 \times 0.2 + 0.1 \times 0.15 + 0.1 \times 0.05 + 0.6 \times 0.15\}/0.97 = 0.21$$

Similarly, for the combined event FC ∪ UC, we get

$$m_{12}(\text{FC} \cup \text{UC}) = \{m_1(\text{FC} \cup \text{UC})m_2(\text{FC} \cup \text{UC}) + m_1(\text{FC} \cup \text{UC})m_2(\text{FC} \cup \text{OC} \cup \text{UC})$$

$$+ m_1(\text{FC} \cup \text{OC} \cup \text{UC})m_2(\text{FC} \cup \text{UC})\}/0.97$$

$$= \{0.1 \times 0.2 + 0.1 \times 0.5 + 0.6 \times 0.2\}/0.97 = 0.20$$

Finally, using the combined evidence measures, m_{12}, we can calculate the combined belief measures (bel_{12}). For example, for OC ∪ FC we have

$$\text{bel}_{12}(\text{OC} \cup \text{FC}) = m_{12}(\text{OC} \cup \text{FC}) + m_{12}(\text{OC}) + m_{12}(\text{FC}) = 0.12 + 0.01 + 0.21 = 0.34$$

The remaining calculated values are shown in Table 15.2.

PROBABILITY MEASURES

When the additional belief axiom (Eq. (15.5)) is replaced with a stronger axiom (illustrated for only two sets, A and B),

$$\text{bel}(A \cup B) = \text{bel}(A) + \text{bel}(B), \quad A \cap B = \emptyset \tag{15.20}$$

we get a *probability measure*. Let us now introduce a formal definition for a probability measure in the context of an evidence theory.

If we have a *bea* for a singleton, x, denoted $m(x) = \text{bel}(x)$, and we have $m(A) = 0$ for all subsets A of the power set, P(X), that are *not* singletons, then $m(x)$ is a probability measure. A probability measure is also a mapping of some function, say $p(x)$, to the unit interval, i.e.,

$$p : x \longrightarrow [0, 1] \tag{15.21}$$

To conform to the literature we will let $m(x) = p(x)$ to denote $p(x)$ as a probability measure. The mapping $p(x)$ then maps evidence only on singletons to the unit interval. The key distinction between a probability measure and either a belief or plausibility measure, as can be seen from Eq. (15.21), is that a probability measure arises when all the evidence is on singletons only, i.e., only on elements x; whereas, when we have some evidence on subsets that are not singletons, we cannot have a probability measure and will have only belief and plausibility measures (both, because they are duals – see Eq. (15.10)). If we have a probability measure, we will then have

$$\text{bel}(A) = \text{pl}(A) = p(A) = \sum_{x \in A} p(x) \qquad \text{for all } A \in P(X) \tag{15.22}$$

where set A is simply a collection of singletons; this would define the probability of set A. Equation (15.22) reveals that the belief, plausibility, and probability of a set A are all equal for a situation involving probability measures. Moreover, Eqs. (15.7) and (15.11) become a manifestation of the excluded middle axioms (see Chapter 2) for a probability measure:

$$\text{pl}(A) = p(A) = \text{bel}(A) \longrightarrow p(A) + p(\overline{A}) = 1 \tag{15.23}$$

Example 15.3. Two quality control experts from PrintLaser Inc. are trying to determine the source of scratches on the media that exit the sheet feeder of a new laser printer already in production. One possible source is the upper arm and the other source is media sliding on top of other media (e.g., paper on paper). We shall denote the following focal elements:

W denotes scratches from wiper arm.

M denotes scratches from other media.

The experts provide their assessments of evidence supporting each of the focal elements as follows:

Focal elements	Expert 1, m_1	Expert 2, m_2
W	0.6	0.3
M	0.4	0.7
W ∪ M	0	0

We want to determine the beliefs, plausibilities, and probabilities for each non-null focal element. We can see that evidence is only available on the singletons, W and M. We find the following relationships for the first expert:

$$\text{bel}_1(W) = m_1(W) = 0.6$$

$$\text{bel}_1(M) = m_1(M) = 0.4$$

$$\text{bel}_1(W \cup M) = m_1(W) + m_1(M) + m_1(W \cup M) = 0.6 + 0.4 + 0 = 1$$

$$\text{pl}_1(W) = m_1(W) + m_1(W \cup M) = 0.6 + 0 = 0.6$$

$$\text{pl}_1(M) = m_1(M) + m_1(W \cup M) = 0.4 + 0 = 0.4$$

$$\text{pl}_1(W \cup M) = m_1(W) + m_1(M) + m_1(W \cup M) = 0.6 + 0.4 + 0 = 1$$

We note that $\text{bel}_1(W) = \text{pl}_1(W)$, $\text{bel}_1(M) = \text{pl}_1(M)$, and $\text{bel}_1(W \cup M) = \text{pl}_1(W \cup M)$. From Eq. (15.23), these are all probabilities. Hence, $p_1(W) = 0.6$, $p_1(M) = 0.4$, and $p_1(W \cup M) = p(W) + p(M) = 0.6 + 0.4 = 1$ (this also follows from the fact that the probability of the union of disjoint events is the sum of their respective probabilities). In a similar fashion for the second expert we find

$$p_2(W) = 0.3, \qquad p_2(M) = 0.7 \qquad \text{and} \qquad p_2(W \cup M) = 0.3 + 0.7 = 1$$

POSSIBILITY AND NECESSITY MEASURES

Suppose we have a collection of some or all of the subsets on the power set of a universe that have the property $A_1 \subset A_2 \subset A_3 \subset \cdots \subset A_n$. With this property these sets are said to be *nested* [Shafer, 1976]. When the elements of a set, or universe, having evidence are nested, then we say that the belief measures, bel(A_i), and the plausibility measures, pl(A_i), represent a *consonant* body of evidence. By consonant we mean that the evidence allocated to the various elements of the set (subsets on the universe) does *not* conflict, i.e., the evidence is free of dissonance.

For a consonant body of evidence, we have the following relationships [Klir and Folger, 1988] for two different sets on the power set of a universe, i.e., for A, B $\in$ P(X):

$$\text{bel}(A \cap B) = \min[\text{bel}(A), \text{bel}(B)] \tag{15.24}$$

$$\text{pl}(A \cup B) = \max[\text{pl}(A), \text{pl}(B)] \tag{15.25}$$

The expressions in Eqs. (15.24)–(15.25) indicate that the belief measure of the intersection of two sets is the smaller of the belief measures of the two sets and the plausibility measure of the union of these two sets is the larger of the plausibility measures of the two sets.

In the literature consonant belief and plausibility measures are referred to as necessity (denoted η) and possibility (denoted π) measures, respectively. Equations (15.24) and (15.25) become, respectively for all A, B $\in$ P(X),

$$\eta(A \cap B) = \min[\eta(A), \eta(B)] \tag{15.26}$$

$$\pi(A \cup B) = \max[\pi(A), \pi(B)] \tag{15.27}$$

For a consonant body of evidence, the dual relationships expressed in Eq. (15.10) then take the forms,

$$\pi(A) = 1 - \eta(\overline{A})$$

$$\eta(A) = 1 - \pi(\overline{A})$$

(15.28)

Since the necessity and possibility measures are dual relationships, the discussion to follow focuses only on one of these, possibility. If necessity measures are desired, they can always be derived with the expressions in Eq. (15.28).

We now define a possibility distribution function as a mapping of the singleton elements, x, in the universe, X, to the unit interval, i.e.,

$$r : X \longrightarrow [0, 1]$$

(15.29)

This mapping will be related to the possibility measure, $\pi(A)$, through the relationship

$$\pi(A) = \max_{x \in A} r(x)$$

(15.30)

for each $A \in P(X)$ [see Klir and Folger, 1988, for a proof]. Now, a possibility distribution can be defined as an ordered sequence of values,

$$\mathbf{r} = (\rho_1, \rho_2, \rho_3, \dots, \rho_n)$$

(15.31)

where $\rho_i = r(x_i)$ and where $\rho_i \geq \rho_j$ for $i < j$. The *length* of the ordered possibility distribution given in Eq. (15.31) is the number n. Every possibility measure also can be characterized by the n-tuple, denoted as a basic distribution [Klir and Folger, 1988],

$$\mathbf{m} = (\mu_1, \mu_2, \mu_3, \dots, \mu_n)$$

(15.32a)

$$\sum_{i=1}^{n} \mu_i = 1$$

(15.32b)

where $\mu_i \in [0, 1]$ and $\mu_i = m(A_i)$. Of course, the sets A_i are nested as is required of all consonant bodies of evidence. From Eq. (15.17) and the relationship

$$\rho_i = r(x_i) = \pi(x_i) = \mathrm{pl}(x_i)$$

(15.33)

it can be shown [Klir and Folger, 1988] that

$$\rho_i = \sum_{k=i}^{n} \mu_k = \sum_{k=i}^{n} m(A_k)$$

(15.34)

or, in a recursive form,

$$\mu_i = \rho_i - \rho_{i+1}$$

(15.35)

where $\rho_{n+1} = 0$ by convention. Equation (15.35) produces a set of equations of the form

$$
\begin{aligned}
\rho_1 &= \mu_1 + \mu_2 + \mu_3 + \cdots + \mu_n \\
\rho_2 &= \quad\quad \mu_2 + \mu_3 + \cdots + \mu_n \\
\rho_3 &= \quad\quad\quad\quad \mu_3 + \cdots + \mu_n \\
&\quad\quad\quad\quad\quad \cdots \\
\rho_n &= \quad\quad\quad\quad\quad\quad\quad\quad \mu_n
\end{aligned}
\tag{15.36}
$$

Nesting of focal elements can be an important physical attribute of a body of evidence. Consider the following example where physical nesting is an important feature of an engineering system.

Example 15.4. Suppose there are seven nodes in a communication network X, labeled $x_1 - x_7$ and represented by boxes. Of these seven nodes, one is causing a problem. The company network expert is asked for an opinion on which node is causing the communications problem. The network expert aggregates these nodes into sets, as given in the accompanying table. The third column in the table represents the expert's basic distribution (basic evidence assignments), and the last column is the possibility distribution found from Eq. (15.34).

Set A	Aggregation of focal elements	$\mu_n = \mu(A_n)$	ρ_i
A_1	x_1	0.4	1
A_2	$x_1 \cup x_2$	0.2	0.6
A_3	$x_1 \cup x_2 \cup x_3$	0	0.4
A_4	$x_1 \cup x_2 \cup x_3 \cup x_4$	0.1	0.4
A_5	$x_1 \cup x_2 \cup x_3 \cup x_4 \cup x_5$	0	0.3
A_6	$x_1 \cup x_2 \cup x_3 \cup x_4 \cup x_5 \cup x_6$	0.2	0.3
A_7	$x_1 \cup x_2 \cup x_3 \cup x_4 \cup x_5 \cup x_6 \cup x_7$	0.1	0.1

The physical significance of this nesting (shown in Fig. 15.4) can be described as follows. In the network expert's belief, node x_1 is causing the problem. This node has new hardware and

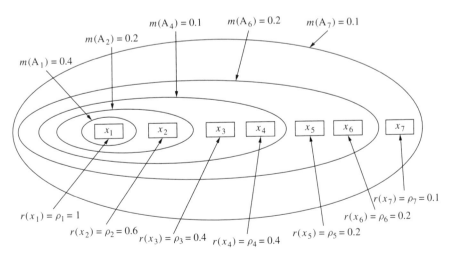

FIGURE 15.4
Nesting diagram, Example 15.4.

is an experimental CPU. For these reasons, the network expert places the highest belief on this set (set A_1). The next set with nonzero belief (supporting evidence), A_2, is comprised of the union, x_1 or x_2. The network expert has less belief that node x_1 or x_2 is causing the problem. Node x_2 has new hardware as well, but has a trusted CPU. The next set with nonzero belief, A_4, is nodes x_1 or x_2 or x_3 or x_4. The network expert has even less belief that the problem is caused by this set. The expert reasons that x_3 or x_4 has trusted hardware and CPUs. The next set with nonzero belief, A_6, is nodes x_1 or x_2 or x_3 or x_4 or x_5 or x_6. The network expert has slightly more belief that this set is the problem than set A_4, but much less than the initial set A_1, the reasons being that there are two new programmers using these nodes for testing communications software. The final set with evidence is the union of all seven nodes. The expert has little belief that this set is the problem because node x_7 is usually turned off.

Note that the first element of any ordered possibility distribution, ρ_1, is always equal to unity, i.e., $\rho_1 = 1$. This fact is guaranteed by Eq. (15.32b). The smallest possibility distribution of length n has the form $\mathbf{r} = (1, 0, 0, 0, \ldots, 0)$, where there are $(n - 1)$ zeros after a value of unity in the distribution. The associated basic distribution would have the form $\mathbf{m} = (1, 0, 0, 0, \ldots, 0)$. In this case there would be only one focal element with evidence, and it would have all the evidence. This situation represents *perfect evidence*; there is no uncertainty involved in this case.

Alternatively, the largest possibility distribution of length n has the form $\mathbf{r} = (1, 1, 1, 1, \ldots, 1)$, where all values are unity in the distribution. The associated basic distribution would have the form $\mathbf{m} = (0, 0, 0, 0, \ldots, 1)$. In this case all the evidence is on the focal element comprising the entire universe, i.e., $A_n = x_1 \cup x_2 \cup \cdots \cup x_n$; hence, we know nothing about any specific focal element in the universe except the universal set. This situation is called *total ignorance*. In general, the larger the possibility distribution, the less specific the evidence and the more ignorant we are of making any conclusions.

Since possibility measures are special cases of plausibility measures and necessity measures are special cases of belief measures, we can relate possibility measures and necessity measures to probability measures. Equation (15.23) shows that, when all the evidence in a universe resides solely on the singletons of the universe, the belief and plausibility measures become probability measures. In a like fashion it can be shown that the plausibility measure approaches the probability measure from an upper bound and that the belief measure approaches the probability measure from a lower bound; the result is a range around the probability measure [Yager and Filev, 1994],

$$\text{bel}(A) \leq p(A) \leq \text{pl}(A) \tag{15.37}$$

Example 15.5. Probabilities can be determined by finding point-valued quantities and then determining the relative frequency of occurrence of these quantities. In determining the salvage value of older computers, the age of the computer is a key variable. This variable is also important in assessing depreciation costs for the equipment. Sometimes there is uncertainty in determining the age of computers if their purchase records are lost or if the equipment was acquired through secondary acquisitions or trade. Suppose the age of five computers is known, and we have no uncertainty; here the ages are point-valued quantities.

Computer	Age (months)
1	26
2	21
3	33
4	24
5	30

With the information provided in the table we could answer the following question: What percentage of the computers have an age in the range of 20 to 25 months, i.e., what percentage of the ages fall in the interval [20, 25]? This is a countable answer of $\frac{2}{5}$, or 40%.

Now suppose that the age of the computers is not known precisely, but rather each age is assessed as an interval. Now the ages are set-valued quantities, as follows:

Computer	Age (months)
1	[22, 26]
2	[20, 22]
3	[30, 35]
4	[20, 24]
5	[28, 30]

With this information we can only assess possible solutions to the question just posed: What percentage of the computers possibly fall in the age range of [20, 25] months? Because the ages of the computers are expressed in terms of ranges (or sets on the input space), the solution space of percentages will also have to be expressed in terms of ranges (or sets on the solution space).

To approach the solution we denote the query range as Q, i.e., Q = [20, 25] months. We denote the age range of the ith computer as D_i. Now, we can determine the certainty and possibility ranges using the following rules:

1. Age (i) is certain if $D_i \subset Q$.
2. Age (i) is possible if $D_i \cap Q \neq \emptyset$.
3. Age (i) is not possible if $D_i \cap Q = \emptyset$.

The first rule simply states that the age is certainly in the query range if the age range of the ith computer is completely contained within the query range. The second rule states that the age is possibly in the query range if the age range of the ith computer and the query range have a non-null intersection, i.e., if they intersect at any age. The third rule states that the age is not possible if the age range of the ith computer and the query range have no age in common, i.e., their intersection is null. We should note here that a solution that is certain is necessarily possible (certainty implies possibility), but the converse is not always true (things that are possible are not always certain). Hence, the set of certain quantities is a subset of the set of possible quantities. In looking at the five computers and using the three rules already given, we determine the following relationships:

Computer	η or π
1	Possible
2	Certain
3	Not possible
4	Certain
5	Not possible

In the table we see that, of the five computers, two have age ranges that are certainly (denoted $\eta(Q)$) in the query interval and three have age ranges that are possibly (denoted $\pi(Q)$) in the query interval (one possible and two certain). We will denote the solution as the response to the query, or resp(Q). This will be an interval-valued quantity as indicated earlier, or

$$\text{resp}(Q) = [\eta(Q), \pi(Q)] = [\tfrac{2}{5}, \tfrac{3}{5}]$$

Hence, we can say that the answer to the query is "Certainly 40% and possibly as high as 60%." We can also see that the range represented by resp(Q) represents a lower bound and an upper bound to the actual point-valued probability (which was determined earlier to be 40%) as indicated by Eq. (15.37).

In Example 15.5 all computer ranges were used in determining the percentages for possibilities and certainties (i.e., all five). In evidence theory null values are *not* counted in the determination of the percentages as seen in the normalization constant, K, expressed in Eq. (15.19). This characteristic can lead to fallacious responses, as illustrated in the following example.

Example 15.6. Suppose again we wish to determine the age range of computers, this time expressed in units of years. In this case we ask people to tell us the age of their own computer (PC). These responses are provided in the accompanying table. In the table the null symbol, Ø, indicates that the person queried has no computer.

Person	Age of PC (years)
1	[3, 4]
2	Ø
3	[2, 3]
4	Ø
5	Ø

Let us now ask the question: What percentage of the computers have an age in the range $Q = [2, 4]$ years? Using the rules given in Example 15.5, we see that we have two certainties (hence, we have two possibilities) and three null values. If we include the null values in our count, the solution is

$$\text{resp}(Q) = [\eta(Q), \pi(Q)] = [\tfrac{2}{5}, \tfrac{2}{5}] = \tfrac{2}{5}$$

In this case, we have not used a normalization process because we have counted the null values. If we decide to neglect the null values (hence, we normalize as Dempster's rule of combination suggests), then the solution is

$$\text{resp}(Q) = [\eta(Q), \pi(Q)] = [\tfrac{2}{2}, \tfrac{2}{2}] = 1$$

We see a decidedly different result when normalization is used.

A graphical interpretation of the evidence theory developed by Dempster and Shafer is provided in what is called the ball–box analogy [Zadeh, 1986].

Example 15.7   [Zadeh, 1984]. Suppose the king of country X believes a submarine, S, is in the territorial waters of X. The king summons n experts, $E_1, E_2, \ldots, E_n$, to give him advice on the location of the submarine, S. The n experts each provide their assessment of the location of S; call these possible locations $L_1, L_2, \ldots, L_m, \ldots, L_n$, where $m \leq n$. Here, L_i are subsets of the territorial waters, X. To be more specific, experts $E_1, E_2, \ldots, E_m$ say that S is in $L_1, L_2, \ldots, L_m$ and experts $E_{m+1}, \ldots, E_{n-1}, E_n$ say that S is *not* located in the territorial waters of X, i.e., $L_{m+1} = L_{m+2} = \cdots = L_n = \emptyset$. So there are $(m - n)$ experts who say that S is not in the territorial waters. Now the king asks, "Is S in a *subset* A of our territorial waters?" Figure 15.5 shows possible location regions, L_i, and the query region of interest, region A.
    If we denote E_i as the location proffered by the ith expert, we have the following two rules:

1. $E_i \subset A$ implies that it is certain that $S \in A$.
2. $E_i \cap A \neq \emptyset$ implies that it is possible that $S \in A$.

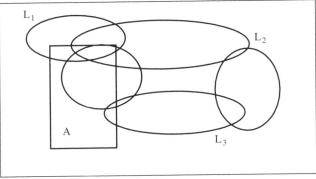

Territorial waters

FIGURE 15.5
Ball–box analogy of the Dempster–Shafer evidence theory.

We note again that a certainty is contained in the set of possibilities, i.e., certainty implies possibility. We further assume that the king aggregates the opinions of his experts by averaging. Thus, if k out of n experts vote for rule number 1, then the average certainty $= k/n$; if l out of n (where $l \geq k$) experts vote for rule number 2, then the average possibility $= l/n$. Finally, if the judgment of those experts who think there is no submarine anywhere in the territorial waters is ignored, the average certainty and possibility will be k/m and l/m, respectively (where $m \leq n$). Ignoring the opinion of those experts whose L_i is the null set corresponds to the normalization (Eq. (15.19)) in the Dempster–Shafer evidence theory.

Dempster's rule of combination may lead to counterintuitive results because of the normalization issue. The reason for this [Zadeh, 1984] is that normalization throws out evidence that asserts the object under consideration does not exist, i.e., is null or empty (Ø). The following example from the medical sciences illustrates this idea very effectively.

Example 15.8   [Zadeh, 1984]. A patient complaining of a severe headache is examined by two doctors (doctor 1 and doctor 2). The diagnosis of doctor 1 is that the patient has either meningitis (M) with probability 0.99 or a brain tumor (BT) with probability 0.01. Doctor 2 agrees with doctor 1 that the probability of a brain tumor (BT) is 0.01, but disagrees with doctor 1 on the meningitis; instead doctor 2 feels that there is a probability of 0.99 that the patient just has a concussion (C). The following table shows the evidence for each of the focal elements in this universe for each of the doctors (m_1 and m_2) as well as the calculated values for the combined evidence measures (m_{12}). In the table there is evidence only on the singletons M, BT, and C; hence, all the measures are probability measures (the doctors provided their opinions in terms of probability).

Focal element	m_1	m_2	m_{12}
M	0.99	0	0
BT	0.01	0.01	1
C	0	0.99	0
M $\cup$ BT	0	0	
M $\cup$ C	0	0	
BT $\cup$ C	0	0	
M $\cup$ BT $\cup$ C	0	0	

The combined evidence measures are calculated as follows. First, by using Eq. (15.19) the normalization constant, K, is calculated. Then use of Eq. (15.18) produces values for m_{12}. For example, $m_{12}(BT)$ for a brain tumor is found as

$$m_{12}(BT) = \frac{0.01(0.01)}{1 - \{0.99[0.01 + 0.99] + 0.01[0 + 0.99] + 0[0 + 0.01]\}} = \frac{0.0001}{1 - 0.9999} = 1$$

One can readily see that the combined measures $m_{12}(C)$ and $m_{12}(M)$ will be zero from the fact that

$$m_1(C) = m_2(M) = 0$$

The table reveals that using the Dempster–Shafer rule of combination (in Shafer [1976], null values are not allowed in the definition of belief functions but do enter in the rule of combination of evidence) results in a combined probability of 1 that the patient has a brain tumor when, in fact, both doctors agreed individually that it was only one chance in a hundred! What is even more confusing is that the same conclusion (i.e., $m_{12}(BT) = 1$) results regardless of the probabilities associated with the other possible diagnoses.

In Example 15.8 it appears that the normalization process *suppressed* expert opinion; but is this omission mathematically allowable? This question leads to the conjecture that the rule of combination cannot be used until it is ascertained that the bodies of evidence are conflict-free; that is, at least one parent relation exists that is absent of conflict. In particular, under this assertion, it is not permissible to combine distinctly different bodies of evidence. In the medical example, the opinions of both doctors reveal some missing information about alternative diagnoses. A possibility theory might suggest an alternative approach to this problem in which the incompleteness of information in the knowledge base propagates to the conclusion and results in an interval-valued, possibilistic answer. This approach addresses rather than finesses (like excluding null values) the problem of incomplete information.

POSSIBILITY DISTRIBUTIONS AS FUZZY SETS

Belief structures that are nested are called consonant. A fundamental property of consonant belief structures is that their plausibility measures are possibility measures. As suggested by Dubois and Prade [1988], possibility measures can be seen to be formally equivalent to fuzzy sets. In this equivalence, the membership grade of an element x corresponds to the plausibility of the singleton consisting of that x; that is, a consonant belief structure is equivalent to a fuzzy set F of X where $F(x) = \text{pl}(\{x\})$.

A problem in equating consonant belief structures with fuzzy sets is that the combination of two consonant belief functions using Dempster's rule of combination in general does not necessarily lead to a consonant result [Yager, 1993]. Hence, since Dempster's rule is essentially a conjunction operation, the intersection of two fuzzy sets interpreted as consonant belief structures may not result in a valid fuzzy set (i.e., a consonant structure).

Example 15.9   [Yager, 1993]. Suppose we have a universe comprised of five singletons, i.e.,

$$X = \{x_1, x_2, x_3, x_4, x_5\}$$

and we have evidence provided by two experts. The accompanying table provides the experts' degrees of belief about specific subsets of the universe, X.

Focal elements	Expert 1, $m_1(A_i)$	Expert 2, $m_2(B_i)$
$A_1 = \{x_1, x_2, x_3\}$	0.7	
$A_2 = X$	0.3	
$B_1 = \{x_3, x_4, x_5\}$		0.8
$B_2 = X$		0.2

Because $A_1 \subset A_2$ and $B_1 \subset B_2$ we have two consonant (nested) belief structures represented by A and B. Using Dempster's rule of combination and applying Eqs. (15.18)–(15.19), we have for any set D on the universe X

$$m(D) = \frac{1}{1-k} \sum_{A_i \cap B_j = D} m_1(A_i) \cdot m_2(B_j)$$

and

$$k = \sum_{A_i \cap B_j = \emptyset} m_1(A_i) \cdot m_2(B_j)$$

Since there are two focal elements in each experts' belief structures, we will have $2^2 = 4$ belief structures in the combined evidence case, which we will denote as m. We note for these data that we get a value of $k = 0$, because there are no intersections between the focal elements of A and B that result in the null set. For example, the intersection between A_1 and B_1 is the singleton, x_3. Then, we get

$$D_1 = A_1 \cap B_1 = \{x_3\} \qquad m(D_1) = 0.56 \quad \text{(i.e., } 0.7 \times 0.8\text{)}$$

$$D_2 = A_1 \cap B_2 = \{x_1, x_2, x_3\} \qquad m(D_2) = 0.14 \quad \text{(i.e., } 0.7 \times 0.2\text{)}$$

$$D_3 = A_2 \cap B_1 = \{x_3, x_4, x_5\} \qquad m(D_3) = 0.24 \quad \text{(i.e., } 0.3 \times 0.8\text{)}$$

$$D_4 = A_2 \cap B_2 = X \qquad m(D_4) = 0.06 \quad \text{(i.e., } 0.3 \times 0.2\text{)}$$

For the focal elements D_i we note that $D_1 \subset D_2 \subset D_4$ and $D_1 \subset D_3 \subset D_4$, but we do not have $D_2 \subset D_3$ or $D_3 \subset D_2$. Hence, the combined case is not consonant (i.e., not completely nested).

Yager [1993] has developed a procedure to prevent the situation illustrated by Example 15.9 from occurring; that is, a method is available to combine consonant possibility measures where the result is also a consonant possibility measure. However, this procedure is very lengthy to describe and is beyond the scope of this text; the reader is referred to the literature [Yager, 1993] to learn this method.

Another interpretation of a possibility distribution as a fuzzy set was proposed by Zadeh [1978]. He defined a possibility distribution as a fuzzy restriction that acts as an elastic constraint on the values that may be assigned to a variable. In this case the possibility distribution represents the degrees of membership for some linguistic variable, but the membership values are strictly monotonic as they are for an ordered possibility distribution. For example, let $\underset{\sim}{A}$ be a fuzzy set on a universe X, and let the membership value, μ, be a variable on X that assigns a "possibility" that an element of x is in $\underset{\sim}{A}$. So we get

$$\pi(x) = \mu_{\underset{\sim}{A}}(x) \tag{15.38}$$

Zadeh points out that the possibility distribution is nonprobabilistic and is used primarily in natural language applications. There is a loose relationship, however, between the two through a possibility/probability consistency principle [Zadeh, 1978]. In sum, what is *possible* may not be *probable*, but what is *impossible* is inevitably *improbable* (see the discussion on this issue later in this chapter).

Example 15.10. Let $\underset{\sim}{A}$ be a fuzzy set defined on the universe of columns needed to support a building. Suppose there are 10 columns altogether, and we start taking columns away until the building collapses; we record the number of columns at the time the building collapses. Let $\underset{\sim}{A}$ be the fuzzy set defined by the number of columns "possibly" needed, out of 10 total, just before the structure fails. The structure most certainly needs at least three columns to stand (imagine a stool). After that the number of columns required for the building to stand is a fuzzy issue (because of the geometric layout of the columns, the weight distribution, etc.), but there is a possibility it may need more than three. The following fuzzy set may represent this possibility:

$$\underset{\sim}{A} = \left\{ \frac{1}{1} + \frac{1}{2} + \frac{1}{3} + \frac{0.9}{4} + \frac{0.6}{5} + \frac{0.3}{6} + \frac{0.1}{7} + \frac{0}{8} + \frac{0}{9} + \frac{0}{10} \right\}$$

A probability distribution on the same universe may look something like the following table, where u is the number of columns prior to collapse, and $p(u)$ is the probability that u is the number of columns at collapse:

u	1	2	3	4	5	6	7	8	9	10
$p(u)$	0	0	0.1	0.5	0.3	0.1	0	0	0	0

As seen, although it is *possible* that one column will sustain the building, it is not *probable*. Hence, a high degree of possibility does not imply a high probability, nor does a low degree of probability imply a low degree of possibility.

Belief measures and plausibility measures overlap when they both become probability measures. However, possibility, necessity, and probability measures do not overlap with one another except for one special measure: the measure of one focal element that is a singleton. These three measures become equal when one element of the universal set is assigned a value of unity, and all other elements in the universe are assigned a value of zero. This measure represents *perfect evidence* [Klir and Folger, 1988].

POSSIBILITY DISTRIBUTIONS DERIVED FROM EMPIRICAL INTERVALS

Analyzing empirical data is an important exercise in any experimental analysis. In practice, it is very common to use probabilistic tools to analyze data from experimental studies. However, there are a number of occasions when it is more appropriate to conduct possibilistic analysis than a probabilistic analysis. For example, in the determination of the residual strength of an existing bridge, one might have only subjective estimates from visual inspections of the bridge or limited information of the strength from nondestructive evaluation. In such cases, data are usually available as a range of numbers or intervals (see Fig. 15.6), and an analyst is required to determine the best possible estimate from

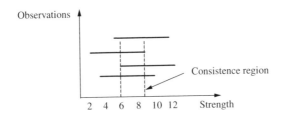

FIGURE 15.6
Residual strength estimates acquired as "consistent" intervals.

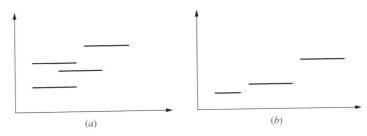

FIGURE 15.7
(a) Overlapping but inconsistent intervals; (b) completely disjoint intervals.

such data. When data are available as a set of intervals, possibilistic analysis captures the true uncertainty in the interval without relying on predetermined distributions and by not requiring an analyst to come up with specific data (i.e., singletons). Possibility distributions capture the imprecision resulting from nonspecificity of the intervals by considering the entire length of the interval. This section describes an approach for developing possibility distributions from such empirical measurements.

As described in the previous section, when the intervals are nested (consonant intervals), possibility measures are plausibility measures as defined by evidence theory. Thus, for nested intervals one can derive a possibility distribution by tracing a contour over all the nested intervals. In practice, however, intervals are seldom nested and are usually available as sets of overlapping and nonoverlapping portions. In such cases, special methods are required to transform the nonconsonant intervals into consonant intervals in accordance with the available evidence. The nonconsonant intervals can be consistent, where at least one common interval exists among all the measurements (Fig. 15.6), partially overlapping, where some intervals overlap (Fig. 15.7a), or disjoint, where there is no overlap among the intervals (Fig. 15.7b). The next section describes a method for developing possibility distributions for more common cases where the intervals are consistent and partially overlapping.

Deriving Possibility Distributions from Overlapping Intervals [Donald, 2003]

Let us consider a system that can be described within a domain $X = \{x_1, x_2, \ldots, x_n\}$, the behavior of which is described by evidence obtained as observations over a collection of sets, F. The set X is called the domain of the system. For example, the domain of the system can be all the possible residual strengths of the bridge under consideration for maintenance. Now, let $F = \{\langle A_j, w_j \rangle\}$ represent the original intervals along with their weights. If M measurements are observed, then the weights w_j of each observation A_j is calculated by frequency analysis as

$$w_j = \frac{n(A_j)}{M} \tag{15.39}$$

where $n(A_j)$ is the frequency count of interval A_j, and $\sum_j w_j = 1, \sum_j n(A_j) = M$.

In conventional probability analysis, the observations A_1 and A_2 are disjoint such that $A_1 \cap A_2 = \emptyset$, and the frequency of occurrence of any event A_j is simply the ratio of the count of a particular event to the total number of occurrences of all the events. However, such disjoint measurements are uncommon and measurements are usually such that F consists of overlapping intervals A_j. In deriving a set of consonant intervals it is assumed that the underlying evidence is coherent such that the observations obtained should reveal intervals that are nested within each other. This is accomplished such that the smallest interval (having the least nonspecificity) is selected as the interval that has the most intersections with the observations and the next smallest one as that having the second-most intersections, and so forth. This process yields Q unique countable intersections and a set $G = \{\langle B_k, v(B_k) \rangle\}, k = 1, 2, \ldots, Q$ of intersections, where B_k is the interval obtained by the kth intersection of original intervals A_i and A_j, and $v(B_k)$ is the weight assigned to the corresponding B_k. Union of all the sets, $S = \bigcup_{A_j \in F} A_j$, where $w(A_j) \neq 0$, forms the support interval and is derived as

$$S = [\min_{A_i \in F, x \in A_i} (x), \max_{A_i \in F, x \in A_i} (x)] \tag{15.40}$$

Weights v_k assigned to each of the Q elements of G are determined by utilizing any of the t-norms (see Chapter 2) for conjunctions as $v(B_k) = T(A_i, A_j)_k$, where $T(A_i, A_j)_k$ is any conjunctive triangular norm operating on the sets A_i and A_j that form the kth intersection. In this section the standard fuzzy intersection (t-norm, minimum operator) is used to determine weights $v(B_k)$ of the conjunction of sets A_i and A_j, and therefore the weights are obtained as

$$v(B_k) = \min(w_i, w_j)_k \tag{15.41}$$

Since a min t-norm is based on a weaker axiom of nonadditivity, the weights do not necessarily add to 1, and hence the weights derived are normalized as

$$\eta(B_k) = \frac{v(B_k)}{\sum_{k=1}^{Q} v(B_k)} \tag{15.42}$$

such that, $\sum_k \eta(B_k) = 1$.

In deriving the elements of G, however, some elements by virtue of their origin from the intersections from parent sets (sets from which the intervals were derived) tend to be consonant with the parent set while they are not necessarily consonant with other parent sets. Therefore, if Q represents the total number of focal elements (intervals with nonzero weights) in the focal set G derived from the intersections of original measurements, there are Q_H elements in the consonant set H, and Q_I elements in the nonconsonant set I, such that

$$\sum_{B_k \in Q_H} \eta(B_k) + \sum_{B_k \in Q_I} \eta(B_k) = 1.0 \qquad (15.43)$$

Consonant intervals can be extracted from the intersection set G as a combination of frequency analysis and expert judgments. The most possible interval is the one that occurs most frequently, and in which an expert has the most confidence. Once the task of selecting consonant intervals is accomplished, the weight from the remaining nonconsonant intervals is redistributed to the consonant intervals in such a manner that the total information from underlying evidence is preserved.

Redistributing Weight from Nonconsonant to Consonant Intervals

Weights are redistributed from nonconsonant to consonant sets according to the dissonance between individual intersecting sets (the conflict between two sets). The logic used here is that the higher the similarity between two sets (or the lower the conflict) the greater is the weight that can be transferred between the two sets. Parameters that are useful in the redistribution of weight are identified as the cardinality of each set, $|H_i|$, and the number of common elements between the sets $|H_i \cap I_j|$. In the case of continuous intervals, the cardinality $|.|$ can be replaced by the length l of the interval defined over a real line, and the set of real numbers comprising the intersecting interval can be determined as the length that is common to both the intervals. The similarity, β, of two sets is given as

$$\beta_{ij} = \frac{|H_i \cap I_j|}{|H_i|} \qquad (15.44)$$

or, for a set of real numbers,

$$\beta_{ij} = \frac{l[\min \omega_i \mid \omega_i \in H_i \cap I_j \neq \emptyset, \max \omega_i \mid \omega_i \in H_i \cap I_j \neq \emptyset]}{l[H_i]} \qquad (15.45)$$

where $l\,[.]$ denotes the length of the interval and is simply given as $l[a, b] = b - a$, with $l[.]$ equal to 1 for singletons and $\beta_{ij} = 1$ when H_i is completely included in I_j.

A redistribution factor κ is then computed as

$$\kappa_{ij} = \frac{\beta_{ij}}{\sum_{i=1}^{Q_H} \beta_{ij}} \qquad (15.46)$$

such that, for any j, $\sum_{i=1}^{Q_H} \kappa_{ij} = 1.0$.

The redistribution factor can be viewed as the fraction of the weight that is transferred from the nonconsonant to the consonant interval. The redistribution weight ρ is then calculated by determining the weight of the nonconsonant interval I_j that is transferred to the corresponding consonant interval H_i. Therefore, for the ith consonant interval and jth nonconsonant interval,

$$\rho_{ij} = \kappa_{ij} * \eta(I_j) \qquad (15.47)$$

From Eq. (15.45) it is clear that when I_j intersects with H_i and no other set then $\beta = 1$, thus assigning the entire weight of I to H, and also, when H does not intersect with any portion of I, $\beta = 0$. The final weights of the initial consonant sets as determined after the redistribution is given over the entire set of nonconsonant intervals as

$$m(H_i) = \eta(H_i) + \sum_{j=1}^{Q_1} \rho_{ij} = \eta(H_i) + \sum_{j=1}^{Q_1} \kappa_{ij}\eta(I_j) \tag{15.48}$$

The above process indicates that the total weight is preserved among the consonant data intervals. It can be proved that $\sum_{i=1}^{Q_H} m(H_i) = 1.0$. From Eq. (15.48) and with the constraint $\sum_{i=1}^{Q_H} \kappa_{ij} = 1.0$, we have

$$\sum_{i=1}^{Q_H} m(H_i) = \sum_{i=1}^{Q_H} \eta(H_i) + \sum_{i=1}^{Q_H}\sum_{j=1}^{Q_1} \kappa_{ij}\eta(I_j)$$

$$= \sum_{i=1}^{Q_H} \eta(H_i) + \sum_{j=1}^{Q_1} \left(\eta(I_j) \sum_{i=1}^{Q_H} \kappa_{ij} \right)$$

$$= \sum_{i=1}^{Q_H} \eta(H_i) + \sum_{j=1}^{Q_1} \eta(I_j)$$

$$= 1.0$$

The possibility distribution from the weights is then obtained as follows:

$$\pi(x) = \sum_{x \in H_i} m(H_i) \tag{15.49}$$

Example 15.11. Suppose it is required to determine the strength of a wooden bridge that is subject to heavy foot traffic and an expert is hired to estimate the residual strength of the bridge. The expert uses various nondestructive techniques at various points of the bridge and offers estimates for the strength as shown in Table 15.3. Based on these estimates, an analyst needs to determine the possible strength of the bridge to decide on the appropriate action.

The first step in the solution is to determine the support of the distribution and the most possible interval based on the intersections of all the measurements. The support is calculated from Eq. (15.40) as $S = [1000, 5000]$, and the intervals obtained by intersections of all the measurements are shown in Table 15.4.

TABLE 15.3
Original data intervals

Estimate	Residual strength (lb/in²)	Weight
1	[1000, 4000]	0.2
2	[2000, 4000]	0.4
3	[3000, 5000]	0.2
4	[2000, 5000]	0.2

TABLE 15.4
Intervals obtained by intersections of original data intervals in Table 15.3

Interval	Weight	Normalized weights
[3000, 4000]	0.2	0.143
[1000, 4000]	0.2	0.143
[2000, 4000]	0.4	0.285
[3000, 5000]	0.2	0.143
[2000, 5000]	0.2	0.143
[1000, 5000]	0.2	0.143

TABLE 15.5
Redistribution of weights between the nonconsonant and consonant intervals

Nonconsonant intervals	Consonant intervals	β	κ	ρ
[3000, 5000]	[3000, 4000]	1	0.316	0.045
	[2000, 4000]	1	0.316	0.045
	[2000, 5000]	0.66	0.209	0.03
	[1000, 5000]	0.5	0.159	.023
[1000, 4000]	[3000, 4000]	1	0.267	0.038
	[2000, 4000]	1	0.267	0.038
	[2000, 5000]	1	0.267	0.038
	[1000, 5000]	0.75	0.199	0.029

From this table, it can be seen that the interval [3000, 4000] is included in all the intervals and, hence, is selected as the most possible interval. The selection of this most possible interval can also be based on other criteria that the expert chooses. The set of *consonant intervals* is then determined from this most possible interval by choosing intervals in the order of increasing cardinality, which include the most possible interval such that the selected intervals are nested. Any intervals that are not part of this nested structure then form the elements for the nonconsonant set. The set of *consonant intervals* forms the intervals for the possibility distribution and the total weight of the evidence is then preserved by transferring the weight from the nonconsonant set to the consonant set (Eqs. (15.44)–(15.46)). Table 15.5 shows this redistribution of weights.

Redistribution parameters β, κ, ρ are determined using Eqs. (15.44)–(15.47) as

$$\beta_{11} = \frac{[3000, 5000] \cap [3000, 4000]}{[3000, 4000]} = 1$$

$$k_{11} = \frac{\beta_{11}}{\beta_{11} + \beta_{21} + \beta_{31} + \beta_{41}} = \frac{1}{3.16} = 0.316$$

$$\rho_{11} = \kappa_{11} * \eta(I_1) = 0.316 * 0.143 = 0.045$$

The final weights are then determined using Eq. (15.48) as

$$m[3000, 4000] = 0.143 + \rho_{11} + \rho_{12}$$

$$= 0.143 + 0.045 + 0.038 = 0.226$$

Similarly, weights for other interval measurements are determined and are shown in Table 15.6.

TABLE 15.6
Possibility intervals and final weights

Interval	Final weight	Possibilistic weight
[3000, 4000]	0.226	1.0
[2000, 4000]	0.368	0.774
[2000, 5000]	0.211	0.406
[1000, 5000]	0.195	0.195

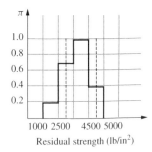

FIGURE 15.8
Possibility distribution for the residual strength of the bridge.

Finally, a possibility distribution is traced over the consonant intervals from Table 15.6 with the corresponding new weights, as shown in Fig. 15.8.

There is significant information in any possibility distribution that is not available when one chooses to represent the same information probabilistically. For example, the possibility distribution in Fig. 15.8 has possibilistic information on residual strength of a wooden bridge. Suppose we choose to focus on a specific interval of strengths, say $A = [2500, 4500]$ lb/in^2 (as shown by the broken lines in Fig. 15.8). The possibility of the actual strength of the bridge being in the interval [2500, 4500] lb/in^2 is unity. This is because the actual interval in Fig. 15.8 with $\pi([3000, 4000]) = 1.0$ is fully contained within the interval of interest, i.e., within [2500, 4500] lb/in^2. Said another way, the possibility of the actual strength not being in the interval [2500, 4500] lb/in^2 is 0.774, i.e., $\pi(\text{not } A) = \max(\text{any possibility value for values outside the interval } A) = \max(0.774, 0.406, 0.195)$. Finally, but most importantly, the necessity (see Eq. (15.28)) is $\eta = 1 - \pi(\text{not } A) = 0.226$. Hence, we can say that the actual strength of the bridge being in the interval [2500, 4500] lb/in^2 is *certainly* 0.226 but could be *possibly* 1.0. The interval of [3000, 4000] lb/in^2 would contain the *most possible* value. A probabilistic assessment of this same question would only provide a confidence level about the interval A, which does not truly represent the kind of evidence available; that is, the information is ambiguous and imprecise and not subject to random variability. In a probabilistic assessment of these data we would get a 95% confidence interval around the *most probable* value, which usually overestimates the interval that would contain the actual strength. Moreover, to get such a confidence interval we have to make assumptions about the data (e.g., the underlying probability distribution) for which we might not have information.

As explained earlier, possibility theory is based on two dual functions, necessity measures (η) and possibility measures (π). The two functions, whose range is [0, 1], can be converted to a single function, C, whose range is $[-1, 1]$, as described below [Klir and Yuan, 1995],

$$C(A) = \eta(A) + \pi(A) - 1 \tag{15.50}$$

Positive values of $C(A)$ indicate the *degree of confirmation* of A by the available evidence, and negative values of $C(A)$ express the *degree of disconfirmation* of A by the evidence. Such a metric adds value to the use of possibility theory in its use in characterizing the forms of uncertainty due to ambiguity, nonspecificity, and imprecision. For instance, in Example 15.11 we compute the *degree of confirmation* for the interval A to be 0.226.

Example 15.12. In the previous example, the intervals were consistent and hence it was relatively straightforward to calculate the most possible interval and subsequent nested intervals. This example illustrates the application of Donald's [2003] method when the intervals are not consistent, i.e, when there is no common interval that spans across all the data intervals. Consider a set of data intervals as shown in Table 15.7.

As there is no interval that is included in all the original intervals, experts can choose the maximally possible interval by relying on their experiences, or they can choose the interval that intersects with the most original data intervals. Due to its objectivity, the latter approach is used in this example. Once the maximally possible interval is identified, the process of choosing the nested intervals is dependent on the degree to which each interval includes a subsequent interval. The following steps illustrate the entire process of generating a possibility distribution. Table 15.8 shows all the intervals produced by taking the intersections of the original data.

From Table 15.8, it can be seen that the interval [11, 13] intersects with the most intervals and thus forms the maximally possible interval. Given this interval, the rest of the consonant intervals are selected according to the nesting structure of the subsequent intervals. Table 15.9 shows one of the series of consonant and nonconsonant intervals that were selected for this example. Once the nesting structure is determined, the weights are then redistributed from the nonconsonant to the consonant intervals as explained in the previous example. Table 15.10 shows the final weights assigned to each interval after this redistribution, and Fig. 15.9 shows the possibility distribution plotted according to the weights on the intervals shown in the Table 15.10.

TABLE 15.7
Original data intervals

Observation	Intervals	Weight
1	[1, 6]	0.25
2	[2, 14]	0.25
3	[7, 13]	0.25
4	[11, 17]	0.25

TABLE 15.8
Intervals produced by taking the intersections of all the original data intervals

Interval	Weight	Normalized weight
[1, 6]	0.25	0.125
[2, 6]	0.25	0.125
[2, 14]	0.25	0.125
[7, 13]	0.25	0.125
[11, 17]	0.25	0.125
[11, 13]	0.25	0.125
[11, 14]	0.25	0.125
[1, 17]	0.25	0.125

TABLE 15.9

Redistribution of weights from the nonconsonant to the consonant intervals

Nonconsonant interval	Consonant interval	β	κ	ρ
[1, 6]	[11, 13]	0	0	0
	[7, 13]	0	0	0
	[2, 14]	0.3333	0.516	0.0645
	[1, 17]	0.3125	0.484	0.0605
[2, 6]	[11, 13]	0	0	0
	[7, 13]	0	0	0
	[2, 14]	0.3333	0.571	0.0714
	[1, 17]	0.2500	0.429	0.0536
[11, 17]	[11, 13]	1	0.512	0.0640
	[7, 13]	0.3333	0.170	0.0213
	[2, 14]	0.2500	0.128	0.0160
	[1, 17]	0.3750	0.190	0.0238
[11, 14]	[11, 13]	1	0.566	0.0708
	[7, 13]	0.3333	0.187	0.0233
	[2, 14]	0.2500	0.141	0.0176
	[1, 17]	0.1875	0.106	0.0133

TABLE 15.10

Final possibilistic weights after the transfer of weight from the nonconsonant to the consonant intervals

Interval	Final weight	Possibilistic weight
[11, 13]	0.26	1.00
[7, 13]	0.17	0.74
[2, 14]	0.29	0.57
[1, 17]	0.28	0.28

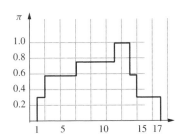

FIGURE 15.9

Possibility distribution generated from nonconsistent intervals.

Comparison of Possibility Theory and Probability Theory

Both possibility theory and probability theory are special branches of evidence theory. This chapter has shown how the two theories relate. Both share similar axiomatic foundations, but there are a few, quite distinct, differences (see Appendix A). Their normalizations are different: probabilities must sum to unity, and possibilities have one or more maximal elements equal to unity. And the way in which each represents ignorance is different. For probability measures, total ignorance is expressed by the uniform probability distribution function

$$p(x) = \frac{1}{|X|} = m(\{x\}) \tag{15.51}$$

for all $x \in X$. This follows from the fact that basic probability assignments are required to focus only on singletons. This choice is justified on several grounds with probability theory, where it is required that every uncertainty situation be characterized by a single probability distribution. But this reasoning represents a paradox: on purely intuitive grounds if no information is available about a situation, then no distribution is supported by any evidence and, hence, a choice of one over the other is arbitrary. Total ignorance should be represented by all possible distribution functions, but this is not a formulation of the theory.

In possibility theory ignorance is represented naturally (as discussed in the section on evidence theory), and is expressed by

$$m(X) = 1 \tag{15.52}$$

Here we have that all the evidence is allocated only to the full universe, X. Hence the information is completely nonspecific; there is no evidence supporting any singleton or subset of the universe, i.e., $m(A) = 0$ for all $A \neq X$.

Although interpretations of possibility theory are less developed than their probabilistic counterparts, it is well established that possibility theory provides a link between fuzzy set theory and probability theory. When information regarding some situation is given in both probabilistic and possibilistic terms, the two interpretations should, in some sense, be consistent. That is, the two measures must satisfy some *consistency condition* (this form of consistency should not be confused with the same term used in the previous section on consistent intervals). Although several such conditions have been reported [Klir and Yuan, 1995] the weakest one acceptable on an intuitive basis is stated as: "an event that is probable to some degree must be possible to at least that same degree," or, the *weak consistency condition* can be expressed formally as

$$p(A) \leq \pi(A) \tag{15.53}$$

for all $A \in P(X)$. The *strongest consistency condition* would require, alternatively, that any event with nonzero probability must be fully possible; formally,

$$p(A) > 0 \Rightarrow \pi(A) = 1 \tag{15.54}$$

All other consistency conditions fall between the extremes specified by Eqs. (15.53) and (15.54).

SUMMARY

This chapter has summarized very briefly a few of the various elements of monotone measures: beliefs, plausibilities, possibilities, necessities (certainties), and probabilities.

The axiomatic expansion of these measures, along with fuzzy set theory as a special case, is known collectively in the literature as generalized information theory (GIT) [Klir and Wierman, 1999]. GIT contains fuzzy set theory, but this development is not explored in this text. Suffice it to say that the current research in GIT is expanding and showing that, collectively, these various theories are very powerful in representing a large suite of uncertainties: fuzziness, vagueness, unknownness, nonspecificity, strife, discord, conflict, randomness, and ignorance.

The engineering community has mostly relied on probabilistic methods to analyze empirical data. These methods are widely used in experimental analysis for studying the variation in model parameters, for determining structural properties of various materials and for image processing. The variables under consideration are assumed to be randomly distributed and hence the analysis depends on methods that satisfy the axioms of probability. However, on many occasions the data available do not necessarily represent complete knowledge and thus do not support probabilistic analysis. When faced with limited data it might be more appropriate to use possibility distributions such that complete knowledge is not assumed. Possibility distributions can capture the imprecision in data and are thus useful in quantifying uncertainty resulting from incomplete knowledge. This chapter presented some of the recent advances in the area of generation of possibility distributions from empirical data that are imprecise, i.e, when data are available as a set of overlapping intervals. In addition to the methods presented in this chapter, other methods for deriving possibility distributions exist; the reader is referred to Donald [2003] and Joslyn [1997] for alternative methods, especially for deriving possibility distributions from disjoint data sets.

There exists a formal relationship between probability and fuzzy logics; this relationship, detailed in Appendix A, illustrates axiomatically that their common features are more substantial than their differences. It should be noted in Appendix A that, whereas the additivity axiom (axiom 9) is common to both a probability and a fuzzy logic, it is rejected in the Dempster–Shafer theory of evidence [Gaines, 1978], and it often presents difficulties to humans in their reasoning. For a probability logic the axiom of the excluded middle (or its dual, the axiom of contradiction, i.e., $p(x \wedge \overline{x}) = 0$) *must* apply; for a fuzzy logic it *may* or *may not* apply.

There are at least two reasons why the axiom of the excluded middle might be inappropriate for some problems. First, people may have a high degree of belief about a number of possibilities in a problem. A proposition should not be "crowded out" just because it has a large number of competing possibilities. The difficulties people have in expressing beliefs consistent with the axioms of a probability logic are sometimes manifested in the rigidity of the axiom of the excluded middle [Wallsten and Budescu, 1983]. Second, the axiom of the excluded middle results in an inverse relationship between the information content of a proposition and its probability. For example, in a universe of n singletons, as more and more evidence becomes available on each of the singletons, the relative amount of evidence on any one diminishes [Blockley, 1983]. This characteristic makes axiom A.1 inappropriate as a measure for modeling uncertainty in many situations.

Finally, rather than debate what is the correct set of axioms to use (i.e., which logic structure) for a given problem, one should look closely at the problem, determine which propositions are vague or imprecise and which ones are statistically independent or mutually exclusive, and use these considerations to apply a proper uncertainty logic, with or without the axiom of the excluded middle. By examining a problem so closely as to determine these relationships, one finds out more about the structure of the problem in the first place. Then the assumption of a strong truth-functionality (for a fuzzy logic) could be viewed as a computational device that simplifies calculations, and the resulting solutions would be presented as ranges of values that most certainly form bounds around the true answer if the assumption is not reasonable. A choice of whether a fuzzy logic is appropriate is, after all, a question of balancing the model with the nature of the uncertainty contained within it. Problems without an underlying physical model, problems involving a complicated weave of technical, social, political, and economic factors, and problems with incomplete, ill-defined, and inconsistent information where conditional probabilities cannot be supplied or rationally formulated perhaps are candidates for fuzzy logic applications. Perhaps, then, with additional algorithms like fuzzy logic, those in the technical and engineering professions will realize that such difficult issues can now be modeled in their designs and analyses.

REFERENCES

Blockley, D. (1983). "Comments on 'Model uncertainty in structural reliability,' by Ove Ditlevsen," *J. Struct. Safety*, vol. 1, pp. 233–235.

Dempster, A. (1967). "Upper and lower probabilities induced by a multivalued mapping," *Ann. Math. Stat.*, vol. 38, pp. 325–339.

Donald, S. (2003). "Development of empirical possibility distributions in risk analysis," PhD dissertation, University of New Mexico, Department of Civil Engineering, Albuquerque, NM.

Dubois, D. and Prade, H. (1988). *Possibility theory*, Plenum Press, New York.

Gaines, B. (1978). "Fuzzy and probability uncertainty logics," *Inf. Control*, vol. 38, pp. 154–169.

Joslyn, C. (1997). "Measurement of possibilistic histograms from interval data," *Int. J. Gen. Syst.*, vol. 26, pp. 9–33.

Klir, G. and Folger, T. (1988). *Fuzzy sets, uncertainty, and information*, Prentice Hall, Englewood Cliffs, NJ.

Klir, G. and Smith, R. (2001). "On measuring uncertainty and uncertainty-based information: Recent developments," *Ann. Math. Artif. Intell.*, vol. 32, pp. 5–33.

Klir, G. and Wierman, M. (1999). *Uncertainty Based Information: Elements of Generalized Information Theory*, Physica-Verlag/Springer, Heidelberg.

Klir, G. and Yuan, B. (1995). *Fuzzy sets and fuzzy logic*, Prentice Hall, Upper Saddle River, NJ.

Shafer, G. (1976). *A mathematical theory of evidence*, Princeton University Press, Princeton, NJ.

Wallsten, T. and Budescu, D. (1983). *Manage. Sci.*, vol. 29, no. 2, p. 167.

Yager, R. (1993). "Aggregating fuzzy sets represented by belief structures," *J. Intell. Fuzzy Syst.*, vol. 1, no. 3, pp. 215–224.

Yager, R. and Filev, D. (1994). "Template-based fuzzy systems modeling," *J. Intell. Fuzzy Syst.*, vol. 2, no. 1, pp. 39–54.

Zadeh, L. (1978). "Fuzzy sets as a basis for a theory of possibility," *Fuzzy Sets Syst.*, vol. 1, pp. 3–28.

Zadeh, L. (1984). "Review of the book *A mathematical theory of evidence*, by Glenn Shafer," *AI Mag.*, Fall, pp. 81–83.

Zadeh, L. (1986). "Simple view of the Dempster–Shafer theory of evidence and its implication for the rule of combination," *AI Mag.*, Summer, pp. 85–90.

PROBLEMS

15.1. In structural dynamics a particular structure that has been subjected to a shock environment may be in either of the fuzzy sets "damaged" or "undamaged," with a certain degree of membership over the magnitude of the shock input. If there are two crisp sets, functional (F) and nonfunctional (NF), then a monotone measure would be the evidence that a particular system that has been subjected to shock loading is a member of functional systems or nonfunctional systems. Given the evidence from two experts shown here for a particular structure, find the beliefs and plausibilities for the focal elements.

Focal elements	m_1	m_2	bel_1	bel_2	pl_1	pl_2
F	0.3	0.2				
NF	0.6	0.6				
F $\cup$ NF	0.1	0.2				

15.2. Suppose you have found an old radio (vacuum tube type) in your grandparents' attic and you are interested in determining its age. The make and model of the radio are unknown to you; without this information you cannot find in a collector's guide the year in which the radio was produced. Here, the year of manufacture is assumed to be within a particular decade. You have asked two antique radio collectors for their opinion on the age. The evidence provided by the collectors is fuzzy. Assume the following questions:

1. Was the radio produced in the 1920s?
2. Was the radio produced in the 1930s?
3. Was the radio produced in the 1940s?

Let R, D, and W denote subsets of our universe set P – the set of radio-producing years called the 1920s (Roaring 20s), the set of radio-producing years called the 1930s (Depression years), and the set of radio-producing years called the 1940s (War years), respectively. The radio collectors provide *beas* as given in the accompanying table.

	Collector 1			Collector 2			Combined evidence		
Focal elements	m_1	bel_1	pl_1	m_2	bel_2	pl_2	m_{12}	bel_{12}	pl_{12}
R	0.05	0.05	0.8	0.15	0.15	0.85	0.1969		
D	0.1	0.1		0.1	0.1				
W	0	0		0	0				
R $\cup$ D	0.2	0.35	1	0.25	0.5	1	0.2677		
R $\cup$ W	0.05			0.05					
D $\cup$ W	0.1			0.05					
R $\cup$ D $\cup$ W	0.5			0.4					

(*a*) Calculate the missing belief values for the two collectors.
(*b*) Calculate the missing plausibility values for the two collectors.
(*c*) Calculate the missing combined evidence values.
(*d*) Calculate the missing combined belief and plausibility values.

15.3. The quality control for welded seams in the hulls of ships is a major problem. Ultrasonic defectoscopy is frequently used to monitor welds, as is x-ray photography. Ultrasonic defectoscopy is faster but less reliable than x-ray photography. Perfect identification of flaws in welds is dependent on the experience of the person reading the signals. An abnormal signal occurs for three possible types of situations. Two of these are flaws in welds: a cavity (C) and a cinder inclusion (I); the former is the more dangerous. Another situation is due to a loose contact of the sensor probe (L), which is not a defect in the welding seams but an error in measuring. Suppose we have two experts, each using a different weld monitoring method, who are asked to identify the defects in an important welded seam. Their responses in terms of beas are given in the table. Calculate the missing portions of the table.

Focal elements	Expert 1			Expert 2			Combined evidence		
	m_1	bel_1	pl_1	m_2	bel_2	pl_2	m_{12}	bel_{12}	pl_{12}
C	0.3	0.3	0.85	0.2	0.2		0.4	0.4	
I	0.05	0.05		0.1	0.1		0.15	0.15	
L	0.05	0.05		0.05	0.05				
C∪I	0.2	0.55		0.15	0.45		0.16	0.71	
C∪L	0.05	0.4	0.95	0.05	0.3				
I∪L	0.05	0.5		0.15	0.3				
C∪I∪L	0.3	1		0.3	1				

15.4. You are an aerospace engineer who wishes to design a bang–bang control system for a particular spacecraft using thruster jets. You know that it is difficult to get a good feel for the amount of thrust that these jets will yield in space. Gains of the control system depend on the amount of the force the thrusters yield. Thus, you pose a region of three crisp sets that are defined with respect to specific gains. Each set will correspond to a different gain of the control system.

You can use an initial estimate of the force you get from the thrusters, but you can refine it in real time utilizing different gains for the control system. You can get a force estimate and a belief measure for that estimate for a specific set. Suppose you define the following regions for the thrust values, where thrust is in pounds:

$$A_1 \text{ applies to a region } 0.8 \leq \text{ thrust value } \leq 0.9.$$

$$A_2 \text{ applies to a region } 0.9 \leq \text{ thrust value } \leq 1.0.$$

$$A_3 \text{ applies to a region } 1.0 \leq \text{ thrust value } \leq 1.1.$$

Two expert aerospace engineers have been asked to provide evidence measures reflecting their degree of belief for the various force estimates. These *beas* along with calculated belief measures are given here. Calculate the combined belief measure for each focal element in the table.

Focal elements	Expert 1		Expert 2	
	m_1	bel_1	m_2	bel_2
A_1	0.1	0.1	0	0
A_2	0.05	0.05	0.05	0.05
A_3	0.05	0.05	0.1	0.1
$A_1 \cup A_2$	0.05	0.2	0.05	0.1
$A_1 \cup A_3$	0.05	0.2	0.15	0.25
$A_2 \cup A_3$	0.1	0.2	0.05	0.2
$A_1 \cup A_2 \cup A_3$	0.6	1	0.6	1

15.5. Consider the DC series generator shown in Fig. P15.5. Let R_a = armature resistance, R_s = field resistance, and R = load resistance. The voltage generated across the terminal is given by $V_t = E_g - (I_s R_s + I_a R_a)$. *Note:* If there is no assignment for R, i.e., if the value of R is infinity, then the generator will not build up because of an open circuit. Also R_a can have a range of values from a low value to a high value. To generate different load voltages required, we can assign values for R, R_a, and R_s in different ways to get the voltage. They are very much interrelated and the generated voltage need not have a unique combination of R, R_a, and R_s. Hence, nesting of focal elements for these resistances does have some physical significance.

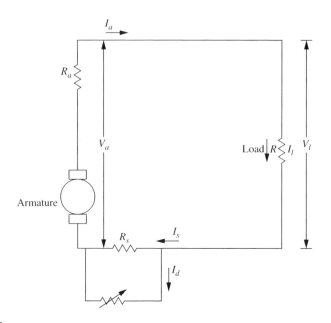

FIGURE P15.5

Let the *basic evidence assignment* for the elements of universe X $\{R, R_a, R_s\}$ be as shown in the accompanying table:

X	m_1	m_2
R_a	0.1	0.1
R	0.1	0
R_s	0.1	0.5
$R_a \cup R_s$	0.3	0.3
$R \cup R_a$	0.1	0
$R_s \cup R$	0.2	0
$R_a \cup R \cup R_s$	0.1	0.1

(a) Does m_1 or m_2 represent a possibility measure?

(b) If either of the evidence measures (or both) is nested, find the possibility distributions.

15.6. A general problem in biophysics is to segment volumetric MRI (Magnetic Resonance Imaging) data of the head given a new set of MRI data. We use a "model head" that has already had the structures in the head (mainly brain and brain substructures) labeled. We can use the model head to help in segmenting the data from the new head by assigning beas to each voxel (a voxel is a three-dimensional pixel) in the new MRI data set, based on what structures contain, or are near, the corresponding voxel in the model head. In this example a nested subset corresponds to the physical containment of a head structure within another structure. We will select a voxel for which the *beas* form a consonant body of evidence.

$$X = \{\text{structures in the MRI data}\}$$

$$H = \text{head}$$

$$B = \text{brain}$$

$$N = \text{neocortex}$$

$$L = \text{occipital lobe}$$

$$C = \text{calcarine fissure}$$

Thus, we have $C \subset L \subset N \subset B \subset H$. For voxel V we have a basic distribution of

$$m = (\mu_C, \mu_L, \mu_N, \mu_B, \mu_H) = (0.1, 0.1, 0.2, 0.3, 0.3)$$

Find the corresponding possibility distribution and draw the nesting diagram.

15.7. A test and diagnostics capability is being developed for a motion control subsystem that consists of the following hardware: a motion control IC (Integrated Circuit), an interconnect between motion control IC, an H-switch current driver, an interconnect between H-switch current driver, a motor, and an optical encoder.

The elements of the motion control subsystem are as follows:

$$x_1 = \text{motion control IC}$$

$$x_2 = \text{interconnect 1}$$

$$x_3 = \text{H-switch current driver}$$

$$x_4 = \text{interconnect 2}$$

$$x_5 = \text{motor}$$

$$x_6 = \text{optical encoder}$$

If a motion control subsystem failure exists, a self-test could describe the failure in the following *bea*: $m = (0.2, 0, 0.3, 0, 0, 0.5)$. This nested structure is based on the level of hardware isolation of the diagnostic software. This isolation is hierarchical in nature. You first identify a motion control subsystem failure $m(A_6)$ that includes a possibility of any component failure $(x_1, x_2, x_3, x_4, x_5, x_6)$. The test then continues and, due to isolation limitations, a determination can be made of the failure possibility consisting of $m(A_3)$, subset (x_1, x_2, x_3), followed by the ability to isolate to an x_1 failure if x_1 is at fault. *Basic evidence assignments* are constructed from empirical data and experience.

Find the associated possibility distribution and draw the nesting diagram.

15.8. Design of a geometric traffic route can be described by four roadway features: a corner, a curve, a U-turn, and a circle. The traffic engineer can use four different evaluation criteria (expert guidance) to use in the design process:

$$m_1 = \text{criteria: fairly fast, short distance, arterial road, low slope points}$$

$$m_2 = \text{criteria: slow, short distance, local road, low slope points}$$

$$m_3 = \text{criteria: fast, long distance, ramp-type road, medium slope points}$$

$$m_4 = \text{criteria: very fast, medium distance, highway, medium slope points}$$

Corner	Curve	U-turn	Circle	m_1	m_2	m_3	m_4
0	0	0	1	0	0	0.2	0.1
0	0	1	0	0.1	0.1	0	0
0	0	1	1	0	0	0	0
0	1	0	0	0.3	0.2	0.3	0.4
0	1	0	1	0	0	0.5	0.5
0	1	1	0	0.1	0.1	0	0
0	1	1	1	0	0	0	0
1	0	0	0	0.2	0.3	0	0
1	0	0	1	0	0	0	0
1	0	1	0	0.1	0.1	0	0
1	0	1	1	0	0	0	0
1	1	0	0	0.1	0.1	0	0
1	1	0	1	0	0	0	0
1	1	1	0	0.1	0.1	0	0
1	1	1	1	0	0	0	0

Using the 15 ($2^4 - 1$) focal elements shown in the accompanying table, determine which, if any, of the four evidence measures (m_1-m_4) results in an ordered possibility distribution.

15.9. Given a communication link with a sender, receiver, and interconnecting link, an error in a message could occur at the sender, receiver, or on the interconnecting link. Combinations such as an error on the link that is not corrected by the receiver are also possible. Let S, R, and L represent sources of error in the sender, receiver, and link, respectively. If E is the universe of error sources, then

$$P(E) = (\emptyset, \{S\}, \{R\}, \{L\}, \{S, R\}, \{S, L\}, \{R, L\}, \{S, R, L\})$$

Now assume each source has its own expert and each of these provides their basic assignment of the actual source of an error as follows:

S	R	L	m_S	m_R	m_L
0	0	0	0	0	0
0	0	1	0.4	0.3	0
0	1	0	0.2	0	0
0	1	1	0.2	0	0
1	0	0	0	0.5	0.5
1	0	1	0.1	0.2	0
1	1	0	0	0	0.4
1	1	1	0.1	0	0.1

Indicate which experts, if any, have evidence that is consonant. For each of these, do the following:

(a) Determine the possibility distribution.

(b) Draw the nesting diagram.

(c) Give the physical significance of the nesting.

15.10 There are a number of hazardous waste sites across the country that pose significant health risk to humans. However, due to high costs involved in exposure analysis only a limited amount of information can be collected from each site to determine the extent of contamination.

Suppose it is determined that one of the sites is contaminated by a new carcinogenic chemical identified as *Tox*. The table shows the results from the chemical analysis of the groundwater samples collected from one of the sites. Given these sparse data, determine the possibility distribution of exposure concentrations for the chemical *Tox*.

Observation	Concentration (mg/L)
1	[0.01, 0.1]
2	[0.03, 0.2]
3	[0.03, 0.15]
4	[0.008, 0.06]

15.11. Due to their excellent self-healing properties, rock salt caverns are used to store nuclear waste from various nuclear plants. One of the properties useful in determining the suitability of a cavern for nuclear waste storage is the creep rate of salt; salt creeps very slowly with time. This creep rate determines the strength of the cavern and the duration that the cavern can be accessible to human operations. The table shows the strain rate results from creep tests conducted on rock salt cores from four locations of the waste repository. Given these data, determine the strain rate interval that is 80% possible (possibilistic weight = 0.8). Also, find the *degree of confirmation*.

Observation	Strain rate (s^{-1})
1	[6.0E-10, 8.5E-10]
2	[8.0E-10, 1.1E-9]
3	[9.0E-10, 2.0E-9]
4	[5.0E-10, 9.0E-10]

15.12. Predicting interest rates is critical for financial portfolio management and other investment decisions. Based on historical variations and other factors, the following interest rates are predicted for the next two months.

Observation	Interest rate
1	[0.75, 1.5]
2	[1.0, 1.25]
3	[0.75, 1.25]
4	[1.5, 2.0]
5	[1.75, 2.25]

(*a*) What is the possibility that the interest rates will be higher than 2%?

(*b*) Give the reason for your choice of consonant intervals.

(*c*) Find the *degree of confirmation*.

APPENDIX A

AXIOMATIC DIFFERENCES BETWEEN FUZZY SET THEORY AND PROBABILITY THEORY

A slight variation in the axioms at the foundation of a theory can result in huge changes at the frontier.

Stanley P. Gudder
Quantum Probability, 1988

In Chapter 1 of this book is a discussion of the relationships and historical confusion between probability theory and fuzzy set theory. It seems fitting that this book should conclude by coming full circle to that same discussion. A paper by Gaines [1978] does an eloquent job of addressing this issue. Historically, probability and fuzzy sets have been presented as distinct theoretical foundations for reasoning and decision making in situations involving uncertainty. Yet when one examines the underlying axioms of both probability and fuzzy set theories, the two theories differ by only one axiom in a total of 16 axioms needed for a complete representation! The material that follows is a brief summary of Gaines's paper, which established a common basis for both forms of logic of uncertainty in which a basic uncertainty logic is defined in terms of valuation on a lattice of propositions. Addition of the axiom of the excluded middle to the basic logic gives a standard probability logic. Alternatively, addition of a requirement for strong truth-functionality gives a fuzzy logic.

In this discussion fuzzy logic is taken to be a multivalued extension of Boolean logic based on fuzzy set theory in which truth values are extended from the endpoints of the

© 2004 John Wiley & Sons, Ltd ISBNs: 0-470-86074-X (HB); 0-470-86075-8 (PB)

interval [0, 1] to a range through the entire interval. The normal logical operations are defined in terms of arithmetic operations on these values; the values are regarded as degrees of membership to truth. The logic operations and associated arithmetic operations are those of conjunction (involving the minimum operator), disjunction (involving the maximum operator), and negation (involving a subtraction from unity). Gaines [1978] points out that the use of the max and min operations in fuzzy logic is not sufficient to distinguish it from that of probability theory – both operators arise naturally in the calculation of the conjunction and disjunction of probabilistic events. Our association of addition and multiplication as natural operations upon probabilities comes from our frequent interest in statistically independent events, not from the logic of probability itself.

In developing a basic uncertainty logic, we begin first by defining a lattice consisting of a universe of discourse, X, a maximal element T, a minimal element F, a conjunction, $\wedge$, and a disjunction, $\vee$. This lattice will be denoted $L(X, T, F, \wedge, \vee)$. For the axioms (or postulates) to follow, lowercase letters x, y, and z denote specific elements of the universe X within the lattice. The following 15 axioms completely specify a basic uncertainty logic.

The basic uncertainty logic begins with the lattice L satisfying idempotency:

1. For all $x \in L$ $x \vee x = x \wedge x = x$ commutativity:

2. For all $x, y \in L$ $x \vee y = y \vee x$ and $x \wedge y = y \wedge x$ associativity:

3. For all $x, y, z \in L$ $x \vee (y \vee z) = (x \vee y) \vee z$ absorption:

4. For all $x, y \in L$ $x \vee (x \wedge y) = x, x \wedge (x \vee y) = x$
   and the definition of the maximal and minimal elements:

5. For all $x \in L$ $x \vee T = T$ and $x \wedge T = x$
   The usual order relation may also be defined:

6. For all $x, y \in L$ $x \leq y$ if there exists a $z \in L$ such that $y = x \vee z$

Now suppose that every element of the lattice L is assigned a truth value (for various applications this truth value would be called a probability, degree of belief, etc.) in the interval [0, 1] by a continuous, order-preserving function, $p : L \to [0, 1]$, with constraints:

7. $p(F) = 0; p(T) = 1$

8. For all $x, y \in L$ $x \leq y$ then $p(x) \leq p(y)$
   and an additivity axiom:

9. For all $x, y \in L$ $p(x \wedge y) + p(x \vee y) = p(x) + p(y)$

We note that for p to exist we have to have

$$p(x \wedge y) \leq \min[p(x), p(y)] \leq \max[p(x), p(y)] \leq p(x \vee y)$$

Now a logical equivalence (or congruence) is defined by

10. For all $x, y \in L$ $x \leftrightarrow y$ if $p(x \wedge y) = p(x \vee y)$

The general structure provided by the first 10 axioms is common to virtually all logics. To finalize Gaines's basic uncertainty logic we now need to define implication and negation, for it is largely the definition of these two operations that distinguishes among various multivalued logics [Gaines, 1978]. We also note that postulate 9 still holds when the outer inequalities become equalities – a further illustration that the additivity of probability-like valuations is completely compatible with, and closely related to, the minimum and maximum operations of fuzzy logic.

To define implication and negation we make use of a metric on the lattice L that measures *distance* between the truth values of two different propositions. This is based on the notion that logically equivalent propositions should have a zero distance between them. So, we define a distance measure:

11. For all $x, y \in$ L $d(x, y) = p(x \vee y) - p(x \wedge y)$

where $d(x, x) = 0, 0 \leq d(x, y) \leq 1$, and $d(x, y) + d(y, z) = d(x, z)$. This axiom is shown schematically in Fig. A.1.

Therefore, a measure of equivalence between two elements can be 1 minus the distance between them, or

12. For all $x, y \in$ L $p(x \leftrightarrow y) = 1 - d(x, y) = 1 - p(x \vee y) + p(x \wedge y)$

Hence, if $d = 0$, the two elements x and y are equivalent, as seen in Fig. A.2.

To measure the strength of an implication, we measure a distance between x and $x \wedge y$, as seen in Fig. A.3.

13. For all $x, y \in$ L $p(x \rightarrow y) = p(x \leftrightarrow x \wedge y) = 1 - d(x, x \wedge y)$
$$= 1 - p(x) + p(x \wedge y)$$
$$= 1 + p(y) - p(x \vee y) = 1 - d(y, x \vee y)$$

Negation can now be defined in terms of equivalence and implication as

14. For all $x \in$ L $p(\overline{x}) = p(x \leftrightarrow F) = 1 - p(x) = 1 - d(x, F)$

We note by combining axioms 9 and 14 that if element y is replaced by element $\overline{x}$, we get

$$p(x \vee \overline{x}) + p(x \wedge \overline{x}) = p(x) + p(\overline{x}) = 1$$

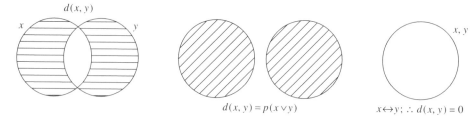

FIGURE A.1
Venn diagrams on distance measure, $d(x, y)$.

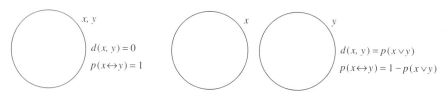

FIGURE A.2
Venn diagram on equivalent and maximally nonequivalent elements.

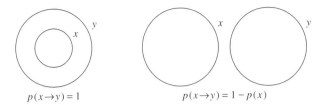

FIGURE A.3
Venn diagram for implication.

This has also been termed the "additivity axiom" in a probability theory.

Finally, we add a postulate of distributivity:

15. For all $x, y, z \in L$ $x \wedge (y \vee z) = (x \wedge y) \vee (x \wedge z)$

which will prove useful for the two specializations of this basic uncertainty logic to be described in the remaining paragraphs.

Axioms 1–15 above provide for a basic distributive uncertainty logic. The addition of a special 16th (denoted 16.1) axiom, known in the literature as the axiom of the excluded middle:

16.1. For all $x \in L$ $p(x \vee \overline{x}) = 1$

leads to Rescher's standard probability logic [Rescher, 1969].

Alternatively, if we add another special 16th (denoted 16.2) axiom to the basic axioms 1–15, we get a special form of fuzzy logic:

16.2. For all $x, y \in L$ $\{p(x \rightarrow y) = 1 \vee p(y \rightarrow x) = 1\}$

known in the literature as the Lukasiewicz infinite-valued logic [Rescher, 1969]. Axiom 16.2 is called the *strong truth-functionality*, or *strict implication*, in the literature.

It should be pointed out that a *weaker* form of axiom 16.2, or

$$\text{For all } x, y \in L \qquad p((x \rightarrow y) \vee (y \rightarrow x)) = 1$$

is embraced by both a probability and a fuzzy logic [Gaines, 1978].

REFERENCES

Gaines, B. (1978). "Fuzzy and probability uncertainty logics," *Inf. Control*, vol. 38, pp. 154–169.
Rescher, N. (1969). *Many-valued logic*, McGraw-Hill, New York.

APPENDIX B

ANSWERS TO SELECTED PROBLEMS

CHAPTER 1

1.9. Crisp: $\chi_{LD_{50}} = 1$, for $0 < LD_{50} \leq 5000$ mg/kg
       $\chi_{LD_{50}} = 0$, for $LD_{50} > 5000$ mg/kg and $LD_{50} \leq 0$
1.11. (b) $A(x) = 0$, $x \leq a$; $1 - e^{-k(x-a)^2}$, $x > a$; (e) $A(x) = e^{-k(x-a)^2}$, $k > 0$
1.16. $P(x) = \{x_1, x_2, \ldots, (x_1, x_2), \ldots, (x_3, x_4), \ldots, (x_1, x_2, x_4), \ldots\}$

CHAPTER 2

2.2. (b) $0.8/1 + 0.4/2 + 0.9/3 + 1/4 + 1/5$
2.4. (c) $0/1.0 + 0.25/1.5 + 0.7/2.0 + 0.85/2.5 + 1.0/3.0$
       (e) $0/1.0 + 0.4/1.5 + 0.3/2.0 + 0.15/2.5 + 0/3.0$
2.7. (a) $1/1 + 1/10 + 1/20 + 0.5/40 + 0.2/80 + 0/100$
       (e) $0/1 + 0/10 + 0/20 + 0.5/40 + 0.8/80 + 1/100$
2.13. Difference: $\underset{\sim}{A}|\underset{\sim}{B} = 0.14/0 + 0.32/1 + 0.61/2 + 0.88/3 + 0.86/4 + 0.68/5 + 0.39/6$
       $+ 0.12/7 + 0/8 + 0.04/9 + 0.01/10$
2.14. (d) $0/1 + 0/2 + 0/3 + 0/4 + 0.2/5 + 0/6 + 0.5/7 + 0/8$
       (f) $1/1 + 1/2 + 1/3 + 1/4 + 0.9/5 + 0.4/6 + 0.5/7 + 0/8$

CHAPTER 3

3.3. (a) $g_{11} = 0.1$, $g_{12} = 0.4$, $g_{22} = 0.9$, $g_{31} = 0.6$;
       (b) $C_{11} = 0.1$, $C_{12} = 0.9$

Fuzzy Logic with Engineering Applications, Second Edition   T. J. Ross
© 2004 John Wiley & Sons, Ltd   ISBNs: 0-470-86074-X (HB); 0-470-86075-8 (PB)

3.4. (a) $R_{11} = 1, R_{13} = 0.25, R_{22} = 0.4, R_{32} = 0.2;$
    (b) $S_{11} = 0.1, S_{23} = 0.25, S_{31} = 1, S_{33} = 0.25$
    (c) $0.3/SRR + 0.3/MRR + 0.25/FRR$
    (d) $0.2/SRR + 0.2/MRR + 0.2/FRR$

3.7. (a) $0.7/1 + 1/2 + 0.7/6$
    (b) $0.56/1 + 1/2 + 0.7/6$

3.11. (a) $R_{11} = R_{12} = R_{13} = R_{14} = R_{15} = R_{16} = R_{21} = R_{31} = R_{41} = R_{51} = 0.1;$
      $R_{52} = R_{53} = R_{54} = R_{55} = R_{56} = R_{26} = R_{36} = R_{46} = 0.2; R_{35} = 0.5;$
      $R_{22} = R_{23} = R_{24} = R_{25} = R_{32} = R_{33} = R_{42} = R_{43} = 0.3; R_{34} = R_{54} =$
      $R_{55} = 0.4;$
    (b) $S_{11} = S_{12} = S_{13} = S_{21} = S_{31} = S_{41} = S_{51} = S_{61} = 0.1; S_{52} = 0.5; S_{42} = 0.4;$
      $S_{62} = S_{63} = 0.2; S_{22} = S_{23} = S_{32} = S_{33} = S_{43} = S_{53} = 0.3;$
    (c) $M_{11} = M_{12} = M_{13} = M_{21} = M_{31} = M_{41} = M_{51} = 0.1; M_{22} = M_{23} = M_{33} =$
      $M_{43} = 0.3; M_{52} = M_{53} = 0.2; M_{32} = 0.5; M_{42} = 0.4$

3.14. (a) $R_{11} = 0.2, R_{23} = 0.5, R_{31} = 0.3, R_{34} = 0.8, R_{43} = 0.9, R_{55} = 0.6$
    (b) $0.3/0.5 + 0.6/1.0 + 0.9/1.5 + 0.9/4 + 0.6/8 + 0.3/20$
    (c) $0.3/0.5 + 0.6/1.0 + 0.81/1.5 + 0.9/4 + 0.6/8 + 0.3/20$

3.17. (a) $R_{11} = 0.1, R_{12} = 0.1, R_{21} = 0.3, R_{22} = 0.7, R_{31} = 0.3, R_{32} = 0.4$
    (b) $S_{11} = S_{21} = 0.1, S_{12} = S_{22} = 0.2, S_{13} = S_{14} = 0.3, S_{23} = 0.8, S_{24} = 0.7$
    (c) $C_{11} = 0.1, C_{22} = 0.2, C_{23} = 0.7, C_{33} = 0.4$

3.18. (a) $R_{11} = 1, R_{12} = 0.4925, R_{13} = 0.2121, R_{55} = 1, R_{45} = 0.0363, R_{34} = 0.1561$
    (b) $R_{11} = 1, R_{21} = 0.8503, R_{13} = 0.3447, R_{55} = 1, R_{45} = 0.0430, R_{34} = 0.3093$

3.21. $R_{11} = 1, R_{12} = 0.538, R_{23} = 0.25, R_{25} = 0.333, R_{35} = 0.176, R_{45} = 0.818, R_{55} = 1$

3.24. (a) Symmetric relation $R_{ii} = 1, \ R_{12} = 0.836, \ R_{13} = 0.914, \ R_{14} = 0.682, \ R_{23} = 0.934, R_{24} = 0.6, R_{34} = 0.441$

3.26. (i) $\mu_T(x_1, z_1) = 0.5, \mu_T(x_1, z_2) = 0.7, \mu_T(x_2, z_2) = 0.7, \mu_T(x_2, z_3) = 0.5$
    (iii) $\mu_T(x_1, z_1) = 0.1, \mu_T(x_1, z_3) = 0.2, \mu_T(x_2, z_1) = 0.1, \mu_T(x_2, z_2) = 0.4$

3.27. Second row of column $\underset{\sim}{B}$: $\mu_B(y_{11}) = 1.0, \mu_B(y_{13}) = 1.0, \mu_B(y_{23}) = 1.0,$
$\mu_B(y_{31}) = 0.7, \mu_B(y_{42}) = 0.9$

CHAPTER 4

4.1. (a) $(\overline{\underset{\sim}{A}})_{0.7} = 1.0/x_1 + 0/x_2 + 0/x_3 + 0/x_4 + 0/x_5 + 1.0/x_6$
    (d) $(\underset{\sim}{A} \cap \underset{\sim}{B})_{0.6} = 0/x_1 + 1.0/x_2 + 0/x_3 + 0/x_4 + 0/x_5 + 0/x_6$

4.3. For $\underset{\sim}{A}$, (i) $\lambda = 0.2, \ x = (0.89, 4.11);$ (ii) $\lambda = 0.4, \ x = (0.09, 4.91);$ (iii) $\lambda = 0.7,$
    $x = (0.007, 4.993);$ (iv) $\lambda = 0.9, x = (0.148, 4.85);$ (v) $\lambda = 1.0, x = (0.00, 1.00).$
    For $\underset{\sim}{B}$, (i) $\lambda = 0.2, \ \lambda = 2.32;$ (ii) $\lambda = 0.4, \ x = 1.32;$ (iii) $\lambda = 0.7, \ x = 0.51;$ (iv)
    $\lambda = 0.9, x = 0.15;$ (v) $\lambda = 1.0, x = 0.00.$
    For $\underset{\sim}{C}$, (i) $\lambda = 0.2, \ x = 0.55;$ (ii) $\lambda = 0.4, \ x = 1.25;$ (iii) $\lambda = 0.7, \ x = 2.69;$ (iv)
    $\lambda = 0.9, x = 4.09;$ (v) $\lambda = 1.0, x = 5.00.$

4.5. (a) $R_{ij} = 1$ (b) $R_{ij} = 1,$ (c) $R_{ij} = 1$ (d) $R_{11} = R_{12} = R_{15} = R_{21} = R_{22} = R_{25} = R_{33} = R_{44} = R_{51} = R_{52} = R_{55} = 1,$ all others equal to 0.

4.9. Max membership, $z^* = 3$; weighted average, $z^* = 3.33$; center of sums, $z^* = 3.43$; center of largest area, $z^* = 2.02$; first of maxima and last of maxima, $z^* = 3$; centroid method, $z^* = 3.56.$

4.12. First maxima, $z^* = 2$; last maxima, $z^* = 3$; center of sums, $z^* = 2.5$; mean max, $z^* = 2.5$; centroid method, $z^* = 2.5$; weighted average methods, $z^* = 2.5$.

4.15. (*ii*) Defuzzified values using two centroids: centroid method, $T^* \approx 80.2°C$; weighted average method, $T^* \approx 79.75°C$.

CHAPTER 5

5.1. If $T(P) = T(Q)$ or P is false and Q is true, then $P \rightarrow Q$ is a tautology (that can be shown in a truth table).

5.4.

P	Q	$\overline{P}$	$\overline{Q}$	$P \wedge Q$	$P \vee Q$	$(P \wedge Q) \wedge (\overline{P} \vee \overline{Q})$	$(P \wedge Q) \wedge (\overline{P} \vee \overline{Q}) \leftrightarrow 0$
0	0	1	1	0	1	0	1
0	1	1	0	0	1	0	1
1	0	0	1	0	1	0	1
1	1	0	0	1	0	0	1

5.8. (*a*) $((P \rightarrow Q) \wedge P) \rightarrow Q$ by contradiction $((P \rightarrow Q) \wedge P) \wedge \overline{Q}$

P	Q	$\overline{P}$	$\overline{Q}$	$\overline{P} \rightarrow \overline{Q} = \overline{P} \vee Q$	$(P \rightarrow Q) \wedge P$	$((P \rightarrow Q) \wedge P) \wedge \overline{Q}$
0	0	1	1	1	0	0
0	1	1	0	1	0	0
1	0	0	1	0	0	0
1	1	0	0	1	1	0

5.12. (*a*) Mamdani: $R_{22} = 0.5, R_{33} = 1, R_{42} = 0.5, R_{54} = R_{35} = 0$
      Product: $R_{22} = 0.25, R_{33} = 1, R_{23} = 0.50, R_{54} = R_{35} = 0$
   (*b*) Mamdani: $0/0 + 0.5/1 + 1/2 + 0.5/3 + 0/4$ (same for product)

5.13. (*b*) $0.3/0 + 0.3/1 + 0.4/2 + 0.6/3 + 0.7/4 + 0.7/5$

5.16. (*b*) $0.4/66 + 0.6/68 + 0.6/70 + 0.6/72 + 0.4/74$

5.18. (*b*) $0.8/10 + 0.8/20 + 0.8/30 + 0.8/40$

5.22. (*a*) (*i*) $0.0/131 + 0.36/132 + 0.64/133 + 0.84/134 + 0.96/135 + 1.0/136$
   (*b*) (*iii*) $0/400 + 0.36/600 + 0.64/700 + 0.84/800 + 0.96/900 + 1.0/1000$

5.27. (*b*) $0/1 + 0.01/2 + 0.04/3 + 0.64/4 + 0/5$

5.29. IF $(x_1 \cap x_2)$, THEN y

5.30. IF $(x_1 \cap x_2)$, THEN t

5.33. No, the response surface obtained using a weighted sum defuzzifier will be different. The weighted average defuzzifier incorporates a denominator composed of the sum of the weights, and due to the manner in which these weights are derived (product, minimum norm) it produces a different response curve.

5.35. Mamdani: $\mu(\text{Re}) = 0.25$, $\mu(\text{Pr}_L) = 0.25$, and $\mu(\text{Pr}_H) = 0.25$, $z = 550$; Sugeno: $\mu(\text{Re}) = 0.25$, $\mu(\text{Pr}_L) = 0.75$, and $\mu(\text{Pr}_H) = 0.25$, $\text{Nu}_1 = 560.0993666$, $\text{Nu}_2 = 559.5643482$, $z = 559.8318574$

CHAPTER 6

6.5. (*a*) Isosceles, 0.92; Right, 0.89; RI, 0.89; Equilateral, 0.69; other, 0.083
      (*c*) Isosceles, 0.75; Right, 0.83; RI, 0.75; Equilateral, 0.83; other, 0.167
6.7.   Rank order: BMW, Mercedes, Infinity, Lexus, Porsche
6.12. For Economy – Class 1; Midsize – Class 2; Luxury – Class 3; $S_1 = 0.387$ ($x = 11$);
       $S_2 = 0.785$ ($x = 14$); $S_3 = 0.284$ ($x = 21$)

CHAPTER 7

7.1. After two cycles, $\hat{\theta} = \{0.3646, 8.1779\}$, Y = 1.4320, 4.0883, 6.4798
7.2. (*a*) After two cycles, $\hat{\theta} = \{1.2976, 7.5519\}$, Y = 1.6316, 3.8452, 6.5230
7.4. B = 0.9212, 6.0292; C = 0.0642, 2.0486; 2.0420, 4.0860; σ = 1.1186, 1.0154; 0.9994,
      1.1291; $(x_{1j}, x_{2j}, y_j) = (0, 2, 1.0398); (2, 4, 5.8662); (3, 6, 6.0288); j = 1, 2, 3$

CHAPTER 8

8.3. (*a*) If 'X$_1$' then 'G$_1$': $R_1 = X_1 \bullet G_1$, $R'_{11} = 1$, $R'_{22} = 0.25$, $R'_{12} = R'_{21} = 0.5$,
       $R'_{91} = 1$, $R'_{82} = 0.25$, $R'_{81} = R'_{92} = 0.5$, all others zero
      (*b*) R(row1) = 1.0, 0.5, 0, 0, 0
       R(row3) = 0, 0.5, 1, 0.5, 1.0
       R(row8) = 0.5, 0.25, 0, 0.25, 0.5
8.5. Discretizing each membership function at integers yields the following:

x	-10	-8	-6	-5	-4	-2	0
y	100	82	26	25	26	2	0

   Note: the function is symmetric.
8.7. (*a*) R$_1$(row 1) = 1, 0.6, 0.2, 0, R$_1$(row 2) = 0.6, 0.6, 0.2, 0, R$_1$(row 4) = 0, 0, 0, 0,
       R$_2$(row 2) = 0, 0.4, 0.4, 0.4, 0.4, 0.4, 0, R$_2$(row 4) = 0, 0.4, 0.8, 1, 0.8, 0.4, 0,
       R$_3$(row 3) = 0.2, 0.4, 0.8, 0.8, 0.8, 0.4, 0, R$_6$(row 6) = 0, 0.4, 0.4, 0.4, 0.4, 0.4, 0
      (*b*)

x	0	0.2	0.4	0.5	0.6	0.8	1.0
y	0	0.1	0.5	0.5	0.5	0.5	0.5

CHAPTER 9

9.1. (*a*) All singular values = 4
9.2. (*a*) $\Sigma_{11} = 19.9126$, $\Sigma_{22} = 5.7015$, $\Sigma_{44} = 0.1033$,
9.4. $d_{ii} = 0.4$, $d_{ij} = 0.2$, where $i \neq j$
9.6. (*c*) $\Sigma_{r_{11}} = 16.3760$, $\Sigma_{r_{22}} = 16.3760$, $E_{r_{33}} = 0$,
       Z_r(row 1) = 1.0404, 2.8772, 0.9494
       Z_r(row 2) = 3.9980, 3.7633, 2.0513
       Z_r(row 3) = 5.0534, 6.2819, 2.7256

9.8. $Z_{URC} = 5.63$

	Low	Medium	High
Input A	0.25*2	0.4*5	0.9*10
Input B	0.63*2	0.37*5	0.1*10
Input C	0.18*2	0.04*5	0.0*10
Accumulator array	2.12	4.05	10.0

CHAPTER 10

10.2. $\mathbf{e} = (0.4, 0.4, 0.3)$

10.4. $\mathbf{e} = (0.2, 0.5, 0.2, 0.2, 0.1)$

10.6. Ranking: x_4, x_3, x_1, x_2 ($C_{12} = 0.83$, $C_{31} = 0.56$, $C_{24} = 0.25$)

10.9. $C(\underset{\sim}{R}) = 0.61$, $m(\underset{\sim}{R}) = 0.53$, distance $= 94\%$

10.12. Average fuzziness, $F(\underset{\sim}{R}) = 0.387$; average certainty, $C(\underset{\sim}{R}) = 0.613$; distance to consensus, $m(\underset{\sim}{R}) = 0.524$; distance to consensus for a Type I relation, $m(\underset{\sim}{R}) = 0.293$; 67% of the way to Type I consensus, Distance $= 0.231$; 48% of the way to Type II consensus, Distance $= 0.524$.

10.15. The second alternative ($\mu = 0.6$).

10.18. $D(pipe) = 0.4$; $D(pond) = 0.5$. The preferred method of construction to mitigate flooding would be the construction of a pond.

10.19. (a) For imperfect information, $P(D_2|M_3) = 0.444$
For perfect information, $P(D_2|M_3) = 0.720$
(b) For perfect information, $E(U_1|M_2) = 3(0.584) + 2(0.4) + (-1)(0.076) = 2.041$, where $P(D_1|M_2) = 0.584$, $P(D_2|M_2) = 0.4$, $P(D_3|M_2) = 0.076$;
For imperfect information, $E(U_1|M_2) = 3(0.371) + 2(0.519) + (-1)(0.11) = 2.476$, where $P(D_1|M_2) = 0.371$, $P(D_2|M_2) = 0.519$, $P(D_3|M_2) = 0.11$

10.21. (a) $V(x) = 1.4169 - 2.6544 = -1.2375$

CHAPTER 11

11.2. 3 classes $\{x_1, x_4\}, x_2, x_3$

11.5. $C_1 = [0.991, 0.994, 0.025, 0.014]$, error $= 0$

11.8. $C_1 = [0.972, 0.816, 0.247, 0.086]$, error $= 0.25$

11.9. $C_1 = [0.438, 0.343, 0.443]$, two cycles

11.12. (b) U(row 1) $\approx 0.893, 0.838, 0.069, 0.143, 0.058$;
$U(row2) \approx 0.107, 0.162, 0.931, 0.857, 0.942$; $F_c(U) = 0.81091$
(d) $R_{11} = 1$, $R_{12} = 0.945$, $R_{13} = 0.176$, $R_{23} = 0.231$, $R_{24} = 0.305$. $R_{35} = 0.989$

11.20. $\max[0.35, 0.6, 0.05, 0] = 0.6$; Resembles Pattern 2 the most.

11.22. $\max[0, 0.25, 0.5, 0.25] = 0.5$; Matches Pattern 3 the most.

11.25. $\max[0.5, 0.52, 0.65, 0.5] = 0.65$; Composed of Black Powder.

11.28. $\max[0, 0.125, 0.7125, 0.1625, 0] = 0.7125$; Heavy Solvent Neutral.

11.31. $\max[0.51, 0.8182, 0.8001, 0.8697] = 0.8697$; The cell is classified under Pattern 4.

11.34. μ for an isosceles trapezoid is $\min[0.8, 1, 0.9, 1] = 0.8$

11.36. $\mu(s) = \mu(a_8)$, for AC, $S \rightarrow I(a_6, a_7)$, for DC, $S \rightarrow I(a_8, A)$

11.38. Scaling the pixel values between 0 and 1 by dividing by 255 we get the following: row 1 $= 0.86, 0.12, 0.04, 0.06, 0.98$; row 2 $= 0.80, 0.90, 0, 0.94, 0.90$; row 3 $= 0.88$,

0.08, 0.88, 0.08, 0.86; row 4 = 0.85, 1, 0.72, 0.04, 0.84; row 5 = 0.86, 0.1, 0.06, 1, 0.92. Using the algorithm presented in the text we get: row 1 = 0.86, 0.12, 0.04, 0.06, 0.98; row 2 = 0.80, 0.25, 0.69, 0.26, 0.90; row 3 = 0.88, 0.92, 0.07, 0.68, 0.86; row 4 = 0.85, 0.29, 0.50, 0.51, 0.84; row 5 = 0.86, 0.1, 0.06, 1, 0.92.

CHAPTER 12

12.1. (a) $[2, 3] + [3, 4] = [5, 7]$
       (c) $[4, 6] \div [1, 2] = [4, 6] * [1, 0.5] = [\min(4, 6, 2, 3), \max(4, 6, 2, 3)] = [2, 6]$
12.3. (a) $Z = 0/-2 + 0.1/1 + 0.6/4 + 0.8/7 + 0.9/10 + 0.7/13 + 0.1/16 + 0/19$
       (d) $Z = 0/0 - 0 + 0.1/1 - 1 + 0.6/2 - 2 + 0.8/3 - 3 + 0.9/4 - 4 + 0.7/5 - 5 +$
       $0.1/6 - 6 + 0.0/7 - 7 = 3.2/0$
       (e) $Z = 0/0 + 0.9/1 + 0.7/2 + 0.5/3 + 0.2/4 + 0.1/5 + 0/6 + 0/7$
12.7. (a) $\underset{\sim}{F} = \underset{\sim}{m} \cdot \underset{\sim}{A} = 0/1 + 0.2/2 + 0.2/3 + 0.5/4 + 0/5 + 0.7/6 + 0.5/8 +$
       $1/9 + 0/10 + 0.5/12 + 0/15 + 0/16 + 0/20$
       (b) (ii) DSW algorithm, $I_0^+ = [0, 20]$, $I_{0.5}^+ = [3.2, 14]$, $I_1^+ = [9, 9]$
12.9. (a) From the DSW algorithm for I_{0+}, $I_{0.5}$, $I_{0.8}$: $y = 0.8/-3.17157 + 0.5/0.8284 + 0/\sqrt{8}$
12.12. (b) $0.1/-9 + 0.1/-6 + 0.3/-4 + 0.1/-3 + 0.3/-2 + 0.7/-1 + 1/0 +$
       $0.7/1 + 0.3/2 + 0.1/3 + 0.3/4 + 0.1/6 + 0.1/9$
12.14. $\underset{\sim}{P} = 0.5/0 + 0.3/10 + 0.9/20 + 1/30 + 0.8/40 + 0.4/50 + 0.4/80 + 0.1/90 +$
       $0.4/120 + 0.4/160 + 0.1/180 + 0.4/200 + 0.1/270 + 0.1/360 + 0.1/450$

CHAPTER 13

13.1. Cycle 1, $x_1(0) = 80°$, $x_2(0) = 85°$, $u(0) = 275.20$; cycle 2, $x_1(1) = 200.2$, $x_2(1) = -5$, $u(1) = 303.10$; cycle 3, $x_1(2) = -102.3$, $x_2(2) = 205.2$, $u(2) = 292.55$; cycle 4, $x_1(3)$ 702.2, $x_2(3) = -307.5$, $u(3) = 303.33$
13.4. Cycle 1, $t^* = -0.67$ N m, $\theta(1) = 0.5$, $\dot{\theta}(1) = -3.97$;
       Cycle 2, $t^* = 0.0$ N m, $\theta(2) = -0.19$, $\dot{\theta}(2) = -12.40$;
       Cycle 3, $t^* = 1.2$ N m, $\theta(3) = -0.5$, $\dot{\theta}(3) = 2.46$;
       Cycle 4, $t^* = 1.4056$ N m, $\theta(4) = -0.19$, $\dot{\theta}(1) = 3.63$
13.7. Cycle 1, $\alpha^*(0) = 47$ mg/cm^3 s
       $W_1(1) = W_1(0) + W_2(0) = 800 - 280 = 520$ mg/cm^3
       $W_2(1) = W_2(0) + \alpha(0) = -280 - 47 = -327$ mg/cm^3 cm
       Cycle 2, $\alpha * (1) = 48$ mg/cm^3 s
       $W_1(2) = W_1(1) + W_2(1) = 193$ mg/cm^3
       $W_2(1) = W_2(0) + \alpha(0) = -375$ mg/cm^3 cm
13.10. (a) The new valve position is approximately 0.6. This is an example of using a "dead band" in a control problem.
       (b) The new valve position should be adjusted to approximately 0.8.
       (c) The new valve position is adjusted to approximately 0.174.

CHAPTER 14

14.2. The moment, M, at an arbitrary distance x from the left support, $-Py = M = EI\,y''$, so $y'' + P/EI\,y = 0$. A possible equation for y in terms of x is $y = C_1 \sin(kx) +$

$C_2 \cos(kx)$. Consider bounding conditions for C_1 and C_2; we have the following: $x = 0$, $y = 0$, so $C_2 = 0$, and $x = 1$, $y = 0$, so $C_1 \sin(kl) = 0$, thus C_1 is not zero and in order for $C_1 \sin(kL) = 0$, $kL = n\pi$, where $n = 1, 2, \ldots, m$. That is, $k^2 = P/EI = (\pi n)^2/L^2$ which gives $P = (n^2\pi^2 EI)/L^2$. Since n is constrained from 0 to 2, $P(2) = (4\pi^2 EI)/L^2$ and $P(0) = 0$; thus $\mu_G(n) = (P(n) - P(0))/(P(2) - P(0)) = n^2/4$. To find the optimal solution n^*, $\mu_{\underset{\sim}{c}}(n) = 1 - n$, $1 \le n \le 2$ and $\mu_{\underset{\sim}{c}}(n) = 0$, $n > 1$; $\mu_D(n) = n^2/4$, $0 \le n \le n^*$ and $\mu_D(n) = 1 - n$, $n > n^*$. $n^2/4 = 1 - n$, $n^* = 0.8284$ and results in a load of $P(n^*) = ((0.8284)^2\pi^2 EI)/L^2$.

14.5. $a_{max} = 1$, $a_{min} = 0$

14.10. $(P_0, C_0) = (0, 0)$ and $(P_1, C_1) = (5000, 0)$

14.12. $[0, 1, 1, 1, -1]$

CHAPTER 15

15.1.

Focal elements	m_1	m_2	bel_1	bel_2	pl_1	pl_2
F	0.3	0.2	0.3	0.2	0.4	0.4
NF	0.6	0.6	0.6	0.6	0.7	0.8
F ∪ NF	0.1	0.2	1.0	1.0	1.0	1.0

15.4.

Focal elements	m_1	bel_1	m_2	bel_2	m_{12}	bel_{12}
A_1	0.1	0.1	0	0	0.09	0.09
A_2	0.05	0.05	0.05	0.05	0.09	0.09
A_3	0.05	0.05	0.1	0.1	0.14	0.14
$A_1 \cup A_2$	0.05	0.2	0.05	0.1	0.07	0.25
$A_1 \cup A_3$	0.05	0.2	0.15	0.25	0.13	0.36
$A_2 \cup A_3$	1	0.2	0.05	0.2	0.10	0.33
$A_1 \cup A_2 \cup A_3$	0.6	1	0.6	1.0	0.38	1.0

15.7. $m = (0.2, 0, 0.3, 0, 0, 0.5)$; $r = (1, 0.8, 0.8, 0.5, 0.5, 0.5)$

15.10. Assuming a uniform weight of 0.25 on the original data intervals, a typical set of intervals and weights might be: [0.008, 0.2], 0.173228; [0.03, 0.2], 0.160663; [0.03, 0.15], 0.175523; [0.03, 0.1], 0.211611; [0.03, 0.06], 0.278975.

15.12. Assuming a uniform weight of 0.20 on the original data intervals, a typical set of intervals and weights might be: [0.75, 2.25], 0.573; [0.75, 1.5], 0.143; [0.75, 1.25], 0.143; [1.0, 1.25], 0.143.

(a) The possibility that the interest rates will be greater than 2% is 0.573.

(b) Generation of consonant intervals depends on the characteristics of the original data intervals and whether the expert desires a pessimistic or an optimistic estimate. In this case, the expert is assumed to lean more toward the pessimistic attitude and hence a more conservative estimate is generated. Therefore, intervals with lower values are assigned more weight.

(c) Degree of confirmation = -0.427

INDEX OF EXAMPLES AND PROBLEMS BY DISCIPLINE

Index

Fuzzy Logic with Engineering Applications, Second Edition   T. J. Ross
© 2004 John Wiley & Sons, Ltd   ISBNs: 0-470-86074-X (HB); 0-470-86075-8 (PB)